AF718022

Fundamentals of General Topology

Mathematics and Its Applications

Managing Editor:

M. HAZEWINKEL
Centre for Mathematics and Computer Science, Amsterdam, The Netherlands

Editorial Board:

A. BIALYNICKI-BIRULDA, *Institute of Mathematics PKIN, Warsaw, Poland*
J. KURZWEIL, *Mathematics Institute, Academy of Sciences, Prague, Czechoslovakia*
L. LEINDLER, *Bolyai Institute, Szeged, Hungary*
L. LOVÁSZ, *Bolyai Institute, Szeged, Hungary*
D. S. MITRINOVIĆ, *University of Belgrade, Yugoslavia*
S. ROLEWICZ, *Polish Academy of Sciences, Warsaw, Poland*
BL. H. SENDOV, *Bulgarian Academy of Sciences, Sofia, Bulgaria*
I. T. TODOROV, *Academy of Sciences, Sofia, Bulgaria*
H. TRIEBEL, *Universität Jena, D.D.R.*

A. V. Arkhangel'skii and V. I. Ponomarev
Faculty of Higher Geometry and Topology, Moscow University, Moscow, U.S.S.R.

Fundamentals of General Topology:
Problems and Exercises

D. Reidel Publishing Company
A MEMBER OF THE KLUWER ACADEMIC PUBLISHERS GROUP
Dordrecht / Boston / Lancaster

Library of Congress Cataloging in Publication Data

CIP

Arkhangel'skii, A. V.
Fundamentals of general topology.

(Mathematics and its applications; v.)
Translation of: Osnovy obscheĭ topologii v zadachakh i uprazhneniiakh.
Bibliography: p.
Includes index.
1. Topology—Problems, exercises, etc. I. Ponomarev, V. I. (Viktor Ivanovich) II. Title. III. Series: Mathematics and its applications (D. Reidel Publishing Company); v.
QA611.19.A7413 1983 514'.076 83-17818

Originally published as: *Osnovy Obsheii Topologii v Zadachakh i Uprajhneniyakh,* by A. V. Arkhangel'skii and V. I. Ponomarev
Izdatel'stvo 'Nauka' Moscow, 1974

Translated from the Russian by V. K. Jain

Published by D. Reidel Publishing Company,
P. O. Box 17, 3300 AA Dordrecht, Holland
in co-edition with Hindustan Publishing Corporation, India.

Distributed in India by
Hindustan Publishing Corporation (India), Delhi-110007.

Sold and distributed in U.S.A. and Canada by
Kluwer Academic Publishers, 190 Old Derby Street, Hingham, MA 02043, U.S.A.

Sold and distributed in all remaining countries by
Kluwer Academic Publishers Group,
P.O. Box 322, 3300 AH Dordrecht, Holland.

All Rights Reserved.
Copyright © 1984 by Hindustan Publishing Corporation, India.
No part of the material protected by this copyright notice may be reproduced or utilized in any form or by any means, electronic or mechanical, including photocopying, recording or by any information storage and retrieval system, without written permission from the copyright owner.

Printed in India by Hindustan Publishing Corporation Press,
C-74, Okhla, Industrial Area, Phase I, New Delhi.

Foreword

Recent years have witnessed the publication of several books in the western literature, which are devoted to the so-called general topology, i.e., are focused mainly on the theory of topological spaces. They are of varying levels. The most notable of them, in my opinion, is Kuratowski's *Topology*, the voluminous two-volume treatise which has to its credit several editions in many languages and was translated into Russian quite some time back. The other remarkable work is from R. Engelking, the Polish mathematician. It was published simultaneously in Polish and English, titled respectively as *Zarya Topologii Ogolnej* and *General Topology*. It is no wonder that both the books should have been contributed by Polish mathematicians, since the predominance of the Polish School of Mathematics in the field of set-theoretical topology for more than half the century is a well-known fact.

During the last fifty years, the Soviet School of Topology has also moved forward to the forefront, though this fact is not fully reflected in the Russian literature of that period. In this connection, a mention may be made of Russian translations of Kuratowski's treatise mentioned above, Hausdorff's celebrated work on set theory (which originally appeared in various editions between 1914 to 1927) and Kelley's *General Topology* (published in 1955, but now more obsolete than both the earlier classical works cited above) as well as my own quite elementary book [3] published in 1948. However, they do not touch the fringe of topology in-as-much as that even the formulation of some general metrization theorems are not made accessible to the reader. It is perhaps against this background of total lack of original Russian publications on general topology that even the new edition of [1] jointly written by me and Urysohn in 1922 failed to augument the stock value of our literature in this field.

The present book bearing the title, *Fundamentals of General Topology : Problems and Exercises*, is addressed to the mathematicians interested in, and students of, topology. It is written by two young mathematicians, who are from among the most outstanding representatives of the Soviet School of Mathematics in the branch of set-theoretical topology. We owe to them many extremely significant results obtained in the theory of topology during the last decade and a half. It is sufficiently borne out when we remark that a completely new and fundamental chapter relating to the general theory of continuous mapping of topological spaces in the present work is basically their creation. Certain conclusions like Ponomarev's theorems on absolute topological spaces or on first-countable spaces as open images of metric spaces, as well as Arkhangel'skii's theorems on the so-called feather spaces introduced by him and, in general, on the relationships between various types of spaces—the relations brought into play under their continuous mappings, form undoubtedly the basic and exceedingly interesting attainments of general topology. As regards the solution of Arkhangel'skill's problem on the cardinality of all compact first-countable Hausdorff spaces, it solves not only a problem five decades old but is also one of the fundamental conclusions in the entire set-theoretic mathematics. It is no wonder that the authors, because of their close association with, and direct contribution to, the development of topology, should be able to reflect extensively in the present comprehensive work its most essential modern directions and also the promising perspectives of its future development. Indeed, the book gives an insight into the attainments and problems which howsoever irritating and even abstract do not detract on the fascination of the subject; this fascination is especially brought out in the manner they are able to demonstrate the method of exposition via the problems (though originality is conceded in their case, it has unfortunately become lately also a fashion), in other words 'through problems and exercises.' In my opinion, this has been more of an impediment than an aid in the attainment of their goal. Nevertheless, *the problems offer to the reader a picturesque description, whose great merit consists in the fact that apart from depicting the vitality of present day general topology,* they are solved here in such a facile manner that one may hardly come across any other work worthy of preference in this respect.

After what has been stated above, it ought to be no surprise to the reader that the entire work reaches its climax on the one hand in the general theory of continuous mappings and on the other in the theory of the so-called cardinal invariants (cardinality, weight, the so-called π-weight, tightness, the Souslin number, etc.). These problems and numerous others connected with them, not only those of closest interest just to the authors but also those which have quite recently arisen objectively in the general topology, are given the foremost place in the book. However, the other basic directions of the theory of topology have also not been left in the background. By way of example is the metrization problem which during the past ten years took an unexpected turn connected to a great extent specifically with the general theory of continuous mappings and, in turn, intimately related to the notion of paracompactness, its extension and reduction. So is the theory of compactifications and in no less degree that of dyadic compacta. The book also does not ignore the relationships to the axiomatic abstract set theory, manifested so conspicuously in the general topology during the recent past.

A concise description of the contents of books is not included in summing up all that has been stated. I wish only to reiterate that the present work initiates the reader into the modern problems of general topology; moreover, the problems have been selected not at random but by those who are really experts in this field.

The book is highly suitable for directing the reader toward rigorous research, giving him at the same time a systematic training to actively analyze other mathematical texts. I have no doubt that it shall receive a warm response from the readers for whom it is designed. It will be of great use to them and season many young readers into contemporary mathematicians, irrespective of the fact whether they continue to be engaged in the general topology or some other branch of mathematics.

P. ALEXANDROFF

Editor's Preface

Approach your problems from the right end and begin with the answers. Then one day, perhaps you will find the final question.

'The Hermit Clad in Crane Feathers' in R. van Gulik's *The Chinese Maze Murders.*

It isn't that they can't see the solution. It is that they can't see the problem.

G. K. Chesterton, The scandal of Father Brown "The Point of a Pin"

Growing specialization and diversification have brought a host of monographs and textbooks on increasingly specialized topics. However, the 'tree' of knowledge of mathematics and related fields does not grow only by putting forth new branches. It also happens, quite often in fact, that branches which were thought to be completely disparate are suddenly seen to be related.

Further, the kind and level of sophistication of mathematics applied in various sciences has changed drastically in recent years: measure theory is used (non-trivially) in regional and theoretical economics; algebraic geometry interacts with physics; the Minkowsky lemma, coding theory and the structure of water meet one another in packing and covering theory; quantum fields, crystal defects and mathematical programming profit from homotopy theory; Lie algebras are relevant to filtering; and prediction and electrical engineering can use Stein spaces. And in addition to this there are such new emerging subdisciplines as 'completely integrable systems', 'chaos, synergetics and large-scale order', which are almost impossible to fit into the existing classification schemes. They draw upon widely different sections of mathematics.

This program, *Mathematics and Its Applications,* is devoted to such (new) interrelations as *exempla gratia*:

- —a central concept which plays an important role in several different mathematical and/or scientific specialized areas;
- —new applications of the results and ideas from one area of scientific endeavor into another;
- —influences which the results, problems and concepts of one field of enquiry have and have had on the development of another.

The *Mathematics and Its Applications* program tries to make available a careful selection of books which fit the philosophy outlined above. With such books, which are stimulating rather than definitive, intriguing rather than encyclopaedic, we hope to contribute something towards better communication among the practitioners in diversified fields.

The best way (if one can find the time) to learn mathematics is by doing it, a statement which can be found in various places in the literature about mathematics. Given that fact, very few high level books have been written (I am excluding programmed courses) which aim to teach the reader a subject in terms of problems and exercises; that is, by letting him develop the subject himself (with enormous help by giving him the right definitions and concepts to work with). Such an approach can significantly speed up the learning process.

General topology provides a large number of concepts which are basic in many mathematical specialisms, but is otherwise not especially known for its vigorous interrelations with other fields. These undoubtedly exist (logic and set theory, $C(X)$, fixed point theory, topological transformation groups) but are not the central theme in this volume. Still, its unique features as a guided-do-it-yourself-tour through an area of which the basic ideas and concepts are used in practically every specialism, justifies, in my view, the inclusion of this volume in the MIA Series.

The unreasonable effectiveness of mathematics in science. . .

Eugene Wigner

Will, if you knows of a better 'ole, go to it.

Bruce Bairnsfather

What is now proved was once only imagined.

Willam Blake

As long as algebra and geometry proceeded along separate paths, their advance was slow and their applications limited.
But when these sciences joined company, they drew from each other fresh vitality and thenceforward marched on at a rapid pace towards perfection.

Joseph Louis Lagrange

Amsterdam
August, 1983

MICHIEL HAZEWINKEL

Table of Contents

Preface

At the time we commenced writing this book, Halmos' *Hilbert Space Problems Book* as well as Glazman and Lubich's *Problems in Finite Dimensional Linear Analysis* were not published. We had therefore an impression that we were perhaps the first to embark on a new approach of exposing the fundamentals of a mathematical discipline by means of problems. Though the above mentioned publications deprive us of the credit of being a pioneer in this direction, they also demonstrate that we are not the only one to opt for a more practical approach rather than relying solely on the so-called indisputable treatises : there are others also who share our viewpoint that the best way to have a deeper grasp of some serious mathematical theorems is to solve them independently. Hence the foremost and basic requirement that we prescribe for our reader is that he should actively, eagerly and patiently seek on his own the solutions of problems offered to him and then compare them with the solutions given in the book. We consider it to be useful that sometimes the reader may even skip over a problem he is unable to solve and pass on to the succeeding ones. Perhaps, he may be able to handle them and thereby gain an insight to endow him with success when he reverts to the problem which had earlier proved too stubborn to solve. The failure by itself must not upset the reader, in many cases the problems set forth here are in reality the theorems tedious to prove, i.e., they involve a serious test of one's capability and are not so conducive to an assured, much less a rapid, progress. In short, a reference to the solutions given in the book must always be the last resort for the reader. This is a different matter and a sensible one if after finding independently the solution of a problem, the reader compares this solution with the one deduced in the book. This is always useful as it provides material for thinking and occassionally also enables the reader to detect an error in his solution. It is impossible to rule out the last eventuality. There is no doubt that the solutions initially deduced by the reader of singularly involved problems would be more often than not inaccurate; this applies also to the process of mathematical calculations.

We must also forewarn the reader that, though the problems are arranged topic-wise, in certain cases we are forced to refer to the solutions of problems which appear subsequently and in a different thematic context. Of course, this gives rise to additional difficulties for the reader. However, it may be assured that even in these not too many cases the reader is never called upon to look beyond the book. It is self-understood that within the frame of each theme we have taken pains to keep the logical sequence of problems.

We have partitioned many difficult problems into comparatively easier ones. Otherwise (for example in the case of Problem 213 of Chap. 3 which is a tedious mathematical problem solved only recently), we have given some hints to solve them. All this, in our opinion, ought to have made the book easier to work upon and enhance its utility to the reader. We sincerely hope and wish that the reader would on his own work his way to the solutions, especially of more cumbersome problems. This is expected to stimulate his creative abilities which may in a series of cases contribute to real scientific progress through the discovery of new and more elegant proofs of some important theorems. We also bring to the notice of the reader that some of the problems listed in this book are still open. The solution of these problems may form the subject matter of a separate publication.

Obviously a passive reader, or a reader who is not directly interested in general topology but whose scientific pursuits necessitate his acquaintance with this subject, may (and the present book affords him this opportunity) treat this work as a handbook or monograph to extract from it the information desired by him. The book is so designed that it can adequately serve the needs of such a reader.

Every chapter, and frequently even a section, is prefaced by a theoretical introduction containing not only the fundamental definitions but also an exposition of the auxiliary concepts requisite for solving the problems of that chapter (or section), provided of course that they have not occurred earlier. It is essential that the reader gets himself acquainted with these definitions and concepts at the very start.

In our analysis of each topic we have strived that our reader is first acquainted with its present status and then led to its modern problems. The proof of many problems rests on theorems which have been proved only recently. There are also those problems which, let us reiterate, are still open.

After all these general and partly even leading remarks, we give below some concrete information about the book which would probably be of use to the reader.

No specialized knowledge in the field of general topology is expected of the reader. However, the reader is assumed to have an excellent grasp of mathematical analysis and analytic geometry, usually covered in first two semesters of the faculty of mathematics and pedagogical institutes. The book is designed for students of third to fifth semesters, to enable them to specialize in the set-theoretical directions of mathematics, graduates and scientists engaged in the field of functional analysis, topology, certain branches of geometry, theory of dynamic systems etc. We suggest that in order to make the best of the book a separate seminar is devoted to it.

The book contains 1587 problems. It is divided into six chapters, each covering a distinct topic. Each chapter, in turn, is divided into sections. The problems are numbered separately in each chapter (for example, Problem 10 of Chap. 1 is referred thus: Prob. 10, Chap. 1).† Having set out the problems, the solutions are derived or hints to solutions are indicated at the end of each chapter. Some problems are not provided with solutions or hints to solution, usually because of their elementary nature or having no bearing on the rest of the text material. In the latter case, the problems are marked with an asterisk which should not be taken to imply that they are more tedious than others. In particular, we have not given as a rule the solutions of problems relating to the theory of metric spaces, which as a matter of fact forms an integral part of many texts on functional analysis and is necessarily covered in the current first two-semester courses. It may also be remarked that for the solutions of some problems the reader is referred either to some book or the original work.

At the end of the book, a small list of references is appended. The list has on no account a pretence for being exhaustive. It includes primarily those books and works which have been cited in the solutions of problems.

One more important feature of the book is that we have nowhere touched upon the historical aspects of a problem nor indicated, save a few exceptions, as to who was the first to prove it or to construct an example. In the formulation of problems, we have also not made any distinction between the difficult and easier ones.

Our deepest gratitude is due to Powel Sergeevich Alexandroff but for whose constant encouragement the book would not have been completed. He carefully watched and guided our progress on the book, quite frequently we drew upon his counsel and recommendations, and he critically went through the manuscript, not once but many times. We express our sincerest

†However, if the problem is from the same chapter, then only the number of problem is mentioned.

thanks to our dear teacher for his multifarious participations and constant help.

We are sincerely grateful to L. A. Tumarkin, the referee of the book, for his valuable comments and suggestions, which have been used to our advantage.

Thanks are due to V. L. Klushin for going through the entire manuscript and painstakingly editing it.

We also thank our students, A. Bashkirov, A. Kombarov, V. Malikhin, C. Shapirovskii, V. Mishkin and others, who carefully checked the solutions of problems and made a number of important corrections.

Moscow University
Faculty of Higher Geometry and Topology
25 June, 1973

A. V. ARKHANGEL'SKII
V. I. PONOMAREV

Chapter 1

Set Theory

We shall take the concept of *set*, as well as the more general notion of *class* (family, collection), of elements to be elementary and intuitively clear. In particular, the meaning of the following statements will be assumed to be understood.

A set consists of *elements* and is defined by its elements. The expression, $a \in A$ is read 'a is an element of A.' *Equality of sets* is understood to mean their coincidence: $A = B$, where A and B are sets, iff $a \in A$ implies $a \in B$ and $b \in B$ implies $b \in A$. There exists (a unique) *empty* set, a set which contains no elements. We denote the empty set by Λ.

A class also consists of elements and is characterized by its elements. All sets are classes. But a class is a set iff it can be an element of some class. This implies that it is forbidden to make use of certain classes as elements for the construction of a new set. An example of such a class is the class of all sets. On the other hand, for each set A there is canonically defined a new set $\exp A$ which, in particular, contains A as an element: $\exp A$ is the set of all *subsets* (see below) of A.

The collection of all conceivable sets into a class not forming a set permits one to avoid the well-known paradoxes that arose in 'naive' set theory at the turn of the century (see, Fraenkel, Bar-Hillel and Levy [1]).

Neither axiomatic set theory nor these paradoxes are discussed in this book.

If, for all a, $a \in A$ implies $a \in B$, where B is a set and A is a class, then we write $A \subset B$ and say that A is a *subset* of B. Moreover, A itself is certainly a set. It is clear that the empty set is a subset of every other set. If $A \subset B$ and $A \neq B$, then A is a *proper* subset (or proper part) of B.

1.1. Notations

To introduce new sets, we frequently use the following notation :

$$A = \{a\colon \ldots F(a) \ldots\}.$$

Let us explain its meaning. A is a definable set, a is a variable ranging over all possible elements and $F(a)$ is a proposition about a (where $F(x)$ is a formula in which x occurs only as a free variable and in which all other variables are bound by quantifiers). In addition, $a \in A$ iff the statement $F(a)$ has a sense and is true (that is, the proposition obtained by substituting an individual variable a in $F(x)$ for all occurrences of x in $F(x)$ is meaningful and correct). By $\{a\}$ we always denote the set (*singleton*) containing the single element a.

1.2. Set Operations

The *sum or union* of two sets A and B, written $A \cup B$, is defined by: $A \cup B = \{a: a \in A \text{ or } a \in B\}$ (here 'or' is used in the sense of 'and/or' implying that at least one of the statements connected by it is satisfied). The *intersection* of two sets A and B, written $A \cap B$, is defined by: $A \cap B = \{a: a \in A \text{ and } a \in B\}$ (the conjunction 'and' always involves the requirement that both of the statements connected by it are true). When $A \cap B = \Lambda$, we say that A *does not intersect* B, or that A and B are *disjoint*. If to each element α of A there corresponds some set P_α, then we put $\bigcup \{P_\alpha : a \in A\} = \{a: \text{there exists } \alpha \in A \text{ for which } a \in P_\alpha\}$ and $\bigcap \{P_\alpha : \alpha \in A\} = \{a: a \in P_\alpha \text{ for every } \alpha \in A\}$. In addition, we call $\bigcup \{P_\alpha : \alpha \in A\}$ the *sum* (or *union*) (of elements) of the collection $\{P_\alpha : \alpha \in A\}$, or the *field* of this collection, and we call $\bigcap \{P_\alpha : \alpha \in A\}$ the *intersection* (of elements) of the collection $\{P_\alpha : \alpha \in A\}$. In axiomatic set theory, it is proved that, under the above conditions, the classes $\bigcup \{P_\alpha : \alpha \in A\}$ and $\bigcap \{P_\alpha : \alpha \in A\}$ are sets. If the union of a family η of sets is equal to X, then η is called a *cover* of X. If $\gamma = \{P_\alpha : \alpha \in A\}$ is a family of sets, and for any $\alpha', \alpha'' \in A$ if $P_{\alpha'} \cap P_{\alpha''} = \Lambda$ whenever $\alpha' \neq \alpha''$, then the family γ is said to be *disjoint*.

The *difference* of an ordered pair of sets A and B is the set $A \setminus B = \{a: a \in A \text{ and } a \notin B\}$ (where the expression '$a \notin B$' is read '*a does not belong to B*'). Sometimes $A \setminus B$ is called the *complement* of B with respect to A, or simply the *complement* of B. Together with $A \setminus B$ and $B \setminus A$ we consider the *symmetric difference* $A \Delta B$ of sets A and B: $A \Delta B = (A \setminus B) \cup (B \setminus A)$. (In this book parentheses play their usual role, in contrast with braces which, as we have seen above, play a special role in set formation.)

The *product* of two sets A and B is the set $A \times B = \{(a, b): a \in A \text{ and } b \in B\}$. Similarly we define the product $A_1 \times A_2 \times \ldots \times A_k$ of any finite collection of sets $A_1, \ldots, A_k$. We also write

$$\underbrace{A \times A \times \ldots \times A}_{k \text{ times}} = A^k.$$

The product of any (not necessarily finite) collection of sets will be defined shortly.

Suppose that, for each $\alpha \in A$, we are given a set X_α. The *free union* (*free sum*, or *disjoint union*) $\bigcup_c \{X_\alpha : \alpha \in A\}$ of the sets X_α over all $\alpha \in A$ has as elements all possible pairs (x, α), where $\alpha \in A$ and $x \in X_\alpha$.

For fixed $\alpha^* \in A$ the set X_{α^*} may be identified with the subset $i_{\alpha^*}(X_{\alpha^*})$ of the free union, which consists of all pairs (x, α) satisfying $\alpha = \alpha^*$. We also put $i_\alpha(x) = (x, \alpha)$ for all $\alpha \in A$ and $x \in X_\alpha$. For distinct $\alpha_1, \alpha_2 \in A$, we may have $X_{\alpha_1} \cap X_{\alpha_2} \neq \Lambda$. But we always have $i_{\alpha_1}(X_{\alpha_1}) \cap i_{\alpha_2}(X_{\alpha_2}) = \Lambda$, provided $\alpha_1 \neq \alpha_2$. This remark reveals the distinguishing characteristic of free unions: in forming them we do not simply unite the terms but arrange them side by side in such a way that they are pairwise disjoint.

1.3. Functions

We assume the concept of *function* (or *mapping*) to be intuitively clear: we say that a function (*mapping*) f of the set X into the set Y is defined (written 'there is defined (a function) $f: X \to Y$') if to every element $x \in X$ there corresponds a uniquely defined element $y \in Y$, which we denote by $f(x)$ (that is, $y = f(x)$). The simplest function on a set X is the *identity function* $e: X \to X$, defined thus: $e(x) = x$ for each $x \in X$. Let the function $f: X \to Y$ be defined. The element $f(x)$ is called the *image of* x (under f). If $A \subset X$, then put $f(A) = \{f(x): x \in A\}$† (this is a shorthand notation for $f(A) = \{y$: there exists $x \in A$, for which $y = f(x)\}$). The set $f(A)$ is called the *image of* A (under f). If ξ is a family of subsets of X, then $f\xi = \{f(A): A \in \xi\}$. Further, if $B \subset Y$, then we put $f^{-1}(B) = \{x: f(x) \in B\}$ and call $f^{-1}(B)$ the *inverse image* of B. The inverse image $f^{-1}\eta$ of a family η of *subsets* of Y is $\{f^{-1}(B): B \in \eta\}$. We call f a function of X *onto* Y, if $f(X) = Y$. If for any x_1 and $x_2 \in X$, $x_1 \neq x_2$ implies $f(x_1) \neq f(x_2)$, then the function f is said to be *one-to-one* (or *inclusion*). Of special interest are functions that are one-to-one and onto. If $f: X \to Y$ is a function and $X' \subset X$, then

†When no confusion is likely, one may dispense with the parentheses, that is, we may write fA for $f(A)$.

the function $f' = f|X'$, $f': X' \to Y$ defined for all $x \in X'$ by $f'(x) = f(x)$ is called the *restriction* of f to X'.

Given the functions $f: X \to Y$ and $g: Y \to Z$, for every $x \in X$, put $\varphi(x) = g(f(x))$. Obviously, φ is a function from the set X into the set Z; it is called the *composition* of f and g and is written $\varphi = gf$.

The *graph* of a function $f: X \to Y$ is the subset $\Gamma(f) = \{(x, f(x)): x \in X\}$ of the product $X \times Y$. The *standard function* φ_f from the domain X of f into the graph $\Gamma(f)$ of f is defined by the following rule: $\varphi_f(x) = (x, f(x)) \in \Gamma(f)$ for all $x \in X$. If $\pi_Y: X \times Y \to Y$ is a *projection*, that is, if $\pi_Y(x, y) = y$ for all $(x, y) \in X \times Y$, then evidently $f = \pi_Y \varphi_f$.

Suppose that to each element $\alpha \in A$ there corresponds some set X_α. A *fiber*† is a function x, defined on A, such that for each $\alpha \in A$ we have $x(\alpha) \in X_\alpha$ ($x(\alpha)$ is usually written x_α). Notice that no extension of the concept of function is involved here: we may regard the fiber x as being a function in the original sense of the set A into the union of the sets X_α. We call x_α the αth *coordinate* of x. The set of all fibers is called the *product* of the X_α over all $\alpha \in A$ and is denoted by $\Pi\{X_\alpha: \alpha \in A\}$ or simply by ΠX_α, if there is no confusion about the identity of A. Put $X = \Pi\{X_\alpha: \alpha \in A\}$. For each $x \in X$ and every $\alpha \in A$ put $\pi_\alpha(x) = x_\alpha$: obviously, π_α is a function from a set X onto the set X_α; it is called the *projection of the product* onto the αth factor, or more simply, the αth *projection*. Let $A' \subset A$. Then the product $\Pi\{X_\alpha: \alpha \in A'\}$ is called the *A'-edge* (or simply an *edge*) of the product $X = \Pi\{X_\alpha: \alpha \in A\}$ and is denoted by $X_{A'}$.

There is a natural projection $\pi_{A'}$ of the product $X = \Pi\{X_\alpha: \alpha \in A\}$ onto the edge $X_{A'}$ (called the *standard function* onto the edge $X_{A'}$) described by the rule : $\pi_{A'}(x) = x|A'$ for each $x \in X$, where $x|A'$ is the element $x' \in X_{A'} = \Pi\{X_\alpha: \alpha \in A'\}$ such that $x'_\alpha = x_\alpha$ for all $\alpha \in A'$. Formally, an edge of a product is not a subset, but for each point $x^* \in X$ there is a *natural inclusion* $i_{x^*}^{A'}$ of $X_{A'}$ into X, defined for each $x' \in X_{A'}$ by putting $i_{x^*}^{A'}(x') = x$, where $x|A' = x'$ and $x|(A \setminus A') = x'|(A \setminus A') = x^*|(A \setminus A')$. In particular, each factor, that is, every X_α with $\alpha \in A$, can be regarded as the edge of X obtained when $A' = \{\alpha\}$. In this case we put $i_{x^*}^{\alpha} = i_{x^*}^{A'}$, obviously $i_{x^*}^{\alpha}$ is an inclusion of X_α into X.

The inclusions $i_{x^*}^{A'}$ and $i_{x^*}^{\alpha}$ are called *standard inclusions* into the product (of edges, factors) *along* x^*.

†In the next chapter, the term 'fiber' is used in a more specialized sense to be made explicit presently.

The following specialization of the notion of product is useful: if A and B are two sets, then by B^A we denote the *set of all functions from A to B.* Obviously, $B^A = \Pi\{B_\alpha: \alpha \in A\}$, where $B_\alpha = B$ for each $\alpha \in A$. If for each $\alpha \in A$ there is defined a function f_α from some set X_α to some set Y_α, then the product $f = \Pi\{f_\alpha: \alpha \in A\}$ of the f_α is the function from $X = \Pi\{X_\alpha: \alpha \in A\}$ to $Y = \Pi\{Y_\alpha: \alpha \in A\}$, defined by the following rule: $f(x) = y \Leftrightarrow f_\alpha(x_\alpha) = y_\alpha$ for all $\alpha \subset A$.†

If $X_\alpha = Z$ for each $\alpha \in A$, then in the product $X = \Pi\{X_\alpha: \alpha \in A\}$ we have the following subset Δ, called the *diagonal*: $\Delta = \{x \in X: x_{\alpha'} = x_{\alpha''}$ for every $\alpha' \in A, \alpha'' \in A\}$. If, in addition, the functions $f_\alpha: X_\alpha = Z \to Y_\alpha$, $\alpha \in A$, are defined, then their *diagonal product* $f_\Delta = \Delta\{f_\alpha: \alpha \in A\}$ is defined to be the restriction of the product f of the functions f_α, $\alpha \in A$, to the diagonal Δ. Since the diagonal is naturally identified with the set Z by means of an arbitrary projection π_α, f_Δ can be interpreted as the mapping of Z into the set $Y = \Pi\{Y_\alpha: \alpha \in A\}$, defined by the condition: $f_\Delta(z) = y \Leftrightarrow f_\Delta(z) = y_\alpha$ for each $\alpha \in A$.

The concept of *number* can be defined in terms of the concept of function. It is of crucial importance that this procedure enables us to define not only the finite (natural) numbers, but also different infinite (cardinal, ordinal) numbers. For details the reader is referred to Sierpinski [1]. A set is called *countable* if either it consists of a finite number of elements (in particular, it may be empty), or if there exists one-to-one function of the set onto the set N of all natural numbers. If there exists a one-to-one function of X onto a set Y we say that X and Y are *equipotent* and write $|X| = |Y|$. The class of all sets, equipotent with a given set X is called the *cardinality* of a set X and is denoted by $|X|$. The cardinality is also called the *cardinal number*. If there exists $Y' \subset Y$, such that $|X| = |Y'|$, then we write $|X| \leqslant |Y|$. If $|X| \leqslant |Y|$, but the inequality $|Y| \leqslant |X|$ does not hold, then we write $|X| < |Y|$. For an arbitrary set X we put $2^{|X|} = |\exp X|$. If τ is a cardinal number, we put $\exp_\tau X = \{A \in \exp X: |A| < \tau\}$. Further, if $Y = \Pi\{Y_\alpha: \alpha \in A\}$ and $y^* \in Y$, then $\Sigma_{\tau, y^*}\{Y_\alpha : \alpha \in A\}$ is the set of all $y \in Y$ such that the cardinality of the set of all those $\alpha \in A$ for which $y^*_\alpha \neq y_\alpha$ is less than τ. $\Sigma_{\tau, y^*}\{Y_\alpha: \alpha \in A\}$ is called the *product of the sets Y_α with respect to all $\alpha \in A$ and with centre at y^*.* This set is often denoted by $\Sigma_\tau(y^*)$ and is called the *τ-hull* of the point y^* in the product $\Pi\{Y_\alpha: \alpha \in A\}$.

†The sign $\Leftrightarrow$ symbolizes the expression 'iff'.

We now introduce the important notion of a (partial) *ordering* of a set. A (partial) ordering of a set or class X is a relation $\prec$ on X such that:

(*a*) if $x \prec y$ and $y \prec z$, then $x \prec z$;
(*b*) $x \prec y$ and $y \prec x$, iff $x = y$.

An ordering $\prec$ is said to be *linear* (or *total*), if the following supplementary condition is satisfied:

(*c*) for every $x, y \in X$, either $x \prec y$ or $y \prec x$.

A set X together with an ordering $\prec$ of it is called an *ordered set* $(X, \prec)$.

Any subset on which $\prec$ is a linear ordering is called a *chain* or *nest* in an ordered set $(X, \prec)$.

An element $b \in X$ is called a *majorant* (*upper bound*) of a set $A \subset X$ with respect to an ordering $\prec$ defined on X if $a \prec b$ for every $a \in A$.

An ordered set $(X, \prec)$ (or its ordering $\prec$) is said to be *inductive* if each chain $A \subset X$ in X has a majorant.

An element $c \in (X, \prec)$ is called a *maximal* element of X if $a \in X$ and $c \prec a$ imply that $a = c$.

The Kuratowski-Zorn Postulate (which we accept) asserts: *every inductive set has at least one maximal element.*

Correspondingly, an element a of a subset A of an ordered set $(X, \prec)$ is called a *minimal* element of A if $a \in A$ and $a' \in A$, $a' \prec a$ imply $a' = a$. A linear ordering of a set X is called a *well-ordering* if every non-empty set $A \subset X$ has a minimal element. We assume without proof the principle: *every set can be well-ordered.*

Let $(x, \prec)$ be an ordered set and let $x \in X$. We agree to denote by $I(x)$ (occasionally $T(x)$ or $I_{<}(x)$) the set of all $y \in X$ for which $y \neq x$, $y \prec x$, and put $\tilde{I}(x) = I(x) \cup \{x\}$. When $\prec$ is a well-ordering, $I(x)$ is called *the interval with end-point* x, and $\tilde{I}(x)$ the *segment with end-point* x.

A well-ordering $\prec$ of X is called *minimal*, if $|I(x)| \neq |X|$ for every $x \in X$.

Put $A(x) = X \setminus \tilde{I}(x)$.

If $A(x) \neq \Lambda$, then by $x + 1$ we denote a minimal (relative to $\prec$) element of $A(x)$. For an arbitrary subset A of a well-ordered set $(X, \prec)$ we denote by $\sup A$ the *first* (or *smallest*) element of the set $\{x \in X: a \prec x$ for all $a \in A\}$, if the latter is non-empty; if not, the expression $\sup A$ becomes meaningless. By $(N^+, <)$ we denote the set of positive integers endowed with the natural well-ordering, and by N the set of all integers with the usual linear ordering. The cardinal number of the set N^+ is denoted by $\aleph_0$. A one-to-one function $f: (X, <) \to (Y, \leqslant)$ from one ordered set into other is called $(+)$-*monotonic*, if for every $x_1, x_2 \in X$, $x_1 < x_2$ implies $f(x_1) \leqslant f(x_2)$.† If, in addition, $f(X) = Y$ and f^{-1} is also a $(+)$-monotonic function, then the function f is called a *similarity*.

Sometimes the very nature of the elements of a set enables us to define a natural ordering of it. This is of prime significance in the following concrete situation. Let $\mathcal{E}$ be some family of subsets of X. For $A_1 \in \mathcal{E}$, $A_2 \in \mathcal{E}$ we put $A_1 < A_2$ iff $A_1 \subset A_2$. The relation $<$ is an order on $\mathcal{E}$. In addition, a chain or nest in $\mathcal{E}$ is any subfamily $\mathcal{E}'$ of $\mathcal{E}$ such that for every $A_1 \in \mathcal{E}'$, $A_2 \in \mathcal{E}'$ either $A_1 \subset A_2$ or $A_2 \subset A_1$. In this case, we say that the *rank* of $\mathcal{E}'$ is unity. In addition to nests, an important role in set theory and general topology is played by *centralized* families of sets (or families of sets having the *finite intersection property*), which are characterized by the fact that the intersection of any finite set of their elements is nonempty.

If X is a set and $\mathcal{E}$ some family of subsets of X, then we shall always denote by $\mathcal{Y}(\mathcal{E})$ the ordered set of all centralized families of sets included in $\mathcal{E}$ ($\mathcal{Y}(\mathcal{E})$ is ordered as a family of sets in the natural way by the above rule). The set $\mathcal{Y}(\mathcal{E})$ is a family of collections of subsets of X. Notice that the ordered set $\mathcal{Y}(\mathcal{E})$ is inductive (see Problem 127), while $\mathcal{E}$ ordered in the natural manner is not always inductive and, in general, X is not even assigned an ordering.

As in any inductive ordered set, there are maximal elements in $\mathcal{Y}(\mathcal{E})$ (see Problem 144). In the present case the maximal elements are the *maximal centralized* families of elements of $\mathcal{E}$, that is, the centralized families, all of whose elements belong to $\mathcal{E}$, and which are not proper subsets of any other centralized families.

A centralized family ξ of sets belonging to $\mathcal{E}$ is called a prefilter in $\mathcal{E}$ if for each $A, B \in \xi$ there is $C \in \xi$ such that $C \subset A \cap B$. A filter base ξ in $\mathcal{E}$ is a *filter* in $\mathcal{E}$, if $A \in \xi$, $B \in \mathcal{E}$ and $B \supset A$ imply that $B \in \xi$. If $\mathcal{E}$ coincides with the set of all subsets of the set X under consideration, we speak simply of a prefilter (on X) or a filter (on X). In this case a maximal centralized family $\mathcal{E}$ is called an *ultrafilter* or *maximal filter* (on X).

A centralized family ξ of sets (whether prefilter, filter or ultrafilter) is

†For denoting specific orderings we use '$<$', '$\leqslant$' and other similar symbols in addition to '$<$'. We agree also to write $x <_c y$ for $x < y$ and $x \neq y$.

called *free* if $\cap \{U: U \in \xi\} = \Lambda$. Otherwise, we say that ξ is *singular*, or *trivial*, or *fixed*.

A family η of nonempty sets is called a *base* for a family ξ of sets, if for each $U \in \xi$ there exists $V \in \eta$ such that $V \subset U$. In this sense, we can speak of a base for a prefilter or ultrafilter.

If a family has a countable base, it is said to be *countably generated.*

1.4. PROBLEMS

1.4.1. *Set Operations. Countable Sets*

1. Let A, B and C be subsets of X. Show that

(*a*) $(A \cap B) \cup C = (A \cup C) \cap (B \cup C)$;
(*b*) $X \setminus (X \setminus A) = A$;
(*c*) $A \subset B$ iff $X \setminus B \subset X \setminus A$;
(*d*) $A \setminus B = A \cap (X \setminus B)$;
(*e*) $A \cap B = X \setminus ((X \setminus A) \cup (X \setminus B))$.

2. Let B be a set and let $\{A_\alpha: \alpha \in M\}$ be a family of sets. Show that

(*a*) $B \cap (\cap \{A_\alpha: \alpha \in M\}) = \cap \{B \cap A_\alpha: \alpha \in M\}$;
(*b*) $B \cap (\cup \{A_\alpha: \alpha \in M\}) = \cup \{B \cap A_\alpha: \alpha \in M\}$;
(*c*) $B \cup (\cap \{A_\alpha: \alpha \in M\}) = \cap \{B \cup A_\alpha: \alpha \in M\}$.

3. Let X be a set and $\{A_\alpha: \alpha \in M\}$ a family of subsets of X. Show that

(*a*) $X \setminus \cap \{A_\alpha: \alpha \in M\} = \cup \{X \setminus A_\alpha: \alpha \in M\}$;
(*b*) $X \setminus \cup \{A_\alpha: \alpha \in M\} = \cap \{X \setminus A_\alpha: \alpha \in M\}$.

These relations are called *de Morgan's laws*.

4. Show that if $A \setminus B$ and $B \setminus A$ are equipotent sets, then $|A| = |B|$ (but, generally speaking, the converse is not true).

5. Show that $|A_1| = |B_1|$ and $|A_2| = |B_2|$ do not impl , in general, that $|A_1 \cap A_2| = |B_1 \cap B_2|$.

6. Let A be a finite set and n its power (cardinal number). What is the power (cardinal number) of the set exp A?

7. Give an example of nonempty set X, each of whose elements is a subset of X.

8. For each integer $n > 0$ construct a set A_n, having exactly n elements such that for each $a \in A_n$, $a' \in A_n$ either $n \in a'$ or $a' \in a$.

9. Show that a set is finite iff every linear ordering on it is a well-ordering.

10. Let $(X, \leqslant)$ be a well-ordered (or order-complete) set. For $x_1, x_2 \in X$ put $x_1 < x_2$ iff $x_2 \leqslant x_1$. Show that $<$ well-orders X iff X is a finite set.

11. Show that a set X is finite iff it is not equipotent with any of its proper subsets.

12. Show that every subset of a countable set is countable.

13. Show that every infinite set has a countably infinite subset.

14. Show that, for every infinite set X, we have $\aleph_0 \leqslant |X|$.

15. Show that each countably infinite set can be represented as the union of two disjoint infinite (also countable) subsets.

16. If A and B are two countable sets, show that $A \cup B$ is also countable.

17. If X is an uncountable set and its subset A is countable, show that $|X \setminus A| = |X|$.

18. Show that the union of a countable collection of finite sets is countable.

19. Show that the union of a countable collection of countable sets is countable.

20. Prove that the product of two (finite) countable sets is countable.

21. Show that image of a countable set (under an arbitrary mapping) is countable.

22. Show that the following sets are countable:

(*a*) the set of all rational numbers on a straight line;

(*b*) the set of all points of the plane (space), all of whose coordinates are rational (the coordinate system is asssumed to be fixed);

(*c*) the set of all finite subsets of a countable set;

(*d*) the set of all intervals on a straight line with rational endpoints;

(*e*) the set of all spheres in space with centres at rational points (that is, all of whose coordinates are rational) and rational radii;

(*f*) the set of all polynomials in a finite number of variables with rational coefficients (where we identify any two polynomials which can be transformed into one another by rearrangement of variables);

(*g*) any set of pairwise disjoint figure eights lying on a fixed surface.

1.4.2. *General Problems on Functions*

23. Let $f: X \to Y$ be a function, $\eta = \{B_\alpha: \alpha \in M\}$ a family of subsets of Y and $\xi = \{A_\beta: \beta \in L\}$ a family of subsets of X. Show that

(*a*) $f^{-1}(\cap \{B_\alpha: \alpha \in M\}) = \cap \{f^{-1} B_\alpha: \alpha \in M\}$;
(*b*) $\cup \{fA_\beta: \beta \in L\} = f(\cup \{A_\beta: \beta \in L\})$;
(*c*) $f(\cap \{A_\beta: \beta \in L\}) \subset \cap \{fA_\beta: \beta \in L\}$.

Show that $f(\cap \{A_\beta: \beta \in L\}) \neq \cap \{fA_\beta: \beta \in L\}$ is possible.

24. Let $f: X \to Y$ be a function, $A \subset X$ and $B \subset Y$. Show that $f^{-1}(f(A) \cap B) \supset A \cap f^{-1}B$ and $f(A \cap f^{-1}B) = f(A) \cap B$. Hence, deduce that $f(A) \cap B = \Lambda$ iff $A \cap f^{-1}B = \Lambda$. Show that $f^{-1}(Y \setminus B) = X \setminus f^{-1}B$.

25. Show that, if there is a function f from a set X onto a set Y, then $|Y| \leqslant |X|$.

26. Let $f: X \to Y$, $g: Y \to X$ be two functions and let $|g(Y)| > 2$. Show that there exist non-empty sets X_1, X_2, Y_1, Y_2 which satisfy $X_1 \cap X_2 = \Lambda$, $Y_1 \cap Y_2 = \Lambda$, $fX_1 = Y_1$ and $gY_2 = X_2$.

27. Prove that, for any nonempty sets A, B and C, $|(A \times B)^C| = |A^C \times B^C|$.

28. For any three nonempty sets A, B and C, such that $B \cap C = \Lambda$, establish the equality: $|A^{B \cup C}| = |A^B \times A^C|$.

29. For any nonempty sets A, B and C, show that $|(A^B)^C| = |A^{B \times C}|$.

30. Show that, for any two sets X and Y, at least one of them is equipotent to some subset of the other.

31. Let $f: X \to Y$ and $g: Y \to X$ be one-to-one functions (not necessarily onto). Construct by means of f and g a one-to-one function from X onto Y.

32. Show that, for any two sets X and Y, either $|X| < |Y|$, or $|X| = |Y|$, or $|Y| < |X|$.

33. Show that the relation $\leqslant$ among cardinal numbers linearly orders the class of all cardinal numbers.

34. Show that we always have $|X| < |\exp X|$.

35. Let $\{S_i: i \in N^+\}$ be a sequence of finite sets S_i and for each $a \in S_{i+1}$ let $\varphi(a)$ be some element of S_i for all $i \in N^+$. Show that it is always possible to choose $a_i \in S_i$ such that for all $i \in N^+$ we have $\varphi(a_{i+1}) = a_i$.

1.4.3. *General Problems on Well-ordered (Order-complete) Sets*

Let $(X, <)$ be a linearly ordered set. A set $A \subset B$ is called a *segment* of this set, if $x \in A$, $y \in X$ and $y < x$ always imply that $y \in A$. If the linearly ordered sets $(A, <)$ and $(B, \preccurlyeq)$ are disjoint, then their *sum*, written $(A, <) + (B, \preccurlyeq)$ is the set $A \cup B$, linearly ordered by the relation $\prec$, defined as follows: if $a \in A$ and $b \in B$, then $a \prec b$; if $a_1 \in A$ and $a_2 \in A$, then $a_1 \prec a_2 \Leftrightarrow a_1 < a_2$; if $b_1 \in B$ and $b_2 \in B$, then $b_1 \prec b_2 \Leftrightarrow b_1 \preccurlyeq b_2$.

More generally: if $(A, \prec)$ is a linearly ordered set and for each $\alpha \in A$ we are given a set B_α, linearly ordered by the relation $<_\alpha$, and $B_{\alpha'} \cap B_{\alpha''} = \Lambda$ for $\alpha' \neq \alpha''$, then the *ordered sum of the sets* $(B_\alpha, <_\alpha)$ *over* $(A, \prec)$, written $\Sigma\{(B_\alpha, <_\alpha): \alpha \in (A, \prec)\}$, consists of the points of the set $X = \bigcup \{B_\alpha: \alpha \in A\}$ ordered by the relation $<$ given by the following rule: if $x, y \in X$ and $x \in B_{x''}$, $y \in B_{\alpha''}$, then for $\alpha < \alpha''$ we put $x < y$ and for $\alpha' = \alpha''$ we put $x < y \Leftrightarrow x <_{\alpha'} y$.

The lexicographic product $(A, <) \times (B, \preccurlyeq)$ of two linearly ordered sets $(A, <)$ and $(B, \preccurlyeq)$, taken in a definite order (this order is usually expressed in the dictionary (lexicographic) ordering of a sequence of symbols, denoting the sets under consideration), is obtained when the product $A \times B$ of the sets A and B is assigned the following relation $\prec$. If $(a_1, b_1) \in A \times B$ and $(a_2, b_2) \in A \times B$, then $(a_1, b_1) \prec (a_2, b_2)$ iff $b_1 \preccurlyeq b_2$ or else $b_1 = b_2$ and $a_1 < a_2$. The ordering $\prec$ is called the *lexicographic ordering*, or *dictionary ordering* (of the product).

We note that multiplication, as opposed to addition, is defined for all (and not only for disjoint) linearly ordered sets. In particular, we may consider a (lexicographically) ordered *square* of a linearly ordered set. In Problem 50, we shall define the Cantor square of a linearly ordered set which is distinct, as we shall see, from the lexicographic square, usually called the *simple square*.

For any two (nonempty) linearly ordered sets $(A, <)$ and $(B, \preccurlyeq)$ where, in addition, $(B, \preccurlyeq)$ has a first element θ, we also define the notion of the power $(B, \preccurlyeq)^{(A, <)}$ as follows. The elements of $(B, \preccurlyeq)^{(A, <)}$ are the functions x from A into B, such that $x_\alpha = \theta$ for all $\alpha \in A \setminus K(x)$, where $K(x)$ is some finite set. Let $x, y \in (B, \preccurlyeq)^{(A, <)}$: since $K(x) \cup K(y)$ is a finite subset of A, it has a largest element which we denote by $\alpha(x, y)$. The ordering $\prec$ on $(B, \preccurlyeq)^{(A, <)}$ is then defined by: $x \prec y$ iff $x_{\alpha*} \preccurlyeq y_{\alpha*}$ for $\alpha^* = \alpha(x, y)$.

In the sequel we shall often come across the ordered set $(D, <)$. It consists of two elements 0 and 1, the ordering being $0 < 1$. We shall sometimes use the symbols 0, 1, 2, . . . to denote the first, second, . . . elements of an arbitrary well-ordered set. In this scction $(N, <)$ always denotes the set of natural numbers with their natural ordering; the two symbols N and $<$ are sometimes replaced by the single symbol ω_0.

36. Show that, if every countable subset of a linearly ordered set is well-ordered, then the entire set is also well-ordered.

37. Show that an arbitrary segment A of a well-ordered set $(X, <)$ either coincides with X or is an interval with end-point at some point of X.

38. (*a*) Show that the union of any set of segments of a linearly ordered set is a segment of this set.

(*b*) Show that, among any two segments A, A' of a linearly ordered set $(X, <)$, one is contained in the other.

39. Show that the identity map is the unique similarity function f of a well-ordered set $(X, <)$ into itself, such that $f(X)$ is a segment of X.

40. Let $(X, <)$ and $(Y, \leqslant)$ be two well-ordered sets, and let $f_1 : (X, <) \to (Y, \leqslant)$, f_2: $(X, <) \to (Y, \leqslant)$ be similarity functions, such that $f_1(X)$ and $f_2(X)$ are segments of the set $(Y, \leqslant)$. Then $f_1 = f_2$.

41. Show that, for any two well-ordered sets $(X, <)$ and, $(Y, \leqslant)$, exactly one of them can be carried onto some segment of the other by a similarity function.

42. For any two well-ordered sets $(X, <)$ and $(Y, \leqslant)$, show that there is exactly one (to within, possibly, the passage to the inverse function) similarity function from one of them onto some segment of the other.

43. Show that, for each infinite set X, there exists a well ordering $\prec$, on X with respect to which X does not have a last element (that is, for each $x \in X$, the set $\{x' \in X : x \prec x' \text{ and } x \neq x'\}$ is non-void).

44. Show that, if A is an infinite set, A' a subset and $A \setminus A'$ consists of exactly one point, then $|A| = |A'|$.

45. A well-ordering of a set X is called *minimal* if it can be mapped into any other well-ordering of X by similarity function. Show that each set has a minimal well-ordering.

46. Show that, if a well-ordered set $(A, <)$ is similar to a well-ordered set $(B, \leqslant)$ and $<$ is minimal, then $\leqslant$ is also minimal.

47. Show that, for any minimal well-orderings $<$ and $\leqslant$ of a set X, there is a unique similarity function from $(X, <)$ onto $(X, \leqslant)$.

48. Show that the natural well-ordering of any set of natural numbers is minimal.

49. Show that, if $<$ is a minimal well-ordering of an infinite set X, then $(X, <)$ has no last element.

50. Let $(A, <)$ be a well-ordered set. Consider the ordering of the set $A \times A$ defined as follows.

Let $(x_1, y_1), (x_2, y_2) \in A \times A$ and let $\max\{x_1, y_1\} \neq \max\{x_2, y_2\}$.† If $\max\{x_1, y_1\} < \max\{x_2, y_2\}$, then put $(x_1, y_1) \lessdot (x_2, y_2)$ (correspondingly, if $\max\{x_2, y_2\} < \max\{x_1, y_1\}$, then put $(x_2, y_2) \lessdot (x_1, y_1)$). If $\max\{x_1, y_1\} = \max(x_2, y_2)$, then put $(x_1, y_1) \lessdot (x_2, y_2)$, provided $x_1 < x_2$ or $[y_1 < y_2$ and $x_1 = x_2]$. In the remaining cases put $(x_2, y_2) \lessdot (x_1, y_1)$.

Show that†† $(A \times A, \lessdot)$ is a well-ordered set.

51. Let $(N, <)$ be the set of natural numbers with its natural well-ordering, and let $(N \times N, \lessdot)$ be its Cantor square (see Problem 50). Show that $(N, <)$ and $(N \times N, \lessdot)$ are similar.

52. If the set A is infinite, show that A and $A \times A$ are equipotent.

53. Show that any two of the following sets are equipotent: (*a*) a straight line; (*b*) a plane; (*c*) an n-dimensional Euclideanspace, $n = 1, 2, 3, \ldots$; (*d*) a segment; (*e*) a square; (*f*) a ball; (*g*) a sphere; (*h*) an interval.

54. Show that, if $(A, <)$ is an infinite well-ordered set and $<$ is minimal, then the well-ordering $\lessdot$ of the Cantor square of $(A, <)$ is also minimal.

55. Show that, if $<$ is a minimal well-ordering of an infinite set A, then $(A, <)$ is similar to its Cantor square $(A \times A, \lessdot)$.

56. Show that, if $X = \prod_{i=1}^{k} X_i$ and $|X_i| \leqslant \tau$ for all $i = 1, \ldots, k$, where τ is some infinite cardinal number, then $|X| \leqslant \tau$.

57. Let X be a set and $\{X_\alpha : \alpha \in M\}$ some family of its subsets, where $|M| \leqslant \tau$ and $|X_\alpha| \leqslant \tau$ for all $\alpha \in M$. Show that $|\bigcup\{X_\alpha : \alpha \in M\}| \leqslant \tau$.

58. Show that the set $\mathcal{K}(A)$ of all finite subsets of any fixed infinite set A is equipotent with A.

59. Let X be an infinite set. Show that there are X_1 and X_2 such that $X = X_1 \cup X_2$, $X_1 \cap X_2 = \Lambda$ and $|X_1| = |X| = |X_2|$.

60. Let X be an infinite set and let $|X| = \tau$. Show that X can be represented as the union of a pairwise disjoint family $\mathcal{F}$ of subsets of X, such that $|\mathcal{F}| = \tau$ and each member of $\mathcal{F}$ has cardinality τ.

†By $\max\{x, y\}$ we denote the largest of the elements x, y of a linearly ordered set.

††We agree to call the set $A \times A$ together with the well-ordering $\lessdot$ the *Cantor square* of the well-ordered set $(A, <)$.

61. Verify that the sum of disjoint well-ordered sets over a well-ordered set (see p. 11) is well-ordered.

62. Show that, if $(A, <)$ and $(B, \preccurlyeq)$ are disjoint infinite well-ordered sets and their sum $(A, <) + (B, \preccurlyeq)$ is a minimal well-ordered set, then $|A| < |B| + \aleph_0$ and $(B, \preccurlyeq)$ is a minimal well-ordered set. Show that the converse also holds.

63. Show that the operation of addition of well-ordered sets is not commutative.

64. Verify that the multiplication of linearly ordered sets is associative: $((A, <) \times (B, \preccurlyeq)) \times (C, \prec)$ is similar to $(A, <) \times ((B, \preccurlyeq) \times (C, \prec))$ for any linearly ordered sets $(A, <), (B, \preccurlyeq)$ and $(C, \prec)$.

65. Let $(A, <)$ and $(B, \preccurlyeq)$ be two well-ordered sets and $(C, \prec) = (A, <) \times (B, \preccurlyeq)$ be their product. Show that $(C, \prec)$ is a well-ordered set.

66. Prove that† the product $(D, <) \times (N, <)$ is similar to $(N, <)$.

67. Prove an assertion similar to the one formulated in Problem 66, replacing $(N, <)$ by an arbitrary well-ordered set having no last element.

68. True or false ? $(N, <) \times (D, <)$ and $(N, <)$ are similar.

69. Show that the multiplication of well-ordered sets is not commutative.

70. Let α, β, γ be ordered sets. Show that we always have $\alpha \times (\beta + \gamma) = (\alpha \times \beta) + (\alpha \times \gamma)$ and that $(\beta + \gamma) \times \alpha \neq (\beta \times \alpha) + (\gamma \times \alpha)$ is possible.

71. Show that the (lexicographic) well-ordering of the square $(A, <) \times (A, <)$ of an infinite well-ordered set $(A, <)$ is never minimal.

72. Show that the Cantor square and the square of a (lexicographically) well-ordered set seldom coincide.

73. Let $(A, <)$ and $(B, \preccurlyeq)$ be two well-ordered sets and let $(C, \prec) = (B, \preccurlyeq)^{(A, <)}$. Show that $(C, \prec)$ is well-ordered.

74. Consider the well-ordered set $(D, <)^{(D, <)}$.

75. Prove that, if the well-ordered sets $(A, <)$ and $(B, \preccurlyeq)$ consist of a finite number of elements, then $(A, <)^{(B, \preccurlyeq)}$ is similar to the well-ordered set of natural numbers less than $|A|^{|B|}$.

76. Show that, if the ordered sets $(A, <)$ and $(B, \preccurlyeq)$ are infinite, then the cardinal number of the set of elements of the power $(A, <)^{(B, \preccurlyeq)}$ does not exceed the larger of the cardinal numbers $|A|, |B|$.

77. Show that, if $(A, \preccurlyeq)$ is linearly ordered, then $(A, \preccurlyeq)^{(D, <)} = (A, \preccurlyeq) \times (A, \preccurlyeq)$.

---

†For notation, see p. 12.

78. Show that for any well-ordered sets α, β and γ, we have $\alpha^{\beta+\gamma} = \alpha^{\beta} \times \alpha^{\gamma}$.

79. Produce a well-ordered set $(A, \leqslant)$ of cardinality $\aleph_0$ such that $(D, <)^{(A, <)}$ is similar to $(A, \leqslant)$.

80. *The principle of transfinite induction.* Let W be a well-ordered set and $Q \subset W$. If for each $x \in W$, $I(x) \subset Q$ implies $x \in Q$, then $Q = W$.

81. *Practical formulation of proof by the principle of transfinite induction.* Let W be a well-ordered set and to each $x \in W$ let there be assigned some proposition $P(x)$. If the following statements hold, then $P(x)$ is true for every $x \in W$:

(1) $P(0)$ is true (where 0 is the first element of W); and
(2) for each $x \in W$, if $P(\alpha)$ is true for every $\alpha \in I(x)$, then $P(x)$ is also true.

82. *Construction by the principle of transfinite recursion.* Let W be a well-ordered set, S some class and for each $x \in W$ suppose we have a function K_x, which assigns to each function f of $I(x)$ into S some element $K_x(f)$ of S. Then there exists a unique function φ: $W \rightarrow S$ such that $\varphi(x) = K_x(\varphi \mid I(x))$ for all $x \in W$.

1.5. Properties of Cardinal Numbers

We define addition and multiplication of cardinal numbers by taking unions and powers in such a way that for natural numbers they are identical to the usual arithmetic operations. The arithmetic of cardinal numbers is highly distinctive, and its value for set-theoretic topology immense. Cardinal numbers enter essentially and non-trivially into many important definitions, structures and formulations of theorems in general topology.

A cardinal number is *infinite* if it is the cardinality (power) of an infinite set. Otherwise it is *finite*. A finite cardinal number is naturally identified with a natural number, which we conceive as the cardinality of the set of natural numbers smaller than it. In particular, the empty set is said to have cardinal number 0. We agree to confine our attention to the ordering $<$ of the class of cardinal numbers introduced at the beginning of this book; no other ordering is to be considered. An infinite countable set N is said to have cardinal number $\aleph_0$. On solving Problem 84, it will be immediately clear that each nonempty set of cardinal numbers has a smallest element. This fact is of crucial importance to the whole of mathematics. In accordance with what has been stated, we denote by τ^+, for an arbitrary cardinal number τ, the smallest of all cardinal numbers, strictly greater than τ.

Let λ and τ be two cardinal numbers. Fix disjoint sets A and B, satisfying $|A| = \lambda$, $|B| = \tau$. Then the cardinal number $\lambda + \tau$ is the *sum of cardinal numbers* λ and τ, and is defined to be the cardinal number of the set $A \cup B$. The *product* $\lambda\tau$ is the cardinal number of the product $A \times B$ of the sets A and B. The cardinal number λ^τ which reads 'λ to the (power of) τ' is, by definition, the cardinal number of the set A^B. It is clear that the arbitrariness in the choice of the sets A and B does not imply any ambiguity in the definitions of the sum, product and power of cardinal numbers.

The *generalized continuum-hypothesis* says that $2^\tau = \tau^+$ for every infinite cardinal number τ. The *continuum-hypothesis* in the classical sense asserts $2^{\aleph_0} = \aleph_1$ (where $\aleph_1 = \aleph_0^+$; this is conventional notation). P. J. Cohen has proved that the continuum-hypothesis (and the generalized continuum-hypothesis) is independent of the usual axioms of set theory.

Given a cardinal number τ, we often consider the *derived* (from τ) *number* $cf(\tau)$. By definition, $cf(\tau)$ is the smallest cardinal number λ, such that τ can be represented as the sum of some set M, $|M| = \lambda$, of cardinal numbers each of which is less than τ (the term 'less than', in the case of cardinal numbers, always implies 'strictly less than'). We ought just to explain what is meant by the sum of an infinite set of cardinal numbers. Let A be some set and let there be given for each $\alpha \in A$ the cardinal number τ_α. Assume that τ_α is the cardinal number of some set X_α; in addition, the sets X_α can be chosen for each $\alpha \in A$ such that for $\alpha' \neq \alpha''$ we always have $X_{\alpha'} \cap X_{\alpha''} = \Lambda$. (It suffices to pass from X_α to their copies $i_\alpha(X_\alpha)$ lying in the free (disjoint) sum of the sets X_α over all $\alpha \in A$.) Put $X = \bigcup \{X_\alpha: \alpha \in A\}$ and call the cardinal number of X the *sum of the cardinal numbers* τ_α over all $\alpha \in A$, written $\Sigma\{\tau_\alpha: \alpha \in A\}$. It is clear that this definition is unambiguous. Similarly, the product of an arbitrary infinite set of cardinal numbers can be defined.

A cardinal number τ is called *isolated* or a *successor*, if $\tau = \lambda^+$ for some cardinal number λ. If τ is not isolated, then it is called a *limit*. A cardinal number τ is called *regular*, if $\tau = cf(\tau)$. We say that τ is a *strong limit* cardinal, if for any λ, $\lambda < \tau$ implies that $2^\lambda < \tau$. A regular strong limit cardinal number is called *strongly inaccessible*.

83. Let λ and τ be two arbitrary cardinal numbers. Show that the following always hold:

(*a*) $\lambda + \tau = \tau + \lambda$;
(*b*) $\lambda \cdot \tau = \tau \cdot \lambda$.

84. Show that each nonempty set of cardinal numbers has a smallest element.

85. Let $(X, <)$ be a well-ordered set of cardinality τ and λ a cardinal number, $\lambda < \tau$. Show that there exists $x \in X$ such that $|I(x)| = \lambda$.

86. Show that the cardinality of the set $\mathcal{E}$ of all cardinal numbers which are less than a given cardinal number τ does not exceed τ.

87. Let τ be the cardinal number of some infinite set. Show that

(*a*) $\tau + \tau = \tau$;
(*b*) $\tau \cdot \tau = \tau$;
(*c*) if $\lambda \neq \tau$, then $\lambda \cdot \tau$ is equal to the larger of the cardinal numbers λ, τ.

88. Show that if $\tau_\alpha \leqslant \tau$ for each $\alpha \in A$ and $|A| \leqslant \tau$, then $\Sigma\{\tau_\alpha : \alpha \in A\} \leqslant \tau$.

89. Show that each cardinal number τ is the sum of all preceding cardinal numbers.

90. Let $\tau = \Sigma\{\tau_\alpha : \alpha \in A\}$. Show that τ is equal to the larger of the two cardinal numbers: $\sup\{\tau_\alpha : \alpha \in A\}$ and $|A|$.

91. In what follows, we shall always denote by $\aleph_1$ the smallest cardinal number which is greater than $\aleph_0$. Let $T_{\aleph_1}$ (or $T(\omega_1)$) be a set of cardinality $\aleph_1$, assigned a minimal well-ordering $<$ (all such sets are similar by virtue of Problem 47). Show that for any countable set A, there is an element $b \in T_{\aleph_1} \setminus A$ such that $a < b$ for all $a \in A$.

92. Prove that the following two definitions of the cardinal number 2^τ are equivalent:

(*a*) 2^τ is the cardinal number of $\exp X$, the set of all subsets of a set X such that $|X| = \tau$;
(*b*) 2^τ is the cardinal number of D^X, the set of all functions from a set X such that $|X| = \tau$ into a set D with exactly 2 elements, zero and unity.

93. Show that $\tau < 2^\tau$ for every cardinal number τ.

94. Show that there exists no largest cardinal number.

95. Let λ and τ be two infinite cardinal numbers. We always have $\lambda^{\tau} = \tau^{\lambda}$. True or false?

96. Let λ, μ and τ be cardinal numbers. Show that, if $\lambda < \mu$, then $\lambda^{\tau} \leqslant \mu^{\tau}$ and $\tau^{\lambda} \leqslant \tau^{\mu}$.

97. For any three cardinal numbers λ, μ, τ, show that the following relations hold: $\lambda^{\mu+\tau} = \lambda^{\mu} \cdot \lambda^{\tau}$, $(\lambda^{\mu})^{\tau} = \lambda^{\mu \cdot \tau}$ and $(\lambda \cdot \mu)^{\tau} = \lambda^{\tau} \cdot \mu^{\tau}$.

98. Show that a set of all countable subsets of (the points of) a straight line has the cardinal number $2^{\aleph_0}$.

99. Show that $\aleph_1^{\aleph_0} = 2^{\aleph_0}$.

100. Show that, if $\mu \geqslant \aleph_0$ and $2 \leqslant \lambda \leqslant \mu$, then $\lambda^{\mu} = 2^{\mu}$.

101. Let λ, μ and τ be cardinal numbers. We always have $(\lambda^{\mu})^{\tau} = \lambda^{(\mu^{\tau})}$. True or false?

102. Show that $\aleph_0$ is strongly inaccessible.

103. Give an example of a strong limit number other than $\aleph_0$.

104. True or false? If a cardinal number λ is less than a cardinal number τ, then $2^{\lambda} < 2^{\tau}$.

105. Show that, if $X = \bigcup \{X_\alpha : \alpha \in A\}$, where $|A| < |X|$ and $|X_\alpha| < |X|$ for each $\alpha \in A$, then $X = \bigcup \{Y_\beta : \beta \in B\}$, where $|B| \leqslant |A|$ and $|A| < |Y_\beta| < |X|$ for all $\beta \in B$, and $|Y_{\beta'}| \neq |Y_{\beta''}|$, whenever $\beta' = \beta''$.

106. Let τ be a cardinal number and let $(X, <)$ be a minimally well-ordered set of cardinality τ. Show that $cf(\tau)$ is the smallest of the cardinal numbers λ such that in $(X, <)$ there is a cofinal subset of cardinality λ. (The subset $M \subset X$ is *cofinal* in $(X, <)$ if for each $x \in X$ there exists $a \in M$ satisfying $x < a$.)

107. Show that, for any cardinal number τ, if $\lambda = cf(\tau)$, then $cf(\lambda) = \lambda$.

108. Compute $cf(\tau^*)$ for the cardinal number τ^*, defined in the solution of Problem 103.

109. Show that every isolated infinite cardinal number is regular.

110. Let A and X_α, where $\alpha \in A$, be two sets with $|A| = \tau$ and $|X_\alpha| \leqslant \lambda$ for each $\alpha \in A$. Show that the cardinal number of the product $X = \Pi\{X_\alpha : \alpha \in A\}$ does not exceed λ^{τ}.

111. Let τ and λ be two cardinal numbers, where $\lambda < cf(\tau)$ and the number τ is a limit. Show that $\tau^{\lambda} = \Sigma\{\mu^{\lambda} : \mu < \tau\}$.

112. Prove the following assertion, assuming the generalized continuum hypothesis: if τ and λ are cardinal numbers, where $1 \leqslant \lambda < cf(\tau)$, then $\tau^{\lambda} = \tau$.

113. Let M be a set and for each $\alpha \in M$ let there be given sets A_α and B_α such that $|A_\alpha| < |B_\alpha|$. Show that the cardinality of the union of all A_α,

$\alpha \in M$, is strictly less than that of the product of all B_α, $\alpha \in M$.

114. Suppose given, for each $\alpha \in M$, a set A_α satisfying the conditions:

(i) if $\alpha' \neq \alpha''$, $\alpha' \in M$ and $\alpha'' \in M$, then $|A_{\alpha'}| \neq |A_{\alpha''}|$;
(ii) the set $\{|A_\alpha|: \alpha \in M\}$ has no largest element.

Show that $|\bigcup\{A_\alpha: \alpha \in M\}| < |\Pi\{A_\alpha: \alpha \in M\}|$.

115. Show that, for any cardinal number τ, we have $\tau < cf(2^\tau)$.

116. Show that a segment in the real line cannot be represented as the union of a countable family of sets of smaller cardinality.

117. Given sets A_1, A_2, B_1 and B_2 such that

(*a*) $A_1 \cap A_2 = \Lambda$; (*b*) $B_1 \cap B_2 = \Lambda$; (*c*) $|A_1 \cup A_2| = |B_1 \cup B_2|$;
(*d*) $|A_1| = |A_2|$; (*e*) $|B_1| = |B_2|$.

Show that $|A_1| = |B_1|$ (and $|A_2| = |B_2|$).

118. Show that there is a family $\mathcal{E}^*$ of subsets of the set N of natural numbers such that: (*a*) if $A \in \mathcal{E}^*$, $B \in \mathcal{E}^*$ and $A \neq B$, then $|A \cap B| < \aleph_0$; (*b*) for each $A \in \mathcal{E}^*$, we have $|A| = \aleph_0$; (*c*) $|\mathcal{E}^*| = 2^{\aleph_0}$; (*d*) if $C \subset N$ and $|C| = \aleph_0$, then $|A \cap C| = \aleph_0$ for some $A \in \mathcal{E}^*$.

119. Let $(B, <)$ be a minimally well-ordered set of cardinality τ^+. Let $\alpha_0 \in B$ be such that $|\{\alpha \in B: \alpha < \alpha_0\}| = \tau$, $C = \{\alpha \in B: \alpha_0 \leqslant \alpha\}$ and let $f: C \to B$ be a function such that $f(\alpha) < \alpha$ for all $\alpha \in C$ (in other words, f is an arbitrary element of the product $\Pi\{B_\alpha: \alpha \in C\}$, where $B_\alpha = \{\beta \in B: \beta < \alpha\}$). Show that there exist $\alpha' \in C$ and $\alpha'' \in C$ satisfying $\alpha' \neq \alpha''$, but $f(\alpha') = f(\alpha'')$.

120. Let τ be a cardinal number, $\tau \geqslant \aleph_0$, $\lambda = 2^\tau$, X a set with $|X| > \lambda$, and M the set of all unordered pairs of distinct elements of X. Let f be a function from M into some set T of cardinality τ. Show that there exists $X^* \subset X$ such that $|X^*| > \tau$ and $f(x, y) = f(x', y')$ for any $x, y, x', y' \in X^*$, where $x \neq y$ and $x' \neq y'$.

121. Let A be a set, τ a cardinal number, $\tau \geqslant \aleph_0$ and $|A| > 2^\tau$. Assume that a set M of all unordered pairs of distinct elements of A is decomposed into a sum of pairwise disjoint sets to be such that for each $a \in A$ the set $\{(a, x): x \in A \setminus \{a\}\}$ is contained in the union of a family of cardinality τ of these summands. Show that then there exists $A^* \subset A$, $|A^*| > \tau$ such that the set M^* of all unordered pairs of distinct elements of A^* lies in one of the summands.

122. Given on a set Y a family $\mathcal{E} = \{\gamma_\alpha: \alpha \in A\}$ of collections γ_α of subsets of Y, satisfying the conditions:

(i) if $\alpha', \alpha'' \in A$ and $\alpha' \neq \alpha''$, then there exist $F^{\alpha'} \in \gamma_{\alpha'}$ and $F^{\alpha''} \in \gamma_{\alpha''}$ such that

$$F^{\alpha'} \cap F^{\alpha''} = \Lambda;$$

(ii) each γ_α contains an intersection of its any two elements;
(iii) $|\gamma_\alpha| \leqslant \tau$, and $|A| > 2^\tau$ for all $\alpha \in A$ and some cardinal number τ.

Show that there can be selected a set $A^* \subset A$, and for each $\alpha \in A^*$ a set $F^\alpha \in \gamma_\alpha$ so that the following conditions are satisfied: (iv) $|A^*| > \tau$, and (v) if $\alpha' \in A^*$, $\alpha'' \in A^*$ and $\alpha' \neq \alpha''$, then $F^{\alpha'} \cap F^{\alpha''} = \Lambda$.

1.6. Prefilters, Filters and Ultrafilters. Centralized and Maximal Centralized Families of Sets

Before we proceed to solve the problems of this section and draw conclusions from them, which are extensively applied in many structures and arguments of general topology, the reader is strongly advised to revert to the theoretical introduction given in the beginning of the chapter and carefully read the concluding portion of Sec. 3. The present section further makes explicit only a restricted number of especial concepts.

An ultrafilter defined on a set X is called a *σ-ultrafilter*, if the intersection of every countable set of its elements is not empty. The cardinality of X is called a *measurable cardinal number*, if on X there is a non-trivial (that is, free) ultrafilter.

Let ξ_1 and ξ_2 be two filters defined on X_1 and X_2, respectively. We say that ξ_1 and ξ_2 are *equivalent*, if there is a one-to-one function f of X_1 onto X_2 such that† $f\xi_1 = \xi_2$. The filters ξ_1 and ξ_2 are called *weakly equivalent* if there exist $A_2 \in \xi_2$ and $A_1 \in \xi_1$ such that the restrictions $\xi_1 | A_1$ and $\xi_2 | A_2$ (considered as the filters defined on A_1 and A_2, respectively) are equivalent.

1.6.1. *Problems*

123. Show that if ξ is a prefilter on X, then $\tilde{\xi} = \{B \subset X$: there exists $A \subset \xi$, satisfying $A \subset B\}$ is a filter. We also say that ξ is a filter base for $\tilde{\xi}$.

124. Every centralised family ξ of subsets of X is contained in some filter on X.

† $f\xi_1 = \{f(A)\colon A \in \xi_1\}$.

125. Let ξ be a prefilter on X and B a subset of X, satisfying $A \cap B \neq \Lambda$, whatever $A \in \xi$. Show that then $\xi' = \xi \cup \{B\}$ is a centralized system of sets (or, a system of sets with finite intersection property).

126. Show that every maximal centralized system of sets is a filter. Verify that the converse is false.

127. Let X be a set and $\mathcal{E}$ some family of nonempty subsets of X. Show that every centralized family of elements of $\mathcal{E}$ is contained in a maximal centralized family of elements of $\mathcal{E}$.

128. Show that the following assertions are equivalent:

(*a*) ξ is a maximal centralized family of subsets of X;
(*b*) ξ is an ultrafilter on X;
(*c*) ξ is a prefilter and for each $A \subset X$, $A \in \xi$ implies that there exists $B \in \xi$, satisfying $A \cap B = \Lambda$;
(*c'*) ξ is a prefilter and for each $A \subset X$, if $A \cap B \neq \Lambda$ for every $B \in \xi$, then $A \in \xi$;
(*d*) for each $A \subset X$ either $A \in \xi$ or $X \setminus A \in \xi$ and ξ is a centralized family of sets;
(*e*) ξ is a centralized family of sets on X, $\xi \ni X$, and $A_1 \cup A_2 \in \xi$ implies that either $A_1 \in \xi$ or $A_2 \in \xi$.

129. Let X be an infinite set and let $\eta(X) = \{X \setminus A : A \subset X \text{ and } |A| < |X|\}$. Show that $\eta(X)$ is a filter, but not an ultrafilter.

130. Let X be a set. Show that every filter ξ on X is contained in some ultrafilter on X.

131. An ultrafilter ξ on X is trivial iff $\{x\} \in \xi$ for some point $x \in X$.

132. Let ξ be a filter on X and $X' \subset X$.

Show that: (*a*) $\xi \mid X'$ is a filter on X' iff $A \cap X' \neq \Lambda$ for all $A \in \xi$; (*b*) if ξ is an ultrafilter, then $\xi \mid X'$ is an ultrafilter on X' iff $X' \in \xi$.

133. The restriction $\xi \mid A$ of the ultrafilter ξ on any $A \in \xi$ is a subfamily of ξ.

134. Let X be a set, x some element of X and A some subset of X. Introduce the following notations :

$$\xi(x) = \{B \subset X : B \ni x\},$$
$$\xi(A) = \{B \subset X : B \supset A\}.$$

Show that

(i) $\xi(x)$ is a singular ultrafilter;
(ii) $\xi(A)$ is a singular filter;

(iii) $\xi(A)$ is an ultrafilter iff A consists of exactly one point;
(iv) X is infinite iff there is a free ultrafilter on X.

135. Let X and Y be two sets, $f: X \to Y$ a function, ξ a family of sets on X and η a family of sets on Y. Show that:

(*a*) if ξ is a prefilter, then $f\xi$ is also so;
(*b*) if ξ is a filter, then $f\xi$ may not be a filter (even for $fX = Y$);
(*c*) if η is a prefilter, then $f^{-1}\eta$ is also so;
(*d*) if η is a filter, then $f^{-1}\eta$ may not be a filter (even for $fX = Y$);
(*e*) if ξ is an ultrafilter and $fX = Y$, then $f\xi$ is also an ultrafilter;
(*f*) if f is a one-to-one function of X onto Y, then ξ is a filter (corresp. prefilter, ultrafilter) iff $f\xi$ is a filter (corresp. prefilter, ultrafilter).

136. The restriction of a σ-ultrafilter ξ defined on a set X by the set $X_0 \subset X$ is a σ-ultrafilter on X_0, if $X_0 \in \xi$.

137. If ξ is an ultrafilter on X and τ a cardinal number, $\tau \geqslant \aleph_0$, where for each family $\xi' \subset \xi$ whose cardinality does not exceed τ, we have $\cap \{A: A \in \xi'\} \neq \Lambda$, then for every ξ', $|\xi'| \leqslant \tau$, we have $\cap \{A: A \in \xi'\} \in \xi$.

138. If $X \subset Y$ and on a set X there exists a non-trivial σ-ultrafilter, then a non-trivial σ-ultrafilter exists also on Y.

139. If $\tau_1 < \tau_2$ and τ_1 is a measurable cardinal number, then the cardinal number τ_2 is also measurable.

140. If ξ is a non-trivial σ-ultrafilter on X and $\xi' \subset \xi$, $|\xi'| \leqslant \tau$, $\cap \{A: A \in \xi'\} = \Lambda$, then the cardinal number τ is measurable.

141. Let A and X be two sets, where $X = \cup \{X_\alpha: \alpha \in A\}$ and $|A| = \tau$, and $|X_\alpha| = \tau_\alpha$ for each $\alpha \in A$ be nonmeasurable cardinal numbers. Then, the cardinal number of X is also non-measurable.

142. If the cardinal number of X is nonmeasurable, then that of exp X is also so.

143. The smallest measurable cardinal number (if existing) is a strongly inaccessible cardinal number other than $\aleph_0$.

144. The cardinal number of a collection of all distinct (that is, different from one another even if there is only a single element) maximal centralised families of subsets of X is equal to $2^{2^{|X|}}$.

145. Show that there exist non-equivalent free ultrafilters on a set of natural numbers.

146. Show that on a set of natural numbers there exist free ultrafilters which are not weakly equivalent (cf. p. 20). What is the cardinality of a family of all pairwise nonweakly equivalent free ultrafilters on a set of natural numbers? (Compare with Prob. 144.)

1.7. SOLUTIONS

6. 2^n, (prove by induction; recall that an empty set is an element of $\exp A$).

7. The appropriate example is $A = \{\Lambda\}$ or $B = \{\Lambda, \{\Lambda\}\}$, and so on.

8. Put $a_1 = \{\Lambda\}$, $a_2 = \{\Lambda, \{\Lambda\}\}$, $a_3 = \{\Lambda, \{\Lambda\}, \{\{\Lambda\}\}\}, \ldots$ Then $a_i \in a_j$ for $i < j$ and $A = \{a_1, \ldots, a_n\}$ is the set we sought.

9. This is left as an excercise for the reader (see next problem).

10. Let $<$ be a well-ordering on X. In $(X, <)$ there is a smallest element, which we denote by x^*. Then x^* is a largest element in $(X, \leqslant)$. If X is infinite, then the set $A = \{y: y \in X \text{ and } \tilde{I}_{\leqslant}(y) \text{ is infinite}\}$ is nonempty (because $x^* \in A$). Let a be a smallest element of A (relative to $\leqslant$). Then $I_{\leqslant}(a)$ is an infinite set which, as is easy to see, has no smallest element relative to the ordering $<$ of elements. A contradiction. The rest is clear.

11. Let X be infinite and $\leqslant$ a well-ordering relation on X. If in X there is a maximal element, then it is only one. In this case, denote this by ξ and put $X' = X \setminus \{\xi\}$. In other words, assume $X' = X$. For each $x \in X'$, the element $x + 1 \in X$ is defined. The function $\varphi : X' \to X$ defined by the rule $\varphi(x) = x + 1$ is one-to-one. If $X' = X$, then $\varphi: X \to X$ is one-to-one correspondence between X and the characteristic subset $\varphi(X)$ of X, because a smallest element η of X does not belong to $\varphi(X)$. If $X' \neq X$, then $X \setminus X' = \{\xi\}$. Put $\varphi(\xi) = \eta$. Then, again $\varphi: X \to X$ is one-to-one. Show that $\varphi(X) \neq X$. The set $I_{\leqslant}(\xi)$ is infinite (because X is infinite, and ξ is a maximal element). Put $B = \{y \in X: I_{\leqslant}(y) \text{ is infinite}\}$. Then $\xi \in B$, and hence $B \neq \Lambda$. We call y_0 a minimal element of B. No matter what $z \in X$ is, surely $y_0 \neq z + 1$: if $y_0 = z + 1$, then $I_{\leqslant}(y_0) = I_{\leqslant}(z) \cup \{z\}$ and again $I_{\leqslant}(y_0)$ is infinite, so that $I_{\leqslant}(z)$ is also infinite, contradicting the choice of y_0, because $z < y_0$ and $z \neq y_0$. However, then, $y_0 \notin \varphi X$, that is, $\varphi(x)$ is the characteristic subset of X. The rest is clear.

12. Let $X' \subset X$. If X is finite, then X' is also so. If X is infinite but countable, then there exists a one-to-one function $\varphi: X \to N^+$ of X into the set N^+ of positive integers. Put $N' = \varphi(X')$, $N' \subset N^+$. It suffices that N' carries onto N^+ in a one-to-one fashion. On N^+ and N' there is defined a natural order $\leqslant$. For an arbitrary $x' \in N'$ put $\psi(x') = |\{y' \in N': y' \leqslant x'\}|$, interpreting now positive integers as the cardinal numbers of finite nonempty sets. It is easy to prove that $\psi: N' \to N^+$ is a one-to-one function of N' onto N^+ (it is necessary to use the fact that $\leqslant$ is a well-ordering).

15. Map X one-to-one onto N, the set of all natural numbers and consider in X the inverse image of a set N_0 of all even numbers (denote it by X_0) and the inverse image of a set N_1 of all odd numbers (denote it by X_1). X_0 and X_1 are the desired sets (cf. Prob. 12).

16. It can be assumed that $A \cap B = \Lambda$. (In the contrary case replace B by $B' = B \setminus A$; by virtue of Prob. 12, B' is countable. In addition, $A \cup B = A \cup B'$.) Analyze just the case when A and B are infinite. Denote by N_0 (resp. N_1) a set of all even (resp. odd) numbers. By Prob. 12 even N_0 and N_1 are carried over one-to-one onto the set N^+ of all natural numbers. Hence, A can be carried over one-to-one onto N_0 and B onto N_1. However, then the union of these mappings (in the usual sense) is a one-to-one function of $A \cup B$ onto N.

17. The set $X \setminus A$ is uncountable (Prob. 16). Hence, in $X \setminus A$ an infinite countable subset C can be selected (Prob. 13). The infinite set $A \cup C$ is countable (Prob. 16);

there exists a one-to-one function of C onto $A \cup C$. Uniting this function with the identity function of the set $X \setminus (A \cup C)$ into itself, we obtain the desired one-to-one function of $X \setminus A$ onto X.

18. We shall assume that to each $n \in N$ there corresponds some finite set A_n (without any risk, if $A_n = \Lambda$). It can be assumed that $A_{n_1} \cap A_{n_2} = \Lambda$ for $n_1 \neq n_2$; otherwise, it would be necessary to pass onto (finite) sets $A_n = A_n \setminus (\bigcup \{A_k : k < n\})$, satisfying

$$\bigcup \{A'_n : n \in N\} = \bigcup \{A_n : n \in N\}.$$

Define on $A = \bigcup \{A_n : n \in N\}$ the order releation $\preccurlyeq$ as follows. Fix some well-ordering $\preccurlyeq_n$ on each A_n. If $x, y \in A_n$, then put $x \preccurlyeq y$ iff $x \preccurlyeq_n y$. If there exists no n such that $x, y \in A_n$, then put $x \preccurlyeq y$ if $n_1 < n_2$, where $A_{n_1} \ni x$, $A_{n_2} \ni y$. It is trivial to verify that $\preccurlyeq$ well-orders A, and for each $x \in A$, $\tilde{I}_{\preccurlyeq}(x)$ is a finite set. Of our interest is only the case when the entire A is infinite. Define a function $\varphi: A \to N$ by the rule: $\varphi(x) = |\tilde{I}_{\preccurlyeq}(x)|$. Using the fact that $(N, <)$ and $(A, \preccurlyeq)$ are well-ordered sets, show that φ is a one-to-one function of A onto N.

19. It suffices to analyze the case (see Probs. 18 and 12) when to each integer $n > 0$ there corresponds a countable infinite set A_n obeying the condition: $A_{n_1} \cap A_{n_2} = \Lambda$ for $n_1 \neq n_2$. Fix for each $n \in N^+$ some one-to-one function φ_n of N^+ onto A_n. Put $\varphi_n(k) = x_{nk}$. Then $A = \bigcup \{A_n : n \in N^+\} = \{x_{nk} : n \in N^+, k \in N^+\}$. Put $B_l = \{x_{nk} : n + k = l\}$ for each $l \in N^+ \setminus \{1\}$. Then $\bigcup \{B_l : l \in N^+, l \neq 1\} = \{x_{nk} : n \in N^+, k \in N^+\} = A$, and every B_l is a finite set. It remains to draw upon the conclusion of Prob. 18.

20. Let A_1, A_2 be countable sets and φ_1, φ_2 be one-to-one functions $\varphi_1: A_1 \to N$, $\varphi_2: A_2 \to N$. Then $\varphi_1 \times \varphi_2: A_1 \times A_2 \to N \times N$ is also a one-to-one function. Hence, it suffices to establish that $N \times N$ is countable. Put $A_n = \{(n, k) : k \in N\}$ for each $n \in N$. Then A_n is countable, because $p_n: A_n \to N$, where $p_n(n, k) = k$ for each $k \in N$, is a one-to-one function of A onto N. Obviously, $\bigcup \{A_n : n \in N\} = N \times N$, that is, $N \times N$ is a countabie set (cf. Prob. 19). Now, by induction, the result is extended to the case of every finite family of countable factors.

21. Let $f: X \to Y$ be a function and X a countable set. For each $y \in f(X)$, fix $x(y) \in f^{-1}(y)$. Put $\varphi(y) = x(y)$; then $\varphi: f(X) \to X$ is a one-to-one function of the set $f(X)$ onto $\varphi f(X)$, which is countable as a subset of a countable set (cf. Prob. 12).

22. (*a*) Consider the product $N^+ \times N^+$ of the set of all positive integers by itself. For $(m, n) \in N^+ \times N^+$ put $\chi(m, n) = m/n$. This χ is a function of a countable (cf. Prob. 20) set $N^+ \times N^+$ onto R^+, a set of positive rational numbers. Consequently, (cf. Prob. 21), R^+ is countable. Thereupon the set of negative rational numbers is also countable (on multiplying by (-1) a one-to-one correspondence is established between R^+ and R^-). Hence, the set $R^+ \cup \{0\} \cup R^-$ is also countable (cf. Prob. 16).

(*b*) It suffices to refer to Probs. 12, 20 and 22(*a*).

(*c*) Suppose that A is countable and $A^* = \{A_i : i = 1, 2, \ldots\}$, where $A_1 = A$; $A_2 = A \times A = A^2$ and, in general, $A_i = A \times A^{i-1}$. A^* is countable (cf. Probs. 19 and 20). To each element of A^* there corresponds naturally some finite subset of A, and thus a collection of all nonempty finite subsets in A is obtained, that is, a natural function of A^* onto a collection of all nonempty finite subsets of A. Consequently, the latter (a complete empty set as an element) is countable (cf. Probs. 21 and 16).

(*d*) The proof is the same as in Prob. 22(*a*).

(e) Cf. Probs. 22(a) and 20.

(f) This is implied by Prob. 22(c).

(g) Choose inside each of the circumferences, 8 such points whose both coordinates are rational. This yields, as is easy to see, a one-to-one function of a set of 8's contained in the family onto a countable set.

27. Let $f: C \to A \times B$ be a function and $\pi_A: A \times B \to A$, $\pi_B: A \times B \to B$ be projections, that is, $\pi_A(a, b) = a$, $\pi_B(a, b) = b$ for all $(a, b) \in A \times B$. Put $f_1 = \pi_A f$, $f_2 = \pi_B f$, $f_1: C \to A$, $f_2: C \to B$. Verify that the rule $\chi(f) = (f_1, f_2)$ establishes a one-to-one correspondence between the sets $(A \times B)^C$ and $A^C \times B^C$.

28. It suffices to note that by virtue of the relation $B \cap C = \Lambda$ a one-to-one correspondence holds between the functions $f: B \cup C \to A$ and a pair (f_1, f_2) of their contractions $f_1 = f \mid B$, $f_1: B \to A$, $f_2 = f \mid C$, $f_2: C \to A$.

29. Let $f \in A^{B \times C}$ (that is, $f: B \times C \to A$) and let $(b, c) \in B \times C$, $b \in B$, $c \in C$. Put $\varphi_c^f(b) = f(b, c) \in A$. For fixed $f \in A^{B \times C}$ and $c \in C$, this rule defines the function $\varphi_c^f: B \to A$ (thus, $\varphi_c^f \in A^B$); putting $\chi_f(c) = \varphi_c^f$, we obtain the function $\chi_f: C \to A^B$. Finally, the rule $\lambda(f) = \chi_f$ defines the function $\lambda: A^{B \times C} \to (A^B)^C$. It is found that λ is a one-to-one and 'onto' function. In fact, let $g: C \to A^B$ be a function. For an arbitrary element $(b, c) \in B \times C$ put $f_g(b, c) = (g(c))(b) \in A$. The rule $\mu(g) = f_g$ is defined by some function $\mu: (A^B)^C \to A^{B \times C}$. Verify that μ and λ are functions, inverse of each other.

30. We denote a triple by the collection $t = (A, f, B)$, where $A \subset X$, $B \subset Y$ and f is a one-to-one function of A onto B. Let T denote a set of all triples. Let $t = (A, f, B)$ and $t' = (A', f', B')$ be two arbitrary triples. Put $t < t'$ if $A \subset A'$, $B \subset B'$ and $f' \mid A = f$; $t = t'$, if $A = A'$, $B = B'$, $f = f'$. Verify that the ordered set $(T, <)$ is inductive. By the Zorn-Kuratowski principle, there exists in $(T, <)$ some maximal element $t^* = (A^*, f^*, B^*)$. If $A^* = X$ or $B^* = Y$, then the problem is solved. Otherwise, one can choose $x^* \in X \setminus A^*$ and $y^* \in Y \setminus B^*$. Putting $\tilde{A} = A^* \cup \{x^*\}$, $\tilde{B} = B^* \cup \{y^*\}$ and $\tilde{f}(x^*) = y^*$, $\tilde{f} \mid A^* = f^*$ yields the triple $\tilde{t} = (\tilde{A}, \tilde{f}, \tilde{B}) \in T$, where $t^* < \tilde{t}$, $t^* \neq \tilde{t}$, contradicting the maximality of t^* in $(T, <)$. Hence, either $A^* = X$ or $B^* = Y$.

31. It can be assumed that X and Y are disjoint sets. Otherwise, it would be necessary to multiply the set $X \cup Y$ by $\{0, 1\}$, a set consisting of two elements 0 and 1; then $X \times \{0\}$, $Y \times \{1\}$, the 'copies' of X and Y, are disjoint subsets of $(X \cup Y) \times \{0, 1\}$. Let ξ denote an object which is not an element of either X or Y (for example, it is appropriate to assume that $\xi = \{X \cup Y\}$ is a set whose only element is a set $X \cup Y$). Put $Z = X \cup Y$. For each $z \in Z$, define by (usual) induction a sequence $S(z) = \{z_i: i = 1, 2, \ldots\}$ of elements of the set $Z \cup \{\xi\}$, as follows. We always have $z_1 = z$. Let $z_i \in S(z)$ be already defined for every $i \leqslant n$. Then either $z_n \in X$, or $z_n \in Y$, or else $z_n = \xi$ (but these relations are not realized simultaneously). If $z_n \in X$ and $z_n \in g(Y)$, then we put $z_{n+1} = g^{-1}(z_n)$. If $z_n \in X$ and $z_n \notin g(Y)$, then we put $z_{n+1} = \xi$. If $z_n \in Y$ and $z_n \in f(X)$, then we let $z_{n+1} = f^{-1}(z_n)$. If $z_n \in Y$ and $z_n \notin f(X)$, then we let $z_{n+1} = \xi$. Finally, if $z_n = \xi$, then also we put $z_{n+1} = \xi$. Put $X_\infty = \{x \in X: S(x)$ does not contain $\xi\}$, $Y_\infty = \{y \in Y: S(y)$ does not contain $\xi\}$, $X_{odd} = \{x \in X \setminus X_\infty: \max\{n: x_n \neq \xi\}$ is an odd number$\}$, $Y_{odd} = \{y \in Y \setminus Y_\infty: \max\{n: y_n \neq \xi\}$ is an odd number$\}$, $X_{even} = \{x \in X \setminus X_\infty: \{\max n: x_n \neq \xi\}$ is an even number$\}$, $Y_{even} = \{y \in Y \setminus Y_\infty: \max\{n: y_n \neq \xi\}$ is an even number$\}$. Then it is easy to see that the following relations are true;

$X = X_{odd} \cup X_{even} \cup X_\infty$, $Y = Y_{odd} \cup Y_{even} \cup Y_\infty$, $fX_\infty = Y_\infty$, $fX_{odd} = Y_{even}$, $gY_{odd} = X_{even}$. Define $\varphi: X \to Y$ thus: $\varphi \mid X_\infty = f \mid X_\infty$, $\varphi \mid X_{odd} = f \mid X_{odd}$, $\varphi \mid X_{even} = g^{-1} X_{even}$. Verify that φ is a one-to-one function of X onto Y.

32. This is implied by Probs. 30 and 31.

34. Put $f(x) = \{x\}$ for each $x \in X$. Then $f(x) \in \exp X$ and f is a one-to-one function of X into $\exp X$. It is now required to establish that if $\varphi: X \to \exp X$ is a one-to-one function, then $\varphi(X) \neq \exp X$. Put $A = \{x \in X: x \notin \varphi(x)\}$. If $A \in \varphi(x)$, then $A = \varphi(x^*)$ for some $x^* \in X$ and either $x^* \in X$, or $x \notin A$. Show that both the alternatives lead to a contradiction. If $x^* \in A$, then $x^* \in \varphi(x^*)$ and by the definition of A, $x^* \notin A$, a contradiction. If $x^* \notin A$, then $x^* \notin \varphi(x)$ and by the definition of A, $x^* \in A$, a contradiction. Consequently, $A \notin \varphi(X)$.

If $X = \Lambda$ be an empty set, then the set $\exp X$ consists of exactly one element, and X has no element. Consequently, it is impossible to establish a one-to-one correspondence between X and $\exp X$.

35. If $k \in N^+$, $i \in N^+$ and $a \in S_{i+k}$, put

$$\varphi_k(a) = \varphi(\varphi_{k-1}(a)) = \underbrace{\varphi(\varphi(\ldots\varphi}_{k \text{ times}}(a))).$$

Further, $\varphi_k(S_{i+k}) = \{\varphi_k(a) : a \in S_{i+k}\}$. Since $S_{i+k} \neq \Lambda$, then we have also $\varphi_k(S_{i+k}) = \Lambda$. Obviously, $\varphi_k(S_{i+k}) \subset S_i$. Put $S_i^* = \bigcap \{\varphi_k(S_{i+k}) : k \in N^+\}$. From $\varphi(S_{i+k+1}) \subset S_{i+k}$ it follows that $\varphi_{k+1}(S_{i+k+1}) \subset \varphi_k(S_{i+k})$. Consequently, the sequence $\xi_i = \{\varphi_k(S_{i+k}): k \in N^+\}$ is monotonically decreasing (but, in general, not strictly). Thus the elements ξ_i form a nonempty finite set. Hence, $S_i^* = \varphi_{k^*}(S_{i+k^*})$ for some $k^* \in N^+$ and, therefore, $S_i^* \neq \Lambda$. Further, $\varphi(S_{i+1}^*) = \varphi(\bigcap\{\varphi_k(S_{i+k+1}): k \in N^+\}) \subset \bigcap\{\varphi_{k+1}(S_{i+k+1}): k \in N^+\} \subset S_i^*$. On the other hand, let $a \in S_i$ and $A = \{b \in S_{i+1}: \varphi(b) = a\}$. If $S_{i+1}^* \cap A = \Lambda$, then since ξ_{i+1} is a monotonic increasing family of finite sets, there exists $k' \in N^+$ satisfying $\varphi_{k'}(S_{i+k'+1}) \cap A = \Lambda$. But then $\varphi_{k'+1}(S_{i+k'+1}) \cap \varphi(A) = \Lambda$ (because A is a pre-image under φ), that is, $a \notin \varphi_{k'+1}(S_{i+k'+1})$ and, furthermore, $a \notin S_i^*$. This shows that $\varphi(S_{i+1}^*) \supset S_i^*$. Hence, $\varphi(S_{i+1}^*) = S_i^*$ for every $i \in N^+$. Now, the desired sequence $\{a_i: i \in N^+\}$ is easy to construct; select an arbitrary $a_1 \in S_1^*$ and, if every a_i is already defined for $i \leqslant n$ to be such that $a_i \in S_i^*$, then every element of the set S_{i+1}^* is taken as a_{n+1} satisfying $\varphi(a_{n+1}) = a_n$.

36. Let there exist a set $A \subset X$, $A \neq \Lambda$, which has no smallest element. Then in an obvious manner we construct by induction the sequence $\{a_n\} \subset A$ such that† $a_{n+1} < a_n$, $a_{n+1} \neq a_n$ for each n. Obviously, $\{a_n : n = 1, 2, \ldots\}$, a countable set which is not well-ordered by the relation $<$, is defined on X.

38. The assertion (*a*) is immediate from the definition of a fragment. This leaves (*b*) to prove. If A' is not contained in A, then $x' \notin A$ for some $x' \in A'$. Let x be an arbitrary element of A. The relation $x' \prec x$ cannot be true, otherwise $x' \in A$. Hence, since $\prec$ is a linear order, $x \prec x'$. But then $x \in A'$, since A' is a fragment permitting us to conclude that $A \subset A'$.

---

†$\{a_n\}$ is here the shorthand notation for $\{a_n: n \in N^+\}$.

39. If $A = \{x \in X: f(x) < x \text{ and } f(x) \neq x\} \neq \Delta$, then let x^* be the first element of A. Then $\tilde{x} = f(x^*) \neq x^*$ and $\tilde{x} < x^*$. Consequently, $\tilde{x} \notin A$. Hence the fact that $f(\tilde{x}) < \tilde{x} = f(x^*)$ is false. But f is a similarity function; therefore, $\tilde{x} < x^*$ implies that $f(\tilde{x}) < f(x^*)$. A contradiction, so that $A = \Delta$. Put $B = \{x \in X: f(x) \neq x\}$. Let $B \neq \Delta$, and let x_0 be a smallest element of B. Consider the set $I(x_0) = \{x \in X: x < x_0 \text{ and } x \neq x_0\}$. Then, $f(I(x_0)) = I(x_0)$, because for each $x \in I(x_0), f(x) = x$. Put $C = X \setminus (I(x_0) \cup \{x_0\})$. We have $f(x_0) \neq x_0$ and $x_0 < f(x_0)$; hence, $f(x_0) \in C$. Further, $f(C) \subset C$ (because $A = \Delta$). Consequently, $f(X) \not\ni x_0$, and $f(X) \ni f(x_0)$, where $x_0 < f(x_0)$. This contradicts that $f(X)$ is a fragment of the set $(X, <)$.

40. Evidently, either $f_1(X) \subset f_2(X)$, or $f_2(X) \subset f_1(X)$ (because $f_1(X)$ and $f_2(X)$ are fragments of the set $(Y, \leqslant)$. For definiteness, let $f_1(x) \subset f_2(X)$. Then $Y_1 = f(X_1)$ and $Y_2 = f(X_2)$ themselves are well-ordered as subsets of a well-ordered set $(Y, \leqslant)$, Y_1 being a fragment of $(Y_2, \leqslant)$. For each $y \in Y_2$ put $\varphi(y) = f_1 f_2^{-1}(y)$. As a superposition of two similarities: $f_2^{-1}: (Y_2, \leqslant) \to (X, <)$ and $f_1: (X, <) \to (Y_1, \leqslant)$, φ is a similarity function of $(Y_2, \leqslant)$ onto its fragment $(Y_1, \leqslant)$. However, then (cf. Prob. 39) $Y_1 = Y_2$ and φ is an identity function of the set Y_2 onto itself, that is, $\varphi(y) = y$ for every $y \in Y_2$. This implies that f_1 and f_2^{-1} are one-to-one; finally, $f_1(x) = f_2(x)$ for each $x \in X$.

41. Call $\mathcal{P}$ a family of all fragments of the set $(X, <)$ which (being endowed with well-ordering in $(X, <)$) admits a similarity mapping onto some fragment of the set $(Y, \leqslant)$. $\mathcal{P}$ has the natural well-ordering $\prec: A \prec B$, if $a \subset B$ for every $A, B \in \mathcal{P}$. Let us show that $(\mathcal{P}, \prec)$ is an inductive set. In the first place, if $A = \bigcup \{A_\alpha : \alpha \in M\}$, where $A_\alpha \in \mathcal{P}$ for each $\alpha \in M$, then A is a fragment of $(X, <)$ (cf. Prob. 38(*a*)). Now, let $f_\alpha : A_\alpha \to (Y, \leqslant)$ be a similarity function of the set $(A_\alpha, <)$ onto a fragment of $(Y, \leqslant)$ for each $\alpha \in M$. The image of a fragment under a similarity mapping is again a fragment (this is trivial). Consider $f_{\alpha'}: (A_{\alpha'}, <) \to (Y, \leqslant)$, $f_{\alpha''}: (A_{\alpha''}, <) \to (Y, \leqslant)$ for any $\alpha', \alpha'' \in M$. Then, either $A_{\alpha'}$ is a fragment in $(A_{\alpha''}, <)$, or $A_{\alpha''}$ is a fragment in $(A_{\alpha'}, <)$. Problem 40 implies that the functions $f_{\alpha'}$ and $f_{\alpha''}$ are identical wherever both are defined. Now, define $f: A \to Y$ by the rule: $f(x) = f_\alpha(x)$ for each $x \in A$ and every $\alpha \in M$ such that $x \in A_\alpha$. By what has been stated above this definition is correct. Since the family $\{A_\alpha : \alpha \in M\}$ is a nest (chain), for any $x_1, x_2 \in A$ there exists $\alpha' \in M$ such that $x_1, x_2 \in A_{\alpha'}$.

From $f \mid A_{\alpha'} = f_{\alpha'}$ and the fact that $f_{\alpha'}$ is a similarity function it now follows that f also is a similarity function. Hence $(\mathcal{P}, \prec)$ is an inductive set. By the Kuratowski-Zorn principle, there exists in $(\mathcal{P}, \prec)$ a maximal element A^*. By the definition of the family $\mathcal{P}$, there exists a similarity function f^* of $(A^*, <)$ onto some fragment Y^* of $(Y, \leqslant)$. If $A^* = X$ or $Y^* = Y$, then a chain is obtained. Otherwise, let x^* be the first element of $X \setminus A^*$ and y^* of $Y \setminus Y^*$. Put $\tilde{A} = A^* \cup \{x^*\}$, $\tilde{Y} = Y^* \cup \{y^*\}$ and define the function $\tilde{f}: \tilde{A} \to \tilde{Y}$ as follows: $\tilde{f}(x^*) = y^*$ and $\tilde{f} \mid X^* = f^*$. Then $\tilde{A}$ is a fragment of $(X, <)$ and $\tilde{f}$ is its similarity function onto the fragment $\tilde{Y}$ of $(Y, \leqslant)$. Hence, $\tilde{A} \in \mathcal{P}$. But $A^* \neq \tilde{A}$ and $A^* \prec \tilde{A}$, contradicting that A^* is a maximal element of $\mathcal{P}$. Hence, either $A = X$ or $Y^* = Y$.

42. It suffices to draw on Probs. 40 and 41.

43. Let $<$ be any well-ordering on X. If it does not satisfy the hypothesis of the problem, then the set $B = \{x \in X: \aleph_0 \leqslant |I_<(x)|\}$ is not empty (because, B then contains at least the last element of $(X, <)$). Call x^* the first element of B. Then $(I(x^*), <)$

is an infinite well-ordered set without a last element. Define on X a new well-ordering $\prec$ by the rule: (a) if $x_1, x_2 \in I(x^*)$ or $x_1, x_2 \in X \setminus I(x^*)$, then $x_1 \prec x_2$ iff $x_1 < x_2$; (b) for any $x_1 \in X \setminus I(x^*)$ and $x_2 \in I(x^*)$ put $x_1 \prec x_2$ (and, as ever, $x \prec x$ for every $x \in X$). There is no reason to doubt that $\prec$ is indeed a well-ordering relation, and this is what we sought.

44. This is immediate from Probs. 17, 16.

45. By the well-ordering principle, X can be well-ordered by some relation $<$. If this ordering is not minimal, then $A = \{x \in X: |\{x' \in X: x' < x \text{ and } x' \neq x\}| = |X|\}$ is not an empty set. (The abridged definition of the set A now becomes: $A = \{x \in X: |I(x)| = |X|\}$). Denote by x^* the first element of A. Then $|I(x^*)| = |X|$, and for any $x \in I(x^*)$ we have $|I(x)| < |X|$. Fix any one-to-one function φ of X onto $I(x^*)$ and put $x_1 \leqslant x_2$ iff $\varphi(x_1) < \varphi(x_2)$. Then φ is a similarity function, whence it is straightforward that $\leqslant$ is a minimal well-ordering on X.

47. Cf. Prob. 42.

49. If x is the last element of $(X, <)$, then $X = I(x) \cup \{x\}$. Prob. 44 implies that $|X| = |I(x)|$. But, then, $<$ is not minimal.

50. For every $a \in A$, put $\Gamma_a = \{(x, y) \in A \times A: \max\{x, y\} = a\}$. Let B be an arbitrary nonempty subset of $A \times A$ and $B_a = B \cap \Gamma_a$. Denote by a^* the first element of $\{a \in A: B_a \neq \Delta\}$. Obviously, $\Gamma_{a^*} = \Gamma^1_{a^*} \cup \Gamma^2_{a^*} \cup \{(a^*, a^*)\}$, where $\Gamma^1_{a^*} = \{(x, y) \in \Gamma_{a^*} \setminus \{(a^*, a^*)\}: \max\{x, y\} = y\}$, $\Gamma^2_{a^*} = \{(x, y) \in \Gamma_{a^*} \setminus \{(a^*, a^*)\}: \max\{x, y\} = x\}$. The projections $\pi_1: (\Gamma^1_{a^*}, \leqslant) \to (A, <)$ and $\pi_2: (\Gamma^2_{a^*}, \leqslant) \to (A, <)$ (where $\pi_1(x, y) = x$, $\pi_2(x, y) = y$) are similarlity functions. Consequently, $(\Gamma^1_{a^*}, \leqslant)$ and $(\Gamma^2_{a^*}, \leqslant)$ are well-ordered sets.

The definition of the relation $\leqslant$ implies that each element of the set $\Gamma^1_{a^*}$ precedes every element of the set $\Gamma^2_{a^*}$ and all elements in $\Gamma^1_{a^*} \cup \Gamma^2_{a^*}$ precede the element (a^*, a^*). From what has been stated the conclusion is immediate that $(\Gamma_{a^*}, \leqslant)$ is a well-ordered set. Hence, there is a first element in B_{a^*}. This is also a first element of B.

51. From the definition of $\leqslant$ (cf. Prob. 50) it follows that $\leqslant$ is a minimal well-ordering on $N \times N$. It now suffices to draw on Probs. 42 and 48.

52. Assume the contrary and let τ^* be the least of cardinal numbers, whose square is different from them. Then $\aleph_0 < \tau^*$ (cf. Probs. 51 and 14). Choose a set A^* satisfying $|A^*| = \tau^*$. Let $<$ be some minimal well-ordering on A^* (cf. Prob. 45) and $(A^* \times A^*, \leqslant)$ be the Cantor square of a well-ordered set $(A^*, <)$ (cf. Prob. 50). From $\tau^* = |A^*| < |A^* \times A^*|$ it follows (cf. Prob. 41 and the definition of well ordering $\leqslant$) that there exists $a^* \in A^*$, satisfying $|(I((a^*, a^*))| = \tau^*$. Since $(A^*, <)$ has no last element, the element $a^* + 1 \in (A^*, <)$ is defined. But $I(a^* + 1) \times I(a^* + 1) \supset I((a^*, a^*))$; hence $\tau^* \leqslant |I(a^* + 1) \times I(a^* + 1)|$ and $\aleph_0 \leqslant |I(a^* + 1)|$. By the minimality of the well-ordering $<$ we have $|I(a^* + 1)| < \tau^*$. The definition of τ^* now implies that $|I(a^* + 1) \times I(a^* + 1)| = |I(a^* + 1)| < \tau^*$, but this contradicts the relation proved earlier.

54. Put $\tau = |A| = |A \times A|$ (cf. Prob. 52) and let there exist $a \in A$, satisfying $|I((a, a))| = \tau$. An element $a + 1 \in A$ is defined by virtue of the minimality of $<$. We have $I((a, a)) \subset I(a + 1) \times I(a + 1)$, and hence $\tau = |I((a, a))| \leqslant |I(a + 1)|$ (cf. the solution of Prob. 52). But $|I(a + 1)| < \tau$ by the minimality of $<$. We obtain a contradiction: $\tau < \tau$.

55. By Prob. 42, there exists a similarity function of the set $(A, <)$ onto some fragment I of the set $(A \times A, \leqslant)$. If I is a characteristic fragment, then well-ordering $\leqslant$

is not minimal (because, $|I| = |A| = |A \times A|$ (cf. Probs. 52 and 54). Hence, $I = A \times A$.

56. Make use of the conclusion of Prob. 52.

57. Fix some set X^* satisfying $|X^*| = \tau$. For each $\alpha \in M$, there is a one-to-one function $i_\alpha : X_\alpha \to X^*$. Put $i_\alpha(X_\alpha) = X_\alpha^*$. By Prob. 56, $|X^* \times M| \leqslant \tau$. Let $Y^* = \bigcup \{X_\alpha^* \times \{\alpha\} : \alpha \in M\}$. From $Y^* \subset X^* \times M$ it follows that $|Y^*| \leqslant \tau$. Put $\varphi(x_\alpha^*, \alpha) = i_\alpha^{-1}(x_\alpha^*) \in X_\alpha$ for each $\alpha \in M$ and every $x_\alpha^* \in X_\alpha^*$. This rule defines the function $\varphi : Y^* \to X$, and $\varphi(Y^*) = \bigcup \{X_\alpha : \alpha \in M\}$. But then (cf. Prob. 25) $|\bigcup \{X_\alpha : \alpha \in M\}| \leqslant |Y^*| \leqslant \tau$.

58. For every natural number n the cardinality of A^n, the product of n copies of the set A, is equal to the cardinality of A by Prob. 56. However, the set A^n admits the obvious natural mapping onto the set B_n of all subsets of A, consisting of not more than n elements. Consequently, $|B_n| \leqslant |A^n| = |A|$. We have $\mathcal{K}(A) = \bigcup \{B_n : n \in N\}$ (where N is the set of all natural numbers). Since $|N| = \aleph_0$ is the least cardinal number and A is an infinite set, $|N| \leqslant |A|$. Problem 57 permits to conclude that $|\mathcal{K}(A)| \leqslant |A|$. Obviously, $|A| \leqslant |\mathcal{K}(A)|$ (because $\{x\} \in \mathcal{K}(A)$ for each $x \in A$). Hence $|\mathcal{K}(A)| = |A|$.

59. Fix $x_1^* \in X$, $x_2^* \in X$, $x_1^* \neq x_2^*$. Consider $X^2 = X \times X = \{(x_1, x_2) : x_1 \in X, x_2 \in X\}$. Put $A_{x_i^*} = \{(x_i^*, x) : x \in X\}$, $i = 1, 2$, $Y_1 = A_{x_1^*}$, $Y_2 = X^2 \setminus A_{x_1^*}$. Then $X^2 = Y_1 \cup Y_2$, $Y_1 \cap Y_2 = \Lambda$, $|Y_1| = |X|$, $|Y_2| = |X|$. The latter implies also that $A_{x_2^*} \subset Y_2 \subset X^2$ and $|X^2| = |X| = |A_{x_2^*}|$. The set X^2 is equipotent with X (cf. Prob. 52). Hence, there exists a one-to-one correspondence between them. The images of sets Y_1 and Y_2 under this correspondence are the desired X_1 and X_2.

60. The set $X \times X$ can be mapped one-to-one onto X (Prob. 52), whence the solution is straightforward.

63. Let α be an infinite minimal well-ordered set and β a (trivially well-ordered) set consisting of a single element not belonging to α. Then $\beta + \alpha$ is similar to α (cf. Prob. 42), but $\alpha + \beta$ is not similar to α, since in $\alpha + \beta$ there is a largest element but in α there is no such element.

65. Leaving out the detailed working to the reader as an exercise, let us only show that in each nonempty subset Q of C, there exists a first or last element relative to $<$. Put $S = \{b \in B : \text{there exists } a \in A, \text{ satisfying } (a, b) \in Q\}$. Evidently, $S \neq \Lambda$. Denote by b_1 the first element of S relative to $\leqslant$. Put $\mathcal{P} = \{a \in A : (a, b_1) \in Q\}$. By the definition of b_1, we have $\mathcal{P} \neq \Lambda$. Denote by a_1 the first element of $\mathcal{P}$. Then (a_1, b_1) is the first element of S (verify this).

66. Show that the well-ordering of a product is minimal and then draw upon Prob. 42.

68. False. The ordering of the product $(N, <) \times (D, <)$ is not minimal, but $(N, <)$ is a minimal ordered set.

69. Cf. Probs. 66 and 68.

70. The first assertion is easy to adduce from the definition. For proof of the second, the following example is appropriate. Let each of the sets β and γ consist of exactly one element and these elements be distinct (and the ordering be trivial, there being no other possibility under this setting). Introduce α as a set of natural numbers with the natural well-ordering. Then $\beta + \gamma$ is an ordered set consisting of two elements, and its product with α is similar to α (cf. Prob. 66). At the same time $(\beta \times \alpha) + (\gamma \times \alpha)$ is the sum of two disjoint well-ordered sets, similar to α. Hence, the well-ordering on

$(\beta \times \alpha) + (\gamma \times \alpha)$ is not minimal, and since α is a minimal well-ordered set, α and $(\beta \times \alpha) + (\gamma \times \alpha)$ are not similar.

71. In fact, if θ is the first element of the set $(A, <)$, then $i_\theta A = \{(a, \theta) : a \in A\}$; evidently, a characteristic fragment of a lexicographic square of $(A, <)$ is equipotent with the whole of this set (because $|A| = |A \times A|$ by Prob. 52).

72. The Cantor square of a minimal well-ordered set is minimal well-ordered. But the well-ordering of a lexicographic square of an infinite well-ordered set is never minimal (Prob. 71). Consequently, they are not similar.

73. First verify that the relation $\prec$ is transitive, and then proceed along the usual reasonings not explicitly stated here.

(i) Let $x, y, z \in C$ and $x \prec y$, $y \prec z$. Denote by a' the largest element of the set A such that $x_{a'} \neq y_{a'}$ and by a'' the largest element of A such that $y_{a''} \neq z_{a''}$. All $x_{a'}, y_{a'}, z_{a'}, x_{a''}, y_{a''}, z_{a''}$ are elements of the set B and from their definition and the definition of $\prec$ it follows that $x_{a'} \leqslant_c y_{a'}$ and $y_{a''} \leqslant_c z_{a''}$.

Consider three cases : (*a*) $a' = a''$. Then $x_a = y_a$, $y_a = z_a$ and, consequently, $x_a = z_a$, if $a' = a'' <_c a$ and $x_{a'} \leqslant_c y_{a'} = y_{a''} \leqslant_c z_{a''}$; hence in this case $x \prec z$. (*b*) $a' <_c a''$. Then, by the definitions of a' and a'', we have $x_{a''} = y_{a''}$, $y_{a''} \leqslant_c z_{a''}$ and, therefore, $x_{a''} \leqslant_c z_{a''}$. In addition, $x_a = y_a$ and $y_a = z_a$ for $a'' <_c a$. Hence $x \prec z$. (*c*) $a'' <_c a'$. Then $y_{a'} = z_{a'}$, $x_{a'} \leqslant_c y_{a'}$ and, hence, $x_{a'} \leqslant_c z_{a'}$. For $a' <_c a$ we have $x_a = y_a$, $y_a = z_a$ and, consequently, $x_a = z_a$. Hence $x \prec z$. This completes the first part of the proof.

(ii) We agree to call 0 the first element of the set $(B, \leqslant)$ and $(A_x, <)$, where $x \in A$, a segment $\tilde{I}(x)$ of the set $(A, <)$ ordered by the relation $<$. Note that if $x = y + 1$, then $(B, \leqslant)^{(A_x, <)}$ is similar to $(B, \leqslant)^{(A_y, <)} \times (B, \leqslant)$ (it suffices to compare the definitions of power and product). The latter set is well-ordered if the set $(B, \leqslant)^{(A_y, <)}$ is so. Now, let $\tilde{A} = \bigcup \{A_y : y \in Q\}$ for some set $Q \subset A$; assume also that no element of the set Q is the last (largest in $(Q, <)$). From the definition of power, it is clear that $(B, \leqslant)^{(\tilde{A}, <)}$ is similar to $\bigcup \{(B, \leqslant)^{(A_y, <)} : y \in Q\}$, where for $y' < y''$ the set $(B, \leqslant)^{(A_{y'}, <)}$ is naturally identified with some segment of the set $(B, \leqslant)^{(A_{y''}, <)}$ and the order on the union is defined as the union of orders. Hence, if $(B, \leqslant)^{(A_y, <)}$ is a well-ordered set for each $y \in Q$, then $(B, \leqslant)^{(\tilde{A}, <)}$ is also a well-ordered set. Since $(B, \leqslant)^{(0, <)}$ is similar to $(B, \leqslant)$ and, consequently, well-ordered, we have all that is necessary to conclude by the principle of transfinite induction (cf. Prob. 81) that $(B, \leqslant)^{(A, <)}$ is well-ordered.

74. From the definition of power it follows that $(D, <)^{(D, <)}$ is $(0, 0) \prec (1, 0) \prec (0, 1) \prec (1, 1)$.

79. Denote by ω a set of natural numbers with the usual order. Put $\varepsilon = \omega + \omega^\omega + \omega^{\omega^\omega} + \ldots$ (Here the *n*th terms has the form $\underbrace{\omega^{\omega^{\cdot^{\cdot^{\cdot^{\omega}}}}}}_{n \text{ times}}$ the parentheses are necessarily arranged thus $\omega^{(\omega^{(\omega^{\cdot^{\cdot^{\cdot^{\omega))}}}}})}$. The addition is effected over the well-ordered set ω). The number ε is unknown. (For details see Sierpinski [1], pp. 326-329.)

80. Evidently, among the elements of W not belonging to the set Q, there cannot be the first element. Hence, there is no such element.

81. Cf. Prob. 80. Evidently, the requirement is to accept Q as a set of all $x \in W$, for which $P(x)$ is true.

82. (*a*) Let $\varphi : W \to S$ and $\varphi' : W \to S$ satisfy the prescriptions of the problem and and let $\varphi' \neq \varphi$. Then, there is a first element $x \in W$, satisfying $\varphi(x) \neq \varphi'(x)$. For this x, evidently $\varphi \mid I(x) = \varphi' \mid I(x)$ and, hence, $K_x(\varphi \mid I(x)) = K_x(\varphi' \mid I(x))$, that is $\varphi(x) = \varphi'(x)$, a contradiction. Thus, there exists only one function $\varphi : W \to S$ with the properties prescribed.

(*b*) Denote by A a set of all $x \in W$, for which there exists on $I(x)$ the desired function φ, which we denote by φ_x. By what has been proved in (*a*), if $x_1 < x_2$ and $\varphi_{x_1}, \varphi_{x_2}$ are defined, then $\varphi_{x_1} \mid I(x_1) = \varphi_{x_2}$. Put $\tilde{A} = \bigcup \{I(x) : x \in A\}$. Then, either $\tilde{A} = W$ or there exists $x \in W$ satisfying $\tilde{A} = I(x)$. The function $\psi : \tilde{A} \to S$ is defined by the rule : $\psi(y) = \varphi_x'(y)$, where $x' \in \tilde{A}$ and $y \in I(x')$. If $\tilde{A} = W$, then ψ is evidently the desired function. If $\tilde{A} \neq W$, then put $\varphi(x) = K_x(\psi)$ and $\varphi(x') = \psi(x')$ for $x' \in \tilde{A}$. The function $\varphi : I(x+1) \to S$ satisfies the hypotheses of the problem; hence, $x \in I(x+1) \subset A = I(x)$, but this contradicts $x \notin I(x)$. Consqeuently, $\tilde{A} \neq W$ is impossible and ψ is the function sought.

84. Let $\tilde{\tau}$ be some element of M and $\tilde{M} = \{\tau \in M : \tau < \tilde{\tau}\} \neq \Lambda$. Denote by $\tilde{X}$ any set with the cardinal number $\tilde{\tau}$. By the well-ordering principle, it can be assumed that $\tilde{X}$ is well-ordered by some relation $\leqslant$. Further, let X_τ for each $\tau \in \tilde{M}$ be a set of cardinality τ and well-ordered by some relation $<_\tau$. By Prob. 41, there exists exactly one similarity function f_τ of the set $(X_\tau, <_\tau)$ onto some fragment $\tilde{X}_\tau$ of the set $(\tilde{X}, \leqslant)$ (it is already known that $\tau < \tilde{\tau}$). From $\tau < \tilde{\tau}$ it follows that $\tilde{X}_\tau \neq \tilde{X}$. Denote by x_τ the first element of $\tilde{X} \setminus \tilde{X}_\tau$. For every $\tau_1, \tau_2 \in \tilde{M}$, either $\tilde{X}_{\tau_1} \subset \tilde{X}_{\tau_2}$, or $\tilde{X}_{\tau_2} \subset \tilde{X}_{\tau_1}$, because $\tilde{X}_{\tau_1}$ and $\tilde{X}_{\tau_2}$ are fragments. However, if $\tau_1 < \tau_2$, then the inclusion $\tilde{X}_{\tau_2} \subset \tilde{X}_{\tau_1}$ is impossible (so also is the equality $\tilde{X}_{\tau_1} = \tilde{X}_{\tau_2}$) so that, then, surely, $\tilde{X}_{\tau_1} \subset \tilde{X}_{\tau_2}$ where $\tilde{X}_{\tau_2} \setminus \tilde{X}_{\tau_1} \neq \Lambda$. Hence, $x_{\tau_1} \in \tilde{X}_{\tau_2} \setminus \tilde{X}_{\tau_1}$ and $x_{\tau_1} \leqslant x_{\tau_2}$. Since $x_{\tau_2} \notin \tilde{X}_{\tau_2}$, the conclusion is that $x_{\tau_1} \neq x_{\tau_2}$. The assertions above imply that the function $\varphi : (\tilde{M}, \leqslant) \to (\tilde{X}, \leqslant)$, defined by the formula $\varphi(\tau) = x_\tau$ for each $\tau \in \tilde{M}$, is similar to a function of the ordered set $(\tilde{M}, \leqslant)$ onto some subset A of a well-ordered set $(\tilde{X}, \leqslant)$. However, in A there is a first element, which we denote by x^*. There exists $\tau^* \in \tilde{M}$, satisfying $\varphi(\tau^*) = x^*$ (because $\varphi(\tilde{M}) = A$). Since φ is a similarity function, τ^* is the first (least) element of $(\tilde{M}, \leqslant)$. From the definition of $\tilde{M}$ it now follows that τ^* is the least element of the set M.

85. Fix a well-ordered set $(Y, \leqslant)$ of cardinality λ. It is mapped similarly onto some characteristic fragment $I(x)$ of the set $(X, <)$ (cf. Prob. 42; it is known that $\lambda < \tau$). Then x is the desired element of the set X.

86. Let $(A, <)$ be any well-ordered set of cardinality τ. Set in correspondence to each $\tau' \in \mathcal{E}$ some element $x_{\tau'} \in A$ such that $\mid I(x_{\tau'}) \mid = \tau'$ (this is possible, cf. Prob. 85). Evidently, $x_{\tau'} \neq x_{\tau''}$ for $\tau' \neq \tau''$. Hence, we obtain a one-to-one function of the set $\mathcal{E}$ into the set A, whence it follows that $\mid \mathcal{E} \mid \leqslant \mid A \mid = \tau$.

87. Cf. Probs. 52 and 57.

88. The conclusion is immediate from the definition of the sum of cardinal numbers and the results of Prob. 57.

89. Denote by $\mathcal{E}$ a set of all cardinal numbers less than τ. Since $|\mathcal{E}| \leqslant \tau$ (cf. Prob. 86) and $\tau' < \tau$ for each $\tau' \in \mathcal{E}$, hence $\tau^* = \Sigma\{\tau' : \tau' \in \mathcal{E}\} \leqslant \tau$ (Prob. 88). If we had $\tau^* < \tau$, then since the integer τ is a limit, we would have a cardinal number $\tilde{\tau}$ satisfying $\tau^* < \tilde{\tau} < \tau$. But then $\tilde{\tau} \in \mathcal{E}$ and $\tilde{\tau} \leqslant \tau^*$, a contradiction. Hence $\tau^* = \tau$.

90. Cf. Prob. 57.

91. From the definition of $\aleph_1$ and minimality of $<$ it follows that $|\tilde{I}(a)| \leqslant \aleph_0$ for each $a \in A$. But then $|\bigcup \{\tilde{I}(a) : a \in A\}| \leqslant \aleph_0$ by Prob. 57. Consequently, $T_{\aleph_1} \setminus (\bigcup \{\tilde{I}(a) : a \in A\}) \neq \Lambda$. The definition of the set $\tilde{I}(a)$ (see p. 6) implies that the element of the set $T_{\aleph_1} \setminus (\bigcup \{\tilde{I}(a) : a \in A\})$ satisfies the hypotheses of the problem.

92. Set in correspondence to each $F \in \exp X$ a function $\varphi_F : X \to D$, defined by the condition $\varphi_F^{-1}(1) = (F)$. Put $\varphi_F = f(F)$. Then f is a one-to-one function of the set $\exp X$ on to the set DX.

93. Cf. Prob. 34 and the definition of the cardinal number 2^τ (see Prob. 92).

94. Cf. Prob. 93.

95. False, the equality is not always true. Put $\lambda = 2$, $\tau = \aleph_0$. Then $\tau^\lambda = \aleph_0^2 = \aleph_0$ (Prob. 87) and $\lambda^\tau = 2^{\aleph_0}$. But $\aleph_0 < 2^{\aleph_0}$ (Prob. 93).

97. It suffices to compare the definitions of sum, product and power of a cardinal number and the results of Probs. 28, 27 and 29.

98. Evidently, the cardinality of a set of all countable subsets of a straight line does not exceed the cardinality of the product of a countable set of straight lines. However, the cardinality of the latter is equal to $(2^{\aleph_0})^{\aleph_0} = 2^{\aleph_0}$ (cf. Prob. 97 and the definition of the power of cardinal numbers).

99. From $2 < \aleph_1$ it follows that $2^{\aleph_0} \leqslant \aleph_1^{\aleph_0}$, and from $\aleph_1 \leqslant 2^{\aleph_0}$ that $\aleph_1^{\aleph_0} \leqslant (2^{\aleph_0})^{\aleph_0} = 2^{\aleph_0 \aleph_0} = 2^{\aleph_0}$ (Probs. 87, 97). Hence, $2^{\aleph_0} = \aleph_1^{\aleph_0}$

100. We have $2^\mu \leqslant \lambda^\mu$, because $2 \leqslant \lambda$. But $2^\mu = 2^{\mu \cdot \mu} = (2^\mu)^\mu$. Comparing $(2^\mu)^\mu$ and λ^μ, we conclude that $\lambda^\mu \leqslant (2^\mu)^\mu$, because $\lambda < 2^\lambda \leqslant 2^\mu$. Thus, $\lambda^\mu \leqslant 2^\mu$ and, consequently, $2^\mu = \lambda^\mu$.

101. False. Let $\lambda = \aleph_0$, $\mu = \aleph_0$ and $\tau = \aleph_0$. Then $(\lambda^\mu)^\tau = (\aleph_0^{\aleph_0})^{\aleph_0} = \aleph_0^{\aleph_0 \cdot \aleph_0} = \aleph_0^{\aleph_0} = 2^{\aleph_0}$ (cf. Probs. 100 and 97). However, $\lambda^{(\mu^\tau)} = \aleph_0^{(\aleph_0^{\aleph_0})} = \aleph_0^{(2^{\aleph_0})} = 2^{(2^{\aleph_0})}$. By Prob. 93, $2^{\aleph_0} < 2^{(2^{\aleph_0})}$.

103. Use the operations of addition and taking power available at our disposal. Put $\tau = \aleph_0$, $\tau_1 = \aleph_0^{\aleph_0}$, $\tau_2 = \aleph_0^{\tau_1}, \ldots, \tau_{n+1} = \aleph_0^{\tau_n}, \ldots$ and $\tau^* = \Sigma\{\tau_i : i = 0, 1, 2, \ldots, n, \ldots\}$ (the sum is taken over all natural numbers). We are convinced of τ^* being inaccessible.

104. This problem is rather subtle. It is solved in the set axiomatic theory (cf. P. Cohen [1]).

105. Form the union of equipotent terms, unite also in a single set all terms of not quite large cardinality (cf. Prob. 57).

106. If A is a cofinal subset in $(X, <)$, then $X = \bigcup \{I(x) : x \in A\}$. Since $|I(x)| < |X| = \tau$ for each $x \in X$ (the order $<$ is minimal), hence $cf(\tau) \leqslant |A|$ (cf. the definition of $cf(\tau)$). On the other hand, let M be a set of cardinality $\mu = cf(\tau)$ and let $X = \bigcup \{X_\alpha : \alpha \in M\}$, where $|X_\alpha| < |X|$ for every $\alpha \in M$. Within this setting of reasonings it would, evidently, suffice to consider the case when $\mu < \tau$. For each $\alpha \in M$, then there is an element $x_\alpha \in X$ such that $|X_\alpha| = |I(x_\alpha)|$ (cf. Prob. 85). Put $\tilde{A} = \{x_\alpha : \alpha \in M\}$. It can be assumed that $|X_{\alpha'}| \neq |X_{\alpha''}|$ for any distinct $\alpha', \alpha'' \in M$ (Prob. 105). Then, we have also $x_{\alpha'} \neq x_{\alpha''}$ for $\alpha' \neq \alpha''$; hence, $|\tilde{A}| = |M| = cf(\tau) = \mu$. The set $\tilde{A}$ is cofinal to $(X, <)$, otherwise $\bigcup \{I(x_\alpha) : \alpha \in M\} = I(x^*)$ for some $x^* \in X$ and $|X_\alpha| = |I(x_\alpha)| \leqslant |I(x^*)| < \tau$. Then, $\tau = |X| = |\bigcup \{X_\alpha : \alpha \in M\}| = \Sigma\{|X_\alpha| : \alpha \in M\} \leqslant \max\{|I(x^*)|, \mu\} < \tau$, a contradiction.

107. Evidently, we always have $cf(\lambda) \leqslant \lambda$. Assume that $\mu = cf(\lambda) < \lambda$. Let A be a set of cardinality λ, B of μ and X of τ. Then, there exist decompositions into the sum of pairwise disjoint sets : $X = \bigcup \{X_\alpha : \alpha \in A\}$, $A = \bigcup \{A_\beta : \beta \in B\}$, where $|X_\alpha| < |X| = \tau$ for each $\alpha \in A$ and $|A_\beta| < |A| = \lambda$ for every $\beta \in B$. Put $Y_\beta = \bigcup \{X_\alpha : \alpha \in A_\beta\}$ for each $\beta \in B$. Then, $\bigcup \{Y_\beta : \beta \in B\} = X$. From $|B| < \lambda = cf(\tau)$ it follows that $|Y_{\beta^*}| = |X| = \tau$ for some $\beta^* \in B$. But $|X_\alpha| < \tau$ for each $\alpha \in A_{\beta^*}$, and $|\bigcup \{X_\alpha : \alpha \in A_{\beta^*}\}| = \tau$. Hence $\lambda = cf(\tau) \leqslant |A_{\beta^*}|$, but this violates the fact that $|A_{\beta^*}| < \lambda$.

108. $cf(\tau^*) = \aleph_0$. The proof is left as an exercise to the reader (cf. Probs. 88 and 93).

109. This assertion is immediate from the definitions and the conclusion of Prob. 88.

110. Let Y be a set of cardinality λ, $Y_\alpha = Y$ and let $f_\alpha : Y_\alpha \to X_\alpha$ be a function of Y_α onto X_α for each $\alpha \in A$. Consider the products $\Pi\{Y_\alpha : \alpha \in A\}$ and $f = \Pi\{f_\alpha : \alpha \in A\}$. Then, $\Pi\{Y_\alpha : \alpha \in A\} = Y^A$ and f is a function of the set Y^A onto X. By definition $\lambda^\tau = |Y|^{|A|} = |Y^A|$. Thereupon, from Prob. 25 it follows that $|X| \leqslant |Y^A|$.

111. Consider two sets A and X of cardinality λ and τ, respectively, and fix on X some minimal well-ordering $<$. Thereupon, if $f \in X^A$, then $f(A) \subset I(x)$ for some $x \in X$ (because $|f(A)| \leqslant |A| < cf(\tau)$, cf. Prob. 106). Consequently, $X^A = \bigcup \{(I(x))^A : x \in X\}$. Choose $x(\mu) \in X$ for each $\mu < \tau$ in order that $|I(x(\mu))| = \mu$ (Prob. 85). Then $X^A = \bigcup \{(I(x))^A : x = x(\mu), \mu < \tau\}$. (Here also a note has been taken of the fact that τ is the limit of cardinal numbers.) Last relation is equivalent to the conclusion of the problem.

112. Problems 97, 87 and 109 imply that it will suffice to consider the case when τ is the limit of cardinal numbers and $\lambda \geqslant \aleph_0$. Then, by Prob. 111, $\tau^\lambda = \Sigma\{\mu^\lambda : \mu < \tau\}$. But $\mu^\lambda \leqslant \max(\mu^\mu, \lambda^\lambda) = \max(2^\mu, 2^\lambda)$ (cf. Probs. 100 and 96). By the generalized continuum-hypothesis $2^\mu = \mu^+$ (or, $\mu < \aleph_0$), $2^\lambda = \lambda^+$. From $\lambda < cf(\tau) \leqslant \tau$ and $\mu < \lambda$ it follows that $\mu^+ \leqslant \tau$, $\lambda^+ \leqslant \tau$. Thus $\mu^\lambda \leqslant \tau$ for each $\mu < \tau$. However, then, $\tau \leqslant \tau^\lambda = \Sigma\{\mu^\lambda : \mu < \tau\} \leqslant \tau$, that is, $\tau^\lambda = \tau$.

113. Put $A = \bigcup \{A_\alpha : \alpha \in M\}$ and $B = \Pi\{B_\alpha : \alpha \in M\}$. It can be assumed that $A_{\alpha'} \cap A_{\alpha''} = \Lambda$ for $\alpha' \neq \alpha''$. Denote by φ an arbitrary function of A into B and put $\varphi_\alpha = \pi_\alpha \varphi$, where $\pi_\alpha : B \to B_\alpha$ is a projection (see p. 4), $C_\alpha = \varphi_\alpha(A_\alpha)$, $\alpha \in M$. Then $|C_\alpha| \leqslant |\varphi_\alpha(A_\alpha)| \leqslant |A_\alpha| < |B_\alpha|$ and, consequently, $D_\alpha = B_\alpha \setminus C_\alpha \neq \Lambda$. Put $D = \Pi\{D_\alpha : \alpha \in M\} \subset B$ and verify that $\varphi(A) \cap D = \Lambda$. Let $b \in \varphi(A)$. Then, for some $\alpha^* \in M$ we have $b = \varphi(a)$, where $a \in A_{\alpha^*}$. We conclude that $\pi_{\alpha^*}(b) = \pi_{\alpha^*}\varphi(a) = \varphi_{\alpha^*}(a) \in C_{\alpha^*}$, implying that $b \notin D$ (because, $\pi_{\alpha^*} D = D_{\alpha^*} = B_{\alpha^*} \setminus C_{\alpha^*}$). Consequently, $\varphi(A) \neq B$ and $|A| < |B|$.

114. For each $\alpha \in M$ define $\varphi(\alpha) \in M$ of the following specifications : (*a*) $|A_\alpha| < |A_{\varphi(\alpha)}|$ and (*b*) if $\beta \in M$ and $|A_\alpha| < |A_\beta|$, then $|A_{\varphi(\alpha)}| \leqslant |A_\beta|$. Since a class of cardinal numbers is well-ordered in the variables, (i) implies that an element $\varphi(\alpha)$ satisfying the conditions (*a*) and (*b*) exists (and is unique) for each $\alpha \in M$. Put $B_\alpha = A_{\varphi(\alpha)}$.

Then, $|A_\alpha| < |B_\alpha|$ for each $\alpha \in M$. By Prob. 113, $|\bigcup\{A_\alpha : \alpha \in M\}| < |\Pi\{B_\alpha : \alpha \in M\}|$. Since the set $\Pi\{B_\alpha : \alpha \in M\} = \Pi\{A_{\varphi(\alpha)} : \alpha \in M\}$ is a bound of the product $\Pi\{A_\alpha : \alpha \in M\}$ (note that φ is a one-to-one function of A into itself), then $|\bigcup\{A_{\varphi(\alpha)} : \alpha \in M\}| \leqslant |\Pi\{A_\alpha : \alpha \in M\}|$.

115. Assume the contrary : let $\lambda = cf(2^\tau) \leqslant \tau$. Fix a minimally well-ordered set $(A, <)$ of cardinality 2^τ, and let $M \subset A$, M cofinal to A, $|M| = \lambda$. For an arbitrary $\alpha \in M$, put $A_\alpha = \{\beta \in A : \beta < \alpha\}$. Then $|A_\alpha| < 2^\tau$ and $\bigcup\{A_\alpha : \alpha \in M\} = A$. Thence and from $|M| \leqslant \tau$ it follows that in the set $\{|A_\alpha| : \alpha \in M\}$ there is no largest element (otherwise this largest element when multiplied by τ would give 2^τ, which is impossible (cf. Prob. 87(*c*)). From similar reasonings it follows that one can find $M^* \subset M$, which is cofinal to M (and, consequently, to A) and satisfies the following condition: if $\alpha' \in M^*$, $\alpha'' \in M^*$ and $\alpha' \neq \alpha''$ then, $|A_{\alpha'}| \neq |A_{\alpha''}|$. Thereupon (Prob. 114), $|\bigcup\{A_\alpha : \alpha \in M^*\}| < |\Pi\{A_\alpha : \alpha \in M^*\}|$. But $|A_\alpha| < 2^\tau$ for each $\alpha \in M^*$ and $|M^*| \leqslant |M| = \lambda \leqslant \tau$. Hence, $|\Pi\{A_\alpha : \alpha \in M^*\}| \leqslant (2^\tau)^\tau = 2^\tau$. On the other hand, M^* is confinal to A and, hence, $A = \bigcup\{A_\alpha : \alpha \in M^*\}$ and $|\bigcup\{A_\alpha : \alpha \in M^*\}| = |A| = 2^\tau$. Thus, $2^\tau < 2^\tau$, a contradiction.

116. Otherwise, we would have $cf(2^{\aleph_0}) \leqslant \aleph_0$, which is impossible (cf. Prob. 115).

118. By Prob. 50 of Chap. 4, there is a family of subsets of N, satisfying the conditions (*a*), (*b*) and (*c*). A family $\mathcal{T}$ of all such collections, partially ordered by the inclusion relation, is inductive (and as just remarked is not empty). Hence, there is a maximal element in $\mathcal{T}$. Evidently this is also the desired maximal element of $\mathcal{E}^*$.

119. Assume that f is one-to-one. Denote by A a least subset of B containing B_{α_0}, satisfying $f^{-1}A = A$ (i.e., $A = B_{\alpha_0} \cup f^{-1}B_{\alpha_0} \cup f^{-1}f^{-1}B_{\alpha_0} \cup f^{-1}f^{-1}f^{-1}B_{\alpha_0} \cup \ldots$). Then $|A| \leqslant \aleph_0 \cdot \tau = \tau$ and, hence, $B \setminus A \neq \Lambda$. From $A \supset B_{\alpha_0}$ it follows that $B \setminus A \subset C$. Hence, the function f is defined on $B \setminus A$ and $f^{-1}A = A$ implies that $f(B \setminus A) \subset B \setminus A$. Fix $\alpha^* \in B \setminus A$. Then $f(\alpha^*) \in B \setminus A$, $ff(\alpha^*) \in B \setminus A, \ldots$, and by virtue of the assumption on f, we have $\alpha^* > f(\alpha^*) > ff(\alpha^*) \ldots$ We have thus constructed an infinite monotonic (strictly) decreasing sequence of the elements of a well-ordered set $(B, <)$, that is, we obtain a contradiction.

120. For any $t \in T$, $x \in X$ and $A \subset X$, put $(A)_{x,t} = \{y \in A : f(x, y) = t\}$ (we identify (x, y) with (y, x) and always assume that $x \neq y$). Evidently, $x \notin (A)_{x,t}$ and $(A_{x,t}) \subset A$. Let $(\mathcal{Y}, <)$ be a minimally well-ordered set of cardinality τ^+. For each $\alpha \in \mathcal{Y}$, put $\mathcal{Y}_\alpha = \{\beta \in \mathcal{Y} : \beta < \alpha\}$. Then $|\mathcal{Y}_\alpha| \leqslant \tau$ for every $\alpha \in \mathcal{Y}$. Further, $\alpha + 1$ is always the first element of $\mathcal{Y}$, greater than α. In every nonempty set $A \subset X$, fix an element $c(A)$. By induction side by side $(\mathcal{Y}, <)$ for each $\alpha \in \mathcal{Y}$ there is now defined some family $\mathcal{P}_\alpha$ of subsets of X. Put $\mathcal{P}_0 = \{X\}$. If $\beta \in \mathcal{Y}$, $\beta \neq 0$ and for every $\alpha < \beta$ the family $\mathcal{P}_\alpha$ is already defined, we denote by Q_β a family of all nonempty $A \subset X$ of the form $A = \bigcap\{P_\alpha : \alpha < \beta\}$, where $P_\alpha \in \mathcal{P}_\alpha$ for every $\alpha < \beta$. Put $\gamma_{\beta,A} = \{(A)_{c(A),t} : t \in T \text{ and } (A)_{c(A),t} \neq \Lambda\} \cup \{\{c(A)\}\}$. Evidently, $\gamma_{\beta,A}$ is the disjoint cover of A, with $|\gamma_{\beta,\alpha}| \leqslant |T| = \tau$. Put $\mathcal{P}_\beta = \bigcup\{\gamma_{\beta,A} : A \in Q_\beta\}$. If $|\mathcal{P}_\alpha| \leqslant \lambda$ for each $\alpha < \beta$ (for $\beta = 1$ this is true, because $|\mathcal{P}_0| = 1$), then $|Q_\beta| \leqslant \lambda^{|\mathcal{Y}_\beta|} = \lambda^\tau = 2^\tau = \lambda$, and, hence, $|\mathcal{P}_\beta| \leqslant \tau \cdot \lambda = \lambda$. If every $\mathcal{P}_\alpha$, where $\alpha < \beta$, covers X, then Q_β also does so and thereby $\mathcal{P}_\beta$ too covers X. In addition $\mathcal{P}_0$ covers X. If all $\mathcal{P}_\alpha$, where $\alpha < \beta$, are disjoint (and $\mathcal{P}_0$ is *a priori* so), then $\mathcal{P}_\beta$ is also disjoint. Clearly, $\mathcal{P}_\beta$ is a refinement of $\mathcal{P}_\alpha$ for $\alpha < \beta$. We regard the family $\mathcal{P}_\alpha$ to be defined, as above, for each $\alpha \in \mathcal{Y}$. The following fact, though trivial, is important in its own right;

(1°) If $\alpha' \in \mathcal{Y}$, $\alpha < \alpha'$, $A' \in \mathcal{Y}_{\alpha'}$, $A \in \mathcal{P}_\alpha$ and $A \cap A' \neq \Lambda_\lambda$, then $A' \subset A$; also, if $|A'| \geqslant 2$, then $c(A) \neq c(A')$ and there exists $t' \in T$, satisfying $A' \subset (A)_{c(A),t'}$.

Put $\mathcal{P} = \cup \{\mathcal{P}_\alpha : \alpha \in \mathcal{Y}\}$, $\mathcal{P}_\lambda = \{A \in \mathcal{P} : |A| \leqslant \lambda\}$ and $X_\lambda = \cup \{A : A \in \mathcal{P}_\lambda\}$. Then $|\mathcal{P}_\lambda| \leqslant |\mathcal{P}| \leqslant \tau^+\lambda = \lambda$ and, hence, $|X_\lambda| \leqslant \lambda$. Hence, $X \setminus X_\lambda \neq \Lambda$. Fix $x_\lambda \in X \setminus X_\lambda$. For each $\alpha \in \mathcal{Y}$ choose $A_\alpha \in \mathcal{P}_\alpha$ such that $x_\lambda \in A_\alpha$. From $x_\lambda \notin X_\lambda$ it follows that $A_\alpha \notin \mathcal{P}_\lambda$, that is, $|A_\alpha| > \lambda > 1$ for every $\alpha \in \mathcal{Y}$. Put $x_\alpha = c(A_\alpha)$. By (1°) $x_\alpha \neq x_{\alpha'}$ for $\alpha \neq \alpha'$ and $A_{\alpha+1} \subset A_\alpha$. Since $|A_{\alpha+1}| > \lambda$, for each $\alpha \in \mathcal{Y}$ there exists $t_\alpha \in T$ such that $A_{\alpha+1} \subset (A_\alpha)_{x_\alpha, t_\alpha}$. For $\alpha < \alpha'$ we have $A_{\alpha'} \subset A_{\alpha+1}$ (see 1°), whence and from $x_{\alpha'} \in A_{\alpha'}$ it follows that $f(x_{\alpha'}, x_\alpha) = t_{\alpha'}$. But $|T| = \tau < \tau^+$. Consequently, there exists $t^* \in T$, for which the set $\mathcal{Y}^* = \{\alpha \in \mathcal{Y} : t_\alpha = t^*\}$ is equipotent with $\mathcal{Y}$. Put $X^* = \{x_\alpha : \alpha \in \mathcal{Y}^*\}$. We have $|X^*| = |\mathcal{Y}^*|_{\tau^+} > \tau$. If $x \in X^*$, $y^* \in X$ and $x \neq y$, then $x = x_\alpha$, $y = x_{\alpha'}$ for some $\alpha, \alpha' \in \mathcal{Y}^*$, where $\alpha \neq \alpha'$. For definiteness, let $\alpha < \alpha'$. Then, $f(x_\alpha, x_{\alpha'}) = t_\alpha = t^*$, that is, $f(x, y) = t^*$ for every unordered pair of distinct elements of X^*. Hence, X^* is the desired set.

121. Fix a set T of cardinality τ. Linearly order a set A by some relation $<$. For each $a \in A$ fix some function φ_a of T onto a set of all terms, which intersect the set $\{(a, x) : x \in A \text{ and } a < x\}$, and also some function ψ_a of T onto a set of all terms, which intersect the set $\{(x, a) : x < a\}$. Now, let $a' \in A$, $a'' \in A$ and $a' \neq a''$. Then, either $a' < a''$, or $a'' < a'$. For definiteness, assume that $a' < a''$. Then, there exist $t' \in T$ and $t'' \in T$ for which $(a', a'') \in \varphi_{\alpha'}(t')$ and $(a', a'') \in \psi_{\alpha''}(t'')$. In conformity with the fact that, $a' < a''$, consider an ordered pair $(t', t'') \in T \times T$ and put $f(a', a'') = (t', t'')$. We have $|T \times T| = \tau$ and $|A| > 2^\tau$. By Prob. 120, there exists $A^* \subset A$ and an ordered pair (t_1, t_2) of elements of T such that $|A^*| > \tau$ and $f(a', a'') = (t_1, t_2)$, for every $a' \in A^*$, $a'' \in A^*$ and $a' \neq a''$. Write $\tilde{A}^* = A^*$ if $(A^*, <)$ has no largest element, and $\tilde{A}^* = A^* \setminus \{m\}$ if m is a largest element in $(A^*, <)$. $\tilde{A}^*$ is the desired set. Indeed, if $a', a'' \in \tilde{A}^*$ and $a' < a''$, then $\varphi_{\alpha'}(t_1) \ni (a', a'')$ and $\psi_{\alpha''}(t_2) \ni (a', a'')$, that is, $\varphi_{\alpha'}(t_1) \cap \psi_{\alpha''}(t_2) \neq \Lambda$, and since by assumption distinct terms do not intersect, $\varphi_{\alpha'}(t_1) = \psi_{\alpha''}(t_2)$. But then $\varphi_a(t_1) = \varphi_b(t_1)$ for all $a, b \in \tilde{A}^*$, because $c \in \tilde{A}^*$ can be chosen such that $a < c$ and $b < c$ and, by what has been proved above, we have $\varphi_a(t_1) = \psi_c(t_2)$, $\varphi_b(t_1) = \psi_c(t_2)$ and, therefore, $\varphi_a(t_1) = \varphi_b(t_1)$. From $\{(a', a''): a', a'' \in \tilde{A}^* \text{ and } a' \neq a''\} \subset \cup \{\varphi_a(t_1): a \in \tilde{A}^*\}$ the conclusion sought is now immediate.

122. Fix a set L of cardinality τ and carry it over into each γ_α. Then $\gamma_a = \{F_l^\alpha : l \in L\}$. Put $T = L \times L$. Then, $|T| = \tau^2 = \tau$. For each pair (α', α'') of distinct elements of A fix $t = (l', l'') \in T$ so that we had : $F_{l'}^{\alpha'} \cap F_{l''}^{\alpha'} \cap F_{l'}^{\alpha''} \cap F_{l''}^{\alpha''} = \Lambda$ (plausible by (i)), and put $f(\alpha', a'') = t$ (the order of elements in the pair (α', α'') is, evidently, not effected by the choice of t). The set-up now admits the assertion of Prob. 120 (on interchanging A with X). The conclusion is that there exist $A^* \subset A$ and $t^* \in T$ such that $|A^*| > \tau$ and $f(\alpha', \alpha'') = t^*$, if $\alpha' \in A^*$, $\alpha'' \in A^*$ and $\alpha' \neq \alpha''$. Let $t^* = (l_1, l_2)$. For each $\alpha \in A^*$ put $F^\alpha = F_{l_1}^\alpha \cap F_{l_2}^\alpha$. By (ii), $F^\alpha \in \gamma_\alpha$. Evidently, F^α, $\alpha \in A^*$, is the desired set.

124. By Prob. 123, it suffices to show that ξ is contained in some prefilter on X. Clearly, such a prefilter is, in particular, a family $\xi^\cap$ of all subsets of X, representable in the form of an intersection of a finite collection of elements of ξ.

125. If we had $B_1, \ldots, B_l \in \xi'$ satisfying $\cap \{B_i : i = 1, \ldots, l\} = \Lambda$, then, surely one of the $B_1, \ldots, B_l$ would coincide with B (because ξ is centralized). Let $B_1 = B$. Then it can be assumed that $B_2, \ldots, B_l \in \xi$. Since ξ is a prefilter, there exists $B' \in \xi$,

$B' \subset \bigcap \{B_i : i = 2, \ldots, l\}$. Then $B' \cap B = B_1 \cap B' \subset \bigcap \{B_i : i = 1, \ldots, l\} = \Lambda$, contradicting the hypothesis.

127. A set $\mathcal{T}(\mathcal{E})$ of all centralized systems of element $\mathcal{E}$ is assigned natural order by the rule: if $\xi_1 \in \mathcal{T}(\mathcal{E})$ and $\xi_2 \in \mathcal{T}(\mathcal{E})$, then $\xi_1 < \xi_2$ iff $\xi_1 \subset \xi_2$. Show that the ordered set $(\mathcal{T}(\mathcal{E}), <)$ is inductive (see p. 6), that is, for every chain $\mathcal{A} \subset (\mathcal{T}(\mathcal{E}), <)$ there exists $\tilde{\xi} \subset \mathcal{T}(\mathcal{E})$, a majorant for $\mathcal{A}$. Recalling that every $\xi \in \mathcal{A}$ (as also, in general, each $\xi \in \mathcal{T}(\mathcal{E})$) is a subset of $\mathcal{E}$, put $\tilde{\xi} = \bigcup \{\xi : \xi \in \mathcal{A}\}$. Verify that $\tilde{\xi}$ is a centralized family of sets. If $A_1, \ldots, A_k$ is an arbitrary finite collection of elements $\tilde{\xi}$ then, by the definition of $\tilde{\xi}$, there exist $\xi_1, \ldots, \xi_k \in \mathcal{A}$ such that $A_i \in \xi_i$, $i = 1, \ldots, k$. Since $\mathcal{A}$ is a chain in $(\mathcal{T}(\mathcal{E}), <)$, there can be found a permutation $i_1, \ldots, i_k$ of the numbers $1, \ldots, k$ for which $\xi_{i_1} < \ldots < \xi_{i_k}$. Thereupon, by the definition of the relation $<$, we obtain $\xi_{i_1} \subset \ldots \subset \xi_{i_k}$ and, hence, $A_i \in \xi_{i_k}$ for all $i = 1, \ldots, k$. But, then, since ξ_{i_k} is a centralized family of sets, $\bigcap \{A_j : j = 1, \ldots, k\} \neq \Lambda$. This proves that $\tilde{\xi}$ is a centralized family. From the definition of $\tilde{\xi}$ it now follows that $\tilde{\xi} \in \mathcal{T}(\mathcal{E})$ and that $\xi \subset \tilde{\xi}$ for every $\xi \in \mathcal{A}$. But the latter signifies that $\xi < \tilde{\xi}$ for every $\xi \in \mathcal{A}$; thus, $\tilde{\xi}$ is a majorant of the set $\mathcal{A}$. On the basis of Kuratowski-Zorn principle, we now conclude that in the ordered set $(\mathcal{T}(\mathcal{E}), <)$ there is a maximal element (see p. 6), which we denote by ξ^*. It is to be shown that ξ^* is a maximal centralized system of elements $\mathcal{E}$ in the sense of the definition given on p. 7. Let $A \in \mathcal{E} \setminus \xi^*$. Put $\xi' = \xi^* \cup \{A\}$. Then, $\xi^* \subset \xi'$ and $\xi^* \neq \xi'$. If ξ' is a centralized family of sets, then we obtain the relations $\xi' \in \mathcal{T}(\mathcal{E})$, $\xi^* < \xi'$ and $\xi^* \neq \xi'$, contradicting the maximality of ξ^* in $(\mathcal{T}(\mathcal{E}), <)$. Consequently, the family ξ' is not centralized, giving the desired proof.

128. *(a)* $\Rightarrow$ *(b)*. From *(a)* it follows by Prob. 126 that ξ is a filter. If $\xi' \supset \xi$ and ξ' is a filter on X, then ξ' is a centralized family of sets and, by virtue of *(a)*, $\xi' = \xi$.

(b) $\Rightarrow$ *(a)*. From *(b)* it follows that ξ is a centralized family of sets. Consider an arbitrary family ξ' centralized on X and containing ξ. By Prob. 124, there exists a filter ξ'' on X, containing ξ'. However, from *(b)* it now follows that $\xi'' = \xi$. Moreover, $\xi' = \xi$.

(a) $\Rightarrow$ *(c)*. From *(a)* it follows that ξ is a filter (furthermore, ξ is a prefilter). If $A \notin \xi$, but $A \cap B \neq \Lambda$ for every $B \in \xi$, then by Prob. 125, $\xi' = \xi \cup \{A\}$ is a centralized family on X, where $\xi' \neq \xi$, $\xi' \supset \xi$, contradicting *(a)*. Evidently, *(c')* is trivially a reformulation of the hypothesis *(c.)*

(c) $\Rightarrow$ *(e)*. If $A_1 \notin \xi$ and $A_2 \notin \xi$, then by virtue of *(c)*, there exist $B_1 \in \xi$ and $B_2 \in \xi$, for which $A_1 \cap B_1 = \Lambda$ and $A_2 \cap B_2 = \Lambda$. Then, $(A_1 \cup A_2) \cap (B_1 \cap B_2) = \Lambda$. However, since ξ is a prefilter, there exists $B' \in \xi$, $B' \subset B_1 \cap B_2$. Then, $(A_1 \cup A_2) \cap B' = \Lambda$, contradicting that ξ is centralized. Obviously, $\xi \ni X$.

(e) $\Rightarrow$ *(d)*. We have $X = (X \setminus A) \cup A$, $X \in \xi$. Now *(e)* implies *(d)*.

(d) $\Rightarrow$ *(a)*. Let ξ' be a family of sets centralised on X, containing ξ. If $A \in \xi'$, then $X \setminus A \notin \xi'$ and, moreover, $X \setminus A \notin \xi$. By *(d)*, then $A \in \xi$. Hence, $\xi' \subset \xi$, that is, $\xi' = \xi$, and *(a)* is proved.

129. By Prob. 90, if the cardinality of each of the sets $A_1, \ldots, A_k$ is less than that of X, then we have also $|\bigcup \{A_i : i = 1, \ldots, k\}| < |X|$. It is now clear that $\eta(X)$ is a filter. In Prob. 59, it has been shown that X can be represented as the union of disjoint sets X_1 and X_2, each of which is equipotent with X. Consequently (cf. Prob. 128*(d)*), $\eta(X)$ is not an ultrafilter.

130. A filter ξ is a centralized family of sets. Every such family by Prob. 127 is contained in some maximal centralized family of sets, the latter by Prob. 128 is an ultrafilter.

135. The assertions (*a*), (*b*), (*c*) and (*d*) are routine to prove. We shall, therefore, establish (*e*). It is to be verified that for $f\xi$ the condition (*c*) of Prob. 128 is satisfied. By (*a*), $f\xi$ is a prefilter. Let $A \notin f\xi$. Then $f^{-1}A \notin \xi$ (because $Y = fX$). Hence, by Prob. 128(*c*), there exists $B \in \xi$, for which $(f^{-1}A) \cap B = \Lambda$. But $f((f^{-1}A) \cap B) = A \cap fB$. Hence, $A \cap fB = \Lambda$. Also, $fB \in f\xi$, because $B \in \xi$. Consequently, the condition (*c*) of Prob. 128 is satisfied for $f\xi$, and $f\xi$ is an ultrafilter by Prob. 128.

137. Put $A' = X \setminus \bigcap \{A : A \in \xi'\}$. Then $A' \cap (\bigcap \{A : A \in \xi'\}) = \Lambda$, which implies owing to the assemption on ξ (and the fact that $|\xi' \cup \{A'\}| < \tau$), that $A' \notin \xi$. Since ξ is an ultrafilter on X, the conclusion is that $X \setminus A' \in \xi$, that is, $\bigcap \{A : A \in \xi'\} \in \xi$.

138. Let ξ be a non-trivial σ-ultrafilter on X. Put $\tilde{\xi} = \{A \subset Y : A \cap X \in \xi\}$. If $A \cup B = Y$, then $(A \cap X) \cup (B \cap X) = X$, and from ξ being an ultrafilter on X it follows (cf. Prob. 123) that either $A \cap X$, or $B \cap X$ is an element of ξ. But, then, either $A \in \tilde{\xi}$, or $B \in \tilde{\xi}$. These reasonings establish (cf. Prob. 128) that $\tilde{\xi}$ is an ultrafilter on If $A_i \in \tilde{\xi}$, where $i = 1, 2, \ldots$, then $\bigcap \{A_i : i = 1, 2, \ldots\} \cap X = \bigcap \{A_i \cap X : i = 1, \ldots\} \in \xi$ because $A_i \cap X \in \xi$ for each $i = 1, 2, \ldots$ and ξ is an σ-ultrafilter. Hence, $\bigcap \{A_i : i = 1, 2, \ldots\} \in \tilde{\xi}$. Evidently, $\xi \subset \tilde{\xi}$ implying that $\bigcap \{A : A \in \tilde{\xi}\} \subset \bigcap \{A : A \in \xi\} = \Lambda$. Thus, it is proved that $\tilde{\xi}$ is a non-trivial σ-ultrafilter on Y.

139. Cf. Prob. 138.

140. Assume the contrary and let τ^* be the smallest cardinal number such that there is $\xi^* \subset \xi$, satisfying $|\xi^*| \leqslant \tau^*$ and $\bigcap \{A : A \in \xi^*\} = \Lambda$. Then $\tau^* \leqslant \tau$ and, hence (Prob. 139), τ^* is non-measurable. Fix a family $\xi^* \subset \xi$, having the properties mentioned. Let there be set in one-to-one correspondence to the elements of ξ^* the elements of some minimal well-ordered set $(M, <)$ of cardinality τ^* : $\xi^* = \{A_\alpha : \alpha \in M\}$. Put (for each $\alpha \in M$) $B_\alpha = \bigcap \{A_{\alpha'} : \alpha' < \alpha\}$. By the minimality of the well-ordernig under consideration, the definition of τ^* and Prob. 137, $B_\alpha \in \xi$ for each $\alpha \in M$. Thence and from the definition of τ^* it follows that $\{B_\alpha : \alpha \in M\}$ is not a standard set: a passage to a suitable cofinal subset of M can be made in order that $B_{\alpha'} \neq B_{\alpha''}$ for $\alpha' \neq \alpha''$. Assume this condition to be already satisfied. Evidently, if $\alpha' < \alpha''$, then $B_{\alpha'} \supset B_{\alpha''}$. Let θ be the first element of $(M, <)$. Put $B_\theta = Z$. Define a function $\pi: Z \to M$ by the rule: $\pi^{-1}(\alpha) = B_\alpha \setminus B_{\alpha+1}$ for each $\alpha \in M$. Since $\pi^{-1}(M) = Z \in \xi$ and the inverse image of the sum is equal to the sum of inverse images, as well as the inverse image of an intersection is equal to the intersection of inverse images, hence the σ-ultrafilter η on M is defined by the condition $L \in \eta$, where $L \subset M$ iff $\pi^{-1}L \in \xi$. Since $B_\alpha = \pi^{-1}\pi(B_\alpha) \in \xi$, $\pi(B_\alpha) \in \eta$ and $\bigcap\{\pi(B_\alpha) : \alpha \in M\} = \pi(\bigcap\{B_\alpha : \alpha \in M\}) = \pi(\Lambda) = \Lambda$. Consequently, η is a non-trivial σ-ultrafilter on M; also $|M| = \tau^*$, contradicting that τ^* is non-measurable.

141. Let ξ be an σ-ultrafilter on X. There are the following two possibilities:

(i) $X_{\alpha_0} \in \xi$ for some $\alpha_0 \in A$. Then the restriction $\xi_{\alpha_0} = \xi \mid X_{\alpha_0}$ of an ultrafilter on the set X_{α_0} is a σ-ultrafilter on X_{α_0} (Prob. 136). Since $\tau_{\alpha_0} = |X_{\alpha_0}|$ is a non-measurable cardinal number, we conclude that the ultrafilter ξ_{α_0} is trivial, that is, $\{x_{\alpha_0}\} \in \xi_{\alpha_0}$ some point $x_{\alpha_0} \in X_{\alpha_0}$. But $\xi_{\alpha_0} \subset \xi$ (cf. Prob. 133); hence $\{x_{\alpha_0}\} \in \xi$, and the utrafilter ξ is trivial.

(ii) $X_\alpha \notin \xi$ for each $\alpha \in A$. Then (Prob. 128) $X \setminus X_\alpha \in \xi$ for each $\alpha \in A$. Since $= \tau$ is a non-measurable cardinal number and ξ is a σ-ultrafilter, hence $\bigcap \{X \setminus X_\alpha;$

$\alpha \in A\} \in \xi$ (cf. Probs. 137, 139 and 140). However, by De Morgan's rules, $\bigcap\{X \setminus X\alpha : \alpha \in A\} = X \setminus (\bigcup \{X\alpha : \alpha \in A\} = \Lambda$. Thus, $\Lambda \in \xi$, contradicting the definition of an ultrafilter.

142. Assume the contrary, and let ξ be a non-trivial σ-ultrafilter on exp X. Put, for each point $x \in X$, $\eta_1(x) = \{A \subset X : A \ni x\}$, $\eta_2(x) = \{A \subset X : A \not\ni x\}$. Further, let $F = \{x \in X : \eta_1(x) \in \xi\}$, $\Phi = X \setminus F$. Then, $\eta_1(x) \cup \eta_2(x) = \exp X$ and $\eta_1(x) \cap \eta_2(x) = \Lambda$. Hence, $\Phi = \{x \in X : \eta_2(x) \in \xi\}$ (since ξ is an ultrafilter). From $|\Phi| \leqslant |X|$ and $|F| \leqslant |X|$, $|X|$ being non-measurable, it follows that (cf. Probs. 137, 139, 140) that $A^* = \bigcap \{\eta_1(x) : x \in F\} \in \xi$ and $B^* = \bigcap \{\eta_2(x) : x \in \Phi\} \in \xi$. Hence, $A^* \cap B^* \in \xi$. It is to be shown that $A^* \cap B^* = \{F\}$, that is, the subset $A^* \cap B^*$ of the set exp X consists of exactly one point, and this point is an element of $F \in \exp X$. From $x \in \Phi = X \setminus F$ and the definition of $\eta_2(x)$ it follows that $F \in \eta_2(x)$ for each $x \in \Phi$. Hence, $F \in B^*$. Exactly in a similar way, from the definition of $\eta_1(x)$ it follows that $F \in \eta_1(x)$ for every $x \in F$. Hence, $F \in A^*$, that is, $F \in A^* \cap B^*$. Furthermore, if $Q \in A^*$ then, evidently, $Q \supset F$, and if $Q \in B^*$ then it is equally obvious that $Q \cap \Phi = \Lambda$, that is, $Q \subset F$. Consequently, if $Q \in A^* \cap B^*$, then $Q = F$ and $\{F\} = A^* \cap B^* \in \xi$. But then the ultrafilter ξ is trivial, it is a collection of all subsets of exp X, containing F (as an element). A contradiction.

143. If τ is the least measurable cardinal number, then τ is regular (by Prob. 141). In Prob. 142, it is established that if λ is non-measurable, then 2^λ is also so. Hence, τ is inaccessible (cf. Prob. 139).

144. Cf. Prob. 53 of Chap. 4 and Prob. 180 of Chap. 6.

145. The solution is routine via Prob. 63 of Chap. 4.

146. See, Arkhangel'skii [6] and Frolick [3].

Chapter 2

Topological Spaces. Metric Spaces. Basic Concepts in Topological and Metric Spaces

2.1. Theoretical Introduction

Suppose that X is a set. A family $\mathcal{T}$ of subsets of X is called a *topology* on X if the following conditions are satisfied:

T–1. The intersection of any finite collection of elements of $\mathcal{T}$ is an element of $\mathcal{T}$.

T–2. The union of any collection of elements of $\mathcal{T}$ again belongs to $\mathcal{T}$.

T–3. The entire set X and the empty subset Λ are elements of $\mathcal{T}$.

The set X together with a topology $\mathcal{T}$ fixed on it is called a *topological space* (or simply a *space*) and denoted by the pair $(X, \mathcal{T})$ or just X. Every subset of X belonging to $\mathcal{T}$ is called an *open set* in the topological space $(X, \mathcal{T})$. A set $F \subset X$ is called *closed* in $(X, \mathcal{T})$ if the set $X \setminus F$ is open in $(X, \mathcal{T})$, that is, if $X \setminus F \subset \mathcal{T}$. A set which is both open and closed is called *clopen.*

All the subsequent notations and concepts relate to points and sets lying in a fixed but arbitrary space X.

Any open set containing A is called a *neighbourhood* (*nhood*, for short) of A. We often consider in particular a *nhood of a point* (that is, of a one point subset) of the space X.

We call x an *adherent point* (or *closure* point) of the set A if every nhood of x has a nonempty intersection with A, and we write $x \delta A$ or $x \in [A]$.

The set of all points of the space X which are adherent points of A is called the *closure* of A and denoted by $[A]$. The *interior* of A is the set $X \setminus [X \setminus A]$, and is denoted by $\operatorname{Int} A$. The *frontier* or *boundary* of the set A is the set $[A] \cap [X \setminus A]$; we denote it by $\operatorname{Fr} A$. If $x \in \operatorname{Int} A$, then x is called an *interior point* of A. A point x is called a *limit point* (or *accumulation point*) of A if every nhood of this point contains some point of A distinct from x. A point x in a space X is said to be *isolated* if the set $\{x\}$ is open in X. The space X is said to be *discrete* if all its points are isolated (the topology of such a space is called the *discrete* topology).

A sequence $\{x_n : n = 1, 2, \ldots\}$ of points of a space X *converges* to a point x, if for each nhood Ox of x there exists an integer n^* such that $n > n^*$ implies $x_n \in Ox$. We express this as $\lim \{x_n\} = x$ and call x the *limit* of the sequence $\{x_n : n = 1, 2, \ldots\}$. An *adherent* point of a *family* ξ of sets is any point that belongs to the closure of each member of this family. The set of all adherent points of ξ is denoted by $[\xi]$; evidently, $[\xi] = \bigcap \{[A] : A \in \xi\}$. We say that ξ *converges* to a set $A \subset X$, if for each nhood A of OA there is an element $Q \in \xi$ included in OA. If $A = \{x\}$, where $x \in X$, and ξ converges to A we say that ξ converges to x; x is called a limit of ξ and we write $x = \lim \xi$. We also denote by $\lim \xi$ the set of all points of X which are limits of ξ. Of special importance in topology is convergence of filters and ultrafilters in the above sense. In keeping with this general approach, it is useful to formulate the notion of convergence of sequences in the following way. We say that a sequence $\{x_n : n = 1, 2, \ldots\}$ of points of X *converges to a set* $A \subset X$, if for each nhood OA of A there is an n^* such that $x_n \in OA$ for all $n > n^*$.

In many cases the notions of *nontrivial* sequence and an *essential* sequence are useful. A sequence $\{x_n : n = 1, 2, \ldots\}$ is called *essential* if the set of points of this sequence is infinite (that is, the set $\{y :$ there exists an index n such that $x_n = y\}$ is infinite), and *nontrivial* if $n' \neq n''$ always implies that $x_{n'} \neq x_{n''}$.

A family σ of sets is called a *net* in X if each open set is the union of a family of elements of σ. If σ is a net in X and all elements of σ are open sets, then σ is called a *base for the space* X (or a base for the topology defined on X). A family γ of open subsets of a topological space X is called a *subbase*, if the collection of all finite intersections of elements of γ forms a base for X.

A family η of nhoods of a subset A of X is called a *defining family of nhoods of* A, if for each nhood U of $A \in X$ there is a $V \in \eta$ such that $V \subset U$. If A consists of a single point x and η is a defining family of nhoods of A, then we say that η is a base for X (or, for the topology on X) *at the point* $x \in X$.

In topology the following essentially set-theoretic notion is quite important. Suppose that γ is a family of subsets of X and $A \subset X$. The set $\bigcup \{U \in \gamma : U \cap A \neq \Lambda\}$ is called the *star* of A with respect to γ and always denoted by $\gamma(A)$. If $A = \{x\}$ we write $\gamma(x)$ for $\gamma(\{x\})$.

If λ is a family of subsets of a space X and each nonempty open set includes some nonempty element of λ, then λ is said to be *dense* in X. A family of open sets dense in X is called a *π-base* for X.

A subset M is said to be *dense* (in X), if M intersects every nonempty open set U (that is, if $U \cap M \neq \Lambda$). A set P is called *nowhere dense* (in X), if for every nonempty open set U there exists a nonempty open set $U' \subset U$, having no common point with $P : U' \cap P = \Lambda$.

A number of the following notions concern fixed but arbitrary function $f: X \to Y$ from a topological space X into a topological space Y.† The function f is called *continuous* if $f^{-1}(G)$ is open in X whenever G is open in Y.

If f is continuous, one-to-one and $f(X) = Y$, f is called a *contraction*.

If both $f: X \to Y$ and $f^{-1}: Y \to X$ are contractions, then we call f a *homeomorphism* (or *topological mapping*) of X onto Y; the spaces X and Y are called *homeomorphic* (or *topologically equivalent*) if there is a homeomorphism of one onto the other.

Let $\mathcal{T}_1$ and $\mathcal{T}_2$ be two topologies on a set X. We say that $\mathcal{T}_1$ is *larger* (or *stronger*, or *finer*) than or equal to $\mathcal{T}_2$, if $\mathcal{T}_2 \subset \mathcal{T}_1$ (not excluding the possibility that $\mathcal{T}_2 = \mathcal{T}_1$).†† If $\mathcal{T}_1 = \mathcal{T}_2$, we say that the topologies $\mathcal{T}_1$ and $\mathcal{T}_2$ coincide or are equal.

Let $(X, \mathcal{T})$ be a topological space and $X_0 \subset X$. The topology $\mathcal{T}$ *induces* (or generates) a topology $\mathcal{T}_0 = \mathcal{T} \mid X_0$ on X_0 in the following manner: $\mathcal{T}_0 = \{U \cap X_0 : U \in \mathcal{T}\}$. Also, $(X_0, \mathcal{T}_0)$ (or, simply X_0) is said to be a *subspace* of X.†††

When considering a space X, a subspace X_0 and a subset A of X_0, the closure of A in X is denoted by $[A]_X$, or as usual by $[A]$, but the closure of A in the subspace X_0 is invariably denoted by $[A]_{X_0}$.

A homeomorphism of X onto some subspace Y_0 of Y is called a homeomorphism of X *into* Y, or an *embedding* of X into Y, and we say that X is *embedded* in Y by f.

A property $\mathcal{P}$ of a topological space is called a *topological* property, or a *topological invariant*, if it is preserved under homeomorphism, that is, if from the fact that $\mathcal{P}$ is possessed by any topological space X it always follows that each space homeomorphic to X has also property $\mathcal{P}$.

---

†When we speak of a function f from a topological space X into a topological space Y we mean a function of the set X into the set Y and topologies fixed on X and Y.

††When $\mathcal{T}_2 \subset \mathcal{T}_1$ and $\mathcal{T}_2 \neq \mathcal{T}_1$, we say that $\mathcal{T}_1$ is *strictly larger* than $\mathcal{T}_2$.

†††If X_0 is a closed (open) set of X, we call $(X_0, \mathcal{T}_0)$ a closed (open) subspace of $(X, \mathcal{T})$.

An important topological invariant is *compactness.*† The credit for developing the concept of sequential compactness† as a basic mathematical concept goes to P. S. Alexandrov.

A *cover* (or *covering*) of a topological space X is just a cover of the underlying point set of this space, that is, every family ξ of subsets of X such that for each $x \in X$ there is an element A of ξ, containing x. A covering ω of X is called *open* (resp. *closed*) if ω consists of open (resp. closed) sets.

If a subfamily ω' of ω is itself a cover of the space X, then ω' is called a *subcover* of ω. A space X is *compact* iff every open cover of X has a finite subcover.

In addition to the notion of a subcover, that of a *refinement* of a cover plays an important role in topology. We say that a cover (family of sets) λ is a refinement of a cover (family of sets) γ, or that λ *refines* γ if for each $V \in \lambda$ there is a set $U \in \gamma$ such that $V \subset U$.

The notion of Hausdorff connectedness is related to a number of topological invariants already introduced. A space X is *connected* if it cannot be represented as the union of two nonempty disjoint open (in X) sets. In the contrary case, the space X is said to be *disconnected.* A subset A of X is called *connected* if the subspace A of X is a connected space in the sense of the above definition.

A property naturally opposed to connectedness is that of 0-*dimensionality*. *A* space is called 0-*dimensional* if it has a base consisting of clopen sets. 0-*dimensionality* is also a topological invariant.

The property of separation is one of the basic topological invariants. It is characterized by the so-called *separation axioms*, generating a hierarchy of requirements.

A topological space X satisfies the T_0-separation axiom, or is a T_0-*space*, if for each pair x_1 and x_2 of distinct points of X there exists a nhood of at least one point which does not contain the other (A. N. Kolmogorov).

A topological space X satisfies the T_1-separation axiom, or is a T_1-*space*, if for each pair x_1 and x_2 of distinct points of X there exist nhoods Ox_1 and Ox_2 of x_1 and x_2 such that $x_2 \notin Ox_1$ and $x_1 \notin Ox_2$.

A topological space X is called a *Hausdorff space*, or T_2-space, if whenever x_1 and x_2 are any two distinct points of X there exist disjoint nhoods of x_1 and x_2 (the topology on X is then called a Hausdorff topology). A compact Hausdorff space is called a *compactum*. A T_1-space X is called a

†In Russian works the terms 'compact' and 'sequentially compact' are usually referred to as 'bicompact' and 'compact'.

regular space (also a T_3-*space*) if for every point $x \in X$ and every closed set $F \subset X$ with $x \notin F$ there are nhoods Ox of x and OF of F such that $Ox \cap OF = \Lambda$.

A T_1-space X is called a *normal* space if every pair of disjoint closed sets in X have disjoint nhoods.

There is one more important nontrivial restrictive type of separability called *complete regularity*, which is weaker than normality but stronger than regularity. A T_1-space X is *completely regular* (or a *Tychonoff space*) if for every point $x_0 \in X$ and every closed set F in X with $x \notin F$ there is a continuous function f defined on X such that $0 \leqslant f(x) \leqslant 1$ for all $x \in X$, $f(x_0) = 0$ and $f(x) = 1$ for all $x \in F$ (see Problems 23 and 13 of Chap. 3).

A number of highly important topological invariants involve cardinal numbers, the latter themselves being the values of these invariants.

The least cardinal of a base for the space X is called the *weight* of X and is denoted by $w(X)$. The least cardinal of a net in X is called the α-*weight* of X and denoted by $\alpha(X)$. The *P-weight* of X denote by $Pw(X)$ is the cardinal of a π-base for X. The least cardinal of a dense subset of X is called the *density* of X and denoted by $s(X)$. We denote by $c(X)$ the least cardinal number τ such that the cardinal of every disjoint collection of nonempty open subsets of X is at most τ. The invariant $c(X)$ is called the *Souslin number* (or cellular number) of the space X.

The *character* $\chi(A, X)$ *of a subset* $A \subset X$ *of a space* X is the least cardinal of a defining family of nhoods of A in X. If $A = \{x\}$, where $x \in X$, then instead of $\chi(\{x\}, X)$ we write $\chi(x, X)$; we also call $\chi(x, X)$ the *character of a point* x in X (or, conversely, *the character* of X *at the point* x). The *pseudo-character* $\psi(A, X)$ of a subset A *of* a T_1-space X is the least cardinal of a family $\mathcal{E}$ of open sets in X such that $A = \cap \{V : V \in \mathcal{E}\}$ (in Problem 41 of Chap. 2 this definition is shown to be correct). If $A = \{x\}$, $x \in X$, then we write $\psi(x, X)$ for $\psi(\{x\}, X)$ and call $\psi(x, X)$ the *pseudocharacter* of X at x. If $\psi(A, X) \leqslant \tau$, we say that A is a G_τ-*set* in X. It is normal practice to call a $G_{\aleph_0}$-set a G_δ-set (in X).

The *tightness* $t(x, X)$ of X at $x \in X$ is the least cardinal number satisfying the condition: if $A \subset X$ and $x \in [A]$, then there exist $A' \subset A$ satisfying $|A'| \leqslant \tau$ and $x \in [A']$. The *character* $\chi(X)$ (resp. the *pseudocharacter* $\psi(X)$, the *tightness* $t(X)$) of X is the least cardinal number τ such that $\chi(x, X) \leqslant \tau$ (resp. $\psi(x, X) \leqslant \tau$, $t(x, X) \leqslant \tau$) for all $x \in X$. The space X satisfies the *second axiom of countability*, or is *second countable* if $w(X) \leqslant \aleph_0$, that is, if X has an at most countable open base. The space X satisfies the *first axiom of countability* or is *first countable*, if $\chi(X) \leqslant \aleph_0$, that is, if X has a countable base at each point. If $s(X) \leqslant \aleph_0$, then we say that X is *separable*; if $c(X) \leqslant \aleph_0$, then X is said to have the *Souslin property*. We call X a *Fréchet-Urysohn space* if $x \in X$, $A \subset X$ and $x \in [A]$ always imply that there exists a sequence $\{x_n : n = 1, 2, \ldots\}$ of points of A which converges to x. We call X a *sequential space* if $A \subset X$ and $A \neq [A]$ always imply that there is a sequence $\{x_n : n = 1, 2, \ldots\}$ of points of A converging to some point $x \in [A] \setminus A$. It differs from the preceding definition in that in the latter case the point $x \in [A] \setminus A$ cannot always be chosen arbitrarily. Finally, the *index of compactness* $ic(X)$ of X is defined to be the least cardinal number τ such that every open cover of X has a subcover whose cardinal number is at most τ. If the index of compactness of X equals $\aleph_0$, that is, if every open cover of this space has a countable subcover, then X is said to be a *Lindelöf space*.† If X is normal and every closed subset is a G_δ-set, then X is said to be *completely normal*.

The *disjoint union* (or *free union*) of the topological spaces X_α, where $\alpha \in A$, is the disjoint union $X = \bigcup_c\{X_\alpha : \alpha \in A\}$ of the sets X_α for each $\alpha \in A$ endowed with the topology uniquely defined by the following requirements: (*a*) each $i_\alpha(X_\alpha)$, as a subspace of the disjoint union X, is homeomorphic to X_α, the homeomorphism being the map $i_\alpha : X_\alpha \to \bigcup_c\{X_\alpha : \alpha \in A\}$; (*b*) $i_\alpha(X_\alpha)$ is open in X for each $\alpha \in A$ (see p. 3).

In what follows i_α will be called the *standard embedding* of X_α into the disjoint union, and $i_\alpha(X_\alpha)$ itself the *standard realization* of the space X_α in the disjoint union.

The notion of the *topological* (or *Tychonoff*) *product* of topological spaces is very important. Suppose we are given a space $(X_\alpha, \mathcal{T}_\alpha)$ for each $\alpha \in A$. Let X be the product $\Pi\{X_\alpha : \alpha \in A\}$ of the sets X_α. We now define a topology $\mathcal{T}$ on X called the *product of the topologies* $\{\mathcal{T}_\alpha : \alpha \in A\}$. We recall that $\mathcal{E}_k(A) = \exp_{\aleph_0}(A)$ is the set of all finite subsets of A. If $U \subset X$ and $U = \Pi\{U_\alpha : \alpha \in A\}$, where $U_\alpha \in \mathcal{T}_\alpha$ for all $\alpha \in A$ and $U_\alpha = X_\alpha$ for all $\alpha \in A \setminus F$, for some finite subset $F \subset A$ (that is, $F \in \mathcal{E}_k(A)$), then we say that U is a *standard* set. We also put $k(U) = \{\alpha \in A : U_\alpha \neq X_\alpha\}$. Then $k(U) \in \mathcal{E}_k(A)$ for each standard set U. The family of all standard sets serves as a base for some topology $\mathcal{T}$ on X; we call this topology the *product topology* (or *Tychonoff topology*) and denote it by $\Pi\{\mathcal{T}_\alpha : \alpha \in A\}$. However, we usually write for short: $X = \Pi\{X_\alpha : \alpha \subset A\}$, it being under-

†A *finally compact* space in Russian terminology.

stood that the set X is endowed with the topology corresponding to the topologies defined on the X_α, which are also not explicitly indicated. The base for a product topology formed by the collection of standard sets will be called the *standard base* and the space X itself together with the product topology is called the *product topological* (or *Tychonoff topological*) *space*, or, simply *product space*.

The construction of the product topology as well as the simpler operations of passage to a subspace and splitting apart a disjoint union of topological spaces enable us to construct new spaces from old. Further constructions of this sort are made possible by the notion of a *quotient map*.

A map $f: X \to Y$ of a topological space X onto a topological space Y is a *quotient* (or *natural map*) if, whenever V is an open set of $Y, f^{-1}(V)$ is open in X and conversely. The notion of a quotient map is closely related to that of a *decomposition* (*partition*) of X (Alexandrov). A *decomposition* of X is a closed disjoint cover of X. In what follows, we shall denote by ω a fixed decomposition of the topological space $(X, \mathcal{T})$. Each point $x \in X$ is contained in a unique element of the decomposition ω, that is, there is defined a *natural map* (or *decomposition map*) $\pi_\omega : X \to \omega$ of the topological space X onto the set ω. We can now assign to ω a *topology*, called the *quotient* or *natural topology on* ω by calling a set $V \subset \omega$ open iff $\pi_\omega^{-1}(V)$ is open in X, that is, if the union of elements of V, regarded as a subset of X, is an open set in X. The natural topology on ω is denoted by $\mathcal{T}_\omega$ ($\mathcal{T}$ being a topology defined on X). We call the topologies $\mathcal{T}' = \mathcal{T}_\omega$ *associated* owing to the fact that $\pi_\omega: (X, \mathcal{T}) \to (\omega, \mathcal{T}')$ is always a quotient map. The set ω endowed with the topology $\mathcal{T}' = \mathcal{T}_\omega$ is called the ω-*quotient space* or the *decomposition space* ω of X, symbolically written as $(\omega, \mathcal{T}') = X | \omega$. The map π_ω is also called the *natural projection of the space* X onto the *decomposition space* ω.

Continuous open maps and continuous closed maps are important special cases of quotient maps. A map is called *open* (*closed*), if the image of each open (closed) set is open (closed). A map that is simultaneously open and closed is called *clopen*. The modifier continuous will hereafter be dropped but always implied, unless stated to the contrary. It should be emphasized that the definitions given above apply to all mappings and not just to 'onto' mappings. A continuous mapping $f: X \to Y$ is said to be *compact* if the subspace $f^{-1}y$ is compact for every point $y \in Y$. A compact closed and continuous mapping is called a *perfect mapping*.

Historically, the notion of a topological space was preceded by that of a metric space introduced by Fréchet in 1906. Let X be a set. A function ρ which assigns to each pair of points x and y of X a non-negative real number $\rho(x, y)$ (called the *distance* from x to y) is said to be a *metric* on X, provided the following axioms are satisfied for all $x, y, z \in X$:

M–1. $\rho(x, y) = \rho(y, x)$ (*symmetry axiom*).
M–2. $\rho(x, x) \geqslant 0$; $\rho(x, y) = 0$ iff $x = y$ (*identity axiom*).
M–3. $\rho(x, z) \leqslant \rho(x, y) + \rho(y, z)$ (*triangle axiom*).

Let $x \in X$ and let $M \subset X$ be any subset of a metric space (X, ρ). The number $\rho(x, M) = \inf\{\rho(x, y): y \in M\}$ is called the *distance from* x to M and the number $\rho(M_1, M_2) = \inf\{\rho(x, M_2): x \in M_1\} = \inf\{\rho(x, y): x \in M_1, y \in M_2\}$ the *distance from the set* M_1 *to the set* M_2.

A subset A of a metric space (X, ρ) is said to be *bounded* if there is some constant K such that $\rho(x, y) < K$ for all $x \in A$, $y \in A$; the *diameter* of the bounded set A (diam A) is the real number $\sup\{\rho(x, y): x \in A, y \in A\}$. Let $A \subset X$ and let ε be a positive number. The set $\{y \in X: \rho(A, y) < \varepsilon\}$ is called an ε-*nhood* of A and denoted by $O_\varepsilon(A)$. If $A = \{x\}$, where $x \in X$, then $O_\varepsilon(x) = O_\varepsilon(\{x\})$ is called an ε-nhood of x or an ε-*ball* (spherical nhood) in (X, ρ) of *radius* ε and *center* x.

A subset S of the metric space (X, ρ) is called an ε-*net* in X if for each point $x \in X$ there is a point $x' \in S$ such that $\rho(x, x') < \varepsilon$, that is, if $\cup\{O_\varepsilon(x): x \in S\} = X$. In symbols: $O_\varepsilon(S) = X$. The metric space (X, ρ) is called *totally bounded* if X has a *finite* ε-net for each $\varepsilon > 0$.

If A is a subset of a metric space (X, ρ), then A inherits from X the metric $\rho' = \rho \mid A$: $\rho'(x, y) = \rho(x, y)$, if $x \in A, y \in A$. This converts A into the metric space (A, ρ') which is called a (metric) *subspace of the metric space* (X, ρ); it is often denoted by (A, ρ) or A for short.

A subset A of a metric space (X, ρ) is said to be *totally bounded*, if the metric space (A, ρ) is totally bounded.

The next notion plays a central role in the theory of metric spaces.

A sequence $\{x_n\}$† of points in a metric space X *converges* to a point x_0, if the sequence of real numbers $c_n = \rho(x_0, x_n)$ converges to zero (that is, if for every $\varepsilon > 0$ there is a positive integer n_0 such that whenever $n > n_0$ we have $\rho(x_n, x_0) < \varepsilon$). A sequence of points $\{(x_n)\}$ in X is said to be a *Cauchy* (or *fundamental*) *sequence* if for each $\varepsilon > 0$ there is a positive

†For short we write $\{x_n\}$ instead of

$$\{x_n : n = 1, 2, \ldots\} \quad \text{or} \quad \{x_n : n \in N^+\}.$$

integer n_0 such that $\rho(x_n, x_m) < \varepsilon$ whenever $n, m > n_0$. The metric space X is said to be *complete* if every Cauchy sequence of points in X converges to a point in X.

A subset U of a metric space (X, ρ) is open if $\rho(x, X \setminus U) > 0$ for each $x \in U$.

We denote by $\mathcal{T}_\rho$ tbe collection of all open sets in a metric space (X, ρ). The system $\mathcal{T}_\rho$ forms a topology on X (this assertion, of course, needs proof) and is called the *topology generated* (or *induced*) by the *metric* ρ on X. Thus every metric space (X, ρ) can be regarded as the topological space $(X, \mathcal{T}_\rho)$.

Let ρ and ρ^* be two metrics on X. We write $\rho = \rho^*$ if $\rho(x, y) = \rho^*(x, y)$ for all $x, y \in X$. It may happen that $\rho \neq \rho^*$ but $\mathcal{T}_\rho = \mathcal{T}_{\rho^*}$. If $\mathcal{T}_\rho = \mathcal{T}_{\rho^*}$, then ρ and ρ^* are said to be *topologically equivalent* metrics.

There are topologies which are not induced by any metric. A topological space $(X, \mathcal{T})$ for which $\mathcal{T} = \mathcal{T}_\rho$, where ρ is some metric on X, is said to be *metrizable*. Metrizability is also a topological invariant.

A one-to-one map $f: (X, \rho) \to (Y, \rho')$ of a metric space (X, ρ) onto a metric space (Y, ρ') is called an *isometry*, or an *isometric mapping*, if for each pair of points x_1 and x_2 of X we have $\rho(x_1, x_2) = \rho'(fx_1, fx_2)$. An isometry is thus a distance-preserving map.

An *isometric embedding* of a metric space (X, ρ) into another metric space (Y, ρ^*) is an isometry from (X, ρ) onto some metric subspace of (Y, ρ^*).

If there is an isometry of (X, ρ) onto (Y, ρ^*), then we say that (X, ρ) and (Y, ρ^*) are *isometric*, or that each of them is an isometric image of the other. Metric spaces that are isometric to one another are often said to be *metrically equivalent*. It should be noted that the inverse of an isometric onto map is an isometry.

If $\mathcal{P}$ is a certain topological property, then we say that a space X has this property *hereditarily*, or that X is $\mathcal{P}$-hereditary iff X and every subspace $X' \subset X$ also has property $\mathcal{P}$.

2.2. Simple Problems in General Topology

A mapping of the family of all subsets of a fixed set X into itself is called an *operation on the family of all subsets of* X. An operation C on the family of all subsets of X is called a *Kuratowski* (*closure*) *operation*, if the following four properties (also called Kuratowski closure axioms) are satisfied:

(i) For each $A \subset X$, $A \subset C(A)$.
(ii) If $A \subset X$, $B \subset X$ and $A \subset B$, then $C(A) \subset C(B)$.

(iii) If $A \subset X$ and $B \subset X$, then $C(A \cup B) = C(A) \cup C(B)$.
(iv) For each $A \subset X$, $C(C(A)) = C(A)$.

Let $(X, \mathcal{T})$ be a a topological space. The operation which assigns to each subset A of X its closure $[A]$ in $(X, \mathcal{T})$ (see p. 39), is called the *closure operation.*

Another important operation defined on the family of all subsets of a topological space is the *interior* (Int) operation which assigns to each set $A \subset X$ its interior Int A, or the interior of A in $(X, \mathcal{T})$. A closed set F (an open set U) is said to be *regular closed* (resp. *regular open*), if $F = [\text{Int}\, F]$ (resp. $\text{Int}\,[U] = U$).

A subset A of a topological space X is called *discrete-in-itself* if A, as a subspace of X, is a discrete space (see p. 39). A closed discrete-in-itself set is *discrete in X* (or *relatively discrete*). A topological space X is called an *indiscrete* space if the topology on X consists of just two open subsets: the empty set and X itself. Each infinite set X has many topologies. It is useful to write $\mathcal{H}(X)$ for the collection of all topologies on X. The collection $\mathcal{H}(X)$ may be naturally regarded as an ordered set. If $\mathcal{T}', \mathcal{T}'' \in \mathcal{H}(X)$, we put $\mathcal{T}' < \mathcal{T}''$ whenever $\mathcal{T}' \subset \mathcal{T}''$. Of course, this is not a linear ordering. Hence it is helpful to consider a *chain* or a *nest* of topologies, that is, a chain (a linearly ordered set) in the ordered set $(\mathcal{H}(X), <)$. In $(\mathcal{H}(X), <)$ there is a maximal and a minimal element, namely the discrete and indiscrete topology, respectively.

The *(topological) union* $\bigvee \mathcal{E}$ of an arbitrary collection of topologies $\mathcal{E}$, on a given set X, is the topology whose subbase is the (set-theoretic) union $\mathcal{P} = \bigcup \{\mathcal{T}: \mathcal{T} \in \mathcal{E}\}$ of the topologies in $\mathcal{E}$ (moreover, each topology is understood to be a subfamily of the set of all subsets of X). The (topological) *intersection* $\bigwedge \mathcal{E}$ of the family $\mathcal{T}$ of topologies is defined simply as their set theoretic intersection $\bigwedge \mathcal{E} = \{\mathcal{T}: \mathcal{T} \in \mathcal{E}\}$.

Suppose that X is a set and B a fixed subset of X. We write $\mathcal{D}(B)$ for the family whose members are X and all subsets of B. The family $\mathcal{D}(B)$ is a topology on X. We call this the *B-topology.* The B-topology is of little interest in itself. However, nontrivial topologies can be obtained if the B-topology is meshed with other topologies in the following way. The *B-modification* $\mathcal{T}_B$ of an arbitrary topology $\mathcal{T}$ on X is the topological union of the topologies $\mathcal{T}$ and $\mathcal{D}(B)$. The B-extension is the topology which has a subbase consisting of the sets in $\mathcal{T} \cup \mathcal{T} \mid B$, $\mathcal{T} \mid B = \{U \cap B: U \in \mathcal{T}\}$.

Linearly ordered spaces form a distinctive class of topological spaces. Let $(X, <)$ be a linearly ordered set. By an *interval* we understand any set of the form $(x', x'') = \{x \in X: x' <_c x <_c x''\}$, where $x', x'' \in X$. The *order topology* $\mathcal{T}_<$ on a set X generated by the linear order $<$ is the topology that has a base consisting of all possible intervals in $(X, <)$. A linearly ordered set $(X, <)$ together with the topology generated on X by the linear order $<$ is called a *linearly ordered* (or just an *ordered*) *space*.

Locally compact spaces form an important class. A space X is *locally compact* if each point $x \in X$ has a nhood Ox such that $[Ox]$ is a compact subspace.

1. Suppose that X is a set. Show that for an arbitrary Kuratowski operation C on X there is a topology $\mathcal{T}$ on X such that $CA = [A]$ for all $A \subset X$.

2. Suppose that X is an infinite set and let $\mathcal{T}_k = \{A \subset X: X \setminus A$ finite, or $A = \Lambda\}$; we agree to call $\mathcal{T}_k$ the *finite topology on X*.

Show that

(*a*) $\mathcal{T}_k$ is a topology on X;
(*b*) $(X, \mathcal{T}_k)$ is a T_1-space;
(*c*) If $(X, \mathcal{T})$ is a topological space satisfying the separation axiom T_1, then $\mathcal{T}_k \subset \mathcal{T}$;
(*d*) any two nonempty open sets in $(X, \mathcal{T}_k)$ intersect;
(*e*) each point $x \in X$ is a limit for any infinite set $A \subset X$;
(*f*) for every infinite subset A of X, $[A] = X$ (in other words, every subset of X is dense in X);
(*g*) X is compact and separable.

3. Let X be an infinite set, ξ a free ultrafilter on X and $\mathcal{T}_\xi = \xi \cup \{\Lambda\}$. Show that

(*a*) $\mathcal{T}_\xi$ is a topology on X;
(*b*) $(X, \mathcal{T}_\xi)$ is a T_1-space;
(*c*) $\mathcal{T}_k \subset \mathcal{T}_\xi$ (see Prob. 2);
(*d*) there are no isolated points in $(X, \mathcal{T}_\xi)$;
(*e*) if $U \in \xi$, then $X \setminus U$ is a discrete subspace of $(X, \mathcal{T}_\xi)$;
(*f*) if $\mathcal{T}$ is a topology on X and $\mathcal{T}_\xi \subset \mathcal{T}$, $\mathcal{T}_\xi \neq \mathcal{T}$, then there is an isolated point in $(X, \mathcal{T})$ [the assertion (*f*) asserts that the topology $\mathcal{T}_\xi$ is maximal with respect to the property (*d*).

4. Suppose that X is a topological space, ξ a prefilter on the set X and let $x \in X$ be an adherent point of ξ. Show that there is an ultrafilter ξ^* containing ξ and converging to x.

5. Let $\mathcal{T}$ and $\mathcal{T}'$ be two topologies on X, $\mathcal{T} \subset \mathcal{T}'$. Show that if $\mathcal{T}$ is a T_1-topology (resp. Hausdorff topology), then $\mathcal{T}'$ is also a T_1-topology (resp. Hausdorff topology).

6. Let $\{\mathcal{T}_\alpha: \alpha \in M\}$ be a family of topologies on X; let $\mathcal{T}' = \bigcap\{\mathcal{T}_\alpha. \alpha \in M\}$, and $\mathcal{T}'' = \bigcup\{\mathcal{T}_\alpha: \alpha \in M\}$. Then :

(i) Is $\mathcal{T}'$ a topology?
(ii) Is $\mathcal{T}''$ a topology?

7. Show that each ordered space satisfies the Hausdorff separation axiom.

8. Show that any finite T_1-space X is discrete.

9. True or false? Every discrete space is 0-dimensional.

10. Give an example of a space X none of whose one-point subsets is closed, (i.e., the space is not T_1).

11. Give an example of a T_0-space in which no one-point subset is closed.

12. Let X be a set, $B \subset X$ and $\mathcal{T} = \mathcal{D}(B)$ be the B-topology on X. Show that

(i) B is a set of all isolated points in $(X, \mathcal{T})$;
(ii) $[B] = X$;
(iii) $\mathcal{T}$ is a discrete topology iff $B = X$;
(iv) $\mathcal{T}$ is an indiscrete topology iff $B = \Lambda$.

13. Suppose that X is a topological space and let σ be some family of open sets in X. Show that σ is a base for X iff for each $x \in X$ and any nhood Ox of x there exists $U \in \sigma$ such that $x \in U \subset Ox$.

14. Let X be a set, $\mathcal{E}$ an arbitrary family of subsets of X, $\tilde{\mathcal{E}}$ the family of all sets representable as finite intersections of elements of $\mathcal{E}$ and let $\tilde{\tilde{\mathcal{E}}}$ be the family of all sets representable as the union of elements of $\tilde{\mathcal{E}}$. Show that $\tilde{\tilde{\mathcal{E}}} \cup \{\Lambda\} \cup \{X\}$ is a topology on X, $\tilde{\mathcal{E}} \cup \{X\}$ is a base and $\mathcal{E} \cup \{X\}$ is a subbase for this topology.

15. Verify that the collection of standard sets of a product space (p. 44) indeed forms a base for some topology.

16. Describe any set containing a topology defined on a plane, which has a subbase consisting of all straight lines parallel to a fixed straight line l.

17. Consider on the plane the topology which has a subbase consisting of the collection of all straight lines. What is this topology?

18. Let $(X, \mathscr{T})$ be a topological space.

(*a*) Show that the intersection of any collection of closed sets is a closed set.
(*b*) Show that a finite union of closed sets is a closed set.
(*c*) Show that the whole space and the empty set are closed sets.

19. Show that a family $\mathscr{T}$ of subsets of X is a topology on X iff the family $\mathscr{F} = \{X \setminus U: U \in \mathscr{T}\}$ satisfies the following conditions:

(*a*) $\{X\} \in \mathscr{F}$ and $\{\Lambda\} \in \mathscr{F}$;
(*b*) every finite union of elements of $\mathscr{F}$ is again an element of $\mathscr{F}$;
(*c*) any interesection of elements of $\mathscr{F}$ belongs to $\mathscr{F}$.

20. Show that a subset F of a topological space $(X, \mathscr{T})$ is closed iff $[F] = F$.

21. Show that an open set U intersects a set Q iff U intersects the closure of Q.

22. Let $(X, \mathscr{T})$ be a topological space. Show that

(*a*) $A \subset [A]$ for any subset $A \subset X$;
(*b*) if $A \subset B$, then $[A] \subset [B]$;
(*c*) $[A_1 \cup A_2] = [A_1] \cup [A_2]$;
(*d*) $[[A]] = [A]$.

23. Show that, if $[A] \subset A$, then A is closed.

24. Let σ be a base for a space X and $A \subset X$. Show that $x \in [A]$ iff for any set $U \in \sigma$ containing the point x, we have $U \cap A \neq \Lambda$.

25. Let X be a topological space and $A \subset X$. Show that, if each subset of A is closed in X, then $A \in X$ has no limit points.

26. Show that the closure of any set A in a space X is the smallest closed set in X containing A.

27. Show that, for each subset A of X, we have $[A] = \cap \{P: P \subset X, P \text{ closed and } P \supset A\}$.

28. Show that, if X is a topological space, $A \subset X$, $x \in X$, and $x \in [A]$, then $x \in U \cap A$ for any open set $U \ni x$.

29. Show that, if A is dense in X and U is open in X, then $[A \cap U] = [U] \supset U$.

30. Show that (finite) intersection of dense open sets in X is dense (and open) in X.

31. Show that every nonempty subset of an indiscrete space (even a singleton) is dense in it.

32. Let $\mathscr{B}$ be a subbase for X, $A \subset X$ and suppose that $A \cap U \neq \Lambda$ or each $U \in \mathscr{B}$. Does this imply that A is dense in X?

33. Show that, if U is an open set and F a closed set, then $U \setminus F$ is open and $F \setminus U$ is closed.

34. Show that a space X satisfies the T_1 separation axiom iff every one-point subset of X is closed.

35. Let X be a T_1-space and fix $x \in X$, $A \subset X$. Show that the following assertions (i) and (ii) are equivalent:

(i) $|Ox \cap A| \geqslant \aleph_0$ for every nhood Ox of x;
(ii) For each nhood Ox of x there is a point $x' \in Ox \cap A$ such that $x \neq x'$.

36. Let X be a topological space and A a subset of X. Show that the following assertions are equivalent:

(*a*) There exists in X a nhood Ox of each point $x \in A$ such that $A \cap Ox$ is closed in Ox;
(*b*) $[A] \setminus A$ is a closed set;
(*c*) A is an intersection of a closed set with an open set.

37. Show that for any two disjoint open sets the closure of one does not intersect the other.

38. Show that, if the sets U and V are open and disjoint, then $(\text{Int}\,[U]) \cap (\text{Int}\,[V]) = \Lambda$.

39. Show that, if there exist disjoint nhoods of subsets A and B of a topological space X, then there are also disjoint nhoods of A and B which are regular open sets.

40. Give an axample of a space in which all one-point (singleton) sets are closed (T_1-axiom) but at the same time any two nonempty open sets intersect (non-Hausdorfness).

41. What spaces have the property that each subset is the intersection of a collection of open sets?

42. True or false? On every infinite set there is a Hausdorff topology† relative to which no point of the set X is isolated.

43. Show that a space X is regular iff for each $x \in X$ and any nhood Ux of x there is a nhood $U'x$ of x such that $[U'x] \subset Ux$.

44. If M is a countable discrete subspace of a regular space X, show

†A Hausdorff topology is one satisfying the Hausdorff separation axiom.

that there is a family $\{Oy: y \in M\}$ of mutually disjoint nhoods in X of the points $y \in M$.

45. Show that a space X is normal iff for each closed set F and every nhood UF of F there is a nhood $U'F$ of F such that $[U'F] \subset UF$.

46. Show that, if $\mathcal{T}$ and $\mathcal{T}'$ are two topologies on X and $\mathcal{T} \subset \mathcal{T}'$, then $[A]_{\mathcal{T}'} \subset [A]_{\mathcal{T}}$ for each $A \subset X$.

47. Let X be a subspace of a topological space Y and G an open set in X. Show that the following assertions are equivalent:

(i) $\tilde{G} = \cup \{V \subset Y: V \text{ open in } Y \text{ and } V \cap X = G\}$;

(ii) $\tilde{G}$ is an open set in Y, $\tilde{G} \cap X = G$, and if $U \subset Y$, U open in Y, $U \cap X = G$, then $U \subset \tilde{G}$.

(iii) $\tilde{G} = Y \setminus [X \setminus G]_Y$;

(iv) $z \in Y \setminus \tilde{G}$ iff each nhood of z intersects the set $X \setminus G$.

48. If P is an open set in X and $Q \subset X$, show that $\text{Int}\,[P \cap Q] = \text{Int}\,[P] \cap \text{Int}\,[Q]$.

49. Show that, if U is an open set in a space X, then $F = [U] \setminus U$ is nowhere dense in X.

50. Show that, if a set $A \subset X$ is nowhere dense in a space in X, then the closure of this set is also nowhere dense in X.

51. Show that a closed subset F in X is nowhere dense iff $X \setminus F$ is dense in X. Does this assertion remain true if the word 'closed' is omitted?

52. Show that, if an open set is nowhere dense, it is empty.

53. If U is an open set in a space X and A is nowhere dense in X, then $A \cap U$ is nowhere dense in U (as a subspace of X).

54. (Example). Denote by Q^2 any countable dense (relative to the usual topology) subset of the plane (see Prob. 173), and by $\mathcal{T}$ the topology generated (see p. 41 and Prob 58) on Q^2 by the usual topology of the plane. Put $\tilde{\mathcal{T}} = \{U \setminus A: U \in \mathcal{T}, A \text{ nowhere dense in } (Q^2, \mathcal{T})\}$. Show that $(Q^2, \tilde{\mathcal{T}})$ is a countable Hausdorff space without isolated points in which there are no nontrivial (that is, consisting of mutually distinct points) convergent sequences. Moreover, $(Q^2, \tilde{\mathcal{T}})$ does not satisfy the first axiom of countability.

55. Show that: (*a*) The closure of an open set is a regular closed set; (*b*) if A is closed, then its interior Int A is a regular open set.

56. Show that the complement of a regular open set is a regular closed set, and the complement of a regular closed set is a regular open set.

57. (*a*) Show that a finite union of regular closed sets is a regular closed set.

(*b*) Determine whether the intersection of two regular closed sets is always regular closed.

(*c*) Show that for every pair of regular closed sets A_1 and A_2 in X, the set $[\mathrm{Int}\, A_1 \cap \mathrm{Int}\, A_2]$ is regular closed.

58. Let $(X, \mathcal{T})$ be a topological space and $X' \subset X$ a subset. Put $\mathcal{T}' = \{U \cap X' : U \in \mathcal{T}\}$. Then:

(*a*) verify that $\mathcal{T}'$ is a topology on X'.
(*b*) show that $A \subset X'$ is closed in X' iff $A = F \cap X'$, F a closed set in X;
(*c*) show that $[A]_{X'} = [A]_X \cap X'$ for every set $A \subset X'$;
(*d*) let σ be a base for the topology $\mathcal{T}$. Show that $\sigma' = \{X' \cap U: U \in \sigma\}$ is a base for the topology $\mathcal{T}'$.

59. Let X be a topological space and X' a closed subset of X. Show that every set closed in the subspace X' is also closed in X.

60. If X' is an open subspace of X, show that every open set in X' is open in X

61. Let X be a topological space and X' a subspace of X. Show that whenever any of the following conditions is satisfied, it holds also for X':

(*a*) X is a T_0-space;
(*b*) X is a T_1-space;
(*c*) X is a Hausdorff space;
(*d*) X is a regular space.

62. Show that a closed subspace of a normal space is a normal space.

63. Which spaces X have the following property: if $A \subset X$ and A is dense in X, then $A = X$?

64. Show that a dense subspace of a T_1-space without isolated points itself has no isolated points.

65. Let $(X, \mathcal{T})$ be a topological space, $B \subset X$ and let $\tilde{\mathcal{T}}$ be the B-extension of the topology $\mathcal{T}$. Show that:

(*a*) $U \in \tilde{\mathcal{T}}$ iff $U = (V \cap B) \cup W$ for some $V \in \mathcal{T}$, $W \in \mathcal{T}$;
(*b*) A nhood of the set $A = X \setminus B$ in $(X, \tilde{\mathcal{T}})$ contains a nhood in $(X, \mathcal{T})$.

66. Give an example of a countable Hausdorff non-regular space.

67. Show that the set-theoretic union of a chain of topologies forms a base for their topological union.

68. Let $\mathcal{E}$ be a chain of topologies on a set X and $\mathcal{T}^*$ the topological union of all topologies belonging to $\mathcal{E}$. Show that if $(X, \mathcal{T})$ for each $\mathcal{T} \in \mathcal{E}$ is a space without isolated points, then $(X, \mathcal{T}^*)$ also has no isolated points.

69. Suppose that $\mathcal{E}$ is a family of topologies on X and $\mathcal{T}^*$ is their topological union. Show that $(X, \mathcal{T}^*)$ is regular if $(X, \mathcal{T})$ is regular for each $\mathcal{T} \in \mathcal{E}$.†

70. Suppose that $\mathcal{E}$ is a family of topologies on X and $\mathcal{T}^*$ is their topological union. Show that $(X, \mathcal{T}^*)$ is naturally homeomorphic to some subspace of the product topology $\Pi\{(X, \mathcal{T}): \mathcal{T} \in \mathcal{E}\}$.

71. Suppose that $(X, \mathcal{T})$ is a Hausdorff space without isolated points. Show that there is a Hausdorff topology $\mathcal{T}^*$ on X such that:

(*a*) $\mathcal{T} \subset \mathcal{T}^*$;
(*b*) there are no isolated points in $(X, \mathcal{T}^*)$;
(*c*) if $\mathcal{T}'$ is a topology on X and $\mathcal{T}' \supset \mathcal{T}$, $\mathcal{T}' \neq \mathcal{T}$, then there are no isolated points in $(X, \mathcal{T}')$.††

72. Give an example of topologies $\mathcal{T}$ and $\mathcal{T}'$ on some set X, the former regular and the latter non-regular, for which $\mathcal{T} \subset \mathcal{T}'$.

73. Suppose that $(X, \mathcal{T})$ is a regular space without isolated points. Show that there is a regular topology $\mathcal{T}^*$ on X, satisfying: (*a*) $\mathcal{T}^* \supset \mathcal{T}$; (*b*) $(X, \mathcal{T}^*)$ has no isolated points; and (*c*) if $(X, \mathcal{T}')$ is a regular space without isolated points and $\mathcal{T}' \supset \mathcal{T}^*$, then $\mathcal{T}' = \mathcal{T}^*$.†††

74. If $\mathcal{T}$ is a maximal or maximally regular topology on X and X_1, X_2 are dense sets in $(X, \mathcal{T})$, then $X_1 \cap X_2 \neq \Lambda$.

2.3. Cardinal-Valued Characteristic Spaces

We say that the *pointwise cardinality* of a family γ of sets in X does not

---

†For a similar assertion about T_1 and Hausdorff spaces, see Prob. 5.

††A Hausdorff topology with properties (*b*) and (*c*) is called a *maximal* topology (see footnote††† below).

†††A regular topology satisfying conditions (*b*) and (*c*) is called a *maximally* regular topology. This should not be confused with the notion of a maximal topology. A maximally regular topology (the one so defined by us) cannot, in general, be maximal. The problem of the existence of maximal topologies that are regular has been dealt with only recently by Malikhin.

exceed a cardinal number τ, if for each x the cardinality of the collection of all members of γ containing x (as a member), does not exceed τ. If the pointwise cardinality of a family does not exceed $\aleph_0$, it is said to be *point-countable*. If every point is contained in at most finitely many elements of the family, it is said to be *point-finite*.

Suppose that X is a space and Y a subspace of X. A family $\mathcal{B}$ of open sets in X is called an *exterior base* for Y in X, if for each $y \in Y$ and each nhood G of a point y in X there exists $V \in \mathcal{B}$ such that $x \in V \subset G$. Note that an external base of Y in X is not necessarily the same as a defining system of nhoods of a set Y in X (see p. 40). The smallest cardinal number of an exterior base for Y in X is called the *exterior weight* of Y in X.

A crucial property conveyed elegantly via the language of cardinal numbers is that of semi-metrizable topology. A *semi-metric* on a set X is a non-negative real-valued function d defined on the set of all unordered pairs of elements of X, such that $d(x, y) = 0$ iff $x = y$.

It is clear that every metric is a semi-metric but the converse does not hold. With every semi-metric d defined on X there is associated uniquely a topology $\mathcal{T}_d$ on X, described as follows: A set $F \subset X$ is closed iff $d(y, F) > 0$ for each $y \in X \setminus F$. Moreover, $d(y, F) = \inf \{d(y, x) : x \in F\}$. Of course, one must verify that $\mathcal{T}_d$ is a topology. The topology $\mathcal{T}_d$ is said to be *induced* by the semi-metric d. A topology on X induced by some semi-metric is called a *semi-metrizable* topology: the space itself also is said to be *semi-metrizable*.

The general notion of the *sequential closure* of a set A in a space X is useful in the study of semi-metrizable spaces. This is defined to be the collection of limits in X of all convergent sequences of elements of A. A set which coincides with its sequential closure is called *sequentially closed*.

75. Show that, if a cover γ of a space X has a refinement of cardinality τ, then γ includes a subcover of X of cardinality $\leqslant \tau$.

76. If ξ is an open cover of X, $\mathcal{B}$ a base for X and $\mathcal{B}_\xi = \{U \in \mathcal{B}$: there is $G \in \xi$ for which $U \subset G\}$, show that $\mathcal{B}_\xi$ is a base for X.

77. Show that each open cover of a space X with a countable base (or even a countable net) has a countable subcover (i.e., X is Lindelöf). Prove a similar assertion for arbitrary cardinalities.

78. Show that in a T_1-space X there is no family of sets which is a base for two distinct points.

79. Let X be a T_1-space, x a non-isolated point of X, $\mathcal{B}$ a base for X

at x, and γ a finite family of open sets. Show that $\mathcal{B}\setminus\gamma$ is also a base at x.

80. Show that, if X is T_1, γ a point-finite family of open sets such that each element of γ contains at least two points and $\mathcal{B}$ is a base for X, then $\mathcal{B}' = \mathcal{B}\setminus\gamma$ is a base for X.

81. Show that, if X is T_1, and a point $x \in X$ has a base in X consisting of a finite number of elements, then x is isolated.

82. Show that, if each element of a base for a point x in X contains not less than two points, then x is a non-isolated point.

83. Are the following conditions equivalent for an arbitrary Hausdorff space?

(*a*) X has a countable dense subset;
(*b*) X has a countable base.

84. Give an example of a countable Hausdorff space without a countable base.

85. Show that the cardinality of an arbitrary disjoint family P of non-empty open sets of a separable space does not exceed $\aleph_0$.

86. Show that every ordered, separable, connected topological space has a countable base.

87. True or false? If every countable subset of a topological space is closed, then the space is discrete.

88. Show that the topological space $(X, \mathcal{T}_\xi)$, where X is an infinite set and ξ is a non-singular ultrafilter on X (see p. 8 and Prob. 3), never has a countable base.

89. Show that $(X, \mathcal{T}_\xi)$ defined in Prob. 3 is a separable space iff there is a countable subset A of X belonging to ξ.

90. Show that, if every nhood of a set A in a space X intersects a fixed set $B \subset X$, then $[B] \cap A \neq \Lambda$.

91. If the character of a set A in X equals τ and each nhood of A intersects a fixed set $B \subset X$, show that there exists $C \subset B$ such that $|C| \leqslant \tau$ and $[C] \cap A \neq \Lambda$.

92. Give an example of a Hausdorff space in which the character of a certain point is not equal to its pseudocharacter (see p. 43).

93. Let X be a topological space, Y a closed subspace of X, $F \subset X$ and γ a defining family of nhoods of a set F in X. Show that $\gamma' = \{U \cap Y: U \in \gamma\}$ is a defining family of nhoods of the set $\Phi = F \cap Y$ in Y.

94. If Y is a closed subspace of X and $F \subset X$, $\Phi = F \cap Y$, then $\chi(\Phi, Y) \leqslant \chi(F, X)$.

95. Let $(x, <)$ be a totally ordered set having a largest element x_m and endowed with the order topology. In addition, let the total order $<$ be minimal on $I(x_m)$. Show that the character of X at x_m is $cf\,|X|$ (see p. 16).

96. Establish the following relations for an arbitrary topological space X and each point $x \in X$:

$$c(X) \leqslant s(X) \leqslant a(X) \leqslant w(X),$$

and

$$t(x, X) \leqslant \chi(x, X) \leqslant w(X).$$

97. Let X be a topological space and X' a subspace of X. What relations connect $s(X)$ and $s(X')$, $a(X)$ and $a(X')$, $c(X)$ and $c(X')$, $t(X)$ and $t(X')$?

98. Show that the weight and the density of a discrete space coincide with its cardinality.

99. If X is a space of weight τ, show that the weight of every subspace of X does not exceed τ.

100. If $\mathcal{E}$ is a subbase for a topology $\mathcal{T}$ on X and $|\mathcal{E}| \geqslant \aleph_0$, show that the weight of $(X, \mathcal{T})$ does not exceed the cardinality of $\mathcal{E}$.

101. Let X be a topological space of weight τ. Show that each base σ for this space contains a base of cardinality τ. Prove that the analogous assertion fails for nets.

102. If $\mathcal{B}$ is a base (or π-base, see p. 40) for X, and U is an open set in X, show that there is a family $\mathcal{E}^* \subset \mathcal{B}$ of mutually disjoint sets such that

$$\cup\,\{V: V \in \mathcal{E}^*\} \subset U \subset [\cup\{V: V \in \mathcal{E}^*\}].$$

103. Give an example of a non-regular Hausdorff space with a countable base.

104. Consider the half-open interval $[0, 1)$ of the real line. Introduce on $[0, 1)$ the topology which has a base consisting of half-open intervals $[\alpha, \beta)$, $0 \leqslant \alpha < 1$, $\alpha < \beta \leqslant 1$. We denote by X^* the resulting topological space. Prove successively the following properties of this space:

(i) X^* is a T_1-space;
(ii) X^* is regular;
(iii) X^* satisfies the first axiom of countability;
(iv) X^* is separable;

(v) X^* is hereditarily Lindelöf, i.e., every subspace of X is Lindelöf (see p. 44);

(vi) X^* is completely normal (see p. 44);

(vii) X^* does not have a countable base;

(viii) X^* is non-metrizable.

105. Let X be hereditarily Lindelöf, and for each $\alpha \in T(\omega_1)$ (see Prob. 91, Chapter 1), suppose given a closed set $F(\alpha)$ in X in such a way that $F(\alpha_1) \subset F(\alpha_2)$ for $\alpha_2 < \alpha_1$. Show that there exists $\alpha_0 \in T(\omega_1)$ satisfying $F(\alpha_0) = F(\alpha)$ for every $\alpha \geqslant \alpha_0$.

106. Let X be a topological space and τ a cardinal number. Show that the following conditions are equivalent:

(*a*) for every $A \subset X$ and each point $x \in [A]$ there exists $A' \subset A$ such that $x \in [A']$ and $|A'| \leqslant \tau$;

(*b*) for every non-closed set $A \subset X$ there exists a point $y \in [A] \setminus A$ and a set $A' \subset A$ such that $y \in [A']$ and $|A'| \leqslant \tau$.

107. Show that a sequential space has countable tightness.

108. Is every Hausdorff space countably tight and sequential?

109. Is every sequential Hausdorff space a Fréchet-Urysohn space?

110. Let a topology be induced on a space X by a semi-metric d and let A be a closed set in X. Show that, if x is a non-isolated point in A, then $d(x, A \setminus \{x\}) = 0$.

111. If a topology is induced on X by a semi-metric d, $x \in X$, and $\xi = \{x_n : n \in N^+\}$ is a sequence of points of the set X such that

$$\lim_{n \to \infty} d(x_n, x) = 0,$$

show that ξ coverges to x.

112. Show that, if a topology of X is induced by a semi-metric d, then X is a sequential space.

113. Let $X = \{x_{nm} : n \in N^+, m \in N^+\}$ be any countably infinite set, whose elements are enumerated in a one-to-one fashion by all possible pairs of positive integers, let $Y = \{y_n : n \in N^+\}$ be a countably infinite set disjoint from X and θ any object not belonging to $X \cup Y$. Put $Z = \{\theta\} \cup Y \cup X$ and define $d(z', z'')$ for all $z', z'' \in Z$ as follows:

(*a*) $z' = z''$ implies $d(z', z'') = 0$;

(*b*) $z' = \theta$ and $z'' \in X$ imply $d(z', z'') = 1$;

(c) $z' = \theta$ and $z'' = y_n$ imply $d(z', z'') = 1/n$;
(d) $z' = y_{n'}$ and $z'' = x_{n''m}$, imply $d(z', z'') = 1$ for $n' \neq n''$ and $d(z', z'') = 1/m$ for $n' = n''$;
(e) $z', z'' \in X$ and $z' \neq z''$ imply $d(z', z'') = 1$;
(f) $z', z'' \in Y$ and $z' \neq z''$ imply $d(z', z'') = 1$;
(g) in all other cases $d(z', z'')$ is defined by the requirement that $d(z', z'') = d(z'', z')$.

Show that

(i) d is a semi-metric on Z;
(ii) Z is a normal space;
(iii) Z is a sequential space, but not a Fréchet-Urysohn space, it is neither locally compact nor first countable;
(iv) The topology of Z is the intersection of two locally compact metrizable topologies with countable bases;
(v) Z is the image of a locally compact metric space with a countable base under a quotient double mapping (that is, a mapping in which the inverse image of every point consists of at most two points).

The assertions (ii) through (v) refer to the topology $\mathcal{T}$ on Z induced by semi-metric d.

114. Let X be a Hausdorff space whose topology is induced by a semi-metric d. Show that the following conditions are equivalent:

(a) X is first countable;
(b) X is a Fréchet-Urysohn space;
(c) $[A] = \{y \in X: d(y, A) = 0\}$ for each $A \subset X$;
(d) for every $\varepsilon > 0$ and each $x \in X$, Int $O_\varepsilon x \ni x$, where $O_\varepsilon x = \{y \in X: d(x, y) < \varepsilon\}$.

115. Is every separable Hausdorff space hereditarily (see p. 47) separable?.

116. Show that, if a space X has a countable π-base $\mathcal{B}$, then every dense subspace of X is separable.

117. Show that, if X is a first countable separable space, then every dense subspace of X is also separable.

118. Let X be a space, τ a cardinal number, Y a dense subset of X and $s(X) \leqslant \tau$, $t(X) \leqslant \tau$. Show that $s(Y) \leqslant \tau$.

119. True or false? The product of every pair of normal hereditarily separable spaces is hereditarily separable.

120. True or false? Any subset of cardinality $2^{\aleph_0}$ of a completely regular (see p. 43) separable space has a limit point.

121. Show that, if X is a normal separable space and $A \subset X$, $|A| \geqslant 2^{\aleph_0}$, then A has a limit point in X.

122. Let X be a Hausdorff space and A a dense subset of X. Show that $|X| \leqslant 2^{2^{|A|}}$.

123. If X is a Hausdorff space of weight $\tau \geqslant \aleph_0$, show that $|X| \leqslant 2^{\tau}$.

124. Suppose that the cardinality of a Hausdorff space X is greater than $2^{\aleph_0}$. Does X contain a discrete subspace of uncountable cardinality?

125. Does every hereditarily separable Hausdorff space have cardinality greater than $2^{\aleph_0}$?

126. Let X be a Hausdorff space and $M \subset X$, $|M| \leqslant 2^{\aleph_0}$. Show that there is a sequentially closed (see p. 56) set $\tilde{M}$ such that $M \subset \tilde{M} \subset [M]$ and $|\tilde{M}| \leqslant 2^{\aleph_0}$.

127. Let X be a regular space. Let $s(X) \leqslant \aleph_0$ and $\psi(x, X) \leqslant \aleph_0$ for each $x \in X$. Show that $|X| \leqslant 2^{\aleph_0}$.

128. Can every Hausdorff (regular) space X of a countable pseudocharacter be contracted to a first countable Hausdorff space?

129. Let X be regular, $\psi(x, X) \leqslant \aleph_0$ for every point of X, $s(X) \leqslant 2^{\aleph_0}$ and $t(X) = \aleph_0$. Show that $|X| \leqslant 2^{\aleph_0}$.

130. Let X be a regular space, $\psi(x, X) \leqslant \aleph_0$ for each $x \in X$ and let $t(X) = \aleph_0$. Show that, for every $M \subseteq X$, $|M| \leqslant 2^{\aleph_0}$ implies that $|[M]| \leqslant 2^{\aleph_0}$.

131. Show that, if a first countable Hausdorff space X has the Souslin property, then $|X| \leqslant 2^{\aleph_0}$.

132. Let $(X', \mathcal{T}')$ be a dense subspace of a regular space $(X, \mathcal{T})$. Show that the family $\tilde{\mathcal{T}}' = \{\mathrm{Int}[U]: U \in \mathcal{T}'\}$ is a base for the topology $\mathcal{T}$.

133. Let X be a space, X' a dense subspace of X and let X' have weight τ. Show that the exterior weight of X' in X is τ (see p. 56).

134. Let X be a regular space and A a dense subset of X. Show that the weight of X does not exceed $2^{|A|}$.

135. Suppose that X is regular and X' is a dense subspace of X. Show that the character of each point x in X' is equal to its character in X.

136. Show that every base for a topological space is a net in that space.

137. In what spaces is every net a base?

138. Show that the family of all one-point subsets of an arbitrary space is a net.

139. If in a space X there is a net of cardinality τ, show that in every subspace of X there is a net of cardinality not exceeding τ.

140. Give an example of a normal space with a countable net but without a countable base.

141. Let X be a space, X_α a subspace of X for each $\alpha \in M$, and $X = \bigcup \{X_\alpha : \alpha \in M\}$. Show that, if S_α is a net in X_α, $\alpha \in M$, then $S = \bigcup \{S_\alpha : \alpha \in M\}$ is a net in X. Also, show that the corresponding assertion for base does not hold in general.

142. If in a regular space there is a net of cardinality τ, show that in this space there is a net of cardinality not exceeding τ, each of whose elements is a closed set.

143. If X has a net of cardinality τ, show that every open cover of X has a subcover of cardinality not exceeding τ.

144. If in a space X there is a net of cardinality τ, each of whose elements is a closed set, show that every closed set in X is the intersection of a family γ of open sets such that $|\gamma| \leqslant \tau$.

145. True or false? In every Hausdorff space with a countable base an arbitrary closed set is a countable intersection of open sets.

146. Does a countable net of closed sets always exist in a Hausdorff space with a countable base?

147. If in a Hausdorff space $(X, \mathcal{T})$ there is a net of cardinality τ, show that there is a Hausdorff topology $\mathcal{T}^*$ on X whose weight does not exceed τ, and for which $\mathcal{T}^* \subset \mathcal{T}$ (cf. Prob. 148).

148. Let $\mathcal{T}$ be a Hausdorff topology on X having a net of cardinality $\tau \geqslant \aleph_0$. Show that there is a (Hausdorff) topology $\mathcal{T}'$ on X whose weight does not exceed τ and for which $\mathcal{T} \subset \mathcal{T}'$.

149. A Hausdorff space has a net of cardinality τ iff some Hausdorff space of weight τ is contractible to it (see p. 41).

150. A Hausdorff space has a net of cardinality τ iff it is the continuous image of some Hausdorff space of weight τ.

2.4. Metric Spaces

A set F in a topological space X is called *perfect* if it is closed in X and *dense in itself*, that is, as a subspace of X it has no isolated point. Thus, F is perfect in X iff F is closed in X and each point $x \in F$ is a limit point of F.

We say that a topological space X has the *Baire property* (or is a *Baire space*) if the intersection of a countable number of dense, open sets is itself dense. A set A in X is said to be of *second category* if the complement of the set $X \setminus A$ is of *first category*, that is, if $X \setminus A$ is the union of a countable number of nowhere dense sets in X (see p. 41). The definition of metric space as well as the basic notions relevant to metric spaces have been presented in the theoretical introduction in Sec. 1 of this chapter.

Closely related to the notion of an isometric mapping is the notion of a *contraction* mapping: a function $f: (X, \rho) \to (Y, \rho')$ is called a *contraction mapping* if there is a real number $\alpha < 1$ such that $\rho'(fx_1, fx_2) \leqslant \alpha\rho(x_1, x_2)$ for all $x_1, x_2 \in X$.

The notion of *continuity* of a function from one metric space to another may be formulated in an entirely classical way. A mapping f of a metric space (X, ρ) into a metric space (Y, ρ') *is continuous at the point* $x_0 \in X$, if for each positive real number ε there exists a positive real number δ such that $\rho'(fx, fx_0) < \varepsilon$ whenever $\rho(x, x_0) < \delta$. If a mapping f of (X, ρ) into (Y, ρ') is continuous at every point $x \in X$, then we say that f is *continuous* on X. Finally, a function f from (X, ρ) into (Y, ρ') is said to be *uniformly continuous* if for each $\varepsilon > 0$ there exists some $\delta > 0$ such that for all $x_1, x_2 \in X$, satisfying $\rho(x_1, x_2) < \delta$, we have $\rho'(fx_1, fx_2) < \varepsilon$.

A *completion* of a metric space (X, ρ) is a complete metric space $(X, \tilde{\rho})$ together with an isometric embedding $i: (X, \rho) \to (Y, \tilde{\rho})$ such that $\tilde{\rho}(y, i(X)) = 0$ for all $y \in Y$ (hereafter we simply say that $i(X)$ is dense in $(Y, \mathcal{T}_{\tilde{\rho}})$). We shall also frequently identify x with $i(x)$, X with $i(X)$, and write $X \subset Y$.

Let (X_i, ρ_i), $i \in N^+$, be a countable family of metric spaces, the diameter of each being not greater than unity (that is, $\sup\{\rho_i(x_i, x_i'): i \in N^+, x_i \in X_i, x_i' \in X_i\} \leqslant 1$).

Consider the product $X = \Pi\{X_i: i \in N^+\}$. We define a metric on the set X as follows: if $x = \{x_i: i \in N^+\} \in X$, $y = \{y_i: i \in N^+\} \in X$, then

$$\rho(x, y) = \sum_{i=1}^{\infty} \frac{\rho_i(x_i, y_i)}{2^i}.$$

It is easy to verify that the function $\rho(x, y)$ is indeed a metric on X. The metric space $(\Pi\{X_i: i \in N^+\}, \rho)$, thus defined is called the *metric product* of the spaces (X_i, ρ_i) with respect to $i \in N^+$. Moreover, $\rho_i'(\pi_i x, \pi_i y) \leqslant \rho(x, y)$, where $x \in \Pi\{X_i: i \in N^+\}$, $y \in \Pi\{X_i: i \in N^+\}$, $\rho_i'(\pi_i x, \pi_i y) =$

$\frac{1}{2^i}\rho_i(\pi_i x, \pi_i y)$ and $\pi_i : \Pi\{X_i : i \in N^+\} \to (X_i, \rho_i')$ is a projection. Hence it is a uniformly continuous mapping.

151. Let (X, ρ) be a metric space. Show that, for each $x \in X$, every $\varepsilon > 0$ and every $x' \in O_\varepsilon x$, there is some $\varepsilon' > 0$ such that

$$x' \in O_{\varepsilon'} x' \subset O_\varepsilon x.$$

152. Let (X, ρ) be a metric space and $\mathcal{T}_\rho$ the family of subsets defined on p. 47. Show that $\mathcal{T}_\rho$ is a topology and $O_\varepsilon x \in \mathcal{T}_\rho$ for each $x \in X$ and every $\varepsilon > 0$.

153. Let X be a set; put $\rho(x, y) = 1$ if $x \in X$, $y \in X$, $x \neq y$, and $\rho(x, x) = 0$ for each $x \in X$. Show that (X, ρ) is a complete metric space, bounded but not totally bounded. Which sets are open in (X, ρ)?

154. Show that any closed interval of the real line (with the natural metric) is totally bounded.

155. Let (X, ρ) be a metric space. Show that $U \subset X$ is an open set in (X, ρ) iff for each $x \in U$ there exists $\varepsilon > 0$ such that $x \in O_\varepsilon x \subset U$.

156. Let (X, ρ) be a metric space. Show that the collection of all *spherical nhoods* (or *balls*) of all possible points in X is a base for the space $(X, \mathcal{T}_\rho)$.

157. Show that the topology of every metric space satisfies the Hausdorff separation axiom.

158. Let (X, ρ) be a metric space, $A \subset X$. Show that $x_0 \in [A]$ iff $\rho(x_0, A) = 0$.

159. Show that in a metric space limits of convergent sequences are unique.

160. Let (X, ρ) be a metric space. Show that x_0 is an adherent point of the set $A \subset X$ iff there is a sequence $\{x_n : n \in N^+\}$ of points of A which converges to x_0.

161. Show that a topology of every metric space (X, ρ) is first countable.

162. Let (X, ρ) be an arbitrary metric space. Show that:

(*a*) $F \subset X$ is a closed set iff for each $x \notin F$, we have $\rho(x, F) > 0$;

(*b*) F is closed iff for each sequence $\{x_n : n \in N^+\}$ of points of F convergent in X, we have $\lim_{n\to\infty} x_n \in F$.

163. Show that if x_0 is a limit point of a set M in a metric space (X, ρ), then we can always choose a sequence $\{x_n: n \in N^+\}$ of mutually distinct points of M which converges to x_0.

164. Let X be a metric space, $A \subset X$, and ε a positive number such that $\rho(x, y) \geqslant \varepsilon$ for all $x, y \in A$. Show that the set A has no limit point in X.

165. Show that every subsequence of a convergent sequence in a metric space is also convergent and has the same limit. Prove that every subsequence of a Cauchy sequence is also a Cauchy sequence.

166. Show that every convergent subsequence in a metric space (X, ρ) is a Cauchy sequence.

167. Give an example of a metric space and a Cauchy sequence in it which has no limit.

168. Show that a Cauchy sequence $\{x_n: n \in N^+\}$ in a metric space (X, ρ) is convergent if it has a convergent subsequence.

169. Show that every bounded infinite subset of a straight line has a limit point on the line.

170. Show that every essential (see p. 40) and bounded sequence $\{x_n: n \in N^+\}$ of points of the real line has a convergent subsequence consisting of mutually distinct points.

171. Show that every bounded sequence $\{x_n: n \in N^+\}$ of real numbers has a convergent subsequence.

172. Show that the real line with the usual metric $\rho(x, y) = |x - y|$ is a complete separable metric space.

173. Let R^2 be the Euclidean plane. Let $x, y \in R^2$, and let $\rho(x, y)$ be the distance from x to y, i.e., the length of the segment joining the points x and y.

(*a*) Show that (R^2, ρ) is a metric space.
(*b*) Show that $\rho(x, y) = \sqrt{(x_1 - y_1)^2 + (x_2 - y_2)^2}$ for $x = \{x_1, x_2\} \in R^2$, $y = \{y_1, y_2\} \in R^2$.
(*c*) Show that (R^2, ρ) is separable.
(*d*) Show that (R^2, ρ) is complete.

Assign to R^2 the Cartesian coordinate system.

(*e*) Show that the following sets in R^2 are open:
  (i) the interiors of all discs;
  (ii) the interiors of all rectangles.
(*f*) Show that the sets described in (i) and (ii) above form a base for the topology of the plane.
(*g*) Put $\rho^*(x, y) = \max\{|x_1 - y_1|, |x_2 - y_2|\}$ for $x = \{x_1, x_2\} \in R^2$,

$y = \{y_1, y_2\} \in R^2$. Show that ρ^* is a metric on R^2 inducing the same topology as the metric ρ.

(*h*) Describe the ε-nhoods of a point $x \in R^2$ with respect to metric ρ^*.

(*i*) Consider also a metric $\rho'(x, y)$ on R^2, defined by the formula

$$\rho'(x, y) = |x_1 - y_1| + |x_2 - y_2|, \quad x = \{x_1, x_2\}, y = \{y_1 y_2\}.$$

Solve the problems analogous to (*g*) and (*h*) for the metric ρ'.

174. Give an example of a set X and two distinct metrics on X which induce the same topology on X. Do every two metrics on a given set X induce the same topology ?

175. Let (X, ρ) be a metric space and $X_0 \subset X$, a subset. The set X_0 naturally becomes a metric space, a subspace of (X, ρ) by putting $\rho_0(x, y) = \rho(x, y)$, where $x, y \in X_0$. Show that (X_0, ρ_0) is a metric space (hereafter the subset X_0 of (X, ρ) with the metric defined above will be referred to as a (metric) subspace of (X, ρ)).

176. Let (X, ρ) be a metric space and (A, ρ') a subspace (that is, $A \subset X$ and $\rho' = \rho \mid A$ is the restriction of the metric ρ to A). Show that $(A, \mathcal{T}_{\rho'})$ is a topological subspace of $(X, \mathcal{T}_\rho)$.

177. Give an example of two disjoint closed sets of a straight line (plane), the distance between which is zero.

178. Show that every family of disjoint open subsets of a straight line is finite or countable. Prove this also for a plane.

179. Show that the union of every set $\mathcal{A}$ of intervals of the real line containing a fixed point is again an interval (possibly, the entire line, or a set of the form $(-\infty, a)$ or $(b, +\infty)$).

180. Show that every open subset of the real line is the union of a finite or countable number of pairwise disjoint intervals, whose endpoints belong to the complementary closed set (among these intervals there may occur one or two infinite intervals of the form $(-\infty, a)$ or $(a, +\infty)$. These intervals are called *adjacent intervals* to the complementary closed set or the *components* of this open set.

181. Show that a closed set F is a perfect straight line iff its adjacent intervals (cf. Prob. 180) are mutually disjoint.

182. Show that every pair of countable dense-in-themselves (see p. 63) subsets of the real line having neither smallest nor largest element are similar to each other (see p. 7).

183. Show that the space of rationals, and also a space of irrationals is 0-dimensional.

184. Show that every nowhere dense subset X of the real line R^1 is 0-dimensional (see p. 42).

185. Let [0, 1] be the closed unit interval in the real line. In this we isolate the two closed intervals, $\Delta_0 = \left[0, \frac{1}{3}\right]$ and $\Delta_1 = \left[\frac{2}{3}, 1\right]$ (which we call closed intervals of type 1, and between which lies the open interval $\delta = \left(\frac{1}{3}, \frac{2}{3}\right)$ called an interval of type 1). From each closed interval of type 1 we isolate two closed intervals of type 2 : $\Delta_{00} = \left[0, \frac{1}{9}\right]$, $\Delta_{01} = \left[\frac{2}{9}, \frac{1}{3}\right]$, and $\Delta_{10} = \left[\frac{2}{3}, \frac{7}{9}\right]$, $\Delta_{11} = \left[\frac{8}{9}, 1\right]$. Between these closed intervals lie the corresponding type 2 open intervals $\delta_0 = \left(\frac{1}{9}, \frac{2}{9}\right)$ and $\delta_1 = \left(\frac{7}{9}, \frac{8}{9}\right)$. Continuing in this way, we obtain, for each n, 2^n closed intervals $\Delta_{i_1 i_2 \ldots i_n}$ of type n, where each of the indices $i_1, \ldots, i_n$ is either 0 or 1. Each closed interval $\Delta_{i_1 \ldots i_n}$ is divided into 3 similar parts: two outer closed intervals $\Delta_{i_1 \ldots i_n 0}$ and $\Delta_{i_1 \ldots i_n 1}$, called closed intervals of type $n + 1$, and the open interval of type $n + 1$ lying between them.

Let Π_n be the union of all the closed intervals of type n, and define $C = \bigcap \{\Pi_n : n \in N^+\}$. The set C is called the *Cantor perfect set*, or the *Cantor discontinuum*, or the *Cantor ternary set* (of countable weight).

Show that:

(*a*) the Π_n are closed subsets of [0, 1];

(*b*) C is a nonempty closed subset of [0, 1];

(*c*) the sum of the lengths of all open intervals removed is equal to the length of the entire interval [0, 1];

(*d*) C is a perfect set;

(*e*) C has the power of the continuum;

(*f*) C is nowhere dense on [0, 1].

186. Show that a closed (metric) subspace of a complete metric space is complete.

187. Show that a metric space (X, ρ) is complete iff every decreasing (nested) countable sequence $\{F_i : i \in N^+\}$ of nonempty closed sets in (X, ρ), for which diam $F_i \leqslant \varepsilon_i$ with $\varepsilon_i \to 0$ as $i \to \infty$, has a nonempty intersection (such an intersection contains exactly one point). Give an example of a complete metric space for which there exists a decreasing (nested) countable sequence of nonempty closed sets with empty intersection.

188. Show that a metric space (X, ρ) is complete iff the intersection of every decreasing sequence of closed balls with radii tending to zero is nonempty.

189. Let (X, ρ) be a metric space and M be a subspace with metric ρ' induced by ρ. Show that M is closed in $(X, \mathcal{T}_\rho)$ provided (M, ρ') is complete.

190. Let (X, ρ) be a metric space. Put $\rho^*(x, y) = \min\{1, \rho(x, y)\}$. Show that:

(*a*) ρ^* is a metric on X, inducing the same topology as the metric ρ does;
(*b*) (X, ρ^*) is of diameter at most 1;
(*c*) (X, ρ^*) is a complete space if (X, ρ) is.

191. Show that in every metric space one can introduce a bounded metric compatible with its topology.

192. Show that a complete metric space (X, ρ) is homeomorphic to a bounded complete metric space.

193. Let (X_i, ρ_i), $i \in N^+$, be metric spaces; diam $X_i \leqslant 1$. Consider the product $X = \Pi\{X_i : i \in N^+\}$ of the X_i. Let $x = \{x_i\} \in X$, $y = \{y_i\} \in X$. Put

$$\rho(x, y) = \sum_{i=1}^{\infty} \frac{\rho_i(x_i, y_i)}{2^i}.$$

Show that ρ is a metric on X and with this metric diam $X \leqslant 1$.

194. Show that the Tychonoff product of a countable set of metrizable spaces is always metrizable.

195. Show that the Tychonoff product of countably many spaces, each homeomorphic to the real line, is metrizable (and has a countable base).

196. Show that the metric product of a countable family of complete bounded (by a constant) metric spaces is complete.

197. Show that the Tychonoff product of a countable family of spaces homeomorphic to a complete metric space is itself homeomorphic to a complete metric space.

198. The points of *Euclidean space* R^n consist of all possible sequences of n real numbers. Let $x = \{x_i\} \in R^n$, $y = \{y_i\} \in R^n$. Put

$$\rho(x, y) = \sqrt{\sum_{i=1}^{n} (x_i - y_i)^2}.$$

Show that (R^n, ρ) is: (*a*) a metric space; (*b*) separable; (*c*) complete.

199. Let $\{x_n\}$ and $\{y_n\}$ be two Cauchy sequences in a metric space (X, ρ). Form a sequence $\{z_m\}$, where $z_m = x_n$ if $m = 2n - 1$ and $z_m = y_n$ if $m = 2n$. Show that $\{z_m\}$ is a Cauchy sequence iff $\lim\limits_{n\to\infty} \rho(x_n, y_n) = 0$.

200. Let (X, ρ) be a metric space and $\{x_n\}$, $\{y_n\}$ be two Cauchy sequences in it. Show that the sequence of real numbers $\alpha_n = \rho(x_n, y_n)$ converges. The limit of this sequence is called the *oscillation* of the given two Cauchy sequences in (X, ρ).

201. Let(X, ρ) be a metric space.

Two Cauchy sequences in (X, ρ) are said to be ρ-*equivalent* if their *oscillation* (see preceding problem) is zero. Show that the ρ-equivalence on the set of all Cauchy sequences in (X, ρ) is an equivalence relation (that is, it is reflexive, symmetric and transitive). This ρ-equivalence relation partitions the set of all Cauchy sequences into pairwise disjoint ρ-equivalence classes; the collection of all such classes is denoted by ρX and the equivalence class of the sequence $\{x_n\}$ by $|\{x_n\}|$.

202. Let (X, ρ) be a complete metric space. Show that for every pair of points $\xi, \eta \in (X, \rho)$ and each of the sequences $\{x_n\}$ and $\{y_n\}$, $x_n, y_n \in (x, \rho)$, $n = 1, 2, \ldots$, for which $\lim x_n = \xi$, $\lim y_n = \eta$, we have

$$\rho(\xi, \eta) = \lim_{n\to\infty} \rho(x_n, y_n).$$

203. Let (X, ρ) be a metric space and ρX the set of all ρ-equivalence classes of its Cauchy sequences. Put $\tilde{\rho}(\xi, \eta) = \tilde{\rho}(\{x_n\}, \{y_n\})$, where $\tilde{\rho}(\{x_n\}, \{y_n\})$ is the oscillation of two Cauchy sequences $\{x_n\}$ and $\{y_n\}$, for which $|\{x_n\}| = \xi, |\{y_n\}| = \eta$. Show that $(\rho X, \tilde{\rho})$ is a complete metric space containing the isometric image of (X, ρ) as a dense subset.

204. Let (X', ρ') and (X'', ρ'') be two completions of a metric space (X, ρ). Show that there exists an isometry $f\colon (X', \rho') \to (X'', \rho'')$, $f(X') = X''$ for which $fx = x$ whenever $x \in X$.

205. Show that the completion (ρX, ρ) of a totally bounded metric space (X, ρ) is also totally bounded.

206. Let X be a metrizable space and ρ_1, ρ_2 be two metrics on X compatible with its topology. Let us say that $\rho_1 \leqslant \rho$ if every Cauchy sequence in X in the metric ρ_1 is also a Cauchy sequence in X in the metric ρ_2. Show that $\rho_1 \leqslant \rho_2$ iff there exists a continuous function $f: \rho_1 X \to \rho_2 X$ from a completion $\rho_1 X$ in the metric ρ_1 into a completion $\rho_2 X$ in the metric ρ_2 such that $fx = x$ for all $x \in X$.

207. Show that a contraction mapping f of a complete metric space (X, ρ) into itself has always a unique *fixed point*, that is, a point $x_0 \in X$ such that $fx_0 = x_0$.

208. Give an example of a complete metric space and an isometry into itself having no fixed point.

209. Show that every contraction map of a metric space is continuous.

210. Give an example of a non-complete metric space and a contraction map into itself that does not have a fixed point.

211. Show that every set A closed in a metrizable space is a G_δ.

212. Show that every metrizable space is normal.

213. Show that for a metrizable space X the following conditions are equivalent:

(i) X is separable;
(ii) X has a base;
(ii) X has a countable π-base;
(iii) X is a Lindelöf space.

214. Show that a metrizable space X is separable iff it satisfies the Souslin hypothesis.

215. Show that the following conditions are equivalent for a metrizable space X:

(i) X has a countable base;
(ii) X is hereditarily separable (that is, every subspace of X is separable).
(iii) every discrete subspace $X_0 \subset X$ is countable;
(iv) every closed discrete subspace $X_0 \subset X$ is countable;
(v) X has the Souslin property.

216. Show that the weight of a metrizable space is always equal to its density, index of compactness and π-weight (see p. 44).

217. Show that the weight of every metrizable space is equal to its Souslin number and also to the supremum of the cardinalities of closed discrete subspaces of X.

218. Let X be a 0-dimensional separable metric space. Show that, for every $\varepsilon > 0$, there exists a countable disjoint cover ω of this space consisting of clopen sets of diameter $< \varepsilon$.

219. Let X_0 be a connected subset of a space X. Show that the set $[X_0]$ is also connected.

220. Let X be a topological space, and $\{M_\alpha : \alpha \in A\}$ the family of all connected subsets, which contain the point $x_0 \in X$. Show that $M = \bigcup \{M_\alpha : \alpha \in A\}$ is a connected closed set and includes every connected subset of X containing x_0.

221. Show that the image of a connected space under a continuous mapping is connected.

222. Show that every interval is connected (relative to the usual topology).

223. Show that a subset A of the real line l is connected iff whenever it contains two points it includes all of the segment between them.

224. Specify, to within homeomorphism, all connected subsets of the real line. How many are there?

225. The only connected and sequentially compact subsets of the real line are the empty set, the singletons, and the intervals.

226. Determine which of the following spaces are homeomorphic, which are not and why:

(i) a segment of the real line;
(ii) a half-open interval of the real line;
(iii) an interval of the real line;
(iv) the real line;
(v) a circle.

227. Show that the image of a closed interval of the real line under an open mapping into a Hausdorff space is either a closed interval or a point.

228. Show that a straight line is never homeomorphic to Euclidean n-space for $n \geqslant 2$.

229. Show that the complement of an arbitrary countable subset of Euclidean n-space, where $n \geqslant 2$, is connected.

230. Show that any interval has a base of connected sets.

231. Find two non-homeomorphic sequentially compact subspaces of the usual plane, whose products with a segment are homeomorphic. Can such sequentially compact subsets be found on a straight line?

232. Show that a complete metric space can be homeomorphic to a non-complete metric space.

233. Let X be a complete meterizable space. Show that every open subset of X is complete metrizable.

234. Let X be a complete metrizable space and $M \subset X$ be a G_δ-set in X. Show that the subspace M is homeomorphic to some complete metric space.

235. Let Y be a dense subspace of a Hausdorff space X. Show that, if Y is complete meterizable, then Y is a G_δ-set in X.

236. Let X be a complete metrizable space. Show that a subspace $M \subset X$ is homeomorphic to a complete metric space iff it is a G_δ-set in X (Alexandroff-Hausdorff theorem).

237. Show that the space of irrational points on the real line is homeomorphic to a complete metric space.

238. Show that every complete metrizable space X is a Baire space (see p. 63).

239. Show that a complete metrizable space X cannot be represented as the union of a finite number of subsets nowhere dense in X.

240. Show that the space of rational numbers on the real line is not homeomorphic to any complete metric space.

241. Show that a σ-discrete (that is, a space which is a countable union of discrete subspaces) complete metric space X has a dense set of isolated points.

242. Can a countable complete metric space (X, ρ) contain no isolated points?

243. In a complete metric space (X, ρ), can a countable (or even finite) subset be of the second category? Under what conditions?

244. Show that every separable metric space X has at most the cardinality of the continuum.

245. The points of the *Baire space* $B(\aleph_0)$ are all sequences $x = \{n_k: k \in N^+\}$ of natural numbers n_k. Let $x = \{n_k: k \in N^+\}$ and $y = \{n'_k: k \in N^+\}$ be two points in $B(\aleph_0)$. Put $\rho(x, y) = 1/i(x, y)$, where $i(x, y) = \min \{k: n_k \neq n'_k\}$ and $\rho(x, y) = 0$, if $n_k = n'_k$ for each $k \in N^+$:

(*a*) Show that $(B(\aleph_0), \rho)$ is a metric space.
(*b*) Show that it is a complete metric space.

(c) Put $U_{n_1^0 n_2^0 \cdots n_k^0} = \{x = \{n_k\} \in B(\aleph_0)\colon n_1 = n_1^0, \ldots, n_k = n_k^0\}$. Show that the set $U_{n_1^0 n_2^0 \cdots n_k^0}$ is a spherical nhood of radius $1/k$ of each of its points, that is, for every point $x = \{n_k\} \in U_{n_1^0 n_2^0 \cdots n_k^0}$ we have $U_{n_1^0 n_2^0 \cdots n_k^0} = \{x' = \{n_k'\} \in B(\aleph_0)\colon \rho(x, x') < 1/k\}$. A set of the form $U_{n_1^0 n_2^0 \cdots n_k^0}$ is called a *Baire interval.*

(d) Show that the Baire interval is not only open but also closed in $B(\aleph_0)$.

(e) Show that the Baire intervals form a base for the topology of the metic space $B(\aleph_0)$.

(f) Show that the weight of $B(\aleph_0)$ is $\aleph_0$.

(g) Show that two distinct Baire intervals are either disjoint or one is contained in the other.

246. Let W be a set of cardinality τ. Let $B(\tau)$ be the set of all sequences $x = \{\xi_1, \ldots, \xi_n, \ldots\}$ of elements $\xi_1 \in W, \ldots, \xi_n \in W, \ldots$ Thus, every sequence $x = \{\xi_1, \ldots, \xi_n, \ldots\}$, $\xi_1 \in W, \ldots, \xi_n \in W, \ldots$, is a point in $B(\tau)$. Let $x = \{\xi_n\} \in B(\tau)$, $y = \{\eta_n\} \in B(\tau)$. Put $\rho(x, y) = 1/k(x, y)$, where $k(x, y) = \min\{n\colon \xi_n \neq \eta_n\}$.

(a) Show that $(B(\tau), \rho)$ is a metric space (this space will be called the *Baire space of weight* τ).

(b) Put $U_{\xi_1^0 \cdots \xi_n^0} = \{x = \{\xi_i\} \in B(\tau)\colon \xi_1 = \xi_1^0, \ldots, \xi_n = \xi_n^0\}$. Show that any set of the form $U_{\xi_1^0 \cdots \xi_n^0}$ is clopen in $B(\tau)$ and is a spherical nhood of every point belonging to it (such a set is called a *Baire interval*).

(c) Show that the family of sets of the form $U_{\xi_1^0 \cdots \xi_n^0}$ form a base for the topology of $B(\tau)$. Show that $w(B(\tau)) = \tau$.

(d) Show that any two Baire intervals are either disjoint or one is contained in the other.

247. Show that $B(\tau)$ (see preceding problem) is a complete space.

248. Show that the Baire space $B(\aleph_0)$ is homeomorphic to the Tychonoff product of a countable family of discrete countable spaces.

249. The points of *Hilbert space* l^2 are the sequences $x = \{x_n\}$ of real numbers, for which the series

$$\sum_{n=1}^{\infty} x_n^2$$

converges.

(*a*) Show that $x + y = \{x_n + y_n\} \in l^2$, if $x = \{x_n\} \in l^2$, $y = \{y_n\} \in l^2$.

(*b*) Put

$$(x, y) = \sum_{n=1}^{\infty} x_n y_n,$$

where $x = \{x_n\} \in l^2$, $y = \{y_n\} \in l^2$. Show that the series

$$\sum_{n=1}^{\infty} x_n y_n$$

converges.

(*c*) Show that the function (x, y), $x, y \in l^2$ satisfies all the axioms of a scalar product, that is, $(x, y) = (y, x)$; $(x, x) \geqslant 0$ and $(x, x) = 0$ iff all the coordinates of the point x are zero; $(x, y + y') = (x, y) + (x, y')$; $(x, \lambda y) = \lambda(x, y)$, λ an arbitrary real number.

(*d*) Show that

$$\rho(x, y) = \sqrt{(x - y, x - y)} = \sqrt{\sum_{n=1}^{\infty} (x_n - y_n)^2},$$

with $x = \{x_n\} \in l^2$, $y = \{y_n\} \in l^2$, is a metric on l^2.

250. Show that Hilbert space is: (*a*) complete; (*b*) separable.

251. Let A be a discrete space of cardinality τ, $P = (0, 1]$ a half-open interval with the usual topology, $P \times A$ the Tychonoff product and θ an object not belonging to $P \times A$. We then denote by $Kw(\tau)$ the set $(P \times A) \cup \{\theta\}$ with the metric satisfying the following conditions:

(i) if $z_1 \in P \times A$ and $z_1 = (r, a)$, then $\rho(z_1, \theta) = r$;

(ii) if $z_1, z_2 \in P \times A$ and $z_1 = (r_1, a_1)$, $z_2 = (r_2, a_2)$, then $\rho(z_1, z_2) = |r_1 - r_2|$ whenever $a_1 = a_2$, but $\rho(z_1, z_2) = r_1 + r_2$ whenever $a_1 \neq a_2$.

Verify that $Kw(\tau)$ is a connected, complete metric space of weight τ (such a space is called a *hedgehog space*).

252. Show that Euclidean space R^n and Hilbert space l^2 have a countable base.

253. Show that the cube Q^n in R^n is totally bounded.

254. Show that a bounded subset of a finite-dimensional Euclidean space is totally bounded.

255. Show that if a metric space (X, ρ) has a finite ε-net even for at most one $\varepsilon > 0$, then it is bounded.

256. Show that a (metric) subspace of totally bounded metric space is itself totally bounded.

257. If a metric space (X, ρ) has a finite ε-net s for some $\varepsilon > 0$, then every subspace A of (X, ρ) has a finite 2ε-net for the same $\varepsilon > 0$.

258. Let (X, ρ) be a metric space and $\{A_i: i \in N^+\}$ a sequence of sets in X, for which a finite $(1/i)$-net exists for each $i \in N^+$ in A_i. Show that $A = \bigcap \{A_i: i \in N^+\}$ is a totally bounded set.

259. Let X be a metric space and $\{A_i: i \in N^+\}$ a monotone non-increasing family of subsets of X (that is, $A_i \subset A_{i+1}$ for all $i \in N^+$). Show that diam $(\bigcup \{A_i: i \in N^+\}) = \sup \{\text{diam } A_i: i \in N\}$.

260. Show that a metric space (X, ρ) is totally bounded iff for each $\varepsilon > 0$ it has a finite cover by closed sets of diameter at most ε.

261. Show that any totally bounded metric space (X, ρ) is separable, has a countable base and is Lindelöf.

262. Show that a metric space (X, ρ) is not totally bounded iff there exist $\varepsilon > 0$ and an infinite set $A \subset X$ such that $\rho(x, y) \geqslant \varepsilon$ for each $x, y \in A$, $x \neq y$.

263. Show that a metric space (X, ρ) is totally bounded iff each sequence has a Cauchy subsequence.

264. Let (X_i, ρ_i) be a totally bounded metric space of diameter at most 1. Show that the metric product (X, ρ) of these spaces is also totally bounded.

265. Give an example of a bounded but not totally bounded set in a Hilbert space.

266. Let (X, ρ) and (Y, μ) be two metric spaces. Show that a function $f: X \to Y$ is continuous in the topologies induced on X and Y by the metrics ρ and μ iff, for any convergent sequence $\{x_n\}$ of points in X, the sequence $\{y_n\} = \{fx_n\}$ also converges and $\lim_{n \to \infty} f(x_n) = f(\lim_{n \to \infty} x_n)$.

267. Let (X, ρ) be a metric space and $A \subset X$ an arbitrary subset. Prove the inequality $\rho(x, A) - \rho(y, A) \leqslant \rho(x, y)$.

268. Let A be an arbitrary subset of a metric space (X, ρ). Show that

(*a*) the function $f(x) = \rho(x, A)$, $x \in X$, is continuous;

(*b*) continuity of the function $f(x, y) = \rho(x, y)$ as a function of two variables $x, y \in X$ means the following: if $x_0, y_0 \in X$ are arbitrary fixed points and $\varepsilon > 0$ is any positive real number, then there exists $\delta(\varepsilon) > 0$ such that $|\rho(x_0, y_0) - \rho(x, y)| < \varepsilon$ whenever $\rho(x_0, x) < \delta(\varepsilon)$ and $\rho(y_0, y) < \delta(\varepsilon)$.

269. Let R^1 be the real line. Construct the graph of the function $f(x) = \rho(x, A)$, when: (i) A is a segment in the line; (ii) A is an interval in the line; (iii) A is the set of all rational points of the line; (iv) A is the set of all irrational points of the line; (v) A is the Cantor perfect set constructed on the interval $[0, 1]$ (see Prob. 185).

270. Show that a function f from (X, ρ) onto (Y, ρ') is continuous iff for each point $x_0 \in X$ and every set $A \subset X$ for which $\rho(x_0, A) = 0$, we have $\rho'(fx_0, fA) = 0$.

271. Show that every isometry of metric spaces is a homeomorphism.

272. Let X be a straight line (a plane) with the natural metric. Does there exist an isometry of X onto a proper subset?

273. Show that the Euclidean space R^2 (that is, the plane) contains an isometric image of every metric space consisting of not more than three points.

274. Show that Hilbert space l^2 is not universal for all possible metric spaces consisting of four points (in the sense of isometric embedding), that is, there is a metric space consisting of four points which is not isometric to any subset of a Hilbert space.

275. Let f be a function from a metric space (X, ρ) into a metric space (Y, ρ') and $\varepsilon > 0$ a number such that for every $\delta > 0$ there exist points x', $x'' \in X$, for which $\rho(x', x'') < \delta$, but $\rho'(fx', fx'') > \varepsilon$. Show that exactly one of the following two assertions is true:

(i) there is $a \in Y$ such that whatever $\delta > 0$ there exist $x', x'' \in X$, for which $\rho(x', x'') < \delta$, $\rho'(fx', x'') > \varepsilon$ and $\rho'(fx', a) < \varepsilon/3$;

(ii) for any finite set $A \subset X$, there exists $\delta > 0$ such that if $\rho(x', x'') < \delta$ and $\rho'(fx', fx'') > \varepsilon$, then $\rho'(fx', A) \geqslant \varepsilon/3$ and $\rho'(fx'', A) \geqslant \varepsilon/3$.

276. Show that a function f from (X, ρ) into (Y, ρ') is uniformly continuous iff whenever we have $A \subset X$, $B \subset X$ and $\rho(A, B) = 0$ we also have the relation $\rho'(fA, fB) = 0$.

277. Let f be a uniformly continuous function from (X, ρ) onto (Y, ρ'). Show that (Y, ρ') is totally bounded if (X, ρ) is.

278. Verify whether a metric space (Y, ρ'), the image of a bounded metric space (X, ρ) under a uniformly continuous mapping $f: (X, \rho) \to (Y, \rho')$, is always bounded (cf. the preceding problem).

279. Let $f: (X, \rho) \to (Y, \rho')$ be a uniformly continuous function from a complete metric space (X, ρ) onto a metric space (Y, ρ'). Is (Y, ρ') necessarily complete?

280. Let $f: X \to Y$ be a uniformly continuous homeomorphism of a metric space X onto a complete metric space Y. Show that X is complete. Is it possible to omit the continuity of the inverse mapping?

2.5. Continuous Functions on Topological Spaces. First Problems

Let ω be a decomposition of a topological space X (see p. 45). We call a subset A of X *saturated relative to a decomposition* ω or a *distinguished set* for ω if A is a union of elements of ω, that is, if there is a subfamily $\omega' \subset \omega$, for which $A = \bigcup\{F : F \in \omega'\}$. The notion of saturated set under a mapping is defined analogously. Let $f: X \to Y$ be a function from a set X onto a set Y. A subset $A \subset X$ is called a *saturated set under* f: $X \to Y$ or a *distinguished set under the map* f, if $A = f^{-1}fA$. Let f be a function from a topological space X into a topological space Y. We shall be interested in open (closed) saturated sets. Of crucial importance is a continuous decomposition of a topological space. A decompositson ω of X is *continuous* if for each member $F \in \omega$ of this decomposition and each nhood UF of F there is a nhood $U'F$ of F such that if $F' \in \omega$ and $F' \cap U'F \neq \Lambda$, then $F' \subset UF$. A decomposition ω of X is continuous iff, for each $F \in \omega$ and each nhood UF of F there exists a saturated nhood $U'F$ for which $U'F \subset UF$ (the proof is left as an exercise to the reader).

Let ξ be a prefilter on a set X and f a function from X into a topological space Y. We say that f is *continuous with respect to* ξ or, what is the same, *along* ξ if there is a point $y \in Y$, each of whose nhoods contains the image of at most one member of ξ. The set of all such points is denoted by $\lim_\xi f$. Thus, f is *continuous along* ξ iff $\lim_\xi f \neq \Lambda$. We also call $y \in \lim_\xi f$ a *limit of f along* ξ. The sequence $\{x_n: n \in N^+\}$ of points of X is said to be *cofinal in a prefilter* ξ if, for each $A \in \xi$, there is an integer k such that $x_n \in A$ for every $n \geqslant k$. We call a function f *sequentially continuous* along ξ if the sequence $\{fx_n: n \in N^+\}$ converges in Y for each sequence $\{x_n: n \in N^+\}$, cofinal in ξ.

A continuous function $f: X \to Y$ is called *pseudo-open* if for each point $y \in Y$ and every nhood $Uf^{-1}y$ of the set $f^{-1}y$ in X we have $y \in \operatorname{Int} fU$ (here Int FU is the set of all interior points of fU relative to Y; see p. 39).

281. Let $f: X \to Y$ be a function from a space X into a space Y. Show that the following conditions are equivalent:

(i) f is continuous;
(ii) the inverse image $f^{-1}V$ of every element V of some base σ of Y is an open set in X;
(iii) for each point $x \in X$ and any nhood V of a point fx in Y, there is a nhood U of x for which $fU \subset V$;
(iv) $f[A] \subset [fA]$ for every set $A \subset X$;
(iv′) if $x \in X$, $A \subset X$ and $x\delta A$, then $f(x)\delta f(A)$;
(v) for every closed set Φ in Y, $f^{-1}\Phi$ is closed in X.

282. Let $f: X \to Y$ be a continuous function from a space X onto a space Y. Show that:

(*a*) if A is dense in X, then its image fA is dense in Y;
(*b*) if S is a net in X, then its image $fS = \{f(s): s \in S\}$ is a net in Y (also, evidently, see Prob. 25 of Chap. 1, $|fS| \leqslant |S)|$;
(*c*) the image of a base in X is not necessarily a base in Y.

283. Let $f: X \to Y$ be a function from a topological space X into a topological space Y, and suppose that the image of every dense set in X is dense in Y. Does this necessarily imply that f is a continuous function?

284. Show that $(X, \mathcal{T})$ is a discrete space iff all its mappings into any topological space are continuous.

285. Show that on each set X there exists a unique topology $\mathcal{T}_A$ such that every function f from every topological space $(Y, \mathcal{T})$ into $(X, \mathcal{T}_A)$ is continuous. Describe this topology.

286. A composition (p. 16) of continuous functions is a continuous function. Prove a similar assertion for quotient, pseudo-open, closed and open mappings.

287. If $f_1: X \to Y$ and $f_2: X \to Y$ are continuous functions from a topological space X into a Hausdorff space Y such that $A = \{x \in X: f_1(x) = f_2(x)\}$ is dense in X, then $f_1 \equiv f_2$ on X.

288. If $Y \subset X$, $[Y] = X$ and f is a continuous function from a space

Y into a regular space Z, then f can be extended to a continuous function from X into Z iff for each point $x \in X$, f can be extended to a continuous function from $Y \cup \{x\}$ into Z.

289. Suppose we are given two families of spaces $\{X_\alpha: \alpha \in A\}$ and $\{Y_\alpha: \alpha \in A\}$ and a continuous function $f_\alpha: X_\alpha \to Y_\alpha$ for each $\alpha \in A$. Show that the product $f: \Pi\{X_\alpha : \alpha \in A\} \to \Pi\{Y_\alpha: \alpha \in A\}$ of the functions f_α is continuous.

290. Given a topological space X and a family of topological spaces $\{Y_\alpha: \alpha \in A\}$, let $f_\alpha: X \to Y_\alpha$ be a continuous function for each $\alpha \in A$. Show that the diagonal product f of the functions f_α is a continuous function into the Tychonoff product $\Pi\{Y_\alpha: \alpha \in A\}$ of the spaces Y_α.

291. Let $f: X \to Y$ be a continuous function from X into Y and $X \times Y$ be the topological product of X and Y. Put $\Gamma_f = \{(x, y) \in X \times Y: y = f(x)\}$. The subset Γ_f of $X \times Y$ is called the *graph* of f. Show that :

(*a*) Γ_f, as a subspace of $X \times Y$, is homeomorphic to X;
(*b*) if Y is a Hausdorff space, then Γ_f is closed in $X \times Y$.

292. Let X be a square (together with its boundary), let Y be one of its sides and let f be a projection of X onto Y. Show that f is a continuous, open and closed function (the topologies on X and Y being the usual ones).

293. Let X_0 be the set of points of the unit square without its boundary, let Y_0 be its side (that is, a segment without end points), and let $\pi: X_0 \to Y_0$ be the natural projection. Then π is a continuous, open but not a closed function (the topologies on X_0 and Y_0 being the usual ones). Compare this problem with the preceding one.

294. Let X be a topological space and let $X' \subset X$. Show that the relative topology on the subspace X' is the smallest of all topologies on X' relative to which the identity function $i: X' \to X$ is continuous.

295. Let $f: X \to Y$ be a function from a topological space X into a topological space Y, and let $Y' \supset f(X)$, Y' a subspace of Y. Show that:

(*a*) if $f: X \to Y$ is continuous, then so is $f: X \to Y'$;
(*b*) if $f: X \to Y$ is open, then so is $f: X \to Y'$;
(*c*) if $f: X \to Y$ is closed, then so is $f: X \to Y'$.

296. Verify that the family of subsets of a decomposition, called a quotient topology (p. 45), is indeed a topology.

297. Show that the quotient topology on decomposition Z of X is the largest (see p. 41) of all topologies on Z relative to which the natural projection $\pi: X \to Z$ is continuous.

298. Show that $f: X \to Y$, where $f(X) = Y$, is a quotient mapping iff for every $F \subset Y$ the following conditions are equivalent:

(*a*) F is closed in Y;
(*b*) $f^{-1}F$ is closed in X.

299. Let $f: X \to Y$ be a continuous function from a topological space X onto a T_1-space Y, and let $Z = \{f^{-1}y: y \in Y\}$. Show that Z, as a decomposition of X, it being endowed with the quotient topology, can be contracted to the space Y by means of the natural function $\varphi: Z \to Y$, described by the formula: $\varphi(f^{-1}y) = y$. Verify that f is a quotient mapping iff φ is a homeomorphism.

300. Let $f: X \to Y$ be a continuous function, let $X' \subset X$, $f' = f|X'$, $f': X' \to Y$ be a quotient mapping, with $f'(X') = f(X') = Y$. Show that f is also a quotient mapping.

301. Let a family of topologies $\{\mathcal{T}_\alpha: \alpha \in M\}$ be given on a set X and let $\mathcal{T} = \bigcap \{\mathcal{T}_\alpha: \alpha \in M\}$ (that is, in $(X, \mathcal{T})$ exactly those sets are open which are open in every $(X, \mathcal{T}_\alpha)$, $\alpha \in M$). Show that the *natural function*† π from the disjoint union X_ξ of the spaces in the family $\xi = \{(X, \mathcal{T}_\alpha): \alpha \in M\}$ onto the space $(X, \mathcal{T})$ is a quotient mapping.

302. Consider the subsets $A = \left\{\left(\frac{1}{n}, k\right): k = 0, 1, 2, \ldots; n = 1, 2, \ldots\right\}$ $B = \{(0, k): k = 1, 2, \ldots\}$, $C = A \cup B$ of the Euclidean plane with some Cartesian coordinate system. The set C is endowed with the topology induced by the plane and S is defined to be a B-trivial decomposition†† of the space C. Show that:

(i) the natural projection $\pi: C \to S$ is closed;
(ii) S is nonmetrizable;
(iii) π is not an open mapping.

303. Let A be the same as in Prob. 302. Denote by $*$ some object not belonging to A and define on a set $\Phi = A \cup \{*\}$ the topologies $\mathcal{T}_1$ and

†If $y \in X_\xi$, then $y \in (i_\alpha X, \mathcal{T}_\alpha)$ for some $\alpha \in M$. We then put $\pi y = i_\alpha^{-1} y$, where i_α is the standard embedding of $(X, \mathcal{T}_\alpha)$ into the disjoint union X_ξ (see p. 44).

††That is, whose unique nontrivial element is the closed set $B \subset C$.

$\mathcal{T}_2$ as follows. All points of A are isolated relative to both $\mathcal{T}_1$ and $\mathcal{T}_2$. Put

$$U_j = \{*\} \cup \left\{\left(\frac{1}{n}, k\right); k = 0, 1, 2, \ldots; n > j\right\}, \quad j = 1, 2, \ldots$$

The family $\{U_j: j = 1, 2, \ldots\}$ forms a base for $\mathcal{T}_1$ at the point $*$. Put

$$P_k = \left\{\left(\frac{1}{n}, k\right): n = 1, 2, \ldots\right\}, k = 0, 1, 2, \ldots$$

A set $W \subset \Phi$, $W \ni *$, belongs to $\mathcal{T}_2$ iff $P_k \setminus W$ is finite for each $k = 0, 1, 2, \ldots$ Show that:

(i) the space $(\Phi, \mathcal{T}_2)$ is homeomorphic to the space S constructed in Prob. 302;

(ii) $\mathcal{T}_1 \neq \mathcal{T}_2$;

(iii) the spaces $(\Phi, \mathcal{T}_1)$ and $(\Phi, \mathcal{T}_2)$ are not compact but are normal.

304. Let $f: X \to Y$ be a continuous function of a topological space X into a topological space Y, let X' be a subspace of X and let f' be the restriction of f to X', $f': X' \to Y$. Verify whether the following are true or false:

(*a*) f' is continuous;

(*b*) if X' is closed in X and f is closed, then f' is also closed;

(*c*) if X' is open in X and f is open, then f' is also open;

(*d*) if f is a quotient map and X' is a clopen in X, then f' is also a quotient map.

(*e*) If f is a quotient map and $fX' = Y$, then f' is also a quotient map.

305. Let $f: X \to Y$ be a quotient map and $Y' \subset Y$, $X' = f^{-1}Y'$, $f' = f | X'$, $f': X' \to Y'$. Show that:

(*a*) if Y' is open in Y, then f' is a quotient map;

(*b*) if Y' is closed in Y, then f' is a quotient map;

(*c*) if Y' is neither open nor closed in Y, then f' cannot be a quotient map.

306. Determine the truth or falsity of the following assertions:

(*a*) the product (see p. 4) of any collection of quotient maps is a quotient map;

(*b*) the product of any two quotient maps is a quotient map.

307. True or false? The product of a quotient map by an identity map on any space is a quotient map.

308. Let $f: X \to Y$ be a continuous function of X onto Y. Show that the following conditions are equivalent:

(*a*) for each $Y' \subset Y$ the restriction f' of f to $X' = f^{-1}Y'$, the inverse image of Y' is a quotient map of X' onto Y';

(*b*) for each $y \in Y$ and every open set U in X containing $f^{-1}y$, the interior $\operatorname{Int} f(U)$ of the image of U contains y (that is, f is pseudo-open).

(*c*) whenever $A \subset Y$ and $y \in Y$ satisfies $y \in [A]$, we have $f^{-1}y \cap [f^{-1}A] \neq \Lambda$.

309. Let $f: X \to Y$ be a continuous function from some subspace X' of X onto Y, a pseudo-open space. Show that f is also a pseudo-open space.

310. Determine the truth or falsity of the following assertions:

(*a*) the product of any collection of pseudo-open functions is pseudo-open;

(*q*) the product of two pseudo-open functions is pseudo-open.

311. Show that the image of a Fréchet-Urysohn space under a continuous pseudo-open map is a Fréchet-Urysohn space.

312. If $f: X \to Y$ is a quotient map and $f(X) = Y$, show that $t(Y) \leqslant t(X)$.

313. If $f: X \to Y$ is a continuous, pseudo-open map, $f(X) = Y, \Phi \subset Y$ and $F = f^{-1}\Phi$, show that for any defining family ξ of nhoods of F in X the family $f \cdot \xi = \{\operatorname{Int} fU: U \in \xi\}$ is a defining family of nhoods of Φ in Y.

314. If $f: X \to Y$ is a continuous, pseudo-open function, $f(X) = Y, \Phi \subset Y$ and $F = f^{-1}\Phi$, prove that the character of the set Φ in Y does not exceed that of F in X.

315. Prove that every continuous closed map is a quotient map.

316. Prove that every continuous closed function is pseudo-open.

317. True or false? Every closed function from one space onto another is open.

318. Show that the natural projection π of X onto a Φ-trivial† decomposition space of X is closed for every closed set Φ in X.

†That is, a decomposition, the unique nontrivial element of which is a closed set Φ in X.

319. If $f: X \to Y$ is a closed map and $Y' \subset Y$, $X' = f^{-1}Y'$, $f' = f|X'$, prove that $f': X' \to Y'$ is also a closed map.

320. Show that a function f from X into Y is closed iff $[f(A)] \subset f([A])$ for each $A \subset X$.

321. Show that a function $f: X \to Y$, where $Y = f(X)$, is closed iff for every open set U in X, the set $f\#U = \{y \in Y: f^{-1}y \subset U\}$ is open in Y.

322. Show that a quotient mapping f of a T_1-space X onto a T_1-space Y is closed iff the decomposition $\alpha = \{f^{-1}y: y \in Y\}$ is continuous.

323. If $f: X \to Y$ is a closed function, $\Phi \subset Y$, $F = f^{-1}\Phi$ and U is any open set in X containing F, show that there exists an open set V in Y for which $\Phi \subset V$ and $f^{-1}V \subset U$.

324. Let $f: X \to Y$ be a continuous closed function with $f(X) = Y$ and $\Phi \subset Y$; let $F = f^{-1}\Phi$ and ξ (resp. η) be a defining family of nhoods of F in X (resp. of Φ in Y). Show that $f^{-1}\eta = \{f^{-1}U: U \in \eta\}$ (resp. $f\cdot\xi = \text{Int}\, fV$: $V \in \xi$) is a defining family of nhoods of F in X (resp. of Φ in Y).

325. Prove that a function $f: X \to Y$ is closed iff for each $y \in Y$ and every nhood U of $f^{-1}y$ there is a nhood V of y such that $f^{-1}(V) \subset U$.

326. Let X be a spece, Φ a closed set in X, Z_Φ a Φ-trivial decomposition space of X, $\pi: X \to Z_\Phi$ the natural projection, $z^* = \pi(\Phi) \in Z_\Phi$. Show that the character of Z_Φ at a point z^* equals the character of Φ in X.

327. If $f: X \to Y$ is a continuous closed function with $f(X) = Y$, show that the character of each point $y \in Y$ equals the character of the set $f^{-1}y$ in X.

328. Show that a function $f: X \to Y$ from a Hausdorff space X into a Hausdorff space Y is continuous and closed iff the projections $p_X: \Gamma_f \to X$ and $p_Y: \Gamma_f \to Y$, with $\Gamma_f = \{(x, f(x)): x \in X\}$, are closed, where the graph of a function carries the subspace topology inherited from the topological product $X \times Y$.

329. Show that the image of a normal space under a continuous, closed map is a normal space.

330. Show that any projection of a Tychonoff product onto a factor is a continuous open function (in general, not closed).

331. Is every open function from one space onto another closed ?

332. Show that every continuous open function is pseudo-open and also a quotient map.

333. Show that a projection π of X onto a decomposition space Z, endowed with the quotient topology is open iff for every open set U in X its *saturation* $\tilde{U}$ under the decomposition Z (that is, the set $\bigcup \{P \in Z: P \cap U \neq \Lambda\}$) is open in X.

334. Show that a function $f: X \to Y$ is open iff the image under f of each element of a base for X is an open set in Y.

335. Show that a function $f: X \to Y$ is open iff for every set $B \subset Y$, we have $f^{-1}[B] \subset [f^{-1}B]$.

336. Show that a function $f: X \to Y$ is open iff $f^{-1}(\mathrm{Fr}(B)) = \mathrm{Fr}\,(f^{-1}(B))$ for all $B \subset Y$, where $\mathrm{Fr}(B) = [B] \setminus \mathrm{Int}\, B$ is the *frontier* of B in Y.

337. Show that, if $f: X \to Y$ is an open function and $Y' \subset Y$, $X' = f^{-1}Y'$, $f' = f \mid X'$, then $f': X' \to Y'$ is also an open function.

338. Show that the product of every set of open maps is an open map.

339. Show that, if $f: X \to Y$ is a continuous open map from X onto Y, then the weight of Y is at most the weight of X.

340. Show that, if $f: X \to Y$ is an open map and $X' \subset X$, and $[X' \cap f^{-1}y] = f^{-1}y$ for all $y \in Y$, then the restriction $f \mid X': X' \to Y$ is also an open map.

341. Let $f: X \to Y$ be a continuous open map, let Y be a T_1-space, $f(X) = Y$, $y \in Y$, $x^* \in f^{-1}y$ and $X' = (X \setminus f^{-1}(y)) \cup \{x^*\}$. Show that $f' = f \mid X'$, $f': X' \to Y$ is also open.

342. Let $f: X \to Y$ be a one-to-one continuous function from X onto Y. Show that the following conditions are equivalent: (*a*) f is a quotient map; (*b*) f is open; (*c*) f is closed; (*d*) f is a homeomorphism.

343. Construct two non-homeomorphic topological spaces X and Y, each of which is homeomorphic to a subspace of the other.

344. Find non-homeomorphic topological spaces X and Y, each of which can be mapped one-to-one and continuously onto the other.

345. Let $[Y] = X$, where X is T_1 and let $\varphi: X \to Z$ be a continuous map whose restriction $\varphi \mid Y: Y \to \varphi(Y)$ is a homeomorphism. Show that $\varphi(X \setminus Y) \subset Z \setminus \varphi(Y)$.

346. Let Z be any continuous decomposition space of a space X. Let Z^1 be the subspace of Z consisting of all elements of the decomposition which contains exactly one point of X. Put $X^1 = \{x \in X: \{x\} \in Z^1\}$. Show that X^1, as a subspace of X, is homeomorphic to Z^1.

347. Does the conclusion of Problem 346 remain true if the requirement that the decomposition be continuous is omitted ?

348. Let $X = X_1 \cup X_2$, where X is T_1, X_1 and X_2 are closed disjoint subspaces of X; let A_1 and A_2 be closed subsets of X_1 and X_2, respectively and let f be a closed function from A_1 onto A_2. Put $Z = Z_1 \cup Z_2 \cup Z_3$, where $Z_1 = \{\{x\}: x \in X_1 \setminus A_1\}$, $Z_2 = \{\{x\}: x \in X_2 \setminus A_2\}$ and $Z_3 = \{F_x: x \in A_2\}$, where $F_x = \{x\} \cup f^{-1}(x)$ for each $x \in A_2$. Show that:

(*a*) Z is a continuous decomposition of X into closed sets;
(*b*) the natural projection π of X onto the decomposition space Z induces a homeomorphism between X_2 and a closed subspace of Z.

349. On the class of completely regular spaces construct a quotient map whose product with some identity function is not a quotient map.

350. Let ξ be a prefilter on a set X and let $f: X \to Y$ be a function into a Hausdorff space Y. Show that $|\lim_\xi f| \leqslant 1$.

351. Give an example of filter ξ and a real-valued function f defined on the set R of all real numbers, such that f is sequentially continuous, but not continuous, along ξ.

352. Let ξ be a countable prefilter (see p. 7) on X and let $f: X \to Y$ be a function into a Hausdorff space Y. Show that f is continuous along ξ iff it is sequentially continuous along ξ.

2.6. The Tychonoff Product

In general topology an important role is played by a whole series of concepts belonging to abstract set theory and involving the notion of product of sets. Fix a product $X = \Pi\{X_\alpha: \alpha \in A\}$ and let $x, x'' \in X$. Put $H(x', x'') = \{\alpha \in A: x'_\alpha \neq x''_\alpha\}$; we call $H(x, x'')$ the *support* (of the points x' and x''). Evidently, $H(x', x'') = H(x'', x')$. Observe that $x \mid B = y \mid B$, where $B \subset A$ and $x \in X$, $y \in X$, implies that $H(x, y) \subset A \setminus B$. Let $y \in X$ and $B \subset A$. The set $\{x \in X: H(x, y) \subset B\}$ is called *the B-hull* of y. We say that $Y \subset X$ is *independent* of $B \subset A$, if $y \in Y$, $x \in X$ and $H(y, x) \subset B$ always imply that $x \in Y$. Let $P \subset Z \subset X = \Pi\{X_\alpha: \alpha \in A\}$ and let f be a function of Z into some set Y. We say that f is independent of $B \subset A$ on P if $x \in P$, $z \in Z$ and $H(x, z) \subset B$ imply that $f(x) = f(z)$. If, in addition, $P = Z = X$ we simply say that f is independent of B. Fix $B \subset A$ and $y \in X$. The function $\langle B, y\rangle: X \to X$, called the *$\langle B, y\rangle$-projection*, is defined as follows: If $x, z \in X$, then $z = \langle B, y\rangle(x)$ if $z_\alpha = y_\alpha$ for each $\alpha \in B$ and $z_\alpha = x_\alpha$ for every $\alpha \in A \setminus B$ (in other words: $z = \langle B, y\rangle(x)$, if $H(x, z) \subset B$ and $H(y, z) \subset A \setminus B$).

A set $P \subset X$ is said to be *B-convex* if $x, y \in P$ implies that $\langle B, y\rangle(x) \in$

P. Let τ be a cardinal number. We say that P is *τ-convex* if P is B-convex for every $B \subset A$ such that $|B| \leqslant \tau$. The notion of the product set Σ_τ centered at a given point is also of crucial importance to us. We shall of course use all the notations and terms already defined for product sets and products of topological spaces (see pp. 4, 5, 44).

Closely related to the notion of product topology is the notion of the topology of pointwise convergence (or, the pointwise topology) on the set $\mathcal{E}(X, Y)$ of all functions from one topological space X into another topological space Y. $\mathcal{E}(X, Y)$ may be identified with Y^X, the product of $|X|$ copies of Y. This product is regarded as a product space with the natural topology (which, evidently, does not depend on the topology of X). A straightforward description of this natural topology $\mathcal{E}(X, Y)$ is as follows. Fix any finite set $K \subset X$. For each $x \in X$ fix a nonempty open set $V_x \subset Y$. The set $\{f \in \mathcal{E}(X, Y): f(x) \in V_x$ for each $x \in K\}$ is an element of base for some topology on $\mathcal{E}(X,Y)$. For obvious reasons, this topology is called the *topology of pointwise convergence* (or, the *pointwise topology*) on $\mathcal{E}(X, Y)$. We denote by $C_\omega(X, Y)$ the subspace of $\mathcal{E}(X, Y)$ consisting of all continuous functions from X into Y. The topology on this as well as other subspaces of $\mathcal{E}(X, Y)$ is also called the pointwise topology.

In this section we first encounter the *generalized Cantor discontinuum*, which will play an important role in the sequel. This is the Tychonoff product of an arbitrary family of discrete spaces, each of which consists of exactly two points. If the number of factors is precisely τ, the resulting generalized Cantor discontinuum is denoted by D^τ. The reason for this notation is obvious. Finally, if $\tau \neq \tau'$, then D^τ is not homeomorphic to $D^{\tau'}$.

In contrast to generalized Cantor discontinua there are *connected* (or *bounded*) products of 2-point spaces. Let D_0 be the two-point connected topological space, which consists of two points (usually denoted by 0 and 1), whose open sets are D_0, the empty set and one of the two singleton subsets. We shall assume that the set $\{0\}$ is open and the set $\{1\}$ is not open (the latter being, of course, closed). The Tychonoff product of τ copies of this two-point connected space is denoted by F^τ.

353. Give an example of an infinite space which cannot be written as the product of spaces, each of which is different from it.

354. Let $X = \Pi\{X_\alpha: \alpha \in A\}$ be the Tychonoff product of a family of topological spaces, U any standard open set (see p. 44), $k(U) = \{\alpha_1, \ldots, \alpha_s\} \subset A$, $U_\alpha = \pi_\alpha(U)$, $\alpha \in A$. Show that $U = \bigcap\{\pi_{\alpha_i}^{-1} U_{\alpha_i}: i = 1, \ldots, s\}$.

355. Let $X = \Pi\{X_\alpha: \alpha \in M\}$ be the Tychonoff product, $M = \cup\{M_\beta: \beta \in L\}$, where $M_{\beta'} \cap M_{\beta''} = \Lambda$ for $\beta' \neq \beta''$, let $Y_\beta = \Pi\{X_\alpha: \alpha \in M_\beta\}$ be the Tychonoff product and let Z_β be a subspace of Y_β for each $\beta \in L$. Show that there is a *natural* homeomorphism from the Tychonoff product $Z = \Pi\{Z_\beta: \beta \in L\}$ onto a subspace of X that coincides with X if for each $\beta \in L$ we have $Z_\beta = Y_\beta$.

356. Show that each member of a family of T_1-spaces is homeomorphic to a closed subspace of the product of the family.

357. Let $\{X_\alpha: \alpha \in M\}$ be a family of topological spaces and, for each $\alpha \in M$, let Y_α be a dense subspace of X_α. Show that the Tychonoff product $Y = \Pi\{Y_\alpha: \alpha \in M\}$ is dense in the Tychonoff product $X = \Pi\{X_\alpha: \alpha \in M\}$.

358. Show that the product of a countable set of separable spaces is separable.

359. Show that the product of a countable set of spaces, each of which has a countable net, itself has a countable net.

360. Show that the weight of a Tychonoff product of spaces is not less than that of each of its factors.

361. Let $X = \Pi\{X_\alpha: \alpha \in M\}$ be the Tychonoff product and Y_α a closed subspace of X_α for each $\alpha \in M$. Show that the Tychonoff product $Y = \Pi\{Y_\alpha: \alpha \in M\}$ is a closed subspace of X.

362. For each $\alpha \in A$, let X_α be a topological space, let Y_α be a subspace of X_α and let γ_α be a family of G_δ-sets in X_α such that $\cup\{U: U \in \gamma_\alpha\} = X_\alpha \setminus Y_\alpha$. Further, let $\tilde{U} = U \times \Pi\{X_{\alpha'}: \alpha' \in A\setminus\{\alpha\}\}$ for each $U \in \gamma_\alpha$ and $\tilde{\gamma}_\alpha = \{\tilde{U}: U \in \gamma_\alpha\}$, $\tilde{\gamma} = \cup\{\tilde{\gamma}_\alpha: \alpha \in A\}$, $Y = \Pi\{Y_\alpha: \alpha \in A\}$ and $X = \Pi\{X_\alpha: \alpha \in A\}$ (X and Y being the Tychonoff products). Show that:

(i) for each $\alpha \in A$ and every $U \in \gamma_\alpha$, $\tilde{U}$ is a G_δ in X:
(ii) $X \setminus Y = \cup\{\tilde{U}: \tilde{U} \in \tilde{\gamma}\}$.

363. Let $X = \Pi\{X_\alpha: \alpha \in M\}$ be the product of the spaces X_α with respect to $\alpha \in M$. Verify that the product topology is the smallest on X for which each $\pi_\alpha: X \to X_\alpha$ is continuous.

364. (*a*) Show that a two-point connected space is a connected T_0-space.

Consider the family $\{D_\alpha, \alpha \in A\}$, $|A| = \tau$, each D_α being a two-point connected space and their Tychonoff product $\Pi\{D_\alpha: \alpha \in A\} = F^\tau$.

(*b*) Show that, for a T_0-space X, $w(X) \leqslant \tau$ iff X is homeomorphic to a subspace of F^τ.

365. Show that, for any topological space X and any nonempty set A, X is homeomorphic to the diagonal $\Delta = \{z \in X^A$: there is $x \in X$ such that $\pi_\alpha(z) = x$ for all $\alpha \in A\}$, regarded as a subspace of the Tychonoff power X^A.

366. Show that the diagonal Δ is closed in any power of a Hausdorff space.

367. Show that, if X is Hausdorff and $\mathcal{E}$ is a family of subspaces of X, then $Y = \bigcap \{U: U \in \mathcal{E}\}$ with the topology induced by X is homeomorphic to a closed subspace of the Tychonoff product $Z = \Pi \{U: U \in \mathcal{E}\}$.

368. Show that a space X is T_1 iff the diagonal Δ in the Tychonoff product $X \times X$ is an intersection of a family of open sets.

369. Let $X^2 = X \times X$ be the product of a set X with itself and let $\Delta = \{(x, y) \in X \times X: x = y\}$. Let $\mathcal{T}$ and $\mathcal{T}'$ be two Hausdorff topologies on X. Show that:

(*a*) Δ is closed in the topological product $(X, \mathcal{T}) \times (X, \mathcal{T})$.
(*b*) if $\mathcal{T} \neq \mathcal{T}'$, then Δ is not necessarily closed in the product topology $(X, \mathcal{T}) \times (X, \mathcal{T}')$ (give an example).

370. Let $\mathcal{T}_1$ and $\mathcal{T}_2$ be topologies on X. Show that the diagonal $\Delta = \{(x, x): x \in X\}$ is closed in $(X, \mathcal{T}_1) \times (X, \mathcal{T}_2)$ iff the following condition is satisfied:

(*a*) for every $x_1, x_2 \in X$, $x_1 \neq x_2$, there exist $Ox_1 \in \mathcal{T}_1$, $Ox_1 \ni x_1$ and $Ox_2 \in \mathcal{T}_2$, $Ox_2 \ni x_2$ such that $Ox_1 \cap Ox_2 = \Lambda$.

371. Show that a space X is Hausdorff iff the diagonal Δ of the square $X \times X$ is closed in $X \times X$.

372. If X is a Lindelöf T_1-space and the diagonal Δ has a countable defining family of nhoods in $X \times X$, show that X has a countable base.

373. Let $X = \Pi\{X_\alpha: \alpha \in A\}$ be a topological product. Show that, whenever a separation axiom: T_0, T_1, T_2 (or, Hausdorff axiom) or T_3 (or, regularity axiom) is satisfied by X_α for each $\alpha \in A$, the same axiom is satisfied by X.

374. Show that the space $D^\tau = \Pi\{X_\alpha: \alpha \in M\}$, where $X_\alpha = \{0, 1\}$ for each $\alpha \in M$ and $|M| = \tau$, contains a discrete subspace of cardinality τ.

375. Show that the space D^τ has cardinality 2^τ and weight τ (here $\tau \geqslant \aleph_0$).

376. Show that the topological product $X = \Pi\{\tilde{D}: \alpha \in A\}$, where $|A| = \tau \geqslant \aleph_0$ and each $\tilde{D}_\alpha$ is a finite discrete space consisting of more than one point, is homeomorphic to the generalized Cantor discontinuum D^{τ}.

377. Let τ be a cardinal number, $\tau \geqslant \aleph_0$ and let A be a set of cardinality 2^{τ}. Show that there exists on A a Hausdorff topology of weight τ.

378. Show that the space D^{τ} can be mapped one-to-one and continuously onto the space F^{τ}.

379. If $|M| = \tau \geqslant \aleph_0$ and the X_α are spaces with more than one point and weight not exceeding τ for each $\alpha \in M$, show that the weight of the Tychonoff product $X = \Pi\{X_\alpha: \alpha \in M\}$ is τ.

380. Let $X = \Pi\{X_\alpha: \alpha \in A\}$ be the Tychonoff product, where $X_\alpha = Y$ for each $\alpha \in A$, Y a discrete space of cardinality $\tau \geqslant \aleph_0$ and $|A| = 2^{\tau}$. Show that X has a dense subset of cardinality τ.

381. Let $\tau \geqslant \aleph_0$ and let $X = \Pi\{X_\alpha: \alpha \in M\}$ be the Tychonoff product, where the density of X_α does not exceed τ for each $\alpha \in M$ and $|M| \leqslant 2^{\tau}$. Show that X has a dense subset of cardinality at most τ.

382. Let $X = \Pi\{X_\alpha: \alpha \in M\}$ be the Tychonoff product, each X_α containing a countable dense subset. Further, let Y be the $\aleph_1$-hull of some point $x^* \in X$ (see pp. 5-6). Show that, for any cardinal $\tau \geqslant \aleph_0$ and any set $A \subset Y$ for which $|A| \leqslant 2^{\tau}$, there is a set $\tilde{A} \subset X$ such that $|\tilde{A}| \leqslant \tau$ and $[\tilde{A}] \supset A$.

383. If $X = \Pi\{X_\alpha: \alpha \in M\}$ is a Tychonoff product and each X_α has a countable dense subset, show that the cardinality of any family of pairwise disjoint, nonempty open sets in X is at most $\aleph_0$ (that is, $c(X) \leqslant \aleph_0$).

384. If $X = \Pi\{X_\alpha: \alpha \in M\}$ is the Tychonoff product and U an element of the standard base for X, show that U is independent of $L = M \setminus C$, for some finite set C.

385. Let $X = \Pi\{X_\alpha: \alpha \in M\}$ be the Tychonoff product, let X_α be separable for each $\alpha \in M$ and $A = [U]$, where U is an open set in X. Show that there exists a countable set $C \subset M$ such that A is independent of $L = M \setminus C$.

386. Let $X = \Pi\{X_\alpha: \alpha \in M\}$ be the Tychonoff product and f a continuous function from X into a regular space Y. Show that if each X_α is separable and Y is a regular space with a countable base, then there exists a countable set $C \subset M$ such that f is independent of $L = M \setminus C$.

387. Let $X = \Pi\{X_\alpha: \alpha \in A\}$ be the Tychonoff product, $P \subset X, B \subset A$. For each $x \in X$, put $\langle B, y\rangle(x) = z$, where $z_\alpha = y_\alpha$ for each $\alpha \in B_\alpha$ and $z_\alpha = x_\alpha$ for each $\alpha \in A \setminus B$. For obvious reasons, the function $\langle B, y\rangle: X \to X$ is called the *$\langle B, y\rangle$-projection*. Show that:

(*a*) $\langle B, x\rangle$ is a continuous function; moreover, if $y', y'' \in X$, then

$y'_\alpha = y''_\alpha$ for all $\alpha \in A \setminus B$, which is equivalent to $\langle B, x\rangle(y') = \langle B, x\rangle(y'')$ for each $x \in X$;

(*b*) if $\langle B, x\rangle(y') = \langle B, x\rangle(y'')$, then $\langle B, x'\rangle(y') = \langle B, x'\rangle(y'')$ for each $x' \in X$;

(*c*) for every $x, y \in X$, $\langle B, y\rangle\langle B, x\rangle = \langle B, y\rangle$;

(*d*) if $y' \in [P]$ and $P' = \langle B, y'\rangle(P)$, then $y' \in [P']$;

(*e*) if $y' \in [P]$ and $y'' \in X$, $y''_\alpha = y'_\alpha$ for all $\alpha \in A \setminus B$, then $y'' \in [P'']$, where $P'' = \langle B, y''\rangle(P)$.

388. Let $X = \Pi\{X_\alpha : \alpha \in M\}$ be the Tychonoff product, $Z \subset X$ a subspace, $f: Z \to Y$ a continuous function into a Hausdorff space Y, $B \subset Z$, $L \subset M$ and Z be L-convex. Show that, if f is independent of L on B, then the same is true on $[B]$.

389. Let Y be a Hausdorff space. Let $X = \Pi\{X_\alpha : \alpha \in M\}$ be the Tychonoff product of separable spaces X_α, let $Z \subset X$ be a subspace and let $f: Z \to Y$ be continuous. Show that, for each $x \in Z$, there exist sets $A_x \subset Z$ and $C_x \subset M$ such that: (*a*) $f(A_x) = f(x)$; (*b*) A_x is independent of $L_x = M \setminus C_x$; (*c*) $| C_x | \leqslant | \mathcal{P} |$, where $\mathcal{P}$ is any fixed family of open sets in Y for which $\{f(x)\} = \cap \{V: V \in \mathcal{P}\}$.

390. Let $X = \Pi\{X_\alpha : \alpha \in M\}$ be the Tychonoff product, where each X_α, $\alpha \in M$, is a space with a countable net (p. 40), let $Z \subset X$ be an $\aleph_0$-convex subspace of X, let f be a continuous function from Z onto a Hausdorff space Y, and let $B \subset Y$ be such that each point $y \in B$ is G_δ in Y (that is, $\psi(y, Y) \leqslant \aleph_0$). Show that, for any $z^* \in Z$, there is then a countable set $C \subset M$ for which the image under f of some countable subset of the C-hull of the points z^* in Z (see p. 85) is included and is dense in B.

391. Let $X = \Pi\{X_\alpha : \alpha \in M\}$ be the product of sequentially compact spaces X_α (that is, of metrizable compact spaces, or, what is the same, of compact spaces with a countable base, see Prob. 257 of Chap. 3) and let X' be either a $\Sigma\aleph_1$-product of the X_α centered at some point $x^* \in X$, or let $X' = X$.

Given a continuous function $f: X' \to Y$, where Y is Hausdorff, and a set $Y' \subset Y$, $f(X') = Y$ and suppose that the character of Y at each point $y \in Y'$ is at most $\aleph_0$. Show that $[Y']$ is compact metrizable.

392. Show that the Tychonoff product of an uncountable set of real lines is not normal.

393. Let X_α, $\alpha \in A$, be a family of complete metrizable spaces and let $X = \Pi\{X_\alpha : \alpha \in A\}$ be their topological product. Show that the $\aleph_1$-hull of each point $x \in X$ in X is a normal space.

394. Let X_α, $\alpha \in A$, be an uncountable family of separable metrizable spaces and let $X = \Pi\{X_\alpha : \alpha \in A\}$ be their topological product. Is the $\aleph_1$-hull of any $x \in X$ in X always a normal space?

395. Let τ be a cardinal number, $\tau \geqslant \aleph_0$ and let $\{X_t : t \in T\}$ be a family of topological spaces such that $c(X_t) \leqslant \tau$ for each $t \in T$. Show that $c(X) \leqslant 2^\tau$, where $X = \Pi\{X: t \in T\}$.

396. Let X be a T_1-space, Y the set of all nonisolated points of X and let $A = X \setminus Y$. Let ψ_x be any minimal linearly ordered set of cardinality $\psi(x, X)$ for $x \in Y$. Furthermore, A and ψ_x, where $x \in Y$, are endowed with a discrete topology.

Show that A is homeomorphic to a closed set in $X^* = X \times \Pi\{\psi_x : x \in Y\}$.

397. Let $C_\omega[0, 1]$ be the set of all continuous functions of the usual unit interval $I = [0, 1]$ into itself, endowed with the pointwise topology (see p. 86). Show that: (*a*) $C_\omega[0, 1]$ is not first countable, (*b*) $C_\omega[0, 1]$ is homeomorphic to a subspace of the Tychonoff product of $2^{\aleph_0}$ copies of $[0, 1]$, (*c*) every point of $C_\omega[0, 1]$ is a G_δ.

398. Let X and Y be spaces, each of which has a countable base, Y being regular. Show that the space $C_\omega(X, Y)$ of all continuous functions of X into Y, endowed with the pointwise topology†, is hereditarily separable, hereditarily Lindelöf and regular.

2.7. Solutions

1. Call $U \subset X$ an open set if $C(X \setminus U) = X \setminus U$. Let $\mathcal{T}$ be the family of open sets thus defined. It is left to the reader to show that $\mathcal{T}$ is the topology sought on X.

2. The assertions (*a*) through (*g*) are trivial to prove. For example, we prove (*e*). Every open set U in X contains all, except perhaps finitely many, points of X. Hence U contains also all points of A, except perhaps finitely many of them. Since A is infinite, it follows that U contains infinitely many points of A. This applies, in particular, to every nhood of x.

3. (*a*) Since $A \in \xi$ and $X \supset B \supset A$ imply that $B \in \xi$, the union of every set of elements of the family $\mathcal{T}_\xi$ is an element of $\mathcal{T}_\xi$. If $U, V \in \mathcal{T}_\xi$, and even either U or V is empty, then $U \cap V = \Lambda \in \mathcal{T}_\xi$. If $U, V \neq \Lambda$, then $U, V \in \xi$, and since ξ is a filter (see p. 7), then $U \cap V \in \xi$, whence $U \cap V \in \mathcal{T}_\xi$. Hence, the intersection of every finite

†The same conclusion is derived along similar lines for the compact open topology on $C[0, 1]$, which we do not consider here (see, Michael [3], [4] and Archangel'skii [1]).

set of elements of $\mathcal{T}_\xi$ belongs to $\mathcal{T}_\xi$. But $\Lambda \notin \xi$ and since ξ is a filter, it follows that $X \in \xi$. Hence $X \in \mathcal{T}_\xi$. Consequently, $\mathcal{T}_\xi$ is a topology on X.

(*b*) Let $x \in X$. From $X = (X \setminus \{x\}) \cup \{x\}$ and $X \in \xi$ it follows, since ξ is an ultrafilter (see Prob. 128 of Chap. 1), that either $X \setminus \{x\} \in \xi$, or $\{x\} \in \xi$. But the latter is impossible, because ξ is a free ultrafilter (otherwise, one would have $x \in A$ for each $A \in \xi$, whence it would follow that $\xi = \{A \subset X : A \ni x\}$). Hence, $X \setminus \{x\} \in \xi$ and furthermore, $X \setminus \{x\} \in \mathcal{T}_\xi$. Consequently, $\{x\} = X \setminus (X \setminus \{x\})$ is a closed set in $(X, \mathcal{T}_\xi)$.

(*c*) By Prob. 2, $\mathcal{T}_k \subset \mathcal{T}_\xi$.

(*d*) Suppose that x^* is an isolated point of $(X, \mathcal{T}_\xi)$. Then $\{x^*\} \in \mathcal{T}_\xi$ and since $\{x^*\} \neq \Lambda$, we have $\{x^*\} \in \xi$. But then ξ is a singular ultrafilter (as seen in (*b*)). This contradiction shows that $(X, \mathcal{T}_\xi)$ has no isolated points.

(*e*) If $U \in \xi$ and $y \in X \setminus U$, then $U \cup \{y\} \in \xi$ (since ξ is a filter) and hence $U \cup \{y\}$ is an open set in $(X, \mathcal{T}_\xi)$. But $(U \cup \{y\}) \cap (X \setminus U) = \{y\}$, which shows that $\{y\}$ is an open set in $X \setminus U$, a subspace of $(X, \mathcal{T}_\xi)$, that is, y is an isolated point in this subspace.

(*f*) Let $U \in \mathcal{T} \setminus \mathcal{T}_\xi$. Then $U \neq \Lambda$ and $U \notin \xi$. Since ξ is an ultrafilter, $U \cap V = \Lambda$ for some $V \in \xi$. Choose $y \in U$. Then $V \cup \{y\} \supset V$ and hence $V \cup \{y\} \in \xi$. But $\xi \subset \mathcal{T}_\xi \subset \mathcal{T}$. Therefore, $V \cup \{y\} \in \mathcal{T}$. Hence $\{y\} = U \cap (V \cup \{y\}) \in \mathcal{T}$, that is, y is an isolated point in $(X, \mathcal{T})$.

4. Let η be the family of all nhoods of x in X. Then $\xi^* = \xi \cup \eta$ is a prefilter on X and x is an adherent point of ξ^* (trivial). Furthermore, since $\xi^* \supset \eta$, ξ^* converges to x. Take $\tilde{\xi}$ to be any ultrafilter on X containing ξ^* (such ultrafilters exist, see Problems 125, 127, 128 of Chap. 1). Since ξ^* converges to x and $\tilde{\xi} \supset \xi^*$, $\tilde{\xi}$ also converges to x.

6. Evidently, $\mathcal{T}'$ is a topology but $\mathcal{T}''$ is, in general, not a topology.

10. Consider any set containing more than one element. Give this the indiscrete topology. The resulting space is the one required.

11. Let N be the set of all integers, $N_k = \{n \in N : n \geqslant k\}$ for each $k \in N$. The family $\tilde{\mathcal{T}} = \{N\} \cup \{\Lambda\} \cup \{N_k : k \in N\}$ is a topology on N. $(N, \tilde{\mathcal{T}})$ is the required space. To see this, let $m, n \in N$ and $m \neq n$. For definiteness assume that $m < n$. If $U \in \tilde{\mathcal{T}}$ and $U \ni m$, then $U \supset N_m \ni n$, and hence $m \in [n]$ (thus $\{n\}$ is not closed in $(N, \tilde{\mathcal{T}})$). But N_n is a nhood of n not containing m. Consequently, $(N, \tilde{\mathcal{T}})$ is a T_0-space.

13. (*a*) Let σ be a base, let $x \in X$ and let Ox be any nhood of x. There exists a subfamily $\sigma' \subset \sigma$, for which $Ox = \bigcup \{U: U \in \sigma'\}$.

Take any $U \in \sigma'$, for which $x \in U$. Then $x \in U \subset Ox$ and $U \in \sigma$.

(*b*) Let $H \subset X$ be any open set. For each $x \in H$ fix any $Ux \in \sigma$ for which $x \in Ux \subset H$. Then $H = \bigcup \{Ux : x \in H\}$. Hence σ is a base for X.

16. Evidently, a set U is open iff together with each point $x \in U$ it contains all straight lines passing through x and parallel to l.

17. Evidently, this is the discrete topology.

18. To solve this problem it is necessary to make use of the definition of a topological space and De Morgan's laws (see Prob. 3 of Chap. 1).

20. (*a*) Suppose $[F] = F$. Then $X \setminus F = X \setminus [F]$ and for each $x \in X \setminus F$ we can fix a nhood Ux such that $Ux \cap F = \Lambda$, that is, $Ux \subset X \setminus F$. Then $X \setminus F = \bigcup\{Ux : x \in X \setminus F\}$ and $X \setminus F$ is an open set, as the union of open sets, so that $F = X \setminus (X \setminus F)$ is a closed set.

(*b*) Let F be a closed set. Then $X \setminus F$ is open and for any $x \notin F$, the set $X \setminus F$ can be regarded as a nhood of x, which does not meet F. Hence, $x \in [F]$ implies that $x \in F$ and, consequently, $[F] = F$.

22. (*a*) This is immediate from the definition of an adherent point of a set A (each $x \in A$ is an adherent point of A).

(*b*) This follows directly from the definition of an adherent point of a set.

(*c*) The inclusion $[A_1] \cup [A_2] \subset [A_1 \cup A_2]$ follows from the obvious inclusion $A_1 \subset A_1 \cup A_2$, $A_2 \subset A_1 \cup A_2$ and property (*b*). Let $x \in [A_1 \cup A_2]$. Assume that $x \notin [A_1] \cup [A_2]$. Then there is a nhood Ux of x, for which $Ux \cap A_1 = \Lambda$ and $Ux \cap A_2 = \Lambda$. Consequently, $Ux \cap (A_1 \cup A_2) = \Lambda$, that is, $x \notin [A_1 \cup A_2]$, a contradiction.

(*d*) The inclusion $[A] \subset [[A]]$ follows from (*a*) and (*b*). Let $x \in [[A]]$. Then every nhood Ux of x has a nonempty intersection with $[A]$. Consider $x' \in Ux \cap [A]$. The set Ux is also a nhood of x', hence $Ux \cap A \neq \Lambda$. Consequently, $x \in [A]$ and (*d*) is proved.

24. Refer to Prob. 13 and the definition of an adherent point of a set.

25. Assume the contrary, namely that $x^* \in X$ is a limit point of A. Put $A^* = A \setminus \{x^*\}$. Then x^* is a limit point also of A^*. Furthermore, $x^* \in [A^*]$. But $x^* \notin A^*$. Hence A^* is not closed, contradicting the hypothesis.

26. If $A \subset B$ and B is closed in X, then $[B] = B$, $[A] \subset [B] = B$, A is a closed set and $A \subset [A]$ (Probs. 2C and 22). Hence $[A]$ is the smallest closed set in X containing A.

27. This follows from Prob. 26.

32. No, it does not. Let X be the real line with the usual topology and let A be the set of all integers (positive or negative). Consider the subbase $\mathscr{B}$ for the usual topology of the line, consisting of all sets of the form $\{x : a < x\}$ and $\{x : x < b\}$, where a and b are any elements of the lines. Then, all the assumptions of the problem are satisfied by A, $\mathscr{B}$ and X but $[A] = A \neq X$.

33. Indeed, $U \setminus F = U \cap (X \setminus F)$ and $F \setminus U = F \cap (X \setminus U)$.

35. Evidently, (i) implies (ii) (whether X is T_1 is immaterial here). If (i) is not satisfied, fix a nhood Ox of x such that $| Ox \cap A | < \aleph_0$. Put $A' = A \setminus \{x\}$ and $P = Ox \cap A'$ and for each $y \in P$ fix a nhood $O_y x$ of x in X, not containing y. Put $V(x) = \bigcap \{O_y x : y \in P\} \cap Ox$. The set $V(x)$ contains x and, as an intersection of a finite family of open sets, is open. We have $V(x) \cap A' = (Ox \cap A') \cap (\bigcap \{O_y x \cap A' : y \in P\}) \subset P \cap (X \setminus P) = \Lambda$. Hence, condition (ii) is not satisfied, showing that (i) and (ii) are equivalent.

36. (*a*) $\Rightarrow$ (*c*). For each $x \in A$ fix a nhood Ox as in (*a*). Put $G = \bigcup \{Ox : x \in A\}$, $F = [A]$ and show that $A = G \cap F$. Evidently, $A \subset G \cap F$. Let $y \in G \cap F$. Fix $x^* \in A$ for which $y \in Ox^*$. Since Ox^* is an open set, $y \in F = [A]$ implies that $y \in [A \cap Ox^*]$. But $[A \cap Ox^*] \cap Ox^* = A \cap Ox^*$ by the choice of Ox^*. Hence $y \in A \cap Ox^* \subset A$.

(*c*) $\Rightarrow$ (*a*). Let $A = \Phi \cap U$, where Φ is closed and U is open in X. Then A is closed in U by the definition of the topology of a sub-space. Hence $Ox = U$ satisfies (*a*) for each $x \in A$.

(*b*) $\Rightarrow$ (*c*). Put $U = X \setminus ([A] \setminus A)$. By (*b*), U is an open set. Evidently, $U \cap [A] = A$.

(*c*) $\Rightarrow$ (*b*). Let $A = \Phi \cap U$, where Φ is closed and U is open in X. A is an open set in Φ. Hence, $[A]_\Phi \setminus A = [A]_\Phi \cap (\Phi \setminus A)$ is a closed set in Φ. But $[A]_\Phi = [A]$, since Φ is closed in X. Thus $[A] \setminus A$ is closed.

38. Use the statement of Prob. 37 twice.

39. This is immediate from Prob. 38.

40. Let X be any infinite set. A subset of X is open if its complement with respect to X is finite. Show that this defines a topology on X and that X together with this topology is the required topological space (see Prob. 2).

41. T_1-spaces and only T_1-spaces. The proof is left as an exercise.

42. The answer is affirmative. Let Y_i, where $i = 1, 2, \ldots, n, \ldots$, be discrete topological spaces equipotent with X, in each of which $y_i^* \in Y_i$ are distinguished points. Consider the subset Y_j^* of the topological product $Y = \prod_{i=1}^{\infty} Y_i$ formed by the points all of whose coordinates with index greater than j coincide with the corresponding y_i^* and put $Y^* = \bigcup_{j=1}^{\infty} Y_j^*$. Then $|Y_j^*| = |Y_1 \times \ldots \times Y_j| = |X|$, and hence $|Y^*| = |X|$. Moreover, it is not difficult to verify that Y^* as a subspace of $Y = \prod_{i=1}^{\infty} Y_i$ has no isolated point. All that now remains is to establish that the mapping of Y^* onto X is one-to-one and the image of an open set under this mapping is open in X. The resulting topology is the one required.

44. The family $\{Oy: y \in M\}$ is constructed by induction in an obvious manner over some index set M.

45. The proposition is proved by taking complements.

47. Evidently, (i) is equivalent to (ii) and (iii) to (iv). If V is open in Y and $V \cap X = G$, then $[X \setminus G]_Y \cap V = \Lambda$. Hence (i) implies that $\tilde{G} \subset Y \setminus [X \setminus G]_Y$. But $(Y \setminus [X \setminus G]_Y) \cap X = G$; hence from (i) it follows that $Y \setminus [X \setminus G]_Y \subset \tilde{G}$. Thus, (i) and (iii) are equivalent.

48. Put $W_1 = \text{Int}\,[P]$, $W_2 = \text{Int}\,[Q]$ and $W = W_1 \cap W_2$. P is a dense set in W_1. Since W is open, $P \cap W$ is dense in W_2. Similarly, Q is dense in W_2 and hence $Q \cap W$ is dense in W. But $P \cap W$ also is open. Hence, $P \cap W \cap Q \cap W = P \cap Q \cap W$ is dense in W and $[P \cap Q] \supset W = \text{Int}\,[P] \cap \text{Int}\,[Q]$. But then $\text{Int}\,[P \cap Q] \supset \text{Int}\,[P] \cap \text{Int}\,[Q]$ (see Prob. 37). On the other hand, $[P \cap Q] \subset [P]$ implies that $\text{Int}\,[P \cap Q] \subset \text{Int}\,[P]$, and $[P \cap Q] \subset [Q]$ implies that $\text{Int}\,[P \cap Q] \subset \text{Int}\,[Q]$. Hence, $\text{Int}\,[P \cap Q] \subset \text{Int}\,[P] \cap \text{Int}\,[Q]$. The inverse inclusion is already proved. Consequently, $\text{Int}\,[P \cap Q] = \text{Int}\,[P] \cap \text{Int}\,[Q]$.

49. If V is a nonempty open set in X and $V \cap F \neq \Lambda$, then $V' = V \cap U$ is also a nonempty open set in X, $V' \subset V$ and $V' \cap F = \Lambda$.

50. Consider an arbitrary nonempty open set O in X. There is a nonempty open set $O' \subset O$, for which $O' \cap A = \Lambda$. Then no $x \in O'$ is an adherent point of A; consequently, $O' \cap [A] = \Lambda$. Hence, $[A]$ is nowhere dense.

51. (*a*) Let $F \subset X$ be a closed nowhere dense set. Then, for every nonempty open set O in X there is a nonempty open set $O' \subset O$ such that $O' \cap F = \Lambda$ and $O \cap (X \setminus F) \supset O' \neq \Lambda$. Hence, $X \setminus F$ is dense in X.

(*b*) Let $X \setminus F$ be dense in X and suppose that F is closed in X. Then, for every open set O in X, certainly $O' = O \cap (X \setminus F) \neq \Lambda$ and O' is open in X. We have $O' \subset O$ and $O' \cap F = \Lambda$. Consequently, F is nowhere dense.

(*c*) The set of rational numbers and its complement, the set of irrational numbers, are both dense in the space of real numbers (where the latter is assigned the usual topology).

54. We have $|Q^2| = \aleph_0$ and $\tilde{\mathcal{T}} \supset \mathcal{T}$. Since $\mathcal{T}$ is a Hausdorff topology, $\tilde{\mathcal{T}}$ is too. If the point x were isolated in $(Q^2, \tilde{\mathcal{T}})$, then for some nhood Ox of x in $(Q^2, \mathcal{T})$ the set $Ox \setminus \{x\}$ would be nowhere dense in $(Q^2, \mathcal{T})$, which is impossible because an open set is nowhere dense iff it is empty (Prob. 52). If $(Q^2, \tilde{\mathcal{T}})$ has a nontrivial convergent sequence (to $x^* \in Q^2$), then this sequence would also converge to this point in $(Q^2, \mathcal{T})$. But, evidently, the set of points of every convergent sequence in $(Q^2, \mathcal{T})$ is nowhere dense in $(Q^2, \mathcal{T})$. Then $U = Q^2 \setminus A$, where A is the set of points of the sequence in question, would be a nhood of x^* in $(Q^2, \tilde{\mathcal{T}})$ which does not meet A. We obtain a contradiction.

57. (*a*) It will suffice to show that the union $A_1 \cup A_2$ of two regular closed subsets A_1 and A_2 of a space X is a canonical closed set. From the definition it follows that $A_1 = [\text{Int } A_1]$, $A_2 = [\text{Int } A_2]$. Then $A_1 \cup A_2 = [\text{Int } A_1] \cup [\text{Int } A_2] = [\text{Int } A_1 \cup \text{Int } A_2]$, that is, $A_1 \cup A_2$ is the closure of the open set $\text{Int } A_1 \cup \text{Int } A_2$ in X. Thus, $A_1 \cup A_2$ is a regular closed set.

(*b*) No, not always. It is easy to find two regular closed sets whose intersection is a closed nowhere dense, hence a non-regular, set. In a straight line or a plane one can find two regular closed sets whose intersection consists of exactly one point.

(*c*) True. By definition it is a regular closed set.

63. X is discrete : if $x \in X$, then $[X \setminus \{x\}] \neq X$ and, consequently, x is an isolated point.

64. Let X be the space in question and let X' be a subspace of X. If $X \supset U \neq \Lambda$ and U is open in X, then $U \cap X'$ is infinite. Indeed, if $P = U \cap X'$ is finite, then it is closed in X since X is a T_1-space and hence $U \setminus P$ is an open set. But $(U \setminus P) \cap X' = \Lambda$. From $[X'] = X$ it now follows that $U \setminus P = \Lambda$, that is, $U = P$. Hence, U is a nonempty finite set. Choosing $x^* \in U$, we conclude that $U \setminus \{x^*\}$ is a finite and, consequently, a closed set in X. Hence $\{x^*\} = U \setminus (U \setminus \{x^*\})$ is an open set, that is, x^* is an isolated point of X. We obtain a contradiction.

65. (*a*) Put $\mathcal{T}^* = \{V \cap B : V \in \mathcal{T}\}$. Since $(V_1 \cap B) \cap (V_2 \cap B) = (V_1 \cap V_2) \cap B$, the intersection of any finite set of elements of $\mathcal{T}^*$ is an element of $\mathcal{T}^*$. The relation: $\bigcup \{V_\alpha \cap B : \alpha \in M\} = B \cap (\bigcup \{V_\alpha : \alpha \in M\})$ implies that the union of any set of elements of $\mathcal{T}^*$ is an element of $\mathcal{T}^*$. Also, if $G \in \mathcal{T}^*$ and $U \in \mathcal{T}$, then $U \cup G \in \mathcal{T}^*$. It follows that $\{U \cup G : U \in \mathcal{T}, G \in \mathcal{T}^*\}$ is a topology on X and that it equals $\tilde{\mathcal{T}}$.

(*b*) Let $U \in \tilde{\mathcal{T}}$ and $U \supset A = X \setminus B$. By (*a*) there exist $W, V \in \mathcal{T}$ for which $U = (V \cap B) \cup W$. But $V \cap B \cap A = V \cap B \cap (X \setminus B) = \Lambda$. Hence, $W \supset A$. W is the desired nhood.

66. Let Q be the set of all rational numbers, $\mathcal{T}$ the usual topology on Q generated by the natural metric, and $\tilde{\mathcal{T}}$ the B-extension of $\mathcal{T}$, where $B = Q \setminus A$ and $A = (1/n : n \in N^+)$. Then A is closed in $(Q, \tilde{\mathcal{T}})$ but every nhood of A in $(Q, \tilde{\mathcal{T}})$ contains some nhood of A in $(Q, \mathcal{T})$. Sets of the form $O_\varepsilon(0) \cap B$, where $\varepsilon > 0$, constitute a base at 0 in $(Q, \tilde{\mathcal{T}})$. Evidently, every element of this base intersects every nhood of A in $(Q, \mathcal{T})$ and, consequently, every nhood of A in $(Q, \tilde{\mathcal{T}})$ (see also Prob. 54).

68. Assume the contrary. Let $x^* \in X$ be an isolated point in $(X, \mathcal{T}^*)$. Then, there exists a finite family of topologies $\mathcal{T}_1, \ldots, \mathcal{T}_k \in \mathcal{E}$ and elements $U_1 \in \mathcal{T}_1, \ldots, U_k \in \mathcal{T}_k$, for which $\bigcap\{U_i : i = 1, \ldots, k\} = \{x^*\}$ (by the definition of a base for the union of topologies, cf. p. 48). But $\mathcal{E}$ is a nest, hence there exists $i_0 \in \{1, \ldots, k\}$ for which $\mathcal{T}_i \subset \mathcal{T}_{i_0}$ for every $i \in \{1, \ldots, k\}$. Then, $U_1, \ldots, U_k \in \mathcal{T}_{i_0}$ and $\{x^*\} = \bigcap\{U_i : i = 1, \ldots, k\}$ is an open set in $(X, \mathcal{T}_{i_0})$, that is, x^* is an isolated point in $(X, \mathcal{T}_{i_0})$. This is a contradiction.

69. Let x be an arbitrary point and O^*x an nhood of x in $(X, \mathcal{T}^*)$. From the definition of a base for the topology $\mathcal{T}^*$ it follows that there exist $U_1 \in \mathcal{T}_1 \in \mathcal{E}, \ldots, U_k \in \mathcal{T}_k \in \mathcal{E}$, for which $x \in \bigcap\{U_i : i = 1, \ldots, k\} \subset O^*x$. Using the regularity of the spaces $(X, \mathcal{T}_i)$, $i = 1, \ldots k$, choose sets $V_i \in \mathcal{T}_i$ for which $x \in V_i \subset [V_i]_{\mathcal{T}_i} \subset U_i$. Put $V = \bigcap\{V_i : i = 1, \ldots, k\}$. Then $[V]_{\mathcal{T}_i} \subset [V]_{\mathcal{T}_i} \subset [V_i]_{\mathcal{T}_i} \subset U_i$ for each $i \in \{1, \ldots, k\}$, since $\mathcal{T}^* \supset \mathcal{T}_i$ for $i = 1, \ldots, k$ (see Prob. 46). Hence, $x \in V \subset [V]_{\mathcal{T}^*} \subset \bigcap\{U_i : i = 1, \ldots, k\} = U \subset O^*x$, so that V is the desired nhood of x.

70. The natural homeomorphism is defined thus: each $x \in X$ is associated with the element of the product $\Pi\{X, \mathcal{T}) : \mathcal{T} \in \mathcal{E}\}$, each of whose coordinates equals x. It is clear that this mapping is one-to-one. Its being a homeomorphism is immediate from the definition of the union of topologies and of the product topology.

71. Consiner the family $\mathcal{E}$ of all Hausdorff topologies $\mathcal{T}'$ on X such that $\mathcal{T}' \supset \mathcal{T}$ and $(X, \mathcal{T}')$ has no isolated points. Clearly, $\mathcal{E} \neq \Lambda$, since $\mathcal{T} \in \mathcal{E}$. $\mathcal{E}$ contains a maximal nest by the maximal element principle (see p. 6). Let $\mathcal{E}^*$ be a maximal nest in $\mathcal{E}$. Then, the union $\mathcal{T}^*$ of the topologies belonging to $\mathcal{E}^*$ is a Hausdorff topology on X (see Prob. 5) and $(X, \mathcal{T}^*)$ has no isolated points (see Prob. 68). The maximality of $\mathcal{E}^*$ implies that $\mathcal{T}^*$ also satisfies hypothesis (*c*).

72. The topologies constructed in Probs. 54 and 66 serve as the desired nonregular topologies relative to the usual (Euclidean) topology for the set considered in these problems. The latter is regular.

73. This problem is solved along lines similar to Prob. 71 using Probs. 69 and 68.

74. Assume the contrary. Put $X_2' = X \setminus X_1$, $\mathcal{T}_1 = \{U \cap X_1 . U \in \mathcal{T}\}$, $\mathcal{T}_2 = \{U \cap X_2' : U \in \mathcal{T}\}$. Let $\mathcal{T}'$ be the topology on X with base the family $\beta = \mathcal{T}_1 \cup \mathcal{T}_2$. Then $\mathcal{T}' \supset \mathcal{T}$ and $\mathcal{T}' \neq \mathcal{T}$, since $X_1 \in \mathcal{T}'$, but $X_1 \notin \mathcal{T}$ because otherwise X_2 would not be dense in $(X, \mathcal{T})$. However, the spaces $(X_1, \mathcal{T}_1)$ and $(X_2', \mathcal{T}_2)$ are without isolated points (see Prob. 64). Since $(X_1, \mathcal{T}_1)$ and $(X_2', \mathcal{T}_2)$ are subspaces of $(X, \mathcal{T}')$ and $X = X_1 \cup X_2'$, it follows that $(X, \mathcal{T}')$ has no isolated points. We thus obtain a contradiction to the maximality of $\mathcal{T}$, because $\mathcal{T}'$ is a Hausdorff (resp. regular) topology if $\mathcal{T}$ is Hausdorff (resp. regular), see footnotes on pp. 55 and 56.

75. Let λ be a refinement of γ and let $|\lambda| \leqslant \tau$. Choose for each $U \in \lambda$ exactly one element of the family γ containing it and denote the latter by $\varphi(U)$. Then $\mu = \{\varphi(U) : U \in \lambda\}$ is a subcover of the cover γ and $|\mu| \leqslant |\lambda| \leqslant \tau$.

78. Recall that by the definition of a base at a point every element of the base must contain this point. If $x \neq y$ and γ is a base at x, then $X \setminus \{y\}$ is a nhood of y. Hence there is $U \in \gamma$ for which $x \in U \subset X \setminus \{y\}$. Then $U \not\ni y$ and, consequently, γ is not a base at y.

79. Since x is not an isolated point, γ is not a base for X at x. This means that there is a nhood Ox of $x \in X$ such that no element of γ both contains x and is included in Ox. Then $\mathcal{B}^* = \{U \in \mathcal{B}: x \in U \subset Ox\} \subset \mathcal{B}\setminus\gamma$ and $\mathcal{B}^*$ is a base for X at x. Hence $\mathcal{B}\setminus\gamma$ is a base for X at x.

80. This is immediate from Probs. 78 and 79.

83. (*b*) always implies (*a*). However, (*a*) generally does not imply (*b*) (Probs. 54, 66 and 84).

84. Let $\varphi = \{X_i : i \in N^+\}$ be a family of pairwise disjoint countable sets. Put $X = \bigcup \{X_i : i \in N^+\}$ and $X^* = X \cup \{x^*\}$, x^* being some object not contained in any element of φ. The points of X are to be regarded as isolated and an arbitrary nhood of x^* is defined to be a set of the form $\{x^*\} \cup (\bigcup \{X_i\setminus A_i) : i \in N^+\}$, where $\{A_i : i \in N^+\}$ is an arbitrary sequence of finite sets, $A_i \subset X_i$. The topological space X^* thus defined is not first countable at x^*. Assume the contrary. Let $B = \{U_i : i \in N^+\}$ be a base at x^*, where

$$U_i = \{x^*\} \cup (\bigcup \{X_j\setminus A_j^i : j \in N^+\}), \quad i \in N^+.$$

Let $\tilde{A}_j^i$ be some finite subset of X_i strictly containing A_j^i. Then

$$\tilde{U} = \{x^*\} \cup (\bigcup \{X_j\setminus \tilde{A}_j^j : j \in N^+\})$$

is a nhood of x^* not containing any U_i:

$$U_i\setminus\tilde{U} \supset \tilde{A}_i^i\setminus A_i^i \neq \Delta, \quad i \in N^+.$$

It can be shown that this space is normal.

An alternative solution. Let $\mathcal{R}$ be the set of rational numbers with the usual topology and N the set of all integers. Consider the decomposition Z of $\mathcal{R}$, whose only element with more than one point is the set N. Show that this decomposition space (see p. 45) is not first countable at the point $\{N\}$ and is normal. See also Prob. 54.

85. Let γ be a disjoint family of open sets in X and $A = \{a_i : i \in N^+\}$ a countable dense set in X. Select for each $U \in \gamma$ a point $a(U) \in U \cap A$. In this way γ is mapped one-to-one into A, so that $|\gamma| \leqslant \aleph_0$.

86. Let A be a dense subset in a topological space X, whose topology is generated by the ordering $<$. Then, the family $\{\langle a_1, a_2\rangle : a_1, a_2 \in A\}$, where $\langle a_1, a_2\rangle = \{x \in X: a_1 <_c x <_c a_2\}$, is a countable base for the topology of X.

87. False. Consider an uncountable set A, in which all points but one (which we denote by $*$) are isolated and the only open sets containing the point $*$ are those whose complements are countable. The resulting topological space is not discrete, but every countable subset is closed.

88. Assume that there is a sequence $\varphi = \{U_n : n \in N^+\}$ of nonempty open sets in X whose elements form a base for $(X, \mathcal{T}_\xi)$. Note that each U_n is infinite, otherwise the ultrafilter ξ would be singular (see Probs. 128 and 131 of Chap. 1). Select some $x_1, y_1 \in U_1$, $x_1 \neq y_1$, and assume $x_i, y_i \in U_i$ to be defined for every $i \leqslant k$. Then, take x_{k+1}, y_{k+1} to be any two distinct points of the set $U_{k+1}\setminus(\bigcup\{\{x_i, y_i\} : i = 1, \ldots, k\}$. Put $A = \{x_i : i \in N^+\}$ and $B = \{y_i : i \in N^+\}$. Then $A \cap B = \Delta$. Since $A \cap U_i \ni x_i$, we have $(X\setminus(X\setminus A)) \cap U_i = A \cap U_i \neq \Delta$, $i \in N^+$, but this means that the set $X\setminus A$ contains no U_i.

Furthermore, $X \setminus A$ if not the union of sets U_i. Hence, $X \setminus A \notin \xi$ (since φ is a base for the topology $\mathcal{T}_\xi$). But $(X \setminus A) \cap U_i \ni y_i$, $i \in N^+$. Consequently, A cannot be written as the union of some family of elements of φ, that is, $A \notin \xi$ (since φ is a base). But then $X = A \cup (X \setminus A) \notin \xi$. This contradicts $X \in \xi$.

89. Let A be a countable subset of X, dense in $(X, \mathcal{T}_\xi)$. Assume that $X \setminus A \neq \Lambda$. From $A \cup (X \setminus A) = X \in \xi$ it follows, since ξ is an ultrafilter (see Prob. 128 of Chap. 1), that if $X \setminus A \notin \xi$, then $A \in \xi$. The necessity of the conditions in question follows. If, however, $X \setminus A \in \xi$, then $X \setminus A$ is a nonempty open set in $(X, \mathcal{T}_\xi)$ containing no point of A, but this contradicts the density of A in $(X, \mathcal{T})$. Hence $X \setminus A \notin \xi$.

For the sufficiency, let A be countable and $A \in \xi$. If $U \neq \Lambda$ and $U \in \mathcal{T}_\xi$, then $U \in \xi$ and hence $U \cap A \neq \Lambda$ (see Prob. 128 of Chap. 1—each ultrafilter is a centralized family of sets). This proves that A is dense in $(X, \mathcal{T}_\xi)$.

90. In the opposite case $U = X \setminus [B]$, a nhood of A, does not intersect B.

91. Fix a defining system $\mathcal{E}$ of nhoods of A in X, for which $|\mathcal{E}| = \tau$. For each $U \in \mathcal{E}$ select $x(U) \in U \cap B$ and put $C = \{x(U) : U \in \mathcal{E}\}$. Then $C \subset B$ and $|C| \leqslant \tau$. From $U \cap C \ni x(U)$ it follows that every nhood of A intersects C. Hence (Prob. 90), $A \cap [C] \neq \Lambda$.

92. An appropriate example is any countable Hausdorff space containing a point at which the first axiom of countability fails. (See, for example, Prob. 84.)

93. Consider an arbitrary nhood G of a set Φ in Y. Then there exists an open set $\tilde{G} \subset X$ in X such that $\tilde{G} \cap Y = G$. Put $G^* = \tilde{G} \cup (X \setminus Y)$. Then G^* is open in X and $G^* \cap Y = \tilde{G} \cap Y = G$. In addition, $F \subset (X \setminus Y) \cup (Y \cap F) = (X \setminus Y) \cup \Phi \subset G \cup (X \setminus Y) \subset G^*$. Consequently, there exists $U^* \in \gamma$ for which $F \subset U^* \subset G^*$. Then, $U^* \cap Y \in \gamma'$ and $\Phi \subset U^* \cap Y \subset G$, giving the desired result.

94. This is immediate from Prob. 93.

95. Make use of the assertions of Prob. 106 of Chap. 1 and the definition of the order topology on p. 49.

96. This follows from the definitions given on pp. 43-44.

97. By definition, we always have $a(X') \leqslant a(X)$, $t(X') \leqslant t(X)$. But there is an example where $s(X) < s(X')$ (Prob. 119) and $c(X) < c(X')$. However, $c(X) \geqslant c(X')$ if X' is dense in X.

99. See Prob. 58.

100. The collection $\tilde{\mathcal{E}}$ of all sets representable as the intersection of a finite set of elements of $\mathcal{E}$ is a base for the topology $\mathcal{T}$. However, by Prob. 58 of Chap. 1, $|\tilde{\mathcal{E}}| = |\mathcal{E}|$.

101. Let σ be an arbitrary base for a space X, and δ a base for a space of cardinality equal to the weight of X: $|\delta| = w(X) = \tau$. Note that $|\delta^2| = \tau$, where $\delta^2 = \delta \times \delta$ is the product of the set δ with itself. A pair of sets $(V, V') \in \delta^2 (V, V' \in \delta)$ is said to be distinguished if there exists $U \in \sigma$ such that $V \subset U \subset V'$.

Let γ be the family of all distinguished pairs $(V, V) \in \delta^2$. We have $\gamma \subset \delta^2$ and, consequently, $\gamma| \leqslant \tau$ (see Prob. 58 of Chap. 1). With every distinguished pair $(V, V') \in \gamma$ associate a set $U_{V,V'} \in \sigma$ for which $V \subset U_{V,V'} \subset V'$. Write

$$\sigma^* = \{U_{V,V'} : (V, V') \in \gamma\}.$$

We have $|\sigma^*| = |\gamma| \leqslant \tau$. We show that σ^* is a base for X. Let $x \in X$ and let Ox be an arbitrary nhood of x. Since δ and σ are bases for X, there exist (Prob. 13) sets $V' \in \delta$, $U \in \sigma$, $V \in \delta$ for which $x \in V \subset U \subset V' \subset Ox$. Then $(V, V') \in \gamma$, that is, it is a distinguished pair. Consequently, with this pair we associate the set $U_{V,V'} \in \sigma^* \subset \sigma$ such that $x \in V \subset U_{V,V'} \subset V' \subset Ox$. It follows that $\sigma^* \subset \sigma$ is a base for X of cardinality at most τ. Since $wX = \tau$, we have $|\sigma^*| = \tau$. The first part of the assertion is proved. To prove the second part it suffices to consider an interval with the usual topology: from the net consisting of all of its one-point subsets, it is impossible to select one that is countable (since the interval is uncountable).

102. Denote by $\mathcal{P}$ the set of all disjoint subfamilies $\mathcal{E}$ of the family $\mathcal{B}$ for which $\bigcup\{V : V \in \mathcal{E}\} \subset U$. Consider the partial ordering $<$ on $\mathcal{P}$ of set incuslon : if $\mathcal{E}', \mathcal{E}'' \in \mathcal{P}$, then $\mathcal{E}' < \mathcal{E}''$ means $\mathcal{E}' \subset \mathcal{E}''$. Evidently, the ordering $<$ on $\mathcal{P}$ is inductive (see p. 6). Hence there exists (p. 6) a maximal element $\mathcal{E}^* \in \mathcal{P}$ relative to $<$. Then, $\bigcup\{V: V \in \mathcal{E}^*\} \subset U$ (by virtue of $\mathcal{E}^* \in \mathcal{P}$) and $[\bigcup\{V: V \in \mathcal{E}^*\}] \supset U$; otherwise $W = U \cap (X \setminus [\bigcup\{V: V \in \mathcal{E}^*\}])$ would be a nonempty open set and there would exist $V^* \in \mathcal{B}$, $V^* \neq \Lambda$ for which $V^* \subset W$. Then, $V^* \cap V = \Lambda$ for each $V \in \mathcal{E}^*$, implying that $V^* \notin \mathcal{E}^*$ and $\mathcal{E}^{**} = \mathcal{E}^* \cup \{V^*\} \in \mathcal{P}$, $\mathcal{E}^* < \mathcal{E}^{**}$, $\mathcal{E}^{**} \neq \mathcal{E}^*$, but this contradicts the maximality of $\mathcal{E}^*$ in $(\mathcal{P}, <)$.

103. Let X be the set of points of the interval $[0, 1]$ and A the set of all points of the form $1/n$, where $n \in N^+$. In addition to the sets open in the usual topology $\mathcal{T}$ on the interval, we declare all sets of the form $U \cup (X \setminus A')$, where $U \in \mathcal{T}$ and $A' \subset A$ to be open. Show that a topology is thus defined on X. This topology $\mathcal{T}^*$ contains more open sets than the usual topology of the interval. Hence $(X, \mathcal{T}^*)$ is a Hausdorff space, A is a closed set in $(X, \mathcal{T}^*)$ and $0 \notin A$, but every nhood of the point 0 in $(X, \mathcal{T}^*)$ intersects every nhood of A in $(X, \mathcal{T}^*)$ (a defining family of nhoods of A in $(X, \mathcal{T}^*)$ is formed by all sets open in the usual topology of an interval and containing A; a base at 0 in $(X, \mathcal{T}^*)$ is formed by all sets of the form $\{a \in [0, 1] : a < \varepsilon \text{ and } a \notin A\}$, where $\varepsilon > 0$).

104. Properties (i) through (iv) are trivial to prove. Properties (ii) and (v) imply (vi) (by Prob. 206 of Chap. 3). Properties (v) and (vii) imply (viii) (by Prob. 213). Thus, it remains to prove properties (v) and (vii).

(v) By virtue of Prob. 206 of Chap. 3 it will suffice to show that from any family γ of basic open sets of a space X^* one can select a countable subfamily γ_0 for which

$$\tilde{\gamma}_0 = \bigcup\{U : U \in \gamma_0\} = \tilde{\gamma} = \bigcup\{U: U \in \gamma\}.$$

Every $U \in \gamma$ is a half-open interval $[x, \beta)$. Hence, we also denote such a U by U_x. Consider the interval $V(U) = U_x \setminus \{x\}$ and the family $\delta = \{V(U): U \in \gamma\}$ of these intervals. Note that every $V(U)$ is also open in X^*. From the family δ of intervals one can select (Probs. 172 and 213; Prob. 206 of Chap. 3) a countable subfamily δ_0 for which

$$\tilde{\delta}_0 = \bigcup\{V(U): V(U) \in \delta_0\} = \tilde{\delta} = \bigcup\{V(U): V(U) \in \delta\}.$$

We claim that $A = \tilde{\gamma} \setminus \tilde{\delta}$ is countable. To see this, if $x_1, x_2 \in A$, $x_1 \neq x_2$ and $U_{x_1}, U_{x_2} \in \gamma$, then evidently, $U_{x_1} \cap U_{x_2} = \Lambda$. Hence, for each point $x \in A$ we take some $U_x \in \gamma$ (x being the left endpoint of the halfopen interval U_x) and denote the family of these half-open intervals by $\gamma_1 : \gamma_1 = \{U_x: x \in A\}$. The family γ_1 is disjoint and hence countable by the separability of X, so that A is countable. Consider now the families $\gamma_2 = \{U: V(U)$

$\in \delta_0\}$ and $\gamma_0 = \gamma_1 \cup \gamma_2$. The family γ_0 is countable, $\gamma_0 \subset \gamma$ and $\tilde{\gamma}_0 = \tilde{\gamma}$, completing the proof.

(vii) We claim that X^* has no countable base. If X^* had a countable base, then some countable family σ of base sets in X^* (i.e., half-open intervals $[a, b)$) would form a base for X^* (see Prob. 101). This would imply that X^* is countable, a contradiction.

REMARK. In the same way one can show that every uncountable subspace X^* has no countable base and, consequently, by property (v) and Prob. 213, is nonmetrizable.

105. Put $F = \bigcap \{F(\alpha) : \alpha \in T(\omega_1)\}$ and $Y = X \setminus F$. For each point $y \in Y$ there is $\alpha \in T(\omega_1)$ such that $y \notin F(\alpha)$. Then one can select a nhood Oy for which $Oy \cap F(\alpha) = \Lambda$ (since $F(\alpha)$ is closed). The open covering $\{Oy: y \in Y\}$ of Y has a countable subcovering (since Y is Lindelöf), which we denote by γ. Then, by the definition of γ, for each $U \in \gamma$ one can find $\alpha(U) \in T(\omega_1)$ such that $U \cap F(\alpha(U)) = \Lambda$. There exists $\alpha^* \in T(\omega_1)$ for which $\alpha(U) < \alpha^*$ for every $U \in \gamma$, since $|\gamma| \leqslant \aleph_0$ (see Prob. 91 of Chap. 1). Then, $F(x^*) \subset \bigcap \{F(\alpha(U)): U \in \gamma\}$. From this and the de Morgan formulas it follows that $F(\alpha^*) \cap (\bigcup \{U: U \in \gamma\}) = \Lambda$, that is, $F(\alpha^*) \cap Y = \Lambda$. Hence $F(\alpha^*) = F = \bigcap \{F(\alpha): \alpha \in T(\omega_1)\}$ and $F(\alpha^*) = F(\alpha)$ for each α greater than α^*.

106. Clearly, *(a)* implies *(b)*. Let us derive assertion *(a)* from *(b)*. Put $\tilde{A} = \{y$: there exists $A' \subset A$ for which $y \in [A']$ and $|A'| \leqslant \tau\}$ and show that $\tilde{A}$ is closed. Assume the contrary. Then, by *(b)*, there exist $y^* \in [\tilde{A}] \setminus A$ and $A^* \subset \tilde{A}$ for which $|A^*| \leqslant \tau$ and $y^* \in [A^*]$. For each $y \in A^*$, choose a set $A(y) \subset A$ such that $y \in [A(y)]$ and $|A(y)| \leqslant \tau$, which is possible by the definition of $\tilde{A}$. Put $A' = \bigcup \{A(y): y \in A^*\}$. Then $A' \subset A$, $|A'| \leqslant \tau$ and $y^* \in [A']$, which implies $y^* \in \tilde{A}$, a contradiction. Thus $\tilde{A}$ is closed, which proves *(a)*.

108. No. Consider any countable Hausdorff space without isolated points and without a nontrivial convergent sequence (Prob. 54).

109. No, see Prob. 113.

110. Put $A' = A \setminus \{x\}$. If $y \notin A'$ and $y \neq x$, then $y \notin A$ and, since A is closed in X, $d(y, A) \neq 0$. Moreover, $d(y, A') \neq 0$. If $d(x, A') \neq 0$, then $d(y, A') \neq 0$ for every $y \notin A'$, that is, A' is closed in X. But then $x \notin [A']$, that is, x is an isolated point in A.

111. Let U be any open set in X containing x. Then $P = X \setminus U$ is closed and $x \notin P$. Hence $d(x, P) > 0$. Select n^* such that $d(x_n, x) < d(x, P)$ for all $n > n^*$. Then, evidently, $x_n \in U$ for every $n > n^*$. Hence ξ converges to x.

112. Let $A \subset X$ and suppose that A is not closed in X. Then there exists $X \in [A] \setminus A$ for which $d(x, A) = 0$. Hence there is a sequence $\xi = \{x_n: n \in N^+\}$ of points of A such that $d(x_n, x) < 1/n$ for every $n \in N^+$. Then $\lim_{n \to \infty} d(x_n, x) = 0$ and by Prob. 111 the sequence ξ converges to x. Hence X is sequential.

113. Assertion(i) is straightforward. Since $|Z| = \aleph_0$, it suffices to show that Z is regular. However, it is easy to see that Z has a base of clopen sets. We have: $d(\theta, X) = 1$ but $d(\theta, Y) = 0$ and $d(y_n, X)$ for every $n \in N^+$. Hence $\theta \in [Y]$, $Y \subset [X]$ and therefore, $\theta \in [X]$. From this and Probs. 112 and 114 we conclude that (iii) holds (actually (iii) is easy to prove directly, which is left as an exercise). To prove (iv) we define two metrics ρ_1 and ρ_2 on Z. Put $\rho_1 | (Y \cup \{\theta\}) = d | (Y \cup \{\theta\})$ and $\rho_1 | (X \cup \{\theta\}) = d | (X \cup \{\theta\})$. If $z' \in Y$ and $z'' \in X$ (or vice-versa), then $\rho_1(z', z'') = 1$. Further put $\rho_2 | (X \cup Y) = d | X \cup Y)$ and $\rho_2(\theta, z) = 1$, $\rho_2(z, \theta) = 1$ for $z \neq \theta$. We verify that ρ_1

and ρ_2 are metrics. Clearly, $d(z', z'') = \min\{\rho_1(z', z''), \rho_2(z', z'')\}$ for every $z', z'' \in Z$. From this it is easy to deduce that the topology $\mathcal{T}$ on Z constructed above is the intersection of topologies $\mathcal{T}_1$ and $\mathcal{T}_2$ generated on Z by the metrics ρ_1 and ρ_2, respectively. Evidently, the spaces $(Z, \mathcal{T}_1)$ and $(Z, \mathcal{T}_2)$ have the properties enumerated in (iv). For the proof of (v), note that the disjoint (free) sum of $(Z, \mathcal{T}_1)$ and $(Z, \mathcal{T}_2)$ is mapped onto $(Z, \mathcal{T})$ by precisely the mapping prescribed (see Prob. 301).

114. (*a*) obviously implies (*b*). Let us derive (*c*) from (*b*). We have always $[A] \supset \{y \in X: d(y, A) = 0\}$. Let $z \in [A] \setminus A$. By (*b*) there is a sequence $\{x_n: n \in N^+\}$ of points of A converging to z. The set S consisting of all points of this sequence and the point z is closed in X (since X is a Hausdorff space). From $z \notin A$ it follows that z is not an isolated point in S. Consequently (Prob. 110), $d(z, S\setminus\{z\}) = 0$. But $S\setminus\{z\} \subset A$. Hence $d(z, A) = 0$ and $[A] = \{y \in X: d(y, A) = 0\}$. (*d*) implies (*c*). If $x \notin \operatorname{Int} O_\varepsilon x$, then $x \in [X\setminus O_\varepsilon x]$. By (*c*) $d(x, X\setminus O_\varepsilon x) = 0$ and hence $O_\varepsilon x \cap (X\setminus O_\varepsilon x) \neq \Lambda$, a contradiction. From (*d*) and the fact that the topology on X is generated by a semimetric d it follows that the family $\{\operatorname{Int} O_\varepsilon x: \varepsilon \text{ a positive real number}\}$ forms a base in X at each $x \in X$. Hence (*d*) implies (*a*).

115. The answer is negative; see Probs. 13, 11, 87 and 115 of Chap. 3 and Prob. 52 of Chap. 4.

116. Let Y be an arbitrary dense subspace of X. Every element U of the π-base $\mathcal{B}$ intersects Y. Choose a point in their intersection and call this $x(U)$. Then $\{x(U): U \in \mathcal{B}\}$ is the desired countable dense set in Y.

117. Let Q be a countable dense set in X. For each $x \in Q$ let γ_x be a countable base at x. Then $\gamma = \bigcup\{\gamma_x: x \in Q\}$ is a countable family of open and dense sets in X, that is, γ is a countable π-base (this is obvious). However, each dense subspace of a space with a countable π-base is separable (Prob. 116).

118. Fix $C \subset X$ such that $|C| \leqslant \tau$ and $[C] = X$. For each $y \in C$, choose $B(y) \subset Y$ such that $[B(y)] \ni y$ and $|B(y)| \leqslant \tau$; possible, since $y \in [Y]$ and $t(X) \leqslant \tau$. Put $B = \bigcup\{B(y): y \in C\}$. Then $B \subset Y$, $|B| \leqslant \tau$, $[B] \supset [C] \supset Y$. Hence $s(Y) \leqslant \tau$.

119. False. The space of Prob. 104 is hereditarily separable, but its square has a closed uncountable discrete-in-itself subspace. See Prob. 11 of Chap. 3.

120. False. It follows from Probs. 100 and 101 of Chap. 5 that $\mathcal{P}$ is a separable space ($\mathcal{P}_0$ is a countable dense subset) and contains a closed discrete subspace $\mathcal{P}_1$ of cardinality $2^{\aleph_0}$. See also Probs. 11 and 13 of Chap. 3.

121. Assume the contrary, namely let the given set A be of cardinality $2^{\aleph_0}$, every subset of which is closed in X. Fix a countable set S, dense in X. Since X is normal, for every $P \subset A$ there is an open set $\varphi(P)$ in X such that $P \subset \varphi(P) \subset [\varphi(P)] \subset X\setminus (A\setminus P)$. Put $\psi(P) = \varphi(P) \cap S$. Since $[S] = X$, $[\psi(p)] \cap A = P$. Hence if $P' \subset A$, $P'' \subset A$ and $P' \neq P''$, then $\psi(P') \neq \psi(P'')$. Thus we have constructed a one-to-one mapping of the set $\exp A$ of all subsets of A into the set $\exp S$ of all subsets of S. But, then, $2^{2^{\aleph_0}} = |\exp A| \leqslant |\exp S| = 2^{\aleph_0}$, a contradiction.

122. For each $x \in X$ put $\varphi(x) = \{P: P \subset A \text{ and } [P] \ni x\}$. Then φ is a one-to-one mapping of the set X into the set $\exp \exp A$. Indeed, if $x_1 \neq x_2$, then there are open sets U_1 and U_2 in X for which $U_1 \cap U_2 = \Lambda$, $x_1 \in U_1$, $x_2 \in U_2$. Put $P_1 = U_1 \cap A$, $P_2 = U_2 \cap A$. Then (see Prob. 29) $[U_1] = [P_1] \supset U_1 \ni x_1$, $[U_2] = [P_2] \supset U_2 \ni x_2$. Consequently, $P_1 \in \varphi(x_1)$, $P_2 \in \varphi(x_2)$. But $P_1 \notin \varphi(x_2)$, because $[P_1] = [U_1] \subset X\setminus U_2$. Hence $\varphi(x_1) \neq$

$\varphi(x_2)$. Thus X is equipotent to some subset of the set exp exp A, so that $|X| \leqslant |\exp \exp A| = 2^{2^{|A|}}$.

123. Fix in X a base $\mathscr{B}$ of cardinality τ. For each $x \in X$ put $\varphi(x) = \{U: U \in \mathscr{B}$ and $U \ni x\}$.† Then $\varphi(x) \in \exp \mathscr{B}$ and φ is a one-to-one map of the set X into the set exp $\mathscr{B}$. For if $x_1 \neq x_2$, then there exist open sets U_1 and U_2 in X such that $x_1 \in U_1$, $x_2 \in U_2$ and and $U_1 \cap U_2 = \Lambda$. Then $U_2 \in \varphi(x_2)$ but $U_2 \notin \varphi(x_1)$ because $x_2 \notin U_2$. Consequently, $\varphi(x_1) \neq \varphi(x_2)$. Hence, X is equipotent with a subset of exp $\mathscr{B}$, that is, $|X| \leqslant |\exp \mathscr{B}| = 2^\tau$.

124. See Prob. 125.

125. Recently, the Hungarian topologist I. Juhasz has constructed, under certain restrictions compatible with the conventional axioms of set theory, a hereditarily separable space X for which $|X| = 2^{2^{\aleph_0}} > 2^{\aleph_0}$.

126. Consider $y(\aleph_1)$, a minimal linearly ordered set of cardinality $\aleph_1$. Put $M_0 = M$. Let $\beta \in y(\aleph_1)$, and for every $\alpha < \beta$ suppose that M_α is already defined. Put $L_\beta = \bigcup \{M_\alpha : \alpha < \beta\}$ and $M_\beta = \{x \in X$: there is a sequence in L_β converging to $x\}$. Evidently, $M \subset M_\alpha \subset M_\beta$ for $\alpha < \beta$. If $|M_\alpha| \leqslant 2^{\aleph_0}$ for every $\alpha < \beta$, then also $|L_\beta| \leqslant 2^{\aleph_0}$. Fixing for each $x \in M_\beta$ a sequence $\xi(x)$ in L_β converging to x, we obtain a one-to-one mapping ξ of M_β into the set of all countable subsets of L_β. Since $(2^{\aleph_0})^{\aleph_0} = 2$, it follows that $|M_\beta| \leqslant 2^{\aleph_0}$. Hence, if for every $\alpha \in y(\aleph_1)$ the set M_α is defined according to the above procedure and $M^* = \bigcup \{M_\alpha : \alpha \in y(\aleph_1)\}$, then $|M^*| \leqslant 2^{\aleph_0}$. Evidently, $M \subset M^* \subset [M]$. We claim that M^* is sequentially closed. Let $x_i \in M^*$, $i \in N^+$, and $x = \lim_{i \to \infty} \{x_i\}$. Then, $x_i \in M_{\alpha_i}$ for some $\alpha_i \in y(\aleph_1)$ for $i \in N^+$. There exists $\alpha^* \in y(\aleph_1)$ such that $\alpha_i < \alpha^*$ for every $i \in N^+$ (Prob. 91 of Chap. 1). The sequence $\{x_i : i \in N^+\}$ is contained in L_{α^*} and hence its limit, the point x, belongs to M_{α^*}. Thus $x \in M^*$ and M^* is sequentially closed.

127. Let M be dense in X and let $|M| \leqslant \aleph_0$. Associate with each point $x \in X$ a countable sequence of nhoods $\theta(x) = \{U_i x : i \in N^+\}$ such that $\{x\} = \bigcap \{U_i x : i \in N^+\}$ and $[U_{i+1}x] \subset U_i x$ (here we have made use of the condition $\psi(x, X) = \aleph_0$ for each $x \in X$ and the regularity of X). Write $M_i x = U_i x \cap M$. Clearly, $M_i x \subset M$, $|M_i x| \leqslant \aleph_0$, $x \in [M_i x]$. Now associate x with the sequence $\gamma(x) = \{M_i x : i \in N^+\}$ of countable sets $M_i x$. Denote by $\Sigma(x)$ the family $\{\gamma(x) : x \in X\}$ of sequences $\gamma(x)$ constructed for each $x \in X$. Since $|M| \leqslant \aleph_0$, $|M^{\aleph_0}| \leqslant c$ and $|(M^{\aleph_0})^{\aleph_0}| \leqslant c$; consequently, $|\Sigma(x)| \leqslant |(M^{\aleph_0})^{\aleph_0}| \leqslant c$.

We claim that the correspondence $\gamma: X \to \Sigma(X)$ is one-to-one. Let $x_1, x_2 \in X$, $x_1 \neq x_2$. Let i_1 be such that $x_2 \notin [U_{i_1}x_1]$, where $U_{i_1}x_1 \in \theta(x_1)$. Take i_2 such that $x_1 \notin [U_{i_2}x_2]$ where $U_{i_2}x_2 \in \theta(x_2)$. Suppose $i_2 \geqslant i_1$. Then, clearly, $x_1 \notin [U_{i_2}x_2]$ and $x_2 \notin [U_{i_2}x_1]$. Consider $M_{i_2}x_2 = U_{i_2}x_2 \cap M$, $M_{i_2}x_1 = U_{i_2}x_1 \cap M$. Since $x_1 \in [M_{i_2}x_1]$, $x_2 \in [M_{i_2}x_2]$ and $x_1 \notin [U_{i_2}x_2]$, $x_2 \notin [U_{i_2}x_1]$, we have $x_1 \notin [M_{i_2}x_2]$, $x_2 \notin [M_{i_2}x_1]$; consequently, $M_{i_2}x_1 \neq M_{i_2}x_2$, that is, $\gamma(x_1)$ and $\gamma(x_2)$ are distinct sequences. Thus, the correspondence $\gamma: X \to \Sigma(X)$ is one-to-one. Hence $|X| \leqslant |\Sigma(X)| \leqslant c$, completing the proof.

128. The answer is not known to the authors.

129. Let $M \subset X$ be such that $[M] = X$, $|M| \leqslant c$. Then for every point $x \in X$ (since

---

†The conjunction 'and' always implies that both the conjuncts are satisfied.

$t(X) = \aleph_0$) there exists $M_x \subset M$ such that $|M_x| \leqslant \aleph_0$ and $x \in [M_x]$. Since $\psi(x', X) \leqslant \aleph_0$, for each $x' \in X$ and $s([M_x]) \leqslant \aleph_0$. $|[M_x]| \leqslant c$ (see Prob. 127). Denote by $\Sigma(M)$ the collection of all finite or countable sets belonging to M. Since $|M| \leqslant c$ and $c^{\aleph_0} = c$, we have $|\Sigma(M)| \leqslant c$. Furthermore, since for each $x \in X$ there exists a countable set $M_x \in \Sigma(M)$ for which $x \in [M_x]$, we have $X = \bigcup\{[N] : N \in \Sigma(M)\}$. But $|\Sigma(M)| \leqslant c$ and $|[N]| \leqslant c$ for every $N \in \Sigma(M)$; hence $|X| \leqslant c$. This completes the proof.

130. Consider the closed set $[M]$ in X. Clearly $s([M]) \leqslant c$ and $\psi(x, [M]) \leqslant \aleph_0$ for each $x \in [M]$. In addition, $t([M]) = \aleph_0$. Hence using the preceding problem, we conclude that $|[M]| \leqslant c$, completing the proof.

131. For each $x \in X$ fix a countable base γ_x for X at x, which contains the intersection of each pair of its elements, and put $\mathcal{E} = \{\gamma_x : x \in X\}$. If $|X| > 2^{\aleph_0}$, then $\mathcal{E}$ satisfies all the conditions of Prob. 122 of Chap. 1. The conclusion of that problem now implies that there is in X a disjoint family of nonempty open sets of uncountable cardinality; but this contradicts the assumption that X has the Souslin property.

132. Let $x \in X$, $V \in \mathcal{T}$ and $x \in V$. Choose $W \in \mathcal{T}$ for which $x \in W \subset [W] \subset V$. Put $U = W \cap X'$. Then $U \in \mathcal{T}'$ and $[U] = [W]$ (see Prob. 29). Consequently, $\text{Int}\,[U] = \text{Int}\,[W] \supset W \ni x$ and $\text{Int}\,[U] = \text{Int}\,[W] \subset [W] \subset V$. Thus, $x \in \text{Int}\,[U] \subset V$. Also, $\text{Int}\,[U] \in \tilde{\mathcal{T}}'$, by definition of $\tilde{\mathcal{T}}'$. Hence, $\tilde{\mathcal{T}}'$ is a base for $\mathcal{T}$.

133. Let $\mathcal{B}'$ be a base for X'. For each $U \in \mathcal{B}'$ fix an open set $\tilde{U}$ such that $\tilde{U} \cap X' = U$ (which is possible by the definition of a subspace). Then $\tilde{U} \subset [U]$ by Prob. 29. Put $\mathcal{B} = \{\tilde{U} : U \in \mathcal{B}'\}$. If $x \in V$, where $x \in X'$ and V is open in X, then there exists an open set W in X for which $x \in W \subset [W] \subset V$ (X being regular). Put $W' = W \cap X'$. The set W' is open in X' and $x \in W'$. Hence there exists $U' \in \mathcal{B}'$ for which $x \in U' \subset W'$. Then, $x \in \tilde{U}' \subset [U'] \subset [W'] \subset [W] \subset V$ and $\tilde{U}' \in \mathcal{B}$. This proves that $\mathcal{B}$ is an exterior base for X' in X. Evidently, $|\mathcal{B}'| = |\mathcal{B}|$.

134. The solution is immediate from Prob. 132.

135. Similar to the solution of Prob. 133.

137. Only in a discrete space. In fact, the collection of all one-point subsets of a space is always a net. If this net is a base, then every point of X is isolated.

140. An appropriate example is a countable normal space without a countable base constructed in Prob. 113.

143. Solve this problem along similar lines to Prob. 75.

145. False. Consider the set of all points of the real line with the topology with a base consisting of all possible intervals and their intersections with the set of irrational numbers. This is a Hausdorff space with a countable base in which the set Q of all rational numbers is closed, but is not a countable intersection of open sets.

146. No : cf. the space described in Prob. 145.

147. Let S be a net in $(X, \mathcal{T})$ of cardinality τ. For an arbitrary pair of elements (s_1, s_2) of S, let us fix a pair of open sets $\varphi(s_1, s_2) = (U_1, U_2)$ in X such that $s_1 \subset U_1$, $s_2 \subset U_2$ and $U_1 \cap U_2 = \Lambda$, if this is possible. Otherwise, put $\varphi(s_1, s_2) = (\Lambda, \Lambda)$. Put $\mathcal{E} = \bigcup\{\varphi(s_1, s_2) : (s_1, s_2) \in S \times S\}$, that is, $U \in \mathcal{E} \Leftrightarrow U \in \varphi(s_1, s_2)$ for some $s_1, s_2 \in S$. Since $|S \times S| = |S| = \tau$ (Prob. 52 of Chap. 1), $|\mathcal{E}| \leqslant \tau$, The family $\mathcal{E} \cup \{X\}$ is a subbase for some topology $\mathcal{T}^*$ on X (see Prob. 14), with $w(X, \mathcal{T}^*) \leqslant |\mathcal{E}| \leqslant \tau$ (Prob. 58 of Chap. 1) and $\mathcal{T}^* \subset \mathcal{T}$, since $\mathcal{E} \cup \{X\} \subset \mathcal{T}$. Let $x_1 \neq x_2$. There exist $U_1', U_2' \in \mathcal{T}$ for which $x_1 \in U_1'$, $x_2 \in U_2'$ and $U_1' \cap U_2' = \Lambda$. Since S is a net in $(X, \mathcal{T})$, $x_1 \in s_1' \subset U_1'$,

$x_2 \in s_2' \subset U_2'$, for some $s_1', s_2' \in S$. But then $\varphi(s_1', s_2') = (U_1, U_2) \neq (\Lambda, \Lambda)$ and $s_1' \subset U_1'$, $s_2' \subset U_2$, $U_1 \cap U_2 = \Lambda$, $U_1 \in \mathcal{E} \subset \mathcal{T}^*$, $U_2 \in \mathcal{E} \subset \mathcal{T}^*$.

Hence $\mathcal{T}^*$ is a Hausdorff topology.

148. Let S be a net of cardinality τ in $(X, \mathcal{T})$. It can be assumed that $\{X\} \in S$ (because $\tau \geqslant \aleph_0$). By Prob. 14, S is a subbase for some topology $\mathcal{T}'$ of weight at most $|S| = \tau$. But, by the definition of net, every element of $\mathcal{T}$ is the union of some set of elements of S. Hence $\mathcal{T} \subset \mathcal{T}'$. From $\mathcal{T}' \supset \mathcal{T}$ it follows that $\mathcal{T}'$ is a Hausdorff topology.

149. See Probs. 148 and 282.

150. See Probs. 149 and 282.

194. The case of a finite number of factors reduces to the case of $\aleph_0$ factors, of which all but finitely many are one-point spaces. Thus we assume that for each $i \in N^+$ a metrizable topological space X_i is given, and that on each X_i a metric ρ_i is fixed which generates a topology $\mathcal{T}_i$ on X_i. We also assume that the ρ_i-diameter of X_i is at most 1 (see Prob. 190). On $X = \Pi\{X_i : i \in N^+\}$ we define a metric ρ by the following rule. If $x' = \{x_i' : i \in N^+\} \in X$ and $x'' = \{x_i'' : i \in N^+\} \in X$, then

$$\rho(x', x'') = \sum_{i=1}^{\infty} \frac{\rho_i(x_i', x_i'')}{2^i}.$$

By Prob. 193, ρ is a metric on X. Now compare the Tychonoff topology $\mathcal{T}$ on X and the topology $\mathcal{T}_\rho$ generated by the metric ρ. Fix $x = \{x_i : i \in N^+\} \in X$ and consider the set $O_\varepsilon x = \{x' \in X : \rho(x, x') < \varepsilon\}$. Choose $n \in N^+$ so that

$$\sum_{i=n+1}^{\infty} \frac{1}{2^i} = \frac{1}{2^n} < \frac{\varepsilon}{2}.$$

In each X_i, $i \leqslant n$, consider the nhood $O_{\varepsilon/2} x_i = \{x_i' \in X_i : \rho_i(x_i, x_i') < \varepsilon/2\}$. Since ρ_i generates the topology $\mathcal{T}_i$ on X_i, $O_{\varepsilon/2}\, x_i \in \mathcal{T}_i$. Consequently, the set $U = \Pi\{O_{\varepsilon/2}\, x_i : 1 \leqslant i \leqslant n\} \times \Pi\{X_i : n+1 \leqslant i < \infty\}$ belongs to the product topology $\mathcal{T}$. But $x \in U$, and if $x' \in U$,then $\rho_i(x_i, x_i') < \varepsilon/2$ for every $i \leqslant n$ and $\rho_i(x_i, x_i') \leqslant 1$ for $i > n$. Hence

$$\rho(x, x') < \sum_{i=1}^{n} \frac{1}{2^i} \frac{\varepsilon}{2} + \sum_{i=n+1}^{\infty} \frac{1}{2^i} < \frac{\varepsilon}{2} + \frac{1}{2^n} < \varepsilon.$$

Consequently, $x \in U \subset O_\varepsilon x$.

This establishes that $O_\varepsilon x \in \mathcal{T}$, that is, $\mathcal{T}_\rho \subset \mathcal{T}$. On the other hand, the projections $\Pi_i : X \to X_i$ are all continuous, and from the definition of ρ it is immediate that $2^i \rho(x, x') \geqslant \rho_i(\pi_i x, \pi_i x')$. But $\mathcal{T}$ is the smallest topology on X relative to which the projection on each factor is continuous (see Prob. 363). Hence, $\mathcal{T} \subset \mathcal{T}_\rho$ and, so $\mathcal{T} = \mathcal{T}_\rho$. This shows that the product topology $\mathcal{T}$ on X is metrizable.

211. The intersection of all spherical nhoods of radius $1/n$, $n \in N^+$, of a closed set A is equal to A, because if $x \notin A$, then $\rho(x, A) > 0$ (and a spherical nhood is an open set).

212. Let X be a set, ρ a metric on X and $\mathcal{T}_\rho$ the topology generated on X by ρ. We have to show that $(X, \mathcal{T}_\rho)$ is a normal space. Consider any disjoint closed sets F and Φ in $(X, \mathcal{T}_\rho)$. Then $\varepsilon_x = \rho(x, \Phi) > 0$ for each $x \in F$ and $\varepsilon_y = \rho(y, F) > 0$ for each $y \in \Phi$. Put $Ox = \{z \in X: \rho(x, z) < \varepsilon_x/2\}$ for $x \in F$, $Oy = \{z \in X : \rho(y, x) < \varepsilon_y/2\}$ for $y \in \Phi$ and $U = \bigcup\{Ox : x \in F\}$, $V = \bigcup\{Oy : y \in \Phi\}$. The sets U and V are open in $(X, \mathcal{T}_\rho)$ and $F \subset U$, $\Phi \subset V$. We claim that $U \cap V = \Lambda$. Assume the contrary, that is, let $z \in U \cap V$. Then, $z \in Ox \cap Oy$ for some $x \in F$ and $y \in \Phi$. We have $\rho(x, y) \leqslant \rho(x, z) + \rho(z, y) < \frac{1}{2}\rho(x, \Phi) + \frac{1}{2}\rho(y, F) \leqslant \frac{1}{2}\rho(x, y) + \frac{1}{2}\rho(y, x) = \rho(x, y)$. Thus $\rho(x, y) < \rho(x, y)$, a contradiction.

213. (*a*) (i) $\Rightarrow$ (ii). Let $A = \{a_i: i \in N^+\}$ be a countable dense set in X and ρ a metric on X compatible with its topology. For each i and each rational number $0 \leqslant r < 1$ consider the nhood $O_r a_i = \{x \in X: \rho(x, a_i) < r\}$ of the point $a_i \in A$. Put $\sigma = \{O_r a_i: 0 \leqslant r \leqslant 1, i \in N^+\}$. The family σ, being the union of a countable family of countable families $\sigma_i = \{O_r a_i: 0 \leqslant r \leqslant 1\}$, is itself countable. We claim that σ is a base for X. Fix a point $x_0 \in X$ and a nhood $O_\varepsilon x_0 = \{x: \rho(x, x_0) < \varepsilon\}$ of x_0. Consider rational numbers r_1, r_2 for which $r_1 < \varepsilon/4 < r_2 < \varepsilon/2$ and a point $a_{i_0} \in A$ for which $\rho(a_{i_0}, x_0) < r_1$. Then $x_0 \in O_{r_2} a_{i_0} \subset O_\varepsilon x_0$, completing the proof.

(*b*) (ii) $\Rightarrow$ (iii) is trivial (see Prob. 77).

(*c*) (iii) $\Rightarrow$ (i). For each point $x \in X$ consider the nhood $O_i x = \{x' \in X: \rho(x', x) < 1/i\}$ and the open cover $\omega_i = \{O_i x : x \in X\}$. Using the Lindelöfness of X, extract from ω_i a countable subcover $\omega_i^0 = \{O_i x_j : j \in N^+\}$. From each set $O_i x_j$ select a point $a_{i_j} \in O_i x_j$. It is easy to see that $A = \{a_{i_j} : j \in N^+, i \in N^+\}$ is a countable dense set in X. Moreover, obviously, (ii′) implies (i) and (ii) follows from (ii′).

214. Clearly, every separable, not necessarily metrizable, space has the Souslin property. We prove the converse for a metrie space. Consider for each $n \in N^+$ an open cover ω_n of a metric space X composed of sets of diameter less than $1/n$. By Prob. 102, there exists a disjoint family ω_n' of open sets in X, which refines ω_n, and is such that $\bigcup\{U: U \in \omega_n'\}$ is an open dense set in X. Since X has the Souslin property, $|\omega_n'| \leqslant \aleph_0$. For each $U \in \omega_n'$ select a point $x(U) \in U$ and consider the set $A_n = \{x(U): U \in \omega_n'\}$. Clearly, $|A_n| \leqslant \aleph_0$. Noting that diam $U < 1/n$ for each $U \in \omega_n'$ it is easy to show that $A = \bigcup\{A_n: n \in N^+\}$ is a countable dense set in X.

215. Since every subspace of a space X with a countable base also has a countable base and every space with a countable base is separable (even without the assumption of metrizability of X), (i) implies (ii). The implications (ii) $\Rightarrow$ (iii) and (iii) $\Rightarrow$ (iv) are trivial (without the assumption of metrizability of X). We show that (iv) $\Rightarrow$ (v). Let $\gamma = \{U_t: t \in T\}$ be a disjoint family of open sets in X. Without loss of generality, we may assume that $U_t = U_{\varepsilon(t)} x_t$, where $U_{\varepsilon(t)} x_t$ is a spherical nhood of radius $\varepsilon(t)$ of some point $x_t \in X$. Denote by A the set of all poinis x_t, $t \in T$ and by A_i the set of all those points x_t for which $\varepsilon(t) > 1/i$. Clearly, $A = \bigcup\{A_i: i \in N^+\}$ and each A_i is a closed discrete subspace of X. By hypothesis, $|A_i| \leqslant \aleph_0$, $i = 1, 2, \ldots$ Then $|A| \leqslant \aleph_0$; consequently, $|\gamma| \leqslant \aleph_0$.

It remains to prove the implication (v) $\Rightarrow$ (i); however, this is immediate from Probs. 214 and 213.

218. With the aid of Prob. 213, it is easy to verify that there is a countable cover $\omega = \{U_1, \ldots, U_i, \ldots\}$ in X of clopen sets of diameter less than ε. Further, assume that

$V_1 = U_1, V_2 = U_2 \setminus V_1, \ldots, V_n = U_n \setminus \bigcup \{V_i : i = 1, \ldots, n-1\}, \ldots$ Evidently, the family $\omega^* = \{V_i : i \in N^+\}$ is disjoint and cnnsists of clopen sets. Since $V_n \subset U_n$ for every $n \in N^+$, diam $V_n < \varepsilon$ for every $n \in N^+$. It is easy to see that ω^* is a cover of X.

227. The problem may be solved, for example, in the following way. Let $f: I \to Y$ be an open continuous function from the segment I onto a Hausdorff space Y. For each point $y \in Y$ cansider $x(y) = \sup f^{-1}y$. By virtue of the fact that $f^{-1}y$ is closed in I, we have $x(y) \in f^{-1}y$. Put $X = \{x(y): y \in Y\} \subset I$. Evidently f, considered only on X, is a one-to-one continuous function from the subspace X onto Y.

It is somewhat tedious to prove that X is closed in I, after which one can assert that $f' = f \mid X$ is a homeomorphism of X onto Y. It then remains to refer to Prob. 225.

233. Let Γ be an arbitrary open subspace of the complete metric space (X, ρ). For each $x \in \Gamma$, put $\rho_x = \rho(x, X \setminus \Gamma)$. This number is positive. For each pair $x, y \in \Gamma$ define

$$\rho^*(x, y) = \frac{\rho(x, y)}{\rho(x, y) + \rho_x + \rho_y}.$$

Put $\rho^{**}(x, y) = \max [\rho(x, y), \rho^*(x, y)]$. It is left as an exercise to the reader to prove that ρ^{**} is the desired metric on Γ (see Hausdorff [2]).

234. If $M = \bigcap \{U_i : i \in N^+\}$, U_i an open subset of a complete metric space (X, ρ), then M, as a subspace of (X, ρ) is homeomorphic to a closed subspace of the product of the spaces U_i over $i \in N^+$ (see Prob. 367). However, each U_i is complete metrizable (Prob. 233). It now remains to refer to Probs. 186 and 197. Another proof can be obtained by making use of Prob. 233.

235. Fix a complete metric ρ on Y which induces on Y the topology inherited from X. For each $n \in N^+$ denote by γ_n the family of all open sets in X which meet Y in a set of diameter at most $1/n$. Put

$$G_n = \bigcup \{U: U \in \gamma_n\} \text{ and } \tilde{Y} = \bigcap \{G_n: n \in N^+\}.$$

Clearly, γ_n covers Y. Hence $G_n \supset Y$ for each $n \in N^+$ and, consequently, $\tilde{Y} \supset Y$. The set G_n is open in X for every $n \in N^+$. We claim that $\tilde{Y} = Y$. Fix $x \in \tilde{Y}$. For each $n \in N^+$ choose $U_n \in \gamma_n$ such that $x \in U_n$. Then $\lambda = \{U_n: n \in N^+\}$ is a centralized family of open sets in X. Put $F_n = [U_n \cap Y]_Y$, $n \in N^+$. Since $[Y] = X$, $\mu = \{F_n: n \in N^+\}$ is a centralized family of closed sets in Y, the diameter of F_n being at most $1/n$. Since (Y, ρ) is complete, there is a point $y \in Y$ such that $\bigcap\{F_n: n \in N^+\} = \{y\}$ (see Prob. 187). Then $y = x$. For if $y \neq x$, then there are disjoint open sets U and V in X for which $y \in U$ and $x \in V$. But $y \in Y$ and $U \cap Y$ is a nhood of y in (Y, ρ). Hence there is $\varepsilon > 0$ such that $O_\varepsilon y = \{y' \in Y: \rho(y, y') < \varepsilon\} \subset U$. Then $[O_\varepsilon y]_X \cap V = \Lambda$ and, moreover, $[O_\varepsilon y]_X \not\ni x$. Choose $n \in N^+$ for which $1/n < \varepsilon$. Then $F_n \ni y$ implies (since the diameter of F_n is at most $1/n$) that $F_n \subset [O_\varepsilon y]$. Hence $[F_n]_X \subset [O_\varepsilon y]$. But $[F_n]_X = [[U_n \cap Y]_Y]_X \supset U_n \ni x$ since Y is dense in X. Consequently, $[O_\varepsilon y]_X \ni x$, which contradicts the earlier conclusion. Thus $y = x$ and $x \in Y$, that is, $Y = \tilde{Y} = \bigcap \{G_n: n \in N^+\}$.

276. (*a*) Let f be uniformly continuous. Fix any $\varepsilon > 0$. Then one can choose $\delta > 0$ such that if $\rho(x', x'') < \delta$, then $\rho'(fx', fx'') < \varepsilon$. But $\rho(A, B) = 0$ implies that there exist $x' \in A$ and $x'' \in B$, for which $\rho(x' x'') < \delta$. Then $\rho'(fA, fB) \leqslant \rho'(fx', fx'') < \varepsilon$. Since this is true for every $\varepsilon > 0$, we conclude that $\rho'(fA, fB) = 0$.

(*b*) Suppose that f is not uniformly continuous and that $\varepsilon > 0$ is chosen so that for every $\delta > 0$ there exist points $x', x'' \in X$ such that $\rho(x', x'') < \delta$ but $\rho(fx', fx'') > \varepsilon$. We now construct in X by (the usual) induction two sequences $\xi = \{x_n: n \in N^+\}$ and $\eta = \{y_n: n \in N^+\}$ such that the following conditions are satisfied: (j_1) $\rho'(fx_m, fy_k) \geqslant \varepsilon/3$ for every $m \in N^+$, $k \in N^+$, and (j_2) $\rho(x_n, y_n) < 1/n$ for every $n \in N^+$. Consider now the two following exhaustive possibilities.

(*i*) In the space (Y, ρ') there is a point $a \in Y$ for which assertion (i) of Prob. 275 holds. In this case, we choose $x_n, y_n \in X$ independently, for each $n \in N^+$, in such a way that condition (j_2) is satisfied together with (t_1): $\rho'(fx_n, fy_n) > \varepsilon$ for every $n \in N^+$ and (t_2): $\rho'(fx_n, a) < \varepsilon/3$. The triangle axiom then implies that condition (j_1) is also fulfilled. Denote by A (resp. B) the set of points of sequence ξ (resp η). Then $\rho(A, B) = 0$ (by (j_1)), but $\rho'(fA, fB) \geqslant \varepsilon/3$ (by (j_2)).

(*ii*) Assertion (ii) of Prob. 275 is satisfied by the triple f, (X, ρ), (Y, ρ') and ε chosen above. In this case, ξ and η are constructed by (the usual) induction. Let $n \in N^+$ and for every $m < n$ suppose that the points $x_m, y_m \in X$ are already defined. Put $A = \{x_m: m < n\} \cup \{y_m: m < n\}$ (if $n = 1$, then $A = \Lambda$). Using (ii) of Prob. 275 we choose x_n and y_n so that $\rho(x_n, y_n) < 1/n$, $\rho'(fx_n, fy_n) > \varepsilon$ and $\rho'(fx_n, A) \geqslant \varepsilon/3$, $\rho'(fy_n, A) \geqslant \varepsilon/3$. Now continue as above.

281. (*a*) (i) $\Rightarrow$ (ii) is trivial.

(*b*) (ii) $\Rightarrow$ (iii). Let $x_0 \in X$ and let V be an arbitrary nhood of fx_0. Consider a $V_1 \in \sigma$ for which $fx_0 \in V_1 \subset V$ (see Prob. 13). By (ii) $f^{-1}V_1$ is an open set in X. In addition, $x_0 \in f^{-1}V_1$; consequently, $f^{-1}V_1$ is a nhood of x_0 for which $f(f^{-1}V_1) = V_1 \subset V$.

(*c*) (iii) $\Rightarrow$ (iv). Let $x_0 \in [A]$. We must show that $fx_0 \in [fA]$. Consider an arbitrary nhood V of fx_0. Using (iii), choose a nhood Ux_0 of x_0 for which $f(Ux_0) \subset V$. Since $x_0 \in [A]$, $Ux_0 \cap A \neq \Lambda$. Then $f\,Ux_0 \cap fA \neq \Lambda$. Consequently, $V \cap fA \neq \Lambda$, that is, $fx_0 \in [fA]$. Assertions (iv) and (iv') are obviously equivalent.

(*d*) (iv) $\Rightarrow$ (v). By virtue of (iv) and Φ being closed, we have

$$f[f^{-1}\Phi] \subset [ff^{-1}\Phi] = [\Phi] = \Phi.$$

It follows that $[f^{-1}\Phi] \subset f^{-1}\Phi$, implying that $f^{-1}\Phi$ is a closed set.

(*e*) (v) $\Rightarrow$ (i). Continuity of the function f follows from the equality $f^{-1}V = X \setminus f^{-1}(Y \setminus V)$ and the fact that this set, a closed set, is the complement of an open set.

283. No. Let X consist of two (disjoint) parts X_1 and X_2, clopen in X, all the points of X_1 being isolated in X, $f(X_1) = Y$, and the restriction of f to X_2 discontinuous. Evidently, to satisfy both conditions is not difficult; in fact, to fix f, it is enough to define it independently on X_1 and X_2. Then any dense set A in X contains X_1, hence $fA = fX_1 = Y$ is a dense set in Y. f is not continuous however, since its restriction to X_2 is not continuous.

287. Let $x \in X$ and $y_1 = f_1(x) \neq y_2 = f_2(x)$. Choose disjoint open sets U_1 and U_2 in Y for which $y_1 \in U_1$, $y_2 \in U_2$. By the continuity of f_i, $i = 1, 2$, there is a nhood V_i of $x \in X$ for which $f_i(V_i) \subset U_i$, $i = 1, 2$. But then $f_1(V_1) \cap f_2(V_2) = \Lambda$. On the other hand, $V_1 \cap V_2$ is a nonempty open set (because $x \in V_1 \cap V_2$) and, by assumption, $A' = V_1 \cap V_2 \cap A$ is a nonempty set. But then $f_1 \setminus A' = f_2 \setminus A'$ and, consequently, $f_1(A') = f_2(A') \neq \Lambda$. However, $f_1(V_1) \supset f_1(A')$ and $f_2(V_2) \supset f_2(A')$. Hence $f_1(V_1) \cap f_2(V_2) \neq \Lambda$, a contradiction.

288. Necessity is clear. We prove sufficiency. Since Z is a Hausdorff space for each $x \in X$ the extension of f to a continuous function $\tilde{f}$ on $Y \cup \{x\}$ is unique (and, by hypothesis, exists). Let $\tilde{f}(x)$ be the image of the point x under this extension. Thus, the function $\tilde{f}: X \to Z$ is defined. We show that this is continuous. Fix $x^* \in X$ and put $z^* = \tilde{f}(x^*)$. Let Oz^* be an arbitrary nhood of $z^* \in Z$ and O^*z^* a nhood of this point such that $[O^*z^*] \subset Oz^*$. Since $\tilde{f} | (Y \cup \{x\}^*)$ is continuous, there is a nhood U of $x^* \in X$ for which $f(U \cap Y) \subset O^*z^*$. Consider any point $x' \in U$. Then $x' \in [U \cap Y]$ (since $[Y] = X$) and the continuity of $\tilde{f} \setminus (Y \cup \{x'\})$ implies that $\tilde{f}(x') \in [\tilde{f}(U \cap Y)] = [\tilde{f}(U \cap Y)] \subset [O^*z^*] \subset Oz^*$. Thus $\tilde{f}(U) \subset Oz^*$, so that $\tilde{f}$ is continuous.

291. (*a*) The function $\varphi: X \to \Gamma_f$ defined by putting $\varphi(x) = (x, f(x))$ for each $x \in X$ is continuous and the inverse of φ is also continuous. It is clear that φ is a one-to-one function of X onto Γ_f. Now let Ox (resp. W) be an arbitrary nhood of the point $x \in X$ (resp. $\varphi(x) = (x, f(x))$ in Γ_f). Then there exist open sets U and V in X and Y, respectively, for which $x \in U, f(x) \in V, (U \times V) \cap \Gamma_f \subset W, U \subset Ox$, and $f(U) \subset V$ (satisfaction of the last two conditions is always possible by shrinking U). Put $W' = (U \times V) \cap \Gamma_f$. Then $(x, f(x)) \in W'$ and $x \in \varphi^{-1}(W') \subset U \subset Ox$, which implies that φ^{-1} is continuous. Furthermore, $\varphi(U) \subset U \times f(U) \subset U \times V$ and $\Phi(U) \subset \Gamma_f$. Hence $\varphi(U) \subset (U \times V) \cap \Gamma_f \subset W$. Consequently, φ is continuous, that is, φ is a homeomorphism.

(*b*) We use the same notation as above. Let $(x, y) \notin \Gamma_f$. Then $y_1 = f(x) \neq y$. By assumption, there exist disjoint open sets V_1 and V in Y for which $y_1 \in V_1$, and $y \in V$. Since f is continuous, there is an open set U in X for which $f(U) \subset V_1, x \in U$. Then $\varphi(U) \subset U \times V_1$. From this and $V_1 \cap V = \Lambda$ it follows that $(U \times V) \cap \Gamma_f = \Lambda$ (and $(x, y) \in U \times V$).

293. π is continuous and open, by the same argument as for the function f in Prob. 292 (see also the general assertion of Prob. 330). But f is not closed. To see this, denote by A and C the extremeties of the side Y_0 and by B the mid-point of the side of the square under consideration opposite to Y_0. Then the set of all points of X_0 lying outside or on the boundary of the triangle ABC is closed in X_0, but its image is $Y_0 \setminus \{D\}$, where D is the mid-point of Y_0. The set $Y_0 \setminus \{D\}$ is not closed in Y_0.

301. If $U \in \mathcal{T}$ then $U \in \mathcal{T}_\alpha$ for every $\alpha \in M$. We have $\pi^{-1} U = \bigcup \{i_\alpha(U): \alpha \in M\}$, which is open in X_ξ. Conversely, if $\pi^{-1}A$ is open in X_ξ for some $A \subset X$, then $\pi^{-1}A \cap i_\alpha(X)$ is open in $i_\alpha(X)$, that is, $A \in \mathcal{T}_\alpha$ for each $\alpha \in M$. Hence $A \in \bigcap \{\mathcal{T}_\alpha: \alpha \in M\} = \mathcal{T}$. This shows that π is a quotient mapping. Note that if $\mathcal{T}_\alpha \neq \mathcal{T}$, then the restriction of π to the clopen subspace $i_\alpha(X)$ of X_ξ is not a quotient mapping, for otherwise this quotient mapping would be a homeomorphism, that is, we would have $\mathcal{T}_\alpha = \mathcal{T}$.

302. Assertion (i) is a special case of Prob. 318. The weight of S at the point $\pi(B)$ equals the character of B in the space C, since f is closed (see Prob. 326). But (Prob. 81 of Chap. 6) the character of B in C is uncountable, since $B = \operatorname{Fr} B$ is a noncompact† subspace of a normal space C (C is a normal, as is every metrizable space, see Prob. 212). Consequently, S is nonmetrizable. Clearly, π is not an open mapping, otherwise S would have a countable base (see Prob. 339).

†See Probs. 259 and 254 of Chap. 3.

303. (i) Put $f(\pi(B)) = *$ and $f(a) = a$ for every $a \in A$. Obviously, f is a one-to-one function of S onto Φ. The topology of $(\Phi, \mathcal{T}_2)$ can be described as the largest topology for which every $P_k \cup \{*\}$ is a compactum with a unique nonisolated point $*$. It is easy to prove that the same holds for the topology of the space S. This also implies (i).

(ii) The point $*$ has in $(\Phi, \mathcal{T}_1)$ a countable base (the same family $\{U_j: j = 1, 2, \ldots\}$), but in $(\Phi, \mathcal{T}_2)$ the character of $*$ is uncountable (Prob. 302). Hence $\mathcal{T}_1 \neq \mathcal{T}_2$.

(iii) The infinite set $\{(1, k): k = 0, 1, 2, \ldots\}$ has no limit point either in $(\Phi, \mathcal{T}_1)$ or in $(\Phi, \mathcal{T}_2)$. Hence $(\Phi, \mathcal{T}_1)$ and $(\Phi, \mathcal{T}_2)$ are not compact. Both spaces are, however, normal, as is every space with a unique nonisolated point.

304. Assertions (*a*), (*b*) and (*c*) are true, but (*d*) and (*e*) are false.

305. Assertions (*a*) and (*b*) are easy to prove. As regards (*c*), see Prob. 308 of this chapter and Prob. 24 of Chap. 6 or the solution of Prob. 347.

306. Both (*a*) and (*b*) are false (see Prob. 307).

307. False. Consider the mappings $\pi: C \to (\Phi, \mathcal{T}_2)$, $i: (\Phi, \mathcal{T}_1) \to (\Phi, \mathcal{T}_1)$, $(\pi \times i): C \times (\Phi, \mathcal{T}_1) \to (\Phi, \mathcal{T}_2) \times (\Phi, \mathcal{T}_1)$, $(\pi \times i)(x, y) = (\pi(x), i(y))$ for every $(x, y) \in C \times (\Phi, \mathcal{T}_1)$, where $C, \Phi, \mathcal{T}_1$ and $\mathcal{T}_2$ are the same as in Probs. 302 and 303. We also write $(\Phi, \mathcal{T}_2)$ in place of S and $\pi: C \to (\Phi, \mathcal{T}_2)$ in place of $\pi: C \to S$, keeping in mind that S is identified with Φ via the natural mapping (see the solution of Prob. 303). π is a quotient and also a closed mapping (Prob. 302) and i is the identity mapping. We must show that $h = \pi \times i$ is not a quotient mapping.

Put (see Prob. 302) $A_{m, l} = \{(1/n, k): 1/m \leqslant 1/n$ and $l \leqslant k\}$ and $(1/r, m) \times A_{m,r} = B_{m,r}$, where l, m, r are any positive integers, $F_r = \bigcup \{B_{r, m}: m = 1, 2, \ldots\}$, $F = \bigcup \{F_r: r = 1, 2, \ldots\}$. F may be regarded as a subset of both $C \times \Phi$ and $\Phi \times \Phi$; moreover, evidently, $(\pi \times i)^{-1}\pi F = F$. However, F is closed in $C \times (\Phi, \mathcal{T}_1)$ but not in $(\Phi, \mathcal{T}_2) \times (\Phi, \mathcal{T}_1)$ and, in particular, at the point $(*, *) \in [F] \setminus F$. Verify all this independently. Another solution is given in Prob. 349.

308. (*a*) $\Rightarrow$ (*b*). Put $A = Y \setminus f(U)$, $Y' = A \cup \{y\}$, $X' = f^{-1}(Y')$ and $B = f^{-1}A$. Then $X' = B \cup f^{-1}y$ and $U \cap B = \Lambda$, $U \supset f^{-1}y$. Hence $f^{-1}y$ is open in X' and (*a*) implies that $\{y\}$ is open in Y', that is, $y \notin [A]$ and $y \in Y \setminus [Y \setminus fU] = \operatorname{Int} f(U)$.

(*b*) $\Rightarrow$ (*c*). If $f^{-1}y \cap [f^{-1}A] = \Lambda$, then $U = X \setminus [f^{-1}A] \supset f^{-1}y$ and U is open in X. Hence $fU \cap A = \Lambda$ and, by (*b*), $\operatorname{Int} fU \ni y$. $\operatorname{Int} U \cap A = \Lambda$ now implies that $y \notin [A]$.

(*c*) $\Rightarrow$ (*a*). Let $A \subset Y'$, $B = f^{-1}A \subset X'$ and let B be closed in X'. Put $\tilde{B} = [B]$. Then $f^{-1}Y' = X'$ implies that $f\tilde{B} \cap Y' = A$. If $y \in [A]$, then by (*c*) $f^{-1}y \cap \tilde{B} \neq \Lambda$ and, consequently, $f\tilde{B} \ni y$. Hence, if $y \in [A] \cap Y'$, then $y \in f\tilde{B} \cap Y' = A$, that is, A is closed in Y', completing the desired proof (see Prob. 298).

309. If U is an open set in X and $U \supset f^{-1}y$, where y is some fixed point of the space Y, and if $f' = f \mid X'$, $U' = U \cap X'$, then $\operatorname{Int} fU \supset \operatorname{Int} fU' = \operatorname{Int} f'U' \ni y$, giving the required proof.

310. Both (*a*) and (*b*) are false. See Prob. 316 and the solution of Prob. 307.

311. Let $f: X \to Y$ be a pseudoopen continuous function, $f(X) = Y$, $y \in Y$, $B \subset Y$ and $y \in [B]$. Put $A = f^{-1}B$, $F = f^{-1}y$. Then (Prob. 302), $[A] \cap F \neq \Lambda$. Fix $x \in [A] \cap F$. If X is a Fréchet-Urysohn space, then there is a sequence $\{x_n: n \in N^+\} \subset A$ converging to x. Put $y_n = fx_n$. Since f is continuous, $\lim_{n \to \infty} y_n = y$.

313. Let G be any nhood of the set Φ. Since f is continuous, there exists a nhood U of F for which $fU \subset G$. Choose $W \in \xi$ for which $F \subset W \subset U$. Then $fW \subset G$ and

$W \supset F \supset f^{-1}y$ for every point $y \in \Phi$. Hence $\text{Int} fW \ni y$ and, consequently, $\text{Int} fW \supset \Phi$. Thus $\Phi \subset \text{Int} fW \subset G$, and $\text{Int} fW \in f\xi$.

314. This is immediate from Prob. 313.

316. This is immediate from Prob. 321.

317. False. If X is the boundary of a square and π is the projection of this square onto one of its sides parallel to the other side, then the restriction of π to X is continuous and closed but is obviously not an open mapping (the side (without vertices) parallel to which a projection is effected, is an open set in X but is mapped into a non-isolated point of the image).

320. If P is closed in X, then $f(P) = f([P])$. $[f(P)] \subset f([P])$ now implies that $f(P) \supset [f(P)]$. Hence $f(P)$ is closed. If f is a closed function and $A \subset X$ is chosen arbitrarily, then $f([A])$ is closed. But $f(A) \subset f[A]$ (since $A \subset [A]$). Hence $[f(A)] \subset [f[A]] = f[A]$.

321. Since $f(X) = Y$, for every $P \subset X$ we have $fP = Y \setminus f^{\#}(X \setminus P)$.

322. Let f be closed, $y \in Y$, $F = f^{-1}y$ and let U be any open set in X containing F. Then $G = f^{-1}(Y \setminus f(X \setminus U)$ is open in X, $F \subset G \subset U$ and $f^{-1}fG = G$. Consequently, if $\Phi \in \alpha$ and $\Phi \cap G \neq \Lambda$, then $\Phi \subset G$. Hence α is a continuous decomposition.

We now proceed from the assumption that the decomposition α is continuous. Let P be closed in X and $y \notin fP$. Then $f^{-1}y \subset U = X \setminus P$. Since U is open in X, $f^{-1}y \in \alpha$ and α is continuous, there exists an open set V in X for which $f^{-1}y \subset V \subset U$, V being such that if $\Phi \in \alpha$ and $\Phi \cap V \neq \Lambda$, then $\Phi \subset V$. But then $f^{-1}fV = V$ and since f is a quotient mapping it follows that fV is open in Y. However, $V \cap P = f^{-1}fV \cap P = \Lambda$ and hence $fV \cap fP = \Lambda$. Since $f^{-1}y \subset V$, $y \in fV$. Thus fV is a nhood of $y \in Y$, disjoint from P. Consequently, P is closed.

323. Put $P = X \setminus U$ and $Q = fP$, $V = Y \setminus Q$. P is closed in X, hence Q is closed in Y but V is open in Y. Evidently, $Q \cap \Phi = \Lambda$, hence $V \supset \Phi$. Also, $f^{-1}V \cap P = \Lambda$, hence, $f^{-1}V \subset U$.

324. Use Probs. 323 and 321.

328. Let p_X and p_Y be closed. We have $p_X^{-1}(x) = (x, f(x))$ and $p_Y p_X^{-1}(x) = p_Y(x, f(x)) = f(x)$ for every $x \in X$. Hence $f = p_Y p_X^{-1}$ (note that the domain of p_X and p_Y is bounded throughout this problem by the set Γ_f, hence p_X is a one-to-one onto mapping and the mapping p_X^{-1} is defined). Let A be closed in X. Then $f(A) = p_Y(p_X^{-1}A)$, and since p_X is continuous and p_Y is closed, $f(A)$ is closed in Y. Hence f is a closed mapping. Let B be closed in Y. Then $f^{-1}B = p_X p_Y^{-1}(B)$. Since p_Y is continuous and p_X is closed, $f^{-1}(B)$ is closed in X. Hence f is continuous.

Assume now that $f: X \to Y$ is continuous and closed. The mapping $p_X: \Gamma_f \to X$ is then a homeomorphism (see Prob. 291), it being only the continuity of f that is relevant here. Furthermore, p_X is closed. Let $\Phi \subset \Gamma_f$ and Φ be closed in Γ_f. Then $p_Y(\Phi) = f(p_X\Phi)$. However, f is closed and $p_X\Phi$ is closed in X. Consequently, $f(p_X\Phi) = p_Y(\Phi)$ is closed in Y and p_Y is a closed mapping.

329. Let $f: X \to Y$ be such a mapping, with $f(X) = Y$ and F_1, F_2 closed sets in Y such that $F_1 \cap F_2 = \Lambda$. Put $\Phi_1 = f^{-1}F_1$, $\Phi_2 = f^{-1}F_2$. Φ_1, Φ_2 are closed sets in X and $\Phi_1 \cap \Phi_2 = \Lambda$. Since X is normal, there are open sets $U_1 \subset X$, $U_2 \subset X$ such that $U_1 \cap U_2 = \Lambda$, $\Phi_1 \subset U_1$, $\Phi_2 \subset U_2$. By Prob 323 there exist open sets V_1 and V_2 in Y for which $F_1 \subset V_1$, $F_2 \subset V_2$ and $f^{-1}V_1 \subset U_1$, $f^{-1}V_2 \subset U_2$. Then $V_1 \cap V_2 = \Lambda$.

331. The answer is negative. Let π be an orthogonal projection of a plane P carrying the Cartesian coordinate system onto the axis of abscissae and let F be the graph of the function $y = 1/x$ (that is, F is a hyporbola). Then F is closed in P but πF is the axis of abscissae without origin, a nonclosed set (the topology being the usual one). (See also Prob. 293).

335. (*a*) If f is open, then $f(X \setminus [f^{-1}B]$ is open and, evidently, $f(X \setminus [f^{-1} B]) \cap B = \Lambda$. Hence $[B] \cap f(X \setminus [f^{-1}B]) = \Lambda$, that is, $f^{-1}[B] \subset [f^{-1}B]$.

(*b*) Let U be open in X. Put $B = Y \setminus fU$. Then $f^{-1}B \cap U = \Lambda$, and since U is open, $[f^{-1} B] \cap U = \Lambda$. From $f^{-1}[B] \subset [f^{-1}B]$ it now follows that $f^{-1}[B] \cap U = \Lambda$, that is, $[B] \cap fU = \Lambda$. But $fU = Y \setminus B$. Hence, $[B] \subset B$ and $B = [B]$.

337. Let $U \subset X'$, where U is open in X'. Then $U = \tilde{U} \cap X'$ for some open set $\tilde{U}$ in X. $X' = f^{-1}Y'$ inplies that $f(X \setminus X') \subset Y \setminus Y'$ and $fU \subset Y'$. Hence $f\tilde{U} \cap Y' = fU = f'U$. But $f\tilde{U}$ is open in Y, since f is an open mapping. Hence $f'U$ is open in Y'.

338. If $f_\alpha : X_\alpha \to Y_\alpha$ is an open mapping for each $\alpha \in M$ and $U = \Pi \{U_\alpha : \alpha \in M\}$ is a standard open set in $X = \Pi \{X_\alpha : \alpha \in M\}$, and if $f = \Pi \{f_\alpha : \alpha \in M\}$ is the product of the mappings f_α, $\alpha \in M$; $f: X \to Y = \Pi \{Y_\alpha : \alpha \in M\}$, then $fU = \Pi \{f_\alpha U_\alpha : \alpha \in M\}$ is obviously a standard open set in Y (for simplicity, we assume that $f_\alpha X_\alpha = Y_\alpha$ for each $\alpha \in M$).. Now use Prob. 334.

339. Let $\mathcal{B}$ be a base for X whose cardinality equals the weight of X. Then $\mathcal{B}' = \{fU: U \in \mathcal{B}\}$ is a base for Y, $| \mathcal{B}' | \leqslant | \mathcal{B} |$; the conclusion is now immediate.

340. Let U' be any open set in X' let U be open in X, and $U \cap X' = U'$. We claim that $fU = fU'$, which will also establish that fU' is open. If $y \in fU$, then $U \cap f^{-1}y \neq \Lambda$. From the hypothesis it follows that $U \cap f^{-1}y \cap X' \neq \Lambda$. Hence, $fU' \ni y$ and $fU' = fU$, that is, $f' = f \setminus X'$ is open.

341. Consider an arbitrary open set U' in X', and an open set U in X such that $X' \cap U = U'$. If $U' \cap f^{-1}y = \Lambda$, then U' can be taken as U (because then $U \subset X \setminus f^{-1}y \subset X'$, where $X \setminus f^{-1}y$ is an open set in X). If $U' \cap f^{-1}y \neq \Lambda$, then $U \setminus U' \subset f^{-1}y$ where $y \in fU \cap fU'$, Hence in both the cases $fU' = fU$, and fU' is an open set in Y, since f is an open mapping.

343. Let X be the real line and Y a closed interval in X. Then Y is a subspace of X, but X is homeomorphic to an interval obtained from Y by deleting the end-points. However, X and Y are not homeomorphic, since Y is compact but X is not (see Prob. 226).

344. Let R be the space of real numbers with the usual topology. Put $X = \{3n: n \in N\} \cup (\cup \{[1 + 3n, 2 + 3n): n \in N\})$ and $Y = \{3n: n \in N\} \cup (\cup\{[1 + 3n, 2 + 3n): n \in N\})$. The spaces X and Y (taken with the topology induced by R) are not homeomorphic. To see this, notice that the components of the point $\frac{3}{2}$ in X form an interval $(1, 2)$ and in Y there is no point whose components would be homeomorphic to an interval. But under a homeomorphism, components are homeomorphically mapped onto components. This shows that there can be no homeomorphism between X and Y.

However, X can be mapped continuously and one-to-one onto Y, say as follows: all points of the intervals carry over onto themselves but the isolated points of one X are divided into two infinite subsets, of which the first is mapped (in any one-to-one manner) onto a set of isolated points of the space Y and the other is mapped one-to-one onto the set of the end-points of the half-open intervals.

The required mapping of X onto Y has thus been constructed. To construct a one-to-one continuous mapping of Y onto X is rather simple. For this, the set $\{[3n+1, 3n+2): n = 0, 1, 2, \ldots\}$ of half-open intervals is arbitrarily partitioned into a countable collection of countable sets A_k, $k = 1, 2, \ldots$, that is, such that each A_k is a countable family of half-open intervals. Each interval $(3n+1, 3n+2)$ contained in X is partitioned into a countable set of pairwise disjoint half-open intervals C_i^n, where

$$C_i^n = \left[(3n+1) + \frac{1}{i+1},\quad 3n+1+\frac{1}{i}\right).$$

Between the families $C^k = \{C_i^k: i = 1, 2, \ldots\}$ and A_k one establishes some one-to-one correspondence and each half-open interval contained in A_k is homeomorphically mapped onto the half-open intervals corresponding to it in C_k. The isolated points of Y are then mapped onto those of X by some one-to-one correspondence, completing the construction of the required mapping of Y onto X.

345. Assume that $x \in X \setminus Y$, $y \in Y$ and $\varphi(x) = \varphi(y) = z \in \varphi(Y)$. Choose disjoint nhoods Ox and Oy of $x, y \in X$. Put $V = Oy \cap Y$. Then $\varphi(Y)$ is a nhood of $z \in \varphi(Y)$, and hence there is an open set U in Y such that $U \cap \varphi(Y) = \varphi(V)$. Since $z \in U$, $\varphi(x) = z$ and φ is continuous, there exists an open set W in X for which $x \in W \subset Ox$ and $\varphi(W) \subset U$. Then $W \cap Y \neq \Lambda$ and $\varphi(W \cap Y) \subset \varphi(Y)$; hence $\varphi(W \cap Y) \subset U \cap \varphi(Y) = \varphi(V)$. Thus, $(W \cap Y) \cap V = \Lambda$ but $\varphi(W \cap Y) \cap \varphi(V) \neq \Lambda$, contradicting the one-to-oneness of $\varphi \mid Y$.

346. Let $\pi: X \to Z$ be the natural mapping. It is closed (Prob. 332). Moreover, $X^1 = \pi^{-1}Z^1$, that is, X^1 is the inverse image under π of a subspace $Z^1 \subset Z$. Consequently, $\pi^1 = \pi \mid X^1$, the restriction of π to X^1 is a one-to-one closed continuous mapping of X^1 onto Z^1 (see Prob. 319), which implies that π^1 is a homeomorphism (Prob. 342).

347. False. Let X be the usual space of nonnegative real numbers and for each positive integer n consider the two-point set $F_n = \{n, 1/n\}$. Let Z be the decomposition space of X, whose nontrivial elements are just the F_n. Let $\pi: X \to Z$ be the natural projection. We shall use the notation employed in the solution of Prob. 346. Evidently, $X^1 = \{0\} \cup Y_1 \cup Y_2$, where $Y_1 = \{x \in X: x < 1 \text{ and } x \neq 1/n \text{ for all } n = 1, 2, \ldots\}$, $Y_2 = \{x \in X: x > 1 \text{ and } x \neq n \text{ for all } n = 1, 2, \ldots\}$. Obviously $0 \in [Y_1]$, but $0 \notin [Y_2]$. However, $[\pi(Y_2)]_Z \supset [\{\pi(n): n = 1, 2, \ldots\}] = [\{\pi(1/n): n = 1, 2, \ldots\}] \ni \pi(0)$. Thus $0 \notin [Y_2]$ but $\pi(0) \in [\pi(Y_2)]_Z$, that is, the restriction of π to X^1 is not a homeomorphism. However, $\pi^{-1}(Z^1) = X^1$ and π is a quotient mapping. The conclusion that the restriction of a quotient mapping to an inverse image is not necessarily a quotient mapping follows (see the related Problem 305).

349. Let X_1 be the interval $(0, 1)$, X_2 the set $\{2, 2+\frac{1}{2}, 2+\frac{1}{3}, \ldots, 2+1/n, \ldots\}$, $X = X_1 \cup X_2$; also let X_1, X_2 and X be endowed with the usual topology. Let Y be the decomposition space whose elements are the pairs $(1/n, 2+1/n)$ and the distinguished points of X. Let Z be the collection of all points of the half-open interval $[0, 1)$, different from $1/n$ for all $n \in N^+$. The product $\pi \times i_Z$ of the natural quotient mapping $\pi: X \to Y$ and the identity mapping i_Z of Z onto itself is not a quotient mapping. For $P = \{(\pi(x), x): x \in Z, x > 0\}$ is not a closed set in $Y \times Z$, because $(\pi(2), 0) \in [P] \setminus P$, although $(\pi \times i_Z)^{-1}P$ is closed in $X \times Z$ (obviously, $(\pi \times i_Z)^{-1}P = \{(x, x): x \in (0, 1), x \neq 1/n, n \in N^+\}$).

351. Let N be the set of all natural numbers and ξ the family of all open (relative to the usual topology) subsets U of R, for which $N \setminus U$ is finite. Define f on R thus: $f(x) = 0$ if $x \in N$, and $f(x) = 1$ if $x \in R \setminus N$. It is now easy to show that if a sequence $\{x_n: n \in N^+\}$ of points of R is cofinal in ξ, then there is an integer k for which $x_n \in N$ for every $n \geqslant k$. But then $\lim \{f(x_n): n \in N^+\} = 0$. Hence the function f, regarded as a mapping of the set R into the usual space R is sequentially continuous along ξ. However, it is not continuous along ξ, since $fU \ni 0$ and $fU \ni 1$ for every $U \in \xi$.

352. Fix a countable prefilter η, which is a base for ξ. Then $\eta \subset \xi$ and each element of ξ contains some element of η. Thus the sequence $\{x_n: n \in N^+\}$ is cofinal in ξ iff it is cofinal in η, so that sequential coutinuity of f along ξ is equivalent to its sequential continuity along η. Similarly, the continuity of f along ξ is equivalent to the continuity of f along η. It is also clear that the continuity of f along η implies the sequential continuity of f along η (this is true for every prefilter and not just for η). Assume now that f is not continuous but is sequentially continuous along η. Index the elements of η: $\eta = \{U_n: n \in N^+\}$. By modifying η, if necessary, we may assume $U_{n+1} \subset U_n$, $n \in N^+$. Choose any point $x_n \in U_n$. Then the sequence $\{x_n: n \in N^+\}$ is cofinal in ξ and there is a point $y \in Y$, each nhood of which contains members of the sequence $\{fx_n: n \in N^+\}$ with sufficienly large index. But $y \notin \lim_\eta f$, since $\lim_\eta f = \Lambda$. Consequently, there is a nhood O^*y for which $f(U_n) \setminus O^*y \neq \Lambda$, $n \in N^+$. Fix $y_n \in U_n$, for which $fy_n \notin O^*y$. The sequence $\{y_n: n \in N^+\}$ is cofinal in η and therefore $\{fy_n: n \in N^+\}$ converges to some point $y' \notin Y$. Obviously, $y' \neq y$. Now put $z_n = x_n$ if n is even and $z_n = y_n$ if n is odd. Evidently, the sequence $\{z_n: n \in N^+\}$ is cofinal in ξ, so that the sequence $\{fz_n: n \in N^+\}$ converges in Y. But this sequence would then contain two subsequences, one converging to y' and the other to y, with $y \neq y'$. Since Y is Hausdorff, this is impossible.

353. It suffices to take a simple countably infinite compact space, with exactly one non-isolated point.

360. The projection of a product space onto a factor is a continuous open mapping. But in such a case the weight of the factor is at most the weight of the product (Prob. 339).

364. (i) This is easy (see p. 86).

(ii) Note that F^τ is a T_0-space of weight τ (and is also compact; see Prob. 75 of Chap. 3), since the Tychonoff product of T_0-spaces is also a T_0-space, and $w(F^\tau) = |A|$ (see Prob. 379). Then every subspace $X \subset F^\tau$ is also a T_0-space and $w(X) \leqslant w(F^\tau) = \tau$. Conversely, let X be a T_0-space and $w(X) \leqslant \tau$. Consider a base $\sigma = \{U_\alpha : \alpha \in A\}$ for X. We may assume that $|A| = \tau$. We now construct a topological mapping f of X into F^τ. Let $p \in X$ be an arbitrary point. For each $\alpha \in A$ put $x_\alpha(p) = 0$ if $p \in U_\alpha$, and $x_\alpha(p) = 1$ if $p \notin U_\alpha$. Set $f(p) = \{x_\alpha(p): \alpha \in A\} \in F^\tau$.

(*a*) We show that f is one-to-one. Let $p, p' \in X$ and $p' \neq p$. Since X is a T_0-space, there is $\alpha \in A$ such that U_α contains either p or p' but not both. For example, let $p \in U_\alpha$, $p' \in X \setminus U_\alpha$. Then $x_\alpha(p) = 0$, $x_\alpha(p') = 1$ and so the points fp, fp' are distinct in F^τ.

(*b*) f is continuous. Let $f(p) = \xi \in F^{\tau}$. Choose an arbitrary basic nhood

$$H^{0\,0\,\ldots\,0}_{\alpha_1\alpha_2\ldots\alpha_s} = \{\xi' \in F^{\tau} : \xi'_{\alpha_1} = 0, \ldots, \xi'_{\alpha_s} = 0\}$$

of the point $f(p) = \xi$ in F^{τ}. Then $f(\bigcap\{U_{\alpha_i}: i = 1, \ldots, s\}) \subset H^{0\,0\,\ldots\,0}_{\alpha_1\alpha_2\ldots\alpha_s}$, which implies that f is continuous at the point $p \in X$.

(*c*) The inverse mapping f^{-1} is continuous. In fact, it suffices to show that fU_α is open in $fX \subset F^{\tau}$ for each $\alpha \in A$ (since $\sigma = \{U_\alpha: \alpha \in A\}$ is a base for X), which however is immediate from the obvious equality $fU_\alpha = H^0_\alpha \cap fX$.

365. The natural homeomorphism φ between X and Δ is given by the rule: $\varphi(x) = z \Leftrightarrow \pi_\alpha(z) = x$ for every $\alpha \in A$.

366. If $z \in X^A \setminus \Delta$, there exist $\alpha_1, \alpha_2 \in A$, for which $x_1 = \pi_{\alpha_1}(z) \neq \pi_{\alpha_2}(z) = x_2$. Choose disjoint nhoods in X of the points x_1 and x_2; corresponding to this pair is a standard nhood of the point z in X^A containing no point of the set Δ.

367. Assign to each point $y \in Y$ the point $\varphi(y) \in Z$, all of whose coordinates are equal to y. Obviously φ is a homeomorphism of Y into Z. That $\varphi(Y)$ is a closed set in Z is straightforward from the Hausdorffness of X, using the fact that no two coordinates of any point in $Z \setminus \varphi(Y)$ are equal (the latter is obvious). (Note that it is necessary to take a standard nhood in Z, which corresponds to some disjoint nhoods of these coordinates.)

368. If X is a T_1-space, then $X \times X$ is a T_1-space and hence every set in $X \times X$ is an intersection of some family of open sets. Suppose that X is not a T_1-space. Choose a pair of points $x, y \in X$ such that every nhood of x in X contains y. Consider an arbitrary open set U in $X \times X$ containing Δ. Then $U \ni (x, x)$ and by the definition of the product topology there are open sets $O'x$ and $O''x$ in X such that $(x, x) \in O'x \times O''x \subset U$. But, by hypothesis, $y \in O'x \cap O''x$. Hence $(x, y) \in U$, that is, the point (x, y) of the product $X \times X$, which is obviously not on the diagonal Δ, belongs to the intersection of all open sets in $X \times X$ containing Δ.

369. The proof of (*a*) is left to the reader. The example sought in (*b*) can be constructed thus: let the set X consist of the elements of some sequence $\{x_n: n \in N^+\}$ and two more points: y and z. The topology $\mathcal{T}$ is so defined that $\{x_n: n \in N^+\}$ converges to y and all points other than y are isolated. Similarly, the topology $\mathcal{T}'$ is so defined that $(X, \mathcal{T}')$ is compact and has only one non-isolated point, namely z. Verify that (y, z) is a limit point of the diagonal $\Delta = \{(x, x): x \in X\}$ in the space $(X, \mathcal{T}) \times (X, \mathcal{T}')$.

370. If condition (*a*) is satisfied, then $Ox_1 \times Ox_2$ is a nhood of the point (x_1, x_2) in the space $(X, \mathcal{T}_1) \times (X, \mathcal{T}_2)$, disjoint from Δ. Hence Δ is closed. Conversely, if Δ is closed, then for any pair $(x_1, x_2) \in (X \times X) \setminus \Delta$ we have $x_1 \neq x_2$ and there is a set $U = Ox_1 \times Ox_2 \ni (x_1, x_2)$ for which $U \cap \Delta = \Lambda$ and $Ox_1 \in \mathcal{T}_1$, $Ox_2 \in \mathcal{T}_2$. Then obviously $x_1 \in Ox_1$, $x_2 \in Ox_2$ and $Ox_1 \cap Ox_2 = \Lambda$.

374. For each $\alpha_0 \in M$ define $\varphi(\alpha_0) \in D^{\tau}$ thus: $x = \varphi(\alpha_0) \Leftrightarrow x_{\alpha_0} = 1$ and $x_\alpha = 0$ for $\alpha \neq \alpha_0$. Further, put $U_{\alpha_0} = \{x \in D^{\tau}: x_{\alpha_0} = 1\}$ and $A = \{\varphi(\alpha): \alpha \in M\}$. Then $\varphi(\alpha') \neq \varphi(\alpha'')$ for $\alpha' \neq \alpha''$ implies that $|A| = |M| = \tau$. Clearly, $U_\alpha \cap A = \{\varphi(\alpha)\}$ for each $\alpha \in M$. Keeping in mind the definition of a subspace and the fact that each U_α is open in D^{τ}, we conclude that A is a discrete subspace of D^{τ}.

375. (*a*) Let $X_\alpha = \{0, 1\}$ for each $\alpha \in M$, where $\{0\}$, $\{1\}$ are open sets in X_α and

$|M| = \tau$. For each $x \in D^\tau = \Pi\{X_\alpha: \alpha \in M\}$ put $\varphi(x) = \{\alpha \in M: x_\alpha = 1\}$. Clearly, φ is a one-to-one function from X onto $\exp M$. Consequently, $|D^\tau| = |\exp M| = 2^\tau$.

(*b*) The notation remains the same as in (*a*). Consider a standard base $\mathcal{B}$ for D^τ. Assign to each element $U \in \mathcal{B}$ a pair $d(U) = (N, \psi)$, where N is a finite subset of M and ψ a function of N into the set $\{0, 1\}$ such that $U = \{x \in D^\tau: x_\alpha = \psi(\alpha)$ for every $\alpha \in N\}$. The pair $d(U)$ exists for each $U \in \mathcal{B}$ by the definition of the standard base $\mathcal{B}$ for a product. We obtain a one-to-one function d of $\mathcal{B}$ into (in fact onto) the set $L \times Q$, where L is the collection of all finite subsets of M, $Q = \bigcup\{Q_N: N \in L\}$, and Q_N is the set of all functions of N into $\{0, 1\}$. But $|L| = |M| = \tau$ (since M is infinite), and $|Q_N| < \aleph_0$ for every $N \in L$, hence $|Q| = |L| = \tau$. Hence $|L \times Q| = \tau$. Since $\mathcal{B}$ is equipotent to a subset of $|L \times Q|$, $|\mathcal{B}| \leqslant \tau$. Consequently, the weight of D^τ is at most τ.

(*c*) The space D^τ contains a discrete subspace of cardinality τ (see Prob. 374). Since the weight of such a subspace equals τ and the weight of a subspace is always at most that of the space, the weight of D^τ is τ.

376. There exists (see Prob. 60 of Chap. 1) a family $\mathcal{P}$ such that $\bigcup\{P: P \in \mathcal{P}\} = A$, $|P| = \aleph_0$ for every $P \in \mathcal{P}$, $|\mathcal{P}| = |A| = \tau$ and $P' \cap P'' = \Lambda$, if $P', P'' \in \mathcal{P}, P' \neq P''$. Then (see Prob. 355) $X = \Pi\{X_P: P \in \mathcal{P}\}$, where $X_P = \Pi\{\tilde{D}_\alpha: \alpha \in P\}$. But every X_P is homeomorphic to the Cantor perfect set (see Prob. 301). Hence X_P is homeomorphic to $\Pi\{D_\alpha: \alpha \in P\}$, where each D_α is a discrete space consisting of two points. Then X is homeomorphic to the product $\Pi\{D_\alpha: \alpha \in A\}$, and the latter is D^τ.

377. Consider D^τ, the Tychonoff product of τ copies of a discrete two-point space. D^τ is Hausdorff, its weight is τ (see Prob. 375) and the cardinality of the set of all points of D^τ is 2^τ. Hence A and D^τ are equipotent, that is, there is a one-to-one function f from D^τ onto A, enabling us to introduce the desired topology in A: we decree all those subsets of A to be open which are the image under f of open sets in D^τ.

379. Fix in each X_α a base $\mathcal{B}_\alpha = \{U_\alpha^\beta: \beta \in M\}$ of cardinality τ. For each finite subset N of M denote by Q_N the set of all mappings of N into M. Obviously $|Q_N| = |M^{|N|}|$. Hence $|Q_N| = \tau$. Put $L = \{N \subset M: |N| < \aleph_0\}$ and $U(N, \psi) = \{x \in X: x_\alpha \in U_\alpha^{\psi(\alpha)}$ for every $\alpha \in N\}$, $N \in L, \psi \in Q_N$. We have $|L| = \tau$ and, consequently, $|\bigcup\{Q_N: N \in L\}| = \tau$. Hence the cardinality of the set $\mathcal{B} = \{U(N, \psi): N \in L, \psi \in Q_N\}$ is at most τ. But $\mathcal{B}$ is obviously a base for X. Thus, the weight of X is at most τ. From Prob. 360 it now follows that the weight of X is equal to τ.

380. Fix on A a Hausdorff topology of weight τ, which can be done by Prob. 377. Let $\mathcal{B}$ be a base for A of cardinality τ. Denote by $\mathcal{E}$ the set of all finite pair-wise disjoint subfamilies of $\mathcal{B}$. Then (see Prob. 58 of Chap. 1) $|\mathcal{E}| = \tau$. For each $S \in \mathcal{E}$ let Q_S be the set of all functions from S into Y. Then $|Q_S| = |Y^{|S|}| = \tau$ (see Prob. 56 of Chap. 1). In each X_α fix some point x_α^*. We have $|\bigcup\{Q_S: S \in \mathcal{E}\}| = \tau$ (Prob. 57 of Chap. 1), hence the set $\mathcal{P} = \{(S, \psi): S \in \mathcal{E}, \psi \in Q_S\}$ is of cardinality τ. Assign to each element $p = (S, \psi) \in \mathcal{P}$ the point $\varphi(p) \in X$, defined as follows: $x = \varphi(p)$ iff $x_\alpha = \psi(U)$, if $\alpha \in U \in S$; and $x_\alpha = x_\alpha^*$ if $\alpha \notin \bigcup\{U: U \in S\}$. This definition makes sense, since distinct elements of S are disjoint. Put $R = \{\varphi(p): p \in \mathcal{P}\}$. Then $|R| \leqslant |\mathcal{P}| = \tau$ and R is dense in X, as is easily verified.

381. Fix a discrete space Z of cardinality τ. In each X_α choose a dense subspace Y_α of cardinality $\leqslant \tau$. Put $Z_\alpha = Z$ for each $\alpha \in M$ and let f_α be a function from Z_α onto Y_α, which is automatically continuous (see Prob. 284).

Consider the Tychonoff product $Z = \Pi\{Z_\alpha : \alpha \in M\}$ and the map $f = \Pi\{f_\alpha : \alpha \in M\}$ of Z into X. This mapping is continuous, since each of its factors is continuous. But Z has a dense subset of cardinality $\leqslant \tau$ (see Prob. 380). Consequently, so does (Zf), a subspace of X. Now $[f(Z)] = X$, as is immediate from the definition of f and the fact that $f_\alpha(Z_\alpha) = Y_\alpha$ is dense in X_α (see Prob. 357).

382. For each $x \in A$, put $M_x = \{\alpha \in M, x_\alpha \neq x_\alpha^*\}$. From $A \subset Y$ and the definition of Y it follows that $|M_x| \leqslant \aleph_0$. Put $M_A = \bigcup\{M_x : x \in A\}$. Then (see Prob. 57 of Chap. 1), $|M_A| \leqslant 2^\tau$. Put $A^* = \{x \in X : x_\alpha = x_\alpha^* \text{ for } \alpha \notin M_A\}$.

As a subspace of X, A^* is homeomorphic to the Tychonoff product $\Pi\{X_\alpha : \alpha \in M_A\}$ (prove this). Also $A \subset A^*$. However, $\Pi\{X_\alpha : \alpha \in M_A\}$ and hence also A^* has a dense subset of cardinality at most τ (see Prob. 381).

383. Let $\mathcal{E} = \{V_\beta : \beta \in L\}$, $V_{\beta'} \cap V_{\beta''} = \Lambda$ for $\beta' \neq \beta''$, V_β a nonempty open set in X for each $\beta \in L$. Assume that $|L| > \aleph_0$ and choose $L_1 \subset L$ for which $|L_1| = \aleph_1$. Now fix some point $x^* \in X$. Its $\aleph_1$-hull $\Sigma(x^*)$ is dense in X and so meets every V_β. For each $\beta \in L_1$, select in $\Sigma(x^*) \cap V_\beta$ a point x^β and put $A = \{x^\beta : \beta \in L\}$. Then $x^{\beta'} \neq x^{\beta''}$ for $\beta' \neq \beta''$ and hence $|A| = |L_1| = \aleph_1$. From $A \subset \Sigma(x^*)$ and $|A| = \aleph_1 \leqslant 2^{\aleph_0}$ it follows (see Prob. 382) that there exists $\tilde{A} \subset X$ for which $|\tilde{A}| \leqslant \aleph_0$ and $|\tilde{A}| \supset A$. But then $[\tilde{A}] \ni x_\beta$ and since $V_\beta \ni x_\beta$, $\tilde{A} \cap V_\beta \neq \Lambda$ for each $\beta \in L_1$. Since $V_{\beta'} \cap V_{\beta''} = \Lambda$ for $\beta' \neq \beta''$ and $|L_1| = \aleph_1$, we conclude that $\tilde{A}$ must be uncountable, a contradiction.

385. Let $\mathcal{B}$ be a standard base for X. By Prob. 102 there is a family $\mathcal{E}$ of pairwise disjoint elements of $\mathcal{B}$ such that $\bigcup\{V : V \in \mathcal{E}\} \subset U \subset [\bigcup\{V : V \in \mathcal{E}\}] = [U] = A$. Since each X_α is separable, the cardinality of every disjoint family of nonempty open subsets of X is at most $\aleph_0$ (Prob. 383). Hence $|\mathcal{E}| \leqslant \aleph_0$. By Prob. 384 for each $V \in \mathcal{E} \subset \mathcal{B}$ there is a finite set $C_V \subset M$ such that V is independent of $L_V = M \setminus C_V$. But, then, the set $W = \bigcup\{V : V \in \mathcal{E}\}$ is independent of $L = \bigcap\{M \setminus C_V : V \in \mathcal{E}\} = M \setminus \bigcup\{C_V : V \in \mathcal{E}\} = M \setminus C$, where $C = \bigcup\{C_V : V \in \mathcal{E}\}$ is a countable set. It is easy to show that $[W]$ is also independent of L (see also Prob. 388). However, $[W] = A$, since $A = [U]$ and $[U] = [W]$ as we see from the inclusion: $W \subset U \subset [W]$.

386. Let $\mathcal{B} = \{V_i : i \in N^+\}$ be a countable base for Y. Put $U_i = f^{-1}V_i$, $A_i = [U_i]$, $i \in N^+$. By Prob. 385 for each $i \in N^+$ there is a countable set $C_i \subset M$ such that A_i is independent of $L_i = M \setminus C_i$. Put $C = \bigcup\{C_i : i \in N^+\} = M \setminus C$; we show that f is independent of L.

Assume that we have $x', x'' \in X$, such that for every $\alpha \in C = M \setminus L$ we have $x'_\alpha = x''_\alpha$ and $f(x') = z' \neq z'' = f(x'')$. Since $\mathcal{B}$ is a base in Y, one can choose $i', i'' \in N^+$ such that $z' \in V_{i'}$, $z'' \in V_{i''}$ and $[V_{i'}] \cap [V_{i''}] = \Lambda$. Then $f^{-1}[V_{i'}] \cap f^{-1}[V_{i''}] = \Lambda$ and by the continuity of f, we obtain $A_{i'} = [f^{-1} V_{i'}] \subset f^{-1}[V_{i'}]$, $A_{i''} = [f^{-1} V_{i''}] \subset f^{-1}[V_{i''}]$. Consequently, $A_{i'} \cap A_{i''} = \Lambda$. From $C \supset (C_{i''} \cup C_{i'})$ and the choice of the points x', x'' it follows that $x'_\alpha = x''_\alpha$ for every $\alpha \in C_{i'}$. However, $x' \in A_{i'}$. Hence (by the choice of $C_{i'}$), $x'' \in A_{i'}$, but this contradicts $x'' \in A_{i''}$ and $A_{i'} \cap A_{i''} = \Lambda$.

387. To prove assertions (*a*), (*b*), (*c*) and (*d*) it suffices to refer to the relevant definitions. Let us prove (*e*). From (*d*) it follows that $y' \in [P']$. By (*c*), $P'' = \langle B, y'' \rangle (P) = \langle B, y'' \rangle \langle B, y' \rangle (P) = \langle B, y'' \rangle (P')$. From (*a*) it follows that $\langle B, y'' \rangle (y') = y''$. Since $\langle B, y'' \rangle$ is a continuous function and, by (*d*), $y' \in [P']$, we have $y'' \in [P'']$.

388. Let $x' \in [B]_Z$, $x'' \in Z$, $x'_\alpha = x''_\alpha$ for every $\alpha \in M \setminus L$ and $f(x') = y' \neq y'' = f(x'')$. In Y choose open sets V' and V'' for which $V' \ni y'$, $V'' \ni y''$ and $V' \cap V'' = \Lambda$. Put $U' = f^{-1}V'$. U' is an open set in X, $U' \ni x'$ and hence $U' \cap B \neq \Lambda$. Put $A = U' \cap B$. Then $x' \in [A]$ and $fA \subset fU' = V' \subset X \setminus V''$ whence $y'' \notin [fA]$.

Now consider the sets $A' = \langle L, x' \rangle (A)$ and $A'' = \langle L, x'' \rangle (A)$ (which are contained in Z, since Z is L-convex). From $A \subset B$ it follows that f is independent of L on A. Hence $fA'' = fA = fA'$. But $x'' \in [A'']$ (see (*e*) of Prob. 378); consequently, $y'' = fx'' \in [fA''] = [fA]$, a contradiction.

389. Put $y = f(x)$. Let $\mathcal{B}$ be a standard base for the topology of X. For each element $V \in \mathcal{P}$ fix an element $\tilde{V} \in \mathcal{B}$ such that $x \in \tilde{V}$ and $f(\tilde{V} \cap Z) \subset V$ (possible, since f is continuous). Put $Q = \{\tilde{V}: V \in \mathcal{P}\}$ and $A_x = \bigcap \{\tilde{V} \cap Z: \tilde{V} \in Q\}$. Then $Q \subset \mathcal{B}$, $|Q| \leqslant |\mathcal{P}|$, $A_x \ni x$ and $f(A_x) \subset \bigcap \{f(\tilde{V} \cap Z): V \in Q\} \subset \bigcap \{V: V \in \mathcal{P}\} = \{y\}$.

The set $\tilde{V}$ is independent of $L_{\tilde{V}} = M \setminus k(\tilde{V})$ (see p. 44). Put $C_x = \bigcup \{k(\tilde{V}): \tilde{V} \in Q\}$. Then $|C_x| \leqslant |\mathcal{P}|$ and $A_x \subset \{\tilde{V}: \tilde{V} \in Q\}$ implies that A_x is independent of $L_x = M \setminus C_x$.

390. Put $A = f^{-1}B$. For each point $x \in A$ fix sets $A_x \subset A$ and $C_x \subset M$ such that (*a*) $f(Ax) = f(x)$, (*b*) $|C_x| \leqslant \aleph_0$, and (*c*) A_x is independent of $M \setminus C_x$ (possible; see Prob. 389). Now define by induction a sequence $\{A_i: i \in N^+\}$ (resp. $\{C_i: i \in N^+\}$) of countable subsets of A (resp. M) such that for each i the following conditions are satisfied: (i) $A_i \subset A_{i+1}$, (ii) $C_i = \bigcup \{C_x: x \in A_i\}$ and (iii) $\pi_{C_i}(A_{i+1})$ is dense in $\pi_{C_i}(A) \subset \Pi\{X_\alpha: \alpha \in C_i\}$. Start with $A_1 = \{x_1\}$, x_1 an arbitrarily chosen point of A, and $C_1 = \{\Lambda\}$. The induction process does not involve any difficulty, because $\Pi\{X\alpha: \alpha \in C_i\}$ is a space with a countable net and, consequently, every subspace has a countable dense subset. Evidently, $A_i \subset A_{i+1}$ implies that $C_i \subset C_{i+1}$ for every i. Put $A_\infty = \bigcup \{A_i: i \in N^+\}$, $C_\infty = \bigcup \{C_i: i \in N^+\}$ and $A^* = \bigcup \{A_x: x \in A_\infty\}$. Note that $|C_\infty| \leqslant \aleph_0$, $|A_\infty| \leqslant \aleph_0$ and $f(A_\infty) = fA^*$. The set A_∞ is obviously independent of $M \setminus C_\infty$, as is the function f on A_∞ (see the definition of C_i). On the other hand, $\pi_{C_\infty}(A^*)$ is dense in $\pi_{C_\infty}(A) \subset \Pi\{X\alpha: \alpha \in C_\infty\}$, as follows immediately from the definition of the product topology, the relations $C_i \subset C_{i+1}$, $C_\infty = \bigcup\{C_i: i \in N^+\}$ and the density of $\pi_{C_i}(A^*)$ in $\pi_{C_i}(A)$ for each i (the last is valid since $A^* \supset A_{i+1}$). But from $\aleph_0$-convexity (which implies C_∞-convexity) of the set Z and independence of A^* of $M \setminus C_\infty$ (and the relation $A^* \subset A$) it then follows that A^* is dense in A. Hence $fA^* = B^*$ is dense in B. But $B^* = fA^* = fA_\infty$ (by the definition of A_x), so that $|B^*| \leqslant |A_\infty| \leqslant \aleph_0$. Finally, put $\langle M \setminus C_\infty, z^* \rangle (A_\infty) = \tilde{A}$. Then $\tilde{A} \subset Z$, since Z is C_∞-convex and $B^* = fA_\infty = f\tilde{A}$, since f is independent of $M \setminus C_\infty$. Evidently, $|\tilde{A}| \leqslant |A_\infty| \leqslant \aleph_0$, that is, C_∞ and $\tilde{A}$ are the required sets.

391. By Prob. 390, for any point $x^* \in X$ there are countable sets $C \subset M$ and Q, contained in the C-hull of the point x^* such that $[f(Q)] \supset Y'$. But from the hypothesis it follows that $[Q]$ is compact. Since f is continuous, $f[Q]$ is also sequentially compact, but Y is a Hausdorff space and so $f[Q]$ is closed in Y. From these assertions it follows that $f[Q] = [fQ] \supset [Y']$; hence, $[Y']$ is sequentially compact. See also Probs. 265, 292, 56 and 92 of Chap. 3.

392. Let X_α be the usual space of real numbers for each $\alpha \in M$, $|M| > \aleph_0$ and Z_α the collection of all non-negative integers, $Z_\alpha \subset X_\alpha$. Then Z_α is a closed discrete subspace of X_α. Hence the Tychonoff product $Z = \Pi\{Z_\alpha: \alpha \in M\}$ is naturally homeomorphic to a closed subspace of $X = \Pi\{X_\alpha: \alpha \in X\}$ (see Prob. 361). Thus (Prob. 62), it suffices to show that Z is not normal. Put $A = \{z \in Z: |\{\alpha \in M: z_\alpha = i\}| \leqslant 1$ for each $i \geqslant 1\}$ and $B = \{z \in Z: |\{\alpha \in M: z_\alpha = i\}| \leqslant 1$ for each $i \neq 1\}$. Then $A \cap B = \Lambda$, and A and B are closed in Z. Suppose now that Z were normal. Then there would be a continuous real-valued function f on Z, such that $f(A) = 0$ and $f(B) = 1$ (see Prob. 25 of Chap. 3). By Prob. 368 there is a countable set $L \subset M$ such that if $z, z' \in Z$ and $z \mid L = z' \mid L$, then $f(z) = f(z')$.

Now consider $z^0, z^1 \in Z$ which send L one-to-one into the set of all integers greater than 1 and satisfy the condition, $z^0 (M \setminus L) = 0$, $z^1 (M \setminus L) = 1$. Then $z^0 \in A$, and hence $f(z^0) = 0$; $z^1 \in B$, and hence $f(z^1) = 1$; $z^0 \mid L = z^1 \mid L$. Thus $f(z^0) = f(z^1)$, that is, $0 = 1$, a contradiction.

393. See Corson [1] and Kombarov [1], [2].

394. The answer is affirmative. This has been shown recently by Kombarov [1] and Mal'khin [1].

395. Consider an arbitrary family γ of nonempty, open, pairwise disjoint sets in X. Assume that $|\gamma| > 2^\tau$. Without loss of generality, let all $U \in \gamma$ have the form $U = \Pi\{U_t: t \in T\}$ where U_t is open in X_t and $U_t \neq X_t$ for only finitely many $t \in T$. Let $U' \in \gamma$, $U'' \in \gamma$ and $U' \neq U''$. Then $U' \cap U'' = \Lambda$ and $U' = \Pi\{U'_t: t \in T\}$, $U'' = \Pi\{U''_t: t \in T\}$ imply that $U'_{t_0} \cap U''_{t_0} = \Lambda$ for some $t_0 \in T$. Fix such a t_0 and put $f(U', U'') = t_0$. The function f is defined on the set of all unordered pairs of distinct elements of γ. For each $U' \in \gamma$, $\{f(U', U''): U'' \in \gamma,\ U'' \neq U'\} \subset \{t \in T: U'_t \neq X_t\} = T'$ is finite. Consequently, the partition $\{f^{-1} t: t \in T\}$ of the set of unordered pairs of distinct elements of γ satisfies the conditions of Prob. 122 of Chap. 1. Hence there are $\gamma^* \subset \gamma$ and $t^* \in T$ such that $|\gamma^*| > \tau$, if $V' \in \gamma^*$, $V'' \in \gamma^*$, $V' \neq V''$, then $f(V', V'') = t^*$. Then $\gamma^*_{t^*} = \{V_{t^*}: V \in \gamma^*\}$ is a family of pairwise disjoint nonempty open sets in X_{t^*} and $|\gamma^*_{t^*}| = |\gamma^*| > \tau$, a contradiction.

396. Let $\{V_x(t): t \in \psi_x\}$ be a family of nhoods of $x \in Y$ such that $\cap\{V_x(t): t \in \psi_x\} = \{x\}$. Write $E_x(t) = \cap\{V_x(s): s < t, s \in \psi_x\} \setminus V_x(t)$ and $U_x(t) = X \setminus E_x(t)$. Evidently, $\{E_x(t): t \in \psi_x\}$ is a family of pairwise disjoint sets; the union of its elements is $X \setminus \{x\}$. In addition, $U_x(t) \supset V_x(t)$ and so $U_x(t)$ is a nhood of x. For $a \in A$ and $x \in Y$ define $t_x(a)$ to be that (unique) element of ψ_x for which $a \in E_x(t_x(a))$. We can now define an inclusion h of the set A in X^* by the following rule: $h(a) = (a, \{t_x(a): x \in Y\})$ for every $a \in A$. Since the points of A are isolated in X, h is indeed an inclusion, that is, a homeomorphim of A onto $h(A)$; $\pi^{-1}(a)$ is a nhood of a for which $\pi^{-1}(a) \cap h(A) = \{h(a)\}$, where $\pi: X^* \to X$ is the natural projection of the product onto the factors. Let π_x be the natural projection $X^* \to \psi_x$. We claim that $h(A)$ is closed in X^*. Let $z \in X^* \setminus h(A)$. There are two possibilities: (i) $\pi(z) \in A$, and (ii) $\pi(z) \in Y$. Consider (i). Put $a = \pi(z)$. There exists $y_0 \in Y$ for which $\pi_{y_0}(z) \neq t_{y_0}(a)$, otherwise we would have $z = h(a) \in h(A)$. But then $\pi^{-1}(a) \cap \pi_{y_0}^{-1}(\pi_{y_0}(z))$ is a nhood of $z \in X^*$ disjoint from $h(A)$.

Consider (ii). Then $\pi(z) = y_0 \in Y$. Put $t_0 = \pi_{y_0}(z)$. The set $\pi^{-1}(U_{y_0}(t)) \subset \pi_{y_0}^{-1}(t_0)$, a nhood of the point $z \in X^*$, has no point in common with the set $h(A)$, since $a \in U_{y_0}(t_0) \cap A$ always implies that $t_{y_0}(a) \neq t_0$, that is, $\pi_{y_0}(h(a)) \neq t_0$. Thus $h(A)$ is closed in X^*.

397. Consider the null function $\theta \in C_\omega[0, 1]$: $\theta(I) = \{0\}$. If the first countability axiom were satisfied in θ, then we would have a countable set $\mathcal{E}$ of elements of a standard subbase of the space C_ω [0, 1] such that the family $\tilde{\mathcal{E}}$ of all finite intersections of members of $\mathcal{E}$ forms a base in θ. By indexing, we may assume that $\mathcal{E} = \{\langle x_i, U_i \rangle : i \in N^+\}$, where $x_i \in I$ and U_i is some nhood of 0 in I. A closed interval is uncountable; fix $x^* \in I \setminus \{x_i : i \in N^+\}$, and put $U^* = [0, \frac{1}{2}) \subset I$ (that is, $U^* = \{x \in I : 0 \leqslant x < \frac{1}{2}\}$). We claim that the nhood $\langle x^*, U^* \rangle$ of θ contains no set belonging to $\mathcal{E}$.

Consider an arbitrary finite collection $\{\langle x_{i_1}, U_{i_1} \rangle, \ldots, \langle x_{i_k}, U_{i_k} \rangle\}$ of elements of $\mathcal{E}$. Since $x^* \notin \{x_{i_1}, \ldots, x_{i_k}\}$, there is a continuous function $f: I \to I$ such that $f(x^*) = 1$ and $f(x_{i_j}) = 0 \in U_{i_j}$ for every $j = 1, \ldots, k$. Then $f \in \bigcap \{\langle x_{i_j}, U_{i_j} \rangle : j = 1, \ldots, k\}$ and $f(x^*) \notin U^*$, that is, $f \notin (x^*, U^*)$. Consequently, $\bigcap \{\langle x_{i_j}, U_{i_j} \rangle : j = 1, \ldots, k\} \notin \langle x^*, U^* \rangle$.

The proof of (*b*) follows from the fact that $|I| = 2^{\aleph_0}$ and the definition of the Tychonoff product of $2^{\aleph_0}$ copies of the interval (the elements of this product are just the functions from a fixed but arbitrary set of cardinality $2^{\aleph_0}$ into I). We may take the closed interval I itself as such a set. The product topology also gets converted exactly into a pointwise topology. In defining $C_\omega [0, 1]$ we only consider the continuous mapping of the interval into itself, that is, we pass to a subspace of the product. Consequently, $C_\omega[0, 1]$ is completely regular (Prob. 38 of Chap. 3). To prove (*c*) it suffices to represent $\{\theta\}$ as an intersection of a countable family of open sets in $C_\omega[0, 1]$.

Let $Q = \{x_i : i \in N^+\}$ be some countable dense set in I. Put $U_j = [0, 1/j], j \in N^+$. Then $\{\theta\} = \bigcap\{U_{i,j} : i, j \in N^+\}$, because if $f \in C_\omega [0, 1]$ and f belongs to the right-hand side, then $f(Q) = \{0\}$ and the continuity of f implies that $f(I) = \{0\}$, that is, $f = \theta$.

398. As in the preceding problem, it can be shown that $C_\omega(X, Y)$ is homeomorphic to a subspace of $Y^{|X|}$. Since the product of every set of regular spaces is regular and a subspace of a regular space is regular, $C_\omega(X, Y)$ and every subspace of it are also regular. Now, endow the set $C[0, 1]$ of all elements of the space $C_\omega(X, Y)$ with the new topology $\mathcal{T}^*$, having a subbase $\mathcal{P}$ consisting of all sets of the form $\langle U, V \rangle$, where $U \in \mathcal{B}_1$ and $V \in \mathcal{B}_2$; $\langle U, V \rangle = \{f \in C[0, 1] : f(U) \subset V\}$ and $\mathcal{B}_1$ and $\mathcal{B}_2$ are arbitrarily fixed countable bases in X and Y, respectively. Then $|\mathcal{P}| \leqslant \aleph_0$ (see Prob. 56 of Chap. 1) and hence $(C[0, 1], \mathcal{T}^*)$ has a countable base. To verify that $\mathcal{T} \subset \mathcal{T}^*$ it suffices to show that $\langle X, V \rangle \in \mathcal{T}^*$ for each $V \in \mathcal{B}_2$. Let $f \in \langle x, V \rangle$. Then $f(x) \in V$. Since $f: X \to Y$ is continuous, there is a nhood Ox of x for which $f(Ox) \subset V$. Then $f \in \langle Ox, V \rangle \subset \langle x, V \rangle$. Hence $\langle x, V \rangle \in \mathcal{T}^*$ and $\mathcal{T}^* \supset \mathcal{T}$. But then $C_\omega(X, Y)$ as a subspace of a space with a countable base also has a countable net. However, every space with a countable net is hereditarily Lindelöf and hereditarily separable (Probs. 139, 143 and 96).

Chapter 3

Compact Spaces and their Subspaces. Concepts Related to Compactness

3.1. Functional Separation. Completely Regular and Normal Spaces

We shall discuss *functional separation*, or as we shall call it, *α-separation*, a form of separation which is of crucial importance in topology and its applications. We say that a subset A of a topological space X is *α-separable* from a subset B of X if there is a continuous real-valued function f on X such that $f(A) = \{0\}$ and $f(B) = \{1\}$. A family ξ of subsets in X is called an *α-family* if for each $C \in \xi$ there exists $B \in \xi$ such that B is α-separable from $A = X \setminus C$ (in addition, evidently, $B \subset C$). If $A \subset B$ and A is α-separable from $X \setminus B$, then B is called an *α-nhood* of A. By an *α-filter* (*α-prefilter*) we shall mean a filter (prefilter) which is also an α-family. An *α-ultrafilter* is an α-filter which is contained in a unique α-filter. An α-ultrafilter is not necessarily an ultrafilter. The definition of the former can be related to a subfamily of a fixed family $\mathcal{E}$ of sets (as on p. 7). Thus, with respect to the family of all open sets in X (playing the role of $\mathcal{E}$) we define *open α-filter*, *open α-ultrafilter* and so on.†

An important role is played by the notion of separation of a point from a set by a collection of functions. We say that a collection $\mathcal{F}$ of functions of X into some topological space *separates* a point $x \in X$ from a set $A \subset X$, if there exists $f \in \mathcal{F}$, $f: X \to Y_f$ such that $f(x) \notin [f(A)]_{Y_f}$.

Furthermore, $\mathcal{F}$ is called *regular* if every $f \in \mathcal{F}$ is continuous and $\mathcal{F}$ separates each point x of the set X from each set P not containing the closure of the point x in X. Also, if every $f \in \mathcal{F}$ is a real-valued function on X, we say that $\mathcal{F}$ is a *regular* (*or partitioning*) collection of functions on X. If on a T_1-space X there exists a collection of regular functions, X is called a *completely regular* space (see also p. 43).

When considering continuous real-valued functions on a topological space, we naturally single out in this space subsets of a certain special

†All these notions are dealt with in Alexandroff [1].

form, namely, the so-called *zero-set* or *cozero-set*. We call $A \subset X$ a *zero-set* if there is a continuous real-valued function f on X such that $A = f^{-1}(0)$. $B \subset X$ is called a *cozero-set* if $X \setminus B$ is a zero-set.

An *exterior pseudobase* for a set A in a topological space X containing A is a family γ of open sets in X, satisfying $\{x\} = \bigcap\{U\colon U \in \gamma \text{ and } x \in U\}$ for each $x \in X$.

Finally, a particularly crucial role is played by the *Tychonoff cube* I^τ, τ being any cardinal number and I^τ the Tychonoff (topological) product of τ copies of an interval. Obviously, when τ is a finite cardinal (that is, a natural number), I^τ is the usual cube in a Euclidean τ-space.

1. Show that functionally separated sets have a disjoint α-nhood.

2. Show that the family ξ_A of all α-nhoods of a set A in X is an α-family on X.

3. Show that each centralized α-family in a space X is contained in some α-ultrafilter on X.

4. Show that every α-filter is a filter.

5. Let β be an α-ultrafilter on X, A a subset of X and ξ the family of all its α-nhoods in X. Then if $A \cap U \neq \Lambda$ for each $U \in \beta$, show that $\xi \subset \beta$.

6. If β_1 and β_2 are α-filters on X and $U_1 \cap U_2 \neq \Lambda$ for every $U_1 \in \beta_1$, $U_2 \in \beta_2$, then $\beta_1 \cup \beta_2$ is also an α-family,

7. A T_1-space X is completely regular iff for each point $x \in X$ the family α of all open sets in X containing this point is an α-filter.

8. Show that each subspace of a completely regular space is completely regular.

9. Give an example of a space that is regular but not completely regular.

10. Let X be a completely regular space, $F \subset X$, $Y = X \setminus F$, β an α-ultrafilter on X and $\beta' = \{U \cap Y\colon U \in \beta\}$. Then, only one of the following assertions holds: (i) β' is an α-ultrafilter on Y; (ii) each element of β intersects F.

11. Show that $X^* \times X^*$ is not normal (here X^* refers to the space mentioned in Problem 104, Chap. 2), but is completely regular.

12. Let X be a completely regular space, F a compact subspace, $F \subset X$ and P a closed set in X not intersecting F. Then there is a continuous real-valued bounded function f on X such that $f(F) = 0$ and $f(P) = 1$.

13. Let X be the set of points of a half-plane bounded by a straight line l and let $Y = X \setminus l$. Let $O(X, r)$, r a positive real number and $x \in l$, be the *interior of the circle* with radius r lying in Y and touching l at α.

Put $\tilde{O}(x, r) = O(x, r) \cup \{x\}$, $\mathcal{L} = \{\tilde{O}(x, r): x \in l, r > 0\}$; let $\mathcal{D}$ be the family of all sets open in the usual topology of the plane and lying in Y. Show that: (i) $\mathcal{B} = \mathcal{L} \cup \mathcal{D}$ is a base for a topology $\mathcal{T}$ on X which includes the usual topology; (ii) l is a discrete subspace of X; (iii) the topology induced on $(X, \mathcal{T})$ by Y is equivalent to the usual topology of the plane induced by Y; (iv) X is the sum of two metrizable subspaces, of which one is closed and the other open, and is an F_σ-set; (v) X is completely regular; (vi) X is not normal (*a Nemitskii space*).

14. Show that every space X possessing a base $\mathcal{B}$ of clopen sets is completely regular.

15. Show that every T_1-space X, with exactly one nonisolated point x_0, is normal.

16. If X is a Hausdorff space and the set of all nonisolated points in X is finite, show that X is normal.

17. True or false? Every countable Hausdorff space is normal.

18. True or false? Every countable regular space is normal.

19. A regular (Lindelöf) space X is normal. Show that if X is regular, $A_1 \subset X$, $A_2 \subset X$ are Lindelöf subspaces, $A_1 \cap [A_2] = \Lambda$, $A_2 \cap [A_1] = \Lambda$, then A_1 and A_2 have disjoint nhoods.

20. Show that a regular space X with a countable base is perfectly normal (see p. 44).

21. If X is a normal space and $\lambda = \{U_1, \ldots, U_k\}$ is an arbitrary finite open cover of X, show that there exist closed sets $F_1, \ldots, F_k$ such that $\bigcup\{F_i: i = 1, \ldots, k\} = X$ and $F_i \subset U_i$, $i = 1, \ldots, k$.

22. If $A = \{x_n: n \in N^+\}$ is a closed countable discrete subspace of a normal space X, and $x_{n'} \neq x_{n''}$ if $n' \neq n''$, show that there exists a family $\{U_n: n \in N^+\}$ of open sets in X such that $x_n \in U_n$ for every $n \in N^+$ and, moreover, each point $x \in X$ has a nhood intersecting not more than one member of this family.

23. If a Tychonoff product $X \times Y$ of spaces X and Y is hereditarily normal, show that either Y is perfectly normal or each countable subset of X is closed.

24. Let X be a normal space and U_0, U_1 two open sets of X such that $[U_0] \subset U_1$.

(*a*) Construct for each binary rational number $r = q/2^n$, $n = 1, 2, \ldots$, $q = 1, 2, \ldots, 2^n$, an open set U_r such that $[U_{r'}] \subset U_r$, whenever $r' < r$.

(*b*) Show that the open sets $U_t = \cup \{U_r: r \leqslant t\}$ if $t \in [0, 1]$, and $U_t = X$ if $t > 1$, have the property: $[U_{t'}] \subset U_t$, whenever $t' < t$.

(*c*) Show that the function $f(x) = \inf \{t: x \in U_t\}$ is continuous on X and vanishes on the set $[U_0]$.

25. Show that X is a normal space iff for every pair of disjoint closed sets F_0 and F_1 in X there is a continuous function $f(x)$ on X such that: $|f(x)| \leqslant 1, f(x) = 0$ if $x \in F_0$, and $f(x) = 1$ if $x \in F_1$.

26. Show that every normal space is completely regular.

27. Show that the following assertions about a space X are equivalent:

(*a*) each subspace of X is normal;

(*b*) each open subspace of X is normal.

28. If X is a normal space and Y an F_σ-subspace of X (that is, $Y = \cup \{F_i: i \in N^+\}$, where each F_i is closed in X), then Y is also normal.

29. Each subspace of a perfectly normal space X is normal.

30. Show that a closed subset F of a normal space X is a G_δ-set iff it is a zero-set.

31. Let X be a normal space, A and B disjoint closed sets in X and c a positive real number. Show that there is a continuous real-valued function defined on X such that $f(A) = \{-c\}, f(B) = \{c\}$, and $f(X) \subset [-c, c]$.

32. Let X be a normal space, A a closed subspace of X, c a positive real number and f_0 a continuous real-valued function defined on A such that $|f_0(x)| \leqslant c$ for every $x \in A$. Show that there exists a real-valued function g, defined and continuous on X such that $|g(x)| < c/3$ for every $x \in X$ and $|f_0(x) - g(x)| \leqslant 2c/3$ for every $x \in A$.

33. Show that every bounded continuous real-valued function f defined on a closed subset F of a normal space X can be extended to a continuous function on X.

34. If X is a normal space and A a closed subspace of X, show that every continuous mapping f of A into R, the usual space of real numbers, can be extended to a continuous mapping $\tilde{f}$ of X into R.

35. Show that the product $X = \Pi\{X_\alpha: \alpha \in A\}$ of completely regular spaces is regular.

36. Show that a completely regular space X of weight τ has a partitioning family of continuous real-valued bounded functions of cardinality not exceeding τ.

37. Given a space X, T_1-spaces $\{Y_\alpha: \alpha \in A\}$, and a continuous mapping $f_\alpha: X \to Y_\alpha$ for each $\alpha \in A$. Show that whenever a system of mappings $f_\alpha : X \to Y_\alpha$ forms a partitioning, the diagonal product $f: X \to \Pi\{Y_\alpha: \alpha \in A\}$ is a homeomorphism.

38. Show that a space X is completely regular iff it is homeomorphic to a subspace of some Tychonoff cube.

39. Show that a regular space with a countable base is metrizable (*Urysohn's lemma*).

40. Let X be a completely regular space and σ a family of cozero-sets in X of cardinality $|\sigma| \leqslant \tau$. Show that there exists a space Y_σ of weight $wY_\sigma \leqslant \tau$, and a continuous map $f_\sigma : X \xrightarrow{\text{onto}} Y_\sigma$, for which each $U \in \sigma$ is saturated, that is, $f_\sigma U$ is open in Y_σ and $f_\sigma^{-1} f_\sigma U = U$ if $U \in \sigma$.

41. Let X be a completely regular space, $M \subset X$ a subset and σ a family of cozero-sets of X forming an exterior pseudobase for M (see p. 122). Let $|\sigma| \leqslant \tau$. Show that there exists a space Y_M, $wY_M \leqslant \tau$, and a continuous mapping $f_M : X \xrightarrow{\text{onto}} Y_M$, satisfying $M = f_M^{-1} f_M M$ and $f_M^{-1} f_M x = x$ if $x \in M$.

3.2 COMPACTNESS

A point x of a space X is called a *complete accumulation point* of a set $A \subset X$, if $|Ox \cap A| = |A|$ for each nhood Ox of x.

Of great interest in the study of compact (and Lindelöf) spaces is the notion of *free sequence*. A well-ordered set $(A, <)$ of points of X is called a *free sequence* if

$$[\{x \in A: x < y\}] \cap [\{x \in A : x \geqslant y\}] = \Lambda, \text{ and}$$
$$|\{x \in A: x < y\}| < |A|$$

for each $y \in A$. The power of A is called the *length of the free sequence* $(A, <)$.

The *dyadic compacta* form a very interesting family of compact spaces. These are Hausdorff spaces which are the continuous images of generalized Cantor discontinua. The reasons they are considered are not only their strange properties but also the fact (to be dealt with in Sec. 4) that *each* sequentially compact space (that is, a metrizable compact space) is a continuous image of Cantor's perfect set, that is, the simplest of all discontinua. Thus, the notion of dyadic compactness is an extension of the concept of sequential compactness.

A point x of a space X is called a *point of local compactness* if there exists an open set U in X and a compact subspace Φ of X such that $x \in U \subset \Phi$. We denote by $G(X)$ the set of all points of locally compactness in a space X and by $R(X)$ the set $X \setminus G(X)$. It is clear that $G(X)$ is an open set and $R(X)$ is a closed set in X. The space X is *locally compact* if $G(X) = X$.

We know that every subset of a topological space becomes a topological space in a quite natural fashion (see p. 41). However, of no less importance and interest is a topological space contained as a subspace of a given space. A space Y is called an *extension* of a space X if X is a dense subspace of Y (or if there is a topological inclusion $f: X \to Y$ such that the image $f(X)$ of X under f is dense in Y). If an extension Y of X is compact, then it is called a *compactification*. An extension of X which is a Hausdorff space is called a *Hausdorff extension*.

A space X is said to be *Čech-complete* if it is a G_δ-set in some Hausdorff compactification. Closely related to the notion of compactness is the notion of *absolute closedness* or *H-closedness*.

A Hausdorff space X is called *H-closed* (*absolutely closed*), if under every topological inclusion f of X into an arbitrary Hausdorff space Y the image $f(X)$ is closed in Y. This may be also expressed thus: A Hausdorff space is H-closed, if it is closed in every extension of it to a Hausdorff space.

Hausdorff's separation axioms are sharpened by Urysohn's separation axiom: we say that X satisfies *Urysohn's separation axiom* if for every pair of distinct points x_1 and x_2 of X there exist nhoods Ux_1 and Ux_2 such that $[Ux_1] \cap [Ux_2] = \Lambda$. This axiom is obviously satisfied by every regular space; the converse is false. Another, weaker, kind of regularity is the so-called *half-regularity* (or *semi-regularity*): A space X is semi-regular, if the regular open sets (see p. 48) in it form a base.

Also closely related to the notion of compactness is the notion of non-contractability. A Hausdorff space X is called a *non-contractible* or *minimal Hausdorff space*, if every one-to-one continuous map from it onto a Hausdorff space is a homeomorphism.

Finally, in this section we introduce the following notion. A space X is said to be (*topologically*) *homogeneous* if for each pair of its points x and y there exists a homeomorphism f of X onto itself such that $f(x) = y$. For example, any generalized Cantor discontinuum (see p. 86), including the usual Cantor perfect set (see Problem 185, Chap. 2), is topologically homogeneous.

42. Let X be an infinite set endowed with the topology $\mathcal{T}_k$: $U \in \mathcal{T}_k$ iff $U \subset X$ and either $U = \Lambda$ or $X \setminus U$ is a finite set (see Problem 2, Chap. 2).

Show that

(*a*) each subspace of $(X, \mathcal{T}_k)$ is compact;
(*b*) each infinite $M \subset X$ is dense in $(X, \mathcal{T}_k)$.

43. If a compact space is discrete, show that it is finite.

44. For X to be a compact space, show that it is (necessary and) sufficient that from every cover of X by elements of some fixed base one can select a finite cover.

45. Show that X is a compact space iff every centralized family of closed sets (a system of sets with finite intersection property) has a nonempty intersection.

46. Show that X is a compact space iff every centralized family of any sets has an adherent (a closure) point (see p. 40).

47. Show that X is compact iff every maximal family of subsets with the finite intersection property has an adherent point in X.

48. Show that X is compact iff every infinite set $M \subset X$ has a complete accumulation point in X.

49. Show that in every infinite compact space X there is a countable nonclosed subset.

50. Let X be a completely regular space, A a subset of X, and suppose that every infinite subset C of A has an accumulation point in X. Is $[A]_X$ compact?

51. Let X be a completely regular space, A a subset of X and suppose that every centralized family ξ of subsets of A has an adherent point in X. Does this necessarily imply that $[A]_X$ is compact (cf. preceding problem)?

52. If a space X has a subbase $\mathcal{B}$ such that every cover of X by elements of $\mathcal{B}$ has a finite subcover, show that X is compact.

53. Let $\mathcal{F}(X)$ be the set of all nonempty closed subsets of a topological space X. If $A, B \in \mathcal{F}(X)$, put $\mathcal{K}(A, B) = \{C \in \mathcal{F}(X): C \cap A \neq \Lambda$ and $C \subset B\}$ and $\mathcal{P} = \{\mathcal{K}(A, B): A \in \mathcal{F}(X), B \in \mathcal{F}(X)\}$.

Show that if X is compact, then $\mathcal{F}(X)$, endowed with the topology for which $\mathcal{P}$ serves† as a subbase for the closed sets, is compact.

54. Show that a closed subspace of a compact space is compact.

55. Let X be a Hausdorff space, F a compact subspace of X and $x_0 \in X$ a point not belonging to F. Show that there exist nhoods UF of F in X and Ux_0 of $x_0 \in X$, such that $UF \cap Ux_0 = \Lambda$.

†A family $\mathcal{P}$ of sets in X forms a subbase for the closed sets in X if the family $\{X \setminus U: U \in \mathcal{P}\}$ is a subbase for the topology of X (see p. 40).

56. Show that every compact subspace of a Hausdorff space X is closed in X.

57. Show that a nonclosed subspace of a Hausdorff space is not compact.

58. Show that for every nonisolated point x of a compact space X, there exists a set $A \subset X$ such that x is the unique accumulation point of A.

59. Show that every regular noncompact space $(X, \mathfrak{I})$ can be represented as a nonclosed subspace of some Hausdorff space.

60. Show that any regular H-closed space is compact.

61. Show that a Hausdorff space X is H-closed iff every open cover ω of X has a finite subfamily $\omega' \subset \omega$, the closure of whose elements covers X.

62. Show that a regular closed subset $[U]_X$ of an H-closed space X is an H-closed subspace.

63. Let X be a Hausdorff space, each of whose closed subspaces is H-closed. Show that X is compact.

64. Show that a compact Hausdorff space is normal.

65. Let X be a topological space, Φ a compact space, $\Phi \subset X$ and $\{U_n: n \in N^+\}$ a defining system of nhoods of Φ in X. Fix any $x_n \in U_n$. Show that any subsequence $\{x_{n_k}: k \in N^+\}$ of $\{x_n: n \in N^+\}$ has a limit point in Φ.

66. Let X be a compact space and ξ a prefilter consisting of closed subsets of X. Show that, for any open set U in X, containing the set $\Phi = \bigcap \{P: P \in \xi\}$, there exists $P_0 \in \xi$, satisfying $P_0 \subset U$.

67. Let X be compact and ξ a centralized family of nonempty closed sets in X. Show that for every nhood OF_0 of $F_0 = \bigcap \{F: F \in \xi\}$ there is a finite collection of elements $F_1, \ldots, F_s \in \xi$ such that $F_0 \subset \bigcap \{F_i: i = 1, \ldots, s\} \subset OF_0$.

68. Show that in a compact space X the pseudocharacter $\psi(F, X)$ of any closed set F is always equal to its character $\chi(F, X)$ (in particular, the pseudocharacter and character of each point $x \in X$ are the same).

69. Let X be compact, Φ be closed in X, $\Phi \neq \Lambda$ and $U \supset \Phi$, U open in X. Show that there exists a compact set F of countable character in X such that $\Phi \subset F \subset U$.

70. Let X be a Hausdorff space, Φ and F compact subspaces of X, τ a cardinal number, $\tau \geqslant \aleph_0$, $F \subset \Phi$, the character of F in Φ does not exceed τ and suppose that the character of Φ in X does not exceed τ (that is, $\chi(F, \Phi) \leqslant \tau$, $\chi(\Phi, X) \leqslant \tau$). Show that the character of F in X also does not exceed τ.

71. Show that, if X is compact, τ is a cardinal number and Q is a G_τ-

set in X, then for each $x \in Q$ there is a compact $\Phi \subset X$ such that $x \in \Phi \subset Q$ and $\chi(\Phi, X) \leqslant \tau$.

72. Show that, if X is first countable and F a compact countable subspace of X, then the character of F in X does not exceed $\aleph_0$.

73. Show that, if X is a compact T_1-space and F a closed set in X, then there exists a family γ of open sets in X such that $|\gamma| \leqslant \tau$, where τ is the weight of X and $F = \cap \{U: U \in \gamma\}$.

74. In every topological space not necessarily Hausdorff, is the intersection of any two compact subspaces a compact subspace?

75. Show that the topological product $X = \Pi\{X_\alpha: \alpha \in A\}$ of a family $\{X_\alpha: \alpha \in A\}$ of compact spaces is also compact (*Tychonoff's first theorem.*)

76. If A and B are compact subspaces of topological spaces X and Y and W is any nhood of a set $A \times B$ in the product $X \times Y$, show that there exist nhoods U of A in X and V of B in Y such that $U \times V \subset W$.

77. Let $\{X_\alpha: \alpha \in A\}$ be a family of topological spaces, Φ_α a compact subspace of X_α for each $\alpha \in A$, $\Phi = \Pi\{\Phi_\alpha: \alpha \in A\}$ a subspace of the product $X = \Pi\{X_\alpha: \alpha \in A\}$ and W a nhood of the set Φ in X. Show that there exist $\alpha_1, \ldots, \alpha_k \in A$ and an open set U_{α_i} in X_{α_i} containing Φ_{α_i} such that $\Pi\{U_{\alpha_i}: i = 1, \ldots, k\} \times \Pi\{X_\alpha: \alpha \in A \setminus \{\alpha_1, \ldots, \alpha_k\}\} \subset W$.

78. Let F be compact, $\mathcal{E} = \{f_\alpha: \alpha \in A\}$ a family of continuous functions from F into the unit segment $I = [0, 1]$ of real numbers, $J = (0,1] = [0, 1] \setminus \{0\}$ a half open interval, $X = \{x \in F: f_\alpha(x) \neq 0$ for every $\alpha \in A\}$ and $f: F \to I^{|A|}$ the function defined by the rule : $f(x) = \{f_\alpha(x): \alpha \in A\}$ for every $x \in F$. Show that

(i) $f^{-1}(J^{|A|}) = X$;
(ii) $f(X)$ is a closed set in $J^{|A|}$;
(iii) the function $\varphi: X \to J^{|A|}$, where $\varphi(x) = f(x)$ for every $x \in X$, is closed (and continuous).

79. Let F be compact and $\{\Phi_\alpha: \alpha \in A\}$ a compact family in F. Assume that $\chi(\Phi_\alpha, F) \leqslant \aleph_0$ for every $\alpha \in A$. Show that the space $X = F \setminus \cup \{\Phi_\alpha: \alpha \in A\}$ is homeomorphic to a closed subspace of the Tychonoff product $R^{|A| \cdot \tau}$ where R is the real line and τ is the weight of X.

80. (*Alexandroff's double circle*). The points of the space X are the points of two concentric circles: S_0 (the inner one) and S_1 (the outer one). A nhood of the point $x \in S_0$ is the union of an interval containing this point of S_0 and a central projection of this interval onto S_1, from which is removed the projection of x. All points of S_1 are declared to be isolated.

Show that X, a compact Hausdorff space without a countable base, is first countable at all points and is the union (not disjoint) of two of its metrizable subspaces.

81. Let X' and X'' be two replicas of the same compact space X. Consider the set $A(X) = X' \cup X''$. A point $x \in X$, considered as an element in X' (resp. X''), we call x' (resp. x''). Similarly, a set $M \subset X$ is denoted by M' if it is considered in the replica X' and by M'' if in X''. Introduce in $A(X)$ a topology along the lines of Problem 80. The singletons $\{x''\}, x'' \in X''$, and sets of the form $(U' \cup U'')\setminus\{x''\}$, where U is any nhood of a point x of the compact space X, form a base for the topology.

Prove the following successively:

(*a*) $A(X)$ is Hausdorff;
(*b*) X' with the topology inherited from $A(X)$ is homeomorphic to X and hence is compact and closed in $A(X)$;
(*c*) X'' is a discrete subspace, open in $A(X)$;
(*d*) $A(X)$ is compact.

82. (*Kelley's space*). Let X be the set of all nondecreasing functions from the unit segment $[0, 1]$ into itself (that is, $f \in X \leftrightarrow f(x) \geqslant f(y)$ for $x \geqslant y$). Endow X with the pointwise topology (see p. 86), which is the same as the topology induced in the topological product $[0, 1]^{|[0,1]|}$.

Show that X is nonmetrizable, separable, compact and first countable. Another example of such a compact space is given in Problem 87.

83. Show that every completely ordered set $(X, <)$, having a largest element x_m and endowed with an order topology, is compact.

84. Let $T(\omega_1)$ be a minimal completely ordered set of cardinality $\aleph_1$ endowed with the order topology, $\alpha \in T(\omega_1)$ and let $\Phi_\alpha = \{x \in T(\omega_1): x \leqslant \alpha\}$. Show that Φ_α, as a subspace of $T(\omega_1)$, is compact.

85. Let $(X, <)$ be a linearly ordered space with the order topology. Show that the following are equivalent:

(*a*) X is compact;
(*b*) every nonempty set B closed in X has a least and a largest element;
(*c*) for each nonempty $A \subset X$, sup A and inf A exist in $(X, <)$.

86. Show that every compact subspace of a star-space (see Problem 104, Chap. 2) is countable.

87. *An example* (*Alexandroff's two-star space*). Consider two half-open intervals $X = [0, 1)$ and $X' = (0, 1]$, one lying under the other. Let X^{**}

be the set of all points of these two intervals. Topologize X^{**} as follows: a base for the topology is composed of all sets of the form

$$\Gamma_1 = [\alpha, \beta) \cup (\alpha', \beta'), \quad \Gamma_2 = (\alpha, \beta) \cup (\alpha', \beta'].$$

Here $[\alpha, \beta)$ (resp. $(\alpha', \beta']$) is a half-open interval in X (resp. X') and (α', β') [resp. (α, β)] is the projection of (α, β) [resp. (α', β')] onto X' (resp. onto X). Show that the resulting topological space X^{**} satisfies the following:

(i) X^{**} is Hausdorff;
(ii) X^{**} is compact;
(iii) X^{**} admits a perfect mapping onto a real line segment;
(iv) X^{**} is hereditarily Lindelöf;
(v) X^{**} is perfectly normal;
(vi) X^{**} is first countable;
(vii) X^{**} is separable;
(viii) X^{**} has no countable base;
(ix) X^{**} is non-metrizable;
(x) every metrizable subspace of X^{**} is at most countable;
(xi) the square of X^{**} is non-hereditarily normal.

88. Show that every compact space has Baire's property (see p. 62).

89. If X is nonempty and compact and $|X| \leqslant \aleph_0$, show that X has an isolated point.

90. Show that every completely regular Čech-complete space X has Baire's property.

91. Let $f: X \to Y$ be a continuous function from a compact space X onto a space Y. Show that Y is compact.

92. Let $f: X \to Y$ be a continuous function from a compact space into a Hausdorff space Y. Show that f is closed. A decomposition of a compact space X is continuous iff a space of this decomposition is Hausdorff (a projection onto a decomposition space is then closed).

93. Show that a one-to-one continuous mapping f of a compactum X onto a Hausdorff Y is a homeomorphism.

94. Let $f: X \to Y$ be a function from X onto Y, X and Y compacta. Show that f is continuous iff the graph of f is closed in the product $X \times Y$.

95. Let $(Y, \mathcal{T})$ be any compact T_1-space. Is there always a topology $\tilde{\mathcal{T}}$ on Y such that $(Y, \tilde{\mathcal{T}})$ is compact Hausdorff and $\mathcal{T} \subset \tilde{\mathcal{T}}$?

96. Is a one-to-one continuous mapping of one compact space onto another always a homeomorphism?

97. (*a*) Show that every continuous real-valued function on a compact space X is bounded and attains its supremum and infimum.

(*b*) Is there a non-compact normal space on which every continuous real-valued function is bounded?

98. If the compact Hausdorff spaces X and Y are equipotent and have precisely one isolated point, show that X and Y are homeomorphic.

99. If X is compact Hausdorff and only one point x_0 is non-isolated, show that a set $U \subset X$ containing x_0 is open iff $X \setminus U$ is finite.

100. Show that the image of an H-closed space under a continuous mapping is H-closed.

101. Show that a Hausdorff non-contractible space X satisfying the first separation axiom is compact.

102. Show that for a Hausdorff space X to be non-contractible it is necessary and sufficient that X be H-closed and semi-regular.

103. Let X be compact Hausdorff and first countable and let $D \subset X$ be any countable subset of X. Is there always a contraction from the subspace $X_0 = X \setminus D$ onto some compact Hausdorff space?

104. Let X consist of all points (x, y) of the Euclidean plane whose coordinates satisfy the condition $0 < x^2 + y^2 \leqslant 1$, together with a pair of points p^+ and p^-. For (x, y) we have the same nhoods as they have in the Euclidean plane. The nhoods of p^+ and p^- are given by

$$U_n p^+ = \{p^+\} \cup \left\{(x, y) \in X \colon 0 < x^2 + y^2 < \frac{1}{n}, y > 0\right\},$$

$$U_n p^- = \{p^-\} \cup \left\{(x, y) \colon 0 < x^2 + y^2 < \frac{1}{n}, y < 0\right\}.$$

By definition all the nhoods considered form a base for a topology on X. Show that X with this topology is Hausdorff, non-contractible and non-compact.

105. Show that a countably compact Hausdorff space has a countable base and hence is metrizable.

106. If in a compact Hausdorff space X there is a net of cardinality not exceeding τ, show that it has also a base of cardinality not greater than τ.

107. Show that the weight of a compact Hausdorff space never exceeds its cardinality.

108. If X is an infinite compact Hausdorff space and $X = X_1 \cup X_2$, show that $wX = \max\{wX_1, wX_2\}$.

109. If X is compact Hausdorff, τ a cardinal number, $\tau \geqslant \aleph_0$ and $X = \bigcup \{X_\alpha: \alpha \in M\}$, where $wX_\alpha \leqslant \tau$ for each $\alpha \in M$ and $|M| \leqslant \tau$, show that $wX \leqslant \tau$.

110. Does there exist a countably compact T_1-space without a countable base?

111. If a compact Hausdorff space Y is the image of a space X with a countable base under a continuous mapping, show that $wY \leqslant \aleph_0$.

112. If $f: X \to Y$ is continuous, $f(X) = Y$ and Y is compact Hausdorff, show that $wY \leqslant wX$.

113. Show that under a continuous mapping of a compact Hausdorff X onto a Hausdorff Y weight is not increased.

114. If X is compact Hausdorff and the diagonal Δ in $X \times X$ is a G_δ-set, show that X has a countable base.

115. Is every separable compactum hereditarily separable?

116. Suppose that X is compact Hausdorff, τ a cardinal number, and $s(F) \leqslant \tau$ for each closed subspace F of X. Show that $s(Y) \leqslant \tau$ for each $Y \subset X$.

117. Can the cardinality of a hereditarily separable compactum exceed $2^{\aleph_0}$?

118. If the cardinality of a compactum exceeds $2^{\aleph_0}$, then does this compactum necessarily have a discrete subspace of non-countable cardinality?

119. Show that there exists a compactum that is separable and not hereditarily normal.

120. If X is a compact T_1-space, τ is a cardinal number, $\tau \geqslant \aleph_0$ and $\mathscr{B}$ is a base for X such that $|\mathscr{B}_x| \leqslant \tau$ for every $x \in X$, where $\mathscr{B}_x = \{U \in \mathscr{B}: U \ni x\}$, show that $|\mathscr{B}| \leqslant \tau$.

121. Let X be compact Hausdorff, τ a cardinal number, $\tau \geqslant \aleph_0$ and $\mathscr{E}$ a family of open sets in X such that for every $x \in X$ we have $|\mathscr{E}_x| \leqslant \tau$ and $\{x\} = \bigcap\{U: U \in \mathscr{E}_x\}$, where $\mathscr{E}_x = \{U \in \mathscr{E}: U \ni x\}$. Show that $|\mathscr{E}| \leqslant \tau$.

122. Let X be compact Hausdorff and $\mathscr{B}$ a pseudo-base for X (that is, $\{x\} = \bigcap\{U \in \mathscr{B}: U \ni x\}$ for every $x \in X$). Show that X also has a base $\mathscr{B}^*$ satisfying $|\mathscr{B}^*| \leqslant |\mathscr{B}|$.

123. If a compact Hausdorff X has a pseudo-base $\mathscr{B}$ (see preceding problem) satisfying $|\mathscr{B}_x| \leqslant \tau$ for every $x \in X$ (τ a cardinal number, $\tau \geqslant \aleph_0$ and $\mathscr{B}_x = \{U \in \mathscr{B}: U \ni x\}$), show that the weight of X does not exceed τ.

124. Show that the weight of any 0-dimensional compactum X is the cardinality of the family of all clopen sets in X.

125. Show that in every compact Hausdorff space F without isolated points there is a pair of disjoint closed subspaces F_0 and F_1 which also have no isolated points.

126. If a compactum X has no isolated points, show that the cardinality of X is not less than $2^{\aleph_0}$.

127. Assuming $\aleph_1 = 2^{\aleph_0}$, show that every compactum of cardinality not exceeding $2^{\aleph_0}$ is first countable at no more than one point.

128. If a compactum X is first countable (at every point), show that either $|X| \leqslant \aleph_0$, or $|X| \geqslant 2^{\aleph_0}$.

129. Show that every compact T_1-space X of weight not exceeding τ, $\tau \geqslant \aleph_0$, has a net of cardinality not exceeding τ, all of whose elements are closed sets.

130. Show that a compact T_1-space X with a countable base is either countable or has cardinality $2^{\aleph_0}$.

131. Give an example of a compactum S with the following properties:

(*a*) S has no isolated points;
(*b*) S has a countable π-base;
(*c*) in S there are no nontrivial convergent sequences.

132. Give an example of a compactum X, in which there are no nontrivial convergent sequences, but there is a sequence $\{A_i: i \in N^+\}$ of finite sets satisfying

(i) $A_{i'} \cap A_{i''} = \Lambda$, $i' \neq i''$, and
(ii) for each nonempty open set V in X there exists $k \in N^+$ such that $V \cap X_i \neq \Lambda$ for every $i > k$.

133. Show that for every cardinal number $\tau > \aleph_0$ there is a sequential (see p. 44) compactum B_τ of cardinality τ which is first countable at all points except one.

134. Is every sequential compactum a Fréchet-Urysohn space (see p. 44)?

135. Is every compactum of countable tightness a sequential space?

136. Does every infinite compactum of countable tightness contain a nontrivial convergent sequence?

137. If a topology on a compactum X is generated by a semi-metric d, show that X is first countable.

138. If X is compact Hausdorff and its topology is generated by a semi-metric d, show that X has a countable net.

139. Show that every semi-metrizable compactum is metrizable.

140. Show that the length of any free sequence in a compactum X does not exceed its tightness $t(X)$.

141. Let X be compact, $A \subset X$, τ a cardinal number and let $A = \bigcup\{[B]: B \subset A \text{ and } |B| \leqslant \tau\} \neq [A]$. Show that

(i) if $x \in [A] \cap Q$ and Q is a G_τ-set in X (see p. 43), then $Q \cap A \neq \Lambda$;

(ii) in X there is a free sequence of length τ^+, all of whose elements are contained in A.

142. Show that the tightness $t(X)$ of an arbitrary compactum X is the least upper bound (lub) of the length of a free sequence in it.

143. If X is sequentially compact and U a nonempty open set in X, show that there is a set F which is closed and G_δ in X, has cardinality not exceeding $2^{\aleph_0}$ and is included in U.

144. Show that every sequentially compact X has character not exceeding $2^{\aleph_0}$ on a dense set of points.

145. If $\aleph_1 = 2^{\aleph_0}$, show that every sequentially compact X is first countable on a dense set of points.

146. Assume that $2^{\aleph_0} = \aleph_1$. Then every compact countable tightness is first countable on a dense set of points. True or false?

147. Assume that $2^{\aleph_0} = \aleph_1$. Then every hereditarily separable compactum has a countable π-base. True or false?

148. If $\mathcal{F}$ is a family of pairwise disjoint and nonempty compacta of countable character in a Hausdorff space X and if X satisfies the Souslin hypothesis, show that $|\mathcal{F}| \leqslant 2^{\aleph_0}$.

149. If a sequential compactum X satisfies the Souslin hypothesis, show that $|X| \leqslant 2^{\aleph_0}$.

150. Let X, a compact Hausdorff space, satisfy the Souslin hypothesis and have countable tightness. Is $|X| \leqslant 2^{\aleph_0}$?

151. If a compactum X is a sequential space and the character of X at every point does not exceed $2^{\aleph_0}$, show that $|X| \leqslant 2^{\aleph_0}$.

152. If a homogeneous compactum X is sequential, show that either it is finite or its cardinality is $2^{\aleph_0}$.

153. If X is compact, τ a cardinal number and $\chi(x, X) \leqslant \tau$ for every $x \in X$, show that $|X| \leqslant 2^{\tau}$.

154. If X is a homogeneous compactum and $\tau = |X|$, show that there is $\tau' < \tau$ satisfying $\tau \leqslant 2^{\tau'}$.

155. By the generalized continuum hypothesis, $2^{\tau} = \tau^{+}$ for every cardinal number $\tau \geqslant \aleph_0$. Show that there is no homogeneous compactum whose cardinality is the limit of τ.

156. The cardinal number τ is strongly inaccessible (that is, $cf(\tau) = \tau$ and $2^{\tau'} < \tau$ whenever $\tau' < \tau$) and X is a compactum such that $\chi(x, X) < \tau$ for every $x \in X$. Show that $|X| \leqslant \tau$.

157. Assuming the validity of generalized continuum hypothesis (see p. 16), show that there exist either a compactum X or a cardinal number $\tau \geqslant \aleph_0$ such that $\chi(x, X) < \tau$ for every $x \in X$, but $|X| > \tau$.

158. Every non-countable compact T_1-space satisfying the first axiom of countability has a cardinality not less than $2^{\aleph_0}$. True or false?

159. Show that the image of a dyadic compactum under a continuous mapping is a dyadic compactum.

160. Show that the product of any sets of a dyadic compactum is a dyadic compactum.

161. Show that every sequential compactum (that is, metrizable compactum) is a dyadic compactum.

162. Show that the product of any sets of a sequential compactum is a dyadic compactum.

163. Show that a Tychonoff cube is a dyadic compactum.

164. Show that a dyadic compactum is the smallest collection of spaces containing all compacta, the product of any sets of their elements and the continuous images of any of their elements. It also satisfies Hausdorff's axiom.

165. Show that every compactum is a homeomorphic closed subspace of some dyadic compactum.

166. Show that every dyadic compactum has the Souslin property.

167. Show that if a dyadic compactum X is first countable at every point, then it has a countable base.

168. Let X be a dyadic compactum, $Y \subset X$, and let the character of X at each point of Y be countable. Show that $[Y]$ is a compactum with a countable base, that is, $[Y]$ is sequentially compact.

169. Give an example of a compactum which is not dyadic.

170. If $2^{\aleph_1} > 2^{\aleph_0}$, show that every dyadic compactum of cardinality $2^{\aleph_0}$ has a countable base.

171. If $2^{\aleph_1} > 2^{\aleph_0}$, show that every sequential dyadic compactum X

has a countable base.

172. Every sequential dyadic compactum is metrizable without the restriction that $2^{\aleph_1} > 2^{\aleph_0}$. True or false?

173. Let X be any dyadic compactum and $D \subset X$ be an arbitrary countable subset of X. A subspace $X_0 = X \setminus D$ is contractible to a compactum (to a dyadic compactum). True or false?

174. A closed set of every dyadic compact space X is a dyadic compactum. True or false?

3.3. Concepts Closely Related to Compactness

A natural weaker form of compactness is Lindelöf compactness (see p. 44). A notion of weak compactness in the opposite direction is that of countable compactness: A space X is *countably compact* if each countable open cover of X has a finite subcover. It is obvious that exactly those spaces are compact as are simultaneously Lindelöf and countably compact.

An important example of a countably compact non-compact space is the space which we shall call $T(\omega_1)$ throughout this book. We fix a minimal well-ordered set of cardinality $\aleph_1$ and endow it with the order topology (see p. 49). The resulting space is $T(\omega_1)$ (see Problem 91 of Chapter 1; Problem 84 of Chapter 3).

It is known (see Problem 97) that every continuous real-valued function on a compactum X is bounded. The converse is, however, false: There exist non-compact normal spaces on which every continuous function is bounded (Problems 188, 187 and 185). A completely regular space X is called *pseudo-compact*, if every real-valued function defined and continuous on X is bounded. On the class of normal spaces, notions of countable compactness and pseudo-compactness coincide (Problem 192 below).

In this section we introduce the important notion of a locally finite family of sets in a given space, which plays a crucial role in Chap. 5. A family ω of subsets of a space X is called *locally finite* in X iff for each point $x \in X$ there exists a nhood Ux of x such that $|\{A \in \omega: Ux \cap A \neq \Lambda\}| < \aleph_0$ (that is, Ux has nonempty intersection with at most a finite number of sets in ω).

A family ξ of subsets of a space X is called *countably centralized* (or is said to *have the countable intersection property*) (cf. the definition of a centralized system of sets or a system of sets that has the finite intersection property, on p. 7), if every countable subset ξ' of ξ has nonempty intersection.

As we remarked earlier, compact spaces and a compact subset of a topological space, play a vital role even in the study of non-compact spaces. There is a very wide class of spaces (called k-spaces), whose topologies are completely characterized by the family of all their compact subsets. A topological space is called a *k-space* or a *compactly generated space*, if the following condition is satisfied: a subset of it is closed iff its intersection with each compact subset is a compact subspace. The class of k-spaces is sufficienty wide: to this belong all compacta, all locally compact spaces, all Čech-complete spaces (see p. 125) and every space satisfying the first axiom of countability. In this section we shall consider only *Hausdorff k-spaces.*

Let $M \subset X$. Let $[M]^1$ be the set of all points x of X for each of which there exists a compact set $F \subset X$ such that $x \in [F \cap M]$. From a cursory inspection we see that $[M]^1$ is closed in X, provided X is a k-space. However, in general, this is not the case.

It is natural to call X a k_1-space, if for each $M \subset X$ the set $[M]^1$ is closed in X. Every k_1-space is also a k-space.

Among the k-spaces an important position is occupied by pointwise countable spaces and spaces of countable type. A Hausdorff space X is called a *pointwise countable* space (a space of *countable type*) if for each $x \in X$ (for every compact $\Phi \subset X$) there exists a compact $F \subset X$ such that $x \in F$ ($\Phi \subset F$) and $\chi(F, X) \leqslant \aleph_0$.

Finally, we shall also need the notion of a pseudo-base for X at the points defined by $A \subset X$, or what is the same, the notion of an exterior pseudo-base for a set A in a space X containing it. This notion is already introduced on p. 121 and needs no repetition here.

$A \subset X$ is a F_σ-set in X if $X \setminus A$ is a G_δ-set for X (see p. 43).

175. (*Example*). Let X be a half-plane bounded by a straight line l and $Y = X \setminus l$. Relative to the usual topology on X select in X a countable dense set Q contained in Y and for each $x \in l$ fix a sequence $q(x) = \{q_n(x): n \in N^+\}$ of points of Q converging to x. Now define on the set $Z = l \cup Q$ a new topology $\mathscr{T}$ as follows: All points of Q are isolated in Z and a base at any $x \in l$ is formed by the set $Q_k(x) = \{q_n(x): n \geqslant k\}$, $k \in N^+$.

Verify that:

(i) $(Z, \mathscr{T})$ is a locally compact, first countable Hausdorff space;

(ii) $(Z, \mathscr{T})$ has a base of clopen sets;

(ii′) $(Z, \mathcal{T})$ is completely regular;
(iii) $(Z, \mathcal{T})$ is locally metrizable and has locally a countable base;
(iv) $(Z, \mathcal{T})$ is the sum of two discrete subspaces of which one is closed and the other is an open F_σ-set, namely $Z = I \cup Q$;
(v) $(Z, \mathcal{T})$ is separable but has no countable base;
(vi) $(Z, \mathcal{T})$ is non-metrizable;
(vii) $(Z, \mathcal{T})$ is not Lindelöf.

176. Show that every locally compact Hausdorff space X has a Hausdorff compactification αX such that $|\alpha X \setminus X| \leqslant 1$ (*Alexandroff compactification*).

177. Show that every locally compact Hausdorff space is completely regular.

178. Show that an open subspace Y of a compact Hausdorff space X is locally compact.

179. Let bX be a Hausdorff compactification of X. Show that X is locally compact iff X is open in bX.

180. Show that a subspace Y of a locally compact Hausdorff space X is locally compact iff Y is open in $[Y]_X$.

181. Show that the set $G(X)$ of all points at which X is locally compact is open in X and forms a locally compact subspace of X. Correspondingly, the set $R(X)$ of all points at which X is not locally compact is closed in X.

182. Show that a product $X = \Pi\{X_\alpha : \alpha \in A\}$ of infinitely many locally compact non-compact T_1-spaces X_α is not locally compact at any point.

183. If X is compact and Y any space, show that the set of points of product $X \times Y$ not locally compact is $X \times R(Y)$, where $R(Y)$ is a set of points of Y at which Y is not locally compact.

184. Show that every (countable) sequence $\varphi = \{x_n : n \in N^+\}$ of points of a space $T(\omega_1)$ contains a convergent subsequence.

185. Show that the space $T(\omega_1)$ is not compact.

186. Show that:

(*a*) each pair of sets F_1, F_2 closed in $T(\omega_1)$ and of cardinalily $\aleph_1$ has nonempty intersection;
(*b*) each continuous real-valued function f on $T(\omega_1)$ is eventually constant, that is, there exists α_0 such that $f(\alpha) = f(\alpha_0)$ for every $\alpha \geqslant \alpha_0$.

187. Show that $T(\omega_1)$ is a normal space.

188. Show that each continuous real-valued function defined on $T(\omega_1)$

is bounded, that is, $T(\omega_1)$ is pseudo-compact.

189. Show that a topological space X is countably compact iff every infinite subset of X has a limit point in X.

190. Give an example of a space that is countably compact, normal and non-compact.

191. Show that Y, the image of a countably compact space X under a continuous mapping, is countably compact.

192. Show that a normal space X is countably compact iff X is pseudo-compact.

193. Show that $T(\omega_1)$ is a countably compact space.

194. Show that for a topological space X the following are equivalent:

(*a*) X is countably compact;
(*b*) every countable centralized family of closed sets in X has nonempty intersection;
(*c*) every countable increasing sequence of closed sets has nonempty intersection.

195. Show that a completely regular space X is countably compact if X itself and every closed subspace of X are pseudo-compact.

196. Show that a completely regular space is pseudo-compact iff every locally finite family γ of open sets in X (see p. 136) is finite.

197. Show that a completely regular space X is pseudo-compact iff for every countable centralized family ξ of open sets in X we have $\bigcap \{[U]: U \in \xi\} \neq \Lambda$.

198. Let X be a normal space, F a closed set in X and $\Phi = \operatorname{Fr} F = F \cap [X \setminus F]$. Show that, if Φ is not countably compact, then the character of F in X is uncountable.

199. Give an example of a space that is Lindelöf but not compact.

200. If X is Lindelöf, show that for every uncountable set $A \subset X$ there is a point $y \in X$ whose every nhood intersects A in an uncountable set (that is, y is a cluster point of A).

201. A hereditarily regular Lindelöf space cannot be first countable. True or false?

202. Show that a closed subset of a Lindelöf space is Lindelöf.

203. Show that a space Y, which is the image of a Lindelöf space X under a continuous mapping, is itself Lindelöf.

204. Give an example of a regular space that has the Souslin property but is not Lindelöf.

205. Give an example of a space that is regular and separable but not Lindelöf.

206. Show that the following conditions on a regular space X are equivalent:

(i) X is Lindelöf and perfectly normal;
(ii) X is Lindelöf and every open subset of X is a Lindelöf space;
(iii) in every family γ of open sets in X there is a countable subfamily $\gamma_0 \subset \gamma$, whose union is the same as the union of γ;
(iv) X is Lindelöf and so is every subspace of X.

207. Show that a perfectly normal Lindelöf space has the Souslin property.

208. Show that a space which is the union of a countable family of Lindelöf subspaces is itself Lindelöf.

209. Let $I = [0, 1]$ be an interval of a real line with the usual topology. Define on I a new topology in the following manner: Suppose that $A \subset I$, a completely imperfect set (see p. 144) in I, has the cardinality of the continuum (for proof of the existence of such a set, see Problem 285 below). The base for the new topology on I (call I^* the new space) is formed by all open sets of the interval with the natural topology and, in addition, all $x \in A$ are isolated points (that is, each one-point set $\{x\}$, if $x \in A$, is open and closed in the new topology). Show that:

(*a*) I^* is a regular topological space;
(*b*) I^* is Lindelöf;
(*b'*) I^* is normal;
(*c*) I^* has a pointwise countable base (see p. 56);
(*d*) I^* has no countable base;
(*e*) I^* is non-metrizable;
(*f*) I^* is not perfectly normal;
(*g*) I^* is not hereditarily Lindelöf.

210. Give an example of a perfectly normal Lindelöf space in which every compact Hausdorff subspace has countable character and is non-homeomorphic to any subspace of a perfectly normal compact Hausdorff space.

211. Let X be Lindelöf and T_1, F a closed set in X, $|F| < 2^{\aleph_0}$ and suppose each $x \in F$ is an intersection of a family γ_x of open sets in X such that $|\gamma_x| \leqslant 2^{\aleph_0}$. Show that F also is an intersection of at most $2^{\aleph_0}$ open sets.

212. Show that if X is Lindelöf and Hausdorff, which (*a*) is sequential and (*b*) has at each point character not exceeding $2^{\aleph_0}$, then $|X| \leqslant 2^{\aleph_0}$.

213. Show that every Hausdorff, Lindelöf and first countable space has cardinality not exceeding $2^{\aleph_0}$.

214. Suppose that X is any regular, Lindelöf and first countable space of countable pseudo-character. Is X contractible to a Hausdorff first-countable space?

215. Let X be a regular Lindelöf space and $\psi(x, X) \leqslant \aleph_0$ for each $x \in X$. Then $|X| \leqslant 2^{\aleph_0}$. True or false?

216. Show that X is Lindelöf iff every countable centralized system of closed sets of X has nonempty intersection.

217. Let X be Lindelöf and let $f: X \to Y$ be a continuous function from X onto a regular space Y of weight $wY = \tau$, where τ is an admissible cardinal number (that is, $\tau^{\aleph_0} = \tau$). Show that there is a closed subset $F \subset X$ such that $fF = Y$ and $s(F) \leqslant wY = \tau$.

218. Suppose that X is Hausdorff and Lindelöf and has the properties:

(*a*) $t(X) \leqslant \aleph_0$;
(*b*) for every $M \subset X$, if $|M| \leqslant \mathfrak{c}$, then $|[M]| \leqslant \mathfrak{c}$.

Let A be a set such that $|A| = \aleph_1$, and A is minimally well-ordered, that is, $A = T(\omega_1)$. For each $\lambda \in A$ suppose we have a space Y_λ, a continuous function $f_\lambda: X \xrightarrow{\text{onto}} Y_\lambda$ and a closed set F_λ in X in such that:

(i) $s(F_\lambda) \leqslant \mathfrak{c}$ for each $\lambda \in A$,
(ii) $f_\lambda F_\lambda = Y_\lambda$ for each $\lambda \in A$,
(iii) $F_\lambda \subset F_{\lambda'}$, if $\lambda < \lambda'$,
(iv) $f_{\lambda'}^{-1} f_{\lambda'} F_\lambda = F_\lambda$, if $\lambda < \lambda'$,
(v) $f_{\lambda'}^{-1} f_{\lambda'} x \subset f_\lambda^{-1} f_\lambda x$ for each point $x \in X$, if $\lambda < \lambda'$.

Show that $|X| \leqslant \mathfrak{c}$.

219. Let X be a regular Lindelöf space satisfying the following conditions:

(*a*) $t(X) \leqslant \aleph_0$;
(*b*) for each $M \subset X$, $|M| \leqslant \mathfrak{c}$, surely $|[M]| \leqslant \mathfrak{c}$.

Show that, if $\psi(x, X) \leqslant \mathfrak{c}$ for each $x \in X$, we have $|X| \leqslant \mathfrak{c}$.

220. Let X be a regular Lindelöf space, $\psi(x, X) \leqslant \aleph_0$ for each point x and $t(X) = \aleph_0$. Show that $|X| \leqslant \mathfrak{c}$,

221. Show that every metrizable subspace Y of a perfectly normal compact Hausdorff space has a countable base.

222. Suppose that Y is a subspace of some perfectly normal compact Hausdorff space and the topology of Y is generated by a semi-metric. Then Y has a countable base. True or false?

223. Show that every perfectly normal compact Hausdorff space is first countable (at every point).

224. Show that every closed set in a perfectly normal compact Hausdorff space has a countable defining family of nhoods.

225. Show that the cardinality of every perfectly normal compact Hausdorff space does not exceed $2^{\aleph_0}$ (*Alexandroff's theorem*).

226. Show that the cardinality of the set of all closed subsets of a perfectly normal compact Hausdorff space X does not exceed $2^{\aleph_0}$ (the proposition is valid if X is perfectly normal, Lindelöf and first countable, in particular if X has a countable base).

227. There is a perfectly normal compact Hausdorff space, which is universal in the sense that every perfectly normal compact Hausdorff space is homeomorphic to a closed subspace of it. True or false? Show that there exists a family of $2^{\aleph_0}$ pairwise non-homeomorphic perfectly normal compact Hausdorff spaces.

228. If X is a perfectly normal compact Hausdorff space and Y a subspace having a net $\mathcal{S}$ of cardinality not exceeding τ, show that Y also has an exterior base (see p. 56) in X of cardinality not greater than τ. Furthermore, the weight of Y does not exceed τ.

229. Show that no non-metrizable subspace X of a perfectly normal compact Hausdorff space can be represented as the union of a countable family of metrizable subspaces.

230. Let $(X, \mathcal{T})$ be compact Hausdorff and first countable, and such that $|U| > \aleph_0$ for each nonempty $U \in \mathcal{T}$ (e.g. $(X, \mathcal{T})$ is an interval with the usual topology). Put

$$\tilde{\mathcal{T}} = \{U \setminus A\colon U \in \mathcal{T},\ A \subset X \text{ and } |A| \leqslant \aleph_0\}.$$

Show that $(X, \tilde{\mathcal{T}})$ is Hausdorff and satisfies the following conditions:

(0) every countable set A in $(X, \tilde{\mathcal{T}})$ is closed in $(X, \tilde{\mathcal{T}})$;
(1) $c(Y) \leqslant \aleph_0$ for each $Y \subset X$ (that is, $cc(X) = \aleph_0$);
(2) $s(Y) > \aleph_0$ if $Y \subset X$ and $|Y| > \aleph_0$;
(3) each subspace Y of X is Lindelöf;

(4) $(X, \tilde{\mathfrak{T}})$ is not first countable at any point but has countable pseudo-character at all points;

(5) $t(X, \tilde{\mathfrak{T}}) > \aleph_0$;

(6) $[A]_{\aleph_0} = \bigcup \{[B]_{(X, \tilde{\mathfrak{T}})}: B \subset A \text{ and } |B| \leqslant \aleph_0\} = A$ for every $A \subset X$;

(7) $[A]^{\aleph_0} = \{y \in X$: if $Q \subset X$, $y \in Q$ and Q is a G_δ-set in $(X, \tilde{\mathfrak{T}})$, then $Q \cap A \neq \Lambda\} = [A]_{(X, \tilde{\mathfrak{T}})}$ and hence $[[A]_{\aleph_0}]^{\aleph_0} = [A]_{(X, \tilde{\mathfrak{T}})}$ for every $A \subset X$ (cf. Problem 141);

(8) $\mathfrak{T} \subset \tilde{\mathfrak{T}}$ but $\mathfrak{T} \neq \tilde{\mathfrak{T}}$, i.e. the identity mapping $(X, \tilde{\mathfrak{T}}) \to (X, \mathfrak{T})$ on X is a nontrivial contraction;

(9) $(X, \tilde{\mathfrak{T}})$ is an H-closed space;

(10) $[V]_{(X, \tilde{\mathfrak{T}})} = [V]_{(X, \mathfrak{T})}$ for every $V \in \tilde{\mathfrak{T}}$.

231. Show that under a continuous mapping into a perfectly normal compact Hausdorff space, weight does not increase.

232. If a Hausdorff space Y is the image of a perfectly normal compact Hausdorff space X under a continuous mapping f, show that Y is also perfectly normal and compact Hausdorff.

233. Show that under a continuous mapping of a perfectly normal compact Hausdorff space X onto a Hausdorff space Y the weight of any subspace of X does not increase.

234. If the product of infinite compact Hausdorff spaces X and Y is hereditarily normal, show that both X and Y are perfectly normal.

235. If X is a compact and $X \times X \times X$ is a hereditarily normal space, show that X has a countable base.

236. If the Tychonoff product of X with itself is perfectly normal and compact Hausdorff, show that X has a countable base.

237. Show that every first countable space is a k-space.

238. Show that every locally compact Hausdorff space is a k-space.

239. Show that a closed subspace of a k-space is a k-space.

240. Show that every Hausdorff space that is pointwise countable is a k-space.

241. Show that every Čech-complete space is pointwise countable.

242. Show that every Čech-complete space is a k-space.

243. If X is a k-space and x is its non-isolated point, show that there exists a compact Hausdorff $\Phi \subset X$ satisfying $x \in [\Phi \setminus \{x\}]$.

244. Is every Hausdorff k-space a k_1-space?

245. Is every Čech-complete space a k_1-space?

246. Show that the image of a k-space under a quotient mapping is a

k-space.

247. Show that a Hausdorff space is a k-space iff it is a quotient of some locally compact Hausdorff space.

248. Show that an open subspace of a Hausdorff k-space is a k-space.

249. Is every subspace of a Hausdorff k-space a k-space?

250. If every subspace of a Hausdorff space X is a k-space, show that X is a Fréchet-Urysohn space.

251. If Y is a k-space and $f: X \to Y$ is a perfect mapping of a completely regular space X onto Y, show that X is also a k-space.

252. Let X be a k-space and Y be locally compact and Hausdorff. Show that $X \times Y$ is also a k-space.

3.4. Compact Metrizable Spaces

The *Hilbert cube* is an important example of a compact metrizable space. By such a cube we mean the set Q^∞ of all points $x = \{x_i: i = 1, 2, \ldots\}$ of Hilbert space l^2 (see Prob. 249, Chap. 2), satisfying $0 \leqslant x_i \leqslant \frac{1}{2^i}$, together with the topology induced from l_2.

In the study of compact metrizable spaces, an important role is played by the notion of condensation point. A point in a space is called a *condensation point*, if every nhood of this point is an uncountable set.

A subset M of a space X is called *completely imperfect* (in X) if it includes no set perfect in X. Also (see p. 62), a set A is called *perfect* in X if it is closed in X and, as a subspace of X, has no isolated points.

In this section a vital role will be played by the Cantor perfect set defined in Chap. 2.4, Prob. 185.

As we shall see later on, the following concept conflicts in a certain sense with the Cantor perfect set. A space X is *σ-discrete,* if it is the union of a countable family of discrete subspaces (the closures of these subspaces may not be discrete).

Finally, the notion of irreducible mapping will be introduced in this section, although it only assumes crucial role in Chap. 6. A mapping f from a space X onto a space Y is *irreducible* if Y is not the image under f of a closed set F in X, other than X. We shall consider only *continuous* irreducible mappings.

The proofs of the assertions in Problems 253—276 below are left as an exercise to the reader.

253. A complete totally bounded metric space (X, ρ) is countably compact (see pp. 46-47, 136).

254. A countably compact metrizable space is compact.

255. A metric space (X, ρ) is compact iff it is complete and totally bounded.

256. A subspace X of Euclidean space R^n is (sequentially) compact iff it is closed and bounded in R^n.

257. A compact Hausdorff space X is metrizable iff it has a countable base.

258. A closed subspace of a compact metrizable space is itself compact metrizable.

259. A metric space (X, ρ) is countably compact iff every sequence $\{x_i: i \in N^+\}$ of its points has a convergent subsequence.

260. A metric space (X, ρ) is compact iff every decreasing sequence of nonempty closed sets has nonempty intersection.

261. A metric space (X, ρ) is compact iff every continuous real-valued function on (X, ρ) is bounded.

262. A metrizable space X is compact iff every metric on X, compatible with its topology, is bounded.

263. The completion of a metric space (X, ρ) is compact iff (X, ρ) is totally bounded.

264. In a metric space, every compact set is closed.

265. The topological product of a countable family of sequentially compact spaces is sequentially compact.

266. The subspace Q^∞ (cube) of Hilbert space l^2 is sequentially compact. Is very closed bounded subspace of l^2 sequentially compact (cf. Prob. 256)?

267. The class of sequentially compact spaces coincides with a class of closed subspaces of the Hilbert cube.

268. Every sequentially compact subspace of the space of irrational points on the real line is nowhere dense in this space.

269. The space of irrational points on the real line or in an interval is not the union of a countable family of compact subspaces.

270. Every open set Γ in a compact metrizable space X is a locally compact space with a countable base.

271. A metric space X with a countable base is locally compact iff it is the union of a countable increasing sequence of compact subsets: $X = \bigcup_{n=1}^{\infty} F_n$, $F_n \subset F_{n+1}$, $n \in N^+$, such that each $x \in X$ is an interior point of some F_n.

272. A locally compact metrizable space X with a countable base can be turned into a compact metrizable space by adding one point.

273. A metrizable locally compact space X has a locally countable base, that is, each $x \in X$, has a nhood Ux which, as a subspace of X, has a countable base.

274. A metrizable space is locally compact and has a countable base iff it is homeomorphic to an open dense set of some compact metrizable space.

275. In the space of irrational points on the real line or in an interval, no point is locally compact.

276. The Hilbert space l^2 has no locally compact point, but a finite-dimensional Euclidean space R^n is locally compact at each of its points.

277. Give an example of a locally compact subspace X of the real line R^1, whose complement $Y = R^1 \setminus X$ is not locally compact.

278. Construct a non-locally compact subspace X of the real line R^1 which is the union of two locally compact subspaces.

279. Show that every 0-dimensional (see p. 42), complete and separable metric space (Y, ρ) without a locally compact point is homeomorphic to the B-space $B(\aleph_0)$ of countable weight (*Alexandroff's theorem*).

280. Show that the space of irrational numbers is homeomorphic to the space $B(\aleph_0)$.

281. Show that every open subspace of $B(\aleph_0)$ is homeomorphic to $B(\aleph_0)$.

282. Show that a 0-dimensional separable metric space X is homeomorphic to a subspace of $B(\aleph_0)$.

283. Show that:

(*a*) the set $X_1 \subset X$ of all points of a sequentially compact space X other than its condensation points, is a countable open set;
(*b*) the set $X_0 = X \setminus X_1$, i.e. the set of all condensation points of X, is a perfect set (nonempty, if X is uncountable).

284. Show that every uncountable compact metrizable space X has the cardinality of the continuum.

285. Show that:

(*a*) on the real line R^1 there is a family γ of cardinality $\mathfrak{c}$ of pairwise non-homeomorphic compacta;
(*b*) on the real line R^1 there is an *absolutely imperfect* set Z having the cardinality of the continuum.

286. If (X, ρ) is a metric space and $F \subset (X, \rho)$ a compact subspace of (X, ρ), show that for each $x \in X$ there is a point $y^* \in F$ such that

$$\rho(x, y^*) = \inf\{\rho(x, y): y \in F\} = \rho(x, F).$$

287. If (X, ρ) is a compact metric space, show that in (X, ρ) there is a pair of points $x_1, x_2 \in X$ satisfying $\rho(x_1, x_2) = \text{diam}\,(X, \rho)$.

288. If (X, ρ) is a metric space and $F_1, F_2 \subset X$ are two closed disjoint subsets, show that $\rho(F_1, F_2) > 0$ provided either F_1 or F_2 is compact.

289. Show that the distance between every pair of disjoint compact subspaces of a metric space (X, ρ) is positive and attained.

290*. Show that for every finite open cover $\omega = \{U_1, \ldots, U_s\}$ of a compact metric space (X, ρ), there is a number $\eta(\omega) > 0$ such that every set $A \subset X$ of diameter $< \eta(\omega)$ is contained in at least one member of ω.

291*. Show that for every finite closed cover α of a compact metric space (X, ρ) there is a number $\delta(\alpha) > 0$ such that for each set $A \subset X$ satisfying diam $A \leqslant \delta(\alpha)$, the family of sets $\alpha_0 = \{F \in \alpha: F \cap A \neq \Lambda\}$ has nonempty intersection.

292. Show that a Hausdorff space Y which is the continuous image of a compact metrizable space X, is itself compact metrizable.

293. Show that every continuous function f from a compact metric space (X, ρ) onto a compact metric space (Y, ρ') is uniformly continuous.

294. Show that every continuous function on (X, ρ) is uniformly continuous.

295. Show that a contraction mapping f of a compact metric space (X, ρ) into itself is a mapping onto a proper subset.

296. Show that every isometry of a compact metric space (X, ρ) into itself is an isometry onto the whole space (X, ρ). Does the same assertion hold when (X, ρ) is a complete metric space?

297. If (X, ρ) is compact and f is a mapping of X into itself such that $\rho(fx, fy) < \rho(x, y)$ for every pair of distinct points $x, y \in X$, show that f has a unique fixed point in X.

298*. If f is a mapping of a compact metric space (X, ρ) onto itself such that $\rho(fx, fy) \leqslant \rho(x, y)$ for every pair of points $x, y \in X$, show that f is an isometry.

299. Show that every compact metrizable space X is the continuous image of the Cantor perfect set (*Alexandroff's theorem*).

300. Show that 0-dimensional compact metrizable space Y without isolated points is homeomorphic to the Cantor perfect set (*Alexandroff's* *theorem*).

301. Show that:

(*a*) the Cantor perfect set is homeomorphic to the product of a countable number of pairs of isolated points;
(*b*) the product of a countable number of finite Hausdorff spaces is also homeomorphic to the Cantor perfect set;
(*c*) the product of a countable number of spaces, each of which is homeomorphic to a Cantor perfect set, is again homeomorphic to the Cantor perfect set.

302. Show that the Cantor perfect set is topologically homogeneous.

303. Show that a compact metrizable space Y is the image of the Cantor perfect set C under an irreducible continuous mapping iff it has no isolated points.

304. Show that a complete metric space without isolated points contains a topological image of the Cantor perfect set. If X is any complete metric space, then either X contains the Cantor perfect set, or X is the union of a countable number of discrete subspaces.

305. If X is a compact metrizable space and $A \subset X$ is a countable subset, show that the subspace $X_0 = X \setminus A$ admits a continuous one-to-one mapping onto some compact metrizable space.

306. Is the space of closed sets (see Problem 53) of intervals on a real line homeomorphic to the Hilbert cube?

307. Is the square also homeomorphic to the space of closed sets of intervals on a real line?

308. Find a family of cardinality $2^{\aleph_0}$ of pairwise non-homeomorphic compact metrizable spaces, each of which is homeomorphic to its space of closed sets.

3.5. Continuous Functions on Compact (Hausdorff) Spaces

As we know, every continuous real-valued function on a compact Hausdorff space X is bounded and attains its supremum and infimum (see Prob. 97). Every continuous function on a compact Hausdorff space X is *uniformly continuous* in the following sense. For every real $\varepsilon > 0$ there is a finite open cover ω of X such that $|fx_1 - fx_2| < \varepsilon$ iff x_1 and x_2 belong to exactly the same member of ω (prove this). The notion of uniform convergence of a sequence of real-valued functions carries over to any topological space. A sequence of functions $\{f_n: n \in N^+\}$, defined on a topological space X, *converges uniformly* to a function f on X, if for each $\varepsilon > 0$ there is a $n_0 \in N^+$ such that $|fx - f_n x| < \varepsilon$ for all $n > n_0$ and each $x \in X$. A uniformly convergent sequence $\{f_n: n \in N^+\}$ of continuous functions on X has a continuous function as its limit (see Prob. 310 below).

We say that a set C_0 of functions on X *separates* (*distinguishes*) *the points of* X if for each pair of distinct points x_1 and x_2 of X there is a function $f \in C_0$ such that $fx_1 \neq fx_2$. A set C_0 of functions on X *strongly separates the points of* X if for every pair of points $x_1, x_2 \in X$, $x_1 \neq x_2$, and each pair of real numbers a and b, $a \neq b$, there is a function $f \in C_0$ such that $fx_1 = a$, $fx_2 = b$.

This section is mainly concerned with the set (denoted by $C(X)$) of all continuous real-valued functions defined on a compact Hausdorff space X. This set becomes in a natural manner a linear commutative ring (i.e., an algebra) by employing the usual operations of (pointwise) addition (multiplication) of two functions (note that the sum (product) of two continuous functions on X is a continuous function). In addition, the multiplication of a function $f \in C(X)$ with an arbitrary real number is defined in a natural way. $C(X)$ contains a (unique) function identically equal to 1 and a (unique) zero-function, which is identically equal to 0. It is easy to verify that $C(X)$ with addition and multiplication by reals so defined, is a linear space. The operation of multiplication of two members in $C(X)$ is commutative. The set $C(X)$ with the operations defined above, which convert it into a linear commutative ring with unity (i.e., into an algebra) is called the (linear) ring of continuous functions on the compact Hausdorff space X (or an algebra of continuous functions over X).

The algebra $C(X)$ (X compact) plays a unique role not only in general topology but in functional analysis as well. We give a precise definition of a linear ring (to be exact, an algebra, see Naimark [1]). A linear ring is a set X, on which are defined operations of addition and multiplication by a real number converting this set into a linear space, together with one more multiplication operation, combining each ordered pair of elements x, y in X into an element $x \circ y$ of X. In addition, the following axioms are required to be satisfied: (*a*) $\alpha(x \circ y) = (\alpha x) \circ y$; (*b*) $(x \circ y) \circ z = x \circ (y \circ z)$; (*c*) $(x + y) \circ z = x \circ z + y \circ z$; (*d*) $x \circ (y + z) = x \circ y + x \circ z$ for every x, y, $z \in X$ and any α. If there is any element $e \in X$ such that $x \circ e = e \circ x = x$ for each $x \in X$, and, moreover, $x \circ y = y \circ x$ for every pair of elements $x, y \in X$, then the linear ring X is called a *commutative linear ring with unity* (or a *commutative algebra with unity*).

All the concepts that follow relate to the algebra $C(X)$ of continuous functions on a compact space X. A subset $C_0 \subset C(X)$ is called a *subring* if $f + g \in C_0, f \cdot g \in C_0, -f \in C_0$ for each $f, g \in C_0$.

A *subring* $C_0 \subset C(X)$ is called an *ideal*, if $g \cdot f \in C_0$ for every $g \in C_0, f \in C(X)$. In other words, a subring $C_0 \subset C(X)$ is an ideal if $C_0 \cdot C(X) \subset C_0$. An ideal $C_0 \subset C(X)$ is said to be *nontrivial* if $C_0 \not\equiv C(X)$ and C_0 consists of more than one function. Finally, a nontrivial ideal $C_0 \subset C(X)$ is called a *maximal ideal* if C_0 is not contained in any nontrivial ideal other than C_0. It is clear that the whole of $C(X)$ is the trivial ideal if it consists of only one function (necessarily equal to zero, that is, the null element in $C(X)$). Note that for us, by definition, a maximal ideal is a nontrivial ideal. We denote by C_x, $x \in X$ the set of all continuous functions on X vanishing at a point x, that is, $C_x = \{f \in C(X): fx = 0\}$.

$C(X)$ carries a standard metric (and even a norm). We put

$$\|f\| = \max \{|fx|: x \in X\},$$
$$\rho(f, g) = \max \{|fx - gx|: x \in X\} = \|f - g\|$$

for every $f, g \in C(X)$, bearing in mind also that every continuous real-valued function on X is bounded and attains its least upper bound. (Note also that $q(x) = |fx|$ is a continuous function if f is.)

The metric so defined on $C(X)$ is called the *metric of uniform convergence* or the *Chebyshev metric*. Hereafter $C(X)$ will be regarded as endowed with this metric. Thus, on the one hand, $C(X)$ is a commutative algebra with unity, on the other $C(X)$ is a metric space. It is straightforward to check that the operations defined on $C(X)$ are continuous in the metric of uniform convergence. This implies that the mapping φ assigning the function $f + g$ to any pair (f, g) of functions $f, g \in C(X)$ and the mapping ψ assigning the function $f \cdot g$ to (f, g) are continuous mappings of the product $C(X) \times C(X)$ onto the space $C(X)$.

The following notions are also useful. A subset $C_0 \subset C(X)$ is *uniformly bounded* on X if there is a real number $K > 0$ such that $|fx| < K$ for every function $f \in C_0$ and each point $x \in X$. A subset $C_0 \subset C(X)$ is *uniformly continuous* if for every $\varepsilon > 0$ there is a finite open cover ω of X such that $|fx_1 - fx_2| < \varepsilon$ for each $f \in C_0$ and every pair of points $x_1, x_2 \in X$ contained in the same member of ω.

We frequently consider continuous functions defined on a compact

Hausdorff space X, which is the topological product of a family $\{X_\alpha: \alpha \in A\}$ of compact Hausdorff spaces X_α, $\alpha \in A$. Suppose that f is a continuous function from the space $X = \Pi\{X_\alpha: \alpha \in A\}$ onto a space Y. We say that f depends only on the coordinates belonging to $A' \subset A$ if it is independent of the coordinates belonging to the set $A \setminus A'$ (see p. 85). In other words, f depends only on $A' \subset A$, if there is a continuous function $g: \Pi\{X_\alpha: \alpha \in A'\} \to Y$ such that $f = g\,\pi_{A'}$, $\pi_{A'}$ the natural projection of all of the product $\Pi\{X_\alpha: \alpha \in A\}$ onto the A'-face of $\Pi\{X_\alpha: \alpha \in A'\}$. We also say that $f: \Pi\{X_\alpha: \alpha \in A\} \to Y$ depends only the coordinates τ (τ a cardinal number), if there is a set $A' \subset A$, $|A'| = \tau$ such that f depends only on the coordinates in $A' \subset A$.

309. Verify that $C(X)$, with the operations defined on pp. 149-150, forms a commutative algebra.

310. Show that a function f_0 on X, which is the limit of a uniformly convergent sequence of functions continuous on X, is also continuous.

311. Let X be compact Hausdorff and $C(X)$ the algebra of continuous functions on X with the metric defined on p. 150. Show that

(*a*) $C(X)$ is a metric space;
(*b*) a sequence $\{f_n: n \in N^+\}$ of functions in $C(X)$ converges to $f_0 \in C(X)$ with respect to the metric in $C(X)$ iff this sequence converges uniformly to f_0;
(*c*) the metric space $C(X)$ is complete.

312. Show that every subring $C_0 \subset C(X)$, separating the points of a compactum X and containing all the identically constant functions, strongly separates the points of X.

313. Let X be compact Hausdorff and $C_0 \subset C(X)$ a subring containing all constants and forming a closed subset of the metric space $C(X)$. Show that the function $g(x) = |f(x)|$ belongs to C_0, if $f \in C_0$.

314. Suppose that f and g is a pair of functions continuous on a compactum X. Define the functions $M_{f,g}$ and $m_{f,g}$ by

$$M_{f,g}(x) = \max\{f(x), g(x)\}, \quad m_{f,g}(x) = \min\{f(x), g(x)\}.$$

Prove the following equalities:

$$m_{f,g}(x) = \tfrac{1}{2}\{f(x) + g(x) - |f(x) - g(x)|\},$$

$$M_{f,g}(x) = \tfrac{1}{2}\{|f(x) - g(x)| + f(x) + g(x)\}.$$

315. Suppose that $C_0 \subset C(X)$ is a closed subring (i.e., the subring C_0 is a closed set in the metric space $C(X)$). Show that $M_{f,g}, m_{f,g} \in C_0$ if f, $g \in C_0$ (cf. the definition of $M_{f,g}$ and $m_{f,g}$ in the preceding problem).

316. Suppose that the family $\mathfrak{A} \subset C(X)$ of continuous functions on a compactum X satisfies the following two conditions:

(i) $\mathfrak{A}$ strongly separates the points in X;
(ii) if $f, g \in \mathfrak{A}$, then also $M_{f,g}, m_{f,g} \in \mathfrak{A}$.

Show that $\mathfrak{A}$ is dense in the metric space $C(X)$, i.e., $[\mathfrak{A}]_{C(X)} = C(X)$.

317. Suppose that $C_0 \subset C(X)$ is a subring and X is compact Hausdorff. Further, assume that C_0 contains all constants and separates the points of X. Show that C_0 is dense in the metric space $C(X)$.

318. Deduce from the preceding problem the following *classical approximation theorem of Weierstrass*:

(i) every continuous function on the segment $[0, 1]$ is the limit of a uniformly convergent sequence of polynomials;
(ii) every continuous function $f(x)$ on the segment $[-\pi, \pi]$, taking on identical values at the endpoints of this segment, is the limit of a uniformly convergent sequence of trigonometric polynomials.

319. Show that the space $C(I)$ of continuous functions on a segment I is separable (and, consequently, has a countable base).

320. Show that a compactum X is metrizable iff $C(X)$ is separable.

321. Show that every continuous function on the product $\Pi\{X_\alpha: \alpha \in A\} = X$ of a family of compacta $\{X_\alpha: \alpha \in A\}$ depends only on a countable number of coordinates.

322. Let $X = \Pi\{X_\alpha: \alpha \in A\}$ be the product of compacta $\{X_\alpha: \alpha \in A\}$ and $f: X \to Y$ a continuous mapping onto a compactum of weight $w(Y) = \tau$, $\tau \leqslant |A|$, $\tau \geqslant \aleph_0$. Show that f depends on τ.

323. Show that every dyadic compactum X of weight $w(X) = \tau$ is the continuous image of the generalized Cantor discontinuum D^τ of weight τ.

324. Let X be a dyadic compactum and $f: X \to Y$ a continuous mapping of X onto a compactum. Show that there is a closed subset X_f of X such that $w(X_f) = w(Y)$ and $fX_f = Y$.

325. If $f: X \to Y$ is a continuous irreducible (see p. 144) mapping of a dyadic compactum X onto a compactum Y, show that $w(X) = w(Y)$.

326. Show that every compactum X, whose π-weight is τ, admits a continuous irreducible mapping onto a compactum of weight $\leqslant \tau$.

327. Show that the π-weight of every dyadic compactum coincides with its weight.

328. Show that every dyadic compactum of a countable π-weight is metrizable.

329. Show that every closed G_δ-subset X of the generalized Cantor discontinuum D^m of weight m, $m > \aleph_0$, is homeomorphic to D^m.

330. Show that for every closed G_δ-set X in the Cantor discontinuum D^m, there exists a retraction of D^m to X, i.e., a continuous mapping g: $D^m \xrightarrow{\text{onto}} X$, which is the identity on X.

331. Show that a closed G_δ-subspace of a dyadic compactum is a dyadic compactum.

332. Let $C(X)$ be the algebra of all functions continuous on a compactum X. Show that:

(*a*) for each $x_0 \in X$ the set of functions $C_{x_0} = \{f \in C(X): f(x_0) = 0\}$ is an ideal in $C(X)$;

(*b*) every nontrivial ideal C_0 in $C(X)$ is contained in an ideal of the form C_x;

(*c*) every maximal ideal in $C(X)$ is an ideal of the form C_x.

333. Let $C(X)$ be the algebra of all continuous functions on a compactum X, and let $A \subset X$. Show that $x_0 \in [A]$ iff $C_{x_0} \supset \bigcap \{C_x: x \in A\}$.

334. Show that two compacta X and Y are homeomorphic iff their algebras $C(X)$ and $C(Y)$ are isomorphic.

335*. Let X and Y be two compacta and $C(X)$ and $C(Y)$ their algebras of continuous functions with metric of uniform convergence. Show that every continuous function f: $X \to Y$ induces an isomorphic inclusion of $C(Y)$ into $C(X)$ as a closed subring and, conversely, every inclusion of $C(Y)$ into $C(X)$ as a closed subring induces a continuous function from X onto Y.

336. Let X and Y be two metric spaces and $C(X, Y)$ be the set of all bounded functions from X into Y. Put $d(f, g) = \sup \{\rho(f(x), g(x)): x \in X\}$ for every pair of functions $f, g \in C(X, Y)$ (ρ a metric in Y). Show that:

(*a*) $C(X, Y)$ with a distance function so defined is a metric space;

(*b*) $C(X, Y)$ is a complete metric space if (Y, ρ) is.

337*. Let $I = [0, 1]$ be the unit segment of a real line and $C(I)$ the space of all real-valued functions continuous on I. Show that a subspace $\Phi \subset C(I)$

is (sequentially) compact iff Φ is closed in $C(I)$, is uniformly bounded and equicontinuous.

3.6. Connectedness

It is not accidental that this section is titled 'connectedness' and not 'connected spaces.' The notion of connectedness serves not only to segregate the connected spaces into a specific class. It has a definite, occasionally a rather crucial, role also in the study of disconnected spaces. The initial observation is that every topological space has maximal connected parts, the so-called *connected components* (or, simply, *components*).

A subset A of a topological space X is called a *component* of X if it is connected and not strictly included in any connected set B in X. Sometimes, we talk about a *component* set C, a part of the space X, regarding C as a subspace of X. A *component at a point* x (in a space, or set) is always that component (of the space or set in question) which contains x.

A *quasi-component at a point* (in a space), another important notion, is the intersection of all clopen sets (in this space), containing the given point. A space is said to be *totally disconnected* if every quasi-component of this space contains only one point.

A compact connected space is also called a *continuum*. A continuum C is called *irreducible* between two of its points x and y if every continuum C' lying in C and containing both x and y coincides with C.

A space X is called *locally connected* if for each point $x \in X$, each nhood of x includes a connected nhood. A space X is said to be *linearly* (or *arcwise*) *connected* if for each pair of distinct points x and y of X there is a subspace C of X, containing x and y and homeomorphic to the usual interval. A continuous function $f: X \to Y$ is said to be *monotone* if $f^{-1}y$ is connected for each $y \in Y$.

338. Show that the space X is not connected iff there is a nonempty clopen subset of X different from X.

339. Show that a space is connected if it cannot be represented as the union of two nonempty disjoint open sets.

340. Give an example of a family of connected sets whose union is not connected.

341. Give an example of a family of connected sets whose intersection is not connected.

342. If a continuous real-valued function f defined on a connected space X takes both positive as well as negative values, show that it vanishes somewhere.

343. Show that a space X is not connected iff it can be continuously mapped onto a Hausdorff space consisting of exactly two points.

344. Is the image of a disconnected space under a continuous map necessarily disconnected?

345. (*Due to Knaster-Kuratowski*). With respect to the Cartesian coordinate system on the plane we define the following set X: $X = \{c^*\} \cup C_0 \cup \{A_c\colon c \in C_0\}$, where $c^* = (\frac{1}{2}, \frac{1}{2})$, C_0 is the set of points of the Cantor perfect set ordered in the usual manner on the closed interval $[(0, 0), (1, 0)]$ of the x-axis and A_c is a subset of the closed interval with endpoints c^* and c; if c is a point of type 1, i.e., if it is an endpoint of some interval adjacent to C_0 (see Problems 185 and 180, Chap. 2), then A_c consists of all points of the interval whose coordinates are rational, while if c is a point of type 2, then A_c consists of all points whose co-ordinates are irrational. Show that the set X (with the topology induced by the usual topology on the plane) is connected but that only one-point subsets of the space $X \setminus \{c^*\}$ are connected.

346. Let X be a topological space and $\{A_\alpha\colon \alpha \in M\}$ be a family of connected subsets of X such that there is $\alpha_0 \in M$ for which $A_{\alpha_0} \cap A_\alpha \neq \Lambda$ for every $\alpha \in M$. Show that $\cup \{A_\alpha\colon \alpha \in M\}$ is a connected set.

347. Suppose that M is a connected subset of a space X. Show that $[M]_X$ is connected. Prove that if $M \subset Y \subset [M]_X$, then Y is connected.

348. If A and B are connected subsets of a space X and $[A] \cap B \neq \Lambda$, show that $A \cup B$ is connected.

349. Let X be a topological space, x a point of X and $\mathcal{E}_x$ the family of all connected subsets of X containing x. Show that $K_x = \cup \{P\colon P \in \mathcal{E}_x\}$ is a component of X.

350. Prove that a component of a space X is closed in X. Show that every two components of a space are either identical or disjoint.

351. Show that the set of all components of a space X is a cover of X.

352. If A and B are closed sets of X, whose union and intersection are connected sets, show that A and B themselves are connected. Illustrate by an example that the assumption of closedness is essential.

353. Give an example of a countable connected Hausdorff space. Does there exist a regular, connected and countable space? Does there exist a metric space that is countable and connected?

354. Show that the intersection of the elements of a prefilter consisting of connected compacta is a connected compactum.

355. Show that the intersection of a decreasing sequence of connected compacta is a connected compactum.

356. Give an example of a compactum that is connected but not locally connected.

357. Show that for every pair of points x and y of a connected compactum X there is an irreducible continuum containing them.

358. Suppose that X is a connected compactum and U an open set of X, different from X. Show that the closure of any component of U intersects $X \setminus U$.

359. A connected compactum X is never the union of a countable family of disjoint closed subsets other than X.

360. Suppose that X is the subspace of the plane R^2:

$$X = \left(\bigcup_{n+1}^{\infty} \left(\left\{ \frac{1}{n} \right\} \times [0, 1] \right) \right) \cup \{0, 0\} \cup \{0, 1\}.$$

Show that:

(*a*) X is homeomorphic to a complete metric space;
(*b*) X is the union of a countably many compacta;
(*c*) X contains only two points, of which one is not locally compact;
(*d*) X is not contractible to any compact space.

361. Let X' be a subspace of X and $x \in X'$. Show that the component at x in X' is contained in the component at x in X.

362. On a plane with a fixed Cartesian coordinate system consider a set X (with the induced topology), which is the union of a straight line $y = 0$ (axis of abscissae), a family of closed intervals I_n, $n \in N^+$ with endpoints $(0, 1)$ and $(n, 1/n + 1)$, and a family of half-lines $x \leqslant n$, $y = 1/(1 + n)$, $n = 1, 2, \ldots$ Show that X is a connected space and that some component in the complement of the point $(0, 1)$ does not have it as an adherent point.

363. Give an example of a space in which a quasi-component at a point is different from its component.

364. Show that a component K_x of a compactum X containing any point x_0 of X, concides with the quasi-component at x_0.

365. If X is compact, show that the partition Z of X into its components is continuous.

366. The quotient space partitioning a compactum X into its components is 0-dimensional compactum.

367. Give an example of a metric space X, whose partition into components (and into quasi-components) is not continuous.

368. If X is a connected compactum and A, B are nonempty disjoint closed sets in X, show that there is a component of the set $X \setminus (A \cup B)$, whose closure intersects both A and B.

369. Show that every locally connected space X is a disjoint (free) sum of connected and locally connected spaces—the components of X.

370. Show that for a space X the following are equivalent:

(*a*) the component of an open set is an open set;

(*b*) for each point x and every nhood Ox of x there is a connected set V such that $x \in \operatorname{Int} V \subset V \subset Ox$.

371. If X is a connected and locally connected space and A, B are nonempty disjoint closed sets in X, then there is a component of the set $X \setminus (A \cup B)$, whose closure intersects both A and B.

372. Show that the image of a connected, locally connected space under a continuous mapping is not necessarily locally connected.

373. Show that a quotient of a locally connected space is locally connected.

374. Show that a Hausdorff space X, which is the image of a closed interval under a continuous mapping, is connected and locally connected.

375. Show that the product of locally connected spaces is locally connected iff all factors, apart from, perhaps, finitely many, are connected.

376. Show that every connected, locally connected compactum is the continuous image of a closed interval of the real line.

3.7. Solutions

1. If f is a continuous real-valued function on X, $A \subset X$, $B \subset X$ and $f(A) = 0$, $f(B) = 1$, then $V = \{x \in X: f(x) < \frac{1}{2}\}$ and $W = \{x \in X: f(x) > \frac{1}{2}\}$ are disjoint α-nhoods of A and B.

2. Let $U \in \xi_A$. Select a continuous non-negative real-valued function f on X, satisfying $f(A) = 0$ and $f(X \setminus U) = 1$. Put $V = \{x \in X: f(x) < \frac{1}{2}\}$, $\varphi(x) = 2f(x)$ if $f(x) < \frac{1}{2}$ and $\varphi(x) = 1$ if $f(x) \geqslant \frac{1}{2}$; $\psi(x) = 0$ if $x \in V$ and $\psi(x) = 2\,(f(x) - \frac{1}{2})$ if $x \notin V$. The function φ separates A from $X \setminus V$ and ψ separates V from $X \setminus U$. Hence $V \in \xi_A$ and ξ_A forms an α-family.

3. Use the Kuratowski-Zorn lemma.

4. To prove that the intersection of elements U_1 and U_2 of an α-filter ξ belongs to ξ, it is necessary to take an intersection of elements V_1 and V_2 of this α-filter, α-separated from $X\setminus U_1$ (resp. $X\setminus U_2$) and consider the product of the separating functions corresponding to U_1, V_1 (resp. U_2, V_2) to see that it separates $X\setminus(U_1 \cap U_2)$ from $V_1 \cap V_2$.

5. By Chap. 3.2, ξ is an α-family on X (possibly, empty). However, $\xi \cup \beta$ is then a centralized α-family. But β is an α-ultrafilter; consequently, $\beta = \xi \cup \beta$, i.e., $\xi \subset \beta$.

9. Given the following sets and points in the plane R^2: $L_n = \{(n, y): 0 \leqslant y < \frac{1}{2}\}$, where n is any integer, $L = \cup \{L_n: n \text{ even}\}$, $p_{n,k} = \left(n, 1 - \frac{1}{k}\right) \in R^2$, and

$$T_{n,k} = \left\{\left(n \pm t, 1 - t - \frac{1}{k}\right): 0 < t \leqslant 1 - \frac{1}{k}\right\} \subset R^2,$$

where n is odd and k an integer, $k \geqslant 2$. Put $S = \{P_{n,k}: n \text{ odd and } k \text{ an integer}, k \geqslant 2\}$, $T = \cup \{T_{n,k}: n \text{ odd and } k \text{ an integer}, k \geqslant 2\}$. Each $T_{n,k}$ consists of two lateral sides of an isosceles right triangle whose base (hypotenuse) lies on the axis of abscissae and whose vertices are at the points $p_{n,k}$. In addition, $p_{n,k} \notin T_{n,k}$. Evidently, $S \cap T = T \cap L = L \cap S = \Lambda$. Add to the set $S \cup T \cup L$ two new points a^* and b^*. Put $X = S \cup T \cup L \cup \{a^*\} \cup \{b^*\}$. Now define a topology on X; we shall indicate its base at each point of X:

(i) Each point of T is isolated in X.

(ii) A basic nhood at each point $p_{n,k}$ consists of $p_{n,k}$ together with all points, apart from finitely many, of the corresponding set $T_{n,k}$.

(iii) A base at the point $(n, y) \in L_n$ is formed by the family of all sets of the form $\{(n, y)\} \cup (A_{n,y}\setminus K)$, where $A_{n,y} = \{(x, y') \in T: y' = y \text{ and } |n - x| < 1\}$, and K is any finite set.

(iv) A base at the point a^* is formed by the sets $U_n = \{(x, y) \in X: x < n\}$, n even.

(v) A base at the point b^* is formed by the sets of the form $V_n = \{(x, y) \in X: x > n\}$, n even.

It is simple to verify that these conditions define a topology on X: if $\mathcal{B}$ is the collection of all the nhoods in X defined above and $x \in V_1 \cap V_2$, $V_1 \in \mathcal{B}$, $V_2 \in \mathcal{B}$, then there exists $V_3 \in \mathcal{B}$ with $x \in V_3 \subset V_1 \cap V_2$; hence, $\mathcal{B}$ is a base for a topology on X (see Problem 14, Chap. 2). Clearly, the family of closures of elements of $\mathcal{B}$ is a net in X. Consequently, X is regular. We now show that any continuous real-valued function f on X takes the same value at a^* and b^*, which implies that X is not completely regular.

(*a*) The set $T^*_{n,k} = \{(x, y) \in T_{n,k}: f(x, y) \neq f(p_{n,k})\}$ is countable. Hence there is y^* such that $0 \leqslant y^* < \frac{1}{2}$ and $X_{y^*} \cap T^* = \Lambda$, where $X_{y^*} = \{(x, y) \in X: y = y^*\}$, $T^* = \cup \{T^*_{n,k}: n \text{ odd and } k \text{ an integer}, k \geqslant 2\}$. Let φ_n be the reflection of the plane R^2 in the line $x = n$, n odd. Then $\varphi_n(X) = X$ and $\varphi_n: X_{y^*} \to X_{y^*}$ is a continuous function. The set $X_{y^*} \cap T_{n,k}$ consists of precisely two points, namely $z'_{n,k}$ whose abscissa is less than n and $z''_{n,k}$ whose abscissa is greater than n. Clearly, $\varphi_n(z'_{n,k}) = z''_{n,k}$. By the choice of y^*, we have $f(z'_{n,k}) = f(z''_{n,k})$, since the points $z'_{n,k}$ and $z''_{n,k}$ belong to the set $T_{n,k}\setminus T^*_{n,k}$. Put $a_n = L_n \cap X_{y^*}$. Then $\varphi_{n+1}(a_n) = a_{n+2}$. In addition, $\lim_{k\to\infty} z'_{n+1,k} = a_n$ and $\lim_{k\to\infty} z''_{n+1,k} = a_{n+2}$. Since $f_{n+1}(z'_{n+1,k}) = f_{n+1}(z''_{n+1,k})$ and f is continuous, we infer that $f(a_n) = f(a_{n+2})$ for every even n. Hence $f(a_n) = f(a_k)$ for every even n and k. But $\lim_{n\to-\infty}$

$a_n = a^*$ and $\lim_{n \to +\infty} a_n = b^*$. Hence $f(a^*) = \lim_{n \to -\infty} f(a_n) = \lim_{n \to \infty} f(a_n) = f(b^*)$.

10. Assume that (ii) is false. Then $\Lambda \notin \beta'$. Fix $V \in \beta$ for which $V \cap F = \Lambda$, and choose $W \in \beta$, α-separated from $X \setminus V$. Fix also some α-ultrafilter β^* on Y, containing β'. Then $V \in \beta' \subset \beta^*$ and $W \in \beta' \subset \beta^*$. Let $U^* \in \beta^*$. Put $V^* = W \cap U^*$ and choose a set $W^* \in \beta^*$, α-separated in Y from $Y \setminus V^*$. Then V (resp. V^*) is an α-nhood in X (resp. V) of the set W (resp. $W^* \subset W$). Hence (see Problem 69, Chap. 4) V^* is an α-nhood of W^* in X. But W^* intersects each member of the family β' (because $\beta^* \ni W^*$ and $\beta^* \supset \beta'$, β^* is an α-ultrafilter). Furthermore, $W^* \cap U \neq \Lambda$ for each $U \in \beta$. But, then, $V^* \in \beta$ (see Problem 5). Consequently, $U^* \in \beta$ (because $V^* \subset U^* \subset Y$). Thus $\beta^* \subset \beta$. Hence $\beta^* \subset \beta'$, i.e., $\beta^* = \beta'$ and β' is an α-ultrafilter on Y.

11. That the space $X^* \times X^* = Y^*$ is completely regular follows from X^* being completely regular and the assertions of Problem 35. We must show that the space $X^* \times X^*$ is not normal. In $X^* \times X^*$ consider the set R of all points $y = (x, 1 - x), x \in X^*$. The subspace $R \subset X^* \times X^*$ is evidently discrete and closed in $X^* \times X^*$. Represent R as the union: $R = R_1 \cup R_2$ of the rational points R_1 and irrational points R_2. R_1 and R_2 are obviously closed in $X^* \times X^*$ and $R_1 \cap R_2 = \Lambda$. We show that it is impossible to separate them by disjoint nhoods in $X^* \times X^*$. Assume the contrary: suppose that there are nhoods VR_1 and VR_2 of these sets for which $VR_1 \cap VR_2 = \Lambda$. For each $y \in R_2$ consider a nhood $V_y = U_x \times U_{1-x}$, where U_x is an half-open interval $[x, x')$ in X^* and U_{1-x} is a half-open interval $[1 - x, x'')$ such that $V_y \subset VR_2$ and U_x and U_{1-x} are of equal lengths. Write $V'R_2 = \bigcup \{V_y : y \in R_2\}$, open in $X^* \times X^*$, $R_2 \subset V'R_2$ and $VR_1 \cap V'R_2 = \Lambda$. For each nhood V_y, $y \in R_2$, $r(y)$ be the length of U_x and U_{1-x}, a real number. Consider the set $A_k = \{y \in R_2 : r(y) \geqslant 1/k\}$. We have $R_2 = \bigcup \{A_k : k \in N^+\}$. If R_2 is considered as being endowed with the usual topology of the real line, then R_2 is a complete metrizable metric space (Prob. 237, Chap. 2) and, consequently, R_2 cannot be represented as the union of a countable number of nowhere dense sets (Problem 238, Chap. 2). Hence some A_{k_0} is dense in a certain interval (a, b) of R, in the usual topology of a real line. This interval contains a rational point $y_0 = (x_0, 1 - x_0) \in R_1$. Now if V_{y_0} is any basic nhood of this point in $X^* \times X^*$, then there is a point $y^* = (x^*, 1 - x^*) \in A_{k_0} \subset R_2$, for which the nhood V_{y^*} constructed above intersects V_{y_0}. (Here use is made of the dense set A_{k_0} in the interval (a, b) and the fact that $r(y) \geqslant 1/k_0$ is continuous for all $y \in A_{k_0}$.) Hence it follows that $V_{y^*} \cap VR_1 \neq \Lambda$ and so, $VR_1 \cap V'R_2 \neq \Lambda$. A contradiction is thus obtained, showing that $X^* \times X^*$ is not a normal space.

12. Since F is compact and X is completely regular, it is easy to construct (finitely many) open sets $U_1, \ldots, U_k$ covering F, and continuous, bounded real-valued functions $f_1, \ldots, f_k$ on X such that $f_i(U_i) = 0$, $f_i(P) = 1$, $i = 1, \ldots, k$. Then $f = f_1 \times f_2 \times \ldots \times f_k$ (the usual product of functions $f_1, \ldots, f_k$) is the required function.

13. Let $V_1, V_2 \in \mathcal{B}$ and $x \in V_1 \cap V_2$. Then there exists $V_3 \in \mathcal{B}$ such that $x \in V_3 \subset V_1 \cap V_2$. Hence $\mathcal{B}$ is a base in X. The second part of assertion (i) is obvious. Since $\{x\} = \tilde{O}(x, r) \cap l$ for all $x \in l$, (ii) holds. Assertion (iii) is trivial. We have $X = l \cup Y$, where l is discrete and hence metrizable as well as closed in X, and Y is metrizable (and has a countable base) by (iii) and is an open F_σ-set. This proves (iv).

The complete regularity of X at the points of Y stems from the complete regularity of Y, because every function continuous on X relative to the usual topology is continuous also relative to $\mathcal{T}$. Now consider any $O(x, r)$. Draw all possible chords in $O(x, r)$, starting from x. Define a real-valued function f on X as follows: $f_{r,x}(X\setminus O(x, r)) = 1$, $f_{r,x}(x) = 0$, and on the chord indicated define f to be compatible with linearity. It is easily verifiable that $f_{r,x}$ is a continuous function (if $r' < r\varepsilon/2$, then diam $(f(\widetilde{O}(x, r'))) < \varepsilon$).

Represent l in the form $l = R_1 \cup R_2$, $R_1 \cap R_2 = \Lambda$, where R_1 are the rational points in l and R_2 the irrational poins in l. It is plain that R_1 and R_2 are disjoint closed sets. Exactly as in Prob. 11, one shows that they cannot he separated by the disjoint nhoods.

15. If F and Φ are closed in X and $F \cap \Phi = \Lambda$, then at least one of F and Φ consists of isolated points. Suppose that F has this property. Put $U_1 = F$ and $U_2 = X\setminus F$; evidently, U_1 and U_2 are disjoint nhoods of F and Φ.

16. Suppose that $\{x_1, \ldots, x_k\}$ is the set of all non-isolated points of X. Since X is Hausdorff, there exist mutually disjoint nhoods Ox_i of x_i, $i = 1, \ldots, k$. If F and Φ are closed disjoint sets in X, then $OF = F \cup (\cup \{Ox_i: x_i \in F\})\setminus\Phi$ and $O\Phi = \Phi \cup (\cup\{Ox_i: x_i \in \Phi\})\setminus F$ are disjoint nhoods of F and Φ.

17. False. Consider a connected countable Hausdorff space X (such a space exists, see Prob. 353). If X were completely regular, then there would exist a continuous real-valued function f on X, not a constant. Put $f(X) = Y$. Then Y is a countable but not a one-point subspace of the real line. Consequently, Y is disconnected (Probs. 221, 224 of Chap. 2), and hence so is X, a contradiction. Thus, X is not normal (see Prob. 26). Prob. 66 of Chap. 2 gives another example.

18. True. A countable space is Lindelöf (Prob. 208), and a regular Lindelöf space is normal (Prob. 19).

19. Suppose that A and B are disjoint closed subsets of X. For each $x \in A$ fix a nhood Ox whose closure does not intersect B; correspondingly, for $x \in B$ choose a nhood Ox such that $[Ox] \cap A = \Lambda$. Put $\gamma = \{Ox: x \in A\}$, $\lambda = \{Ox: x \in B\}$. Since $\gamma \cup \{X\setminus A\}$ and $\lambda \cup \{X\setminus B\}$ is an open cover of X and $A \cap (X\setminus A) = \Lambda$, $B \cap (X\setminus B) = \Lambda$, there exist countable families $\gamma^* \subset \gamma$ and $\lambda^* \subset \lambda$ for which $\cup \{U: U \in \gamma^*\} \supset A$ and $\cup\{V: V \in \lambda^*\} \supset B$. Evaluate the elements γ^* and λ^*: $\gamma^* = \{U_i: i \in N^+\}$, $\lambda^* = \{V_i: i \in N^+\}$ (assuming that $U_i = U_j$ for $i \neq j$). Put $G_i = V_i\setminus\cup \{[U_j]: j \leqslant i\}$ and $H_i = U_i\setminus\cup \{[V_j]: j \leqslant i\}$. By the choice of the nhood Ox, we have: $A \subset \cup \{G_i: i \in N^+\} = G$, $B \subset \cup \{H_i: i \in N^+\} = H$. Evidently $G_i \cap H_j = \Lambda$ for every i and j. Hence $G \cap H = \Lambda$. Thus G and H are disjoint nhoods of A and B. The second assertion is proved exactly in the same way.

20. See Prob. 77 of Chap. 2 and Probs. 19, 206 of this chapter.

21. If $k = 1$, the assertion is trivial. Assume that the assertion holds for all normal spaces X and for all open covers λ with less than n elements, and let $k = n$, that is, $\lambda = \{U_1, \ldots, U_n\}$. Put $P_n = X\setminus\cup \{U_i: i = 1, \ldots, n-1\}$. Then P_n is closed in X and $P_n \subset U_n$; hence there is an open set V in X for which $P_n \subset V \subset [V] \subset U_n$. Put $X' = X\setminus V$ and $U_i' = U_i \cap X'$, $i = 1, \ldots, n-1$. Then X' is closed in X and hence, together with the topology induced from X, forms a normal space; also $X' = \cup \{U_i': i = 1, \ldots, n-1\}$, i.e., $\{U_1', \ldots, U_{n-1}'\}$ is an open cover of X'. By inductive hypothesis, there exists in X' a closed set $F_i' \subset X'$ such that $F_i' \subset U_i'$ and $\cup \{F_i': i = 1, \ldots, n-1\} = X'$. The sets F_i', $i = 1, \ldots, n-1$ are also closed in X (since X' is closed in X) and $F_i' \subset U_i$. Put $F_i = F_i'$ for $i = 1, \ldots, n-1$ and $F_n = [V]$. Evidently, $\{F_1, \ldots, F_n\}$ is the desired family.

22. Since X is regular, there is a family $\lambda = \{V_n: n \in N^+\}$ of mutually disjoint open sets, for which $x_n \in V_n$ for all $n \in N^+$ (Prob. 44, Chap. 2). Then $A \subset G = \bigcup\{V_n: n \in N^+\}$ and G is open in X. Since X is normal, there is an open set W in X for which $A \subset W \subset [W] \subset G$. Put $U_n = V_n \cap W$. Evidently, $x_n \in U_n$ and U_n is open in X for all $n \in N^+$. We show that the family $\gamma = \{U_n: n \in N^+\}$ is discrete. Evidently, $\bigcup\{U_n: n \in N^+\} = W$. Let $z \notin [W]$. Then $X \setminus [W]$ is a nhood of z, which intersects no element of γ. Assume now that $z \in [W]$. Then $z \in G$ and hence $z \in V_k$ for some $V_k \in \lambda$. Then V_k is a nhood of z, intersecting just one element of γ, namely U_k. Indeed, if $n \neq k$, then $U_n \subset V_n$ and $V_k \cap V_n = \Lambda$ implies that $U_n \cap V_k = \Lambda$.

23. Since Y is normal, it is homeomorphic to a closed subspace of $X \times Y$. Assume that there is in Y a closed set M, not the intersection of countable family of open sets in Y, and a countable set $N \subset X$ not closed in X. Fix $x^* \in [N] \setminus N$ and consider the sets $A = N \times M$ and $B = \{x^*\} \times (Y \setminus M)$. By the definition of the Tychonoff topology on $X \times Y$ it is clear that A contains no limit point of B, and B contains no limit point of A. Hence, A and B are closed disjoint subsets of the space $(X \times Y) \setminus F$, where $F = [A] \cap [B]$ (closure in $X \times Y$). Since $X \times Y$ is hereditarily normal, there exist (in $(X \times Y) \setminus F$, and hence also in $X \times Y$) disjoint open sets U and V, for which $A \subset U$, $B \subset V$. Now consider the set $M_x = \{(x', y): x' = x, y \in M\}$, where $x \in N$. Each M_x is closed in $X \times Y$ (since M is closed in Y) and $M_x \subset A \subset U$. From the definition of product topology it follows that for each $x \in N$ there is an open set U_x in Y containing M such that $M_x \subset \{x\} \times U_x \subset U$. The family $\{U_x: x \in N\}$ is countable. Hence (by the choice of M) $\bigcap\{U_x: x \in N\} \setminus M \neq \Lambda$. Fix $y^* \in \bigcap\{U_x: x \in N\} \setminus M$. Then, $(x^*, y^*) \in [N \times \{y^*\}]$. Since $(x^*, y^*) \in \{x\} \times U_x \subset U$ for all $x \in N$, it follows that $N \times \{y^*\} \subset U$ and hence $(x^*, y^*) \in [U]$. However, $(x^*, y^*) \in B$, but $y^* \notin M$ and $B = \{x^*\} \times (Y \setminus M)$. Hence $(x^*, y^*) \in V$ and $V \cap [U] \neq \Lambda$. Since V is open in X, we conclude that $V \cap U \neq \Lambda$, a contradiction.

24. (*a*) Construct the open sets U_r, $r = q/2^n$, $n \in N^+$, $q = 1, 2, \ldots, 2^n$, by induction on Urysohn's lemmas (Prob. 45, Chap. 2). Suppose we have constructed the sets $U_{q/2^{n-1}}$, $q = 1, 2, \ldots, 2^{n-1}$, with the property that $[U_{q_1/2^{n-1}}] \subset U_{q_2/2^{n-1}}$, whenever $(q_1/2^{n-1}) < (q_2/2^{n-1})$, then the set $U_{q/2^n}$, $q = 1, 2, \ldots, 2^n$ is constructed as follows: for every pair $(q/2^{n-1})$, $((q+1)/2^{n-1})$, by Urysohn's lemma (Prob. 45, Chap. 2), we take an open set U_r ($r = \frac{1}{2}[(q/2^{n-1}) + ((q+1)/2^{n-1})] = (2q+1)/2^n$ for which $U_{q/2^{n-1}} \subset [U_{q/2^{n-1}}] \subset U_r \subset [U_r] \subset U_{q+1/2^{n-1}}$.

(*b*) For the proof of assertion (*b*), use the fact that between every pair of real numbers there are many binary rational numbers, i.e., numbers of the form $q/2^n$, and also use assertion (*a*) of this problem.

(*c*) The fact that the function $f(x) = \inf\{t: x \in U_t\}$ vanishes on the set $[U_0]$ follows immediately from its definition. It remains to show that $f(x)$ is continuous at an arbitrary point $x_0 \in X$. Let $\varepsilon > 0$. Consider the sets $U_{f(x_0)+\varepsilon}$, $U_{f(x_0)-\varepsilon}$ and $Ux_0 = U_{f(x_0)+\varepsilon} \setminus [U_{f(x_0)-\varepsilon}]$, all open. It is easy to verify that $x_0 \in Ux_0$ and that for every point $x \in U_{x_0}$, we have $|f(x) - f(x_0)| < \varepsilon$ which implies the continuity of $f(x)$ at x_0. The proof is complete.

25. Suppose that X is normal, F_0, F_1 are closed in X and $F_0 \cap F_1 = \Lambda$. Consider $U_1 = X \setminus F_1$ and an open set U_0 for which $F_0 \subset U_0 \subset [U_0] \subset U_1$ (Prob. 45, Chap. 2). Using the preceding problem, construct a continuous function $f: X \to [0, 1]$, for which $f(x) = 0$ if $x \in [U_0]$, and $f(x) = 1$ if $x \in X \setminus U_1 = F_1$. Then $f(x)$ is the desired function.

(*b*) Suppose that for every pair of disjoint closed sets F_0 and F_1 in X there exists a continuous function $f(x)$ on X such that

$$f(x) = \begin{cases} 0, \text{ if } x \in F_0, \\ 1, \text{ if } x \in F_1, \\ 0 \leqslant f(x) \leqslant 1. \end{cases}$$

Consider the sets $UF_0 = \{x \in X: f(x) < \frac{1}{2}\}$, and $UF_1 = \{x \in X: f(x) > \frac{1}{2}\}$. The sets UF_0 and UF_1 are obviously open in X, $F_0 \subset UF_0$, $F_1 \subset UF_1$ and $UF_0 \cap UF_1 = \Lambda$.

26. The complete regularity of any normal space is immediate from the preceding problem.

28. Suppose that A and B are disjoint closed sets in Y. Each F_i, being closed, is a normal space. By induction it is easy to construct the open sets U_i and V_i in X, where $i \in N^+$, such that the following conditions hold:

(i) $U_i \supset (A \cap F_i) \cup [U_{i-1}]$;
(ii) $[U_i] \cap (B \cup [V_{i-1}]) = \Lambda$;
(iii) $V_i \supset (B \cap F_i) \cup [V_{i-1}]$;
(iv) $[V_i] \cap (A \cup [U_i]) = \Lambda$

(at every step of this construction, we make use of the normality of X). Consider the sets $U = \bigcup \{U_i: i \in N^+\}$ and $V = \bigcup \{V_i: i \in N^+\}$; then $U \supset A$, $V \supset B$, U and V are open in X and $U \cap V = \Lambda$. $U \cap Y$, $V \cap Y$ are the desired sets.

29. Every open subspace of a perfectly normal space is an F_σ-set and hence is normal (see preceding problem). Consider now an arbitrary subspace Y of X and the disjoint closed sets A, B in Y. Then $[A]_X \cap B = \Lambda$ and $[B]_X \cap A = \Lambda$; otherwise A and B would not be closed in $A \cup B$ and moreover, in $Y \supset A \cup B$. Put $F = [A]_X \cap [B]_X$ and $\tilde{Y} = X \setminus F$. From what has been stated above it follows that $\tilde{Y} \supset A \cup B$. Put $\tilde{A} = [A]_{\tilde{Y}}$ and $\tilde{B} = [B]_{\tilde{Y}}$. From $\tilde{A} \cap \tilde{B} = [A]_{\tilde{Y}} \cap [B]_{\tilde{Y}} \subset [A] \cap [B] = F = X \setminus \tilde{Y}$ and $[A]_{\tilde{Y}} \cap [B]_{\tilde{Y}} \subset \tilde{Y}$ it follows that $\tilde{A} \cap \tilde{B} = \Lambda$. Thus $\tilde{A}$ and $\tilde{B}$ are disjoint closed subsets of the space $\tilde{Y}$, which is open in X and hence normal. Consequently, there exists in $\tilde{Y}$ open sets U and V, for which $U \supset \tilde{A} \supset A$, $V \supset \tilde{B} \supset B$ and $U \cap V = \Lambda$. Then U and V are open in X and $U \cap Y$, $V \cap Y$ are the desired nhoods of A and B. Incidentally, this also proves the assertions of Prob. 27.

30. (*a*) Suppose that X is a normal space and $F \subset X$ is a closed G_δ-set: $F = \bigcap \{U_i: i \in N^+\}$, where U_i, $i \in N^+$, is open in X. By Urysohn's theorem (Prob. 25), for each i there exists a function $f_i(x)$ continuous on X such that $0 \leqslant f_i(x) \leqslant 1$, $x \in X$, and $f_i(x) = 0$ if $x \in F$, and $f_i(x) = 1$ if $x \in X \setminus U_i$. Consider the function

$$f(x) = \sum_{i=1}^{\infty} \frac{1}{2^i} f_i(x)$$

on X. The function $f(x)$ is continuous on X and $0 \leqslant f(x) \leqslant 1$ for every $x \in X$. Also, clearly, $f(x) = 0$ for any point $x \in F$. Let $x_0 \notin F$. Then $x_0 \notin U_{i_0}$ for some index i_0; consequently, $f_{i_0}(x_0) = 1$. But then $f(x_0) \neq 0$. Thus $F = f^{-1}(0)$.

(*b*) Suppose now that X is normal and for $F \subset X$ there exists a continuous function $f(x)$ on X for which $F = f^{-1}(0)$. Consider $U_i = \{x \in X: |f(x)| < 1/i\}$. Then the set

U_i, $i \in N^+$, is open in X and $F = \bigcap\{U_i: i \in N^+\}$.

31. In fact, there exists on X a continuous real-valued function g, for which $g(A) = 0$, $g(B) = 1$ and $g(X) \subset [0, 1]$. Put $f(x) = 2c\,(g(x) - \frac{1}{2})$ for all $x \in X$. f is the desired function.

32. Put $A_1 = \{x \in A: -c \leqslant f_0(x) \leqslant -\frac{1}{3}c\}$, $A_2 = \{x \in A: \frac{1}{3}c \leqslant f_0(x) \leqslant c\}$. The sets A_1 and A_2 are disjoint; they are closed in A and, consequently, in X. Hence there exists a continuous real-valued function g on X such that $g(A_1) = \{-\frac{1}{3}c\}$, $g(A_2) = \{\frac{1}{3}c\}$ and $g(X) \subset [-\frac{1}{3}c, \frac{1}{3}c]$. g is, evidently, the desired function.

33. Assume for definiteness that $f(A) \subset [-1, 1]$. Apply to f the assertions of the preceding problem (where $c = 1$ and f plays the role of f_0). Let g_1 be the function obtained. Put $f_1(x) = f(x) - g_1(x)$ for all $x \in A$. Then $f_1(A) \subset [-\frac{2}{3}, \frac{2}{3}]$. Now again apply the preceding problem to f_1 in the role of f_0 and with $c = \frac{2}{3}$, to obtain the function g_2. Continue this process inductively: thus having defined $g_1, \ldots, g_n$, take

$$f - \left(\sum_{i=1}^{n} g_i \,|\, A\right)$$

as f_0 and $(\frac{2}{3})^n$ as c and choose g_{n+1} according to the conclusions of the preceding problem. Then $|g_n(x)| \leqslant \frac{1}{3}(\frac{2}{3})^{n-1}$ for all $x \in X$ and

$$\left| f(x) - \sum_{i=1}^{n} g_i(x) \right| \leqslant \left(\frac{2}{3}\right)^n$$

for all $x \in A$ (and every n). Hence the series

$$\sum_{i=1}^{\infty} g_i(x)$$

uniformly converges to some function g on X; since all the terms of this series are continuous, g is also a continuous function; this is established exactly in the same way as the corresponding theorem of analysis. Clearly, $g(x) = f(x)$ for all $x \in A$ and $-1 \leqslant g(x) \leqslant 1$ for all $x \in X$.

34. Define a mapping i of the space R onto the interval $J = (-1, 1)$ by the formula $i(x) = x/(1 + |x|)$ for all $x \in R$. Obviously, i is a homeomorphism of R onto J. Then $g = if$ can be regarded as a continuous map of the space A into the interval $[-1, 1]$. By Prob. 33, there exists a continuous map F_1 of the space X into $[-1, 1]$, which is an extension of the mapping g. Since $g(A) \subset (-1, 1)$, $B = F_1^{-1}(\{1, -1\})$ is a closed set in X, disjoint from A. Select a continuous real-valued function $k(x)$ on X, for which $k(A) = 1$, $k(B) = 0$, $k(X) \subset [0, 1]$. Put $F(x) = k(x) \cdot F_1(x)$ for all $x \in X$. On A, the functions F and F_1 coincide with $g = if$. In addition, $F(X) \subset (-1, 1)$. Then $\tilde{f} = i^{-1}F$, continuous from X into R, coincides on A with $i^{-1}if = f$.

35. Consider any point $x^* \in X$, $x^* = \{x_\alpha: \alpha \in A\}$, and any nhood Ox^* of x. There exist $\alpha_1, \ldots, \alpha_k \in A$ and an open set U_{α_i} in X_{α_i}, $i = 1, \ldots, k$, for which $x^* \in \bigcap\{\pi_{\alpha_i}^{-1}$

(U_{α_i}): $i = 1, \ldots, k\} \subset Ox^*$ where π_α is the natural projection of X onto X_α. Since $X_{\alpha i}$ is completety regular, there exists a continuous mapping f_i: $X_{\alpha_i} \to [0, 1]$, for which $f_i(x_{\alpha_i}) = 0$, $f_i(X_{\alpha_i} \setminus U_{\alpha_i}) = \{1\}$ for all $i = 1, \ldots, k$. Then $f_i^* = f_i \pi_{\alpha_i}: X \to [0, 1]$ is also a continuous mapping. Clearly f^* defined by the rule: $f^*(x) = \max \{f_i^*(x): i = 1, \ldots, k\}$, is the desired function: $f(X \setminus Ox^*) = \{1\}$ and $f(x^*) = 0$.

36. Suppose that $\sigma = \{U_\lambda : \lambda \in T\}$ is a base of cardinality τ, equal to the weight of a given space X. The $(U_\lambda, U_{\lambda'})$, $U_\lambda \in \sigma$, $U_{\lambda'} \in \sigma$, is said to be distinguished if there exists a continuous function $f: X \to [0, 1]$, which is equal to 0 at all points of the set $[U_\lambda]$ and to 1 at all points of the set $X \setminus U_{\lambda'}$. Evidently $[U_\lambda] \subset U_{\lambda'}$. Denote by δ the set of all distinguished pairs. Evidently, $|\delta| \leqslant |\delta^2| = |\sigma| = \tau$. With every distinguished pair $(U_\lambda, U_{\lambda'}) \in \delta$ associate some definite continuous function $f_{\lambda\lambda'}: X \to [0, 1]$, for which $f_{\lambda\lambda'}(x) = 0$ if $x \in [U_\lambda]$, and $f_{\lambda\lambda'}(x) = 1$ if $x \in X \setminus U_{\lambda'}$. Denote by Ξ the set of all continuous functions so chosen. Obviously, $|\Xi| = |\delta| \leqslant \tau$, and Ξ is a partitioning family of functions. Indeed, let $x_0 \in X$ and Ux_0 be an arbitrary nhood. Take $U_{\lambda'} \in \sigma$ such that $x_0 \in U_{\lambda'} \subset Ux_0$. By the complete regularity of X, there exists a continuous function $f: X \to [0, 1]$, for which $f(x_0) = 0$ and $f(x) = 1$, if $x \in X \setminus U_{\lambda'}$. Consider a nhood $U'x_0 = \{x \in X: f(x) < \frac{1}{2}\}$ of the point x_0. Evidently, $[U'x_0] \subset U_{\lambda'}$. It now remains to take $U_\lambda \in \sigma$, for which $x_0 \in U_\lambda \subset U'x_0$. We show that the pair $(U_\lambda, U_{\lambda'})$ is distinguished. Construct the function

$$f_1(x) = \begin{cases} 1, \text{ if } f(x) = 1, \\ 0, \text{ if } f(x) \leqslant \frac{1}{2} \\ 2\,(f(x) - \frac{1}{2}), \text{ if } \frac{1}{2} < f(x) < 1. \end{cases}$$

The function $f_1(x)$ is continuous, $0 \leqslant f_1(x) \leqslant 1$ for all $x \in X$; $f_1(x) = 0$ if $x \in [U_\lambda]$; $f_1(x) = 1$ if $x \in X \setminus U_{\lambda'}$. Hence the pair $(U_\lambda, U_{\lambda'})$ is distinguished. Then by the construction of Ξ there exists $f_{\lambda\lambda'} \in \Xi$, for which $f_{\lambda\lambda'}(x) = 0$ if $x \in [U_\lambda]$, and $f_{\lambda\lambda'}(x) = 1$ if $x \in X \setminus U_{\lambda'}$. The proof is complete.

37. (*a*) The diagonal f is a one-to-one map. Indeed, if $x_1, x_2 \in X$, $x_1 \neq x_2$, then there is $\alpha \in A$ such that $f_\alpha x_1 \neq f_\alpha x_2$. But then $fx_1 \neq fx_2$. Also, f is continuous (Prob. 290 of Chap. 2).

(*b*) We must show that the inverse mapping $f^{-1}: fX \to X$ is continuous. Suppose that F is a closed set in X. We must verify that fF is a closed set in $fX \subset \Pi\{Y_\alpha: \alpha \in A\}$. Suppose that $y^0 = fx_0$ does not belong to fF. Then $x_0 \notin F$ and there exists $\alpha_0 \in A$ such that $f_{\alpha_0}x_0 \notin [f_{\alpha_0}F]_{Y_{\alpha_0}}$. Consider the projection $\pi_{\alpha_0}: fX \to Y_{\alpha_0}$. We have $f_{\alpha_0} = \pi_{\alpha_0} f$. Then $\pi_{\alpha_0}[fF] \subset [\pi_{\alpha_0} fF] = [f_{\alpha_0}F]$ (by the continuity of π_{α_0}). Since $f_{\alpha_0}x_0 \notin [f_{\alpha_0}F]_{Y_{\alpha_0}}$, it follows that $f_{\alpha_0}x_0 \notin \pi_{\alpha_0}[fF]_{fX}$, but then $fx_0 \notin [fF]_{fX}$. Thus, $[fF]_{fX} = fF$ for every closed set F in X.

38. The space X is completely regular iff there exists a partitioning family of continuous real-valued bounded functions $f_\lambda : X \to I_\lambda$, $\lambda \in T$. Hence a completely regular space X is mapped homeomorphically via a diagonal map f (see Problem 37) into a Tychonoff cube $Y = \Pi\{I_\lambda : \lambda \in T\}$ (and even into $[fX]_Y$). The rest of the assertion follows from Probs. 35 and 8.

39. Any regular space X with a countable base $\sigma = \{U_i: i \in N+\}$ is perfectly normal (Prob. 20). Drawing on Prob. 30, construct for each i a continuous function $f_i(x)$ on X: $0 \leqslant f_i(x) \leqslant 1$, for which $X \setminus U_i = f_i^{-1}(0)$. The set $\{f_i: i \in N+\}$ of these functions is obviously a partition. Then the diagonal map f maps X homeomorphically into the product of a countable set of intervals : $\prod_{i=1}^{\infty} I_i$, i.e., into a metrizable space (Prob. 194 of Chap. 2). Thus, X itself is also metrizable.

40. Note first that every cozero-set of a space X is open in X. For each $U \in \sigma$ consider a function $f_U: X \to I_U$ continuous on X (here I_U is the interval [0, 1] of the real line endowed with the index U), for which $f_U(X \setminus U) = 0$ and $f_U x \neq 0$ if $x \in U$. Consider the product topology $\Pi\{I_U: U \in \sigma\} = Y$, the Tychonoff cube of weight $|\sigma| \aleph_0$. Clearly, $wY = |\sigma| \leqslant \tau$ (see Prob. 379 of Chap. 2). Furthermore, consider the diagonal product $f_\sigma : X \to \Pi\{I_U: U \in \sigma\}$ of the family of mappings $f_U: X \to I_U$, $U \in \sigma$. It is obvious that $w(f_\sigma X) \leqslant \tau$. As Y_σ take $f_\sigma(X) \subset \Pi\{I_U: U \in \sigma\}$. The mapping $f_\sigma: X \to Y_\sigma$, as the diagonal product of a family of continuous maps, is continuous. It remains to show that $f_\sigma U$ is open in Y_σ and $f_\sigma^{-1} f_\sigma U = U$ if $U \in \sigma$. This is, however, trivial.

41. Consider (drawing upon Prob. 40) a space Y_σ, $wY_\sigma \leqslant \tau$ and a continuous map $f_\sigma: X \xrightarrow{\text{onto}} Y_\sigma$, for which $f_\sigma U$ is open in Y_σ and $f_\sigma^{-1} f_\sigma U = U$, if $U \in \sigma$. Since σ is an exterior pseudobase of the set $M \subset X$, $f_\sigma^{-1} f_\sigma(x) = x$, if $x \in M$. In particular $f_\sigma^{-1} f_\sigma M = M$. The proof is complete.

44. Suppose that $\mathscr{B}$ is a base for X and γ is an open cover. Evidently, from γ one can extract a cover consisting of the elements of $\mathscr{B}$. If from the latter a finite cover can be extracted, then it is apparent that the original cover has a finite subcover.

45. (*a*) Suppose that X is compact, and $\xi = \{F_\lambda: \lambda \in A\}$ is an arbitrary centralized family of closed sets in X. If ξ had empty intersection, then the family $\omega = \{X \setminus F_\alpha: \lambda \in A\}$ of open sets in X would form an open cover of X. Hence, because of the compactness of X, ω has a finite subcover $\omega_0 = \{X \setminus F_{\lambda_1}, \ldots, X \setminus F_{\lambda_s}\}$ of X. This would imply

$$\prod_{i=1}^{s} F_{\lambda_i} = X \setminus \bigcup_{i=1}^{s} (X \setminus F_{\lambda_i}) = X \setminus X = \Lambda,$$

which contradicts the assumption that ξ is a centralized family.

(*b*) Suppose that every centralized family of closed sets in X has a nonempty intersection. We show that X is compact. Let $\omega = \{U_\lambda: \lambda \in A\}$ be an arbitrary open cover of X. The requirement is to extract from it a finite subcover. Assume the contrary, and suppose every finite subfamily of ω fails to be a cover of X. Then $\xi = \{X \setminus U_\lambda: \lambda \in A\}$ is a centralized system and consists of closed sets; consequently, $\bigcap\{(X \setminus U_\lambda): \lambda \in A\} \neq \Lambda$. But then $X \setminus (\bigcap\{(X \setminus U_\lambda): \lambda \in A\}) = \bigcup\{(X \setminus (X \setminus U_\lambda)): \lambda \in A\} = \bigcup\{U_\lambda: \lambda \in A\} \neq X$, which contradicts the assumption that ω is a cover of X.

46. To solve this problem, one need only note that an adherent point of a centralized family of closed sets belongs to all sets of this family, that if a family of sets is centralized, then the family of closures of members of this family is centralized, and make use of Prob. 45.

47. To solve this problem, it is necessary to recall that every centralized family of sets is contained in a maximal centralized family (Prob. 127 of Chap. 1), and use the preceding problem.

48. (*a*) Suppose X is compact. Denote by A the set of all points of X which are not total accumulation points of M. For each $x \in A$, fix a nhood Ox (in X) such that

$|Ox \cap M| < |M|$. If $A = X$, then $\gamma = \{Ox: x \in A\}$ is an open cover of X. Select from this a finite subcover $\lambda = \{Ox_1, \ldots, Ox_k\} \subset \gamma$. Then $M \subset \bigcup \{Ox_i \cap M: i = 1, \ldots, k\}$ and $|M| = \max \{|Ox_i \cap M|: i = 1, \ldots, k\}$ (see Prob. 90 of Chap. 1), a contradiction. Consequently, $X \setminus A \neq \Lambda$, i.e., M has a complete accumulation point in X.

(*b*) Suppose that there is an open cover γ of X, which has no finite subcover. Put $\tau^* = \min \{|\lambda|: \lambda \subset \gamma$ and λ is a cover of $X\}$. Fix $\lambda^* \subset \gamma$, for which $|\lambda^*| = \tau^*$ and $\bigcup \{U: U \in \lambda^*\} = X$. Then, by assumption $\tau^* \geqslant \aleph_0$. Well-order set λ^* by some minimal well-ordering $<$. Then, by the choice of λ^* and the relation $<$, we have $\bigcup \{V: V \in \lambda^*$ and $V < U\} \neq X$ for each $U \in \lambda^*$. Choose a point $x(U) \in X \setminus \bigcup \{V \cup \{x(V)\}: V \in \lambda^*$ and $V < U\}$ for each $U \in \lambda^*$. This is possible, because otherwise one could select from λ^* a cover of smaller cardinality. Put $M = \{x(U): U \in \lambda^*\}$. Let $y \in X$. We show that y is not a complete accumulation point in M. Certainly $y \in U^*$ for some $U^* \in \lambda^*$. since λ^* covers X. But $x(U) \in U^*$ implies by the choice of $x(U)$ that $U < U^*$. Hence $\{U: U \in \lambda^*$ and $x(U) \in U^*\} \subset \{U: U \in \lambda^*$ and $U < U^*\}$. But $|\{U: U \in \lambda^*$ and $U < U^*\}| < |\lambda^*|$, by the minimality of $<$. Consequently, $|M \cap U^*| < |\lambda^*|$. But $|M| = |\lambda^*| \geqslant \aleph_0$ because $V' \neq V''$, and $V', V'' \in \lambda^*$, imply that $x(V') \neq x(V'')$. Thus M has no complete accumulation point in X.

49. Choose an arbitrary countably infinite set A in X. In X, there is a complete accumulation point of A, say $a' \in X$. Put $A' = A \setminus \{a'\}$. A' is the desired set. Another solution can be given which does not rely on the preceding problem.

50. No. Take as X the space $\mathscr{P}$ defined in Problem 100 of Chap. 5, and as A its subspace $\mathscr{P}_0$. Then every infinite subset C of A has a limit point in $\mathscr{P}$, but the latter is also a complete accumulation point of C, because $|C| \leqslant |A| = \aleph_0$. However, $[A]_X = X$, but X is non-compact.

51. Yes. Suppose that $Y = [A]_X$ is not compact. From $X \supset Y$ it follows that Y is completely regular. Consider some compact Hausdorff extension bY of Y (see, for example, Problem 14, Chap. 4). Choose $z \in bY \setminus Y$ and put $\xi = \{Oz \cap A: Oz$ is a nhood of point z in $Y\}$. Since $[A]_{bY} \supset Y, bY = [A]_{bY}$ and, consequently, ξ is a centralized family. Obviously, z is the unique adherent point of ξ in bY, implying that there is no adherent point of ξ in Y, a contradiction.

52. Denote by $\mathscr{K}$ the collection of all families of open sets in X, no finite subfamily of which covers X. $\mathscr{K}$ can be partially ordered by inclusion. Obviously, $\mathscr{K}$ is inductive with respect to this partial ordering. In accordance with the Kuratowski-Zorn principle (see, p. 6) there is a maximal element in $\mathscr{K}$, containing any preassigned element of $\mathscr{K}$, say γ^*. Thus no finite family γ^* of elements forms a cover of X, but if we add to γ^* some suitable open set not belonging to it, then the family obtained does contain a finite cover of X. It suffices to show that γ^* does not cover X. Consider $\lambda = \gamma^* \cap \mathscr{B}$. Since $\lambda \subset \gamma^*$ and $\lambda \subset \mathscr{B}$, the properties of γ^* and $\mathscr{B}$ imply that λ does not cover X. We show that $\bigcup \{U: U \in \gamma^*\} \subset \bigcup \{V: V \in \lambda\}$. If $x \in U \in \gamma^*$, then there are $V_1, \ldots, V_k \in \mathscr{B}$ such that $x \in V_1 \cap \ldots \cap V_k \subset U$ (because $\mathscr{B}$ is a sub-base). If $V_i \notin \gamma^*$ for all $i = 1, \ldots, k$, then for every i, we can choose from $\{V_i\} \cup \gamma^*$, a finite cover of X. But then a finite cover of X can also obviously be selected from the cover $\gamma^* \cup \{\bigcap \{V_i: i = 1, \ldots, k\}\}$. Furthermore, a finite cover of X can be extracted from $\gamma^* \cup \{U\} = \gamma^*$, a contradiction. Hence for some i^*, where $1 \leqslant i^* \leqslant k$, we have $V_{i^*} \in \gamma^*$. But then $V_{i^*} \in \lambda$ and $x \in \bigcup \{V: V \in \lambda\}$. The relation $\bigcup \{U: U \in \gamma^*\} \subset \bigcup \{V: V \in \lambda\}$

is thus proved. Since λ does not cover X, γ^* also does not cover X.

53. By virtue of Problems 45 and 52, it suffices to verify that every centralized family ξ of elements of $\mathcal{P}$ has nonempty intersection. Let $\xi = \{\mathcal{K}(A_\alpha, B_\alpha): \alpha \in M\}$. We show that the family $\eta_\beta = \{A_\beta\} \cup \{B_\alpha: \alpha \in M\}$ of closed subsets of X is centralized for any $\beta \in M$. Let $\alpha_1, \ldots, \alpha_k \in M$. Then $\mathcal{K}(A_\beta, B_\beta), \mathcal{K}(A_{\alpha_1}, B_{\alpha_1}), \ldots, \mathcal{K}(A_{\alpha_k}, B_{\alpha_k})$ $\in \xi$ and hence $\mathcal{K}(A_\beta, B_\beta) \cap \left(\bigcap_{i=1}^{k} \mathcal{K}(A_{\alpha_i}, B_{\alpha_i})\right) \neq \Lambda$; there exists $C \in \mathcal{F}(X)$ for which $C \subset \bigcap\{B_{\alpha_i}: i = 1, \ldots, k\}$, $C \cap A_\beta \neq \Lambda$. But then $A_\beta \cap (\bigcap\{B_{\alpha_i}: i = 1, \ldots, k\}) \supset C \cap A_\beta \neq \Lambda$. Thus every family η_β is centralized. Hence $\bigcap\{B_\alpha: \alpha \in M\} = B^* \neq \Lambda$, $A_\beta \cap B^* \neq \Lambda$ for each $\beta \in M$. But then $B^* \in \bigcap\{\mathcal{K}(A_\alpha, B_\alpha): \alpha \in M\}$, i.e., ξ has nonempty intersection.

55. X is a Hausdorff space, $x_0 \in X$, F is a compact subspace of X and $x_0 \notin F$. Since X is Hausdorff, for each point $x \in F$ and a point x_0 we may choose nhoods Ux of x and $U_x x_0$ of x_0, for which $Ux \cap U_x x_0 = \Lambda$. Consider the open cover $\{Ux: x \in F\}$ of the compact subspace F and choose from it a finite subcover $\{Ux_i: i = 1, 2, \ldots, s\}$: $\bigcup_{i=1}^{s} Ux_i \supset F$. Consider further the corresponding nhoods $U_{x_1}x_0, \ldots, U_{x_s}x_0$ of x_0. We have $Ux_i \cap U_{x_i}x_0 = \Lambda$, $i = 1, 2, \ldots, s$, and so

$$\bigcup_{i=1}^{s} Ux_i \cap \bigcap_{i=1}^{s} U_{x_i}x_0 = \Lambda.$$

Consequently, if

$$UF = \bigcup_{i=1}^{s} Ux_i, \quad Ux_0 = \bigcap_{i=1}^{s} U_{x_i}x_0,$$

we obtain disjoint nhoods UF of these F and Ux_0 of x, completing the proof.

56. This is implied by the preceding problem.

58. The set $X \setminus \{x\}$ is not closed in the Hausdorff space X. Consequently, $Y = X \setminus \{x\}$, as a subspace of X, is not compact (Problem 57), and Y contains an infinite set A, which has no complete accumulation points in Y (Problem 48). But in X there is a complete accumulation point for A, since X is compact (Problem 48) and, by what has been proved, this point can be only x.

59. Fix some open cover γ of X, from which it is impossible to extract a finite subcover. For every point x choose a nhood Ox such that $[Ox] \subset G$ for some $G \in \gamma$. Put $\lambda = \{[Ox]: x \in X\}$. Then λ, a cover of X, is a refinement of γ, and hence no finite set of elements of λ covers X. Consequently, $\xi = \{X \setminus P: P \in \lambda\}$ is a centralized family of open sets in X. Let x^* be any element not belonging to X; put $X^* = X \cup \{x^*\}$ and $\mathcal{B} = \mathcal{T} \cup \xi^*$, where $\xi^* = \{U \cup \{x^*\}: U \in \xi\}$. Obviously, $\mathcal{B}$ is a sub-base for some topology $\mathcal{T}^*$ on X^* and $(X, \mathcal{T})$ is a subspace of $(X^*, \mathcal{T}^*)$. The family ξ^* is a sub-base for the topology $\mathcal{T}^*$ at the point x^*. Then $U^* \cap X \in \xi$ for each $U^* \in \xi^*$. Since every element of ξ is nonempty, we conclude that x^* belongs to the closure of X in $(X^*, \mathcal{T}^*)$. Furthermore, if x' and $x'' \in X$, the points x', x'' have disjoint nhoods in the space $(X^*, \mathcal{T}^*)$ (the same will be disjoint nhoods of these points in $(X, \mathcal{T})$). If $x \in X$, then

$X \setminus [Ox] = Ox^*$, a nhood of x^* in $(X^*, \mathcal{T}^*)$, is disjoint from Ox, which is a nhood in $(X^*, \mathcal{T}^*)$ of x and $[Ox] \in \lambda$. Hence $(X^*, \mathcal{T}^*)$ is a Hausdroff space.

60. This is implied by the preceding problem.

61. Add to X one more point ξ, to obtain the set $\tilde{X} = X \cup \{\xi\}$. Now suppose that $\omega = \{U_\alpha : \alpha \in A\}$ is an arbitrary open cover of X. Define on $\tilde{X}$ a topology $\mathcal{T}(\omega)$ in the following way: every open set in X is also open in $\tilde{X}$; moreover, every set of the form $\tilde{X} \setminus [U_{\alpha_1} \cup \ldots \cup U_{\alpha_s}]_X$, where $\alpha_1 \in A, \ldots, \alpha_s \in A$, is open in $\tilde{X}$. It is not difficult to see that $\tilde{X}$, with the topology $\mathcal{T}(\omega)$, with a base described above, is a Hausdorff space. Here ξ is an isolated point in $\tilde{X}$ iff a finite subfamily can be extracted from ω, the closures of whose elements cover X. Hence the desired assertion follows.

62. Suppose that ω is an arbitrary open cover of a subset $[U]_X$. Consider the open cover Ω of X, consisting of all elements of the system ω and also the set $X \setminus [U]_X$. Using the H-closedness of X and Problem 61, choose a finite subsystem $\Omega_0 \subset \Omega$, the closures of whose elements cover X. The latter signifies that $\tilde{\Omega}_0 = \cup (U' : U' \in \Omega_0\}$ is a dense set in X. It is clear that $\tilde{\Omega}_0 \cap U$ is dense in U, and hence also in $[U]_X$. Therefore, the closures of elements of the family $\omega_0 = \{U' : U' \in \Omega_0$ and $U' \cap [U]_X \neq \Lambda\}$ cover the subspace $[U]_X$. It is obvious that $\omega_0 \subset \omega$, $|\omega_0| < \aleph_0$. By the preceding problem, the subspace $[U]_X$ is H-closed.

63. See, for example, Iliadis and Fomin [1].

64. See the solution of Problem 55. Argue similarly. (A solution can also be obtained from Problem 19.)

65. Assume the contrary. Then, for each point $x \in \Phi$, a nhood Ox can be found such that $Ox \cap \{x_{n_k} : k \in N^+\} = \Lambda$. From the cover $\{Ox : x \in \Phi\}$ of the set Φ extract a finite subcover $\{Ox_1, \ldots, Ox_l\}$. Then $V = \cup \{Ox_i : i = 1, \ldots, l\}$ is a nhood of Φ in X, disjoint from $\{x_{n_k} : k \in N^+\}$. But there exists n^* for which $U_{n^*} \subset V$. Hence $x_{n^*} \in U_{n^*} \subset V$, a contradiction.

66. Put $\eta = \{(X \setminus U) \cap P : P \in \xi\}$. The family η consists of closed sets in X, and $\cap \{Q : Q \in \eta\} \subset (X \setminus U) \cap \Phi = \Lambda$ by virtue of the definition of η and Φ. Since X is compact, we conclude that η is not centralized. Hence there exist $P_1, \ldots, P_k \in \xi$, for which $\bigcap_{i=1}^{k} (P_i \cap (X \setminus U)) = \Lambda$. But ξ is a prefilter, hence there is a $P^* \in \xi$ such that

$$P^* \subset \bigcap_{i=1}^{k} P_i.$$

Then,

$$P^* \cap (X \setminus U) \subset \bigcap_{i=1}^{k} (P_i \cap (X \setminus U)) = \Lambda,$$

i.e., $\quad P^* \subset U.$

67. Let $\tilde{\xi}$ be the family of all finite intersections of elements of the family ξ. Then $\tilde{\xi}$ is a prefilter, all its elements are closed and $F_0 = \cap \{F : F \in \tilde{\xi}\}$. It remains to refer to the preceding problem.

68. Let $F = \bigcap\{U: U \in \gamma\}$, where γ is a family of open sets in X and F is closed in X. For each $U \in \gamma$ fix an open set $V(U)$ such that $F \subset V(U) \subset [V(U)] \subset U$; this is possible since X is normal. Then $F = \bigcap\{[V(U)]: U \in \gamma\}$. By the preceding problem, for any open set OF in X containing F there is a finite collection $U_1, \ldots, U_k$ of elements of γ, for which $\bigcap\{[V(U_i)]: i = 1, \ldots, k\} \subset OF$. Moreover, $\bigcap\{V(U_i): i = 1, \ldots, k\} \subset OF$. Hence, by taking for λ the family $\{V(U): U \in \lambda\}$ and for $\tilde{\lambda}$ the family of all finite intersections of elements of λ, we have that (*a*) $|\tilde{\lambda}| = |\lambda| \leqslant |\gamma|$ (Problem 58, Chap. 1) and (*b*) $\tilde{\lambda}$ forms a defining system of nhood of F in X. It can be further asserted that $|\gamma|$ is the pseudo-character of F in X.

69. Set $U_1 = U$ and assume that the open sets U_i are defined for all $i = 1, \ldots, k$ where k is some positive integer. In addition, let $U_k \supset \Phi$ and $[U_{i+1}] \subset U_i$ for all $i < k$. Then take as U_{k+1} any open set such that $\Phi \subset U_{k+1} \subset [U_{k+1}] \subset U_k$. If the sequence $\{U_i: i \in N^+\}$ is constructed in accordance with the above procedure, then $\bigcap\{U_i: i \in N^+\} = \bigcap\{[U_i]: i \in N^+\}$ is a compactum of countable character in X (because it is a G_δ-set in X; see also Problem 68), containing Φ and included in $U_1 = U$.

70. Suppose that $\varphi = \{U_\alpha: \alpha \in A\}$ is a defining system of nhoods of Φ in X, $\psi = \{V_\beta: \beta \in B\}$ is a defining system of nhoods of F in Φ and $|A| \leqslant \tau$, $|B| < \tau$. For each point $x \in \Phi \diagdown F$ fix a nhood Ox in X for which $[Ox] \cap F = \Lambda$. Then $\gamma_\beta = \{V_\beta\} \cup \{Ox: x \in \Phi \diagdown F\}$ includes finite cover of the compactum Φ; let this be $\omega_\beta = \{V_\beta, Ox_1^\beta, \ldots, Ox_{n(\beta)}^\beta\}$. Set $M_\beta = \bigcup\{Ox_i^\beta: i = 1, \ldots, n(\beta)\}$. Note that M_β is an open set in X and $[M_\beta] \cap F = \Lambda$. For each element (α, β) of $C = A \times B$ put $W_{(\alpha,\beta)} = U_\alpha \diagdown [M_\beta]$. We must show that $\{W_{(\alpha,\beta)}: (\alpha, \beta) \in C\}$ is a defining family of nhoods of F in X. Suppose that G is any open set in X containing F. Fix $V_{\beta^*} \in \psi$ for which $V_{\beta^*} \subset G$. Then, $M_{\beta^*} \cup G$ is a nhood of Φ in X. Fix $\alpha^* \in A$, for which $U_{\alpha^*} \subset M_{\beta^*} \cup G$. Then $F \subset U_{\alpha^*} \diagdown [M_{\beta^*}] \subset G$, i.e., $F \subset W_{(\alpha^*,\beta^*)} \subset G$. The goal is achieved; in fact, $|C| = |A \times B| \leqslant \tau$.

71. Let $Q = \bigcap\{\Gamma_\alpha: \alpha \in A\}$, where $|A| \leqslant \tau$, Γ_α is open in X. For each $\alpha \in A$, fix a compactum F_α of countable character in X such that $x \in F_\alpha \subset \Gamma_\alpha$ (see Problem 69). Then $\Phi = \bigcap\{F_\alpha: \alpha \in A\}$ is the required compactum (see the preceding problem).

72. For each point $x \in F$ fix some countable base γ_x at the point x in X. Set $\gamma = \bigcup\{\gamma_x: x \in F\}$. Then $|\gamma| \leqslant \aleph_0$. Obviously, the family γ^* of all finite unions of members of γ is countable and contains a defining system of nhoods of F in X.

73. By Problem 129, there is a net $\mathcal{S}$ of closed sets in X such that $|\mathcal{S}| \leqslant \tau$. Put $\mathcal{S}_F = \{P \in \mathcal{S}: P \cap F = \Lambda\}$. Then, obviously, $|\mathcal{S}_F| \leqslant |\mathcal{S}| \leqslant \tau$ and $F = \bigcap\{X \diagdown P: P \in \mathcal{S}_F\}$. Thus $\gamma = \{X \diagdown P: P \in \mathcal{S}_F\}$ is the desired family of sets.

74. No. Suppose that X is an arbitrary infinite set and y, z are any two distinct points, not in X. Let any subset of X be open in $\tilde{X} = X \cup \{y\} \cup \{z\}$. Assume further that the complements of finite sets in X, and the empty set, are open in $\tilde{X}$. This transforms $\tilde{X}$ into a topological space. Obviously, $Y = X \cup \{y\}$ and $Z = X \cup \{z\}$ are compact subspaces. But $Y \cap Z = X$ and X, as a subspace of $\tilde{X}$, is discrete and hence noncompact (in fact, $|X| \geqslant \aleph_0$).

75. By virtue of Problem 47, it suffices to show that every maximal centralized family ξ of subsets of the space X has an adherent point. For each $\alpha \in A$ consider the family $\xi_\alpha = \{\pi_\alpha M: M \in \xi\}$ of sets of X_α (here $\pi_\alpha: \Pi\{X_\alpha: \alpha \in A\} \to X_\alpha$ is the natural projection). It is clear that ξ_α is a centralized family; consequently, it has (in view of the space X_α being compact, see Problem 46) an adherent point $x_\alpha \in X_\alpha$. Consider the point $x = \{x_\alpha : \alpha \in A\} \in X$. We show that x is an adherent point of ξ. Let

$$Ox = \bigcap_{i=1}^{s} \pi_{\alpha_i}^{-1} U^{\alpha_i}$$

be an arbitrary nhood of x in the standard Tychonoff base (see Problem 354, Chap. 2). Then $x_{\alpha_1} \in U^{\alpha_1}, \ldots, x_{\alpha_s} \in U^{\alpha_s}$; consequently,

$$\pi_{\alpha_1} M \cap U^{\alpha_1} \neq \Lambda, \ldots, \pi_{\alpha_s} M \cap U^{\alpha_s} \neq \Lambda \tag{1}$$

for any $M \in \xi$; but then

$$\pi_{\alpha_1}^{-1} U^{\alpha_1} \cap M \neq \Lambda, \ldots, \pi_{\alpha_s}^{-1} U^{\alpha_s} \cap M \neq \Lambda \tag{2}$$

for any $M \in \xi$. Using the maximality of the centralized family ξ (see Problem 128, Chap. 1), (2) implies that $\pi_{\alpha_i}^{-1} U^{\alpha_i} \in \xi$, $i = 1, 2, \ldots, s$, whence

$$Ox = \bigcap_{i=1}^{s} \pi_{\alpha_i}^{-1} U^{\alpha_i} \in \xi$$

and $x = \{x_\alpha : \alpha \in A\} \in Ox \cap M$ for any $M \in \xi$, i.e., $Ox \cap M \neq \Lambda$ for any $M \in \xi$. Noticing that Ox is an arbitrary basic nhood of x in X, we get $x \in [M]_X$ for any $M \in \xi$, i.e., $x = \{x_\alpha : \alpha \in A\}$ is an adherent point of ξ.

76. For each pair (x, y) of points $x \in A$, $y \in B$, there are open sets $O^y x$ and $O^x y$ (in X and Y respectively), satisfying $O^y x \times O^x y \subset W$. The family $\gamma^x = \{Ox_y : y \in B\}$ covers B; hence it has a finite subcover λ (because B is compact). Let $\lambda = \{O^x y_1, \ldots, O^x y_k\}$. The set $Ux = \bigcap \{O^{y_i} x : i = 1, \ldots, k\}$ is a nhood of x in X. Put $V^x = \bigcup \{O^x y_i : i = 1, \ldots, k\}$; V^x is a nhood of B in Y. We have, in addition, $Ux \times V^x \subset W$. Fix such Ux and V^x, for each point $x \in A$. Then $\gamma = \{Ux : x \in A\}$ is an open cover of the compact set A. Extract from it a finite subcover $\mu = \{Ux_i : i = 1, \ldots, l\}$. Set $V = \bigcap \{V^{x_i} : i = 1, \ldots, l\}$ and $U = \bigcup \{Ux_i : i = 1, \ldots, l\}$. Then U is a nhood of A in X, V is a nhood of B in Y and $U \times Y \subset W$.

77. By Problem 75, Φ is a compact subspace of X. Each point p of Φ has a standard nhood Op included in W. Denote by Ap, the finite subset of A, corresponding to Op (see p. 44). From the cover $\{Op : p \in \Phi\}$ of Φ extract a finite cover $\{Op_1, \ldots, Op_k\}$. Put $A^* = \bigcup \{Ap_i : i = 1, \ldots, k\}$. The set $W^* = \bigcup \{Op_i : i = 1, \ldots, k\}$ is a nhood of Φ in X, contained in W and independent of the elements of the set $A \setminus A^*$ (see p. 85 and Problem 384, Chap. 2). Consider $\tilde{X} = \Pi \{X_\alpha : \alpha \in A^*\}$, $\tilde{\Phi} = \Pi \{\Phi_\alpha : \alpha \in A^*\}$ and $\tilde{W} = \pi_{A^*} W^*$, where $\tilde{W}$ is the projection of W^* onto the boundary $\tilde{X}$ of the product X. Then $\tilde{W}$ is a nhood of $\tilde{\Phi}$ in $\tilde{X}$ and $\Phi \subset \tilde{W} \times \Pi\{X_\alpha : \alpha \in A \setminus A^*\} \subset W^* \subset W$. This argument reduces the argument to the case when A is a finite set. When $|A| \leqslant 1$, the statement is trivial. For $|A| = 2$ the conclusion is already provided in the preceding problem. Suppose the statement holds when $|A| \leqslant k$. Assume that $|A| = k + 1$. Fix $\alpha' \in A$ and set $A' = A \setminus \{\alpha'\}$, $X' = \Pi \{X_\alpha : \alpha \in A'\}$, $\Phi' = \Pi \{\Phi_\alpha : \alpha \in A'\}$. Then, $X = X' \times X_{\alpha'}$, $\Phi = \Phi' \times \Phi_{\alpha'}$ and W is a nhood of Φ in X. By Problem 76, there are sets $W' \subset X'$, $U_{\alpha'} \subset X_{\alpha'}$, open in X' and in $X_{\alpha'}$ respectively, satisfying $W' \supset \Phi'$, $U_{\alpha'} \supset \Phi_{\alpha'}$ and $W' \times U_{\alpha'}$

$\subset W$. But $|A'| = k$. By the inductive assumption, there exist in X_α open sets U_α, containing Φ_α and satisfying $\Pi\{U_\alpha: \alpha \in A'\} \subset W'$. Then $\Pi\{U_\alpha: \alpha \in A\} \subset W' \times U_{\alpha'} \subset W$. The problem is solved.

78. As a diagonal product of continuous mappings, f is continuous. Since F is compact (and $I^{|A|}$ is a Hausdorff space), f is closed (Problem 92). If (i) holds, then from Problem 319 of Chap. 2 it follows that $\varphi = f|X, \varphi: X \to J^{|A|}$ is a closed mapping and hence $f(X) = \varphi(X)$ is a closed set in $J^{|A|}$. (Simple: $f(X) = J^{|A|} \cap f([X]_F)$.) Thus it suffices to prove (i). But (i) indirectly follows from the definition of X.

79. Since the character of Φ_α in F is countable, each compactum Φ_α can be represented as the set of all zeros of some continuous real-valued function f_α on F (see Problem 30). Further, since the weight of X is τ and F is normal, there is a family $\{\psi_\beta : \beta \in B\}$ of continuous real-valued functions on F such that (i′) $\psi_\beta(F) \not\ni 0$ for all $\beta \in B$; (ii′) $|B| = \tau$; (iii′) there exists $\beta_1 \in B$ satisfying $\psi_{\beta_1}(F) = \{1\}$; (iv′) for any $x_1, x_2 \in X$ there is $\beta' \in B$ such that $\psi_{\beta'}(x_1) \neq \psi_{\beta'}(x_2)$. Substitute $A' = A \times B$ and for each $\alpha' = (\alpha, \beta) \in A'$ put $f_{\alpha'} = f_\alpha \cdot \psi_\beta$. It is clear that $|A'| = |A| \cdot \tau$. It is also obvious that $X = \{x \in F: f_{\alpha'}(x) \neq 0 \text{ for all } \alpha' \in A'\}$.

From the preceding problem it follows that the restriction φ to X of f, the diagonal product of the mappings $f_{\alpha'}$, for $\alpha' \in A'$, is a continuous closed mapping of X into the space $J^{|A'|}$, where $J = (0, 1]$. But by (iii′) and (iv′) the family $\{f_{\alpha'}: \alpha' \in A'\}$ separates the points of X (for any $x_1, x_2 \in X$ there is $\alpha' \in A'$ such that $f_{\alpha'}(x_1) \neq f_{\alpha'}(x_2)$). Hence φ is one-to-one and continuous; therefore, we conclude that φ is a homeomorphism of X onto a closed subspace of $J^{|A'|} = J^{|A| \cdot \tau}$. But the half-interval J is homeomorphic to a closed ray, a closed subspace of the real line R. Hence $J^{|A| \cdot \tau}$ is homeomorphic to a closed subspace of the product $R^{|A| \cdot \tau}$ and, consequently, X is homeomorphic to a closed subspace of the latter.

80. As a subspace of X, S_0 is the usual circumference (a circumference with the usual topology). Hence, any open cover of X with the elements of its base indicated in the definition of X, includes a finite family $\{Ox_1, \ldots, Ox_k\}$ which covers S_0. But then automatically, all of S_1 save for a finite point set is covered by the sets $Ox_1, \ldots, Ox_k$. It now suffices to adjoin to $\{Ox_1, \ldots, Ox_k\}$ a finite number of elements of the cover under consideration in order to cover the excluded points, and the desired finite subcover of X is obtained. Hence X is compact (Problem 44). It is clear that X is a Hausdorff space and that it is first countable at all its points. In addition, $X = S_0 \cup S_1$ and S_1 is metrizable, being a discrete space, but S_0 is metrizable as a subspace of the usual plane. But X has no countable base, for otherwise S_1 would also have a countable base, which is impossible because S_1 is uncountable and discrete.

81. Strictly speaking, we need only prove just the statement (*d*). Suppose that ω is an open cover of the space $A(X)$ by basic open sets (defined in the hypotheses of the problem). Regard ω as a cover of the subspace X' by open sets in $A(X)$. The subspace X' is homeomorphic to the compactum X. Hence, from ω a finite subfamily $\omega' \subset \omega$ can be extracted which covers the compact subspace X'. Since ω' is made up of basic sets, is finite and covers X', there remains uncovered (by definition of a nhood of a point $x' \in X'$ in $A(X)$) at most a finite number of points $x'' \in X''$. Hence by adding, if necessary, to ω' a further finite number of elements of ω covering this finite number of points, we obtain a finite subcover $\omega_0 \subset \omega$ of $A(X)$.

82. Under the natural interpretation of X as a subspace of the space $[0, 1]^{|[0,1]|}$, X is closed in $[0, 1]^{|[0,1]|}$, and a function, which is a limit of a set of non-decreasing functions, is itself non-decreasing. Consequently, X is compact (see Problems 54 and 75). To prove that X is first countable, we rely on the following statement: every non-decreasing real-valued function defined on $[0, 1]$ has at most countably many points of discontinuity. Finally, a countable dense set in X is constructed in exactly the same way as in the solution of Problem 380 of Chap. 2, taking those mappings in X, which assume just a finite set of values, each of these values being a rational number. It is further required that all pre-images of the points under the given mapping are either rational, or an interval with rational endpoints.

83. Let $M \subset X$, $|M| \geqslant \aleph_0$. Set $A = \{x \in X: |M \cap \tilde{I}(x)| = |M|\}$ (see p. 6). Then $x_m \in A$, and hence $A \neq \Lambda$. Let $x^* = \min A$, $x' < x^*$, $x' \neq x^*$ and $U = \{y \in X: x' < y < x^*\}$. Then $|\tilde{I}(x') \cap M| < |M|$ and $|\tilde{I}(x^*) \cap M| = |M|$. But $\tilde{I}(x^*) \cap M = (\tilde{I}(x') \cap M) \cup (U \cap M)$. Hence $|U \cap M| = |M|$ (see Problem 90, Chap. 1). By the definition of order topology, this means that x^* is a complete accumulation point for M. Consequently, X is a compact space (see Problem 48). It is obviously Hausdorff (Problem 7, Chap. 2).

84. This statement is easily deduced from the preceding problem.

85. $(a) \Rightarrow (b)$. Set $F(\beta) = \{\beta': \beta \leqslant \beta' \text{ and } \beta' \in B\}$ for each $\beta \in B$. For any $\beta_1, \beta_2 \in B$ either $F(\beta_1) \subset F(\beta_2)$, or $F(\beta_2) \subset F(\beta_1)$. In addition, $F(\beta) \neq \Lambda$ and $F(\beta)$ is closed in X for all $\beta \in B$. Consequently, $\{F(\beta): \beta \in B\}$ is a centralized family of nonempty closed sets in X and the set $\Phi = \bigcap\{F(\beta): \beta \in B\}$ is nonempty. Fix any element $\beta^* \in \Phi$ and show that $\beta^* \geqslant \beta$ for any $\beta \in B$ (note that $\beta^* \in \Phi \subset B$ by the definition of $F(\beta)$). Assume the contrary: suppose that we have $\tilde{\beta} \in B$ satisfying $\beta^* < \tilde{\beta}$. Then $F(\tilde{\beta}) \not\ni \beta^*$ and $\Phi \subset F(\tilde{\beta})$ implies that $\Phi \not\ni \beta^*$, contradicting the choice of β^*.

Hence β^* is the greatest element of B. Similarly, the least element in B is determined.

$(b) \Rightarrow (c)$. Set $B = [A]$. By (b) B has a greatest element β^*. We show that $\beta^* = \sup A$. From $A \subset B$ it follows that $a \leqslant \beta^*$ for all $a \in A$. Let $c_1, c_2 \in X$ and $c_1 < \beta^* < c_2$. From $\beta^* \in B = [A]$ and the definition of the topology on X it follows that there exists $a \in A$ such that $c_1 < a < c_2$. Then, surely, $c_1 < a < \beta^*$. This shows that $\beta^* = \sup A$. Similarly, the existence of $\inf A$ in $(X, <)$ is established.

$(c) \Rightarrow (a)$. Consider an arbitrary infinite set $A \subset X$; we show that A has a complete accumulation point in X. For each $x \in X$, set $F(x) = \{x' \in A: x \leqslant x'\}$. Consider the set $C = \{x \in X: |F(x)| = |A|\}$. Trivially, $\inf A \in C$; hence $C \neq \Lambda$. By (c), $c^* = \sup C$ exists. We show that c^* is a complete accumulation point for A. Let $\beta_1, \beta_2 \in X$, $\beta_1 < c^* < \beta_2$. Then there is $\tilde{c} \in C$, satisfying $\beta_1 < \tilde{c} < \beta_2$. On the other hand, $\beta_2 \notin C$ by the definition of c^*. Hence $|\{x \in A: x \geqslant \tilde{c}\}| = |A|$, but $|\{x \in A: x \geqslant \beta_2\}| < |A|$. Consequently, $|\{x \in A: \tilde{c} \leqslant x < \beta_2\}| = |A|$. This establishes that the interval with endpoints in β_1 and β_2 intersects A in a set equipotent to A. By the definition of the topology in X, this implies that c^* is a complete accumulation point for A.

86. We identify a set of elements of an 'arrow', as usual, with the half-interval $[0, 1)$ of a set of real numbers. We shall need the following auxiliary statement: (*) for every uncountable subset $A \subset [0, 1)$ there is $x \in A$, for which the set $A_x = \{y \in A: x < y\}$ is uncountable. It is straightforward to show that the set $\tilde{A}$ of condensation points of A (see p. 144) (in relation to the topology on A induced by the usual topology of an interval) is uncountable. Fix $x \in \tilde{A}$, $z \in \tilde{A}$, $x \neq z$. Then either $x < z$, or $z < x$. Let $x < z$. In this case the set A_x is obviously uncountable. The statement (*) is thus proved. With the aid of (*) it is routine to construct in each uncountable set $A \subset [0,1)$ a (strictly) monotonically increasing sequence of elements $\{x_n: n \in N^+\}$. Obviously, with respect to the 'arrow' topology for the sequence $\{x_n: n \in N^+\}$ there is no limit point in $[0, 1)$ and, even more, not even in A. Hence A, as a subspace of an 'arrow', is not compact.

87. (i) and (vii) are established directly, (ii) is implied by (iii) because a perfect pre-image of a compact space is compact (Problem 36, Chap. 6).

(iii) is established in the following way: a closed interval is drawn under our two half-intervals with the usual topology (also of length 1). The desired mapping is the projection of our space onto this closed interval. It is easily verified that the indicated mapping is continuous and perfect (see definition on p. 45).

For the rest note that the half-intervals X and X' with the inherited topology are homeomorphic to an 'arrow' (Problem 104, Chap. 2). Hence properties (iv), (v), (viii), (ix) and (x) are corollaries to the corresponding properties of 'arrow' (Problem 104, Chap. 2) and some propositions (see Problems 208, 206, Chap. 3; Problem 213, Chap. 2). Property (vi) is a corollary to X^{**} being a perfectly normal compactum (see Problem 68), and property (xi) is obtained by observing that the square of X^{**} topologically contains the square of 'arrow', which is not a normal space (Problem 11).

88. Suppose that the open sets U_i, $i \in N^+$ are dense in X. Define by induction a sequence $\{V_i: i \in N^+\}$ of nonempty open sets in X such that the following conditions are satisfied: (α) $[V_i] \subset U_i$ for all values of i, and (β) $[V_i] \subset V_j$ for all $i > j \geqslant 1$. For the construction of V_i we essentially rely on the fact that X is regular and U_i is dense in X. Fix any nonempty open set $W \subset X$. As V_1 take a nonempty open set such that $[V_1] \subset W \cap U_1 \neq \Lambda$. Assume that the V_i are defined for all $i \leqslant k$. Then denote by V_{k+1} any nonempty open set such that $[V_{k+1}] \subset V_k \cap U_{k+1}$ ($V_k \cap U_{k+1}$ is a nonempty set, because U_{k+1} is dense in X). Suppose that V_i are defined by the above rule for all $i \in N^+$. Then we have

$$W \cap (\cap \{U_i: i \in N^+\}) \supset \cap \{V_i: i \in N^+\} = \cap \{[V_i]: i \in N^+\}.$$

But the set on the right-hand side is nonempty, because X is a compactum and $\{V_i: i \in N^+\}$ is a centralized family (see Problem 45). Hence $W \cap (\cap \{U_i: i \in N^+\}) \neq \Lambda$ for any open set W in X, i.e., $\cap \{U_i: i \in N^+\}$ is dense in X.

89. A stronger statement is true, see Problem 126.

90. Suppose that $\{U_i: i \in N^+\}$ is a family of dense open sets in a complete (Čech-complete) completely regular space X; βX is a compactification of the space (for example, the Stone-Čech compactification, see Problem 14, Chap. 4). Then $X = \cap\{\Gamma_i: i \in N^+\}$, where Γ_i, $i \in N^+$ is open in βX. For each i, consider in βX an open set $O(U_i)$, satisfying $O(U_i) \cap X = U_i$. Set $O^*(U_i) = O(U_i) \cap \Gamma_i$. Obviously, $O^*(U_i) \cap X = U_i$. However, already $O^*(U_i) \subset \Gamma_i$. We now show that $O^*(U_i)$ is dense in βX. Suppose that Γ is an arbitrary open set in βX. Then, since X is dense in βX and U_i is dense in X, we have $\Gamma \cap U_i \neq \Lambda$, implying, furthermore, $\Gamma \cap O^*(U_i) \neq \Lambda$, i.e., $O^*(U_i)$ is dense in βX. Then, $\cap \{O^*(U_i): i \in N^+\} \neq \Lambda$ is also dense in βX (Problem 88). But $\cap \{O^*(U_i): i \in N^+\} \subset \cap \{\Gamma_i; i \in N^+\} = X$; consequently, $\cap \{O^*(U_i): i \in N^+\} = X \subset \cap \{O^*(U_i):$

$i \in N^+\} = \bigcap\{U_i: i \in N^+\}$; hence $\bigcap\{U_i: i \in N^+\} = \bigcap\{O^*(U_i): i \in N^+\}$ is a dense set in X. The proof is complete.

92. Suppose that $F \subset X$ is an arbitrary closed subset of a compact space X. The subspace F is compact (see Problem 54). Then fF is also a compact subspace of a Hausdorff space Y (Problem 91) and is, consequently, closed in Y (see Problem 56). The second statement follows from the first and Problem 322, Chap. 2.

93. From the preceding problem it follows that the mapping $f: X \to Y$ is closed. Hence f is a homeomorphism.

94. (*a*) If $f: X \to Y$ is continuous, then the graph of f is closed in $X \times Y$ (see Problem 291, Chap. 2), it being immaterial that X and Y are compact spaces.

(*b*) Suppose that the graph D of the mapping $f: X \to Y$ is closed in the product of the compacta X and Y. Consider the mappings $p_X: D \to X$, $p_Y: D \to Y$. $p_X: D \to X$ is one-to-one and continuous. The fact that D is a compact set implies that the continuous mapping $p_Y: D \to Y$ is closed, and $p_X: D \to X$ is a homeomorphism. Moreover, $f: X \to Y$ is a composition of the mappings p_X^{-1} and p_Y, i.e., $f = p_Y p_X^{-1}$. Suppose that V is an arbitrary open set in Y. Then $f^{-1}V = p_X p_Y^{-1} V$, $p_Y^{-1}V$ is open in D, $p_X p_Y^{-1} V = f^{-1}V$ is open in X. The continuity of f is proved. Note that to prove the continuity of $f: X \to Y$, we just make use of the fact that the projection of the graph of this mapping on X is a topological mapping.

95. This is not always true. Suppose that X is the space of rational numbers with the usual topology, and $Y = bX = \xi \cup X$ is the compact T_1-extension of X, defined in Problem 1 of Chap. 4. A base of the topology of bX at the points of X consists of the nhoods of these points in X (i.e., X is open in bX), and an arbitrary open nhood of the point ξ is a complement to some compact (consequently, closed) subspace of X. Let $\mathcal{T}$ be the topology of bX and consider an arbitrary topology $\tilde{\mathcal{T}}$ on the set of points of bX, for which $\mathcal{T} \subset \tilde{\mathcal{T}}$ and $(bX, \tilde{\mathcal{T}})$ is a compact space. Let $U \in \tilde{\mathcal{T}}$. Then, $bX \setminus U$ is closed in $(bX, \tilde{\mathcal{T}})$ and hence is compact. But then $bX \setminus U$ is a compact subspace of $(bX, \mathcal{T})$ (because an image of a compact space under a continuous mapping is a compact space, see Problem 91). We now show that every compact subspace F of $(bX, \mathcal{T})$ is closed. If $F \not\ni \xi$, then $F \subset X$ is a compact subspace of X. Then, by the definition of $\mathcal{T}$, F is closed in $(bX, \mathcal{T})$. If $F \ni \xi$ and F is not closed in $(bX, \mathcal{T})$, then there is a sequence $\{x_n: n \in N^+\}$ of points of $F \setminus \xi$, which converges to some point $x^* \in X \setminus F$ (in the sense of the topology $\mathcal{T}$, i.e., the usual topology of the rational numbers). Then $\Phi = \{x^*\} \cup \{x_n: n \in N^+\}$ is a compact subspace of X and, consequently, $O\xi = bX \setminus \Phi$ is a nhood of ξ. But $O\xi$ contains no element of the sequence $\{x_n: n \in N^+\}$. Hence, $\{x_n: n \in N^+\}$ has no limit point in F, a contradiction. Consequently, F is closed in $(bX, \mathcal{T})$. It follows that $bX \setminus U$ is closed in $(bX, \mathcal{T})$, and hence $U \in \mathcal{T}$. Thus $\tilde{\mathcal{T}} \subset \mathcal{T}$, and so $\mathcal{T} = \tilde{\mathcal{T}}$. But $(bX, \mathcal{T})$ does not satisfy the Hausdorff separability axiom (since X is not locally compact; see Problem 179). Hence $(bX, \tilde{\mathcal{T}})$ is also non-Hausdorff.

96. No. Take the set of all points of the interval $[0, 1]$ with the topology $\mathcal{T}_k$ (see Prob. 2, Chap. 2 and Prob. 42 of the present chapter). Then the natural (identity) mapping of $[0, 1]$, taken with the usual topology, onto the space $([0, 1], \mathcal{T}_k)$ is continuous, one-to-one and yet not a homeomorphism.

97. (*a*) Suppose that f is a continuous function on a compactum X: $f: X \to R$, where R is the real line. The subspace $fX \subset R$ is compact (Prob. 91) and, consequently (Prob. 256), the set fX is closed and bounded in R. Then $a = \inf fx$ and $b = \sup fx$ exist and belong to fX, i.e., f is bounded: $a \leqslant f(x) \leqslant b$ for all $x \in X$ and $fx_1 = a$, $fx_2 = b$ for some $x_1 \in X$, $x_2 \in X$.

(*b*) Yes. There do exist normal non-compact spaces (see Probs. 188 and 185). Such spaces are called pseudo-compact.

98. Suppose that x_0 is a non-isolated point of X and y_0 of Y. Then $X \setminus \{x_0\}$ and $Y \setminus \{y_0\}$ are equipotent sets. Let φ be any one-to-one mapping of $X \setminus \{x_0\}$ onto $Y \setminus \{y_0\}$. Define $\tilde{\varphi}: X \to Y$ thus: $\tilde{\varphi}(x_0) = y_0$, $\tilde{\varphi}(x) = \varphi(x)$ when $x \in X \setminus \{x_0\}$. We show that $\tilde{\varphi}$ is a homeomorphism. Since $\tilde{\varphi}$ is one-to-one and $\tilde{\varphi}(X) = Y$, it suffices to show that $\tilde{\varphi}$ is continuous. Suppose that U is an arbitrary open set in Y. If $U \not\ni y_0$, then $\tilde{\varphi}^{-1}U \not\ni x_0$, and hence $\tilde{\varphi}^{-1}U$ is open in X_0 (in fact, $X_0 = X \setminus \{x_0\}$ is discrete and open in X). If $U \ni y_0$, then y_0 is not a complete accumulation point of $Y \setminus U$. Any other point of Y also fails to be a complete accumulation point for $Y \setminus U$ (since all points of $Y \setminus \{y_0\}$ are isolated). Hence $Y \setminus U$ is finite. Consequently, $X \setminus \tilde{\varphi}^{-1}U$ is also finite. Hence $X \setminus \tilde{\varphi}^{-1}U$ is closed, $\tilde{\varphi}^{-1}U$ is open in X.

100. Use Problem 61 of the present chapter and 281 of Chap. 2.

101. In consequence of Prob. 102, our space X is H-closed and semi-regular since it is non-contractible. We now show, using the second hypothesis, that it is regular. Suppose that $x_0 \in X$. Take an arbitrary nhood Ox_0 of x_0. Since X is semi-regular, hence there is a nhood $O'x_0 \subset Ox_0$ which is a regular open set. Then $X \setminus O'x_0$ is a regular closed set (see Prob. 56, Chap. 2). Hence $X \setminus O'x_0$ is an H-closed subspace (see Prob. 62). For each point $x \in X \setminus O'x_0$ and for x_0 take nhoods Ux and $U_x x_0$ such that $[U_x] \cap [U_x x_0] = \Lambda$. Consider an open cover $\omega = \{Ux: x \in X \setminus O'x_0\}$ of $X \setminus O'x_0$. Using Problem 61, we select a finite subfamily $\omega_0 = \{Ux_1, Ux_2, \ldots, Ux_s\}$ such that

$$X \setminus O'x_0 \subset \bigcup_{i=1}^{s} [Ux_i].$$

Consider the corresponding nhoods of the point x_0: $U_{x_1}x_0, U_{x_2}x_0, \ldots, U_{x_s}x_0$. We have $[U_{x_i}x_0] \cap [Ux_i] = \Lambda$, $i = 1, 2, \ldots, s$. Set

$$O''x_0 = \bigcap_{i=1}^{s} U_{x_i}x_0.$$

It is easy to see that $x_0 \in O''x_0 \subset [O''x_0]_X \subset O'x_0 \subset Ox_0$. Thus X is regular and H-closed, and hence compact (see Prob. 60). The proof is complete.

102. (*a*) Suppose that the Hausdorff space X is H-closed and semi-regular. Let f: $X \to Y$ be a one-to-one continuous mapping onto a Hausdorff space Y. We show that f is a homeomorphism. Since X is semi-regular, the regular open sets form a base in X. To show that f is a homeomorphism it is necessary to show that the image fU of a regular open set U in X is an open set in X. The set $X \setminus U$ is regular closed in X, hence $X \setminus U$ is an H-closed subspace (see Prob. 62). Then $f(X \setminus U)$ is a H-closed subspace of a Hausdorff space Y (see Problem 100); in particular, $f(X \setminus U)$ is closed in Y; consequently, $fU = Y \setminus f(X \setminus U)$ is open in Y (recalling that f is one-to-one).

(*b*) Suppose that X is non-contractible.

(i) We first show that X is semi-regular. Let $\mathcal{T}$ be the topology on X with a base consisting of the regular open subsets of the Hausdorff space X. It is clear that $(X, \mathcal{T})$ is a Hausdorff space, and $f: X \to (X, \mathcal{T})$ is a one-to-one continuous mapping. Since X is not contractible, f is a homeomorphism, which implies that the regular open sets in X form a base in X, i.e., X is semi-regular.

(ii) We now show that X is H-closed. Assume that X is not H-closed. Then by adjoining to X a non-isolated point ξ, we can extend X to a Hausdorff space $\tilde{X}$. Consider a decomposition of $\tilde{X}$, all of whose members are singletons, except one which consists of two points, namely ξ and another point $x_0 \in X$. This partition is continuous. The resulting decomposition space we denote by Y, and the natural mapping of $\tilde{X}$ onto Y by $\tilde{f}$. Since the decomposition defined above is continuous, the mapping $\tilde{f}: \tilde{X} \to Y$ is continuous and closed (see Problem 322, Chap. 2). It is now easy to see that Y is a Hausdorff space. Consider the restriction $\tilde{f} \mid X = f$. It is clear that $fX = Y, f: X \to Y$ is one-to-one. Moreover, f is not a closed mapping, which is easily seen using the fact that ξ is a non-isolated point in $\tilde{X}$; hence X is dense in $\tilde{X}$. Thus we have constructed a one-to-one continuous non-topological mapping of X onto a Hausdorff space Y, which contradicts the non-contractibility of X.

103. The answer is not known to the authors.

104. The Hausdorff axiom is straightforward to verify. It is also routine to show that the the closures of any two nhoods of the points p^+ and p^- intersect, i.e., X does not satisfy Uryhson's separation axiom. It is further trivial to show that X is semi-regular. It therefore remains to show that X is H-closed. For this, we use the H-closedness criterion (Problem 61). Suppose that ω is an arbitrary open cover of X by basic open sets. Consider two points p^+ and p^-, $p^+ \in U_{n_1}(p^+) \in \omega, p^- \in U_{n_2}(p^-) \in \omega$. Without loss of generality, we may assume that $n_1 = n_2 = n$. Then the subspace

$$X \setminus \left(U_n p^+ \cup U_n p^- \cup \left\{ (x, y): x^2 + y^2 < \frac{1}{n}, \quad y = 0 \right\} \right)$$
$$= \left\{ (x, y): \frac{1}{n} \leqslant x^2 + y^2 \leqslant 1 \right\}$$

is compact, and ω has a finite subfamily ω_0 which covers this compact subspace. Consider the finite family ω_1 of all elements of ω_0 and two more sets $U_n p^+$ and $U_n p^-$ contained in ω. It is clear that $\omega_1 \subset \omega, |\omega_1| \leqslant \aleph_0$. Since $\{(x, y): 0 < x^2 + y^2 < (1/n), y = 0\}$ is nowhere dense in X, the closures of the elements of ω_1 cover X. Thus X is H-closed, semi-regular and does not satisfy Uryhson's separation axiom. Hence X is non-contractible (in consequence of Problem 102) and is non-compact by Problem 64. The proof is complete.

105. See Problems 107 and 39.

106. By Problem 147 of Chap. 2, a Hausdorff space with a net of cardinality less than or equal to τ, can be contracted to a Hausdorff space with a base of cardinality less than or equal to τ. But every contraction of a compactum to a Hausdorff space is a homeomorphism (Problem 93). Hence, X itself has a base of cardinality less than or equal to τ. It is clear that we have used just the non-contractibility of X.

107. The set of all one-point subsets of a compactum is a net. The conclusion is now immediate from the preceding problem.

108. Suppose that $\mathcal{B}_1$ is a base for X_1 and $\mathcal{B}_2$ is a base for X_2, where $|\mathcal{B}_1| \leqslant \tau = \max(wX_1, wX_2)$ and $|\mathcal{B}_2| \leqslant \tau$. Then $\mathcal{S} = \mathcal{B}_1 \cup \mathcal{B}_2$ is a net in X (Problem 141, Chap. 2) and $|\mathcal{S}| \leqslant \tau$ (since $\tau \geqslant \aleph_0$). Hence (see Problem 106) X also has a base of cardinality less than or equal to τ. Clearly, the weight of X cannot be less than the weight of X_1 or X_2.

109. The reasoning is similar to the one in the solution of the preceding problem.

110. Yes, there does exist a countably compact T_1-space without a countable base. Suppose that $(X, \mathcal{T})$ is an arbitrary non-countably compact T_1-space without a countable base (see Problems 84 and 113, Chap. 2). Then $(\tilde{X}, \tilde{\mathcal{T}})$ defined in Problem 1 of Chap. 4, is a countably compact T_1-space without a countable base: if the latter were false, then a countable base would also exist in $(X, \mathcal{T})$, it being a subspace of $(\tilde{X}, \tilde{\mathcal{T}})$.

111. Under a continuous mapping, a space with a countable net is mapped onto a space with a countable net. Hence there is a countable net in Y. But then there is also a countable base in Y (Problem 106).

114. Let $U_i \supset \Delta$, U_i open in $X \times X$ and $\bigcap\{U_i: i \in N^+\} = \Delta$. One may also assume that $[U_j] \subset U_i$ for $j > i$. Then for each open set G in $X \times X$ containing Δ there is i_0 such that $U_{i_0} \subset G$ (see Problem 67). For each point $x \in X$ and every value of i fix a nhood $O^i x$ of x in X such that $O^i x \times O^i x \subset U_i$. We show that the family $\gamma = \{O^i x: i \in N^+, x \in X\}$ is a base for X. Consider an arbitrary nhood Ox^* of fixed but arbitrary point x^*. Then $U^* = (Ox^* \times Ox^*) \cup ((X \times X) \setminus (X \times x^*))$ is a nhood of Δ in $X \times X$. There exists i^* such that $U_{i^*} \subset U^*$. We show that $O^{i^*}x^* \subset Ox^*$. By definition, $O^{i^*}x^* \times O^{i^*}x^* \subset U_{i^*} \subset U^*$ and $O^{i^*}x^* \ni x^*$. Consequently, whatever $y \in O^{i^*}x^*$, we have $(y \times x^*) \in U^* = (Ox^* \times Ox^*) \cup ((X \times X) \setminus (X \times x^*))$. But, then $(y \times x^*) \in Ox^* \times Ox^*$ and hence $y \in Ox^*$. Thus $O^{i^*}x^* \subset Ox^*$. In addition, $O^{i^*}x^* \in \gamma$. Consequently, γ is a base for X. Now, for each $i \in N^+$ consider the cover $\gamma_i = \{O^i x: x \in X\}$ of X and extract from it a finite subcover $\gamma_i^0 = \{O^i x_j: j = 1, \ldots, s(i)\}$. Then $\gamma^0 = \bigcup\{\gamma_i^0: i \in N^+\}$ is also a base for X, $\gamma^0 \subset \gamma$ and $|\gamma^0| \leqslant \aleph_0$. The proof is complete.

115. No. As an example, consider the family $\{D_\alpha: \alpha \in A\}$, $|A| = \mathfrak{c}$, of isolated double-points: $D_\alpha = \{0_\alpha, 1_\alpha\}$. As we know, $D^{\mathfrak{c}} = \Pi\{D_\alpha: \alpha \in A\}$ is a separable compactum (see Problem 381, Chap. 2). For each point $x \in \Pi\{D_\alpha: \alpha \in A\}$, let $k(x) = \{\alpha \in A: x_\alpha \neq 0\}$. Consider the set $M = \{x \in \Pi\{D_\alpha: \alpha \in A\}: |k(x)| < \aleph_0\}$. We show that the subspace $M \subset \Pi\{D_\alpha: \alpha \in A\}$ is not seperable. Suppose that M' is a countable subset of M. Put $k(M') = \bigcup\{k(x): x \in M'\}$. It is clear that $k(x_0) \subset k(M')$ if $x_0 \in [M']_M$. But $|k(M')| \leqslant \aleph_0$, and $|A| = \mathfrak{c} > \aleph_0$. Hence, by taking $x_0 \in M$ such that $k(x_0) \cap k(M') = \Lambda$, we obtain a point M, which does not belong to the set $[M']_M$. Note that M is dense in $D^{\mathfrak{c}}$.

116. The set A of points of any free sequence forms an open discrete subspace in $F = [A]_X$. By hypothesis, there is $C \subset F$ such that $[C] = F$ and $|C| \leqslant \tau$. From the isolatedness of the points of A in F and $[C] \supset A$ it follows that $C \supset A$. Hence $|A| \leqslant \tau$ and the upper bound of the cardinality of a free sequence in X does not exceed τ. Then, we have also $t(X) \leqslant \tau$ (Problem 142). In $\Phi = [Y]$ select a dense set S such that $|S| \leqslant \tau$. For each $y \in S$ fix a set $B(y) \subset Y$ with the properties: $|B(y)| \leqslant \tau$ and $[B(y)]$

$\ni y$ (possible, since $t(X) \leqslant \tau$). Put $B = \bigcup \{B(y): y \in S\}$. Then, $|B| \leqslant \tau$, $B \subset Y$ and $[B] \supset [S] = \Phi \supset Y$. Consequently, $s(Y) \leqslant \tau$.

117. The answer is not known to the authors.

118. The answer is not known to the authors.

119. There are completely regular, non-normal, separable spaces. An example is the Nemitskii space $(Z, \mathcal{T})$ in Problem 13. $(Z, \mathcal{T})$ has a Hausdorff compactification (Prob. 14, Chap. 4) and the latter is obviously also the desired compactum. Another example is the product of $2^{\aleph_0}$ copies of the usual interval, which is a separable compactum (Problem 381, Chap. 2). It contains a homeomorphic image of any completely regular space of weight less than or equal to $2^{\aleph_0}$ (Problems 36 and 37), and among them there is also a non-normal one (Problems 13 and 11).

120. A cover λ of X is said to be minimal if $\gamma \subset \lambda$ and $\gamma \neq \lambda$ imply that γ does not cover X. Denote by $\mathcal{S}$ the set of all minimal covers of X, consisting of elements of the base $\mathcal{B}$. It is clear that every minimal open cover of a compact space is finite. Hence $\mathcal{S} = \bigcup \{\mathcal{S}_n: n \in N^+\}$ where $\mathcal{S}_n = \{\lambda \in \mathcal{S}: |\lambda| = n\}$. It is routine to show $|\mathcal{S}| = |\mathcal{B}|$ (see Problem 58, Chap. 1; we note that each $\lambda \in \mathcal{S}$ is finite and that X is a T_1-space). Assume that $|\mathcal{B}| > \tau$. Then $|\mathcal{S}| > \tau$ and hence $|\mathcal{S}_{n_0}| > \tau$ for some n_0 (Problem 90, Chap. 1). Put (for an arbitrary $\gamma \subset \mathcal{B}$) $\mathcal{S}_{n_0}^{\gamma} = \{\lambda \in \mathcal{S}_{n_0}: \gamma \subset \lambda\}$. We now show by induction that for each $k = 0, 1, \ldots, n_0$ there is $\gamma_k \subset \mathcal{B}$ with the properties: (α) $|\gamma_k| = k$, (β) $|\mathcal{S}_{n_0}^{\gamma_k}| > \tau$. Let $k = 0$; assuming $\gamma_0 = \Lambda$, the condition (β) is satisfied, since $\mathcal{S}_{n_0}^{\gamma_0} = \mathcal{S}_{n_0}$ and $|\mathcal{S}_{n_0}| > \tau$. Assume that γ_l is already defined for certain $l < n_0$ where $|\mathcal{S}_{n_0}^{\gamma_l}| > \tau$ and $|\gamma_l| = l < n_0$. Then, $\bigcup \{U: U \in \gamma_l\} \neq X$, otherwise by the definition of $\mathcal{S}_{n_0}^{\gamma_l} \subset \mathcal{S}_{n_0}$, the family $\mathcal{S}_{n_0}^{\gamma_l}$ would be empty. Fix $x^* \in X \setminus \bigcup \{U: U \in \gamma_l\}$. Since $\mathcal{S}_{n_0}^{\gamma_l} \subset \mathcal{S}_{n_0}$, $\mathcal{S}_{n_0}^{\gamma_l} = \bigcup \{\mathcal{S}_{n_0}^{\gamma_l \cup \{V\}}: V \in \mathcal{B}_{x^*}\}$. But $|\mathcal{B}_{x^*}| \leqslant \tau$ and $|\mathcal{S}_{n_0}^{\gamma_l}| > \tau$. Hence (Problem 88, Chap. 1), there is $V^* \in \mathcal{B}_{x^*}$ such that $|\mathcal{S}_{n_0}^{\gamma_l \cup \{V\}}| > \tau$. Obviously, $V^* \notin \gamma_l$ (in fact $x^* \in V^*$ and $x^* \notin \bigcup \{U: U \in \gamma_l\}$. Consequently, $|\gamma_l \cup \{V^*\}| = l + 1$, i.e., $\gamma_{l+1} = \gamma_l \cup \{V^*\} \subset \mathcal{B}$ is the desired family. This procedure leads us, finally, to the family $\gamma_{n_0} \subset \mathcal{B}$, satisfying $|\gamma_{n_0}| = n_0$ and $|\mathcal{S}_{n_0}^{\gamma_{n_0}}| > \tau$. But since $\gamma_{n_0} \subset \gamma$ and $|\gamma| = n_0$, for any $\gamma \in \mathcal{S}_{n_0}^{\gamma_{n_0}}$ we conclude that $\gamma = \gamma_{n_0}$ for all $\gamma \in \mathcal{S}_{n_0}^{\gamma_{n_0}}$, i.e., $\mathcal{S}_{n_0}^{\gamma_{n_0}} = \{\gamma_{n_0}\}$ and $|\mathcal{S}_{n_0}^{\gamma_{n_0}}| = 1$, which contradicts $|\mathcal{S}_{n_0}^{\gamma_{n_0}}| > \tau \geqslant \aleph_0 > 1$.

121. Denote by $\tilde{\mathcal{E}}$ the collection of all finite intersections of members of the family $\mathcal{E}$. $\tilde{\mathcal{E}}$ serves as a base for some topology $\tilde{\mathcal{T}}$ on X (Problem 14, Chap. 2). Since $\{x\} = \bigcap \{U: U \in \mathcal{E}_x\}$ for all $x \in X$, $(X, \tilde{\mathcal{T}})$ is a T_1-space. By Problem 58 of Chap. 1, $|\tilde{\mathcal{E}}| = |\mathcal{E}|$ and $\tilde{\mathcal{T}} \subset \mathcal{T}$, where $\mathcal{T}$ is the original topology on X. The latter means that $(X, \tilde{\mathcal{T}})$ is compact. Obviously, $|\tilde{\mathcal{E}}_x| \leqslant \tau$ for all $x \in X$, where $\tilde{\mathcal{E}}_x = \{V \in \tilde{\mathcal{E}}: V \ni x\}$. We conclude (Problem 120) that $|\tilde{\mathcal{E}}| \leqslant \tau$. Consequently, $|\mathcal{E}| \leqslant \tau$.

122. Assume that $|X| \geqslant \aleph_0$. Put $\mathcal{F} = \{X \setminus U: U \in \mathcal{B}\}$. Obviously, for each $x \in X$, we have $\{x\} = \bigcap \{F \in \mathcal{F}: F \ni x\}$. Consider the family $\tilde{\mathcal{F}}$ of all finite intersections of elements of $\mathcal{F}$. Then $|\tilde{\mathcal{F}}| = |\mathcal{F}| = |\mathcal{B}|$. All elements of $\tilde{\mathcal{F}}$ are closed sets. We show that $\tilde{\mathcal{F}}$ is a net in X. Let $x \in X$ and let Ox be any nhood of x. Consider $\xi = \{P \cap (X \setminus Ox): P \in \tilde{\mathcal{F}}$ and $P \ni x\}$. Obviously, $\cap \{Q: Q \in \xi\} \subset \cap \{P \in \tilde{\mathcal{F}}: P \ni x\} = \cap \{P' \in \mathcal{F}: P' \ni x\} \subset \{x\}$. On the other hand, $\cap \{Q: Q \in \xi\} \subset X \setminus Ox \subset Ox \subset X \setminus \{x\}$. Consequently, $\cap \{Q: Q \in \xi\} = \Lambda$. Since all the elements of ξ are closed, ξ is not centralized (Problem 45). Hence there are $P_1, \ldots, P_k \in \tilde{\mathcal{F}}$ such that $\cap \{P_i: i = 1, \ldots, k\} \cap (X \setminus Ox) = \Lambda$. But then $x \in \cap \{P_i: i = 1, \ldots, k\} \subset Ox$ and $\cap (P_i: i = 1, \ldots, k\} \in \tilde{\mathcal{F}}$. Thus $\tilde{\mathcal{F}}$ is a net in X. Hence (Problem 106) there is a base $\mathcal{B}^*$ for X such that $|\mathcal{B}^*| \leqslant |\mathcal{B}|$.

124. Let δ be the family of all clopen sets in X. Since X is zero-dimensional, δ is a base for X. Take a sub-family $\delta_0 \subset \delta$ of cardinality $|\delta_0| = w(X)$ such that δ_0 is also a base for X (see Problem 101, Chap. 2), and let $\bar{\delta}_0$ be the family of all finite unions of members of δ_0. Obviously, $\bar{\delta}_0$ is made up of clopen sets, i.e., $\bar{\delta}_0 \subset \delta$; $\bar{\delta}_0$ forms a base for the compactum X and $|\bar{\delta}_0| = |\delta_0| = w(X)$. To show that $\bar{\delta}_0 = \delta$, suppose that $V \in \delta$. For each $x \in V$ take $U_x \in \delta_0$ such that $x \in U_x \subset V$ and consider the open cover $\gamma(V) = \{U_x: x \in V\}$ of V. Since V is clopen in the compactum X, V is a compactum; consequently, there is a finite subcover $\{U_{x_1}, \ldots, U_{x_s}\}$ in $\gamma(V)$ of the subspace V. Then

$$V = \cup \{U_x: U_x \in \gamma(V)\} = \bigcup_{i=1}^{s} U_{x_i}.$$

Since each $Ux_i \in \delta_0$, $i = 1, 2, \ldots, s$, hence

$$\bigcup_{i=1}^{s} Ux_i = V \in \bar{\delta}_0.$$

Thus $\delta = \bar{\delta}_0$. Hence $|\delta| = |\bar{\delta}_0| = |\delta_0| = w(X)$. The proof is complete.

126. Using Problem 125, select disjoint closed sets $F_0 \subset X$ and $F_1 \subset X$, which in the topology inherited from X have no isolated points. Again apply Problem 125 separately to F_0 and F_1, to obtain closed disjoint sets $F_{00} \subset F_0$, $F_{01} \subset F_0$, $F_{10} \subset F_1$, $F_{11} \subset F_1$, each of which as a subspace of X has no isolated points. Continuing this process inductively, put each ordered collection $i_0, i_1, \ldots, i_k$ of zeros and ones in correspondence with a closed set in X, which in the topology inherited from X has no isolated points. Here the following conditions are satisfied: (α) $F_{i_0 i_1, \ldots, i_k} \subset F_{i_0, i_1 \ldots i_{k-1}}$ and (β) $F_{i_0 i_1 \ldots i_{k-1} i_k} \cap F_{i_0 i_1 \ldots i_{k-1} i'_k} = \Lambda$, if $i'_k \neq i_k$, where $i_0, i_1, \ldots, i_{k-1}, i_k, i'_k$ can be arbitrarily replaced by zeros and ones and $k = 0, 1, 2. \ldots$ From (α) it follows that for any sequence $\xi = \{i_0, i_1, \ldots, i_l, \ldots\}$ of zeros and ones, the set $\Phi_\xi = \cap \{F_{i_0, \ldots, i_k}: k = 0, 1, 2, \ldots\}$ is nonempty (since X is a compactum). By virtue of (β), if $\xi \neq \xi'$, then $\Phi_{\xi'} \cap \Phi_\xi = \Lambda$. Since the set of all distinct sequences of zeros and ones has cardinality $2^{\aleph_0}$ (see Problem 92, Chap. 1: this set can be identified with $D^{\aleph_0}$, the product of $\aleph_0$ doubletons), $|X| \geqslant |\cup \{\Phi_\xi: \xi$ is any sequence of zeros and ones$\}| \geqslant 2^{\aleph_0}$.

127. Observe the following:

(i) If the cardinality of a compactum is greater than unity, then there are two disjoint nonempty closed G_δ-sets in it (Problem 69 can be used.)

(ii) The intersection of a countable centralized family of closed G_δ-sets in a compactum is a nonempty closed G_δ-set in this compactum.

(iii) If $\Phi \subset F \subset X$, F is a G_δ-set in X, and Φ is a G_δ-set in F, then Φ is a G_δ-set in X (see Problem 70).

(iv) If $F = \{x\}$ is a G_δ-set in a compactum X, then X is first countable at the point x (see Problem 68).

Now consider a minimal well-ordered set $(A, <)$ of cardinality $\aleph_1$. For each $x \in A$ denote by $Q(x)$ the set of all functions with values 0 or 1, defined on $I(x) = \{y \in A: y < x, y \neq x\}$. If $f \in Q(x)$ and $y < x$, $y \neq x$, then denote by $\langle f, y \rangle$ the restriction of the function f to the set $I(y)$, and by f^{+0} (respectively f^{+1}) the function, which is defined on $\tilde{I}(x) = I(x) \cup \{x\} = I(x+1)$ and coincides with f on $I(x)$, but on x it equals 0 (respectively 1). Now assume that X is not first countable at any point. By induction on $(A, <)$ we construct a mapping φ of the set $Q = \bigcup \{Q(x): x \in A\}$ into the set $\mathcal{E}$ of all nonempty closed G_δ-sets in X such that for all $f \in Q$ the following conditions are satisfied: (α): $\varphi(f^{+0}) \cap \varphi(f^{+1}) = \Lambda$; ($\beta$): $\varphi\langle f, y\rangle \supset \varphi(f)$; ($\gamma$): $|\varphi(f)| > 1$; further, one must make use of remarks (i)-(iv) and the assumptions made. Carrying out the construction is left to the reader.

Consider now the set $\tilde{Q}$ of all mappings of A into the set $D = \{0, 1\}$. As before, we agree to denote by $\langle g, x\rangle$, for each $g \in \tilde{Q}$ and each $x \in A$, the restriction of g to $I(x)$ and define a mapping $\tilde{\varphi}$ of $\tilde{Q}$ into the set $\tilde{\mathcal{E}}$ of nonempty closed subsets of X in the following way: $\tilde{\varphi}(g) = \bigcap \{\varphi\langle g, x\rangle: x \in A\}$. From ($\beta$) it follows that $\tilde{\varphi}(g) \neq \Lambda$ for all $g \in Q$, and (α) implies that $\tilde{\varphi}(g) \cap \tilde{\varphi}(g') = \Lambda$, whenever $g \neq g'$. But $\tilde{Q} = D^A$ and $|\tilde{Q}| = |D|^{|A|} = 2^{\aleph_1}$. From $\aleph_1 = 2^{\aleph_0}$ it follows that $2^{\aleph_1} > \aleph_1 = 2^{\aleph_0}$. We obtain $|\bigcup \{\tilde{\varphi}(g): g \in \tilde{Q}\}| \geqslant |\tilde{Q}| > 2^{\aleph_0}$, a contradiction.

128. We show that if in the hypothesis of the problem X is uncountable, then X has a closed subspace without isolated points.

Denote by X^* the set of all points $x \in X$, having a nhood Ox such that $|Ox| \leqslant \aleph_0$. We show that $|X \setminus X^*| > 1$. Each closed set F in X, which is included in X^*, is countable. If $X \setminus X^* = \{x^*\}$, then $\{x^*\} = \bigcap\{U_n: n = 1, 2, \ldots\}$, where all U_n are open in X and, consequently, $X^* = X \setminus \{x^*\} = \bigcup \{X \setminus U_n: n = 1, 2, \ldots\}$. By what has been stated above, each $X \setminus U_n$ is countable; consequently, X^* and X are also countable. From the contradiction thus obtained, it follows that $|X \setminus X^*| > 1$. It is clear that the subspace $X \setminus X^*$ is closed and has no isolated points. Now refer to Problem 126.

129. Suppose that $\mathcal{B}$ is a base for the space X and $|\mathcal{B}| \leqslant \tau$. Let $\tilde{\mathcal{B}}$ be the family of all finite unions of elements of $\mathcal{B}$. Then (Problem 58, Chap. 1), $|\tilde{\mathcal{B}}| \leqslant \tau$, and since X is a compact T_1-space, for any $x \in X$ and $F \subset X$, where $x \notin F$ and F is closed in X, there is $V \in \tilde{\mathcal{B}}$ such that $x \notin V$, $F \subset V$. Consequently, $\mathcal{S} = \{X \setminus V: V \in \tilde{\mathcal{B}}\}$ is a net in X. Obviously, the elements of $\mathcal{S}$ are closed in X and $|\mathcal{S}| = |\mathcal{B}| \leqslant \tau$.

130. X has a countable net $\mathcal{S}$ of closed sets (preceding problem). Assume that X is uncountable.

$A = \bigcup \{P: P \in \mathcal{S}$ and $|P| \leqslant \aleph_0\}$ is a countable set. Choose $y \in X \setminus A$, $z \in X \setminus A$, $y \neq z$. Since X is a T_1-space, $X \setminus \{y\}$ is an open set that contains z. Hence there exists $P_0 \in \mathcal{S}$ such that $z \in P_0 \subset X \setminus \{y\}$. $X \setminus P_0$ is also an open set in X and contains the point y. Hence there is $P_1 \in \mathcal{S}$ such that $y \in P_1 \subset X \setminus P_0$. From $y \notin A$, $y \in P_1 \in \mathcal{S}$ and $z \notin A$, $z \in P_0 \in \mathcal{S}$ it follows that $\aleph_0 < |P_0|$ and $\aleph_0 < |P_1|$.

We have thus proved the statement (*): if a compact T_1-space with a countable base is uncountable, then it has two closed disioint uncountable subspaces. The latter are also compact, have countable bases and satisfy the T_1-axiom. The statement (*) can be applied again to them. Continuing the construction and arguing exactly in the same way as in the solution of Problem 126, we conclude that X contains a subset of cardinality $2^{\aleph_0}$. The cardinality of a T_1-space with a countable base never exceeds $2^{\aleph_0}$, as is easily shown. Hence $|X| = 2^{\aleph_0}$.

131. An absolute Cantor perfect set is a suitable example, see Problems 221, 239 and 167, Chap. 6.

132. Take as X the space S of Problem 131, which is an absolute Cantor perfect set. Fix a countable π-base $\xi = \{U_n: n \in N^+\}$ in X, this can be done by Problem 239, Chap. 6. Then $U_n \neq \Lambda$ for all $n \in N^+$. Since there is no isolated point in X (see Problem 239, Chap. 6), $|U_n| \geqslant \aleph_0$ for all $n \in N^+$. Put $A_1 = \{y_1^1\}$, where y_1^1 is any point of U_1. Suppose that the finite sets A_n are already defined for all $n \leqslant j$, where j is some element of N^+. Choose a point y_{j+1}^i in $U_i \setminus \bigcup \{A_n: n \leqslant j\}$ and set $A_{j+1} = \{y_{j+1}^n: n = 1, \ldots, j+1\}$. $\{A_i: i \in N^+\}$, which has been constructed inductively, is the desired sequence: if $V \neq \Lambda$ and V is open in X, then $U_{n*} \subset V$ for some $n^* \in N^+$ (since ξ is a π-base), and hence $A_n \cap U_{n*} \neq \Lambda$ for all $n > n^*$. Moreover, $A_n \cap V \neq \Lambda$ for all $n > n^*$. It is clear that the remaining requirements are also met.

133. Denote by B_τ the Alexandroff compactification (see Problem 176) of a discrete space X_τ of cardinality τ. Put $\{\xi_\tau\} = B_\tau \setminus X_\tau$. Then all the points of X_τ are isolated in B_τ. A nhood of ξ_τ in B_τ is the complement in B_τ of a finite subset of X_τ. Since $|X_\tau| = \tau > \aleph_0$, there is no countable base either for B_τ or ξ_τ. But if $\xi_\tau \in [A]$, where $A \subset X_\tau$, then $|A| \geqslant \aleph_0$. Choosing any $A' \subset A$ satisfying $|A'| = \aleph_0$, we have $\xi_\tau \in [A']$. Hence B_τ is sequential, and is also a Fréchet-Urysohn space (see p. 44).

134. No. Consider the space $\mathcal{P}$ constructed in Problem 100 of Chap. 5. It is Hausdorff and locally compact. Suppose that $B = \mathcal{P} \cup \{\xi^*\}$ is a Hausdorff compactification of $\mathcal{P}$ in the sense of Alexandroff (one-point compactification; see Problem 176). We show that B is a sequential, but not a Fréchet-Urysohn space. Since $\mathcal{P}_0$ is dense in $\mathcal{P}$, $\xi^* \in [\mathcal{P}_0]$. But no sequence η of points of $\mathcal{P}_0$ converges to ξ^*, because in η there is always a subsequence converging to some point in $\mathcal{P}$. If $A \subset B$ and A is not closed in B, then either $([A] \setminus A) \cap \mathcal{P} \neq \Lambda$, or $|A \cap \mathcal{P}_1| = \aleph_0$ and $[A] \setminus A = \{\xi^*\}$. In the former case, we have used the fact that $\mathcal{P}$ is a locally metrizable space and, consequently, a Fréchet-Urysohn space. In the latter case, appeal has been made to the fact that any countably infinite set of elements of $\mathcal{P}_1$, as a sequence, converges to ξ^* (which follows immediately from the the definition of the topology in $\mathcal{P} \cup \{\xi^*\}$ and the fact that $\mathcal{P}_1$ is a discrete set in $\mathcal{P}$).

135. The answer is so far not known to the authors.

136. The answer is not known to the authors.

137. It suffices to show (see Problem 68) that each point $x_0 \in X$ is the intersection of a countable family of open sets in X. X is a regular space, hence for each point $x \in X \setminus \{x_0\}$ a nhood Ox can be chosen, whose closure is included in $X \setminus \{x_0\}$. The family $\xi = \{Ox: x \in X \setminus \{x_0\}\}$ covers the set $X \setminus \{x_0\}$. Suppose that $<$ is some minimal well-ordering on ξ (see p. 6). For each $U \in \xi$, set $\tilde{U} = U \setminus \bigcup \{V \in \xi: V < U\}$. If $U_1 \in \xi$, $U_2 \in \xi$ and $U_1 \neq U_2$, then obviously, $\tilde{U}_1 \cap \tilde{U}_2 = \Lambda$. Moreover, $\bigcup \{\tilde{U}: U \in \xi\} = X \setminus \{x_0\}$. (Of course, $U \neq \Lambda$ is possible, but $\tilde{U} = \Lambda$.) Fix any $\xi > 0$. If $U \in \xi$ and there is a point $x \in \tilde{U}$, satisfying $d(x, X \setminus U) > \varepsilon$, then fix some such point and denote it by $x_\varepsilon(U)$. $A_\varepsilon = \{x_\varepsilon(U): U \in \xi$ and $x_\varepsilon(U)$ defined$\}$, is clearly a finite set. It is obvious that if $x' \in A_\varepsilon$, $x'' \in A_\varepsilon$ and $x' \neq x''$, then $d(x', x'') > \varepsilon$. We first show that A_ε is closed. Assume the contrary. Then there is $x^* \in X \setminus A_\varepsilon$ satisfying $d(x^*, A_\varepsilon) = 0$. If $x_\varepsilon(U)$ is defined, then $x_\varepsilon(U) \in U \subset X \setminus \{x_0\}$ implies that $d(x_\varepsilon(U), x_0) > \varepsilon$. Hence $d(x_0, A_\varepsilon) \geqslant \varepsilon$ and $x^* \neq x_0$. Therefore, the family $\xi^* = \{U \in \xi: x^* \in U\}$ is nonempty. Let U^* be its first element. $F^* = X \setminus U^*$ is a closed set and $x^* \notin F^*$. Consequently, there is $\delta > 0$, $\delta < \varepsilon$ such that $d(x^*, F^*) \geqslant \delta$. Since $d(x^*, A_\varepsilon) = 0$, there are U_1 and $U_2 \in \xi$ such that $U_1 < U_2$, $x_\varepsilon(U_1) \in A_\varepsilon$, $x_\varepsilon(U_2) \in A_\varepsilon$ and $d(x^*, x_\varepsilon(U_1)) < \delta < \varepsilon$, $d(x^*, x_\varepsilon(U_2)) < \delta < \varepsilon$. Then the relation $U^* < U_2$ cannot hold, otherwise $x_\varepsilon(U_2) \notin U^*$ and $d(x^*, x_\varepsilon(U_2)) \geqslant \delta$, a contradiction. Hence $U_1 < U_2 \leqslant U^*$. From $d(x_\varepsilon(U_1), X \setminus U_1) > \varepsilon$ and $d(x^*, x_\varepsilon(U_1)) < \varepsilon$ it follows that $x^* \in U_1$, but this contradicts the definition of U^*. Consequently, A_ε is a closed set. Regard A_ε as a subspace of the topological space X. Its topology is generated by the restriction of semi-metric d to A_ε (since A_ε is closed in X, see p. 56). If some point x' is not isolated in A_ε, then $d(x', A_\varepsilon \setminus \{x'\}) = 0$ by Problem 110 of Chap. 2 and there exists $x'' \in A_\varepsilon \setminus \{x'\}$ satisfying $d(x', x'') < \varepsilon$, which contradicts the definition of A_ε. Consequently, A_ε is discrete. Since X is compact, we conclude that A_ε is finite.

Put $\tilde{U}_n = \{x \in \tilde{U}: d(x, X \setminus U) > (1/n)\}$ for each $U \in \xi$. Then $\tilde{U} = \bigcup \{\tilde{U}_n: n \in N^+\}$ and hence $\bigcup \{\tilde{U}_n: U \in \xi$ and $n \in N^+\} = X \setminus \{x_0\}$. But $\gamma_n = \{U \in \xi: \tilde{U}_n \neq \Lambda\}$ is a finite set, as is implied by the above arguments when $\varepsilon = 1/n$. Consequently, $\{\tilde{U}_n: U \in \xi$, $n \in N^+$ and $\tilde{U}_n \neq \Lambda\}$ is a countable family. Since it covers $X \setminus \{x_0\}$ and the closures of its elements do not contain the point x_0, we conclude that $\{x_0\}$ is G_δ-set in X.

138. For each $n \in N^+$ and every $x \in X$ fix an open set $U_n x$ in X such that

$$x \in U_n x \subset [U_n x] \subset O_{1/n} x = \{y \in X: d(y, x) < 1/n\}; \tag{1}$$

this is possible by virtue of X being regular and the statement of Problem 137. From an open cover $\{U_n x: x \in X\}$ of the compactum X choose a finite cover $\gamma_n = \{U_n x_i: i = 1, \ldots, k_n\}$. Set $F^n x_i = [U_n x_i]$, and consider the family $\varphi = \{F^n x_i: i = 1, \ldots, k_n;$ $n \in N^+\}$. Add to φ all possible disjoint finite subfamilies of φ and denote the resulting family by $\tilde{\varphi}$. Obviously, $|\tilde{\varphi}| \leqslant \aleph_0$. We show that $\tilde{\varphi}$ is a net in X. Let $x^* \in X$. Fix for each $n \in N^+$ a set $F^n x_{i(n)} \in \gamma_n \subset \tilde{\varphi}$ such that $x^* \in F^n x_{i(n)}$ (possible, since each γ_n covers X). Then, $x^* \in \Phi = \bigcap \{F^n x_{i(n)}: n \in N^+\}$. We show that $\Phi = \{x^*\}$. Let $x' \in \Phi$. Then $x' \in F^n x_{i(n)}$, and hence $d(x', x_{i(n)}) < (1/n)$. Consequently, $\lim\limits_{n \to \infty} d(x_{i(n)}, x') = 0$ and hence (Problem 111, Chap. 2) $\lim\limits_{n \to \infty} x_{i(n)} = x'$. In particular, $x^* = \lim\limits_{n \to \infty} x_{i(n)}$. But X is a Hausdorff space. Hence each convergent sequence in X has at most one limit. Therefore $x^* = x'$ for all $x' \in \Phi$, i. e., $\Phi = \{x^*\}$. Consider an arbitrary open nhood V of x^*. $\{F^n x_{i(n)} \setminus V: n \in N^+\}$ is not a centralized family (otherwise, $\Phi \setminus V \neq \Lambda$). Hence there exist $n_1, \ldots, n_l \in N^+$ such that $F = \bigcap \{F^{n_j} x_{i(n_j)}: j = 1, \ldots, l\} \subset V$. But $F \in \tilde{\varphi}$. Consequenty, $\tilde{\varphi}$ is a net in X.

139. This follows from Problems 138, 106 and 39.

140. Assume that $\xi = \{x_\alpha: \alpha \in J(\tau)\}$ is a free sequence in X and $t(X) < \tau$ (here

$J(\tau)$ is a minimal well-ordered set of cardinality τ). Then there is a regular cardinal number τ^*, satisfying $t(X) < \tau^* \leqslant \tau$ (see Problem 109, Chap. 1). Let A be a set of points of the free sequence $\xi^* = \{x_\alpha : \alpha \in J(\tau^*)\}$ (which is a part of the free sequence ξ). Then $|A| = \tau^* > t(X)$ and some point $x^* \in X$ is a complete accumulation point of A (see Problem 48). Since $t(x) < \tau^*$, there is a set $A^* \subset A$ satisfying $|A^*| < \tau^*$ and $[A^*] \ni x^*$. But τ^* is regular, hence there is $\beta^* \in J(\tau^*)$ such that $A^* \subset \{x_\alpha : \alpha < \beta^*\}$. Then $x^* \in [A^*] \subset [\{x_\alpha : \alpha < \beta^*\}] \subset X \setminus [\{x_\alpha : \beta^* \leqslant \alpha\}]$ and, consequently, $U^* = X \setminus [\{x_\alpha : \beta^* \leqslant \alpha\}]$ is a nhood of x^* containing less than τ^* elements of A, which contradicts the fact that x^* is a complete accumulation point of A.

141. Since X is a compactum and Q is a G_τ-set in it, there is $\Phi \subset X$, satisfying $x \in \Phi \subset Q$ and $\chi(\Phi, X) \leqslant \tau$ (Problem 71). From $x \in \Phi$ and $x \in [A]$ it follows that every nhood of Φ intersects A. Hence (Problem 91, Chap. 2) there is $C \subset A$ such that $|C| \leqslant \tau$ and $[C] \cap \Phi \neq \Lambda$. But then, by the assumptions on A, $[C] \subset A$, and hence $Q \cap A \supset [C] \cap \Phi \neq \Lambda$. This proves (i). To deduce (ii) from (i), fix $x \in [A] \setminus A$. Put $V_0 = X$ and select any $x_0 \in A$. Let $\beta \in J(\tau^+)$ and suppose that for all $\alpha < \beta$ an open set $V_\alpha \ni x$ and a point $x_\alpha \in A$ are already defined. Set $P_\beta = \{x_\alpha : \alpha < \beta\}$. From $|P_\beta| < \tau^+$ and $P_\beta \subset A$ it follows that $x \notin [P_\beta]$. As V_β take any open set such that $x \in V_\beta \subset [V_\beta] \subset X \setminus [P_\beta]$. $Q_\beta = \bigcap \{V_\alpha : \alpha \leqslant \beta\}$ is a G_τ-set in X and $x \in Q_\beta$; hence by virtue of (i), $Q_\beta \cap A \neq \Lambda$. Take x_β as an arbitrary point of $Q_\beta \cap A$. Thus we have defined $V_\alpha \subset X$ and $x_\alpha \in A$ for all $\alpha \in J(\tau^+)$. From $x_\alpha \in Q_\alpha \subset Q_\beta$ when $\beta \leqslant \alpha$ it follows that $[\{x_\alpha : \beta \leqslant \alpha\}] \subset [Q_\beta] \subset [V_\beta] \subset X \setminus [P_\beta] \subset X \setminus [\{x_\alpha : \alpha < \beta\}]$. Thus $[\{x_\alpha : \alpha < \beta\}] \cap [\{x_\alpha : \beta \leqslant \alpha\}] = \Lambda$ and for every $\beta \in J(\tau^+)$, $\{x_\alpha : \alpha \in J(\tau^+)\}$ is a free sequence in X of length τ^+, all of whose elements are contained in A, proving (ii).

142. Let τ^* be the least upper bound of the length of free sequences in X. Then $\tau^* \leqslant t(X)$ by Problem 140. Suppose now that τ is an arbitrary cardinal number less than $t(X)$. Then there is $A \subset X$ such that $A = \bigcup\{[B] : B \subset A \text{ and } |B| \leqslant \tau\} \neq [A]$. Hence (Problem 141) there is a free sequence of length τ^+ in X, which obviously gives the desired conclusion.

143. Fix a minimal well-ordered set $(A, <)$ of cardinality $\aleph_1$. Assume now that the statement to be proved is not true. By induction on the relation $<$, we can define a mapping f of A into X and a mapping φ of A into the set $\mathcal{E}$ of all closed nonempty G_δ-subsets of X, so that the following conditions are satisfied: (i) $[fI(x)] \cap \varphi(x) = \Lambda$ for all $x \in A$; (ii) $\varphi(x) \ni f(x')$ for $x < x'$; (iii) $\varphi(x) \supset \varphi(x')$, $\varphi(x) \neq \varphi(x')$ for all $x < x'$, $x \neq x'$. Assume f and φ have been constructed with the prescribed properties. Then the conditions (i), (ii) and the fact that $\varphi(x)$ is closed imply that $[fI(x)] \cap [f(A \setminus I(x))] = \Lambda$ for all $x \in A$. The set fA is of cardinality $\aleph_1$, because $|A| = \aleph_1$ and $f(x') \neq f(x'')$ when $x' \neq x''$ (see (i) and (ii)). If x^* is a complete accumulation point of fA, then there is a countable set $A' \subset A$ which satisfies $x^* \in [fA']$, since X is a sequential space (and has countable tightness—see Problem 107 of Chap. 2). But there is $a^* \in A$ such that $a < a^*$ for all $a \in A'$ (see Problem 91, Chap. 1). Then $[fI(a^*)] \ni x^*$ and $[f(A \setminus I(a^*))] \not\ni x^*$, which implies that x^* is not a complete accumulation point of fA. This is a contradiction. Let us show how to construct the f and φ above. A central role is played by the following statements: (α) the cardinality of the closure of a countable set in a Hausdorff sequential space does not exceed $2^{\aleph_0}$ (Problem 126, Chap. 2); (β) the intersection of a countable family of G_δ-sets is a G_δ-set; (γ) for any $x \in X$, $y \in U$, there is a nonempty closed G_δ-set, which contains y but not x and is included in U (see Problem 69). Suppose that $a_0 \in A$ is given and for all $a < a_0$ we have defined $f(a)$, $\varphi(a) \subset U$, where $\varphi(a)$ is a nonempty closed G_δ-set and $\varphi(a) \supset \varphi(a')$ when $a < a' < a_0$. Set $\Phi(a_0) = \bigcap\{\varphi(a'): a' < a_0\}$, $F(a_0) = [\{f(a'): a' < a_0\}]$. We have $|F(a_0)| \leqslant 2^{\aleph_0}$ (Problem 126, Chap. 2). Since X is a compactum, $\Phi(a_0)$ is a nonempty closed G_δ-set. Then, by definition, $|\Phi(a_0)| > 2^{\aleph_0}$. Hence $\Phi(a_0) \setminus F(a_0) \neq \Lambda$. Select in $\Phi(a_0)\setminus F(a_0)$ a nonempty closed G_δ-set and take it as $\varphi(a_0)$ (here we have again applied Problems 69, 70). As $f(a_0)$ we take any point of $\varphi(a_0)$.

144. If $U \subset X$ is open and nonempty, then we find a nonempty closed G_δ-set F included in U such that $|F| \leqslant 2^{\aleph_0}$ (Problem 143). Then $\{y\} = F \cap (\bigcap\{X\setminus\{x\}: x \in F$ and $x \neq y\})$ for all $y \in F$. Hence every point of F is an intersection of not more than $2^{\aleph_0}$ open sets in X. But then (Problem 68), $\chi(y, X) \leqslant 2^{\aleph_0}$ for all $y \in F$.

145. For any nonempty open set $U \subset X$, there is a closed G_δ-set F of cadinality not greater than $2^{\aleph_0}$ and included in U (Problem 143). But every compactum of cardinality not greater than $2^{\aleph_0}$ is first countable at least at one point, assuming that $\aleph_1 = 2^{\aleph_0}$ (Problem 127). Consequently, F contains a point y of G_δ-type in F. But since F is a G_δ-set in X, $\{y\}$ is as well. Hence (Problem 68), X is first countable at y. In addition, $y \in F \subset U$.

146. The answer is not known to the authors.

147. True, even without invoking any auxiliary set-theoretic hypotheses. This has been proved recently by B. Shapirovskii.

148. It is necessary to apply the statement of Problem 122 of Chap. 1, in the same way as we did in the solution of Problem 131 of Chap. 2.

149. For every nonempty open set U in X, there is a compactum $F \subset U$ such that $1 \leqslant |F| \leqslant 2^{\aleph_0}$ and $\chi(F, X) \leqslant \aleph_0$ (Problem 143). By means of the Kuratowski-Zorn postulate (p. 6) and this statement, it is routine to construct a disjoint family $\mathcal{F}$ of nonempty compacta of countable character in X such that $Y = \bigcup\{F: F \in \mathcal{F}\}$ is dense in X and $|F| \leqslant 2^{\aleph_0}$ for all $F \in \mathcal{F}$. But since X satisfies the Souslin hypothesis, $|\mathcal{F}| \leqslant 2^{\aleph_0}$ (see Problem 148). Consequently, $|Y| \leqslant 2^{\aleph_0} \cdot 2^{\aleph_0} = 2^{\aleph_0}$ and since $[Y] = X$ and X is sequential, it follows that $|X| \leqslant 2^{\aleph_0}$ (Problem 126, Chap. 2).

150. The answer is not known to the authors.

151. It follows from either Problem 212 or Problem 219.

152. By Problem 144, there is a point $y \in X$ at which the character of X does not exceed $2^{\aleph_0}$. Since X is homogeneous, we infer that the character of X does not exceed $2^{\aleph_0}$ at any point. But it is shown in Problem 151 that every sequential compactum with these properties has cardinality not greater than $2^{\aleph_0}$. If there is an isolated point in X, then X is discrete and hence $|X| < \aleph_0$. If there is no isolated point in X, then $|X| \geqslant 2^{\aleph_0}$ (see Problem 126). This, in conjunction with what is proved above, enables us to conclude that $|X| = 2^{\aleph_0}$.

153. The argument carries over without any difficulty to the scheme of solution of either Problem 212 or Problem 219, subject to the obvious replacement of τ^+ by $\aleph_1$. See also Arkhangel'skii [8].

154. It suffices to consider the case when $\tau \geqslant \aleph_0$. If $\chi(x, X) = \tau$ for all $x \in X$, then (see Problem 128) $|X| \geqslant 2^\tau > \tau$, a contradiction. Hence $\chi(x^*, X) < \tau$ for some point $x^* \in X$. Put $\tau' = \chi(x_0, X)$. By virtue of the homogeneity of X, then $\chi(x, X) = \tau'$ for all $x \in X$. Hence (Problem 153), $|X| \leqslant 2^{\tau'}$, i.e., $\tau \leqslant 2^{\tau'}$.

155. This follows from the preceding problem.

156. The reasoning is carried out along the lines of the solution of Problem 212 or 219. As the set $(A, <)$ relative to which the construction is carried out by induction, we take any minimal well-ordered set of cardinality τ.

157. The statment is true. Recently this problem has been solved by several authors (see, for example, Juhasz [2]).

158. No matter what the answer to this problem may be, it is not known to the authors.

160. A product of the generalized Cantor discontinua is a generalized Cantor discontinuum and a product of continuous mappings is a continuous mapping.

161. Every compact metrizable space is a continuous image of the Cantor perfect set (Problem 299), and the latter is homeomorphic to $D^{\aleph_0}$ (Problem 301).

162. See Problems 161 and 160.

163. See Problem 162.

165. A Tychonoff cube of weight τ is a dyadic compactum (Problem 163), every compactum is homeomorphically embedded in such a cube (see Problem 38).

166. Every generalized Cantor discontinuum D^τ has the Souslin property by Problem 383 of Chap. 2. Obviously, the image of a space with the Souslin property under a continuous mapping also has the Souslin property.

167. By Problem 391 of Chap. 2, X is the continuous image of some compactum with a countable base. Hence (Problem 111) X also has a countable base.

168. From Problem 391 of Chap. 2 it follows that $[Y]$ is a continuous image of some compactum with a countable base. Consequently, $[Y]$ also has a countable base (Problem 111).

169. By virtue of Problems 166 and 167, it suffices to take any non-metrizable first countable compactum (for example, Alexandroff's two circles) or any compactum, which does not possess the Souslin property, (say), $T(\omega_1) \cup \{\omega_1\}$ (see Problems 80, 84, 184-188 of this chapter and Problem 24 of Chap. 4).

170. If $2^{\aleph_1} > 2^{\aleph_0}$, then every dyadic compactum of cardinality $2^{\aleph_0}$ is first countable on a dense point set (Problem 127). Consequently, it is metrizable by Problem 168.

171. By Problem 145, X is first countable at the points of a dense set. Hence (Problem 168), X has a countable base.

172. True. See Arkhangel'skii and Ponomarev [1] or Arkhangel'skii [7].

173. True. The condition is possible. The problem was solved recently by Belugin.

174. The answer is not known to the authors.

175. Statements (i)-(iv) are completely obvious, in addition (ii′) follows from (ii), see Problem 14. It remains to prove (v)-(vii). $(Z, \mathcal{T})$ has an uncountabe discrete subspace, namely I. Hence (Problem 99, Chap. 2), $(Z, \mathcal{T})$ has no countable base. But $(Z, \mathcal{T})$ is separable, because $[Q]_{(Z, \mathcal{T})} = Z$, $|Q| = \aleph_0$. The space $(Z, \mathcal{T})$ is non-metrizable, because every separable metrizable space has a countable base. If $(Z, \mathcal{T})$ were Lindelöf, then from its cover by open sets possessing a countable base, a countable subcover could be extracted, which yields a countable base in $(Z, \mathcal{T})$, but this contradicts (vi).

176. It will suffice to consider the case when X is not compact. With respect to X a T_1-space $\tilde{X}$ can be constructed, containing X as an open subspace, such that $\tilde{X} \setminus X = \{\xi\}$, the nhoods of ξ being the complements in $\tilde{X}$ of compact subsets of X (all compact subsets of X are closed in it, since X is Hausdorff). From our assumptions on X we deduce that $\tilde{X}$ is Hausdorff. Let $x, y \in \tilde{X}$. If x, y belong to X, then they have disjoint nhoods, which are also nhoods of these points in $\tilde{X}$, since X is open in $\tilde{X}$. If $y = \xi$, $x \neq \xi$, then $x \in X$ and x has a nhood Ox in X with a compact closure. Hence $O\xi = \tilde{X} \setminus [Ox]_X$ is a nhood of ξ disjoint from Ox. It is easy to show also that $\tilde{X}$ is compact.

177. We know (Problem 176) that every locally compact Hausdorff space has a compact Hausdorff extension. But every subspace of a compactum is completely regular (Problem 8).

178. X is a regular space; hence for every point $x \in Y$, there is a nhood Ox of x whose closure (in X) is contained in Y. $[Ox]$ is a compact set, i.e., Ox is the desired nhood.

179. If X is open in bX, then X is locally compact (see Problem 178). Let us prove the converse. Put $F = bx \setminus X$; take an arbitrary point $x \in X$. It has a nhood Ox for which $\Phi = [Ox]_X$ is compact. Φ is a closed set in X and in bX, because these spaces are Hausdorff. Put $G = X \setminus \Phi$. Then $G \cup \Phi = X$. Hence $[G]_{bX} \cup [\Phi]_{bX} = [X]_{bX} = bX$. But $[\Phi]_{bX} = \Phi$ and $[G]_{bX} \cap X = [G]_X \subset X \setminus Ox \subset X \setminus \{x\}$. Hence $x \notin [G]_{bX}$ and $[G]_{bX} \cup \Phi = bX$, i.e., $x \in bX \setminus [G]_{bX} \subset \Phi \subset X$ and X is an open set in bX.

182. Every nonempty open set U in X contains a standard set $V = \Pi\{U_\alpha : \alpha \in A\}$ open in X; here $U_\alpha \neq \Lambda$ for all $\alpha \in A$ and $U_\alpha = X_\alpha$ for $\alpha \in A \setminus A^*$, where A^* is some finite set. Fix $\alpha^* \in A \setminus A^*$ and choose $y_\alpha \in U_\alpha$ for all $\alpha \in A$. Then U contains a closed subspace of X homeomorphic to X_{α^*}, namely $Y^* = X_{\alpha^*} \times \Pi\{\{y_\alpha\}: \alpha \in A \setminus \{\alpha^*\}\}$ (that Y^* is closed in X follows from the fact that every X_α is a T_1-space). But Y^* is not compact; hence U is not compact.

183. It is necessary to compare the definitions and note that under the projection of a product onto a factor a compact subspace is mapped onto a compact subspace (see Problem 330 of Chap. 2 and Problem 91 of this chapter).

184. There is a $a \in T(\omega_1)$ such that $\varphi \subset \tilde{I}(a)$. But $\tilde{I}(a)$ is countably compact (Problem 84). This means that $\tilde{I}(a)$ is compact metrizable so that it has a countable base (Problem 105), and hence the conclusion follows (see also Problem 259).

185. The open cover $\gamma = \{I(x): x \in T(\omega_1)\}$ of X has no finite subcover. Another solution is: there is not a single complete accumulation point for $T(\omega_1)$.

186. (*a*) From $|F_1| = |F_2| = \aleph_1$ it follows that F_1 is cofinal in $T(\omega_1)$ and so is F_2. Since $F_1 \cap F_2 = \Lambda$, a sequence $\{a_n: n \in N^+\}$ can be chosen from $T(\omega_1)$ such that $a_1 < a_2 < a_3 < \ldots < a_i < a_{i+1} < \ldots$ (in the sense of the relation $<$, defined on $T(\omega_1)$) and $a_{2n} \in F_2$, $a_{2n-1} \in F_1$ for all $n \in N^+$. Suppose that a^* is the least element of $T(\omega_1)$ such that $a_i < a^*$ for all $i \in N^+$ (see Problem 91, Chap. 1). Then $a^* \in [\{a_{2n}: n \in N^+\}] \subset [F_2] = F_2$ and $a^* \in [\{a_{2n-1}: n \in N^+\}] \subset [F_1] = F_1$. Consequently, $a^* \in F_1 \cap F_2$, i.e., $F_1 \cap F_2 \neq \Lambda$, a contradiction.

(*b*) For each $\alpha \in T(\omega_1)$ set $F_\alpha = \{\beta \in T(\omega_1): \alpha < \beta\}$. From the definition of the topology on $T(\omega_1)$ it follows that every F_α is closed. Since the real line l is a Hausdorff space of countable tightness, $f(F_\alpha)$ is closed in l for each $\alpha \in T(\omega_1)$ (see Problem 79, Chap. 6). Obviously, if $\alpha' < \alpha''$, then $f(F_{\alpha''}) \subset f(F_{\alpha'})$. Consequently (Problem 105, Chap. 2), there is an $\alpha^* \in T(\omega_1)$ such that $f(F_{\alpha^*}) = f(F_\alpha)$ for all α greater than α^*. Then $\bigcap \{f(F_\alpha): \alpha \in T(\omega_1)\} = f(F_{\alpha^*}) \neq \Lambda$. We show that $|f(F_{\alpha^*})| = 1$. If $x \neq y$, $x \in f(F_{\alpha^*})$, $y \in f(F_{\alpha^*})$, then $P_x = f^{-1}(x)$, $P_y = f^{-1}(y)$ are disjoint closed sets cofinal in $T(\omega_1)$, contradicting (*a*).

187. Suppose that F_1 and F_2 are closed in $T(\omega_1)$ and let $F_1 \cap F_2 = \Lambda$. By the preceding problem, every pair of closed sets in $T(\omega_1)$ of cardinality $\aleph_1$ intersect. Hence we may assume that $|F_1| \leqslant \aleph_0$. Choose an $\alpha_0 \in T(\omega_1)$, satisfying $F_1 \subset \tilde{I}(\alpha_0) = \{x \in T(\omega_1): x \leqslant \alpha_0\}$ (such α_0 exists, see Problem 91 of Chap. 1 and Problem 84 of this chapter). Set $F_2' = F_2 \cap \tilde{I}(\alpha_0)$. The subspace $\tilde{I}(\alpha_0)$ is clopen in $T(\omega_1)$ and is compact. Consequently, $\tilde{I}(\alpha_0)$ is normal and there exist open sets U_1 and U_2' in $\tilde{I}(\alpha_0)$ and hence also in $T(\omega_1)$, such that $\tilde{I}(\alpha_0) \supset U_1 \supset F_1$, $U_2' \supset F_2'$ and $U_1 \cap U_2' = \Lambda$. Set $U_2 = U_2' \cup \{x \in T(\omega_1): \alpha_0 < x\}$. Then U_2 is open in $T(\omega_1)$, $U_2 \supset F_2$ and $U_1 \cap U_2 = \Lambda$.

188. By Problem 186, if f is such a function, then we can find $\alpha_0 \in T(\omega_1)$, satisfying $f(\alpha) = f(\alpha_0)$ for all $\alpha > \alpha_0$. But $\tilde{I}(\alpha_0) = \{x \in T(\omega_1): x \leqslant \alpha_0\}$ is a compact subspace of $T(\omega_1)$ (see Problem 84). Hence $f | \tilde{I}(\alpha_0)$ is a bounded function. Thus f is also bounded.

189. (*a*) Suppose that X is a countably compact space, and $D \subset X$ is an infinite subset. If D has no limit point in X, then D is closed in X and for each point $d \in D$ there is a nhood Ud such that $Ud \cap D = \{d\}$. Then the countable open cover, consisting of the nhoods $Ud, d \in D_0$, where $D_0 \subset D$, $|D_0| = \aleph_0$ together with the open set $X \setminus D_0$ in X, contains no finite subcover. This is a contradiction.

(*b*) Conversely, suppose that every infinite set of X has a limit point in X. Let $\omega = \{U_i: i \in N^+\}$ be a countable open cover of X, which has no finite subcover. Without loss of generality, we may assume that $U_i \not\subset \bigcup \{U_j: j < i\}$ for every $i \in N^+$. For every i take a point $x_i \in U_i \setminus \bigcup \{U_j: j < i\}$. $D = \{x_i: i \in N^+\}$ is an infinite set; consequently, it has a limit point x_0 in X. Consider first i_0 such that $x_0 \in U_i$. Then $x_{i_0} \notin U_{i_0}$ for all $i > i_0$ (by the choice of x_i) and, consequently, x_0 is not a limit of D, a contradiction.

190. $T(\omega_1)$ is such a space. This follows from Problems 189, 185 and 184.

192. (*a*) Suppose that X is countably compact, and $f(x)$ is a countinuous real-valued function on X. The image $Y = fX$ of X under the mapping f is countably compact (see Problem 191). Thus $Y = fX$ is a countably compact subspace of the metric space R, and hence is compact and bounded in R (see Problems 254 and 256). Consequently, $\sup Y < \infty$ and $\inf Y > -\infty$, i.e., f is a bounded function.

(*b*) Conversely, suppose that a normal space X is not countably compact. Then there exists in X a countably infinite subset $D \subset X$ without a limit point (see Problem 189), i.e., D is a closed and discrete subspace. Hence D can be mapped onto the natural numbers N by means of some mapping $f: D \to N$. This mapping is a continuous mapping of the closed set D in X into the real line. Consequently, since X is normal, there is a continuous mapping $\tilde{f}: X \to R$ such that $\tilde{f} | D = f$ (Problem 34). The function $\tilde{f}$ is not bounded.

194. (*a*) $\Rightarrow$ (*b*). Suppose that $\xi = \{F_i: i \in N^+\}$ is a countable centralized family of closed sets in X. If $\bigcap_{i=1}^{\infty} F_i = \Lambda$, then the family $\{X \setminus F_i : i \in N^+\}$ is a countable open cover from which it is impossible (by virtue of ξ being centralized) to select a finite subcover.

$(b) \Rightarrow (c)$ is obvious

$(c) \Rightarrow (a)$. Suppose that $\omega = \{U_i : i \in N^+\}$ is an arbitrary countable open cover. Consider in X the closed sets

$$F_i = X \setminus \bigcup_{j=1}^{i} U_j, \; i \in N^+.$$

Obviously, $F_{i+1} \subset F_i$, $i \in N^+$. By virtue of the fact that ω covers X, we have

$$\bigcap_{i=1}^{\infty} F_i = \bigcap_{i=1}^{\infty} \left(X \setminus \bigcup_{j=1}^{i} U_j \right) = \Lambda.$$

But then for some i_0 the set

$$F_{i_0} = X \setminus \bigcup_{j=1}^{i_0} U_j = \Lambda,$$

i.e., $\{U_1, \ldots, U_{i_0}\}$ is a finite subcover of ω.

196. (*a*) Suppose that X is pseudocompact, and $\gamma = \{U_\alpha : \alpha \in A\}$ is a locally finite family of open sets in X. Suppose that γ is infinite. Without loss of generality, we may assume that γ is countable: $\gamma = \{U_i : i \in N^+\}$. For each $i \in N^+$ choose a point $x_i \in U_i$ in such a manner that $x_i \neq x_j$ when $i \neq j$. Consider continuous functions f_i, $i \in N^+$, on X such that $0 \leqslant f_i(x) \leqslant i$ for every point $x \in X$, $f_i(x_i) = i$, $f_i(x) = 0$, if $x \notin U_i$ (such functions exist because of the complete regularity of X). Further, consider the function

$$f(x) = \sum_{i=1}^{\infty} f_i(x).$$

Let $x_0 \in X$. Since γ is locally finite, there is a nhood Ux_0 of x_0 in X, which intersects only a finite number of elements of γ. Hence $f(x_0) < \infty$. Let $U_{i_1}, \ldots, U_{i_k}$ be all the elements of γ, which intersect U_{x_0}. Then $f_j(x_0) = 0$, if $j \neq i_1, i_2, \ldots, i_k$ and $f(x_0) = f_{i_1}(x_0) + f_{i_2}(x_0) + \ldots + f_{i_k}(x_0)$. Moreover, obviously, $f(x) = f_{i_0}(x) + \ldots + f_{i_k}(x)$ for every point $x \in Ux_0$. Hence if $\varepsilon > 0$ is arbitrary, then using the continuity of the function $g_{i_1 i_2 \ldots i_k}(x) = f_{i_1}(x) + f_{i_2}(x) + \ldots + f_{i_k}(x)$, choose a nhood $U'x_0$ of x_0, satisfying $U'x_0 \subset Ux_0$ and $|g_{i_1 i_2 \ldots i_k}(x) - g_{i_1 i_2 \ldots i_k}(x_0)| < \varepsilon$, if $x \in U'x_0$. Then also $|f(x) - f(x_0)| < \varepsilon$, if $x \in U'x_0$. The continuity of the function $f(x)$ on X follows. Moreover, it is obvious that $\lim_{i \to \infty} f(x_i) = \infty$, because $f_i(x_i) = i$. This contradicts the pseudocompactness of X.

(*b*) Suppose that f is a continuous function on the completely regular space X. For each $n \in N^+$ consider an interval $(n, n + 1)$ of the real line. Obviously, the family of all such intervals is locally finite on the real line. Consider the sets $U_n = f^{-1}((n, n + 1)) = \{x \in X: fx \in (n, n + 1)\}$, open in X (since f is continuous). The family $\gamma = \{U_n: n \in N^+\}$ of open sets in X is obviously locally finite in X. Then, by hypothesis, γ is finite. This implies that, starting from a certain natural number $n \in N^+$ we always have $fX \cap (n, n + 1) = \Lambda$, i.e., f is a bounded function. The proof is complete.

197. Assume that X is pseudocompact and $\cap \{[U]: U \in \xi\} = \Lambda$. Enumerate the elements of the family ξ: $\xi = \{U_1, \ldots, U_i, \ldots\}$. Without loss of generality, we may assume that $U_{i+1} \subset U_i$, $i \in N^+$. Since $\cap \{[U_i]: i \in N^+\} = \Lambda$, $X = \cup \{V_i: i \in N^+\}$, where $V_i = X \setminus [U_i]$. Moreover, $V_{i+1} \supset V_i$, $i \in N^+$, because $U_{i+1} \subset U_i$, $i \in N^+$. Let $x_0 \in X$. Consider first $i_0 \in N^+$ such that $x_0 \in V_{i_0}$. Then $V_{i_0} \cap U_i = \Lambda$ for all $i \geqslant i_0$; hence $V_{i_0} \cap U_i \neq \Lambda$ is possible only for $i < i_0$. Hence ξ is a locally finite family in X, and is therefore finite (see Problem 196). This contradicts that ξ is centralized.

Conversely, suppose that for any countable centralized family ξ of open sets in X, $\cap \{[U]: U \in \xi\} \neq \Lambda$. We must show that X is pseudocompact. If not, then in X there is an infinite locally finite family γ of open sets (see Problem 196). Hence there is in X a countably infinite locally finite family γ of open sets $\gamma = \{V_1, \ldots, V_i, \ldots\}$, satisfying $V_i \subset V_{i+1}$, $i \in N^+$, and $X = \cup \{V_i: i \in N^+\}$. Because of the local finiteness of γ in X, only a finite number $V_i \in \gamma$ can have the property $[V_i] = X$. By removing such V_i from γ, we may assume that $[V_i] \neq X$ for every $i \in N^+$. Then $U_i = X \setminus [V_i] \neq \Lambda$. Since $V_i \subset V_{i+1}$, $i \in N^+$, $U_i \supset U_{i+1}$, $i \in N^+$, so that $\xi = \{U_i: i \in N^+\}$ is an infinite centralized family. Since $X \cup \{V_i: i \in N^+\}$, $\cap \{[U_i]: i \in N^+\} = \Lambda$. We have a contradiction.

198. Suppose that Φ is not countably compact. Then Φ contains a countable closed set $A = \{a_n: n \in N^+\}$, which forms a discrete subspace of X. By induction, it is now easy to construct a family $\{V_n: n \in N^+\}$ of mutually disjoint nhoods of the points of A: $V_n \ni a_n$ (see Problem 22) and such that this family is discrete in X. The latter means that for every point $x \in X$ there is a nhood Ox, which intersects not more than one set in $\{V_n: n \in N^+\}$, see Problem 22. Hence, for any choice of $x_n \in V_n$, $\{x_n: n \in N^+\}$ is a discrete and closed set. Assume now that $\{G^k: k \in N^+\}$ is a countable family of nhoods of the set F in X. Put $V_n^k = G^k \cap V_n$. Since $a_n \in \operatorname{Fr} F$ and V_n^k is a nhood of a_n, $V_n^n \cap (X \setminus F) \neq \Lambda$. For each $n \in N^+$ choose $b_n \in V_n^n \cap (X \setminus F)$ and set $B = \{b_n: n \in N^+\}$. B is closed (this is explained above), and $B \cap F = \Lambda$. Consequently, $G = X \setminus B$ is a nhood of F. But $G^n \cap (X \setminus G) = G^n \cap B \supset V_n^n \cap B \ni b_n$, i.e., $G^n \cap (X \setminus G) \neq \Lambda$ for all $n \in N^+$ and, consequently, $\{G^n: n \in N^+\}$ is not a defining system of nhoods of F.

199. Such, in particular, is every noncompact space with a countable base.

200. Assume the contrary. Fix for each $x \in X$ a nhood Ox such that $|Ox \cap A| < \aleph_0$. The cover $\gamma = \{Ox: x \in X\}$ contains a countable subcover λ. Then $A \subset \cup \{U \cap A: U \in \lambda\}$ and $|\lambda| \leqslant \aleph_0$, $|U \cap A| < \aleph_0$. Consequently, $|A| \leqslant |\cup \{U \cap A: U \in \lambda\}| \leqslant \aleph_0$, a contradiction.

201. True. An appropriate example is any countable regular space not satisfying the first axiom of countability, see Problem 113 of Chap. 2.

204. Since every separable space satisfies the Souslin property, the space $(Z, \mathcal{T})$ in Problem 175 is an appropriate example.

205. Such a space is $(Z, \mathcal{T})$ in Problem 175.

206. (*a*) (i) → (ii). Let $U \subset X$ be open. Since X is perfectly normal, $U = \bigcup_{i=1}^{\infty} F_i$, where the F_i are closed in X, and hence Lindelöf. This implies (Problem 208) that the subspace U is Lindelöf.

(*b*) (ii) → (iii). Let γ be any family of open sets in X. Consider in X the open set $\tilde{\gamma} = \bigcup \{U: U \in \gamma\}$, which by (ii) is a Lindelöf subspace. Hence γ has a countable subfamily γ_0 such that $\tilde{\gamma}_0 = \bigcup \{U: U \in \gamma_0\} = \tilde{\gamma}$.

(*c*) (iii) → (iv). Suppose that $A \subset X$ is any subspace of X. Consider an arbitrary cover γ of A by open sets in X. By virtue of condition (iii), γ has a countable subfamily $\gamma_0 \subset \gamma$ such that $\tilde{\gamma}_0 = \bigcup \{U: U \in \gamma_0\} = \tilde{\gamma} = \bigcup \{U: U \in \gamma\}$. Then γ_0 is a countable subcover of the cover γ of A.

(*d*) (iv) → (i). Suppose that $\Gamma \subset X$ is an arbitrary open set in X. Consider a cover γ of Γ by open sets in X such that $[U] \subset \Gamma$ for every $U \in \gamma$ (such cover of Γ exists, because X is regular and Γ is open in X). Using the Lindelöf property of Γ (condition (iv)), select from γ a countable subcover γ_0 of this subspace. Then $|\gamma_0| \leqslant \aleph_0$ and $\Gamma = \bigcup \{[U]: U \in \gamma_0\}$, i.e. Γ is F_σ. Consequently, X is a perfectly normal space.

207. This follows from Problem 206.

208. Let

$$X = \bigcup_{i=1}^{\infty} X_i,$$

where X_i, $i \in N^+$, are Lindelöf subspaces. Take an arbitrary open cover ω of X. Consider the family $\omega_i = \{U \in \omega: U \cap X_i \neq \Lambda\}$, which covers X_i. Choose from the cover ω_i of X_i a countable subcover ω_i^0. Then,

$$\omega^0 = \bigcup_{i=1}^{\infty} \omega_i^0$$

covers the whole space, $\omega^0 \subset \omega$ and $|\omega^0| \leqslant \aleph_0$.

209. (*a*) Note that A is open in I^*, but $I^* \setminus A$ is closed in I^*. Also, A is a discrete space; all this follows from the definition of the space. If U is an interval or half-interval on the segment I, then its closure $[U]_{I^*}$ in the topology of I^* is automatically contained in the segment $[U]_I$ (this fact is made transparent by the definition of the topology of I^*). From all that has been stated it follows that I^* is a regular space.

(*b*) Before we show that I^* is Lindelöf, note that in I^* a base for the topology on I^* is formed by the family consisting of all possible intervals of the segment [0, 1], half-intervals of the form $[0, \alpha)$ and $(\beta, 1]$, $\alpha \in I$, $\beta \in I$, and one-point sets $\{x\}$, $x \in A$. Denote this base by σ. Hence, to show that I^* is Lindelöf, it suffices to choose a countable subcover from any cover of I^* by elements of σ. Suppose that ω is any cover of I^* by elements of σ. Consider the subspace $I \setminus A$ of I^*. This subspace has the same topology possessed by $I \setminus A$ as a subspace of the real line; consequently, $I \setminus A$ as a subspace of I^* has a countable base, implying that it is Lindelöf (see Problem 77, Chap. 2). Let $\omega_1 \subset \omega$ be a countable subfamily covering $I \setminus A$. $V = \bigcup \{U: U \in \omega_1\}$ is an open set both in the usual topology of I and in I^*. $F = I \setminus V$ is compact in the topology of the segment and obviously $F \subset A$. Since A is completely imperfect, $|F| \leqslant \aleph_0$. We enlarge our family ω_1 to the family ω' by adding $\{x\}$, $x \in F$. Obviously $|\omega'| \leqslant \aleph_0$, $\omega' \subset \omega$ and ω' is a cover of I^*, showing that I^* is Lindelöf.

(b') The normality of I^* follows from its regularity and Lindelöf property (Problem 19).

(c) In order to construct a pointwise countable base for I^* it is necessary to take a countable base σ_1 for I in the natural topology, and consider the family σ_2 of one-point sets $\{x\}$, $x \in A$. Then $\sigma_0 = \sigma_1 \cup \sigma_2$ is the desired base for I^*.

(d), (e), (f), and (g) follow from the fact that I^* has an uncountable discrete subspace, namely A, open in I^* (in this connection, see Problems 99 and 213 of Chap. 2 and Problem 206 of the present chapter).

210. On the plane consider a circle K. Let C be its circumference, and let $O = K \setminus C$. Define a new topology on K as follows. A base at the points of O consists of sets open in the usual topology. We obtain an arbitrary nhood of any point $x \in C$ by drawing any two chords xy' and xy'' from $x(y', y'' \in C)$, taking W as the complement in K with respect to the set of all its points lying inside or on the boundary of the angle $y'xy''$ and passing on to the intersection of the set $W \cup \{x\}$ with an arbitrary nhood of x in the usual topology. It is easy to verify (directly) that a regular space is obtained. The topologies induced on C and O by this space coincide with the usual topology for C and O, as is clear. Consequently (Problem 141, Chap. 2), the space K has a countable net. Hence it is normal (Problem 19), Lindelöf (Problem 143, Chap. 2) and also perfectly normal. Now let C_0 be any completely imperfect subset in C such that $|C_0| > \aleph_0$ (see Problem 285). Also take a countable dense set $O_0 \subset O$ in O. Consider the set $X = C_0 \cup O_0$, endowed with the topology induced in K. The space X has a countable net, is first countable, and every compactum in it is countable, but X itself is uncountable and does not have a countable base—the latter follows from an examination of the topology at the points of C_0 and the fact that C_0 is uncountable. As a subspace of a completely regular space, X is completely regular. Since a countable compactum in any first countable space has countable character (Problem 72), the character of every compactum in X is countable. However, X, as a space with a countable net but without a countable base, is not among the perfectly normal compacta (Problem 228).

211. Put $\gamma = \bigcup \{\gamma_x : x \in F\}$. For every $y \in X \setminus F$ we find $\lambda \subset \gamma$ such that $F \subset \bigcup \{U : U \in \lambda\} \subset X \setminus \{y\}$, in fact X is a T_1-space. But F itself is Lindelöf, being a closed subspace of a Lindelöf space. Consequently, there is $\lambda_y \subset \lambda$, satisfying $F \subset \bigcup \{U : U \in \lambda_y\} \subset X \setminus \{y\}$ and $|\lambda_y| \leqslant \aleph_0$. Fix such a λ_y for each $y \in X \setminus F$ and put $G_y = \bigcup \{U : U \in \lambda_y\}$. Then obviously $\bigcap \{G_y : y \in X \setminus F\} = F$. On the other hand, λ_y is a countable subfamily of γ, and the cardinality of γ does not exceed $2^{\aleph_0}$. Since the cardinality of the set of all countable subfamilies of a family of cardinality $2^{\aleph_0}$ is also $2^{\aleph_0}$, the number of distinct λ_y is not greater than $2^{\aleph_0}$. The proof is complete.

212. Fix a minimal well-ordered set $(A, <)$ of cardinality $\aleph_1$, and a set M of cardinality $2^{\aleph_0}$. Suppose that, for each $a \in A$, $Q(a)$ is the set of all mappings of $\tilde{I}(a) = \{a' \in A : a' < a\}$ into M. If $f' \in Q(a')$ and $f'' \in Q(a'')$, where $a' < a''$, $a' \neq a''$ and the restriction f'' to $\tilde{I}(a')$ equals f', then write $f' \prec f''$ and also $f' = \langle f'', a' \rangle$. Put $Q = \bigcup \{Q(a) : a \in A\}$. If $f \in Q(a)$, then we say that the length of f equals a, and we write

$l(f) = a$. Denote by $\mathcal{E}$ the set of all mappings of A into M. If $\varphi \in \mathcal{E}$ and $a \in A$, then $\langle \varphi, a \rangle$ is the restriction of φ to $\tilde{I}(a)$, so that $\langle \varphi, a \rangle \in Q(a)$. Let $\mathcal{P}$ be the family of all closed sets in X and ξ some choice function on $\mathcal{P}$ which chooses from every nonempty $P \subset X$ a one-point subset $\xi(P) = \{x\} \subset P$. Set also $\xi(\Lambda) = \Lambda$. For all those $R \subset X$ for which this is possible, fix closed sets R_m in X, where $m \in M$, such that (I): $R = \bigcup \{R_m: m \in M\}$. The set of all $R \subset X$, for which (I) is defined, we denote by $\mathcal{R}$. It is possible to assume here that in (I), if $|R| > 1$, then $R_m \neq R$ for all $m \in M$ (invoke the Hausdorf separation axiom).

REMARK (II): If F is closed in X and Φ is closed in X, with $|\Phi| \leqslant 2^{\aleph_0}$ (for example, $\Phi = \Lambda$), then $F \setminus \Phi \in \mathcal{R}$ (this follows from Problem 211, by virtue of the proposition on the character of points in X). Consider a mapping $\psi: Q \to \mathcal{P}$ with the following properties. (III) If $f' < f''$, then $\psi(f') \supset \psi(f'')$, and if, in addition, $|\psi(f')| > 1$, then $\psi(f') \neq \psi(f'')$. (IV) For all $f \in Q$, we have $[\eta(f)] \cap \psi(f) = \Lambda$, here $\eta(f) = \bigcup \{\xi(\psi f' \setminus \psi(f'')): f' < f'' < f\}$. (V) For each $a \in A$, we have $X = \bigcup \{[\eta(f)] \cup \psi(f): f \in Q(a)\}$. Note that $|Q(a)| \leqslant 2^{\aleph_0}$ and $|\eta(f)| \leqslant \aleph_0$. Hence $|Q| \leqslant 2^{\aleph_0}$ and $|[\eta(f)]| \leqslant 2^{\aleph_0}$ (see Problem 126, Chap. 2). Consequently, $|X \setminus \bigcup \{\psi f: f \in Q\}| \leqslant 2^{\aleph_0}$. The relation (IV) can be enlarged thus: (IV′) for every $f \in Q$ we have $\psi(f) \setminus [\eta(f)] = \bigcup \{\psi(f'): f < f'$ and $l(f') = l(f) + 1\}$. The mapping ψ is easy to construct by induction in $(A, <)$, with respect to the length of $f \in Q$. If 0 is the first element of A, then for all $f \in Q(0)$ we set $\psi(f) = X$. Now let ψ be defined on the set $\bigcup \{Q(a): a < a^*, a \neq a^*\}$. For any $f \in Q(a^*)$ let $F^f = \bigcap \{\psi(f'): f' < f\}$, $R^f = F^f \setminus [\eta(f)]$ (where $\eta(f)$ is defined according to (IV)). Since $|[\eta(f)]| \leqslant 2^{\aleph_0}$, $R^f \subset \mathcal{R}$. Put $\psi(f) = R^f$. The definition of ψ is complete.

The reader should not have any difficulty in verifying that the relations (III), (IV), (IV′) and (V) are satisfied. Assume that $|X| > 2^{\aleph_0}$. Then there exists $x^* \in X \setminus (\bigcup \{[\eta(f)]: f \in Q\} \cup \{\psi(f): |\psi(f)| \leqslant 1$ and $f \in Q\})$. From (IV′) it follows that it is possible (by induction in $(A, <)$) to select $f_a \in Q(a)$ for each $a \in A$, such that $f_{a'} < f_{a''}$ if $a' < a''$, $a' \neq a''$ and $\psi(f_a) \ni x^*$ for all $a \in A$. Put $\psi(f_a) = F_a$. By the choice of x^*, we have $|F_a| > 1$ for all $a \in A$, and (III) now implies that $F_{a'} \supset F_{a''}$, $F_{a'} \neq F_{a''}$, when $a' < a''$, $a' \neq a''$. Hence for each $a \in A$, the element $x_a = \xi(F_a \setminus F_{a+1}) \in F_a$ is defined (here, as usual, $a + 1$ is the first element of A greater than a). For each $a_0 \in A$ either $\{x_a: a < a_0\} = \eta(f_{a_0})$ if a_0 has no immediate predecessor, or $\{x_a: a < a_0\} = \eta(f_{a_0}) \cup x_{a'}$, if a_0 has an immediate predecessor, i.e. if $a_0 = a' + 1$ for some $a' \in A$. But, by (IV), $[\eta(f_{a_0})] \cap F_{a_0} = \Lambda$. Since $x_{a'} \notin F_{a_0}$, we infer that we always have $[\{x_a: a < a_0\}] \cap F_{a_0} = \Lambda$. But if $a_0 \leqslant a$, then $x_a \in F_{a_0}$ and F_{a_0} is closed in X. Hence, for all $a_0 \in A$, we have: (VI) $[\{x_a: a < a_0\}] \cap [\{x_a: a_0 \leqslant a\}] = \Lambda$. Obviously, $|\{x_a: a \in A\}| = \aleph_1$. Since X is Lindelöf, there is in X a complete accumulation point for the set $\{x_a: a \in A\}$ (Problem 200). Let y^* be such a point. Then, since X is sequential, there is a countable set $A' \subset A$ such that $y^* \in [\{x_a: a \in A'\}]$ (see Problem 107, Chap. 2). Then there is (Problem 91, Chap. 1) $a^* \in A$ such that $a < a^*$ for all $a \in A'$. We now have $y^* \in [\{x_a: a \in A'\}] \subset [\{x_a: a < a^*\}]$, and hence (see (VI)), $y^* \notin [\{x_a: a^* < a\}]$, but this means, since $\{x_a: a < a^*\}$ is a countable set, that y^* is not a complete accumulation point for $\{x_a: a \in A\}$, a contradiction.

If X is regular, then another solution of this problem can be derived from Problem 219. The requirement $\chi(x, X) \leqslant 2^{\aleph_0}$ can be replaced by the weaker restriction $\psi(x, X) \leqslant 2^{\aleph_0}$.

213. This is a special case of the statement of Problem 212.

214-215. The answer is not known to the authors.

216. This is solved on similar lines to Problem 45.

217. Suppose that σ is a base for Y, satisfying $|\sigma| = wY = \tau$. Let Σ be the family of countable intersections of elements of σ. Since $|\sigma| = \tau$ is an admissible cardinal number, $|\Sigma| = \tau^{\aleph_0} = \tau$. Now, for every nonempty set $B \in \Sigma$ take some point $x(B) \in f^{-1}B$. Set $M = \{x(B): B \in \Sigma, B \neq \Lambda\}$. It is obvious that $|M| = \tau$. Consider the set $F = [M]_X$, closed in X. Clearly, $s(F) \leqslant \tau$. To show that $fF = Y$, first note that the set fM is dense in Y, so that the set fF is also dense in Y. Suppose now that $y_0 \in Y$ is any point and $\sigma_{y_0} = \{V \in \sigma: y_0 \in V\}$. It is obvious that $\sigma_{y_0} \subset \sigma$, $|\sigma_{y_0}| \leqslant \tau$ and σ_{y_0} is a base at y_0. Moreover, it is natural that $f^{-1}y_0 = \bigcap\{f^{-1}V: V \in \sigma_{y_0}\}$. Since Y is regular and the mapping f is continuous.

$$f^{-1}y_0 = \bigcap\{f^{-1}V: V \in \sigma_{y_0}\} = \bigcap\{[f^{-1}V]_X: V \in \sigma_{y_0}\}. \tag{1}$$

To prove the equality $fF = Y$, it is necessary to show that $f^{-1}y_0 \cap F \neq \Lambda$ for every point $y_0 \in Y$. For this, we shall show that the family of closed sets $\{([f^{-1}V]_X \cap F): V \in \sigma_{y_0}\}$ is countably centralized. Indeed, consider an arbitrary countable sub-family $\gamma = \{[f^{-1}V_i]_X: i \in N^+\} \subset \{[f^{-1}V]_X: V \in \sigma_{y_0}\}$. Since

$$\bigcap_{i=1}^{\infty} V_i \neq \Lambda \left(\text{because } y_0 \in \bigcap_{i=1}^{\infty} V_i,\ V_i \in \sigma,\ i \in N^+\right)$$

there is a point

$$x \in f^{-1}\left(\bigcap_{i=1}^{\infty} V_i\right) \cap M = \left(\bigcap_{i=1}^{\infty} f^{-1}V_i\right) \cap M$$

which implies that

$$\bigcap_{i=1}^{\infty} [f^{-1}V_i]_X \cap F \neq \Lambda \tag{2}$$

Thus, the family of closed sets $\{([f^{-1}V]_X \cap F): V \in \sigma_{y_0}\}$ is countably centralized. Since X is Lindelöf (see Problem 216), we have

$$\bigcap\{([f^{-1}V]_X \cap F): V \in \sigma_{y_0}\} \neq \Lambda. \tag{3}$$

From (3) and (1) it then follows that $f^{-1}y_0 \cap F \neq \Lambda$.

218. Since $s(F_\lambda) \leqslant \mathfrak{r}$, and X satisfies condition (b), $|F_\lambda| \leqslant \mathfrak{r}$ for each $\lambda \in A$. Consider the set

$$\Phi = \bigcup\{F_\lambda: \lambda \in A\}. \tag{1}$$

It is clear that $|\Phi| \leqslant \mathfrak{r}$. To show that Φ is closed in X, let $x_0 \in [\Phi]_X$. Since X satisfies condition (a), we find $\Phi' \subset \Phi$, satisfying $|\Phi'| \leqslant \aleph_0$ and $x_0 \in [\Phi']_X$. Then there is $\lambda_0 \in A$ such that $\Phi' \subset F_{\lambda_0}$ and, consequently, since F_{λ_0} is closed, $x_0 \in F_{\lambda_0} \subset \Phi$. Thus

Φ is closed in X. To show that

$$\Phi = X, \tag{2}$$

let $x_0 \in X$. From condition (ii) of the problem it follows that

$$f_\lambda^{-1} f_\lambda x_0 \cap F_\lambda \neq \Lambda \text{ for every } \lambda \in A. \tag{3}$$

Then, a fortiori we have also

$$\Phi_\lambda = f_\lambda^{-1} f_\lambda x_0 \cap \Phi \neq \Lambda \text{ for every } \lambda \in A. \tag{4}$$

From (v) it follows that $\Phi_{\lambda'} \subset \Phi_\lambda$ if $\lambda < \lambda'$. In particular, we find that the family $\{\Phi_\lambda: \lambda \in A\}$ of nonempty closed sets in X is countably centralized. Since X is Lindelöf (see Problem 216), we have

$$\cap\{\Phi_\lambda : \lambda \in A\} = \cap\{(f_\lambda^{-1} f_\lambda x_0 \cap \Phi) : \lambda \in A\} \neq \Lambda. \tag{5}$$

Let $x' \in \cap\{\Phi_\lambda : \lambda \in A\}$. Then $x' \in \Phi$ and

$$f_\lambda x' = f_\lambda x_0 \tag{6}$$

for every $\lambda \in A$. Hence $x' \in F_{\lambda_0}$ for some $\lambda_0 \in A$. Take $\lambda_1 > \lambda_0$. In consequence of condition (iv) we obtain $f_{\lambda_1}^{-1} f_{\lambda_1} F_{\lambda_0} = F_{\lambda_0}$; consequently, $f_{\lambda_1}^{-1} f_{\lambda_1} x' \subset F_{\lambda_0}$ because $x' \in F_{\lambda_0}$. Since (by equality (6)) $f_{\lambda_1} x' = f_{\lambda_1} x_0$, $x_0 \in f_{\lambda_1}^{-1} f_{\lambda_1} x' \subset F_{\lambda_0} \subset \Phi$ because $f_{\lambda_1}^{-1} f_{\lambda_1} x' = f_{\lambda_1}^{-1} f_{\lambda_1} x_0$. Thus $x_0 \in \Phi$, and hence equality (2) is proved. Since $|\Phi| \leqslant \mathfrak{c}$, then also $|X| \leqslant \mathfrak{c}$. The proof is complete.

219. Reduce this statement to the statement of the preceding problem. By utilizing the complete reqularity of X (our space is normal too, see Problem 19), set in correspondence with each point $x \in X$ a family $\theta(x)$ of open sets in x such that $|\theta(X)| \leqslant \mathfrak{c}$, $\{x\} = \cap\{U: U \in \theta(x)\}$ and $X \setminus U$ is a null-set for every $U \in \theta(x)$. Further, fix some minimally well-ordered set A of cardinality $|A| = \aleph_1$. Let 1 be the first element of this set. Consider a space Y_1 such that $wY_1 \leqslant \mathfrak{c}$ and a continuous mapping $f_1: X \xrightarrow{\text{onto}} Y_1$ (for example f_1 can be taken as any continuous real-valued function on X, where Y_1 is a subspace of the real line). Utilizing the fact that $\mathfrak{c}^{\aleph_0} = \mathfrak{c}$ and Problem 217, take a set F_1, closed in X, such that $s(F_1) \leqslant \mathfrak{c}$ and $f_1(F_1) = Y_1$. Assume now that for each $a \in A$, $a < a^*$, $a^* \in A$, we have constructed a space Y_a, $wY_a \leqslant \mathfrak{c}$, a continuous mapping $f_a: X \xrightarrow{\text{onto}} Y_a$, a closed set F_a in X, such that conditions (i) through (v) of Problem 218 are satisfied. Consider the set $M_{a*} = \cup\{F_a: a < a^*\}$ and the family $\theta(M_{a*}) = \cup\{\theta(x): x \in M_{a*}\}$. Since $s(F_a) \leqslant \mathfrak{c}$, by condition (*b*) $|F_a| \leqslant \mathfrak{c}$ for every $a < a^*$. Then $|M_{a*}| \leqslant \mathfrak{c}$, because the set M_{a*} is a union of the countable family $\{F_a: a < a^*\}$ of sets M_a of cardinality $|M_a| \leqslant \mathfrak{c}$. But then $|\theta(M_{a*})| \leqslant \mathfrak{c}$, because $|M_{a*}| \leqslant \mathfrak{c}$ and $|\theta(x)| \leqslant \mathfrak{c}$ for every point $x \in M_{a*}$. For each $U \in \theta(M_{a*})$ fix on X a continuous function

$g_U: X \to I_U$, satisfying $X \setminus U = g_U^{-1}(0)$. It is clear that $g_U(U)$ is open in $g_U X$ and $U = g_U^{-1} g_U U$. Consider the diagonal mapping f_{a*} on X of the family of functions $\{g_U: U \in \theta(M_{a*})\}$ and the family of mappings $\{f_a: a < a^*\}$. Since $|\theta(M_{a*})| \leqslant \mathfrak{c}$, but $|\{f_a: a < a^*\}| \leqslant \aleph_0$ and $\omega Y_a \leqslant \mathfrak{c}$, hence $w(\Pi\{I_U: U \in \theta(M_{a*})\} \times \Pi\{Y_a: a < a^*\}) \leqslant \mathfrak{c}$; consequently, $wY_{a*} \leqslant \mathfrak{c}$, where $Y_{a*} = f_{a*}(X)$. Moreover, $f_{a*}U$ is open in Y_{a*} and $f_{a*}^{-1} f_{a*} U = U$ for every $U \in \theta(M_{a*})$. Hence $f_{a*}^{-1} f_{a*}\, x = x$ if $x \in M_{a*}$. Exactly in the same manner, $f_{a*}^{-1} f_{a*} x \subset f_a^{-1} f_a x$ for every point $x \in X$, if $a < a^*$. Thus the following conditions are satisfied: (iv) $f_{a*}^{-1} f_{a*} F_a = F_a$, if $a < a^*$, and (v) $f_{a*}^{-1} f_{a*} x \subset f_a^{-1} f_a x$ for every point $x \in X$ and any $a < a^*$. Making use of Problem 217, choose F_{a*}, closed in X, which satisfies (i) $s(F_{a*}) \leqslant \mathfrak{c}$, (ii) $f_{a*}(F_{a*}) = Y_{a*}$. We verify that condition (iii) is satisfied. Indeed, if $x \in M_{a*}$, then $f_{a*}^{-1} f_{a*}\, x = x$; hence if $f_{a*}(F_{a*}) = Y_{a*}$, then $M_{a*} \subset F_{a*}$. This implies (iii) $F_a \subset F_{a*}$ if $a < a^*$. Thus for every $a \in (A, <)$ we have constructed a space Y_a, $wY_a \leqslant \mathfrak{c}$, a continuous mapping $f_a: X \xrightarrow{\text{onto}} Y_a$ and a closed set F_a in X in such a way that all the hypotheses of Problem 218 are satisfied; hence, in consequence of the statements of this problem, we obtain $|X| \leqslant \mathfrak{c}$. The proof is complete.

220. This follows from Problem 130 of Chap. 2 and Problem 219 or 212 of this chapter.

221. Every subspace of a perfectly normal compactum is Lindelöf (Problem 206). But every metrizable Lindelöf space has a countable base (Problem 213, Chap. 2).

222. The answer is not known to authors.

223. The character and pseudo-character of a closed set in a compactum always coincide (Problem 68).

225. This statement is a special case of the similar assertion about first countable compacta (see Problems 213, 220). A direct (and original) proof is given in Alexandroff and Urysohn [1].

226. We have $|X| \leqslant 2^{\aleph_0}$ (see Problem 225). Hence the weight of X also does not exceed $2^{\aleph_0}$ (Problem 107). Suppose that $\mathcal{B}$ is a base for X satisfying $|\mathcal{B}| \leqslant 2^{\aleph_0}$. Without loss of generality, we may assume that the union of every countable set of elements of $\mathcal{B}$ is again an element of $\mathcal{B}$. Then every set F closed in X can be set in correspondence with some countable family γ_F of elements of the base $\mathcal{B}$ such that $F = \bigcap\{U: U \in \gamma_F\}$. In addition, $F_1 \neq F_2$ implies that $\gamma_{F_1} \neq \gamma_{F_2}$. This means that the set of all closed subsets of X is mapped one-to-one into a set of all countable subsets of a set of cardinality $2^{\aleph_0}$. But the latter itself is of cardinality $2^{\aleph_0}$ (see Problem 98 of Chap. 1).

227. False. Filippov [4] has constructed $2^{\mathfrak{c}}$ mutually non-homeomorphic perfectly normal compacta. But the cardinality of the set of all closed subsets of any fixed perfectly normal space does not exceed $2^{\aleph_0}$ (Problem 226).

228. We may assume that $\tau \geqslant \aleph_0$. Put $\tilde{\mathcal{S}} = \{[P]: P \in \mathcal{S}\}$. Since X is regular, for each $x \in Y$ and every set U open in X, containing x, there is $F \in \tilde{\mathcal{S}}$ such that $x \in F \subset U$. Fix now for each $F \in \tilde{\mathcal{S}}$ a countable defining family γ_F of nhoods in X (see p. 40) and set $\gamma = \bigcup\{\gamma_F: F \in \tilde{\mathcal{S}}\}$. Then $|\gamma| \leqslant \max\{\aleph_0, \tau\} = \tau$ and there is $V \in \gamma_F \subset \gamma$ such that $x \in F \subset V \subset U$. Since the point and its nhood U were chosen arbitrarily, γ is an exterior base for Y in X.

229. In fact, every subspace of a perfectly normal compactum is Lindelöf (Problem 206). Consequently, all metrizable subspaces of perfectly normal compacta have countable bases. But the sum of a countable set of spaces, each with a countable base, has a countable net. Consequently, X has a countable net. But then X has also a countable base (Problem 228).

230. It is clear that $\tilde{\mathcal{T}}$ is a topology on X and that $(X, \tilde{\mathcal{T}})$ satisfies the Hausdorff separation axiom (in fact, $\mathcal{T} \subset \tilde{\mathcal{T}}$). Obviously, every countable set is closed in $(X, \tilde{\mathcal{T}})$. Consequently, there is no convergent sequence in $(X, \tilde{\mathcal{T}})$. On the other hand, no point of $(X, \tilde{\mathcal{T}})$ is isolated (since $|U| > \aleph_0$ for all nonempty $U \in \mathcal{T}$). Consequently, $(X, \tilde{\mathcal{T}})$ is not first countable. But the pseudo-character of $(X, \tilde{\mathcal{T}})$ equals $\aleph_0$, because $\mathcal{T} \subset \tilde{\mathcal{T}}$ and $(X, \mathcal{T})$ is first countable. From (0) and this argument it also follows that $t(X) > \aleph_0$. In addition, (0) and (iv) directly imply the relations (vi), (vii) and (viii). The statement (i) follows from (iii) (see Problems 206 and 207), which leaves (ii), (iii), (ix) and (x) to be proved. But (ii) follows directly from (0): every countable subset A of Y is closed in Y (in the topology inherited from $(X, \tilde{\mathcal{T}})$) and the set $Y \setminus A$, open in Y, is nonempty (since $|Y| > \aleph_0 = |A|$) and does not contain any point of A. We pass on to (iii).

Given an arbitrary family of elements of $\tilde{\mathcal{T}}$, i.e., a family of the form $\xi = \{U_\alpha \setminus A_\alpha: \alpha \in M\}$. Put $Y = \bigcup\{V: V \in \xi\}$. Consider the family $\eta = \{U_\alpha: \alpha \in M\}$ and set $\tilde{Y} = \bigcup \{V: V \in \eta\}$. Then $\tilde{Y} \supset Y$ and, in addition, since $U_\alpha \in \mathcal{T}$ for all $\alpha \in M$, there is a countable set $M' \subset M$ such that $\bigcup\{U_\alpha: \alpha \in M'\} = \tilde{Y}$. Then, obviously, $\bigcup\{U_\alpha \setminus A_\alpha: \alpha \in M'\} \supset \tilde{Y} \setminus B \supset Y \setminus B$, where $B = \bigcup\{A_\alpha: \alpha \in M'\}$ is a countable set. But every element y of B is covered by some element of ξ; fix the latter and denote it by V_y. Then $\xi' = \{U_\alpha \setminus A_\alpha: \alpha \in M'\} \cup \{V_y: y \in B\}$ is a countable subfamily of ξ and $\bigcup \{V: V \in \xi'\} = Y = \bigcup \{V: V \in \xi\}$. Hence (see Problem 206) $(X, \tilde{\mathcal{T}})$ is hereditarily Lindelöf.

To prove (ix), i.e., that $(X, \tilde{\mathcal{T}})$ is H-closed, assume that (x) is already verified. Let $\gamma = \{U_\alpha \setminus A_\alpha: \alpha \in M\}$ be an arbitrary cover of the set X by nonempty elements of $\tilde{\mathcal{T}}$. Then $\tilde{\gamma} = \{U_\alpha: \alpha \in M\}$ is an open cover of the compactum $(X, \mathcal{T})$ and, consequently, there is a finite set $M^* \subset M$ such that $\bigcup \{U_\alpha: \alpha \in M^*\} = X$. Then $\bigcup\{U_\alpha \setminus A_\alpha: \alpha \in M^*\}$ is dense in $(X, \mathcal{T})$, and hence

$$\begin{aligned} X &= [\bigcup\{U_\alpha \setminus A_\alpha: \alpha \in M^*\}]_{(X, \mathcal{T})} = \bigcup \{[U_\alpha \setminus A_\alpha]_{(X, \mathcal{T})}: \alpha \in M^*\} \\ &= \bigcup \{[U_\alpha \setminus A_\alpha]_{(X, \tilde{\mathcal{T}})}: \alpha \in M^*\} \end{aligned}$$

(the last equality is derived using (x)). Consequently (see Problem 61), $(X, \tilde{\mathcal{T}})$ is H-closed.

We now verify (x). Given nonempty $U_1, U_2 \in \mathcal{T}$, $A_1, A_2 \subset X$; let

$$x \in [U_1 \setminus A_1]_{(X, \tilde{\mathcal{T}})}, \; x \in U_2 \setminus A_2.$$

Then $U_1 \cap U_2 \neq \Lambda$, $U_1 \cap U_2 \in \mathcal{T}$, and hence $(U_1 \setminus A_1) \cap (U_2 \setminus A_2) \supset (U_1 \cap U_2) \setminus (A_1 \cup A_2) \neq \Lambda$ (since $|A_1 \cup A_2| \leqslant \aleph_0$). Therefore,

$$x \in [U_1 \setminus A_1]_{(X, \mathcal{T})} \text{ and } [U_1 \setminus A_1]_{(X, \mathcal{T})} = [U_1 \setminus A_1]_{(X, \tilde{\mathcal{T}})}.$$

231. If $f: X \to Y$ is a continuous mapping, where Y is a perfectly normal compactum, then $f(X)$ has a net $\mathcal{S}$ whose cardinality does not exceed the weight of X. But the weight of $f(X)$ does not exceed the cardinality of $\mathcal{S}$, because $f(X)$ is included in a perfectly normal compactum (Problem 228). Hence the weight of $f(X)$ is not greater than the weight of X.

232. Y is a compact space (Problem 91), f is a closed mapping (Problem 92) and $fX = Y$; hence for every $A \subset Y$ we have $\chi(f^{-1}A, X) = \chi(A, Y)$ (see Problem 324, Chap. 2). If A is closed in Y, then $f^{-1}A$ is closed in X and $\chi(f^{-1}A, X) \leqslant \aleph_0$, so that $\chi(A, Y) \leqslant \aleph_0$.

233. See Problems 231 and 232.

234. Every infinite compactum contains a countable non-closed set (Problem 49). If we now apply the statement of Problem 23 twice (the second time the roles of X and Y are interchanged), we obtain the desired result.

235. By Problems 234 and 23 (and the fact that $X \times X \times X = (X \times X) \times X$), either the product $X \times X$ is perfectly normal, or every countable subset of X is closed. We ignore the latter, because X is a compactum (we consider only the non-trivial case when X is infinite). Hence the former is true, i.e., $X \times X$ is perfectly normal. But then X has a countable base, see Problem 114.

236. See Problem 114.

237. If $x \in [M] \setminus M$ and $\{U_n: n \in N^+\}$ is a base at x, where $U_j \subset U_i$ for $j > i$, then choosing $x_n \in U_n \cap M$ for each $n = 1, 2, \ldots$, we obtain a compact subspace $\Phi = \{x\} \cup \{x_n: n \in N^+\}$, where $\Phi \cap M = \{x_n: n = 1, 2, \ldots\}$ and hence $x \in [\Phi \cap M]$. Note that we have incidentally proved a rather stronger statement than the required one: X is a k_1-space (see p. 137).

238. Let $x \in [M]$. Take Ox to be a nhood of x with compact closure $[Ox]$. Then $x \in [M \cap Ox] \subset [M \cap [Ox]]$, which implies by definition that the space in question is a k-space (even a k_1-space).

240. Suppose that X is a pointwise countable space and $P \subset X$. Assume that $P \cap \Phi$ is a compactum for every compactum Φ included in X. Let $x_0 \in [P]$. By hypothesis, there is a compactum $\Phi_0 \subset X$, which contains x_0 and whose character in X is countable. Suppose that U is an arbitrary nhood of x_0. $U_0 = U \cap \Phi_0$ is an open set in Φ_0. Hence (Problem 69) in U_0 we find a compactum F_0 of countable character in Φ_0 containing x_0. Then the character of F_0 in X is also countable (Problem 70). Let $\gamma = \{U_i: i \in N^+\}$ be some countable defining family of nhoods of F_0 in X such that $U_j \subset U_i$ for $j > i$. Since $x_0 \in [P]$ and $U_i \ni x_0$ for each i, there is a point $x_i \in U_i \cap P$. The sequence $\{x_i: i \in N^+\}$ obtained has at least one limit point in F_0 which, exactly in the same way, also holds for every subsequence (Problem 65). Hence $F_1 = F_0 \cup \{x_n: n = 1, 2, \ldots\}$ is compact. But then by assumption $F_1 \cap P$ is closed. Hence $F_1 \cap P \supset [\{x_n: n = 1, 2, \ldots\}$ is compact. But then by assumption $F_1 \cap P$ is closed. Hence $F_1 \cap P \supset [x_n: n = 1, 2, \ldots\}]$, and since $[\{x_n: n = 1, 2, \ldots\}] \cap F_0 \neq \Delta$, it follows that $F_0 \cap P \neq \Delta$. By the same token, for every nhood U of x_0 there is a point in $U \cap P \cap \Phi_0$. But this also implies that $x_0 \in [P \cap \Phi_0]$. By assumption on the set P, we have $[P \cap \Phi_0] = P \cap \Phi_0$. Hence $x_0 \in P$ and P is closed.

241. Let $X \subset bX$, where bX is a compactum and $X = \cap \{G_i: i \in N^+\}$, where every G_i is an open set in bX. Fix $x \in X$ and put $U_0 = bX$. Assume that the sets U_i, containing x and open in bX, are defined for all $i = 0, 1, \ldots, k$. Then take U_{k+1} as any open set in bX such that $x \in U_{k+1} \subset [U_{k+1}] \subset U_k \cap G_{k+1}$. Suppose that the sequence $\{U_i: i = 0, 1, \ldots\}$ is defined according to the above procedure. Set $F = \bigcap\{[U_i]: i = 0, 1, \ldots\}$. Then

$$F \subset \bigcap\{[G_i]: i \in N^+\} \subset X \quad \text{and} \quad F = \bigcap\{[U_i]: i = 0, 1, \ldots\},$$

i.e., F is a a compactum of countable character in bX (see Problem 70) and $x \in F \subset X$. Consequently, X is a pointwise countable space.

244. No. To show this, we construct a Hausdorff k-space X, which contains a point x and a set M such that $x \in [M] \setminus M$, but $x \notin [\Phi \cap M]$ for any compact subspace Φ of X. This condition is satisfied by the space considered in Problem 113 of Chap. 2. As M we take the set of all its isolated points and as x the point θ (in this connection, see Problem 250). Another example is given in Problem 60 of Chap. 4 (in this connection see Problem 59 of Chap. 4 and Problem 242 of the present chapter).

245. No. See Problems 59 and 60 of Chap. 4.

246. Suppose that $f: X \to Y$ is the mapping in question, where $f(X) = Y$. If $M \subset Y$ and $M \cap \Phi$ is closed in Φ for every compact subspace Φ of Y, then $f^{-1}M \cap F$ must be closed in F for every compact subspace F of X. Indeed, $f^{-1}M \cap F = f_*^{-1}(M \cap fF)$, where f_* is the restriction of f to F. Since f_* is continuous, fF is compact, $M \cap fF$ is closed in fF and $f^{-1}M \cap F$ is closed in F. But X is a k-space. Hence $f^{-1}M$ is closed in X. Since f is a quotient mapping, M is closed in Y. Consequently, Y is a k-space.

247. It is necessary to show just that every k-space is a quotient space of some locally compact Hausdorff space; the converse follows from the following two assertions:

(i) every locally compact Hausdorff space is a k-space (Problem 238);
(ii) a quotient space of a k-space is a k-space (Problem 246).

Suppose that $\mathcal{B} = \{F_\alpha: \alpha \in M\}$ is the family of all compact subsets of X. Consider the disjoint (free) union $Z = \Sigma\{F_\alpha: \alpha \in M\}$ of the topological spaces F_α (see p. 44). By definition, the elements of Z are all possible pairs (x, α), where $x \in F_\alpha$, $\alpha \in M$. The set $F'_\alpha = \{(x, \alpha): x \in F_\alpha\}$ is naturally mapped onto F_α, by the rule: $i_\alpha(x, \alpha) = x$ for all $x \in F_\alpha$. This one-to-one mapping allows us to transfer the topology from F_α to F'_α— the topology thus arising on F'_α we denote by $\mathcal{T}_\alpha$. Then $\bigcup\{\mathcal{T}_\alpha: \alpha \in M\}$ is a base for Z. Let f be the natural mapping of Z onto X whose restriction to each F'_α coincides with i_α. f is a continuous mapping: if U is open in X, then $U \cap F_\alpha$ is open in F_α, hence $i_\alpha^{-1}(U \cap F_\alpha)$ is also open in Z. But obviously $f^{-1}(U) = \bigcup\{i_\alpha^{-1}(U \cap F_\alpha): \alpha \in M\}$. Hence $f^{-1}(U)$ is an open set. Assume now that $P \subset X$ and $f^{-1}P$ is closed in Z. Then $f^{-1}P \cap F'_\alpha$ is closed. Hence (since i_α is a homeomorphism) $i_\alpha(f^{-1}P \cap F'_\alpha)$ is closed in F_α. But $i_\alpha(f^{-1}P \cap F'_\alpha) = P \cap F_\alpha$. Thus $P \cap F_\alpha$ is closed for every compact subspace of X. Since X is a k-space, P is closed in X. Taken together, this implies that f is a quotient mapping.

248. Suppose that G is an open subspace of a k-space X. By the preceding problem, there is a quotient mapping $f: Z \to X$ of some locally compact Hausdorff space Z onto X. Set $Y = f^{-1}G$ and $f' = f \mid Y, f': Y \to G$. Since f is continuous, Y is open in Z; hence Y is a locally compact (Problem 180) Hausdorff space. Moreover, f' is a quotient mapping, because Y is a complete pre-image of an open set (see Problem 305, Chap. 2). Hence, G is also a k-space (Problem 247).

249. No. In particular, Fréchet-Urysohn spaces are spaces not satisfiying this condition (see Problem 250). See example in Problem 113, Chap. 2.

250. Thus, let $x \in X$, $M \subset X$ and $x \in [M] \setminus M$. It is possible to find a set $L \subset M$ such that the following conditions are satisfied: (α) $x \in [L]$; (β) if $L' \subset M$ and $x \in [L']$, then $|L| \leqslant |L'|$. Since $L \cup \{x\}$ is a k-space, there is a compactum $\Phi \subset L \cup \{x\}$ such that $x \in [\Phi \setminus \{x\}]$ (Problem 243). By (β) it is obvious that $|\Phi| = |L|$. Put $|\Phi| = \tau$ and show that $\tau = \aleph_0$. This will prove all that is required, because then Φ would be a metrizable compactum (see Problems 39 and 105). x is not an isolated point in Φ; its character in Φ equals τ, which follows from $|\Phi| = \tau$ and Problem 68. Fix any base $\mathscr{B} = \{U_\alpha : \alpha \in A\}$ at x in Φ, whose cardinality equals τ. Assume that A is minimally well-ordered by some relation $<$. We now perform a construction by induction along $(A, <)$. Let $1 = \min (A, <)$. Fix a nhood $O_1 x$ of x such that $[O_1 x] \subset U_1$. Choose $x_1 \in O_1 x \setminus \{x\}$. Assume that $\beta \in A$ and for all $\alpha < \beta$, $\alpha \in A$, the nhoods $O_\alpha x$ of x are defined and the points $x_\alpha \in \Phi \setminus \{x\}$. Then, by ($\beta$), $[\{x_\alpha : \alpha < \beta\}] \not\ni x$ (because $|\{x_\alpha : \alpha < \beta\}| < \tau$). The space Φ is regular; choose a nhood $O_\beta x$ of x such that $[\{x_\alpha : \alpha < \beta\}] \cap [O_\beta x] = \Lambda$ and $[O_\beta x] \subset U_\beta$. Since $|\{\alpha \in A : \alpha \leqslant \beta\}| < \tau$ and the character of x in Φ equals τ, $\bigcap \{O_\alpha x : \alpha \leqslant \beta\} \setminus \{x\} \neq \Lambda$. Take as x_β any element of the set $\bigcap \{O_\alpha x : \alpha \leqslant \beta\} \setminus \{x\}$. Thus, x_α and O_α can be defined for all $\alpha \in A$. Consider the subspace $X^* = \{x_\alpha : \alpha \in A\} \cup \{x\}$ of X. x is not an isolated point in X^*. On the other hand, for every $\beta \in A$, the set $X \setminus ([\{x_\alpha : \alpha < \beta, \alpha \neq \beta\}] \cup [O_{\beta+1} x])$ is a nhood of the point x_β which does not contain points of $X^* \setminus \{x_\beta\}$. Hence all $x' \subset X^* \setminus \{x\}$ are isolated in X^*. But X^* is a k-space. Hence (Problem 243) there is a compactum $F \subset X^*$ such that $x \in [F \setminus \{x\}]$. But $F \setminus \{x\} \subset M$. From the definition of τ it follows that $|F| = \tau$. Choose in $F \setminus \{x\}$ an infinite countable subset P. P has a limit point in F (because of the compactness of F), but no point in $F \setminus \{x\}$ can be such a point. Hence $[P] \ni x$. But $P \subset M$. Consequently, by condition (β), $\tau = \aleph_0$.

252. The space X can be represented as a quotient space of some locally compact Hausdorff space $\tilde{X}$ (Problem 247). Let f be the corresponding quotient mapping and let j be the identity mapping of the space Y onto itself. Then $f \times j : \tilde{X} \times Y \to X \times Y$ is a quotient mapping (Problem 26, Chap. 6). But $\tilde{X} \times Y$ is a locally compact Hausdorff space, hence (see Problem 247) $X \times Y$ is a k-space.

279. Consider $\varepsilon_1 = \frac{1}{2}$. By Problem 218 of Chap. 2, there is a countable disjoint cover ω_1 of the space (Y, ρ) by clopen sets of diameter not exceeding $\varepsilon_1 = \frac{1}{2}$. Since (Y, ρ) is not compact, ω_1 may be assumed to be infinite. Thus, suppose that $\omega_1 = \{U_{i_1} : i_1 \in N^+\}$ is a countably infinite disjoint cover of the space (Y, ρ) by clopen sets, diam $U_{i_1} \leqslant \frac{1}{2}$, $U_{i_1} \neq \Lambda$ for every $i_1 \in N^+$ and $U_{i_1} \cap U_{i_1'} = \Lambda$, if $i_1 \neq i_1'$. Now for each $U_{i_1} \in \omega_1$ and $\varepsilon_2 = 1/2^2$ consider the countable family $\{U_{i_1 i_2} : i_2 \in N^+\}$ of clopen sets in U_{i_1} (hence also in X), satisfying diam $U_{i_1 i_2} \leqslant \frac{1}{4}$ for every $i_2 \in N^+$, $U_{i_1 i_2} \cap U_{i_1 i_2'} = \Lambda$ if $i_2 \neq i_2'$ and $U_{i_1} = \bigcup \{U_{i_1 i_2} : i_2 \in N^+\}$. Since U_{i_1} is non-compact, $\{U_{i_1 i_2} : i_2 \in N^+\}$ can be so chosen that $U_{i_1 i_2} \neq \Lambda$ for every $i_2 \in N^+$. Note that $Y = \bigcup \{U_{i_1 i_2} : i_1 \in N^+, i_2 \in N^+\}$. Perform the same construction for every $U_{i_1 i_2}$, $i_1 \in N^+$, $i_2 \in N^+$ and $\varepsilon_3 = 1/2^3$, etc. As a result, for every finite sequence of natural numbers $i_1, \ldots, i_n$ we obtain the sets

$U_{i_1 \ldots i_n}$, clopen in X. Also the following conditions are satisfied:

$$U_{i_1 i_2 \ldots i_{n-1} i_n} \cap U_{i_1 i_2 \ldots i_{n-1} i'_n} = \Delta, \text{ if } i_n \neq i'_n; \tag{1}$$

$$U_{i_1 \ldots i_{n-1}} = \bigcup \{U_{i_1 \ldots i_{n-1} i_n} : i_n \in N^+\}; \tag{2}$$

$$Y = \bigcup \{U_{i_1} : i_1 \in N^+\}; \tag{3}$$

$$\operatorname{diam} U_{i_1 \ldots i_n} \leqslant 1/2^n; \tag{4}$$

$$U_{i_1 \ldots i_n} \neq \Delta \text{ for every } i_1 \in N^+, \ldots, i_n \in N^+. \tag{5}$$

The last, i.e., condition (5), is realized because of the fact that at every step a non-compact set is obtained. Consider now the countably infinite family $\{X_n : n \in N^+\}$ of countably infinite discrete spaces (i.e., each X_n is homeomorphic to the space N of natural numbers). Then each point x of the topological product $\Pi\{X_n : n \in N^+\}$ is a countable sequence of natural numbers: $x = \{i_1, i_2, \ldots, i_n, \ldots\}$. Put

$$fx = U_{i_1} \cap U_{i_1 i_2} \cap \ldots \cap U_{i_1 i_2 \ldots i_n} \cap \ldots = \bigcap_{n=1}^{\infty} U_{i_1 \ldots i_n}.$$

Because of conditions (1) — (5) and the completeness of (Y, ρ), for each point $x \in \Pi\{X_n : n \in N^+\}$ the set fx is nonempty and consists exactly of one point $y \in Y$. Thus we have constructed the mapping $f: B(\aleph_0) \to Y$ (recall that $B(\aleph_0) = \Pi\{X_n : n \in N^+\}$). From conditions (iii) and (ii) it follows that f is a mapping onto Y, and from (i) that it is one-to-one. In the space $B(\aleph_0) = \Pi\{X_n : n \in N^+\}$ the sets $V_{i_1^0 \ldots i_n^0} = \{x = \{i_n : n \in N^+\} \in \Pi\{X_n : n \in N^+\} : i_1 = i_1^0, \ldots, i_n = i_n^0\}$ form a standard open base, and the sets $U_{i_1 \ldots i_n}$, $i_1 \in N^+, \ldots, i_n \in N^+$ form a base for X (as is trivial to verify). That the mapping $f: B(\aleph_0) \to Y$ is a homeomorphism now follows from the equality $fV_{i_1^0 \ldots i_n^0} = U_{i_1^0 \ldots i_n^0}$, which is easy to establish by using the conditions (1) — (5).

282. See the solution of Problem 279.

285. (*a*) Suppose that $\mathcal{S} = \{F_n : n \in N^+\}$ is a countably infinite family of compacta in the real line, where $|F| > \aleph_0$ for each $F \in \mathcal{S}$ and each pair of sets from $\mathcal{S}$ are non-homeomorphic (such a family obviously exists). For every countable sequence $\xi = \{n_k : k \in N^+\}$ of natural numbers consider a free union A_ξ (see p. 44) of sequential compacta $F_{n_k} \in \mathcal{S}$ and the Alexandroff compactification (Problem 176) Φ_ξ of the subspace A_ξ. It is easy to see that Φ_ξ is a compactum, contained in the real line. Let γ be the family of all compacta thus constructed over the entire sequence ξ of natural numbers. γ is the required family.

(*b*) Suppose that μ is the family of all compact subsets $F \subset R^1$, $|F| > \aleph_0$. (Note by inspection of Problem 284 that for any compact metrizable space F, $|F| > \aleph_0$, we have $|F| = \mathfrak{c}$.) By statement (*a*), it follows that $|\mu| = \mathfrak{c}$. Order the system μ of compacta $F \in \mu$ (where, as usual, we mean a minimal ordering):

$$\mu = \{F_1, F_2, \ldots, F_\lambda, \ldots\}. \tag{1}$$

Similarly, order the set of of all points of the real line by

$$R^1 = \{x_1, x_2, \ldots, x_\lambda, \ldots\}. \tag{2}$$

Let p_1 be the first point in the ordering (2) such that $p_1 \in F_1$, and let q_1 be the first point in (2) such that $q_1 \in F_1 \setminus p_1$. Moreover, if we have constructed the points p_λ and q_λ for all $\lambda < \lambda^*$, then we take as p_{λ^*} the first point in (2) such that $p_{\lambda^*} \in F_{\lambda^*} \setminus (\{p_\lambda: \lambda < \lambda^*\} \cup \{q_\lambda: \lambda < \lambda^*\})$, and q_{λ^*} as the first point in (2) satisfying $q_{\lambda^*} \in F_{\lambda^*} \setminus (\{p_\lambda: \lambda \leqslant \lambda^*\} \cup \{q_\lambda: \lambda < \lambda^*\})$. By using $|F_\lambda| = \mathfrak{c}$ for every $F_\lambda \in \mu$, and the fact that (1) and (2) are minimal orderings of R^1 and μ of cardinality $\mathfrak{c}$, it is easy to see that the inductive construction has been carried out correctly. As a result, we obtain the set Z consisting of all the constructed points p_λ. In addition, obviously $|Z| = |\mu| = \mathfrak{c}$. Suppose now that F is an uncountable compactum included in R^1; then $F \in \mu$; consequently, by the construction of Z, we surely have $(R^1 \setminus Z) \cap F \neq \Lambda$, i.e., $F \notin Z$. Thus every compactum contained in Z is no more than countable, i.e., Z is completely imperfect set.

299. Suppose that ρ is a metric on a compactum X. Using Problem 260 of Chap. 2, represent X as a union of closed subsets $\Phi_1, \ldots, \Phi_{s_1}$ of diameter less than $\frac{1}{2}$ (in addition, we may assume that s_1 is a power of two, i.e. $s_1 = 2^{n_1}$, provided some equal summands are allowed in the representation

$$X = \bigcup_{i=1}^{s_1} \Phi_i).$$

We call $\Phi_1, \ldots, \Phi_{s_1}$ compacta of type 1. Each compactum Φ_{i_1} of type 1 we divide into compacta $\Phi_{i_1 1}, \ldots, \Phi_{i_1 2^{n_2}}$ of diameter $< 1/2^2$: $\Phi_{i_1} = \Phi_{i_1 1} \cup \ldots \cup \Phi_{i_1 2^{n_2}}$. In addition, each compactum of type 1 is divided into the same (explicitly 2^{n_2}) number of closed subsets, admitting, of course, equal members. Thus, the entire compactum X is representable as a union of $s_2 = 2^{n_1+n_2}$ subcompacta of type 2 (all compacta of type 2 have diameter not greater than $1/2^2$). Repeat the procedure to obtain for each natural number m a representation of the compactum X as a union $s_m = 2^{n_1+\cdots+n_m}$ of subcompacta $\Phi_{i_1 \ldots i_m}$ of type m, each compactum $\Phi_{i_1 \ldots i_{m-1}}$ of type $m-1$ being representable as the union of 2^{n_m} compacta of type m:

$$\Phi_{i_1 \cdots i_{m-1}} = \bigcup_{i=1}^{2^{n_m}} \Phi_{i_1 \cdots i_{m-1} i},$$

here the diameter of every compactum of type m is strictly less than $1/2^m$.

Recall now Problem 185 of Chap. 2, in which the construction of the Cantor perfect set is carried out. For constructing a Cantor perfect set, we have segments of types $1, 2, \ldots, k, \ldots$: two segments of type 1, four of type $2, \ldots, 2^n$ segments of type n, and so on. We denote these segments by $\Delta_{j_1 \ldots j_n}$, where $j_1, j_2, \ldots, j_n$ can take either the value 0 or 1.

Consider now for each natural number m the segments $\Delta_{j_1 \ldots j_{r_m}}$ of type r_m as $s_m = 2^{n_1+\cdots+n_m} = 2^{r_m}$ (they are all of length $1/3^{r_m}$). For convenience, each such segment we denote by $\Delta_{i_1 \ldots i_m}$, where i_k takes the values $1, 2, 3, \ldots, 2^{n_k}$ (the meaning of this notation is easy to understand). Now every point t of the Cantor perfect set C coincides with the intersection of the uniquely defined sequence $\Delta_{i_1} \supset \Delta_{i_1 i_2} \supset \Delta_{i_1 i_2 \cdots i_m} \supset \ldots$ of segments containing it, and this sequence in turn defines the sequence of subcompacta (of a compactum X) $\Phi_{i_1} \supset \Phi_{i_1 i_2} \supset \cdots \supset \Phi_{i_1 i_2 \cdots i_m} \supset \cdots$ and, consequently, $x \in X$ is the point of intersection of these compacta. A mapping f: $C \to X$ of C into X is thus defined. It is trivial to show that this mapping is onto the entire compactum. Indeed, suppose that $x \in X$ is an arbitrary point. This point belongs to some sub-compactum Φ_{i_1} of type 1 (though, in general, not of a compactum of type 1 only), to some sub-compactum $\Phi_{i_1 i_2} \subset \Phi_{i_1}$ of type 2, and so on. Then we consider the corresponding sequences $\Delta_{i_1} \supset \Delta_{i_1 i_2} \supset \cdots \supset \Delta_{i_1 \cdots i_m} \supset \cdots$ of segments and the point t of the Cantor perfect set is obtained as an intersection of the segments in this sequence. Obviously, $ft = x$. It is to be shown that f is a continuous mapping. For any $\varepsilon > 0$ take a natural number m such that $(1/2^m) < \varepsilon$. Define $\delta = 1/3^{r_m}$. Then if $|t - t'| < \delta$, the points t and t' of the Cantor perfect set C belong to a certain segment $\Delta_{i_1} \cdots {}_{i_m}$ of rank r_m and, consequently, the points ft and ft' belong to the same compactum $\Phi_{i_1} \cdots {}_{i_m}$ of type m. Thus $\rho(ft, ft') \leqslant \operatorname{diam} \Phi_{i_1} \cdots {}_{i_m} \leqslant \Phi(1/2^m) < \varepsilon$. $f: C \to X$ is a continuous mapping. The proof is complete.

300. See the solution of the preceding problem. If in the construction (given there) it were possible to choose the compacta $\Phi_{i_1} \cdots {}_{i_m}$ in such away that any two compacta of the same type had empty intersection, then each point $y \in Y$ would have been uniquely defined by a sequence $\Phi_{i_1} \supset \Phi_{i_1 i_2} \supset \cdots \supset \Phi_{i_1} \cdots {}_{i_m} \supset \cdots$ of compacta of different types, containing y. Pursuing this reasoning, the mapping f constructed in the above solution would be one-to-one, hence a homeomorphism (because f is continuous, and the Cantor perfect set and the space Y are compacta). Thus, for every $\varepsilon > 0$, it should be possible to represent any 0-dimensional compactum Y without isolated points as a union of mutually disjoint perfect clopen subsets of diameter less than ε, in such a way that the number of members (without repetitions) is a power of two. This is routine to accomplish and is left as an exercise for the reader.

303. (*a*) Suppose that $f: C \to Y$ is a continuous irreducible mapping of the Cantor perfect set C onto the compactum Y. C has no isolated points (see Problem 185, Chap. 2); hence its irreducible image also has no isolated points (see Problem 111, Chap. 6).

(*b*) Suppose that the compactum Y has no isolated points. By Problem 299, there is a continuous mapping $f: C \to Y$ of the Cantor perfect set C onto Y. There is also a closed set $C_0 \subset C$ such that $fC_0 = Y$ and $f \mid C_0: C_0 \to Y$ is irreducible (see Problem 32 of Chap. 6). If C_0 is mapped irreducibly onto Y having no isolated points, then C_0 also has no isolated points (see Problem 111 of Chap. 6). Hence C_0 satisfies the hypothesis of Problem 300, and thus C_0 is homeomorphic to C. Therefore, the composition $(f \mid C_0) \cdot h$ (where $h: C \to C_0$ is a homeomorphism) is an irreducible continuous mapping of C onto Y.

304. Suppose that (X, ρ) is a complete metric space without isolated points. Exactly as in the solution of Problem 126, construct for each finite sequence of 0's and 1's, $i_1, \ldots, i_k, k \in N^+$, nonempty closed sets $F_{i_1} \cdots {}_{i_k}$ without isolated points in such a way that (*a*) each $F_{i_1} \cdots {}_{i_k}$ is the closure of an open set in X, (*b*) $F_{i_1} \cdots {}_{i_k i_{k+1}} \subset F_{i_1} \cdots {}_{i_k}$ for every $i_1, \ldots, i_k$ and $k \in N^+$, and (*c*) $F_{i_1} \cdots {}_{i_k} \cap F_{i'_1} \cdots {}_{i'_k} = \Lambda$, if $\{i_1, \ldots, i_k\} \neq \{i'_1, \ldots, i'_k\}$. In addition, (*d*) in the case at hand we can set $\operatorname{diam} F_{i_1} \cdots {}_{i_k} < 1/2^k$, $k \in N^+$. Then, for every infinite sequence $\xi = \{i_1, \ldots, i_k, \ldots\}$ of 0's and 1's, $\bigcap \{F_{i_1} \cdots {}_{i_k}: k \in N^+\}$ is nonempty and consists of exactly one point. The set Ξ of all possible sequences ξ of 0's and 1's can be regarded as the product of a countable number of isolated double-points, i.e. as the Cantor perfect set, see Problem 301, and it is easy to show that the subspace $P = \bigcup \{\bigcap \{F_{i_1 \ldots i_k}: k \in N^+\}: \xi \in \Xi\}$ is homeomorphic to the Cantor perfect set.

The second statement of our problem easily reduces to the first one. Proceed as follows: $x \in X$ is called a point of *σ-discreteness* of X, if there is a nhood OX of this point such that the subspace Ox is the union of a countable family of discrete subspaces. Assume that X is not representable in the form of a union of a countable family of discrete subspaces and denote by X^* the set of all points in X which are not σ-discrete. It is easy to show that subject to this restriction X^* is a nonempty closed subspace in X and has no isolated points. Thus the second statement of our problem reduces to the first one.

305. Suppose that ρ is a metric on a metrizable compactum X. It may be assumed without loss of generality that no point $a \in A$ is isolated in X. Enumerate A as a sequence: $A = \{a_1, a_2, \ldots, a_n, \ldots\}$. Since a_n is not isolated in X, the open set $\Gamma_n = X \setminus \{a_n\}$ is dense in X. Hence, so is

$$X_0 = \bigcap_{n=1}^{\infty} \Gamma_n = X \setminus A$$

(see Problem 88). Therefore, for each $n \in N^+$ we can choose a point $x_n \in X_0$ such that $\rho(x_n, a_n) < 1/n$. Define a partition of X as follows: its elements are two-point sets of the form (x_n, a_n) and the remaining elements of the partition are one-point sets. To show the continuity of this partition, suppose that $\{x_0\}$ is a one-point element of the partition. Suppose that x_0 is a limit point for $\{x_n: n \in N^+\}$. Consider an arbitrary ε-nhood $U_\varepsilon x_0$ of this point. Take n to be such that $(2/n) < \varepsilon$. Then, if $\rho(x_n, x_0) < 1/n$, we have $\rho(a_n, x_0) \leqslant \rho(a_n, x_n) + \rho(x_n, x_0) < (2/n) < \varepsilon$, i.e., every element of the partition which intersects the nhood $U_{1/n}x_0$, is contained in $U_\varepsilon x_0$. Suppose now that (x_{n_0}, a_{n_0}) is some two-point element of the partition. Consider the nhoods $U_\varepsilon x_{n_0}$ and $U_\varepsilon a_{n_0}$. Take n to be such that $(2/n) < \varepsilon$ and consider the nhood $U_{1/n}x_{n_0} \cup U_{1/n}a_{n_0}$. It is easy to see that every element of the partition intersecting the nhood $U_{1/n}x_{n_0} \cup U_{1/n}a_{n_0}$ of the element (x_{n_0}, a_{n_0}) of the partition, is contained in the nhood $U_\varepsilon x_{n_0} \cup U_\varepsilon a_{n_0}$ of this element of the partition. Thus the partition of X defined above is continuous. The space Y of this partition is Hausdorff (see Problem 92). Since X has a countable base and is compact, Y also has a countable base (as a compactum, which is a continuous image of the metrizable compactum X, see Problem 292).

Consider the restriction $f \mid X_0 = f_0$ of the mapping f to the subspace X_0. It is obvious that $fx_0 = Y$ and $f_0: X_0 \to Y$ is a one-to-one continuous mapping onto Y. The proof is complete.

306. The answer is not known to the authors.

307. The answer is not known to the authors.

308. See, Pelczyński [1].

312. Let $x_1 \in X$, $x_2 \in X$, $x_1 \neq x_2$ and let a, b be two real numbers, $a \neq b$. Take a function $f \in C_0$ satisfying $fx_1 \neq fx_2$. We shall seek the desired function in the form $g = \alpha F + \beta$, where α and β are constants to be determined (precisely, the function is determined by the constants α and β). Obviously, every function g thus defined belongs to C_0. To determine α and β, we have the equalities $g(x_1) = \alpha f(x_1) + \beta = a$, $g(x_2) = \alpha f(x_2) + \beta = b$. Since $f(x_1) \neq f(x_2)$, α and β are uniquely determined.

313. Let $f \in C_0$. Set $\|f\| = \max_{x \in X} |f(x)| = c$. Furthermore,

$$|f(x)| = \sqrt{f^2(x)} = \sqrt{c^2 - c^2 + f^2(x)} = c\sqrt{1 - \left(1 - \frac{f^2(x)}{c^2}\right)}.$$

Put $g(x) = 1 - (f^2(x)/c^2)$. We have $g \in C_0$ and $0 \leqslant g(x) \leqslant 1$ for every point $x \in X$. Moreover,

$$|f(x)| = c\sqrt{1 - g(x)} = c \sum_{n=1}^{\infty} \frac{1 \cdot 1 \cdot 3 \ldots (2n-3)}{2 \cdot 4 \cdot 6 \ldots 2n} g^n(x)$$

(we have used a power series expansion of the function $\sqrt{1-y}$ in y). Note that all the terms of the series belong to C_0, and the series itself converges uniformly to the function $|f(x)|$ because $0 \leqslant g(x) \leqslant 1$. Hence, since C_0 is closed in $C(X)$, it follows that $|f(x)| \in C_0$.

315. This follows from Problems 313 and 314.

316. Suppose that $f \in C(X)$ is a continuous function on X and $\varepsilon > 0$ is any positive number. It is required to find a function $g \in \mathfrak{A}$ such that $\rho(f, g) = \max_{x \in X} |f(x) - g(x)| < \varepsilon$. Using condition (i), for every x, y of points in X take the function $g_{xy} \in \mathfrak{A}$ to be such that $g_{xy}(x) = f(x)$, $g_{xy}(y) = f(y)$. By the continuity of g_{xy} and f at the points x and y respectively, we find nhoods $U_{xy}(x)$ and $V_{xy}(y)$ of x and y, respectively such that

$$g_{xy}(t) < f(t) + \varepsilon \quad \text{for all } t \in U_{xy}(x); \tag{1}$$

$$g_{xy}(t) > f(t) - \varepsilon \quad \text{for all } t \in V_{xy}(y). \tag{2}$$

Fix $y \in X$, and let x run through the whole compactum X. We obtain an open cover $\{U_{xy}(x) \colon x \in X\}$ of X, from which we extract a finite subcover

$$U_{x_1 y}(x_1), \ldots, U_{x_s y}(x_s). \tag{3}$$

Consider the function

$$g_y(t) = \min\{g_{x_1 y}(t), \ldots, g_{x_s y}(t)\} = m_{g_{x_1 y}}, \ldots, g_{x_s y}(t). \tag{4}$$

By condition (ii) we have $g_y \in \mathfrak{A}$; also, by (1), (4) and the fact that (3) is a cover of X, we obtain the inequality

$$g_y(t) < f(t) + \varepsilon \text{ for all } t \in X. \tag{5}$$

For each $y \in X$ consider the nhood

$$V_y(y) = \bigcap_{i=1}^{s} V_{x_i y}(y). \tag{6}$$

Using (2), (4) and (5), we conclude that

$$g_y(t) > f(t) - \varepsilon \text{ for every } t \in V_y(y). \tag{7}$$

Consider an open cover $\{V_y(y): y \in X\}$ of X. Select from it a finite subcover

$$V_{y_1}(y_1), \ldots, V_{y_k}(y_k). \tag{8}$$

Consider the function

$$g(t) = M_{g_{y_1}, \ldots, g_{y_k}}(t). \tag{9}$$

By condition (ii) $g \in \mathfrak{A}$. Making use of (7), (9) and the fact that (8) is a cover, we obtain the inequality $g(t) > f(t) - \varepsilon$ for all $t \in X$. Moreover, from (5) and (9) it follows that $g(t) < f(t) + \varepsilon$ for all $t \in X$. Thus $g \in \mathfrak{A}$ and $| f(t) - g(t) | < \varepsilon$ for all $t \in X$, i.e., $\rho(f, g) < \varepsilon$. The problem is solved.

317. By Problem 312, C_0 strongly separates the points of X. Consider $\mathfrak{A} = [C_0]_{C(X)}$. The family $\mathfrak{A}$ satisfies all hypotheses of the preceding problem (actually $\mathfrak{A}$ is also a closed subring). Then $[\mathfrak{A}]_{C(X)} = [[C_0]_{C(X)}]_{C(X)} = C(X)$. The problem is solved.

318. The solution is straightforward. One needs only to verify that the set of all polynomials, as well as the set of all trigonometric polynomials, satisfy the hypotheses of the preceding problem. The most important thing is to check that these sets of functions separate the points of the corresponding intervals (it is easy to see that in the former case the polynomial $y = x$ separates the points of the interval $[0, 1]$ and in the latter case two trigonometric polynomials $y = \sin x$ and $y = \cos x$ separate the points of the interval $[-\pi, \pi]$).

319. For the proof, consider the subring $C_0 \subset C(X)$ of all polynomials with rational coefficients and show that C_0 separates the points in I, and then apply Problem 317.

321. Let $C(X)$ be the set of all continuous real-valued functions on $X = \Pi\{X_\alpha: \alpha \in A\}$, and let $C_0 \subset C(X)$ be the set of all continuous real-valued functions depending only on a countabe number of coordinates. First note that C_0 contains all constants. In addition, it is routine to establish that C_0 is a subring of $C(X)$. Thus, for example, if $f_1 \in C_0$ depends on a countable number of coordinates, (say) the coordinates in $A_1 \subset A$, $| A_1 | \leqslant \aleph_0$ and $f_2 \in C_0$ on the coordinates in $A_2 \subset A$, $| A_2 | \leqslant \aleph_0$, then the function $f_1 + f_2$ depends only on $A_1 \cup A_2$, i.e., it also depends on a countable number of coordinates, implying $(f_1 + f_2) \in C_0$. In addition, C_0 is a closed subring, i.e. if the sequence $\{f_n: n \in N^+\}$, $f_n \in C_0$ (f_n depends on the coordinates of $A_n \subset A$, $| A_n | \leqslant \aleph_0$) converges uniformly to the function $f_0(x)$, then $f_0(x)$ is a continuous function, depending on the coordinates of

$$A_0 = \bigcup_{n=1}^{\infty} A_n, \ | A_0 | \leqslant \aleph_0,$$

and this implies that $f_0 \in C_0$. Thus C_0 is a closed subring of $C(X)$ containing all constants which separate the points of X (which is trivial to verify). Hence by the Weierstrass-Stone theorem (see Problem 317) $C_0 = C(X)$. The problem is solved.

322. By the Tychonoff theorem (see Problems 38 and 75) any compactum Y can be imbedded in the Tychonoff cube $I^\tau = \Pi\{I_\beta: \beta \in B\}$, $|B| = w(Y)$. Then the mapping $f: X \to Y$, $Y \subset I^\tau$ can be represented as a family $\{f_\beta: \beta \in B\}$ of functions continuous on X. Each function f_β, $\beta \in B$ depends on a countable number of coordinates (Problem 321). Consequently, f depends on $\tau \cdot \aleph_0 = \tau$ coordinates. The Problem is solved.

323. X is an image of some generalized discontinuum $D^m = \Pi\{D_\alpha: \alpha \in A\}$, $|A| = m$ (of some weight m). From Problem 113 it follows that $m \geqslant \tau$. By the preceding problem, f depends on τ coordinates, where $w(X) = \tau$, i.e., there is $A_0 \subset A$, $|A_0| \leqslant \tau$ and a continuous mapping $g: \Pi\{D_\alpha: \alpha \in A_0\} \to X$ such that $f = g\pi_{A_0}$, where $\pi_{A_0}: \Pi\{D_\alpha: \alpha \in A\} \to \Pi\{D_\alpha: \alpha \in A_0\}$ is a projection. Since $|A_0| \leqslant w(X) = \tau$, $w(\Pi\{D_\alpha: \alpha \in A_0\}) \leqslant \tau$ but since the weight of a compactum under a continuous mapping does not increase, $w(\Pi\{D_\alpha: \alpha \in A_0\}) = \tau$. The problem is solved.

324. Let $w(X) = m$ and $w(Y) = n$. Obviously (see Problem 113), $m \geqslant n$. Suppose further that $g: D^m \to X$ is a continuous mapping of the generalized Cantor discontinuum $D^m = \Pi\{D_\alpha: \alpha \in A\}$, $|A| = m$ onto our dyadic compactum X (see the preceding problem). Denote by $h = gf$ the composition of mappings f and g. By Problem 322, there exist $A_0 \subset A$, $|A_0| = n$ and a continuous mapping $h_0: \Pi\{D_\alpha: \alpha \in A_0\} \to Y$ such that $h = h_0\pi_{A_0}$ (here $\pi_{A_0}: \Pi\{D_\alpha: \alpha \in A\} \to \Pi\{D_\alpha: \alpha \in A_0\}$ is a projection). Write $X_f = g(\Pi\{D^*_\alpha: \alpha \in A\})$, where $D^*_\alpha = D_\alpha$ if $\alpha \in A_0$, $D^*_\alpha = \{o\}$ if $\alpha \in A \setminus A_0$ (D_α is an isolated double-point). Obviously, $\Pi\{D^*_\alpha: \alpha \in A\}$ is homeomorphic to $\Pi\{D_\alpha: \alpha \in A_0\}$. Consequently, $w(\Pi\{D_\alpha: \alpha \in A_0\}) = w(\Pi\{D^*_\alpha: \alpha \in A\}) = |A_0| = n$; hence $w(X_f) \leqslant n$ (under a continuous mapping between compacta weight does not increase), implying $w(X_f) = n = w(Y)$. Obviously, $fX_f = Y$. The problem is solved.

325. By Problem 113, the weight of compacta does not increase under continuous mappings. Recall the definition of the irreducible mapping and make use of the preceding problem.

326. Suppose that $\theta = \{U_\alpha: \alpha \in A\}$ is a π-base for X such that $|\theta| = \tau$ (see p. 41). A pair $(U_\alpha, U_{\alpha'})$ of elements of this π-base is said to be distinguished if $[U_\alpha] \subset U_{\alpha'}$. Obviously, the cardinality of the set σ of distinguished pairs $(U_\alpha, U_{\alpha'})$ of elements $U_\alpha \in \theta$, $U_{\alpha'} \in \theta$ does not exceed τ. Put each distinguished pair $(U_\alpha, U_{\alpha'}) \in \sigma$ in correspondence with (using Problem 25 and the normality of X, see Problem 64) a continuous function $f_{\alpha\alpha'}$, satisfying $0 \leqslant f_{\alpha\alpha'}(x) \leqslant 1$ for all $x \in X$, and $f_{\alpha\alpha'}(x) = 0$ if $x \in [U_\alpha]$, and $f_{\alpha\alpha'}(x) = 1$ if $x \in X \setminus U_{\alpha'}$. The set of functions so chosen has cardinality less than or equal to τ. A diagonal mapping f for this set of functions (see Problem 290, Chap. 2) is a mapping of X into a compactum of weight not exceeding τ, It is routine to verify that the mapping f so constructed is irreducible (here it is necessary to make use of the definition of irreducible mapping, see p. 144, the definition of a diagonal mapping, see p. 5, the above construction and the density of θ in X).

327. Suppose that X is a dyadic compactum. Its π-weight does not exceed its weight (in fact, this is true for every topological space). On the other hand, by the preceding problem X admits an irreducible mapping onto a compactum Y, for which $w(Y)$ does not exceed the π-weight of X. Hence, by Problem 325, the weight of the dyadic compactum X equals the weight of Y. From all that has been stated, we obtain the inequalities $w(Y) \leqslant \pi w(X) \leqslant w(X)$ and $w(X) = w(Y)$, whence $w(X) = \pi w(X)$($\pi w(X)$ denotes the π-weight of X).

328. This follows from the preceding problem and the Urysohn metrization theorem (Problem 39).

329. Suppose that $X \subset \Pi\{D_\alpha: \alpha \in A\}$ is a closed G_δ-set and let $|A| = m > \aleph_0$. Consider a continuous real-valued function f on $\Pi\{D_\alpha: \alpha \in A\}$, which vanishes at exactly the points of X (see Problem 30). f depends only on a countable number of coordinates (see Problem 321), i.e., there exist $A_0 \subset A$, $|A_0| = \aleph_0$ and a continuous function f_1 on $\Pi\{D_\alpha: \alpha \in A_0\}$ such that $f = f_1 \pi_{A_0}$, where π_{A_0} is as usual the natural projection. Consider the closed set $X_0 = f_1^{-1}(0)$ in $\Pi\{D_\alpha: \alpha \in A_0\}$. It is easy to show that $X = X_0 \times \Pi\{D_\alpha: \alpha \in A \setminus A_0\}$. Notice that the product $X_0 \times D^{\aleph_0}$ of any 0-dimensional metric compactum X_0 by the Cantor perfect set is again homeomorphic to the Cantor perfect set. Then we have $X = X_0 \times D^m \approx X_0 \times D^{\aleph_0} \times D^m \approx D^{\aleph_0} \times D^m \approx D^m$. The problem is solved.

330. Let $D^m = \Pi\{D_\alpha: \alpha \in A\}$, $|A| = m$. Exactly as in the solution of the preceding problem, let $A_0 \subset A$, $|A_0| = \aleph_0$ be such that $X = X_0 \times \Pi\{D_\alpha: \alpha \in A \setminus A_0\}$, where X_0 is a closed set in the Cantor perfect set $D^{\aleph_0} = \Pi\{D_\alpha: \alpha \in A_0\}$. (Evidently X_0 is G_δ). Utilizing Problem 77 of Chap. 6, consider a retraction $r: \Pi\{D_\alpha: \alpha \in A_0\} \to X_0$. Note that $D^m = \Pi\{D_\alpha: \alpha \in A\} = \Pi\{D_\alpha: \alpha \in A_0\} \times \Pi\{D_\alpha: \alpha \in A \setminus A_0\}$. Then each point ξ is a pair (x, y), where $x \in \Pi\{D_\alpha: \alpha \in A_0\}$, $y \in \Pi\{D_\alpha: \alpha \in A \setminus A_0\}$. Define $g(\xi)$ by the rule: $g(\xi) = g(x, y) = (r(x), y)$. The pair $(r(x), y)$ is again a point of D^m. It is clear that $(r(x), y) \in X_0 \times \Pi\{D_\alpha: \alpha \in A \setminus A_0\} = X$ and $g(\xi) = \xi$ if $\xi \in X$. Thus g is a mapping of D^m onto X, its restriction to X being the identity. It is also obvious that g is continuous.

331. This follows from Problem 329.

332. (*a*) is established directly by verifying that C_{x_0} is a subring of $C(X)$. It is also trivial to show that C_{x_0} is an ideal.

(*b*) Suppose that $C_0 \subset C(X)$ is a non-trivial ideal. If C_0 is not an ideal of the form C_x, then for every point $x \in X$ there is a function $f_x \in C_0$ satisfying $f_x(x) \neq 0$. Then, using the continuity of f_x, take a nhood Ux of x such that $f_x(x') \neq 0$ for every point $x' \in Ux$. From the infinite open cover $\{Ux: x \in X\}$ extract a finite subcover $\{Ux_1, \ldots, Ux_s\}$. Consider the continuous function $g = f_{x_1}^2 + f_{x_2}^2 + \ldots + f_{x_s}^2$. Evidently, $g \in C_0$ and $g(x) > 0$ for all $x \in X$. Suppose now that h is an arbitrary continuous function. It can be put in the form $h = (h/g) \cdot g$. Since $g \in C_0$ and C_0 is an ideal, $h \in C_0$. Hence $C_0 \equiv C(X)$. This contradicts the fact that C_0 is a non-trivial ideal.

(*c*) We show first that every ideal of the form C_x is maximal. If C_x is not maximal, then there is a non-trivial ideal C_0, properly containing the ideal C_x. According to (*b*), C_0 is contained in an ideal of the form C_{x}': $C_0 \subset C_{x}'$. It is required to show that $x = x'$. If this were not so then, owing to the complete regularity of X, our function f would be equal to 0 at x but 1 at x'. This contradicts the inclusion $C_x \subset C_0 \subset C_{x}'$. Thus every ideal of the form C_x is maximal. Suppose now that C_0 is an arbitrary maximal ideal. According to (*b*), C_0 is contained in some ideal of the form C_x. Hence (because of maximality) C_0 is of the form C_x.

333. (*a*) Let $x_0 \in [A]$ and $f \in \bigcap\{C_x: x \in A\}$. Then $f(x) = 0$ for every point $x \in A$. Hence (by continuity) $f(x_0) = 0$ if $x_0 \in [A]$.

(*b*) Suppose that x_0 is a point such that $C_{x_0} \supset \bigcap\{C_x: x \in A\}$. Assume that $x_0 \notin [A]$. Using the complete regularity of X, take a continuous function f such that $f(x_0) = 1$ and $f(x) = 0$ if $x \in [A]$. Then $f \in \bigcap\{C_x: x \in A\}$ but $f \notin C_{x_0}$, a contradiction.

334. Suppose that $i: C(X) \to C(Y)$ is an isomorphism of the algebra $C(X)$ onto the algebra $C(Y)$. Evidently, under this isomorphim the image of every ideal (maximal ideal) is an ideal (maximal ideal). Note (by Problem 332) that all maximal ideals in $C(X)$ and $C(Y)$ are ideals of the form C_x, $x \in X$; C_y, $y \in Y$. For each point $x \in X$, consider the maximal ideal C_x. An isomorphic image (under the isomorphism i) of the maximal ideal C_x is a maximal ideal C_y. Put $y = fx$. Thus the isomorphism $i: C(X) \to C(Y)$ obviously generates a one-to-one mapping $f: X \to Y$ (onto Y). To show that f is a homeomorphism, let $x_0 \in [A]$, $A \subset X$. Use Problem 333 to obtain $C_{x_0} \supset \bigcap \{C_x : x \in A\}$. Under i this inclusion goes over to the inclusion $C_{fx_0} \supset \bigcap \{C_y : y \in fA\}$ which yields (Problem 333) $fx_0 \in [fA]$. Thus f is continnous, one-to-one and hence a homeomorphism. The converse is trivial.

337. See Kolmogorov and Fomin ([1], p. 105).

342. $f(X)$ is a connected subset of the real line (Problem 221, Chap. 2). By hypothesis, there exist a negative number a and a positive number b belonging to $f(X)$. But then by Problem 223 of Chap. 2, the entire interval joining the points a and b (which contains 0), is included in $f(X)$; hence $0 \in f(X)$, i.e., there is a point $x_0 \in X$ such that $0 = f(x_0)$

344. False. A Hausdorff space consisting of two points can be mapped onto a space consisting of just one point.

345. Suppose that P is a clopen subset of X containing the point c^*. Since P is closed, $\tilde{C} = \{c \in C_0,\ A_c \subset P\}$ is closed. Hence $C' = C_0 \setminus \tilde{C}$ is open in the compactum C_0. Consider $C'' = \{c \in C' : c$ is a point of type 2 in $C_0\}$. Then $C'' \subset C'$ and $C' \setminus C''$ is a countable dense set in C'. Hence C'' is a nowwhere locally compact G_δ-set in the compactum. Note that if $\tilde{C} \neq C_0$, then C'' is a nonempty set (this follows from the fact that every point of a Cantor perfect set is a condensation point). We now represent the space C' as a union of countable family of compacta included in it, which also contradicts the assumption $\tilde{C} \neq C_0$ (see Problem 238, Chap. 2).

A point of the interval I_c, which joins c and c^*, is called *critical* if every nhood of it contains points both of $A_c \cap P$ and of $A_c \setminus P$. Each I_c, for $c \in C''$ has a critical point in view of the connectedness of I_c and the definition of C''. Denote such a point by $\varphi(c)$. From the definition of C'' and A_c and the fact that P is open, it follows that the coordinate $y(\varphi(c))$ of $\varphi(c)$ is a rational number. For every rational number r, set $C^r = \{c \in C'' : y(\varphi(c)) = r\}$. Since P is clopen, it follows that C^r is a closed set in C_0, i.e., it is a compactum. Obviously, $C'' = \bigcup \{C^r : r$ is a rational number$\}$, a contradiction.

To prove that only only one-point subsets of $X \setminus \{c^*\}$ are connected, one first observes that every $A_c \cap (X \setminus \{c^*\})$ is the intersection of clopen sets in $X \setminus \{c^*\}$ (this is obviously implied by the fact that every point of a Cantor set is an intersection of clopen sets). Then note that in each A_c the connected subsets are just the points. It ought to be further pointed out that the quasi-components (see p. 154) of $X \setminus \{c^*\}$ are many-point sets.

346. Put $B_\alpha = A_\alpha \cup A_{\alpha_0}$. Then, by Problem 220 of Chap. 2, B_α is connected for each $\alpha \in M$. In addition, $\bigcap \{B_\alpha : \alpha \in M\} \supset A_{\alpha_0} \neq \Lambda$. Hence from Problem 220 of Chap. 2 it follows that $\bigcup \{B_\alpha : \alpha \in M\}$ is connected. But, obviously, $\bigcup \{B_\alpha : \alpha \subset M\} = \bigcup \{A_\alpha : \alpha \in M\}$.

349. Since $\{x\}$ is connected and $x \in \{x\}$, $\mathcal{E}_x \ni \{x\}$ and $\mathcal{E}_x \neq \Lambda$. Obviously $K_x \ni x$. By Problem 220 of Chap. 2 K_x is connected (because $x \in \bigcap\{P: P \in \mathcal{E}_x\}$ and each $P \in \mathcal{E}_x$ is connected). If $Q \supset K_x$ and Q is connected, then $Q \ni x$ and hence $Q \in \mathcal{E}_x$. Consequently, $Q \subset K_x$. Thus K_x is a maximal connected set in X.

350. (*a*) If K is a component of X, i.e., it is a maximal connected set in X, then $[K]$ is also connected (Problem 219, Chap. 2); consequently, $[K] = K$.

(*b*) The second assertion is a consequence of Problem 220 of Chap. 2 and the definition of a component of a space.

351. This follows from Problems 349 and 350.

352. Suppose that A is not connected. Then $A = A_1 \cup A_2$, where $A_1, A_2 \neq \Lambda$, $A_1 \cap A_2 = \Lambda$ and A_1, A_2 are closed in A and, consequently, in X. Either $A_1 \cap B = \Lambda$ or $A_2 \cap B = \Lambda$, otherwise $A \cap B = (A_1 \cup A_2) \cap B$ would be disconnected. Assume that $A_1 \cap B = \Lambda$. Then A_1 and $C = B \cup A_2$ are disjoint nonempty closed sets in X and in $A \cup B$ as well; but $A_1 \cup C = A \cup B$, which contradicts the connectedness of $A \cup B$. Example: $A = \{x: 0 \leqslant x < 1\}$, $B = \{0, 1\}$. Then B is not connected, but $A \cap B = \{0\}$ is a connected set and $A \cup B = \{x : 0 \leqslant x \leqslant 1\}$ is a connected set.

353. Consider the subspace Z of the plane π (equipped with a Cartesian coordinate system, consisting of all the points whose first coordinate is rational and whose second coordinate is rational and negative. Z is a countable set. Endow it with the following unconventional topology $\mathcal{T}^*$. Fix an irrational number α (for definiteness, say $\alpha = \sqrt{3}$). Put $Z' = \{(x, 0) \in \pi: x \text{ is rational}\}$. All sets which are open in Z' relative to the usual topology inherited from the plane, we declare to be open in Z, i.e., to be elements of $\mathcal{T}^*$. Let $\mathcal{E}$ be the collection of all intervals lying on the x-axis. For each $U \in \mathcal{E}$, put $U^+ = \{(x', y') \in Z \setminus Z'$: the line $(y - y') = \alpha(x - x')$ intersects $U\}$ and $U^- = \{(x', y') \in Z \setminus Z'$: the line $(y - y') = -\alpha(x - x')$ intersects $U\}$. All sets of the form $\{z\} \cup (U \cap Z') \cup (V \cap Z')$, where $U, V \in \mathcal{E}$ and $z \in U^+ \cap V^-$, we declare to be open in Z. All sets so far declared open in Z form a base for a topology on Z (verify this yourself). This is the desired topology. To check that Z is Hausdorff, let $z_1, z_2 \in Z$ and $z_1 \neq z_2$. The cases $z_1 \in Z'$ and $z_2 \in Z'$ are obvious. If $z_1 \notin Z'$, and $z_2 \in Z'$, then the lines l and l' passing through z_1 and forming with the x-axis angles whose tangents are $+\alpha$ and $-\alpha$ respectively, intersect the x-axis at the points $(x, 0)$ and $(x', 0)$ different from z_2 (since α is an irrational number!) Choosing disjoint intervals $U \ni (x, 0)$, $V \ni (x', 0)$ and $W \ni z_2$, we conclude that $\{z_1\} \cup (U \cap Z') \cup (V \cap Z')$ and W are the required nhoods.

Similar reasoning applies in the cases when $z_1, z_2 \in Z \setminus Z'$.

To show that Z is connected, let $Z = Z_1 \cup Z_2$, where Z_1 and Z_2 are nonempty disjoint open subsets of Z. Since $[Z'] = Z$ (this immediately follows from the definition of the topology $\mathcal{T}^*$), $Z' \cap Z_1 \neq \Lambda$ and $Z' \cap Z_2 \neq \Lambda$, where $Z' \cap Z_1$ and $Z' \cap Z_2$ are open sets. On x-axis take intervals U and V satisfying $U \cap Z' \subset Z' \cap Z_1$ and $V \cap Z' \subset Z' \cap Z_2$. Then, on the one hand, $[U \cap Z') \subset [Z_1] = Z_1$ and $[V \cap Z'] \subset [Z_2] = Z_2$, and hence $[U \cap Z'] \cap [V \cap Z'] = \Lambda$. On the other hand, it is clear that $[U \cap Z'] \supset U^+ \cup U^-$ and $[V \cap Z'] \supset V^+ \cup V^-$. It now suffices to consider the sets U^+, U^-, V^+ and V^-, in order to see that for any intervals U and V we certainly have $(U^+ \cup U^-) \cap (V^+ \cup V^-) \neq \Lambda$, a contradiction. It has thus been actually shown that the closures of any two open sets in Z intersect.

There does not exist a countable regular connected space X. Indeed, if X is a countable regular space, then it is normal (Problems 18 and 19) and, consequently, there is a continuous real-valued function on X, which is not a constant. Then $f(X)$ is a many-point countable subset of the real line and, consequently, is disconnected (Problem 223, Chap. 2). It now remains to refer to Problem 221 of Chap. 2.

354. Let ξ be the prefilter under consideration. Fix some element $X_0 \in \xi$. Put $\xi_0 = \{\Phi \in \xi : \Phi \subset X_0\}$. Then ξ_0 is obviously a prefilter on X_0 (see p. 7) and $\bigcap \{\Phi : \Phi \in \xi_0\} = \bigcap \{\Phi : \Phi \in \xi\}$, which is immediate from the basic properties of a prefilter. But X_0 is a normal space (as is every compactum see Problem 64) and $\Phi \in \xi_0$ is closed in X_0 (see Problem 56). Set $F = \bigcap \{\Phi : \Phi \in \xi_0\}$. F is a closed set in X_0 and is a compactum. Suppose that F is not connected. Then $F = F_1 \cup F_2$, where $F_1, F_2 \neq \Lambda$. F_1 and F_2 are closed in F and hence in X_0, also $F_1 \cap F_2 = \Lambda$. There exist (Problems 55 and 19) open sets U_1 and U_2 in X_0 such that $F_1 \subset U_1, F_2 \subset U_2$ and $[U_1] \cap [U_2] = \Lambda$. Put $U = U_1 \cup U_2$. Then $U \supset F$. From Problem 66 it follows that there is $\Phi \in \xi_0$, such that $F \subset \Phi \subset U$. Put $\Phi_1 = \Phi \cap [U_1]$ and $\Phi_2 = \Phi \cap [U_2]$. Since $\Phi \subset [U_1] \cup [U_2]$, $\Phi_1 \cup \Phi_2 = \Phi$. From $[U_1] \cap [U_2] = \Lambda$ it follows that $\Phi_1 \cap \Phi_2 = \Lambda$. Φ_1 and Φ_2 are closed in X_0 since the sets Φ, $[U_1]$ and $[U_2]$ are closed in X_0. Finally $\Phi_1 \neq \Lambda$, because $\Phi_1 \supset F_1 \neq \Lambda$; $\Phi_2 \neq \Lambda$, because $\Phi_2 \supset F_2 \neq \Lambda$. Consequently, Φ is not connected, a contradiction (which stems from our assumption that F is disconnected).

355. The elements of a decreasing sequence of connected compacta form a prefilter. Hence the assertion follows from the preceding problem.

356. Suppose that π is a plane with a fixed Cartesian (x, y)-coordinate system. Put $\mathcal{T} = \{(x, y) \in \pi : 0 < x \leqslant 1, y = \sin(1/x)\}$, $S = \{(x, y) \in \pi : x = 0, -1 \leqslant y \leqslant 1\}$ and $\Phi = \mathcal{T} \cup S$. Φ is a closed bounded subset of π; consequently, Φ is compact. Φ is a connected space. In fact, the space $\mathcal{T}$ is a continuous image (under the mapping $x \to (x, \sin(1/x))$ of the half-interval $(0, 1]$, i.e., of a connected set and hence it is connected. But $\Phi = [\mathcal{T}]$. Consequently (Problem 219, Chap. 2; Problem 347 of the present chapter), Φ is connected. But Φ is not locally connected at any point of S. For instance, consider the open subset U of Φ, which is the intersection of the strip $\Pi = \{(x, y) \in \pi : -\infty < x < +\infty, -\frac{1}{2} < y < \frac{1}{2}\}$ with Φ. Obviously, the component at the point $(0, 0) \in U$ in U is the set $\{(x, y) : x = 0, -\frac{1}{2} < y < \frac{1}{2}\}$, which is not open in U. Hence Φ is not locally connected (see Problem 370).

357. Consider the family $\mathcal{E}$ of all connected compact subsets of X, containing both the points x, y. $\mathcal{E} \neq \Lambda$ since $X \in \mathcal{E}$. In the class of all nests in $\mathcal{E}$, there is a maximal nest (this follows from the Kuratowski-Zorn principle, see p. 6). Suppose that ξ is such a maximal nest. This imples that (a) if $A, B \in \xi$, then either $A \subset B$, or $B \subset A$; (b) $\xi \subset \mathcal{E}$ and (c) if $\{C\} \cup \xi$ is a nest and $C \notin \xi$, then $C \notin \mathcal{E}$. Put $\Phi = \bigcap\{F : F \in \xi\}$. By Problem 354, Φ is a connected compactum and $\Phi \ni x$, $\Phi \ni y$, because all elements of ξ contain x and y. Consider an arbitrary compactum C, included in Φ and different from Φ. Then $\xi \cup \{C\}$ is a nest (it is not verified to be in $\mathcal{E}$) and $C \notin \xi$. Hence $C \notin \mathcal{E}$, i.e., either C is not connected, or $C \not\supset \{x, y\}$. Consequently, Φ is irreducible between x and y.

358. Suppose that P is a component of U and $[P] \subset U$. Then $P = [P]$ (see Problem 347). There is a compactum $V \subset U$, whose open kernel V^0 contains P (i.e. $V^0 \supset P$). Consider a point $x^* \in P$ and its component C_{x^*} in V. Then $C_{x^*} \subset P \subset V^0$ (see

Problem 349) and hence $C_{x^*} = P$. Consider the decomposition space Z of V into components and the natural mapping π: $V \to Z$ (Problem 365). Put $F = V \setminus V^0$, $\Phi = \pi(F)$, $z^* = \pi(C_{x^*})$. From $C_{x^*} \cap F = \Lambda$ it follows that $\Phi \not\ni z^*$. In addition, Z is a compactum with a base of clopen sets (see Problem 366) and Φ is closed in Z as a compact subspce of Z (Φ is compact as a continuous image of the compactum F). Consequently, there is a clopen set $Oz^* \ni z^*$ in Z such that $\Phi \cap Oz^* = \Lambda$. Put $W = \pi^{-1} Oz^*$. Then $P \subset W \subset V^0$ and W is open in V. Moreover, W is open in V^0. But V^0 is open in X. On the other hand, W is closed in V as a pre-image of the closed set Oz^* under the continuous mapping π. Since V is closed in X, W is closed in X. Hence W is clopen in X, $W \supset P \ni x^*$, i.e., $W \neq \Lambda$ and $X \setminus W \supset X \setminus U \neq \Lambda$, i.e., $W \neq X$, contradicting the connectedness of X.

359. Let

$$X = \bigcup_{i=1}^{\infty} X_i.$$

It suffices to consider the case when each X_i is nonempty. Since $X_2 \cap X_1 = \Lambda$, there is an open set U in X such that

$$X_2 \subset U \subset [U] \subset X \setminus X_1. \tag{1}$$

Choose a point $x \in X_2$ and consider a component C of U containing x. Put $\Phi_1 = [C]$. Then, by (1), $\Phi_1 \cap X_1 = \Lambda$, by Problem 358, $\Phi_1 \cap (X \setminus U) \neq \Lambda$ and hence $\Phi_1 \not\subset X_2$ but $\Phi_1 \cap X_2 \neq \Lambda$, since $x \in \Phi_1 \cap X_2$. It is clear that Φ_1 is a connected compactum. Assume that the connected compactum Φ_i is already defined such that

$$\Phi_i \cap \left(\bigcup_{j=1}^{i} X_j\right) = \Lambda \text{ and } \Phi_i \not\subset X_j \text{ for } j > i.$$

Put $j_0 = \min\{j: \Phi_i \cap X_j \neq \Lambda\}$. Then there is a $j' \neq j_0$ such that $\Phi_i \cap X_{j'} \neq \Lambda$. Let V be an open set, containing the set $\Phi_i \cap X_{j'}$ and such that $[V] \cap X_{j_0} = \Lambda$. Then Φ_{i+1} is defined as the closure in X of the components of any point $x \in \Phi_i \cap X_{j'}$, in the space $V \cap \Phi_i$. Then Φ_{i+1} is a connected compact set, satisfying

$$\left(\bigcup_{j=1}^{i+1} X_j\right) \cap \Phi_{i+1} = \Lambda,\ \Phi_{i+1} \cap X_{j'} \neq \Lambda,$$

$$\Phi_{i+1} \cap (\Phi_i \setminus V) \subset \Phi_{i+1} \cap (\Phi_i \setminus X_{j'}),\ \Phi_{i+1} \cap (\Phi_i \setminus X_{j'}) \supset \Phi_{i+1} \cap (\Phi_i \setminus V) \neq \Lambda$$

and $\quad \Phi_{i+1} \subset \Phi_i$.

The above construction yields inductively a decreasing sequence $\{\Phi_i: i \in N^+\}$ of nonempty closed sets, for which $\bigcap_{i=1}^{\infty} \Phi_i$ is a nonempty set, disjoint from the set $\bigcup_{i=1}^{\infty} X_i$, a contradiction.

360. (*a*) Adjoin to X the entire interval $I_0 = \{0\} \times [0, 1]$, to obtain a compactum $\tilde{X}$. In this compactum the interval $I_0 \setminus (\{(0, 0)\} \cup \{(0, 1)\})$ is a F_σ-set. Then, in this compactum X is G_δ. Hence by Problem 234 of Chap. 2, X is homeomorphic to a complete metric space.

(*b*) Write

$$X_0 = \bigcup_{n=1}^{\infty} \left(\left\{ \frac{1}{n} \right\} \times [0, 1] \right).$$

Clearly, X_0 is the union of a countable number of intervals, and the whole space X is the union of a countable number of intervals and one-point compacta $\{(0, 0)\}$ and $\{(0, 1)\}$.

(*c*) Consider the points $(0, 0)$ and $(0, 1)$ of X. It is easy to see that, for every nhood U of $(0, 0)$ or $(0, 1)$, the closure $[U]_X$ is not compact. However, for any other point there is a nhood whose closure is compact.

(*d*) To show now that it is impossible to contract X to a compactum, assume that there is a compactum Y and a one-to-one continuous mapping $f: X \to Y$ onto Y. Clearly, Y is metrizable. Write $I_0 = \{1/n\} \times [0, 1]$, the points $(0, 0) = a_0$ and $(0, 1) = a_1$. The mapping f, restricted to each interval I_n, is a homeomorphism, since f is continuous and I_n is a compact subspace. Hence

$$Y = \bigcup_{n=1}^{\infty} fI_n \cup fa_0 \cup fa_1,$$

where fI_n is homeomorphic to a segment of the real line. Let B be the set of all points $y \in Y$, every nhood of which has nonempty intersection with an infinite number of intervals fI_n, $n = 1, 2, \ldots$ Using the continuity of the mapping, it is easily seen that $fa_0, fa_1 \in B$, so that $B \neq \Lambda$. It is also clear that B is closed in Y. To show that B is a connected set, assume the contrary and let $B = B_1 \cup B_2$, $B_1 \cap B_2 = \Lambda$ where $B_1, B_2 \neq \Lambda$, B_1 and B_2 are closed sets. Using normality, consider open sets V_1 and V_2 in Y satisfying $B_1 \subset V_1$, $B_2 \subset V_2$, $V_1 \cap V_2 = \Lambda$. Consider also the nhood $V(B) = V_1 \cup V_2$ of the closed set B. We show that the nhood $V(B)$ (indeed, every nhood of B) contains the whole interval fI_n, for all $n \geqslant$ some index n_0. If this were not so, then we would have a sequence of indices $n_1 < n_2 < \ldots < n_k \ldots$ and points $\{y_{n_k}: k \in N^+\}$ satisfying $y_{n_k} \in fI_{n_k} \cap (Y \setminus V(B))$. Then the limit point y_0 of the sequence $\{y_{n_k}: k \in N^+\}$ which exists because of the sequential compactness of Y (see Problem 252) does not belong to B, contrary to the definition of this space.

Thus the whole interval fI_n, for all $n \geqslant$ some index n_0, is contained in $V(B)$. Moreover, every nhood of both the points fa_0 and fa_1 intersects every interval fI_n, for $n \geqslant$ some index. In consequence of all this, there is a $n^* \in N^+$ such that $fI_{n^*} \subset V(B)$ and $fI_{n^*} \cap V_1 \neq \Lambda$, $fI_{n^*} \cap V_2 \neq \Lambda$. Noting that $V_1 \cap V_2 = \Lambda$, we obtain a contradiction with the connectedness of fI_{n^*} (see Problem 222, Chap. 2). Thus B is a nonempty, connected and closed set in Y. Moreover, $B \cap \bigcup_{n=1}^{\infty} fI_n = \Lambda$, because otherwise B could be represented as a union of a countable family of mutually disjoint closed sets, contradicting Problem 359.

362. Use the following assertions: a space homeomorphic to a straight line is connected; the union of connected sets, whose intersection is nonempty, is connected (Problem 346); the closure of a connected set is connected (Problem 219, Chap. 2; Problem 347, this chapter). The desired component of the complement with respect to (0, 1) is the line $y = 0$.

363. Take the subspace $X = \{(0,0), (1,0), (x,y): 0 \leqslant x \leqslant 1, y = 1/n, n = 1,2,\ldots\}$ of the plane with Cartesian (x, y)-coordinate system. The component at (0, 0) in X consists of only one point in it (prove this). But a quasi-component at (0, 0) in X consists of two points (0, 0) and (1, 0). It remains to verify that if U is a clopen set in X which contains the point (0, 0), then also $(1, 0) \in U$. Since $U \ni (0, 0)$ and U is open in X, there is an integer n_0 such that $U \cap I_n \neq \Lambda$ for all $n > n_0$, where $I_n = \{(x, y): 0 \leqslant x \leqslant 1, y = 1/n\}$. The clopenness of U and connectedness of I_n (see Problem 222, Chap. 2) imply that then $U \supset I_n$ for all $n > n_0$. But U is closed and $[\bigcup\{I_n: n > n_0\}] \ni (1, 0)$. Hence $U \ni (1, 0)$. Another example is the space $X \setminus \{c^*\} = X_0$ of Problem 345.

364. Let σ_{x_0} be the family of all clopen sets in X containing the point x_0. We show first that the closed set $P_{x_0} = \bigcap\{U: U \in \sigma_{x_0}\}$ is connected. Assume the contrary; then $P_{x_0} = F_1 \cup F_2$, where $F_1 \cap F_2 = \Lambda$, $F_1, F_2 \neq \Lambda$ and F_1, F_2 are closed in P_{x_0} (hence, also in X). X being a normal compactum, there exist nhoods $U(F_1)$ and $U(F_2)$ of these sets (in X), satisfying $U(F_1) \cap U(F_2) = \Lambda$. Consider the nhood $U(P_{x_0}) = U(F_1) \cup U(F_2)$ of P_{x_0}. Using Problems 66 and 67, we find a $U \in \sigma_{x_0}$ such that $U \subset U(P_{x_0})$. Then $U' = U(F_1) \cap U$ and $U'' = U(F_2) \cap U$ are open in X, $U = U' \cup U''$ and $U' \cap U'' = \Lambda$. Consequently, U' and U'' are clopen in X, $U', U'' \neq \Lambda$. Moreover, $F_1 \subset U'$, $F_2 \subset U''$. Hence if $x_0 \in F_1 \subset U'$, then $P_{x_0} = F_1 \cup F_2 \not\subset U'$, since $U' \cap U'' = \Lambda$. This contradicts the definition of P_{x_0}. Thus P_{x_0} is connected. Since K_{x_0} is a connected component of x_0, $P_{x_0} \subset K_{x_0}$.

It remains to prove the reverse inclusion $K_{x_0} \subset P_{x_0} = \bigcap\{U: U \in \sigma_{x_0}\}$. If $K_{x_0} \not\subset U$ for some $U \in \sigma_{x_0}$, then we would obtain a proper nonempty clopen set $U \cap K_{x_0}$ in K_{x_0}, contradicting the connectedness of K_{x_0}. It should be noted that there exist metrizable spaces with a countable base, for which our assumption fails (see Problem 363).

365. Let $A \in Z$ and U be an open set in X, $U \supset A$. Consider the family $\mathcal{E}$ of clopen sets in X containing A. By Problem 364, $A = \bigcap\{Q: Q \in \mathcal{E}\}$. From the last relation and the fact that $\mathcal{E}$ is a family of closed sets in X, containing the intersection of every finite set of its elements, it follows (see Problem 67) that $Q_0 \subset U$ for some $Q_0 \in \mathcal{E}$. However, if $A' \in Z$ and $A' \cap Q_0 \neq \Lambda$, then $A' \subset Q_0$, which follows from the connectedness of A' and clopenness of Q_0. This shows that the partition Z is continuous.

366. The partition under consideration is continuous (see Problem 365) and hence the decomposition space Z is a compactum (Problems 91 and 92). Let $z \in Z$. Regard z as a subset $\tilde{z}$ of X. Then an arbitrary nhood Oz of z in Z corresponds to an open set $O\tilde{z}$ in X containing $\tilde{z}$. There is (see, solution of Problem 365) a clopen set $\tilde{U}$ in X such that $\tilde{z} \subset \tilde{U} \subset O\tilde{z}$. $\tilde{U}$ is the union of elements of the given partition (i.e., it is distinguished, see p. 77). Hence the set U of those points of Z whose pre-image is contained in $\tilde{U}$ (under the natural projection $\pi: X \to Z$, see p. 45) is clopen in Z (see the definition of quotient mapping). Obviously $z \in U \subset Oz$.

367. Such a space has been indicated in the solution of Problem 363.

368. See Bourbaki ([2], p, 225).

369. By Problem 350 different components of an arbitrary space do not intersect. In the case in question, these components are open sets (Problem 370). But an open subspace of a locally connected space is locally connected. We further note that every point of the space belongs to a component and that all components are connected. Finally, it is clear that the space is a free union of the elements of every disjoint open cover.

370. *(a)* ⇒ *(b)*. Consider a component V of the open set Ox, containing the point x. By condition *(a)*, V is open and $x \in \operatorname{Int} V = V \subset Ox$.

(b) ⇒ *(a)*. Let $x \in W$, where W is a component of the open set U. Take U as Ox. From condition *(b)* it then follows that there is a connected set V such that $x \in \operatorname{Int} V \subset V \subset U$. The set $W \cup V$ is connected as the union of two intersecting connected sets. But $W \cup V \subset U$. Since W is a component of U, we conclude that $W \cup V = W$, i.e., that $V \subset W$. Finally, $x \in \operatorname{Int} V \subset W$, which implies that W is an open set.

371. See Bourbaki ([2], p. 183).

372. Use the notation of the solution of Problem 356. Consider the half-interval $X = \{x: 0 < x \leqslant 4\}$ with the usual topology. Put $I_1 = \{(x, y): x = 1, -2 \leqslant y \leqslant \sin 1\}$, $I_2 = \{(x, y): 0 \leqslant x \leqslant 1, y = -2\}$, $I_3 = \{(x, y): x = 0, -2 \leqslant y \leqslant +1\}$. Map the interval $I_0 = \{x: 1 \leqslant x \leqslant 4\}$ onto the space $I_1 \cup I_2 \cup I_3$ by means of some homeomorphism φ, satisfying $\varphi(1) = (1, \sin 1)$, $\varphi(4) = (0, +1)$. Map the half-interval $P_0 = \{x: 0 < x \leqslant 1\}$ onto a set $\mathcal{T}$ by the rule $\psi(x) = (x, \sin(1/x))$. Then φ and ψ are continuous and agree on $P_0 \cap I_0 = \{1\}$. Hence $f: X \to Y = \mathcal{T} \cup I_1 \cup I_2 \cup I_3$, where $f(x) = \varphi(x)$ when $1 \leqslant x \leqslant 4$ and $f(x) = \psi(x)$ when $0 < x \leqslant 1$ is a continuous mapping. X is connected and locally connected, but Y is not locally connected, which is proved exactly in the same way as in the solution of Problem 356 where it was shown that Φ is not locally connected.

373. Suppose that X is a locally connected space, $f: X \to Y$ is a quotient mapping, $f(X) = Y$ and V is an open set in Y. Let C be an arbitrary component of V. It is to be shown that $f^{-1}(C)$ is an open set in X. If $x \in f^{-1}(C)$, then $x \in f^{-1}(V) = U$. The set U is open, hence the component K of U containing x is also an open set (this follows from the fact that X is locally connected). But then $f(K)$ is a connected set and $U \supset f(K) \ni f(x)$. From $f(x) \in C$ we infer that $f(K) \cap C \neq \Lambda$. Consequently (see Problem 346), $f(K) \cup C$ is a connected subset of V, which implies that $f(K) \cup C = C$, i.e., $f(K) \subset C$. Hence $x \in K \subset f^{-1}(C)$, i.e., $f^{-1}(C)$ is an open set. Since f is a quotient mapping, C is also an open set. This implies that Y is a locally connected space (see Problem 370).

374. The interval is connected (Problem 222, Chap. 2). Hence X also is connected (Problem 221, Chap. 2). Further, f is closed (Problem 92) and, consequently, is a quotient mapping (Problem 315, Chap. 2). But the interval is locally connected (Problem 230, Chap. 2). Hence the space X is also locally connected (the preceding problem).

376. See Hausdorff ([1], p. 186), or Kuratowski ([1], Vol. 2, p. 261).

Chapter 4

Compactifications

4.1. Theoretical Introduction

A space Y is called an extension of a topological space X, if Y contains X as a dense subspace. Sometimes, it is convenient to define the notion thus: Y is extension of a space X, if there is a homeomorphism of X onto a dense subspace of Y. These definitions are essentially equivalent. If, in addition, Y is a T_1-space, then Y is called a T_1-extension of X. Correspondingly, if Y is a Hausdorff space, then we say that Y is a Hausdorff extension of X. Of great importance are Hausdorff *compactifications* of a space X, that is, those compacta which contain X as a dense subspace. It is customary to use the symbol bX with modifications to denote a Hausdorff compactification of a space X, and to denote by $\tilde{X}$ or rX (with modifications) an arbitrary extension of X.

The *growth over* X of an extension $\tilde{X}$, or simply the 'growth' of the extension $\tilde{X}$ (keeping in mind, 'relative to given X'), is defined to be the space $\tilde{X} \setminus X$, endowed with the topology inherited from $\tilde{X}$.

Extending of continuous functions from a topological space to an extension constitutes an important general problem. Suppose that f is a continuous function from a topological space X into a topological space Y and $\tilde{X}$, $\tilde{Y}$ are extensions of these spaces. Then any *continuous* function $\tilde{f}$: $\tilde{X} \to \tilde{Y}$ such that $\tilde{f} \mid X = f$ (that is, $\tilde{f}(x) = f(x)$ for every $x \in X$) is called an *extension* of f to $\tilde{X}$. Thus, the assumption of continuity is implicit in the term 'extension.' Extensions r_1X and r_2X of the same space X are called *equivalent*, if there is a homeomorphism φ of the space r_1X onto the space r_2X such that $\varphi(x) = x$ for every $x \in X$ (that is, $\varphi \mid X$ is the identity mapping). The notion of the equivalence of extensions is quite important and so is its generalization presented below.

An extension r_1X is *subordinate* to extension r_2X or is less than r_2X (written $r_1X < r_2X$), if there is a continuous function ψ from r_2X onto r_1X

such that $\psi^{-1}(x) = x$ for every $x \in X$. The term 'less than' does not mean here 'strictly less than'.

Before attempting to solve the problems of this chapter, the reader is recommended to review the theoretical introduction to Sec. 1 of Chap. 3 on functional separability (α-separability) which plays a vital role in the following.

4.2. General Constructions and Basic Problems

1. Show that any T_1-space can be completed to a compact T_1-space (cf. Problem 176 of Chap. 3) by adding one point.

2. Suppose that X is a T_1-space. Show that:

(*a*) every centralized family of closed sets in X is contained in some maximal centralized family of closed sets.

Suppose ξ is a maximal centralized family of closed sets in X. Show that

(*b*) if $F_1, F_2 \in \xi$, then $F_1 \cap F_2 \in \xi$;
(*c*) if $F_1 \in \xi$, and $F_2 \supset F_1$, then $F_2 \in \xi$;
(*d*) if $F_0 \cap F \neq \Lambda$ for every $F \in \xi$, then $F_0 \in \xi$.

3. Suppose that X is a T_1-space, ωX is the collection of all maximal centralized families of closed sets in X. For each closed set F of X, put

$$\Phi_F = \{\xi \in \omega X \colon F \in \xi\}.$$

For every open set U of X, put $W_U = \omega X \setminus \Phi_{X \setminus U}$.

(*a*) Show that for every finite family $\{F_1, \ldots, F_s\}$ of closed sets of X, we always have

$$\bigcap_{i=1}^{s} \Phi_{F_i} = \Phi_{\bigcap_{i=1}^{s} F_i}.$$

(*b*) Show that $W_U = \{\xi \in \omega X$: there is $F_0 \in \xi$ such that $F_0 \subset U\}$.

(*c*) The family of all sets of the form W_U, where U is open in X, is a base for some topology on ωX. Show that ωX is a compact T_1-space relative to this topology. The set ωX is always regarded as being endowed with the indicated topology.

(*d*) Show that the family (x) of all closed sets of X, containing the point $x \in X$, is a maximal centralized family, and the correspondence $x \to (x)$ is a topological mapping of X onto a dense subspace of ωX. Later, we shall identify the space X with this topological image and call ωX the *Wallman compactification* of X.

4. If T_1-spaces X and Y are homeomorphic, then their Wallman compactifications ωX and ωY are also homeomorphic.

5. (*a*) Show that for every closed set F_0 of X, we have $[F_0]_{\omega X} = \Phi_{F_0}$ (cf. Problem 3), where ωX is the Wallman compactification of X.

(*b*) Show also that for any disjoint closed sets F_1 and F_2 in X, we have $[F_1]_{\omega X} \cap [F_2]_{\omega X} = \Lambda$.

6. Show that the Wallman compactification of a T_1-space X is Hausdorff iff X is normal.

7. Every Hausdorff space with a countable base can be represented as a subspace of a compact T_1-space with a countable base. True or false?

8. Show that every completely regular space of weight τ has a compact Hausdorff completion of weight τ.

9. Show that a space is completely regular iff it is homeomorphic to a subspace of a compact space.

10. Show that two Hausdorff compactifications b_1X and b_2X of a completely regular space X are equivalent iff the relations $[P]_{b_1X} \cap [P']_{b_1X} \neq \Lambda$ and $[P]_{b_2X} \cap [P']_{b_2X} \neq \Lambda$ are equivalent for all $P, P' \subset X$.

11. Suppose that X is a completely regular space and f is a continuous mapping from X into some compact space Y. In that case, show that there is a Hausdorff compactification bX of X (having weight not exceeding that of X and Y) and a continuous mapping $\tilde{f}\colon bX \to Y$, satisfying $\tilde{f} \mid X = f$.

12. Show that each completely regular space X has a Hausdorff compactification bX such that every continuous mapping f from X into a compactum can be extended to a continuous mapping $\tilde{f}$ of bX into this compactum.

13. If b_1X and b_2X are Hausdorff compactifications of X, each having the property that any continuous bounded real-valued function defined on X can be extended to a continuous function, on b_1X (correspondingly, on b_2X), show that b_1X and b_2X are equivalent (see p. 215). Accordingly, any Hausdorff compactification, having the indicated extension property is called the *Stone-Čech compactification* of X and is usually denoted by βX.

14. Let X be a completely regular space and $\mathcal{F} = \{f_\alpha: \alpha \in A\}$ the family of all continuous real-valued functions on X, all whose values are contained in the segment $I = [0, 1]$. Put $I_\alpha = I = [0, 1]$ for each $\alpha \in A$ and consider the diagonal product φ of all elements of the family $\mathcal{F}$, $\varphi: X \to \Pi\{I_\alpha: \alpha \in A\} = I^{|\mathcal{F}|}$. Show that φ is a homeomorphism 'into', i.e., X is homeomorphic to $\varphi(X)$, where $\varphi(X)$ is regarded as a subspace of $I^{|\mathcal{F}|}$, and $[\varphi(x)]_{I^{|\mathcal{F}|}}$ is the Stone-Čech compactification of $\varphi(x)$ (or of the space X, if X is identified with $\varphi(X)$ via φ).

15. Show that a Hausdorff compactification bX of X is the Stone-Čech compactification of X iff for every pair of functionally separable sets P and P' of X the condition $[P]_{bX} \cap [P']_{bX} = \Lambda$ holds

16. Show that a Hausdorff compactification bX of a normal space X is the Stone-Čech compactification of X iff the closures in bX of every pair of closed disjoint sets of X do not intersect.

17. Show that the Wallman compactification ωX of a normal space X coincides with the Stone-Čech compactification βX.

18. If a Hausdorff compactification bX of X is the Stone-Čech compactification, then every continuous mapping of X into a compactum can be extended to a continuous mapping of all of bX into this compactum.

19. If βX is the Stone-Čech compactification of X, $P \subset X$ and P is closed in X, then $[P]_{\beta X}$ is the Stone-Čech compactification of the space P iff every continuous bounded real-valued function on P can be extended to a continuous bounded real-valued function, defined on the whole of X. Show that the last condition is not always satisfied.

20. If X is a normal space and A is closed in X, show that $[A]_{\beta X}$ coincides with βA (here βX is the Stone-Čech compactification of X and βA that of A).

21. If βX is the Stone-Čech compactification of a completely regular space X and $X \subset Y \subset \beta X$, show that βX is also the Stone-Čech compactification of Y.

22. If bX is any Hausdorff compactification and βX is the Stone-Čech compactification of X, show that there is a unique continuous mapping f: $\beta X \to bX$ such that $f(x) = x$ for every $x \in X$. This f will be called the *standard* (or *natural*) *mapping* of βX onto bX.

23. Show that a completely regular space X has more than one (up to equivalence) Hausdorff compactification iff there exist functionally separable closed sets P_1 and P_2 of X, neither of which is compact.

24. Show that the space $T(\omega_1)$ (see p. 136) has a unique Hausdorff compactification (namely, $T(\omega_1) \cup \{\omega_1\}$).

25. If X is a normal space, βX the Stone-Čech compactification of X and P a closed G_δ-set of X, show that there is a G_δ-set $\tilde{P}$ in βX such that $\tilde{P} \cap X = P$ and $\tilde{P} \supset [P]_{\beta X}$.

26. If a completely regular space X has only one Hausdorff compactification (up to equivalence), show that X is locally compact and pseudocompact.

27. Let b_1X and b_2X be Hausdorff compactifications of X and let f: $b_1X \to b_2X$ be the natural continuous mapping. Show that $f^{-1}X = X$.

28. Suppose that a space X is a G_δ-set in some Hausdorff compactification vX. Show that X is G_δ in every Hausdorff compactification.

29. If f: $bX \to \beta X$ is the natural continuous mapping of a Hausdorff compactification bX of X onto the Stone-Čech compactification βX of X, show that f is a homeomorphism.

30. Show that the Stone-Čech compactification of a completely regular space X is the same (up to equivalence) as any Hausdorff compactification which, when the points of X are fixed, can be continuously mapped onto every Hausdorff compactification of X.

31. Suppose that f: $X \to Y$ is a continuous mapping of one completely regular space into another. Show that there is a unique continuous mapping $\tilde{f}$: $\beta X \to bY$ (where βX is the Stone-Čech compactification of X and bY is an arbitrary Hausdorff compactification of Y) such that $\tilde{f} \mid X = f$.

32. If two completely regular spaces X and Y are homeomorphic, then their Stone--Čech compactifications βX and βY are also homeomorphic.

33. If βX is the Stone-Čech compactification of X and $\beta X \setminus X \neq \Lambda$, show that at no point $x \in \beta X \setminus X$ is the space βX first countable.

34. Show that two completely regular spaces X and Y, each of which satisfies the first axiom of countability at all points, are homeomorphic iff their Stone-Čech compactifications βX and βY are homeomorphic.

35. Suppose that X is a regular Lindelöf space and βX is its Stone-Čech extension. Then, show that no countably infinite subset of the set $\beta X \setminus X$ is closed in βX.

36. No countably infinite set, lying in $\beta N \setminus N$, is closed in βN (where N is a countably infinite discrete space).

37. Suppose that $\{X_\alpha : \alpha \in A\}$ is a family of metrizable compacta, $\Pi\{X_\alpha : \alpha \in A\} = X$ is their topological product, $p = \{p_\alpha : \alpha \in A\} \in X$ is an arbitrary point in this product and $\Sigma(p)$ is the $\aleph_1$-hull of the point p in X (see p. 5). Show that the Stone-Čech compactification $\beta(\Sigma(p))$ of the subspace $\Sigma(p)$ of X coincides with X.

38. If X is a completely regular first countable space and bX is its Hausdorff compactification, which is a dyadic compactum, show that X and bX both have countable bases.

39. If X is a space with a countable net but without a countable base, show that no extension of X is a completely normal compactum.

40. If X is a metrizable space without a countable base, show that no extension of X is a completely normal compactum.

41. Formulate in terms of the topology of a given space a criterion for the existence of an extension which is a completely normal compactum.

42. If X is completely regular, show that there exists a Hausdorff compactification bX such that $bX \setminus X$ is a Lindelöf space iff X is a space of countable type (cf. p. 137).

43. Suppose that X is a completely regular space, bX is its Hausdorff compactification and $R(X)$ is the set of all points at which X is not locally compact. Show that $R(X) = [bX \setminus X]_{bX} \setminus (bX \setminus X)$.

44. Suppose that X is a completely regular space and bX is a Hausdorff compactification of X. Show that $bX \setminus X$ is pointwise countable iff the set of all points at which X is not locally compact forms a Lindelöf subspace of X.

45. Suppose that every Hausdorff compactification of a completely regular space X has the following property (cf. also Problems 42 and 44):

> (*) for any point $x \in bX \setminus X$ there exists a compactum $\Phi \subset bX \setminus X$ such that $x \in \Phi$ and $\chi(\Phi, bX) \leqslant \aleph_0$.

Show that X is Lindelöf.

46. Show that X is a Lindelöf space iff some and hence also every compactification bX of X has the following property :

> (**) for any compactum $\Phi \subset bX \setminus X$ there exists a compactum $F \subset bX \setminus X$ such that $\Phi \subset F$ and $\chi(F, bX) \leqslant \aleph_0$.

47. Show that, for a completely regular space X to be Lindelöf, it is necessary and sufficient that each Hausdörff compactification of X satisfies the condition (*) of Problem 45 (cf. also Problem 46).

4.3 Problems on βN, the Stone-Čech Compactification of a Countable Discrete Space

Throughout the present section, we denote by N the set of all non-negative integers, endowed with the discrete topology and thus regarded as a topological space. The Stone-Čech compactification of N is denoted by βN.

48. Show that every separable compactum Φ is a continuous image of the space βN.

49. Show that there exists a family $\mathcal{E}$ of subsets of the set R of all rational numbers such that

(i) $|\mathcal{E}| = 2^{\aleph_0}$,
(ii) if $A \in \mathcal{E}$, $B \in \mathcal{E}$ and $A \neq B$, then $|A \cap B| < \aleph_0$, and
(iii) $|A| = \aleph_0$ for all $A \in \mathcal{E}$.

50. Show that there exists a family $\mathcal{E}$ of infinite subsets of the set N such that

(i) $|\mathcal{E}| = 2^{\aleph_0}$, and
(ii) if $A \in \mathcal{E}$, $B \in \mathcal{E}$ and $A \neq B$, then $|A \cap B| < \aleph_0$.†

51. Let $A \subset N$ and $B \subset N$. Show that:

(i) $[A]_{\beta N}$ is an open set in βN,
(ii) if $|A \cap B| < \aleph_0$, then $[A]_{\beta N} \cap [B]_{\beta N} \cap (\beta N \setminus N) = \Lambda$, and
(iii) if $|A| = \aleph_0$, then $[A]_{\beta N} \cap (\beta N \setminus N) \neq \Lambda$.

52. Show that the space $\beta N \setminus N$ has a family of mutually disjoint open sets of cardinality $2^{\aleph_0}$ in $\beta N \setminus N$ (i.e., $\beta N \setminus N$ does not satisfy the Souslin hypothesis).

53. What is the weight of the space βN? What is the cardinality of βN?

54. Let X be the space $\beta N \setminus N$, equipped with the induced topology. If γ is a countable family of open sets in X, for which $\cap \{U: U \in \gamma\} \neq \Lambda$, show that $\cap \{U: U \in \gamma\}$ contains a nonempty open set.

55. Show that, if $\aleph_1 = 2^{\aleph_0}$, then in the space $X = \beta N \setminus N$ there is a point x^* such that the intersection of any countable set of nhoods of x^* is again a nhood of x^*.

†A family of sets satisfying condition (ii) is called *almost disjoint*.

56. Show that the space βN is not hereditarily normal.

57. Let $p \in \beta N \setminus N$ and $X = N \cup \{p\}$ be endowed with the topology inherited from βN. Show that any Hausdorff space Y. containing X as a dense supspace, is non-sequential.

58. Suppose that $y \in \beta N \setminus N$ and Y is the subspace $\beta N \setminus \{y\}$ of βN. Show that Y is locally compact, countably compact but not compact, and βN is its Stone-Čech compactification.

59. Suppose that X is a subspace of βN, where $N \subset X \subset \beta N$ and X is a k_1-space (see p. 137). Show that the set X is open in βN and, consequently (Problem 178 of Chap. 3), X is locally compact.

60. Construct in βN a subspace, which has the Čech property, contains N and is not locally compact.

61. Suppose that R is the usual space of all real numbers, and N is the subspace of all natural numbers, and βR and $\beta N = [N]_{\beta R}$ (cf. Problem 20) are the Stone-Čech compactification of R and N, respectively. Show that the space $X = \beta R \setminus (\beta N \setminus N)$is pseudo-compact, but not countably compact.

62. Let $N^+ = \{1, 2, \ldots\}$ and each $n \in N^+$ be set in correspondence to an infinite set $E_n \subset N^+$ such that $E_{n'} \cap E_{n''} = \Lambda$ for $n' \neq n''$, and n' and $n'' \in N^+$. Show that there exist infinite sets $N^* \subset N^+$, $N^{**} \subset N^+$, where $N^{**} = N^+ \setminus N^*$ and

$$| N^* \cap E_n | < \aleph_0 \text{ for each } n \in N^*,$$
$$| N^{**} \cap E_n | < \aleph_0 \text{ for each } n \in N^{**}.$$

63. Suppose that M is a discrete subspace of the space $\beta N \setminus N$, $| M | = \aleph_0$ and x^* is a limit point of M. Then, show that the spaces $X = \{x^*\} \cup N$ and $Y = \{x^*\} \cup M$ (endowed with the topology from βN) are non-homeomorphic.

64. Show that the tightness of βN equals $2^{\aleph_0}$ (see p. 43).

64′. If $\aleph_1 = 2^{\aleph_0}$, show that any non-metrizable dyadic compactum (see p. 124) contains a topological image of βN.

65. Let $f: N \to N$ be such that f^2 is the identity. Let $C_f(N)$ be the space resulting from the partition by Z of the Stone-Čech compactification βN of the discrete space N: $Z = \{\{x, \tilde{f}(x)\}: x \in BN \setminus N\} \cup \{\{x\}: x \in N\}$ (where $\tilde{f}$ is an extension of f to βN). Show that one may regard $C_f(N)$ as a Hausdorff compactification of the space N. Are $C_f(N)$ and βN always homeomorphic?

66. Suppose that f_1 and f_2 are two mappings of the set N onto itself, whose squares are identity mappings. Assume also that f_1 has no fixed point, and f_2 has exactly one fixed point a. The spaces $C_{f_1}(N)$ and $C_{f_2}(N)$ are defined with respect to f_1 and f_2 in accordance with the preceding problem. Show that the extensions $C_{f_1}(N)$ and $C_{f_2}(N)$ of N are non-equivalent.

67. Suppose that f is the mapping of N onto itself, described by the rule: if n is even, then $f(n) = n - 1$, and if n is odd, then $f(n) = n + 1$. Denote by C the space $C_f(N)$ (see Problem 65), and by C^* the space obtained by adjoining to C an isolated point ξ, which does not belong to C. Show that C is not homeomorphic to C^* (even if C contains infinitely many isolated points).

68. Show that if two new isolated points are added to the space C, defined in the preceding problem, then the resulting space is homeomorphic to C.

4.4. Compactifications and α-Filters

We recall here certain notions from Chap. 3.1, and further supplement them.

The subsets A and B of a topological space X are called *α-separated* (in X), if there is a continuous real-valued function f on X such that $f(A) = \{0\}$ and $f(B) = \{1\}$. Correspondingly, a set $U \subset X$ is called *α-nhood* of A, if A and $B = X \setminus U$ are α-separated. A family ξ of subsets of X is called an *α-family* if for each $U \in \xi$ there is $A \in \xi$ such that U is an α-nhood of A. A filter on X is called an *α-filter*, if it is an α-family and is made up of open sets in X. If an α-filter is not properly contained as a subfamily in any α-filter on X, then it is called an *α-ultrafilter*.

69. Suppose that X is a space, $P \subset X$, $P^* \subset P$, V is an α-nhood of a set P in X and $V^* \subset V$ is an α-nood of a set P^* in V. Show that V^* is an α-nhood of P^* in X.

70. Suppose that X is a completely regular space, bX is a Hausdorff compactification of X, $z \in bX$ and $b_z = \{U \cap X: U$ is open in bX and $U \ni z\}$. Verify that b_z is an α-family on X.

71. Let X be a completely regular space, let βX be its Stone-Čech compactification and let β be an α-filter on X. Show that, for each $U \in \beta$, there is $V \in \beta$ such that $[V]_{\beta X} \cap [X \setminus U]_{\beta X} = \Lambda$.

72. Suppose that X is a completely regular space, βX is its Stone-Čech compactification, $z \in \beta X$ and $\beta_z = \{U \cap X: U$ is open in βX and $U \ni z\}$. Show that β_z is an α-ultrafilter on X.

73. Suppose that X is a completely regular space, βX is its Stone-Čech compactification and β is an arbitrary α-ultrafilter on X. Show that $\beta = \beta_z$ (see the preceding problem) for some point $z \in \beta X$, where $z = \bigcap \{[G]_{\beta X}: G \in \beta\}$.

74. Suppose that X is a completely regular space, βX is its Stone-Čech compactification, G is an open set in X. Let $\tilde{G}$ be the maximum open set in βX, which yields the set G when intersected with X (see Problem 149). Show that an α-ultrafilter β on X contains G iff $\beta = \beta_z$ (see Problem 72) for some $z \in \tilde{G}$ (see Problems 149, 72).

75. Suppose that X is a completely regular space and bX is a Hausdorff compactification of X. For each $z \in bX$, suppose that the family b_z is an α-ultrafilter on X. Show that then bX is the Stone-Čech compactification of X.

76. Suppose that X is a completely regular space, bX a Hausdorff compactification of X and $z \in bX$. Show that b_z is an α-ultrafilter iff in X there are no functionally separated sets $P \subset X$ and $Q \subset X$ such that $z \in [P]_{bX} \cap [Q]_{bX}$.

77. Suppose that X is a completely regular space, F is a compact subspace and let $Y = X \setminus F$. Further, suppose that $z \in \beta X \setminus X$ and $\beta_z^* = \{U \cap Y: U \text{ is open in } \beta X \text{ and } U \ni z\}$. Then, show that β_z^* is an α-ultrafilter on Y.

78. Suppose that X is completely regular. Put $B = \{\beta_z: z \in \beta X\}$ and define the mapping $\varphi: \beta X \to B$ by the rule: $\varphi(z) = \beta_z$ for each $z \in \beta X$. Further, let $\mathcal{T} = \{\varphi U: U \subset \beta X \text{ and } U \text{ is open in } \beta X\}$, $\mathcal{E} = \{V \subset \beta X: V = \bigcup \{W \subset \beta X: W \text{ is open in } \beta X \text{ and } W \cap X = V \cap X\}\}$ and $\mathcal{B} = \{\varphi V: V \in \mathcal{E}\}$. In addition, for each open set G in X denote by $\psi(G)$ the set of all α-ultrafilters on X, which contain G.

Prove the following aseertions:

(i) The mapping φ is one-to-one and $\varphi(\beta X) = B$.
(ii) $\varphi(V) = \psi(X \cap V)$ for each $V \in \mathcal{E}$.
(iii) $\mathcal{B} = \{\psi(G): G \text{ is an open set in } X\}$.
(iv) B is the the set of all α-ultrafilters on X.
(v) $\mathcal{T}$ is a topology on B, $\mathcal{B}$ is a base and $(B, \mathcal{T})$ is compact, where $\varphi: \beta X \to (B, \mathcal{T})$ is a homeomorphism. Under this homeomorphism an arbitrary point of X is mapped to the family of all open sets in X which contain it.

4.5. Q-SPACES AND HEWITT EXTENSIONS

A completetely regular space X is called a ***Q-space or real-compact space*** (*functionally closed space*), if for each completely regular space $Y \neq X$ containing X as a dense subspace, there is a continuous real-valued function f on X which cannot be extended to a continuous real-valued function on Y.

The following concepts occupy a central position in the theory of Q-spaces.

A filter ξ is called $\aleph_0$-*prime*, if there is $\eta \subset \xi$, for which $|\eta| \leqslant \aleph_0$ and $\bigcap \{A: A \in \eta\} = \Lambda$.

Suppose that $(X, \mathcal{T})$ is a topological space. The family $\mathcal{B}_{\aleph_0}$ of all G_δ-sets in $(X, \mathcal{T})$ forms a base for some (unique) topology $\mathcal{T}_{\aleph_0}$ on X. The topology $\mathcal{T}_{\aleph_0}$ is called the $\aleph_0$-*modification* of the topology $\mathcal{T}$. Analogously, the space $(X, \mathcal{T}_{\aleph_0})$ is called the $\aleph_0$-modification of the space $(X, \mathcal{T})$. A $\aleph_0$-modification of an arbitrary space X is usually denoted briefly by $(X)_{\aleph_0}$.

For an arbitrary completely regular space X, we denote by $Q(X)$ the set of all its Hausdorff compactifications bX such that X is closed in $(bX)_{\aleph_0}$. If $Q(X) \neq \Lambda$, then the lower bound of the weights of the spaces belonging to $Q(X)$ is called the *Q-weight* of X. Elements of $Q(X)$ are called ***Q-extensions*** of X.

For an arbitrary Hausdorff compactification bX of a space X, $\mathcal{D}(bX, X)$ denotes the set of all compact Hausdorff subspaces of $bX \setminus X$, whose character in bX is countable. If $\mathcal{D}(bX, X)$ covers $bX \setminus X$, then the lower bound of the cardinalities of subcovers of $bX \setminus X$ contained in $\mathcal{D}(bX, X)$ is called the *Q-defect* of X in the compactification bX and is denoted by Q-def (X, bX). In the theory of Q-spaces, an important role is played by certain special concepts pertaining to cardinal numbers.

The cardinal number τ is called *measurable* if on a set X of cardinality τ there is a non-singular σ-ultrafilter (i.e., an ultrafilter with the property that the intersection of any countable set of its elements is nonempty). In the opposite case, τ is called *non-measurable*. A cardinal number λ is called *strongly inaccessible*, if $\lambda > \aleph_0$, λ is regular, i.e., $cf(\lambda) = \lambda$, and for each τ such that $\tau < \lambda$, we have $2^\tau < \lambda$. Finally, for a cardinal number τ, $Q(\tau)$ denotes the Q-weight of the discrete space of cardinality τ, if this Q-weight is defined (otherwise, the symbol $Q(\tau)$ is meaningless).

79. Show that every compactum is a Q-space.

80. Show that each psedudo-compact Q-space X is compact.

81. Show that a completely regular space X is a Q-space iff for every completely regular extension $\tilde{X}$ of X differing from X by one point (i.e., such that $|\tilde{X} \setminus X| = 1$), there is a continuous real-valued function f on X, which cannot be extended to $\tilde{X}$.

82. Show that a completely regular space X is a Q-space iff for each point $x^* \in \beta X \setminus X$ (where βX is the Stone-Čech compactification of X), there is a compactum $F \subset \beta X \setminus X$, containing x^*, whose character in βX is countable.

83. Show that the space $T(\omega_1)$ (see p. 136) is not a Q-space.

84. Show that a space X is a Q-space iff the set X is closed in the $\aleph_0$-modification of the Stone-Čech compactification βX of X.

85. Show that a completely regular space X is a Q-space iff each free α-ultrafilter ξ on X is $\aleph_0$-prime (see p. 225).

86. Show that a discrete space X is a Q-space iff $|X|$ is a non-measurable cardinal number (see p. (225).

87. If X is a Q-space and βX its Stone-Čech compactification, show that $\beta X \setminus X$ is a pointwise countable space.

88. For a completely regular space X to be a Q-space, show that it is necessary and sufficient that there is a Hausdorff compactification bX of X satisfying the following condition:

(*) for each point $x \in bX \setminus X$ there is a compactum F such that:
(i) $x \in F \subset bX \setminus X$, and
(ii) $\chi(F, bX) = \aleph_0$.

89. For a completely regular space X to be a Q-space, show that it is necessary and sufficient that the growth $bX \setminus X$ of this space in some Hausdorff compactification bX is the union of G_δ-sets in bX.

90. Show that every regular space X with a countable base is a Q-space.

91. Show that every regular Lindelöf space X is a Q-space.

92. If a space X has a Hausdorff compactification which satisfies the first axiom of countability at all its points, show that X is a Q-space.

93. If X is a completely regular, first countable space and bX is a Hausdorff compactification, show that $Y = bX \setminus X$ is a Q-space.

94. If X is a completely regular space without points of local compactness, bX is a Hausdorff compactification of X and $bX \setminus X$ is pointwise countable, show that X is a Q-space.

95. If X is pointwise countable and bX is a Hausdorff compactification of X, show that $Y = bX \setminus X$ is a Q-space.

96. Show that a closed subspace Y of a Q-space X is a Q-space.

97. Show that the product of any set $\{X_\alpha: \alpha \in A\}$ of Q-spaces is a Q-space.

98. Show that a space X is a Q-space iff it is homeomorphic to a closed subspace of the Tychonoff product of subsets of the real line.

99. Suppose that X is a completely regular space, βX is its Stone-Čech compactification and $\mathcal{E}$ is the family of all compacta included in $\beta X \setminus X$ of countable character in βX. Put $\bar{\mathcal{E}} = \cup \{F: F \in \mathcal{E}\}$ and $\nu X = \beta X \setminus \bar{\mathcal{E}}$. Show that:

(0) $X \subset \nu X$;
(i) νX is a Q-space;
(ii) if Y is a Q-space and $X \subset Y \subset \beta X$, then $\nu X \subset Y$.

(The extension νX of X constructed above is called the *Hewitt extension* of X, or the *real compactification* of X).

100. Suppose that X is a completely regular space, νX is its Hewitt extension, and βX is the Stone-Čech compactification. Then, show that νX is also a Hewitt extension of any space Y such that $X \subset Y \subset \nu X$.

101. Show that a completely regular space is a Q-space iff it coincides with its Hewitt extension.

102. If $f: X \to Y$ is a continuous 'onto' mapping, $B \subset Y$, $A = f^{-1}B$, X and B being Q-spaces, show that A is also a Q-space.

103. Show that a pre-image of a Q-space under a perfect mapping is a Q-space.

104. Show that every continuous mapping f of a completely regular space X into a completely regular space Y can be extended (uniquely) to a continuous mapping $\hat{f}$ of the Hewitt extension νX of X into the Hewitt extension νY of Y.

105. Show that every continuous mapping of a completely regular space X into a Q-space Y can be extended to a continuous mapping of νX into Y (see the preceding problem).

106. Show that the space νX (see Problem 99) is the largest of all subspaces of the space βX containing X, to which any continuous real-valued function on X can be continuously extended.

107. Suppose that X is a completely regular space and νX is its Hewitt extension (see Problem 99). Show that νX is made up of exactly all those points y of the Stone-Čech compactification βX of X, for which every continuous real-valued function defined on X can be extended to a continuous real-valued function on $X \cup \{y\}$.

108. Show that for every completely regular space X, the following statements are equivalent:

(i) every subspace of X is a Q-space;
(ii) every open subspace of X is a Q-space;
(iii) for each point $x \in X$, $X \setminus \{x\}$ is a Q-space.

109. Any locally compact Hausdorff space of cardinality not exceeding $2^{\aleph_0}$ is a Q-space. True or false?

110. If X is a Q-space and $A \subset X$ is the set of all points at which some continuous real-valued function f on X vanishes, show that $X \setminus A$ is a Q-space.

111. If X is a Q-space and each point $x \in X$ is a G_δ-set in X, show that every subspace of X is a Q-space.

112. If every subspace of a completely regular space X is a Q-space, then each point in X is a G_δ-set in X (see the preceding problem). True or false?

113. If the pseudo-character of a regular Lindelöf space is countable at all points, show that every subspace is a Q-space.

114. If X is a compeletely regular space, F a compact subspace, $X \supset Y \supset X \setminus F$ and Y is a Q-space, show that X is also a Q-space.

115. If $f: X \to Y$ is a continuous compact mapping of a completely regular space X onto a space Y, every subspace of Y being a Q-space, show that X also is a Q-space.

116. If a completely regular space X is contracted to a space Y, each $Z \subset Y$ being a Q-space, show that every subspace of X is also a Q-space.

117. Show that every discrete space of cardinality less than or equal to $2^{\aleph_0}$ is a Q-space.

118. Show that a Nemitskii space (see Problem 13 of Chap. 3), and all its subspaces, are Q-spaces.

119. If X is a regular Lindelöf space, whose pseudo-character at every point is countable, show that $|X|$ is a non-measurable cardinal number.

120. Show that the union of two discrete Q-spaces, of which one is open and countable and the other is closed, may not be a Q-space (even if it is completely regular).

121*. If a normal space X is the union of a countable family of closed subspaces X_i, where $i \in N$, which are Q-spaces, show that X itself is a Q-space.

122*. Does the concluding part of the statement of the preceding problem remain valid if the assumption that X is normal is replaced by the weaker assumption that it is completely regular.

123*. If a completely regular space X is a union of two closed subspaces, each of which is a Q-space, then X also is a Q-space. True or false?

124. If X is closed in a Q-space Y and bY is a Q-extension of Y, show that $bX = [X]_{bY}$ is also a Q-extension of X, and the Q-defect of X in bX does not exceed the Q-defect of Y in bY.

125. Show that the Q-weight of a closed subspace Y of a Q-space X does not exceed the Q-weight of X.

126. Show that the Q-weight of any complete metric space X with a countable base is $\aleph_0$.

127. Show that the Q-weight of a Q-space X does not exceed the cardinal number τ, where $\tau > \aleph_0$, iff the weight of X is not greater than τ and there is a Hausdorff compactification bX of X such that the Q-defect of X in bX does not exceed τ.

128*. If $X = \Pi\{X_\alpha : \alpha \in A\}$ is a Tychonoff product of Q-spaces X_α, the Q-weight of each of which does not exceed τ, where $\tau \geqslant \aleph_0$, show that the Q-weight of X is not greater than $\max\{\tau, |A|\}$.

129*. Show that the Q-weight of a Q-space X does not exceed τ, where $\tau \geqslant \aleph_0$, iff X is homeomorphic to a closed subspace of a Tychonoff product of τ copies of the real line R.

130*. If $X = \Sigma\{X_t : t \in T\}$ is a disjoint (free) topological union (see p. 44) of completely regular spaces X_t, show that the Q-weight of X is not greater than $\max\{\sup\{Q\text{-weight of } X_t : t \in T\}, Q\text{-weight of the discrete space } T\}$.

131. Show that every 0-dimensional (see p. 42) compactum X, whose weight does not exceed τ, is homeomorphic to a closed subset of D^τ.

132*. Suppose that τ is a cardinal number, $\tau \geqslant \aleph_0$, A is a discrete space of cardinality $\lambda = 2^\tau$, and B is a discrete space of cardinality τ. Show that A is homeomorphic to a closed subspace of B^λ(B^λ is the Tychonoff product of λ copies of B).

133*. If a discrete space A of cardinality τ is a Q-space, show that any discrete space B of cardinality 2^τ is also a Q-space.

134*. If the cardinality of a discrete space A is less than the cardinality of the first strongly inaccessible (see p. 225) cardinal number τ^*, show that the Q-weight of A does not exceed $\| A \| = \sup \{2^\tau : \tau < | A |\}$.

135*. If τ is non-measurable, show that the Q-weight of a discrete space of cardinality τ does not exceed 2^τ.

136. Deduce from the statements of Problems 133 and 86 that, if τ is a non-measurable cardinal number, then 2^τ is also non-measurable (this has been proved before, but differently, in Problem 142 of Chap. 1).

137*. If $Q(\tau) = \lambda$, show that $Q(\tau + 1) \leqslant (\tau + 1) \cdot \lambda$. Here $Q(\tau)$ is the Q-weight of a discrete space of cardinality τ.

138*. If τ^* is the first strongly inaccessible cardinal number and for each $\tau < \tau^*$ we have $\tau^+ = 2^\tau$, show that the Q-weight of any discrete space A of cardinality less than τ^* is $| A |$ (i.e., $Q(\tau) = \tau$ for all $\tau < \tau^*$).

139*. Suppose that θ is a least regular limit (see p. 16) cardinal number. Show that if the cardinality of the discrete space A is less than θ, then the Q-weight of A is $| A |$.

4.6. Subordination

We say that a completely regular space X carries a subordination relation v (or a *subordination* v) if for all pairs of the form (F, H), where F is a closed and H is an open set in X, it is determined that F is v-subordinate to H or not (in the former case we write $F <_v H$, and in the latter $F \nless_v H$). For this, the following axioms (subordination axioms) are required to be satisfied:

K–1. If $F <_v H$, then $X \setminus H <_v X \setminus F$.
K–2. If $F <_v H$, then $F \subset H$.
K–3. If $F_1 \subset F <_v H \subset H_1$, then $F_1 <_v H_1$.
K–4. If $F_1 <_v H_1$ and $F_2 <_v H_2$, then $F_1 \cup F_2 <_v H_1 \cup H_2$.
K–5. If $F <_v H$, then there is a nhood OF of F in X such that $F <_v OF$ and $[OF] <_v H$.
K–6. $\Lambda <_v \Lambda$.
K–7. For every point $x \in X$ in its any nhood Ux, we have $\{x\} <_v Ux$.

Note that from the axioms K–1 and K–6 it follows that $X <_v X$ and that $F_1 <_v H_1$ and $F_2 <_v H_2$ imply that $F_1 \cap F_2 <_v H_1 \cap H_2$.

Recall that two sets A and B of a space X are called functionally separable, if there is a continuous function $f\colon X \to [0, 1]$ such that $f(x) = 0$ if $x \in A$, and $f(x) = 1$ if $x \in B$. We now say that a closed set F is *β-subord-*

inate to an open set H, if the sets F and $X \setminus H$ are functionally separable. The β-subordination relation (the subordination axioms for β are verified in Problem 141) is called the *elementary subordination*.

Now suppose that v is some subordination relation on a completely regular space X. A family ξ of open sets in X is said to be ***v-regular***, if for every $H_1 \in \xi$ there is $H_2 \in \xi$ such that $[H_2] <_v H_1$. A family ξ of open sets in X is called a *v-end* of a space X (or a *v-filter*), if ξ is a v-regular centralized family and is maximal with respect to v-regularity and centralization, i.e., ξ is a v-regular centralized family which is contained in no v-regular centralized family of open sets in X other than ξ. If X is a completely regular space, and v is a subordination defined on X, then we denote by vX the set of all v-ends of X. Further, if H_0 is an open set in X, we put $O_{H_0} = \{\xi \in vX: H_0 \in \xi\}$ (the set O_{H_0} is sometimes denoted by $O^v_{H_0}$). On the set vX a topology is defined, whose base consists of all possible sets of the form O_H, where H is open in X. If F is a closed set in X, then by Φ_F we denote the set of all $\xi \in vX$, for which $H \cap F \neq \Lambda$ for every $H \in \xi$.

The space X is topologically contained in vX and is a dense set in it (see Problem 147), if each point $x_0 \in X$ is set in correspondence to the v-end (x_0) consisting of all nhoods of the point x_0 in X. In the following each point $x \in X$ is identified with the v-end (x). In Problem 152, it is shown that vX is a Hausdorff compactification of the completely regular space X. We say that vX is the *compactification of the space X generated by the subordination v* on X.

If v is a subordination relation on a space X, then the sets $A_1 \subset X$, $A_2 \subset X$ are called *v-distant* (symbolically, $A_1 \bar{v} A_2$) if $[A_1] <_v X \setminus [A_2]$. If the sets $A_1 \subset X$, $A_2 \subset X$ are not v-distant, then they are said to be ***v-close*** (in this case, we write $A_1 v A_2$). Frequently, we have to consider two or more subordination relations on a given completely regular space X. Suppose that on a given space X, the subordination relations v_1 and v_2 are defined. We say that the *subordination relation v_2 is not less than v_1* (and in this case we write $v_2 > v_1$), if for any closed set $F \subset X$ and open set $H \subset X$ such that $F <_{v_1} H$ (i.e., F is subordinate to H in the sense of v_1), we have $F <_{v_2} H$ (i.e., F is subordinate to H also in the sense of v_2). We denote by $\mathcal{P}(X)$ the family of all subordination relations on a given completely regular space X. We shall regard the family $\mathcal{P}(X)$ as being endowed with the ordering indicated above. It is trivial to see that ordering is a partial order on $\mathcal{P}(X)$.

Suppose we are given a completely regular space X together with a subordination v defined on it and a completely regular space Y with a subordination w. A mapping $f: X \xrightarrow{\text{onto}} Y$ is called *(v, w)-uniformly continuous,* if for every v-close sets $A_1 \subset X$, $A_2 \subset X$, the sets fA_1 and fA_2 are w-close to Y. In Problem 159 it is shown that the mapping $f: X \rightarrow Y$ is (v, w)-uniformly continuous iff it is continuous and $\Phi <_w O$ ($\Phi \subset Y$ is closed in Y and O is open in Y) implies that $f^{-1}\Phi <_v f^{-1}O$ in X. Hence, if v_1 and v_2 are two subordination relations on X, then $v_2 > v_1$ iff the identity mapping $e: X \xrightarrow{\text{onto}} X$ is (v_2, v_1)-uniformly continuous. The mapping $f: X \rightarrow Y$ is called *(v, w)-open*, if $F <_v H$ in X always implies $[fF]_Y <_w <_w \operatorname{Int}[fH]_Y$ in Y.

Suppose now that X^* is a Hausdorff compactification of a completely regular space. If $H \subset X$ is an open set in X, then by $O(H)$ (or by $O_{X^*}(H)$) we denote a *maximal open set in* X^* satisfying $O(H) \cap X = H$. Suppose that X^* is a compactification of a space X, generated by some subordination v on X, i.e., $X^* \equiv vX$ for some subordination v. Then in Problem 149 it is shown that $O^v_H = O_{X^*}(H)$ for any open set H in X; accordingly, we do not make a distinction between the sets O^v_H and $O_{X^*}(H)$.

The following definition is important: a compactification X^* of a completely regular space X is called *perfect with respect to an open set H in X,* if the equality $[\operatorname{Fr}H]_{X^*} = \operatorname{Fr}_{X^*}O_{X^*}(H)$ holds. However, if X^* is perfect with respect to every open set in X, then it is called a *perfect compactification* of the space. The following notions are also important.

A base σ of a space X is called *algebraically closed* if for every pair of elements $U_1 \in \sigma$, $U_2 \in \sigma$, we have $U_1 \cup U_2 \in \sigma$ and $X \setminus [U] \in \sigma$ for each $U \in \sigma$.

A base σ of a space X is called *peripherally compact* (or, simply, *π-compact*) if the subspace $\operatorname{Fr}U$ is compact for every $U \in \sigma$. A space having a π-compact base is called a *π-compact space.*

140. Let $\mathscr{B}(X)$ be the family of all (Hausdorff) compactifications of a completely regular space X. Show that the succession ordering on Hausdorff compactifications (for definition see p. 215) is a partial ordering on $\mathscr{B}(X)$.

141. Show that the elementary subordination β on a completely regular space X satisfies all the subordination axioms (see p. 230). Show also that if X is a normal space, then $F <_\beta H$ (here F is closed and H is open in X) iff $F \subset H$.

142. Let (X, ρ) be a metric space. Put $F <_\rho H$ iff $\rho(F, X \setminus H) > 0$.

(*a*) Show that ρ is a subordination on X.

(*b*) Show that on a completely regular space different subordination relations may exist.

143. Show that subordination relations can be introduced only on completely regular spaces. Accordingly, all spaces in this section will be assumed completely regular.

144. Show that on a compactum X the only subordination is the elementary subordination.

145. Suppose that X is a completely regular space, and $\mathcal{P}(X)$ is the family of all subordinations on X. Show that $\prec$ in $\mathcal{P}(X)$ (see p. 231) is a partial order.

146. Suppose that (X, v) is a topological space together with subordination v.

(*a*) Show that every v-regular family is contained in some (not necessarily unique) v-end.

(*b*) Let ξ be a v-end. Then, show that:
(i) $H_1, H_2 \in \xi$ implies $H_1 \cap H_2 \in \xi$;
(ii) $H_1 \in \xi$ and $H \supset H_1$ imply $H \in \xi$.

147. Suppose that v is a subordination on a space X, and vX is the family of all v-ends. Show that:

(*a*) $O_{H_1} \cap O_{H_2} = O_{H_1 \cap H_2}$, where H_1 and H_2 are arbitrary open sets in X;

(*b*) if $H_1 \subset H_2$, then $O_{H_1} \subset O_{H_2}$;

(*c*) vX is a T_1-space.

(*d*) if $x_0 \in X$, then the family (x_0) of all nhoods of x_0 in X is a v-end;

(*e*) the mapping h: $X \to vX$ defined by the equality $hx = (x)$, where $x \in X$, is a topological mapping of X onto a dense subspace of vX.

In the following, $h(X)$ is identified with X.

148. Show that for any open set Γ in vX, we always have

(i) $[O_{X \cap \Gamma}]_{vX} = [\Gamma]_{vX}$, and

(ii) $O_{X \cap \Gamma} \subset [\Gamma]_{vX}$.

149. Suppose that X^* is any extension of a space X. Let H be an open and F a closed set in X. Show that:

(*a*) $O(H)$ (see p. 232) is the union of all open sets Γ in X^*, satisfying $\Gamma \cap X = H$;
(*b*) $X^* \setminus O(H) = [X \setminus H]_{X^*}$ and $[F]_{X^*} = X^* \setminus O(X \setminus F)$;
(*c*) if $H_1 \neq H_2$, then $O(H_1) \neq O(H_2)$;
(*d*) $H_1 \subset H_2$ iff $O(H_1) \subset O(H_2)$;
(*e*) $F_2 \subset F_1$ (where F_1 and F_2 are closed sets in X) iff $[F_2]_{X^*} \subset [F_1]_{X^*}$;
(*f*) $O(H_1 \cap H_2) = O(H_1) \cap O(H_2)$;
(*g*) if $X^* = vX$, where v is some subordination on X, then $O(H) = O_H$ for any open set $H \subset X$ and $\Phi_F = [F]_{vX}$.

150. Suppose that (X, v) is a space with a subordination v and vX is the space constructed in Problem 147. Show that:

(*a*) $[O_{H_0}]_{vX} = \{\xi \in vX \colon H \cap H_0 \neq \Lambda \text{ for every } H \in \xi\}$;
(*b*) $F <_v H_0$ (F is closed and H_0 is open in X) implies $\Phi_F \subset O_{H_0}$ (or, equivalently, $[F]_{vX} \subset O(H_0)$; see Problem 149);
(*c*) vX is a regular space.

151. Suppose that v is a subordination on X and F is any closed set in X. Show that $F = \cap\{UF \colon F <_v UF\}$ and $[F]_{vX} = \cap\{O_{UF} \colon F <_v UF\}$.

152. Show that vX is a compact space.

153. Show that $F <_v H$ iff $[F]_{vX} \subset O_H$.

154. Show that for distinct subordinations v_1 and v_2 on a given space X, the compactifications v_1X and v_2X are distinct.

155. Suppose that X is a completely regular space and X^* is an arbitrary compactification of X. Put $F <_v H$ (F is closed and H is open in X) iff $[F]_{X^*} \subset O(H)$. Show that:

(*a*) v is a subordination satisfying all axioms K–1 to K–7 (see p. 230).
(*b*) The compactification vX coincides with X^*.

156. Show that the operation v, associating each subordination $v \in \mathscr{P}(X)$ on X with the compactification vX of this space, is a one-to-one mapping of the family $\mathscr{P}(X)$ of all subordinations on X onto the family $\mathscr{B}(X)$ of all Hausdorff compactifications of X.

157. Show that the compactification vX, with respect to a given subordination v on X, is maximal (i.e., is the Stone-Čech compactification) iff v is the elementary subordination on X.

158. Suppose that v is a subordination relation on X. Show that the sets $A \subset X$, $B \subset X$ are v-distant (for definition see p. 231) iff $[A]_{vX} \cap [B]_{vX} = \Lambda$.

159. Suppose that (X, v) and (Y, w) are two spaces with the subordinations v and w, respectively. Show that the mapping $f: X \to Y$ onto Y is (v, w)-uniformly continuous iff it is continuous and $\Phi <_w O$ in Y (Φ is closed and O is open in Y) always implies $f^{-1}\Phi <_v f^{-1}O$.

160. Suppose that (X, v) and (Y, w) are two spaces, with subordinations v and w, respectively, and $f: (X, v) \to (Y, w)$ is a (v, w)-uniformly continuous mapping on Y. Put $(*)\, \bar{f}\xi = \cap \{[\bar{f}U]_{wY}: U \in \xi\}$ for each point $\xi \in vX$. Show that:

(*a*) $\bar{f}\xi \neq \Lambda$ for every $\xi \in vX$;
(*b*) a point $\eta \in wY$ belongs to $\bar{f}\xi$ iff $\sigma = \{f^{-1}V: V \in \eta\} \subset \xi$;
(*c*) the set $\bar{f}\xi$ consists of one point for every $\xi \in vX$;
(*d*) $\bar{f}$ is a continuous mapping of the compactification vX onto the compactification wY;
(*e*) $\bar{f}$ is an extension of the mapping $f: (X, v) \to (Y, w)$.

161. Suppose that (X, ρ) and (Y, μ) are two metric spaces with metrics ρ and μ, respectively. Let ρ be the subordination on X, associated with metric ρ and μ the subordination on Y, associated with the metric μ (see Problem 142). Show that $f: (X, \rho) \to (Y, \mu)$ is (ρ, μ)-uniformly continuous iff for every $\varepsilon > 0$ there is a $\delta > 0$ such that $\mu(fx, fy) < \varepsilon$ whenever $\rho(x, y) < \delta$.

162. Show that a mapping $f: (X, v) \xrightarrow{\text{onto}} (Y, w)$ is (v, w)-uniformly continuous iff there is a continuous mapping $\bar{f}: vX \to wY$ of the compactification vX onto the compactification wY of the space Y, which is an extension of the mapping f.

163. Let $h: vX \to wX$ be the natural continuous mapping of a compactification vX of X onto a compactification wX of the same space X. Show that the equality $h^{\#}O_v(U) = O_w(U)$ holds for every open set U in X. As always, $h^{\#}O_v(U) = \{\eta \in wX: h^{-1}\eta \subset O_v(U)\} = wX \setminus h(vX \setminus O_v(U))$.

164. Show that the operation v, which associates (see Problem 155) each subordination $v \in \mathcal{P}(X)$ on a space X with the compactification $vX \in \mathcal{B}(X)$, is an isomorphism between the partially ordered sets $\mathcal{P}(X)$ and $\mathcal{B}(X)$.

165. Let $f: X \to Y$ be a continuous mapping of a space X onto a space Y.

(i) Show that f is (β, β')-uniform, where β and β' are the elementary subordinations on X and Y, respectively.
(ii) Deduce from (i) that every continuous mapping $f: X \to Y$ admits a continuous extension $\bar{f}: \beta X \to \beta Y$.

166. Suppose that $f: X \to Y$ is a continuous mapping of a normal space X onto a normal space Y, and $\bar{f}: \beta X \to \beta Y$ is the continuous extension to the Stone-Čech compactifications of these spaces. Show that f is a closed mapping iff $\bar{f}^{-1}y = [f^{-1}y]_{\beta X}$ for every point $y \in Y$.

167. Suppose that $f: X \to Y$ is a (v, w)-uniformly continuous mapping of a space X with a subordination v onto a space Y with a subordination w. Show that f is (v, w)-open iff the unique (see Problem 162) continuous extension $\bar{f}: vX \to wY$ of the mapping f to the corresponding compactifications vX and wY generated by v and w is open.

168. If $f: X \to Y$ is a clopen mapping of a space X onto a space Y, then the unique continuous extension $\bar{f}: \beta X \to \beta Y$ of f to the Stone-Čech compactifications of these spaces is open.

169. If $f: X \to Y$ is a closed monotonic (see p. 154) mapping of a space X onto a space Y, then the unique continuous extension $\bar{f}: \beta X \to \beta Y$ of f to the Stone-Čech compactifications of these spaces is also monotonic.

170. Show that the compactification vX of a space X generated by a subordination v, is perfectly related to an open set U in X iff for any closed set F in X such that $F \subset U$ and $F <_v X \setminus \mathrm{Fr} U$, we have $F <_v U$.

171. Show that the Stone-Čech compactification βX of any (completely regular) space X is perfect.

172. Show that a compactification vX of a space X is perfect iff for any two disjoint open sets U_1 and U_2 in X, we have $O(U_1 \cup U_2) = O(U_1) \cup O(U_2)$.

173. Let v be a subordination on X and vX the associated compactification of X. Show that the equality

$$O(U \cup V) = O(U) \cup O(V) \tag{1}$$

is satisfied for every pair of open sets U and V in X iff X is normal, and v coincides with the elementary subordination β on X (and, consequently, $vX \equiv \beta X$ is the Stone-Čech compactification of X).

174. Show that a compactification vX of a space X is perfect iff the natural mapping $h: \beta X \to vX$ of the Stone-Čech compactification βX of this space onto vX is monotonic (see p. 154).

175. Let σ be an algebraically closed π-compact base for a regular space X. Put $F <_{\sigma} H$ (where F is closed and H is open in X) when there is a $U \in \sigma$ such that $F \subset U \subset [U] \subset H$.

Show that the relation σ is a subordination relation. The extension σX, generated by this kind of subordination (i.e., a subordination associated with some algebraically closed π-compact base σ for X) is called a *π-extension* of X.

176. Show that a π-compact Hausdorff space is completely regular.

177. Suppose that σX is a π-extension of X generated by an algebraically closed π-compact base σ (see Problem 175). Show that σX is perfect with respect to every $U \in \sigma$.

178. Suppose that X is a π-compact space, σ is a base consisting of all open sets in X with compact boundary. Show that the base σ is algebraically closed and the compactification σX generated by σ (see Problems 175 and 176) is perfect.

179. Prove that in the hypotheses of Problem 178 no compactification vX of a space X, preceding the extension σX, is perfect.

180. Suppose that X is a π-compact space, σ is an algebraically closed π-base of the space, σ is the subordination associated by the base σ (see Problem 175) and σX is the compactification generated by this subordination. Show that the famlly $\Sigma = \{O_{\sigma}(U): U \in \sigma\}$ is a base of the compactum σX, where $O_{\sigma}(U) \cap (\sigma X \setminus X)$ is a clopen set in the subspace $\sigma X \setminus X$. Hence deduce that every connected subset of $\sigma X \setminus X$ is a singleton.

181. Suppose that X is a π-compact space, σ is a base for X consisting of all open sets in X with compact boundary, and σX is the compactification of X generated by σ (see Problem 175). Show that every perfect compactification vX of X admits a natural mapping onto σX.

182. Show that for every π-compact completely regular space X there is a unique compactification γX of X such that :

(i) γX is perfect;
(ii) every compactification of X preceding the extension γX is not perfect;
(iii) every perfect compactification of X other than γX follows γX.

183. Every compactification vX of a space X which follows a perfect extension θX, is also perfect. Is this true?

184. Let R be an open interval on the real line with the natural topology. Show that the subspace $\beta R \setminus R$ has two connected components.

4.7. SOLUTIONS

1. Suppose that X is a T_1-space and ξ is some object not in X. Consider the set $\bar{X} = X \cup \{\xi\}$. We define a topology on $\bar{X}$ as follows. On X, the previous topology stands, i.e., $U \subset X$ is considered an open set in $\bar{X}$ iff U is open in X. As nhoods of ξ we take all sets $V \subset \bar{X}$, $\xi \in V$, for which $\bar{X} \setminus V$ is finite. It is easily verified that $\bar{X}$ is a T_1-space, containing X as an open set. To show that $\bar{X}$ is compact, let ω be an open cover of $\bar{X}$. Let $U_{\alpha_0} \in \omega$ be such that $\xi \in U_{\alpha_0}$. Then $\bar{X} \setminus U_{\alpha_0}$ is finite: $\bar{X} \setminus U_{\alpha_0} = \{x_1, \ldots, x_s\}$. Consequently, there are $U_{\alpha_1}, \ldots, U_{\alpha_s}$ in ω such that $x_1 \in U_{\alpha_1}, \ldots, x_s \in U_{\alpha_s}$. Hence $\omega_0 = \{U_{\alpha_1}, \ldots, U_{\alpha_s}, U_{\alpha_0}\}$ is a finite subcover of the cover ω of $\bar{X}$. The proof is complete.

2. (*a*) is proved using the maximum principle (see p. 6). Let Ξ be the family of all centralized families of closed sets, containing the centralized family ξ of closed sets. Ξ is partially ordered thus : $\xi_2 > \xi_1, \xi_1 \in \Xi, \xi_2 \in \Xi$, if $\xi_2 \supset \xi_1$. It is routine to show that every nest (see p. 6) in Ξ with this order has an upper bound. Then, by the maximum principle (p. 6), Ξ has a maximal element, i.e., there is a maximal centralized family of closed sets in X, which contains the given family ξ.

(*b*) Suppose that ξ is a maximal centralized family of closed sets. Consider the family $\bar{\xi} = \xi \cup \{F_1 \cap F_2 : F_1, F_2 \in \xi\}$. It is easy to verify that $\bar{\xi}$ is centralized. Hence $\xi = \bar{\xi}$. Consequently, if $F_1, F_2 \in \xi$, then $F_1 \cap F_2 \in \xi$.

(*c*) is proved along the same lines as (*b*).

To prove (*d*), suppose that ξ is a maximal centralized family and $F_0 \cap F \neq \Lambda$ for every $F \in \xi$. Consider the family $\bar{\xi} = \xi \cup \{F_0 \cap F : F \in \xi\}$. We show that $\bar{\xi}$ is centralized. Let $F_1, \ldots, F_k \in \xi$ and consider $F_0 \cap F_{k+1}, \ldots, F_0 \cap F_{k+l}$. Then $F_0' = F_1 \cap \ldots \cap F_k \in \xi$, $F_{k+1} \cap \ldots \cap F_{k+l} = F'' \in \xi$ (by virtue of (*b*)). Hence $F_0 \cap F_1 \ldots \cap F_k \cap (F_0 \cap F_{k+1}) \cap \ldots \cap (F_0 \cap F_{k+l}) = F' \cap F_0 \cap F'' = F''' \cap F_0 \neq \Lambda$, where $F''' = F' \cap F''$ (by (*b*)) belongs to ξ. Thus $\bar{\xi}$ is centralized and consists of closed sets in X; consequently, since ξ is maximal, we have $\bar{\xi} = \xi$, so that $F_0 \in \xi$.

3. (*a*) If ξ is such that

$$\xi \in \bigcap_{i=1}^{s} \Phi_{F_i},$$

then $F_i \in \xi$, $i = 1, \ldots, s$. Hence, by Problem 2,

$$\bigcap_{i=1}^{s} F_i \in \xi;$$

Consequently, $\xi \in \Phi_{\bigcap_{i=1}^{s} F_i}$. Conversely, if $\xi \in \Phi_{\bigcap_{i=1}^{s} F_i}$, then $\bigcap_{i=1}^{s} F_i \in \xi$. Therefore, since

$$\bigcap_{i=1}^{s} F_i \subset F_1, \ldots, \bigcap_{i=1}^{s} F_i \subset F_s,$$

by Problem 2 we obtain $F_1 \in \xi, \ldots, F_s \in \xi$, i.e.,

$$\xi \in \bigcap_{i=1}^{s} \Phi_{F_i}.$$

(*b*) Since $W_U = \omega X \setminus \Phi_{X \setminus U}$, $W_U = \{\xi \in \omega X\colon F \cap U \neq \Lambda \text{ for every } F \in \xi\}$. We show that there is a $F_0 \in \xi$ such that $F_0 \subset U$. If $F \not\subset U$ for every $F \in \xi$, then $F \cap (X \setminus U) \neq \Lambda$ for every $F \in \xi$, but then by (*d*) of Problem 2 we would have $X \setminus U \in \xi$. This is a contradiction. Hence $W_U \subset \{\xi \in \omega X : \text{there is } F_0 \in \xi \text{ such that } F_0 \subset U\}$. The reverse inclusion is obvious.

(*c*) It is trivial to show that ωX becomes a T_1-space, if all sets of the form W_U (where U is open in X) are taken as a base. Note that sets of the form Φ_F, where F is closed in X, are closed in ωX (they form a closed base in X). The sets which are intersections of sets of the form Φ_F are also closed in ωX. To prove the compactness of ωX, it suffices to show that every centralized family of sets of the form Φ_F has non-empty intersection. Suppose that $\{\Phi_{F_\alpha} : \alpha \in A\}$ is an arbitrary centralized family of such sets. Using statement (*a*) of this problem, we infer that $\{F_\alpha\colon \alpha \in A\}$ is centralized. Expand it (see Problem 2) to a maximal centralized family ξ of closed sets in X, to obtain

$$\xi \in \bigcap \{\Phi_{F_\alpha}\colon \alpha \in A\}, \text{ i.e., } \bigcap \{\Phi_{F_\alpha}\colon \alpha \in A\} \neq \Lambda.$$

(*d*) The family (x) of all closed sets in X containing a point $x \in X$ is obviously centralized and maximal with respect to this property, because $\{x\} \in (x)$. Then (x) is a point of ωX. The rest is left as a routine exercise for the reader.

5. To begin with note that for every closed set F in X, we have $\Phi_F \cap X = F$.

(*a*) (i) Let $\xi_0 \in \Phi_{F_0}$. This implies that $F_0 \in \xi_0$. Let W_U, where U is open in X, be a basic nhood of the point ξ_0 in ωX, $\xi_0 \in W_U$. This means (see Problem 3) that there is a $F_1^0 \in \xi_0$ such that $F_1^0 \subset U$. But since $F_0 \in \xi_0$, $F_1^0 \cap F_0 \neq \Lambda$. Hence $U \cap F_0 \neq \Lambda$. Therefore, $W_U \cap F_0 \neq \Lambda$, so that $\xi_0 \in [F_0]_{\omega X}$.

(ii) Since $F_0 \subset \Phi_{F_0}$, $[F_0]_{\omega X} \subset [\Phi_{F_0}] = \Phi_{F_0}$, i.e., if $\xi_0 \in [F_0]_{\omega X}$, then $\xi_0 \subset \Phi_{F_0}$.

From (i) and (ii) we obtain $[F_0]_{\omega X} = \Phi_{F_0}$.

(*b*) Suppose that F_1 and F_2 are two closed sets in X such that $F_1 \cap F_2 = \Lambda$. We have $[F_1]_{\omega X} = \Phi_{F_1} = \{\xi \in \omega X\colon F_1 \in \xi\}$, $[F_2]_{\omega X} = \Phi_{F_2} = \{\xi \in \omega X\colon F_2 \in \xi\}$. Hence if $F_1 \cap F_2 = \Lambda$, then also $\Phi_{F_1} \cap \Phi_{F_2} = \Lambda$, i.e., $[F_1]_{\omega X} \cap [F_2]_{\omega X} = \Lambda$.

6. (*a*) Suppose that X is normal and ξ_1, ξ_2 are two distinct points in ωX. Since ξ_1 and ξ_2 are distinct, there are $F_0^1 \in \xi_1$ and $F_0^2 \in \xi_2$ such that $F_0^1 \cap F_0^2 = \Lambda$. Since X is normal, we can choose nhoods UF_0^1 and UF_0^2 of F_0^1 and F_0^2 in X, satisfying $UF_0^1 \cap UF_0^2 =$

Δ. Consider the nhoods $W_{UF_0^1}$ and $W_{UF_0^2}$ of the points ξ_1 and ξ_2 in ωX. We assert that $W_{UF_0^1} \cap W_{UF_0^2} = \Delta$: if some $\xi \in \omega X$ were to belong to their intersection, then for a certain $F_0 \in \xi$ we would have $F_0 \subset UF_0^1$, $F_0 \subset UF_0^2$, which would contradict the fact that the intersection of UF_2^1 and UF_0^2 is empty.

(*b*) Conversely, suppose that ωX is a Hausdorff space. Then it is compact. Assume that F_1 and F_2 are two disjoint closed sets in X, to obtain $[F_1]_{\omega X} \cap [F_2]_{\omega X} = \Delta$ (see Problem 5). Since ωX is normal, we can take open sets W_1 and W_2 in ωX in such a way that we would have

$$[F_1]_{\omega X} \subset W_1, \quad [F_2]_{\omega X} \subset W_2, \tag{1}$$

and

$$W_1 \cap W_2 = \Delta. \tag{2}$$

Then $U_1 = W_1 \cap X$ and $U_2 = W_2 \cap X$ are open and disjoint in X and $F_1 = [F_1]_{\omega X} \cap X \subset U_1$, $F_2 = [F_2]_{\omega X} \cap X \subset U_2$.

7. False. Any compact T_1-space with a countable base has a countable net of closed sets (see Problem 129, Chap. 3). But the properties of such a net are retained under passage to every subspace. It suffices now to take any Hausdorff space without a countable net of closed sets (see, for example, Problem 146 of Chap. 2).

8. If τ is the weight of a completely regular space X, then there exists in X a partitioning family (see Problem 36 of Chap. 3) $\{f_\alpha : \alpha \in A\}$ of functions $f_\alpha : X \to I_\alpha$ of cardinality τ. The diagonal product f of this family is a homeomorphism to a compactum $Y = \Pi\{I_\alpha : \alpha \in A\}$ of weight τ (see Problems 37 of Chap. 3 and 379 of Chap. 2). Then $[fX]_Y$ is a compactum, also of weight τ, which has a dense subspace homeomorphic to X.

9. Every compactum is normal (see Problem 64 of Chap. 3) and hence (Problem 26 of Chap. 3) is completely regular. But a subspace of a completely regular space is completely regular (Problem 8 of Chap. 3). Consequently, every subspace of a compactum is completely regular. The converse statement follows from Problems 38 and 75 of Chap. 3, or from Problem 8 of the present chapter.

10. Necessity is obvious. To prove sufficiency, for each $x \in b_1X$ denote by ξ_x the family of all sets of the form $X \cap Ox$, where Ox is a nhood of x in b_1X. Put $\varphi(x) = \bigcap\{[P]_{b_2X} : P \in \xi_x\}$. ξ_x is centralized, hence $\varphi(x) \neq \Delta$. We show that $|\varphi(x)| = 1$. If $y, z \in \varphi(x)$ and $y \neq z$, then for y and z we can find nhoods Oy and Oz in b_2X, whose closures in b_2X are disjoint. Then $P_1 = [Oy]_{b_2X} \cap X$ and $P_2 = [Oz]_{b_2X} \cap X$ are disjoint closed sets in X, where $P_i \cap P \neq \Delta$ for every $P \in \xi_x$ and $i = 1, 2$. From the definition of ξ_x it now follows that $x \in [P_1]_{b_1X}$ and $x \in [P_2]_{b_1X}$, i.e., $[P_1]_{b_1X} \cap [P_2]_{b_1X} \neq \Delta$, a contradiction. Thus $|\varphi(x)| = 1$. If $x \in X$, then obviously $\varphi(x) \ni x$ and hence $\varphi(x) = \{x\}$. We agree to denote by $\tilde{\varphi}(x)$ the uniqe element of the set $\varphi(x)$. Then $\tilde{\varphi}$ is a mapping of the space b_1X into the space b_2X whose restriction to X is the identity mapping.

To show that $\tilde{\varphi}$ is continuous, let $x \in b_1X$ and $y = \tilde{\varphi}(x)$. Fix a nhood Oy of y in b_2X. Since $\bigcap\{[P]_{b_2X} : P \in \xi_x\} = \{y\}$ and ξ_x is a prefilter (but b_2X is compact), there is $P_0 \in \xi_x$ such that $[P_0]_{b_2X} \subset Oy$. But $P_0 = O_0x \cap X$, where O_0x is a certain nhood of x in b_1X. For any $x' \in O_0x$, surely $P_0 \in \xi_{x'}$ and hence $\tilde{\varphi}(x') \in [P_0]_{b_2X} \subset Oy$. Thus $\tilde{\varphi}(O_0x) \subset Oy$, proving the continuity of $\tilde{\varphi}$. We verify that $\tilde{\varphi}$ is one-to-one. If $x_1 \neq x_2$, $x_1, x_2 \in b_1X$, then $[Ox_1]_{b_1X} \cap [Ox_2]_{b_1X} = \Lambda$ for certain nhoods of points x_1 and x_2 in b_1X. Put $Ox_1 \cap X = P_1$ and $Ox_2 \cap X = P_2$. Then $[P_1]_{b_1X} \cap [P_2]_{b_1X} = \Lambda$, and hence $[P_1]_{b_2X} \cap [P_2]_{b_2X} = \Lambda$. But $P_1 \in \xi_{x_1}$, $P_2 \in \xi_{x_2}$; consequently, $\tilde{\varphi}(x_1) \in [P_1]_{b_2X}$ and $\tilde{\varphi}(x_2) \in [P_2]_{b_2X}$, so that $\tilde{\varphi}(x_1) \neq \tilde{\varphi}(x_2)$. From $\tilde{\varphi}(x) = x$ for all $x \in X$ it follows that $\tilde{\varphi}(b_1x) \supset X$. But $\tilde{\varphi}(b_1X)$ is compact and hence (Problem 56 of Chap. 3) is a closed subspace of b_2X. Consequently, $\tilde{\varphi}(b_1X) = b_2X$. As a one-to-one continuous mapping of the compactum b_1X onto the compactum b_2X, $\tilde{\varphi}$ is a homeomorphism.

11. Let $b'X$ be an arbitrary Hausdorff compactification of a space X (whose weight is equal to that of X, see Problem 8). Consider the graph Γ_f of f. Then $\Gamma_f = \{(x,f(x)): x \in X\}$ is a subspace of the product $X \times Y \subset b'X \times Y$, where the map $\varphi(x) = (x, f(x))$ is a homeomorphism between X and Γ_f (Problem 291 of Chap. 2).

Note that $f(x) = \pi\varphi(x)$ for all $x \in X$, where π: $b'X \times Y \to Y$ is the projection onto the second factor. Put $Z = [\Gamma_f]_{b'X \times Y}$. Then Z is a Hausdorff compactification of the space Γ_f and $\pi \mid Z$ is a continuous mapping of Z into Y, the weight of Z not exceeding the maximum weight of X and Y. If we replace in Z each point $(x, f(x))$ of the set Γ_f by x, then the topology is preserved. Z is thus transformed into a Hausdorff compactification of the space X, the mapping $\pi \mid Z$ into the continuous mapping $\tilde{f}$: $fX \to Y$, whose restriction to X coincides with f.

12. Let $\mathcal{F}$ be any family of continuous mappings of the space X into a compactum, satisfying the condition:

(*) if $f: X \to Y$ is a continuous mapping and Y is a compactum, then there is an $f' \in \mathcal{F}, f': X \to Y'$ and a homeomorphism φ of a compactum Y' into Y such that $\varphi f' = f$.

Of course, one would like $\mathcal{F}$ to be the family of all continuous mappings of X into a compactum; however, the latter is not a set. For us it is important that $\mathcal{F}$, satisfying the condition (*), can be so chosen that it is a set (see p. 1).

For each $f \in \mathcal{F}$ fix some Hausdorff compactification b_fX of X, on which f is extended to a continuous mapping $\tilde{f}$ (see Problem 11). Consider the product ψ of the mappings $\tilde{f}$ with respect to all $f \in \mathcal{F}$. Denote by Y_f the image of $\tilde{f}$, by Y the product $\Pi\{Y_f: f \in \mathcal{F}\}$, and by F the product $\Pi(b_fX: f \in \mathcal{F}\}$. Also, π_f is the projection of Y onto Y_f, and the set $\Delta \subset F$ consists of all those points $i(x)$, all of whose coordinates coincide with some $x \in X$. The correspondence $x \leftrightarrow i(x)$ is a homeomorphism between Δ as a subsace of F and X. By means of this homeomorphism X is identified with Δ. Then $f = (\pi_f\psi) \mid X$. But the mapping $\pi_f\psi$ is defined and is continuous on $[X]_F$. Thus each $f \in \mathcal{F}$ has a continuous extension (namely, $(\pi_f\psi) \mid [X]_F$) onto the Hausdorff compactification $[X]_F$ of X. The condition (*) now guarantees that every continuous mapping of X into a compactum can be extended to a continuous mapping of the space $bX = [X]_F$.

13. This is immediate from Problems 15 and 10.

14. $\mathcal{F}$ is a partitioning family, since X is a completely regular space. Consequently, φ is a homeomorphism of X onto its image Y (Problem 37 of Chap. 3). It now suffices to verify that every continuous real-valued function $g(y)$, defined on Y and such that $g(\varphi(X)) \subset [0,1]$ can be extended to a continuous real-valued function on $F = [\varphi(X)]_{I^{|\mathcal{F}|}}$ (see Problem 13; clearly F is a Hausdorff compactification of the space $\varphi(X)$). Put $\psi = g\varphi$. Then ψ is a continuous real-valued function on X and $\psi(X) \subset [0, 1]$. Hence $\psi \in \mathcal{F}$.

Let $\alpha^* \in A$ be such that $f_{\alpha^*} = \psi$. Consider $\pi_{\alpha^*}: I^{|\mathcal{F}|} \to I_{\alpha^*}$, the projection onto the factor I_{α^*}. By the definition of the diagonal product of mappings, $\pi_{\alpha^*}\varphi = f_{\alpha^*}$. Hence $\pi_{\alpha^*}\varphi = g\varphi$, implying $\pi_{\alpha^*}(y) = g(y)$ for all $y \in \varphi(X)$. Thus g is the restriction of π_{α^*} to $\varphi(X)$. But π_{α^*} is defined and continuous on $I^{|\mathcal{F}|}$ and, in particular, also on $F = [\varphi(X)]_{I^{|\mathcal{F}|}}$. Consequently, $\pi_{\alpha^*} \mid F$ is the required extension of the mapping g.

15. If P and P' are functionally seperable and f is a continuous real-valued function on X, satisfying $f(P) = \{0\}, f(P') = \{1\}$ and $f(X) \subset [0, 1]$, then for a continuous extension $\tilde{f}$ of f to bX the following relations hold: $\tilde{f}([P]_{bX}) = [\{0\}] = \{0\}$ and $\tilde{f}([P']_{bX}) = [\{1\}] = \{1\}$. Hence $[P]_{bX} \cap [P']_{bX} = \Lambda$. Thus necessity is proved.

To prove sufficiency, there is (see Problem 12) a Hausdorff compactification $b'X$, such that every (bounded) continuous mapping of X into a compactum can be extended to $b'X$. By the necessity proved above $[P']_{b'X} \cap [P]_{b'X} = \Lambda$ iff P and P' are functionally separable in X, i.e., (in terms of the proof of sufficiency we give) iff $[P']_{bX} \cap [P]_{bX} = \Lambda$. Hence (Problem 10) the compactifications $b'X$ and bX are equivalent, and therefore every continuous mapping of X into a compactum can be extended to bX.

16. In a normal space, P and P' are functionally separable sets iff they have disjoint closures (Problem 25 of Chap. 3). The conclusion now follows from the preceding problem.

17. If X is normal, ωX is a Hausdorff compactification of X (see Problem 6) and, by Problem 5, for every pair of disjoint closed sets in X their closures in ωX are also disjoint; consequently (see Problem 16), ωX coincides with βX.

19. *Necessity.* If $[P]_{\beta X}$ is the Stone-Čech compactification of the space P and f is a bounded continuous real-valued function on P, then there is a continuous real-valued function $\tilde{f}$ defined on $[P]_{\beta X}$ and coinciding with on P. But βX is a normal space, hence $\tilde{f}$ can be extended from the closed set $[P]_{\beta X}$ in βX to a continuous real-valued function $\tilde{\varphi}$ on βX (see Problem 34 of Chap. 3). The latter is bounded. Then $\tilde{\varphi} \mid X$ is the desired extension of the function f on X.

Sufficiency. Fix a bounded, continuous real-valued function f on P. By hypothesis, it can be extended to some continuous, bounded real-valued function f on X. The latter can be continuously extended to some function φ on βX. Then $\varphi \mid [P]_{\beta X}$ is an extension of f to a continuous function on $[P]_{\beta X}$. The existence of such extension, in view of the arbitrary choice of f, implies that $[P]_{\beta X}$ is the Stone-Čech compactification of P (see Problem 13).

20. This follows from Problems 33 of Chap. 3 and 19 of the present chapter.

21. In order that a bounded continuous real-valued function defined on Y extend to βX, it suffices to take its restriction to X and extend the latter to βX.

22. The identity mapping of X into itself, regarded as a mapping into the compactum bX, can be extended to a continuous mapping $f: \beta X \to bX$ (Problems 12 and 13). The uniqueness of f follows from the fact that every pair of continuous mappings in a Hausdorff space, which coincide on a dense set, coincide everywhere (Problem 287 of Chapt. 2).

23. Consider βX, the Stone-Čech compactification of X. Put $F_1 = [P_1]_{\beta X}$ and $F_2 = [P_2]_{\beta X}$. Then $F_1 \cap F_2 = \Lambda$ (see Problem 15) and $F_1 \setminus P_1 \neq \Lambda$, $F_2 \setminus P_2 \neq \Lambda$. Consequently, $|\beta X \setminus X| \geqslant 2$. Fix $z_1 \in F_1 \setminus P_1$ and $z_2 \in F_2 \setminus P_2$. Consider the partition of βX, whose unique non-trivial (i.e., with more than one point) element is the set $A = \{z_1, z_2\}$. This decomposition space can be regarded (see Problems 318, 329 of Chap. 2 and 92 of Chap. 3) as a Hausdorff compactification of X. Denote it by b^*X. Then $A \in [P_1]_{b^*X}$ and $A \in [P_2]_{b^*X}$ by the definition of the topology in b^*X. Hence $[P_1]_{b^*X} \cap [P_2]_{b^*X} \neq \Lambda$, and the extensions βX and b^*X are not equivalent. This proves sufficiency of the hypothesis. To prove its necessity, suppose that b_1X and b_2X are non-equivalent Hausdorff compactifications of X. By definition, this implies that there are sets $P_1 \subset X$, $P_2 \subset X$ such that $[P_1]_{b_1X} \cap [P_2]_{b_1X} = \Lambda$, but $[P_1]_{b_2X} \cap [P_2]_{b_2X} \neq \Lambda$. Obviously, P_1 and P_2 can be regarded as closed sets in X. From our assumption it follows that $[P_1]_{b_2X} \setminus P_1 \neq \Lambda$ and $[P_2]_{b_2X} \setminus P_2 \neq \Lambda$, i,e., P_1 and P_2 are not closed in b_2X and, consequently, neither of them is compact. The sets $[P_1]_{b_1X}$ and $[P_2]_{b_1X}$ are functionally separable in b_1X, as disjoint closed subsets of the normal space b_1X. Hence P_1 and P_2 are functionally separable in X.

24. Every continuous real-valued function on $T(\omega_1)$ is eventually constant, i e., for some $a \in T(\omega_1)$ and for all x, y greater than a, $f(x) = f(y)$. Consequently, if f is extended to the space $T(\omega_1) \cup \{\omega_1\}$, which is the one-point Alexandroff compactification of $T(\omega_1)$ (see Problem 176 of Chap. 3), and if we put $\tilde{f}(\omega_1) = f(x)$, where x is greater than a, $\tilde{f} \mid T(\omega_1) = f$, then we obtain a continuous function $\tilde{f}$. Hence $T(\omega_1) \cup \{\omega_1\}$ is the Stone-Čech compactification of $T(\omega_1)$. If $b(T(\omega_1))$ is any Hausdorff compactification of $T(\omega_1)$, then there is on $T(\omega_1)$ a continuous map φ: $T(\omega_1) \cup \{\omega_1\} \to b(T(\omega_1))$ which is the identity on $T(\omega_1)$. We have $\varphi(\{\omega_1\}) = b(T(\omega_1)) \setminus T(\omega_1)$ (Problem 27), i.e., φ is a one-to-one mapping. Consequently, φ is a homeomorphism, i.e., an equivalence. A more succinct solution of this problem can be given, based on Problem 23 of this chapter, Problems 186 and 184 of Chap. 3 and the obvious fact that a countably compact countable space is compact.

25. Let $P = \bigcap\{U_i: U_i \in N^+\}$, where each U_i is an open set in X. Then $[P]_{\beta X} \cap [X \setminus U_i]_{\beta X} = \Lambda$ because X is normal; and P and $X \setminus U_i$ are disjoint closed sets in X. Hence $\tilde{P} = \bigcap\{\beta X \setminus [X \setminus U_i]_{\beta X}: i \in N^+\}$ is obviously the desired set.

26. Suppose that bX is a Hausdorff compactification of a space X and $|bX \setminus X| \geqslant 2$. Fix $z_1, z_2 \in bX \setminus X$, $z_1 \neq z_2$, and consider the space Z which continuously partitions the space bX, whose unique element with more than one point is the set $A = \{z_1, z_2\}$ (see Problems 318 and 322 of Chap. 2).

Fix the nhoods Oz_1 and Oz_2 of z_1 and z_2, satisfying $[Oz_1]_{bX} \cap [Oz_2]_{bX} = \Lambda$. Put $P_1 = X \cap [Oz_1]_{bX}$, $P_2 = [Oz_2]_{bX} \cap X$. Then, P_1 and P_2 are disjoint nonempty closed sets in X, where $z_1 \in [P_1]_{bX}$ and $z_2 \in [P_2]_{bX}$. Arguing along the same lines as in the solution of Problem 23, we arrive at the conclusion that Z can be regarded as a Hausdorff compactification b^*X of X, where

$$[P_1]_{b^*X} \cap [P_2]_{b^*X} \neq \Lambda. \tag{A}$$

Consequently, b^*X and bX are non-equivalent extensions of X (Problem 10), i.e., X has at least two Hausdorff compactifications, which contradicts the hypothesis. Hence $|bX \setminus X| \leqslant 1$. Therefore $bX \setminus X$ is a closed set, and the set $X = bX \setminus (bX \setminus X)$ is open in bX. Consequently, X is locally compact. Assume now that X is not pseudo-compact. Then there is an unbounded continuous real-valued function f on X. This allows us to choose in $fX \subset R$ two disjoint subsets A and B closed in R (where R is the usual space of all real numbers), neither of which is compact (or empty). Then $P = f^{-1}A$ and $Q = f^{-1}B$ are disjoint closed sets in X, neither of which is compact (because the image of a compact space under a continuous mapping is compact). P and Q are functionally separable. It therefore suffices to take the composition of the functions f and φ on fX, which separates A from B. This is a contradiction (see Problem 23).

27. Assume that $y \in b_1X \setminus X$ and $f(y) = x \in X$. Then, $y \neq x$ and there are nhoods Oy and Ox of x and y in b_1X such that $[Oy]_{b_1X} \cap [Ox]_{b_1X} = \Lambda$. Set

$$P_1 = [Oy]_{b_1X} \cap X \text{ and } P_2 = [Ox]_{b_1X} \cap X. \tag{B}$$

We have $y \in [P_1]_{b_1X}$, $x \notin P_1$. Hence, by the continuity of f, we obtain $x = fy \in [fP_1]_{b_2X}$. But $fP_1 = P_1$, $x \in X$, and since P_1 is closed in X, $x \in P_1$, a contradiction. See also Problem 345 of Chap. 2.

28. Consider the Stone-Čech compactification βX of a space X and its natural continuous mapping f onto υX. Suppose that bX is an arbitrary Hausdorff compactification of X and $g: \beta X \to bX$ is the natural mapping. Fix a countable family γ of open sets in υX, satisfying $X = \bigcap\{V: V \in \gamma\}$. Put $\lambda = \{f^{-1}V: V \in \gamma\}$. Since f is continuous, λ is a (countable) family of open sets in βX. We have $\bigcap\{U: U \in \lambda\} = \bigcap\{f^{-1}V: V \in \gamma\} = f^{-1}(\bigcap\{V: V \in \gamma\}) = f^{-1}X$. But $f^{-1}X = X$ (see Problem 27). Hence $\bigcap\{U: U \in \lambda\} = X$. Put $\mu = \{bX \setminus g(\beta X \setminus U): U \in \lambda\}$. The mapping g is closed, hence μ consists of open sets in bX. Obviously, $|\mu| \leqslant |\lambda| \leqslant \aleph_0$. We have $g^{-1}(bX \setminus g(\beta X \setminus U)) \subset U$ for all $U \in \lambda$. Hence $g^{-1}(\bigcap\{W: W \in \mu\}) = \bigcap\{g^{-1}W: W \in \mu\} \subset \bigcap\{U: U \in \lambda\} = X$. But $g^{-1}X = X$ (see Problem 27) implies $W \supset X$ for all $W \in \mu$. Hence $\bigcap\{W: W \in \mu\} = X$, i.e., X is a G_δ-set in bX.

29. By Problem 18 there is a continuous mapping $\varphi: \beta X \to bX$, satisfying $\varphi(x) = x$ for all $x \in X$. Put $g = \varphi f$, $g: bX \to bX$. Then $g(x) = \varphi f(x) = x$ for all $x \in X$ and g is a continuous mapping. Consequently, g is the identity mapping of bX onto itself (Problem 287 of Chap. 2). Since f is a mapping on βX, we conclude that f is a homeomorphism; more precisely, f effects an equivalence between the compactifications βX and bX.

30. See Problems 12, 13, 18 and 29.

31. The uniqueness of $\tilde{f}$ follows from the fact that two continuous mappings which coincide on dense sets, coincide everywhere (Problem 287 of Chap. 2). The existence of $\tilde{f}$ follows from Problem 18.

33. Assume the contrary. Let $x^* \in \beta X \setminus X$ and suppose $\{x^*\}$ is a G_δ-set in βX. Then there is a continuous real-valued function f on βX, satisfying $f^{-1}(0) = \{x^*\}$ (see Problem 30 of Chap. 3). Regard $R^* = fX$ as a subspace of the usual space R of all real numbers. From $x^* \in [X]$, $f^{-1}(0) = \{x^*\}$ and the continuity of f it follows that $[R^*]_R \setminus R^* \ni 0$. Then there exist nonempty closed disjoint sets P and Q in R^* such that $0 \in [P]_R$, $0 \in [Q]_R$. Put $f_0 = f|X$. We regard f_0 as a mapping of X into R^*. It is continuous. Hence $A = f_0^{-1}P$ and $B = f_0^{-1}Q$ are nonempty, closed (disjoint) sets in X. The space R^* is normal (as is every metric space), so there is a bounded continuous real-valued function φ on R^* such that $\varphi(P) = \{0\}$, $\varphi(Q) = \{1\}$ and $\varphi(R^*) \subset [0, 1]$. Then $g = \varphi f_0$ is a bounded continuous real-valued function on X, satisfying $g(A) = \{0\}$ and $g(B) = \{1\}$. Hence A and B are functionally separable in X. But $[A]_{\beta X} \ni x^*$, and $[B]_{\beta X} \ni x^*$. Indeed, if, say, $[A]_{\beta X} \not\ni x^*$, then $f[A]_{\beta X} \not\ni 0$ (since $f^{-1}(0) = \{x^*\}$), and since $fA = P$ and f is a closed mapping, $f[A]_{\beta X} \supset [P] \ni 0$. The contradiction obtained shows that $[A]_{\beta X} \cap [B]_{\beta X} \supset \{x\} \neq \Lambda$. But this is impossible. because A and B are functionally separable in X and βX is the Stone-Čech compactification of X.

34. The necessity is immediate as a special case of the general statement (see Problem 32). To prove sufficiency, suppose that $\varphi: \beta X \to \beta Y$ is a homeomorphism, $\varphi(\beta X) = \beta Y$. By Problem 135 of Chap. 2, the space βX (corresp. βY) is first countable at each point $x \in X$ (corresp. $y \in Y$). Furthermore, from the preceding problem it follows that at no point $x \in \beta X \setminus X$ (corresp. $y \in \beta Y \setminus Y$) is the space βX (corresp. βY) first countable. It now suffices to note that under every homeomorphism of a space X onto another space Y points at which X is first countable are mapped to points at which Y is first countable.

35. The following stronger statement holds : every Lindelöf subspace of $\beta X \setminus X$ is not closed in βX, if X is Lindelöf. To prove this (by contradiction), make use of the arguments of Problem 19 of Chap. 3.

37. It is obvious that $\Sigma(p)$ is dense in X. It is required to show (see Problem 13) that every bounded (real-valued) function f defined and continuous on $\Sigma(p)$ admits a continuous extension to the compactum X. But by using Problem 389 of Chap. 2 and arguing along similar lines as in the solution of Problem 390 of Chap. 2, it can be shown that f depends only on a countable set of coordinates, i.e., there is a countable $A_0 \subset A$ such that for every pair of points $x, y \in X$ satisfying $x_\alpha = y_\alpha$ if $\alpha \in A_0$, we have $fx = fy$. Consider the projection $\pi_{A_0}: X \to \Pi\{X_\alpha: \alpha \in A_0\} = Y$. Clearly, $\pi_{A_0}(\Sigma(p)) = Y$. Denote by $\pi^0_{A_0}$ the restriction of π_{A_0} to $\Sigma(p)$. It is easily seen that $\pi^0_{A_0}$ is a continuous open mapping. The fact that f on $\Sigma(p)$ depends only on the coordinates with indices in A_0 implies that there is a real-valued function f_0 on Y such that $f = f_0\pi^0_{A_0}$.

To show that $f_0 : Y \to R$ (where R is the usual real line) is continuous, suppose that U is an open set in R and let $V = f_0^{-1}U$. Then $\pi^0_{A_0}(\Sigma(p)) = Y$ and $f = f_0\pi_{A_0}$ imply $\pi^0_{A_0}f^{-1}(U) = V$. But $f: \Sigma(p) \to R$ is continuous and hence $f^{-1}(U)$ is open. Consequently, V is also open in Y, since $\pi^0_{A_0}$ is an open mapping.

Consider the function $\overline{f} = f_0\pi^0_{A_0}$. Clearly, the function $\overline{f}$ is defined and continuous on X and extends f. Thus, every continuous bounded function defined on $\Sigma(p)$ extends to a continuous function on the compactum X, $\beta\Sigma(p) = X$.

38. This is immediate from Problem 168 of Chap. 3.

39. This is a reformulation of the statement of Problem 228 of Chap. 3.

40. Indeed, every metrizable subspace of a perfect normal compactum has a countable base (Problem 221 of Chap. 3).

41. A satisfactory solution of this problem is not known to the authors.

42. *Necessity.* Fix a compactum $F \subset X$. For each point $p \in bX \setminus X$ choose an open F_σ-set Up in bX, containing p but disjoint from F (plausible, by the complete regularity of bX). The open cover $\{Up : p \in bX \setminus X\}$ of $bX \setminus X$ has a countable subcover λ. Put $U = \bigcup\{G: G \in \lambda\}$. Then U is an open F_σ-set in bX, disjoint from F but containing $bX \setminus X$. Hence $\Phi = bX \setminus U$ is a closed G_δ-set in bX, included in X and containing F. Then Φ is a compactum and its character in bX, and hence also in X, is countable.

Sufficiency. Consider an arbitrary cover γ of the space $bX \setminus X$ by open sets in bX. Put $F = bX \setminus \bigcup\{U : U \in \gamma\}$. F is obviously a compactum included in X. By hypothesis, there is a compactum $\Phi \subset X$ containing F and whose character in X is countable. It is then easy to see that the character of Φ in bX is also countable (see Problem 135 of Chap. 2 and Problem 56 of Chap. 3). Hence $bX \setminus \Phi$ is the union of a countable set of compacta and is therefore Lindelöf (see Problem 208 of Chap. 3). Moreover, $bX \setminus X \subset bX \setminus \Phi \subset \bigcup \{U: U \in \gamma\}$. The cover γ contains a countable cover λ of $bX \setminus \Phi$. Then λ is a countable cover of $bX \setminus X$, which is a subcover of the original cover.

43. Let $y \notin [bX \setminus X]_{bX} \setminus (bX \setminus X)$ and $y \in X$. We have $[bX \setminus X]_{bX} \setminus (bX \setminus X) \subset X$. Hence $y \notin bX \setminus X$, but then also $y \notin [bX \setminus X]_{bX}$. There is a nhood Oy of y in bX such that $[Oy]_{bX} \subset X$. Consequently, X is locally compact at y and $y \notin R(X)$.

Suppose now that $y \notin R(X)$. In other words, X is locally compact at y. Fix an open set U in X and a compactum $\Phi \subset X$ satisfying $y \in U \subset \Phi$. Then $[U]_{bX} \subset [\Phi]_{bX} = \Phi \subset X$, and hence $\widetilde{U} = \operatorname{Int}[U]_{bX}$ is an open set in bX, containing y and included in X. Thus, $\widetilde{U} \cap (bX \setminus X) = \Lambda$ and $y \notin [bX \setminus X]_{bX}$.

44. *Necessity.* Put $X^* = bX \setminus X$ and $bX^* = [X^*]_{bX}$, $X^{**} = bX^* \setminus X^*$. bX^* may be regarded as a Hausdorff compactification of X^*. Since X^* is pointwise countable, $X^{**} = bX^* \setminus X^*$ is Lindelöf by Problem 42. Obviously, however, $X = X^{**} \cup (bX \setminus [X^*]_{bX})$. By virtue of Problem 43, $bX \setminus [X^*]_{bX}$ is the set of all points at which X is locally compact.

Sufficiency. This follows from Problem 42.

45. Consider the Stone-Čech compactification βX of X and an arbitrary compactum Φ included in $\beta X \setminus X$. Further, consider the continuous partition of the space βX, whose unique non-trivial element is Φ. This decomposition space is compact (see Problem 92 of Chap. 3), and can be naturally regarded as a Hausdorff compactification bX of X. Let $f: \beta X \to bX$ be the natural mapping of βX onto bX. Then $\Phi = f^{-1}y^*$ for some point $y^* \in bX \setminus X$. By assumption, bX satisfies condition (*), i.e., there is a compactum $\Phi^* \subset bX \setminus X$ such that $y^* \in \Phi^*$ and $\chi(\Phi^*, bX) < \aleph_0$. Put $F = f^{-1}\Phi^*$, to obtain $\Phi \subset F \subset \beta X \setminus X$ (see Problem 27) and since f is a continuous closed mapping (see Problem 92, Chap. 3), $\chi(F, \beta X) \leqslant \chi(\Phi^*, bX) \leqslant \aleph_0$ (see Problem 324, Chap. 2). But then (see Problem 46) X is Lindelöf.

46. *Necessity.* Fix for each point $x \in X$ a nhood Ox in the space bX such that $[Ox] \cap \Phi = \Lambda$. The family $\gamma = \{Ox : x \in X\}$ covers X and $Ox \cap X$ is an open set in X. Since X is Lindelöf, there is a countable family $\lambda \subset \gamma$, satisfying $\bigcup \{U: U \in \lambda\} \supset X$. Put $F = \bigcap \{bX \setminus [U] : U \in \lambda\}$. We may assume that F is closed in bX. Then all that has been stated implies that $\Phi \subset F \subset bX \setminus X$. Since bX is a compactum and $|\lambda| \leqslant \aleph_0$, we have $\chi(F, bX) \leqslant \aleph_0$ (see Prob. 68, Chap. 3), and F is the required compactum.

Sufficiency. Suppose that γ is an arbitrary open cover of X. For each $U \in \gamma$, let $O(U)$ be an open set in bX such that $O(U) \cap X = U$. By hypothesis, for the compactum $\Phi = bX \setminus \bigcup \{O(U): U \in \gamma\}$ there is a compactum $F \subset bX \setminus X$, satisfying $\Phi \subset F$ and $\chi(F, bX) < \aleph_0$. Hence $bX \setminus F$ in bX is F_σ and, consequently, Lindelöf (Problem 208, Chap. 3) and $X \subset bX \setminus F$, $\{O(U): U \in \gamma\}$ is an open cover of $bX \setminus F$, which has a countable subcover λ. Then $\{V \cap X: V \in \lambda\}$ is a countable subcover of the cover γ of X.

47. It suffices to amalgamate the statements of Problems 46 and 45.

48. Fix a countable dense set Q in Φ and a mapping f of N onto Q. The mapping f is continuous and hence extends to a continuous mapping $\tilde{f}: \beta N \to \Phi$ (Problem 31).

Since Φ is a Hausdorff space and βN is compact, $\tilde{f}$ is closed. Therefore, the relations $Q \subset \tilde{f}(\beta N) \subset \Phi$, $[Q] = \Phi$ also imply that $\tilde{f}(\beta N) = \Phi$.

49. Fix for each irrational number m a sequence of rational numbers converging to it in the usual topology; let $\varphi(m)$ be the set of members of this sequence. Then $|\varphi(m) \cap \varphi(m')| < \aleph_0$ for $m \neq m'$ and $|\varphi(m)| = \aleph_0$, $\varphi(m) \subset R$ for all m. Since the cardinality of the set of all irrational numbers is $2^{\aleph_0}$, $\mathcal{E} = \{\varphi(m): m$ is an irrational number$\}$ is the desired family.

50. This follows from Problem 49.

51. (i) We have $[A]_{\beta N} \cup [N \setminus A]_{\beta N} = [A \cup (N \setminus A)]_{\beta N} = [N]_{\beta N} = \beta N$. Since N is normal and A and $N \setminus A$ are disjoint closed sets in N, $[A]_{\beta N} \cap [N \setminus A]_{\beta N} = \Lambda$ (Problem 16). Consequently, $[N]_{\beta N} = \beta N \setminus [N \setminus A]_{\beta N}$ is an open set in βN.

(ii) This is easy to prove via Problems 16, 17 and 5.

(iii) $[A]_{\beta N} \not\subset N$, because $[A]_{\beta N}$ is an infinite compactum, whereas N is not.

52. This follows trivially from Problems 50 and 51.

53. The family $\mathcal{B} = \{[M]\}_{\beta N} : M \subset N\}$ is a base for a space βN by the definition of the topology of ωN, the Wallman compactification of a space N, and the fact that $\beta N = \omega N$ (we have substituted the sign of equality for the equivalence sign, which is clearly admissible). But $|\mathcal{B}| = 2^{\aleph_0}$. On the other hand, βN has no base of cardinality less than $2^{\aleph_0}$, because in $\beta N \setminus N$ there is a disjoint family of nonempty open sets of cardinality $2^{\aleph_0}$ (Problem 52). Hence the weight of βN is $2^{\aleph_0}$. It is not quite easy to estimate the cardinality of βN. From $\beta N = \omega N$ (see Problems 3 and 17) and the definition of ωN it follows that $|\beta N| \leqslant 2^{\mathfrak{c}}$. But βN can be continuously mapped onto any separable compactum, in particular $D^{\mathfrak{c}}$, the product of continuum (discrete) many double points (Problem 381 of Chap. 2, Problem 48 of present chapter). Hence $|\beta N| \geqslant |D^{\mathfrak{c}}| = 2^{\mathfrak{c}} = 2^{2^{\aleph_0}}$ (see Problem 375, Chap. 2). Consequently, $|\beta N| = 2^{\mathfrak{c}}$.

54. In βN, the family of all sets of the form $[M]_{\beta N}$, where $M \subset N$, forms a base. This follows, for example, from the fact that the Wallman compactification ωN of N coincides with its Stone-Čech compactification and the definition of the topology of ωN (Problems 3 and 17). Hence the family $\mathcal{B} = \{[M]_{\beta N} \cap (\beta N \setminus N): M \subset N$ and $|M| = \aleph_0\}$ is a base for the space $X = \beta N \setminus N$. Put $[M]_{\beta N} \cap (\beta N \setminus N) = \overline{M}$. Since $\cap\{U: U \in \gamma\} \neq \Lambda$, it is possible to construct a family $\lambda = \{\overline{M}_n: n = 1, 2, \ldots\} \subset \mathcal{B}$, satisfying $\Lambda \neq \cap\{\overline{M}_n: n \in N^+\} \subset \cap\{U: U \in \gamma\} \subset \overline{M}_{n+1} \subset \overline{M}_n$. Put $\Phi = \cap\{\overline{M}_n : n \in N^+\}$. It suffices to show that Φ contains a nonempty open set in X. From $\overline{M}_{n+1} \subset \overline{M}_n$ it follows (Problem 51) that $M_{n+1} \setminus M_n$ is a finite (possibly empty) set. But then $M'_{n+1} = M_{n+1} \cap M_n$ implies $\overline{M}'_{n+1} = \overline{M}_{n+1}$ (Problem 51). Hence, we may assume that $M_{n+1} \subset M_n$ for all n. Since $|M_n| = \aleph_0$, $n = 1, 2, \ldots$, it is routine to construct by induction a sequence of elements i_n of N such that: (a) $i_n \in M_n$ and (b) $i_{n+1} \notin \{i_1, \ldots, i_n\}$. Put $M^* = \{i_n: n = 1, 2, \ldots\}$. Then $|M^*| = \aleph_0$ and $|M^* \setminus M_n| < \aleph_0$ for all $n = 1, 2, \ldots$ Hence $\overline{M}^* \neq \Lambda$ and $\overline{M}^* \subset \overline{M}_n$ for all $n = 1, 2, \ldots$, i.e., $\overline{M}^* \subset \cap\{\overline{M}_n: n = 1, 2, \ldots\} \subset \cap\{U: U \in \gamma\}$. The set $\overline{M}^* = [M^*]_{\beta N} \cap (\beta N \setminus N)$ is open in X (Problem 51).

55. The weight of X is $2^{\aleph_0}$, i.e., $\aleph_1$. Let $(T, <)$ be a minimally well-ordered set of cardinality $\aleph_1$ and fix a base $\mathcal{B}$ of X, whose elements are set in one-to-one correspondence with the elements of a set T: $\mathcal{B} = \{U_t: t \in T\}$. Define now by induction a decreasing sequence $\{V_t: t \in T\}$ of open sets in X. Put $V_0 = X$ (where 0 is the first element of T). If $t_0 \in T$ and for all $t \in T$, where $t < t_0$, the nonempty open set V_t in X is defined, then put $W_{t_0} = \bigcap \{V_t : t \in T, t < t_0\}$. If $W_{t_0} \neq \Delta$, then choose a nonempty open set V'_{t_0} whose closure is contained in W_{t_0}, which is possible by Problem 54. Furthermore, if $V'_{t_0} \cap U_{t_0} \neq \Delta$, then put $V_{t_0} = V'_{t_0} \cap U_{t_0}$. Otherwise, set $V_{t_0} = V'_{t_0}$. Suppose that V_t is defined for all $t \in T$. Note that $[V_{t'}] \subset V_{t''}$ for $t'' < t'$ and $V_t \neq \Delta$ for all $t \in T$. Hence $P = \bigcap\{[V_t]: t \in T\} = \bigcap\{V_t: t \in T\} \neq \Delta$. Fix $x^* \in P$, we show that the family $Q = \{V_t: t \in T\}$ forms a base at the point x^*.

Consider an arbitrary nhood Ox^* of x^*. There is (since $\mathscr{B}$ is a base for X) a $t^* \in T$ such that $x^* \in U_{t^*} \subset Ox^*$. From $W_{t^*} \supset V_{t^*} \supset P \ni x^*$ it follows that

$$U_{t^*} \cap W_{t^*} \neq \Delta.$$

Hence $V_{t^*} \subset U_{t^*}$ (by the definition of V_{t^*}). Thus $x^* \in V_{t^*} \subset U_{t^*} \subset Ox^*$, i.e., Q is a base at the point x^*. If R is a countable set of nhoods of x^* in X, then for each $G \in R$ fix $t(G)$ so that $x^* \in V_{t(G)} \subset G$ and choose $\tilde{t} \in T$ to satisfy $t(G) < \tilde{t}$ for all $G \in R$ (possible, since R is countable). Then $\bigcap\{G: G \in R\} \supset \bigcap\{V_{t(G)}: G \in R\} \supset V_{\tilde{t}} \ni x^*$, as required, i.e., x^* is the desired point.

56. There is a seperable compactum Φ, no subspace Z of which is normal (Problem 119, Chap. 3). In view of Problem 48 there is a continuous mapping $f: \beta N \to \Phi$, $f(\beta N) = \Phi$. As a continuous mapping between compacta, f is a perfect mapping. Then βN cannot be hereditarily normal, because then Φ would also be hereditarily normal. The latter is easy to derive by using Problems 319 and 329 of Chap. 2.

57. Assume the contrary. Then, in $Y \setminus \{p\}$ there is a sequence $\{y_n: n \in N\}$ of points converging to p. Now construct by induction a sequence $\xi = \{U_n: n \in N\}$ of nonempty mutually disjoint open sets in Y such that: (a) $[U_n] \not\ni p$ for all $n \in N$ and (b) for any subsequence $\eta = \{U_{nk} : k \in N\}$ of ξ, we have $[\bigcup\{U_{nk} : k \ni N\}] \ni p$. First define for each open set G in Y containing p, and every $n \in N$, a nonempty open set $\varphi(G, n)$ and a number $k(G, n) \in N$ with the properties: (i) $\varphi(G, n) \subset G$; (ii) $[\varphi(G, n)] \not\ni p$; (iii) $k(G, n) > n$; (iv) $\varphi(G, n) \ni y_{k(G, n)}$. To show how this is done, observe that there exist $n', n'' \in N$ such that $n', n'' > n$, $y_{n'} \neq y_{n''}$ and $y_{n'}, y_{n''} \in G$. Choose for the points $y_{n'}$ and $y_{n''}$ non-disjoint nhoods V' and V'' in G. Put $P' = V' \cap N$ and $P'' = V'' \cap N$. Then $P' \cap P'' = \Delta$; since N is dense in Y, $[P'] = [V']$, $[P''] = [V'']$. But, from $P' \cap P'' = \Delta$ and the fact that $N \cup \{p\}$ is a subspace of βN it follows that either $p \notin [P'']$ or $p \notin [P']$. Let $p \notin [P']$. Then $p \notin [V']$, and we write

$$k(G, n) = n', \ \varphi(G, n) = V'.$$

Now take U_0 as any nhood of the point y_0 whose closure does not contain the point p. Let $l \in N$ and suppose that for all $n \leqslant l$ the sets U_n are so defined that condition (a) is satisfied. Then put $G = Y \setminus \bigcup\{[U_n]: n \leqslant l\}$ and $U_{l+1} = \varphi(G, l)$. By (ii), $[U_{l+1}] \not\ni p$. Thus the sets U_i are defined for all $i \in N$. They are nonempty, open and mutually disjoint, and clearly satisfy condition (a). The construction implies that for each $k \in N$ there is a $n_k \in N$, $n_k > n$, such that $U_k \ni y_{nk}$. Since every subsequence of the sequence $\{y_{n_k}: k \in N\}$ converges to p, condition (b) is also satisfied.

Set $A_1 = (\bigcup\{U_i: i \in N, \ i \text{ is even}\}) \cap N$ and $A_2 = (\bigcup\{U_i : i \in N, \ i \text{ is odd}\}) \cap N$. Then, obviously, $A_1 \cap A_2 = \Delta$, but $[A_1] \ni p$ and $[A_2] \ni p$, a contradiction.

58. From Problems 15, 21 and 23 it follows that $\beta N = \beta Y$ is the unique compactification of the space Y. Then Y is locally compact and pseudo-compact (see Problem 26). Establish the countable compactness of Y by recalling its definition.

59. Put $Y = X \setminus N$. For each point $y \in Y$ fix a compactum $\Phi(y)$ in X such that $y \in [\Phi(y) \cap N]$ (obviously, $y \in [N] \setminus N$ for all $y \in Y$, so this is plausible). Set $V(y) = [\Phi(y) \cap N]$. By the definition of the topology in βN, $V(y)$ is a clopen compact subspace of βN (Problem 51). In addition, $V(y) = [\Phi(y) \cap N] \subset \Phi(y) \subset X$. We have $X = Y \cup N = \bigcup\{V(y): y \in Y\} \cup N$. All sets appearing on the right-hand side are open in βN; consequently, X is also open in βN.

60. In $\beta N \setminus N$ choose an arbitrary countably infinite subset M. It is not closed in βN (Problem 36). Let x^* be a limit point of M in $\beta N \setminus M$. Put $Q = \bigcap\{\beta N \setminus \{x\}: x \in M\}$. Then Q has the Čech property by definition, and $Q \supset N$ because $M \cap N = \Delta$. Obviously, $Q \ni x^*$ and Q is dense in βN. The space Q is not locally compact, because otherwise it would be open in βN (see Problem 179, Chap. 3) and would therefore contain a nhood of x^* and hence some point of M, which is impossible because $M \cap Q = \Delta$.

61. The set N is contained in X and has no limit point there. Consequently, X is not countably compact (see Problem 189, Chap. 3). Assume that X is not pseudo-compact. Then there is an unbounded continuous real-valued function f on X. Since $R \setminus N$ is dense in X, the set $f(R \setminus N)$ is a non-compact subspace of R. Hence there is an infinite set $A \subset f(R \setminus N)$, having no limit point in R. For each $a \in A$ fix $x(a) \in R \setminus N$ such that $f(x(a)) = a$. Then $P = \{x(a): a \in A\}$ is an infinite set, contained in X and having no limit point there; the latter is implied by the continuity of f and the properties of A. We have $A \subset R \setminus N$, and A and N are closed sets in R. Consequently (Problem 16), $[A]_{\beta R} \cap [N]_{\beta R} = \Delta$ (since R is a normal space). Hence $[A]_{\beta R} \subset X$. But A is closed in X. Hence $A = [A]_{\beta X}$, but this contradicts the fact that A is an infinite discrete subspace of the space X. Hence X is pseudo-compact.

62. Considering the elements of N^+ in order, we decide for each $n \in N^+$ whether to assign it to N^* or N^{**}. Put $1 \in N^*$ and $E_1 \setminus \{1\} \subset N^{**}$. Let $i > 1$; suppose that for all $j < i$ it is already settled which of the relations $j \in N^*$ or $j \in N^{**}$ holds and that all elements of the set $\bigcup\{E_j : j < i\}$ are distributed in N^* and N^{**}. We determine whether $i \in N^*$ or $i \in N^{**}$ and also give the corresponding rule with respect to the elements of the set E_i as follows. In the first place, it may happen that $i \in \bigcup\{E_j: j < i\}$. Two cases then arise : (i) $i \in N^{**}$. (ii) $i \in N^*$. In the former case, assume that if $n \in E_i$ and $n > i$, then $n \in N^*$ (obviously, for $n \in E_i$, when $n > i$ the question of the membership in N^* or N^{**} has not already been resolved). In case (ii), we agree that if $n > i$ and $n \in E_i$, then $n \in N^{**}$. If however $i \notin \bigcup\{E_j : j < i\}$, then we agree that $i \in N^*$, and for all $n \in E_i$, greater than i, as in case (ii), we set $n \in N^{**}$. Obviously, the above procedure defines a partition of the set N^+ into two infinite disjoint sets with the requisite property—the latter trivially follows from the definitions of N^* and N^{**}.

63. Suppose that there is a homeomorphism of X onto Y. The image of the point x^* under this homeomorphism is surely x^*, since x^* is the unique non-isolated point in both spaces. Let us denote by x_n the image of the elements $n \in N = \{0, 1, 2, \ldots\}$ under this homeomorphism. Thus $M = \{x_n: n = 0, 1, 2 \ldots\}$. Now fix a family of mutually disjoint nhoods Ox_n of the points x_n of the set M in the space βN, taken from a standard base for the topology of βN. Thus $Ox_n = [N_n]_{\beta N}$, where N_n is some infinite subset of the natural numbers, with $N_{n'} \cap N_{n''} = \Lambda$ for $n' \neq n''$.

We use the result of Problem 62. There exist disjoint infinite sets $N^* \subset N$, $N^{**} \subset N$, satisfying $N^* \cup N^{**} = N$ and $|N^* \cap N_n| < \aleph_0$ if $n \in N^*$, $|N^{**} \cap N_n| < \aleph_0$ if $n \in N^{**}$. Then either $[N^*] \ni x^*$, or $[N^{**}] \ni x^*$. For definiteness, let $[N^*] \ni x^*$. Then, by the choice of the set M, we obtain $[\{x_n : n \in N^*\}] \ni x^*$. But $[N^*]$ is an open set in βN (Problem 51), i.e., it is a nhood of the point x^*. Consequently, there is $n^* \in N^*$ such that $x_{n^*} \in [N^*]_{\beta N}$. On the other hand, $(x_{n^*}) \in O(x_{n^*}) = [N_{n^*}]_{\beta N}$. Consequently, $[N^*]_{\beta N} \cap [N_{n^*}]_{\beta N} \cap (\beta N \setminus N) \neq \Lambda$. Hence $|(N_{n^*}) \cap N^*| = \aleph_0$ (Problem 51), but this contradicts the fact that $n^* \in N^*$.

64. There exists a continuous mapping f: $\beta N \to D^{\mathfrak{c}}$ onto the Cantor discontinuum $D^{\mathfrak{c}}$ of weight $\mathfrak{c}$, because $D^{\mathfrak{c}}$ is separable (see Problems 381 of Chap. 2 and 48 of the present chapter). But the tightness of any dyadic compactum is equal to its weight (see Arkhangel'skii and Ponomarev [1]). Hence the tightness of $D^{\mathfrak{c}}$ is $2^{\aleph_0}$, and the tightness under a continuous closed mapping does not increase. On the other hand, obviously, $t(\beta N) \leqslant 2^{\aleph_0}$.

64. See Efimov [2], [3].

67. This follows easily from Problems 65 and 66. See also Bereznitskii [1].

68. This follows easily from the fact that $C_f(N)$ and $C_{f'}(N)$ are homeomorphic, if f and f' have the same number of fixed points or if f and f' coincide everywhere, except, possibly, at a finite set of points.

69. Suppose that φ and ψ are continuous real-valued functions on X and V respectively, satisfying $\varphi(P) = \{1\}$, $\varphi(X \setminus V) = \{0\}$, $\psi(P^*) = \{1\}$, $\psi(V \setminus V^*) = \{0\}$, $0 \leqslant \psi(x) \leqslant 1$ for all $x \in V$. Put $f(x) = \psi(x) \cdot \varphi(x)$ if $x \in V$, and $f(x) = \{0\}$ if $x \in X \setminus V$. The reader should verify that f is continuous on X and $f(P^*) = \{1\}$, $f(X \setminus V^*) = \{0\}$.

70. This is immediate from the complete regularity of bX.

71. By the definition of α-filter, we find a $V \in \beta$, for which there exists a continuous bounded real-valued function f on X such that $f(X \setminus U) = \{0\}$ and $f(V) = \{1\}$. It remains to refer to Problem 15.

72. Since βX is completely regular, it is enough to prove that β_z is maximal. Assume that there is an α-filter β, strictly containing β_z. Consider any $U \in \beta \setminus \beta_z$. Choose $V \in \beta$ for which there exists a continuous bounded real-valued function on X such that $f(X \setminus U) = \{1\}$ and $f(V) = \{0\}$. Every nhood of the point z intersects $X \setminus U$; otherwise, $U \in \beta_z$, contradicting $U \in \beta \setminus \beta_z$. Hence $z \in [X \setminus U)_{\beta X}$. But $[V]_{\beta X} \cap [X \setminus U]_{\beta X} = \Lambda$ (see Problem 15). Consequently, $z \notin [V]_{\beta X}$. Then $Oz = \beta X \setminus [V]_{\beta X}$ is a nhood of the point z in βX, and hence $Oz \cap X \in \beta_z \subset \beta$. But obviously $(Oz \cap X) \cap V = \Lambda$, contradicting the fact that β is a filter.

73. The set $\Phi = \cap\{[G]_{\beta X}: G \in \beta\} \neq \Lambda$, because $\{[G]_{\beta X}: G \in \beta\}$ is centralized and consists of closed sets in the compactum βX. Choose $z \in \Phi$. Then, obviously, $\beta_z \cup \beta$ is an α-ultrafilter. But β_z is also an α-ultrafilter (see Problem 72). From $\beta_z \cup \beta \supset \beta_z$ and $\beta_z \cup \beta \supset \beta$ it now follows, by the definition of α-ultrafilter, that $\beta = \beta_z \cup \beta = \beta_z$.

74. By the preceding problem, we have $\beta = \beta_z$, where $\{z\} = \cap\{[U]_{\beta X}: U \in \beta\}$. It is to be shown that $z \in \widetilde{G}$. If $z \notin \widetilde{G}$, then (see Problem 149) $z \in [X \setminus G]_{\beta X}$. From $G \in \beta$ and the fact that β is an α-ultrafilter (see p. 223) it now follows that $[V]_{\beta X}$

$\cap [X\setminus G]_{\beta X} = \Lambda$ for some $V \in \beta$ (see Problem 15). But $\tilde{G} = \beta X\setminus[X\setminus G]_{\beta X}$ (see Problem 149) and $z \in [V]_{\beta X}$. Consequently, $z \in \tilde{G}$. Obviously, if $z \in G$, then $\tilde{G} \in \beta_z$.

75. Assume the contrary. Then there are (see Problem 15) functionally separable closed sets P and Q in X such that $\Lambda \neq]P]_{bX} \cap [Q]_{bX}$. The set $X\setminus P$ (resp. $X\setminus Q$) belongs to the family ξ_Q (resp. ξ_P) of all α-nhoods of Q (resp. P) in X. Fix $z \in [P]_{bX} \cap [Q]_{bX}$. Since ξ_P is an α-family and a filter and since each element of the α-family b_z which also is a filter, intersects each element of ξ_P, $b_z \cup \xi_P$ is also an α-family. But b_z is an α-ultrafilter. Hence $b_z \cup \xi_P = b_z$, i.e., $b_z \supset \xi_P$. Similarly, it can be shown that $b_z \supset \xi_Q$. But functionally separable sets always have disjoint α-nhoods (see Problem 1 of Chap. 3). Hence some element of ξ_P does not intersect some element of ξ_Q, contradicting the fact that b_z is an α-ultrafilter and $b_z \supset \xi_P \cup \xi_Q$.

76. This has already been proved in the solutions of Problems 73 and 75.

77. Consider $\beta_z = \{U \cap X\colon U$ is open in βX and $U \ni z\}$. By Problem 72, β_z is an α-ultrafilter on X. It remains to observe that $Y = (\beta X\setminus F) \cap X$, $z \in \beta X\setminus F$. Consequently, $Y \in \beta_z$. Reference may now be made to Problem 10 of Chap. 3.

78. Statement (i) holds because X is a T_1-space. By Problem 72, every β_z is an α-ultrafilter on X, and from Problem 73 it follows that B is the set of all α-ultrafilters on X, proving (iv). For each open set G in X there is $\tilde{G} \in \mathcal{E}$ such that $\tilde{G} \cap X = G$ (see Problem 149). Hence (ii) implies (iii). From (i), the definitions of $\mathcal{T}$ and $\mathcal{B}$ and the fact that $\mathcal{E}$ is a base for βX, statement (v) is immediate. Finally, (ii) holds by Problem 74.

80. Every continuous real-valued function f on X is bounded and hence can be extended to a continuous function on βX. By definition, this means that either X is not a Q-space or $X = \beta X$, i.e., X is a compactum.

82. *Necessity.* Suppose that X is a Q-space. Then, there is a continuous real-valued function f on X, which does not extend to the space $X \cup \{x^*\}$. For any $k \in N$, put $X_k^- = \{x \in X\colon |f(x)| < k\}$ and $X_k^+ = \{x \in X\colon |f(x)| \geq k+1\}$. The sets X_k^- and X_k^+ are closed in X and functionally separable (by the function f); hence (see Problem 15) $[X_k^-]_{\beta X} \cap [X_k^+]_{\beta X} = \Lambda$ for every $k \in N$. To show that $x^* \notin [X_k^-]_{\beta X}$, assume the contrary. Then $x^* \notin [X_k^+]_{\beta X}$. Consider the function f' on X, defined as follows: $f'(x) = f(x)$ if $|f(x)| < k+1$; $f'(x) = k+1$ if $f(x) \geq k+1$; $f'(x) = -(k+1)$ if $f(x) \leq -(k+1)$. Obviously, f' is a continuous bounded real-valued function on X. Hence it extends to βX and even to a continuous function $\tilde{f}'$ on $X \cup \{x^*\}$. Put $\tilde{f}(x^*) = \tilde{f}'(x^*)$ and $\tilde{f}(x) = f(x)$ for every $x \in X$. From $x^* \notin [X_k^+]_{\beta X}$ it follows that f' and f coincide on the intersection of some nhood of x^* with X. Hence $\tilde{f}$ is a continuous function on $X \cup \{x^*\}$ extending f, a contradiction. Hence $[X_k^-]_{\beta X} \not\ni x^*$ for every $k \in N$. The set $U_k = \beta X \setminus[X_k^-]_{\beta X}$ is open and contains the point x^*. There is (see Problem 69 of Chap. 3) a compactum Φ_k of countable character in βX, contained in U_k and containing x^*. Then, $F = \cap\{\Phi_k\colon k \in N\}$ is a compactum of countable character in βX and $x^* \in F \subset \cap\{U_k\colon k \in N\} \subset \beta X\setminus X$, the latter inclusion follows from the fact that $\cup\{[X_k^-]_{\beta X}\colon k \in N\} \supset \cup\{X_k^-\colon k \in N\} = X$. Necessity follows.

Sufficiency. Let $F \subset \beta X\setminus X$, $\chi(F, \beta X) = \aleph_0$. Then, there is a continuous real-valued function φ on βX such that $\varphi^{-1}(0) = F$ (see Problem 30 of Chap. 3). Put $f(x) = 1/\varphi(x)$ for every $x \in X$ (note that $\varphi(x) \neq 0$ for $x \in X$, because $F \cap X = \Lambda$). For any point $y \in F$ and every $\varepsilon > 0$, there is a nhood Oy of y in βX such that $|\varphi(x)| < \varepsilon$. Then, $|f(x)|$

$> 1/\varepsilon$ for every $x \in Oy \cap X$. Since we always have $Oy \cap X \neq \Lambda$ and ε is arbitrary, we conclude that it is impossible to extend f to a continuous function on $X \cup \{y\}$, no matter what $y \in F$ we choose. Now let $\tilde{X} = \{x^*\} \cup X$, $[X]_{\tilde{X}} = \tilde{X}$, $x^* \notin X$; $\tilde{X}$ is a completely regular space. Consider the natural mapping g of the Stone-Čech compactification βX of X onto the Stone-Čech compactification $\beta \tilde{X}$ of $\tilde{X}$ (such mapping exists, because $\beta \tilde{X}$ is, obviously, a Hausdorff compactification of the space $X \subset \tilde{X}$). Choose $y^* \in g^{-1}(x^*)$. From $x^* \notin X$ it follows that $y^* \in \beta X \setminus X$ (see Problem 27). As seen above, there is a continuous real-valued function f on X, which does not extend to $X \cup \{y^*\}$. Since the restriction of g to $X \cup \{y^*\}$ is a contraction of the space $X \cup \{y^*\}$ to the space $\tilde{X} = X \cup \{x^*\}$, f has no extension to $\tilde{X}$.

83. See Problems 188, 185 of Chap. 3 and Problem 80 of the present chapter.

84. The collection of all G_δ-sets in βX is a base for the $\aleph_0$-modification of βX. Hence our assertion is immediate from Problem 82 and the following remarks (see Problem 71 of Chap. 3): if A is a G_δ-set in a compactum F and $x \in A$, then there is a compactum Φ such that $x \in \Phi \subset A$ and $\chi(\Phi, F) \leqslant \aleph_0$.

85. Suppose that X is a Q-space and ξ is an arbitrary free α-ultrafilter on X. Then, by Problem 73, there is a point $z \in \beta X \setminus X$ such that $\xi = \beta_z$. Since X is a Q-space, there is a countable family $\lambda = \{U_n: n \in N^+\}$ of open sets in βX such that $z \in \bigcap \{U_n: n \in N^+\} \subset \beta X \setminus X$ (Problem 82). Then, $\eta = \{U_n \cap X: n \in N^+\} \subset \xi$, $|\eta| = \aleph_0$ and $\bigcap \{V: V \in \eta\} = \Lambda$, proving the necessity of the condition.

Conversely, if y is any point of $\beta X \setminus X$, then β_y is a free α-ultrafilter on X (see Problem 72). By hypothesis, it is $\aleph_0$-prime, hence there is a countable family Θ of nhoods of y in βX such that $\bigcap \{U \cap X: U \in \Theta\} = \Lambda$. But, then, $y \in \bigcap \{U: U \in \Theta\} \subset \beta X \setminus X$, proving (see Problem 84) the sufficiency of the condition.

86. If X is discrete, then every α-ultrafilter on X is an ultrafilter on X and, conversely, every ultrafilter on X is an α-ultrafilter on X. The condition that an α-ultrafilter be $\aleph_0$-prime is here identical to the requirement that the ultrafilter not be an σ-ultrafilter (see pp. 225, 20). It now remains to compare the statement of Problem 85 with the definition of a non-measurable cardinal (see p. 225).

87. This is immediate from Problem 82 and the following trivial statement: if $F \subset Y \subset X$ and $\chi(F, X) = \aleph_0$, then $\chi(F, Y) \leqslant \aleph_0$.

88. Necessity follows from Problem 82: if X is a Q-space, then βX has the property (*). It is now to be shown that if some Hausdorff compactification bX of X has the property (*), then the Stone-Čech compactification βX of this space also has this property. Consider the natural mapping $f: \beta X \to bX$ (see Problem 22). Fix $x^* \in \beta X \setminus X$ and put $y^* = f(x^*)$. Choose a compactum $\Phi \subset bX \setminus X$, satisfying $y^* \in \Phi$ and $\chi(\Phi, bX) = \aleph_0$, and put $F = f^{-1}\Phi$. Then F is a compactum, and from $f(\beta X \setminus X) \subset bX \setminus X$ (see Problem 27) it follows that $F \subset \beta X \setminus X$. Since f is a closed mapping, from Problem 324 of Chap. 2 it follows that $\chi(F, \beta X) = \aleph_0$. In addition, $x^* \in F$. Thus, βX has property (*). Hence (Problem 82) X is a Q-space.

89. Necessity follows from Problem 84. Sufficiency follows from Problem 88 and the fact that a G_δ-set in a compactum is always representable as the union of some set of compacta of countable character in this compactum (see Problem 71 of Chap. 3).

91. This follows from Problems 46 and 88.

92. This follows directly from Problem 88 (the desired compacta of countable character are one-point sets),

93. Every point of X satisfies the first axiom of countability in the space bX. Hence $[Y]_{bX}$ is a Hausdorff compactification of Y, satisfying condition (*) of Problem 88. Consequently (Problem 88), Y is a Q-space (see also Problem 89).

94. Since X has no point of local compactness, $bX \setminus X$ is dense in bX (see Problem 43). Hence if $F \subset bX \setminus X$ and F is a compactum of countable character in $bX \setminus X$, then we have also $\chi(F, bX) = \aleph_0$. Since $bX \setminus X$ is pointwise countable, it follows that bX has the property (*) with respect to X (see Problem 88). Consequently (Problem 88), X is a Q-space.

95. Set $bY = [Y]_{bX}$. Then $bY \setminus Y$ is a closed subspace of X. Fix an arbitrary $z \in bY \setminus Y$. Since $z \in X$, there is a compactum $\Phi \subset X$ such that $\chi(\Phi, X) \leqslant \aleph_0$. From $[X]_{bX} = bX$ it follows that $\chi(\Phi, bX) \leqslant \aleph_0$, too. Put $F = (bY \setminus Y) \cap \Phi$. Then, F is a compactum of countable character in bY, where $z \in F \subset bY \setminus Y$ (we know that F is a closed G_δ-set in the compactum bY). Hence, bY has the property (*) and Y is a Q-space (see Problem 88).

96. Use Problem 88.

97. For every $\alpha \in A$ fix a Hausdorff compactification $b_\alpha X_\alpha$ of the space X_α, having the property (*) in Problem 88; for example, $b_\alpha X_\alpha$ can be taken as βX_α, the Stone-Čech compactification of X_α. The Tychonoff product $\Pi\{b_\alpha X_\alpha : \alpha \in A\}$ is a Hausdorff compactification (denote it by bX) of the Tychonoff product $X = \Pi\{X_\alpha : \alpha \in A\}$. From Problem 362 of Chap. 2 or Problem 77 of Chap. 3 it follows that bX has the property (*) with respect to X, i.e., it is a Q-extension of the space X. Hence (Problem 88) X is a Q-space.

98. The real line is a Q-space (Problem 90), the product of Q-spaces is a Q-space (Problem 97) and a closed subspace of a Q-space is a Q-space (Problem 96). Sufficiency is thus immediate. Let us prove the necessity. By Problem 82, if X is a Q-space, then in the Stone-Čech compactification βX of X there is a family $\gamma = \{\Phi_\alpha : \alpha \in A\}$ of compacta such that $\chi(\Phi_\alpha, \beta X) \leqslant \aleph_0$ and $\bigcup\{\Phi_\alpha : \alpha \in A\} = \beta X \setminus X$. From Problem 79 of Chap. 3 it now follows that X is homeomorphic to a closed subspace of the product $|A| \cdot \tau$ of copies of the real line, where τ is the weight of X.

99. If $F \in \mathcal{E}$, then $F \cap X = \Lambda$. Hence $\bar{\mathcal{E}} \cap X = \Lambda$ and $X \subset vX$, i.e., the relation (0) holds. From (0) it follows that βX is also a Hausdorff compactification of the space vX. But the growth $\beta X \setminus vX = \tilde{\mathcal{E}} = \bigcup\{F: F \in \mathcal{E}\}$ is the union of compacta of countable character in βX. Hence (see Problem 82), vX is a Q-space. This proves (i). Let us establish (ii). Using the fact that βX is the Stone-Čech compactification of Y (this is true, since $X \subset Y \subset \beta X$, see Problem 21), Thus Y is a Q-space. Hence (Problem 82), there is a family $\mathcal{E}'$ of compacta contained in $\beta X \setminus Y$ and having countable character in βX such that $\bigcup\{\Phi : \Phi \subset \mathcal{E}'\} = \beta X \setminus Y$. But then (since $X \subset Y$) $\mathcal{E}' \subset \mathcal{E}$, implying $\beta X \setminus Y \subset \bar{\mathcal{E}}$. Hence $Y \supset \beta X \setminus \bar{\mathcal{E}} = vX$, giving the desired result.

100. We have $X \subset Y \subset vX \subset \beta X$. Hence βX is the Stone-Čech compactification of Y (Problem 21), and vX is the least Q-space contained in $\beta X = \beta Y$ and containing Y (see Problem 99). Consequently (recalling Problem 99), $vY = vX$.

102. Using Problem 291 of Chap. 2, it can be shown that A is homeomorphic to a closed subspace of the product $X \times B$. But, then, since a product of Q-spaces is always a Q-space and a closed subspace of a Q-space is a Q-space, A is also a Q-space.

103. Use Problem 47 of Chap. 6 and Problems 102, 79 of the present chapter.

104. There exists (Problem 31) a continuous mapping $\tilde{f}: \beta X \to \beta Y$ such that $\tilde{f} | X = f$. We have (see Problem 99) $X \subset \nu X \subset \beta X$, $Y \subset \nu Y \subset \beta Y$. Put $Z = \tilde{f}^{-1}(\nu Y)$. Then $X \subset Z \subset \beta X$ and Z is a Q-space (Problems 102, 103). Hence $\nu X \subset Z$ (Problem 99) and, consequently, $\tilde{f}(\nu x) \subset \nu Y$. Hence $\hat{f} = \tilde{f} | \nu X$ is the required mapping.

106. The real line is a Q-space (see, for example, Problem 90). Hence every continuous real-valued function on X extends to a continuous function on νX (Problem 105). Let $y \in \beta X \setminus \nu X$. Since νX is a Q-space, there is a continuous real-valued function g on νX, which cannot be extended to a continuous function on $\nu X \cup \{y\}$. Put $f_0 = g | X$. Then f_0 does not extend to $X \cup \{y\}$ (see Problem 288 of Chap. 2); furthermore, f_0 extends to no space containing $X \cup \{y\}$. This obviously implies that if $X \subset Z \subset \beta X$ and f_0 extends to a continuous function on Z, then $Z \not\ni y$ and hence $Z \subset \nu X$.

107. This follows directly from Problem 106.

108. Clearly (i) implies (ii) and (iii) follows from (ii). It remains to deduce (i) from (iii). Suppose that Y is an arbitrary subspace of X and put $A = X \setminus Y$. We have $Y = \bigcap \{X \setminus (x): x \in A\}$.

By Problem 367 of Chap. 2, Y is homeomorphic to a closed subspace of the Tychonoff product $Z = \Pi\{X \setminus \{x\}: x \in A\}$. But Z is a Q-space, because every $X \setminus \{x\}$ is a Q-space (Problem 97). Hence Y is also a Q-space.

120. The space $\mathcal{P}$ in Problems 100 and 101 of Chap. 5 is such a space: $\mathcal{P}_0$ and $\mathcal{P}_1$ are Q-spaces, because $| \mathcal{P}_0 | = \aleph_0$ and $| \mathcal{P}_1 | = 2^{\aleph_0}$ (see Problem 117).

121. See Mrowka [1] (actually, it is routine to prove this independently).

122. No. Cf. the space $\mathcal{P}$ in Problems 100 and 101 of Chap. 5.

123. False. For example, see Mrowka [1].

131. Suppose that $\sigma = \{U_\alpha: \alpha \in A\}$ is a base for the space X of cardinality $| \sigma | \leqslant wX$ and consisting of clopen sets. Assign to each $\alpha \in A$ an isolated two-point set $D_\alpha = \{0_\alpha, 1_\alpha\}$. Let $x \in X$. Put $f_\alpha x = 0_\alpha$ if $x \in U_\alpha$, and $f_\alpha x = 1_\alpha$ if $x \notin U_\alpha$. Clearly, $f_\alpha: X \to D_\alpha$ is a continuous mapping. Consider the diagonal product $f: X \to D^{|A|}$ of the family of mappings $\{f_\alpha: \alpha \in A\}$. The mapping f is continuous (see Problem 290 of Chap. 2), hence (and in consequence of the compactness of X) the set fX is compact and therefore closed in $D^{|A|}$. Furthermore, $f: X \to fX$ is a closed mapping. To show that f is a homeomorphism, it is necessary to show that f is one-to-one. Let $x_1 \in X$, $x_2 \in X$, $x_1 \neq x_2$. Take $\alpha_0 \in A$ to be such that $x_1 \in U_{\alpha_0}$ and $x_2 \notin U_{\alpha_0}$. Then, $f_{\alpha_0} x_1 \neq f_{\alpha_0} x_2$, so that $fx_1 \neq fx_2$.

140. It is necessary to prove the following assertion: if each of the two compactifications bX and cX follows the other, then they coincide. Suppose that $f: bX \xrightarrow{\text{onto}} cX$ and $g: cX \xrightarrow{\text{onto}} bX$ are natural continuous mappings. Consider the mapping $h = gf$. The assertion now follows from Problem 288 of Chap. 2.

141. (*a*) It is to be shown that all axioms are satisfied.

K-1. Let $F <_\beta H$. Then there exists a continuous function $f: X \to [0, 1]$ such that $fF = \{0\}$, and $f(X \setminus H) = \{1\}$. Consider the function $g(x) = 1 - f(x)$. It is obviously continuous: $0 \leqslant g(x) \leqslant 1$. Furthermore, $g(F) = \{1\}$, and $g(X \setminus H) = \{0\}$. Hence $X \setminus H <_\beta X \setminus F$. The axioms K-2, K-3 and K-6 are obviously fulfilled.

To show that K-4 is satisfied, let $F_1 <_\beta H_1$; $F_2 <_\beta H_2$. This implies that there exist continuous functions $f_1: X \to [0, 1]$; $f_2: X \to [0, 1]$ satisfying $f_1(F_1) = \{0\}, f_1(X \setminus H_1) = \{1\}, f_2(F_2) = \{0\}, f_2(X \setminus H_2) = \{1\}$. Consider the function $f(x) = f_1(x)f_2(x)$ continuous on X. Obviously, $0 \leqslant fx \leqslant 1$ for all $x \in X$. Let $x_0 \in F_1 \cup F_2$. Then, evidently, $f(x_0) = f_1(x_0)f_2(x_0) = 0$. Let $x_0 \in X \setminus (H_1 \cup H_2)$; then $x_0 \in X \setminus H_1$ and $x_0 \in X \setminus H_2$; consequently, $f_1(x_0) = 1$ and $f_2(x_0) = 1$, i.e., $fx_0 = 1$. Thus, $F_1 \cup F_2 <_\beta H_1 \cup H_2$.

To establish axiom K-5, let $F <_\beta H$; $f: X \to [0, 1]$ is a continuous function on X, satisfying $fF = \{0\}$ and $f(X \setminus H) = \{1\}$. Consider the functions

$$f_1(x) = \begin{cases} 1, & \text{if } f(x) = 1, \\ 0, & \text{if } f(x) \leqslant \tfrac{1}{2}, \\ 2(f(x) - \tfrac{1}{2}), & \text{if } \tfrac{1}{2} < f(x) < 1, \end{cases}$$

$$f_2(x) = \begin{cases} 0, & \text{if } f(x) = 0, \\ 1, & \text{if } f(x) \geqslant \tfrac{1}{2}, \\ 2f(x), & \text{if } 0 < fx < \tfrac{1}{2}, \end{cases}$$

continuous on X and a nhood $OF = (x \in X: 0 \leqslant fx < \tfrac{1}{2}\}$ of F. Then $f_1([OF]) = \{0\}$, $f_1(X \setminus H) = \{1\}, f_2(F) = \{0\}, f_2(X \setminus [OF]) = \{1\}$. Consequently, $F <_\beta OF$; $[OF] < H$. The fulfilment of K-7 follows from the complete regularity of the space X.

(*b*) By the 'large Urysohn lemma' (see Problem 25 of Chap. 3), in a normal space X $F <_\beta H$ iff $F \subset H$.

142. (*a*) It is necessary to verify that all subordination axioms (see p. 230) are satisfied.

(*b*) Suppose that (X, ρ) is the plane with the natural metric.

It suffices to verify that the elementary subordination (see p. 231 and Problem 141) in (X, ρ) does not coincide with the subordination ρ. For this, in X we take F_1 (resp. F_2) as the set of all points of the line $y = 0$ (resp. hyperbola $xy = 1$), closed in X. Clearly $F_1 <_\beta (X \setminus F_2)$, but F_1 is not subordinate to the set $X \setminus F_2$ in the sense of the subordination associated with ρ, because $\rho(F_1, F_2) = 0$.

143. Any completely regular space carries a subordination, say, an elementary subordination (see Problem 141). Conversely, suppose that the T_1-space X carries a subordination v, satisfying all axioms K-1 through K-7.

We show that X is completely regular. From axioms K-7, K-5 and K-2 it first follows that X is a regular space. The complete regularity of X is, however, proved along the same lines as Urysohn's large lemma (see Problem 25 of Chap. 3); the requirement is that we constantly make use of axiom K-5 for the subordination v.

144. In the first place, a compactum X carries subordination, (say) an elementary subordination (see Problem 141). In order to show that it is unique, suppose that v is any subordination. By axiom K-2, $F <_v H$ implies $F \subset H$. It is required to prove the converse of this assertion. Thus, let $F \subset H$. By axiom K-7, for each point $x \in F$, we certainly have $\{x\} <_v H$; using axiom K-5, take a nhood Ux of x, satisfying $[Ux] <_v H$. Consider an open cover $\omega_F = \{Ux: x \in F\}$ of a closed set F in the compactum X.

Select from this cover a finite subcover $\omega_F^0 = \{Ux_1, \ldots, Ux_s\}$. We have $[Ux_i] <_v H$, $i = 1, 2, \ldots, s$. Then

$$F \subset \bigcup_{i=1}^{s} Ux_i,$$

but by axiom K-4 we obtain

$$[\bigcup_{i=1}^{s} Ux_i] <_v H;$$

consequently (by axiom K-3), $F <_v H$, completing the proof.

146. (*a*) The proof of the fact that every v-regular family of open sets of space X with a subordination v is contained in some v-end is a routine exercise using the maximum principle.

(*b*) Suppose that ξ is a v-end in (X, v).

(i) Consider the family $\bar{\xi} = \xi\{H' \cap H'': H' \in \xi, H'' \in \xi\}$. It is easily seen that $\bar{\xi}$ is centralized. Let $H_0 \in \bar{\xi}$. If $H_0 \in \xi$, then since the family ξ is v-regular, there is $H_1 \in \xi \subset \bar{\xi}$ such that $[H_1] <_v H_0$. If $H_0 = H_1 \cap H_2$, $H_1 \in \xi$, $H_2 \in \xi$, then we take $H_1' \in \xi$ and $H_2' \in \xi$ to be such that $[H_1'] <_v H_1$, $[H_2'] <_v H_2$. This gives (see the subordination axioms on p. 230)

$$[H_1' \cap H_2'] = [H_1'] \cap [H_2'] <_v H_1 \cap H_2 = H_0 \text{ and } H_1' \cap H_2' \in \bar{\xi}.$$

Thus, the family $\bar{\xi}$ is v-regular. Then, since ξ is a maximal v-regular family, i.e., it is a v-end, we obtain $\xi = \bar{\xi}$. Hence, $H_1 \cap H_2 \in \xi$.

(ii) Enlarge our v-end ξ to the family $\bar{\xi}$ by adding all open sets in X, containing at least one $H \in \xi$. It is routine to verify that $\bar{\xi}$ is a v-regular centralized family. Hence $\bar{\xi} = \xi$, so that $H \in \xi$, if $H \supset H_1$ and $H_1 \in \xi$.

147. (*a*) Let $\xi \in O_{H_1} \cap O_{H_2}$. Then $H_1, H_2 \in \xi$. Since ξ is a v-end, $H_1 \cap H_2 \in \xi$ (see Problem 146); consequently, $\xi \in O_{H_1 \cap H_2}$. Conversely, let $\xi \in O_{H_1 \cap H_2}$. This implies that $H_1 \cap H_2 \in \xi$. But $H_1 \supset H_1 \cap H_2$ and $H_2 \supset H_1 \cap H_2$. Hence, since ξ is a v-end, $H_1, H_2 \in \xi$ (see Problem 146), i.e., $\xi \in O_{H_1} \cap O_{H_2}$.

(*b*) Let $H_1 \subset H_2$ and $\xi \in O_{H_1}$. Then, $H_1 \in \xi$, and since ξ is a v-end (see Problem 146), $H_2 \in \xi$, i.e., $\xi \in O_{H_2}$.

(*c*) This is trivial.

(*d*) Let (x_0) be the family of all open sets in X. containing the point $x_0 \in X$. Obviously, (x_0) is centralized and v-regular (the latter follows from the subordination axioms K-5 and K-7). It is to be shown that (x_0) is a v-end. Assume that (x_0) is not a v-end. Then (x_0) is contained in some v-end ξ (see Problem 146). Since the space X is regular and ξ is centralized, we have $[H] \ni x_0$ for every $H \in \xi$. Take $H \in \xi$ to be such that $H \notin (x_0)$, i.e., $x_0 \notin H$. Use the v-regularity of ξ (axiom K-5), take $H' \in \xi$ to be such that $[H'] <_v H$. But, then, by axiom K-2, $[H'] \subset H$; consequently, $x_0 \notin [H']$, a contradiction.

(*e*) It is straightforward to show that $hx_1 \neq hx_2$, if $x_1, x_2 \in X$, $x_1 \neq x_2$, because X is a T_1-space, Suppose now that H is an arbitrary open set in X; denote by (H) the image of H under the mapping h. Then, it is clear that

$$O_H \cap h(X) = (H),$$

and $h: X \to vX$ is a homeomorphism onto a dense subset $h(X)$ of the space vX. On identifying X with its topological image $h(X)$, the above equality assumes the form

$$O_H \cap X = H,$$

where H is an arbitrary open set in X.

148. First note that the set $X \cap \Gamma$ is dense in Γ, where Γ is open in vX. This implies that X is dense in vX (see Problem 147). In particular, $X \cap O_H = H$ is dense in O_H. It is to be shown that the equality $[O_{X \cap \Gamma}]_{vX} = [\Gamma]_{vX}$ holds. Since $X \cap \Gamma$ is dense in $O_{X \cap \Gamma}$, $[X \cap \Gamma]_{vX} = [O_{X \cap \Gamma}]_{vX}$, and since $X \cap \Gamma$ is dense in Γ, $[X \cap \Gamma]_{vX} = [\Gamma]_{vX}$. Hence the equality holds. Finally, this equality obviously implies the inclusion $O_{X \cap \Gamma} \subset [\Gamma]_{vX}$.

149. (*a*) This is obvious.

(*b*) If $O(H)$ is the union of all open sets in X^* whose intersection with X is H, then $X^* \setminus O(H)$ is the intersection of all closed subsets in X^*, whose intersection with X is the set $X \setminus H$, i.e., $X^* \setminus O(H) = [X \setminus H]_{X^*}$. The second equality in (*b*) follows from the first equality.

(*c*) This is obvious.

(*d*) Let $H_1 \subset H_2$. Then, by considering in X^* the open set $\Gamma = O(H_1) \cup O(H_2)$, we have $\Gamma \cap X = H_2$. Hence, by the definition of the set $O(H_2)$, we obtain $\Gamma = O(H_2)$. It is then clear that $O(H_1) \subset O(H_2)$. The converse statement is obvious.

(*e*) This follows from the statement (*d*); it is necessary to use the statement (*b*).

(*f*) $O(H_1) \cap O(H_2) \subset O(H_1 \cap H_2)$, because $O(H_1) \cap O(H_2) \cap X = H_1 \cap H_2$, where $O(H_1 \cap H_2)$ is a maximal open set in X^*, satisfying $O(H_1 \cap H_2) \cap X = H_1 \cap H_2$. The inclusion $O(H_1 \cap H_2) \subset O(H_1) \cap O(H_2)$ follows from the statement (*d*) and from the obvious inclusions $H_1 \supset H_1 \cap H_2$, $H_2 \supset H_1 \cap H_2$.

(*g*) It is required to prove the equality $O(H) = O_H$ for any open set H in X, when $X^* = vX$. The inclusion $O_H \subset O(H)$ follows from the fact that $O_H \cap X = H$ (see Problem 147), and $O(H)$ is the maximal open set in vX, whose intersection with X is H. To establish the converse inclusion $O(H) \subset O_H$, suppose that a given v-end ξ belongs to $O(H)$. Since the family of sets of the form O_H (H is open in X) forms a base for the topology in vX, there is an $H' \in \xi$ such that $\xi \in O_{H'} \subset O(H)$. But, then, $O_{H'} \cap X \subset O(H) \cap X$, i.e., $H' \subset H$. However, $H' \in \xi$; consequently, $H \in \xi$ (see Problem 146), which implies $\xi \in O_H$.

Further, consider $vX \setminus O_H = vX \setminus O(H) = [X \setminus H]_{vX}$. If $\xi \in vX \setminus O_H$, then $H' \not\subset H$ for any $H' \in \xi$; consequently, $H' \cap (X \setminus H) \neq \Lambda$ for any $H' \in \xi$. If $\xi \in vX$ is such that $H' \cap (X \setminus H) \neq \Lambda$ for any $H' \in \xi$, then $\xi \notin O_H$; consequently, $\xi \in vX \setminus O_H$. Thus, $[F]_{vX} = vX \setminus O(X \setminus F) = vX \setminus O_{X \setminus F} = \Phi_F = \{\xi \in vX: H \cap F \neq \Lambda \text{ for any } H \in \xi\}$.

150. (*a*) Let $\xi \in [O_{H_0}]$. Then, for any $H \in \xi$ we have $O_H \cap O_{H_0} = O_{H \cap H_0} \neq \Lambda$; consequently, $H \cap H_0 \neq \Lambda$ for any $H \in \xi$ (here we use Problem 147 and the fact that the family of sets of the form O_H forms a base in vX). Conversely, let $\xi \in vX$ be such that $H \cap H_0 \neq \Lambda$ for any $H \in \xi$. This implies that for $H \in \xi$ we have $O_H \cap O_{H_0} = O_{H \cap H_0} \neq \Lambda$. But, since $\{O_H: H \in \xi\}$ forms a base at the point ξ in vX, we obtain $\xi \in [O_{H_0}]_{vX}$.

(*b*) Let $F <_v H_0$ in X; it is required to show that $[F]_{vX} \subset O_{H_0}$. Let $\xi \in [F]_{vX}$. But (see Problem 149) $[F]_{vX} = \Phi_F = \{\xi \in vX: H \cap F \neq \Lambda \text{ for any } H \in \xi\}$, hence $H \cap F \neq \Lambda$

for any $H \in \xi$. To show that $\xi \in O_{H_0}$, i.e., $H_0 \in \xi$, consider the family $\bar{\xi}$ consisting of all elements of the v-end ξ and all open sets H' in X such that $F <_v H' \subset [H'] <_v H_0$. It is trivial to show that $\bar{\xi}$ is v-regular and centralized. Since ξ is a v-end and $\bar{\xi} \supset \xi$, $\bar{\xi} = \xi$; hence $H_0 \in \xi$.

(*c*) The regularity of the space vX follows directly from (*b*).

151. (*a*) Let $x_0 \notin F$. Then $\{x_0\} <_v X \setminus F$ (axiom K-7). By axiom K-5, there is a nhood Ux_0 satisfying $[Ux_0] <_v X \setminus F$. Hence (by axiom K-1), $F <_v X \setminus [Ux_0]$. Thus, for every point $x_0 \notin F$, we have a nhood $UF = X \setminus [Ux_0]$ of the set F such that $F <_v UF$ and $x_0 \notin UF$. Hence $F = \bigcap \{UF: F <_v UF\}$.

(*b*) By Problem 150, we have the inclusion $[F]_{vX} \subset \bigcap \{O_{UF}: F < vUF\}$, and by Problem 149 the equality $[F]_{vX} = \{\xi \in vX: H \cap F \neq \Lambda \text{ for any } H \in \xi\}$. If $\xi \in \bigcap \{O_{UF}: F <_v UF\}$, then $UF \in \xi$ for any $F <_v UF$. Let $H \cap F = \Lambda$ for some $H \in \xi$. Take $H' \in \xi$ to be such that $[H'] <_v H$. Then $F \subset X \setminus H <_v X \setminus [H']$ and $X \setminus [H'] \notin \xi$, but this contradicts the fact that $\xi \in \bigcap \{O_{UF}: F <_v UF\}$. Thus, $H \cap F \neq \Lambda$ for any $H \in \xi$. Hence $\xi \in [F]_{vX}$. Thus the second equality is also proved.

152. To show that vX is compact, it suffices to show that every centralized family of sets of the form $\Phi_F = [F]_{vX}$, where F is closed in X, has nonempty intersection (indeed, the sets Φ_F, where F is closed in X, form a closed base in vX, because $vX \setminus \Phi_F = O_{X \setminus F}$, and the sets O_H, where H is open in X, form a base for the topology in vX). Thus, suppose that $\Xi = \{\Phi_{F_t}: t \in T\}$ is an arbitrary centralized family of such sets. For each $t \in T$, consider the family $\sigma_t = \{UF_t: F_t <_v UF_t\}$ of open sets in X. From Problem 151 we obtain the equality

$$F_t = \bigcap \{UF_t: UF_t \in \sigma_t\}. \tag{*}$$

Consider the family $\sigma = \bigcup \{\sigma_t: t \in T\}$.

(i) σ is centralized. Indeed, let $U_1 \in \sigma, \ldots, U_s \in \sigma$. Then $U_1 \in \sigma_{t1}, \ldots, U_s \in \sigma_{ts}$. By the definition of the family $\sigma_{t1}, \ldots, \sigma_{ts}$, and the equality (*) we obtain the inclusions $F_{t1} \subset U_1, \ldots, F_{ts} \subset U_s$. Since

$$\bigcap_{i=1}^{s} F_{t_i} \neq \Lambda$$

hence

$$\bigcap_{i=1}^{s} U_i \neq \Lambda$$

also.

(ii) The family σ is v-regular. Let $U \in \sigma$. Then $U \in \sigma_t$ for some $t \in T$; hence (by the definition of the family σ_t and from equality (*)) $F_t <_v U$. Consequently, there is an open set U' in X such that $F <_v U' \subset (U') <_v U$. Then $U' \in \sigma_t \subset \sigma$ and $[U'] <_v U$.

Thus σ is a v-regular centralized family and hence it is contained in some v-end of ξ. Since (see Problem 151) the equalities

$$[F_t]_{vX} = \Phi_{F_t} = \bigcap \{O_{UF_t}: U_{F_t} \in \sigma_t\} \tag{**}$$

hold, $\xi \in \Phi_{F_t}$ for every $t \in T$, whence $\bigcap \{\Phi_{F_t}: t \in T\} \neq \Lambda$.

153. If $F <_v H$, then $[F]_v X \subset O_H$ (see Problem 150). Conversely, let $[F]_{vX} \subset O_H$. Consider a family of open sets in X: $\sigma(F) = \{U \subset X\colon F <_v U\}$. In consequence of Problem 151, we obtain the equalities: $F = \bigcap \{U\colon U \in \sigma(F)\}$, $[F]_{vX} = \bigcap \{O_U\colon U \in \sigma(F)\}$. Note that if $U \in \sigma(F)$, then there is $U' \in \sigma(F)$ such that $[U']_X <_v U$ (axiom K-5), and if $U_1 \in \sigma(F), \ldots, U_s \in \sigma(F)$, then

$$U_0 = \bigcap_{i=1}^{s} U_i \in \sigma(F)$$

(axioms K-1 and K-4). Since, in addition, $[U']_X <_v U$ implies $[[U']_X]_{vX} = [O_{U'}]_{vX} \subset O_U$ (see Problems 148, 150), hence

$$\bigcap \{O_U\colon U \in \sigma(F)\} = \bigcap \{[O_U]_{vX}\colon U \in \sigma(F)\} = [F]_{vX}.$$

Then since vX is compact, there are $U_1 \in \sigma(F), \ldots, U_s \in \sigma(F)$ such that

$$[F]_{vX} \subset \bigcap_{i=1}^{s} [O_{U_i}]_{vX} \subset O_H.$$

But

$$U_0 = \bigcap_{i=1}^{s} U_i \in \sigma(F)$$

and

$$O_{U_0} = O_{\bigcap_{i=1}^{s} U_i} = \bigcap_{i=1}^{s} O_{U_i};$$

consequently, $F <_v U_0$ and

$$O_{U_0} \subset \bigcap_{i=1}^{s} [O_{U_i}] \subset O_H.$$

Hence $F <_v U_0 \subset H$. So $F <_v H$. The proof is complete.

154. Since v_1 and v_2 are distinct subordinations on the space X, there are a closed set F in X and an open set H such that, for example, $F <_{v_1} H$, but $F \not< _{v_2} H$. Assume that $v_1 X \equiv v_2 X$, i.e., there is a natural homeomorphism between them which enables $v_1 X$ to be identified with $v_2 X$. Since $F <_{v_1} H$, (see Problem 150) $[F]_{v_1 X} \subset O_H^{v_1}$. But since $v_1 X \equiv v_2 X$, $[F]_{v_1 X} = [F]_{v_2 X}$ and $O_H^{v_1} = O_H^{v_2}$; consequently, $[F]_{v_2 X} \subset O_H^{v_2}$. By Problem 153, this implies that $F <_{v_2} H$, a contradiction. Hence if $v_1 \neq v_2$, then $v_1 X \not\equiv v_2 X$.

155. (*a*) We verify the subordination axioms (see p. 230).

K-1. Supose $F <_v H$. This means that $[F]_{X^*} \subset O(H)$. Hence it follows that $X^* \setminus O(H) = [X \setminus H]_{X^*} \subset X^* \setminus [F]_{X^*} = O(X \setminus F)$ (see Problem 149). Consequently,

$$X \setminus H <_v X \setminus F.$$

K-2. This follows from the equalities $[F]_{X^*} \cap X = F$, $O(H) \cap X = H$.

K-3. This is obviously satisfied.

K-4. Let $F_1 <_v H_1$ and $F_2 <_v H_2$. This implies $[F_1]_{X^*} \subset O(H_1)$, $[F_2)_{X^*} \subset O(H_2)$. But $[F_1 \cup F_2]_{X^*} = [F_1]_{X^*} \cup [F_2]_{X^*} \subset O(H_1) \cup O(H_2) = O(H_1 \cup H_2)$; consequently, $F_1 \cup F_2 <_v H_1 \cup H_2$.

K-5. Let $F <_v H$. Hence $[F]_{X^*} \subset O(H)$. Since the compactum X^* is normal, we can take an open set Γ in X^* such that $[F]_{X^*} \subset \Gamma \subset [\Gamma] \subset O(H)$ and $\Gamma = O(\Gamma \cap X)$. Then $F <_v (\Gamma \cap X)$ and $[\Gamma \cap X]_X <_v H$.

K-6 and K-7. These are obviously satisfied.

(*b*) Let $\xi \in X^*$. Consider the family $\hat{\xi}$ of all nhoods of the point ξ in X^* and the family $h\xi = \{\Gamma\xi \cap X: \Gamma\xi \in \hat{\xi}\}$ of open sets in X. Clearly, $h\xi$ is a centralized v-regular family (the latter follows from the definition of the subordination v, the regularity of X^* and the obvious equality $[\Gamma]_{X^*} = [O(\Gamma \cap X)]_{X^*} = [[\Gamma \cap X]_X]_{X^*}$, which holds by virtue of the space X being dense in X^*). Since X^* is Hausdorff, $h\xi_1 \neq h\xi_2$, whenever ξ_1 and ξ_2 are distinct points in X^*.

It is to be shown now that $h\xi$ is a v-end. Assume the contrary. Then, $h\xi$ is contained in some v-end of η. Let $V \in \eta \setminus h\xi$. Then $\xi \notin OX^*(V)$, because $OX^*(V) \notin \hat{\xi}$. Let $V' \in \eta$ be such that $[V']_X <_v V$. Then $(O_{X^*}(V'))]_{X^*} \subset O_{X^*}(V)$ and, consequently, $\xi \notin [O_{X^*}(V')]_{X^*}$. Consider the nhood $\Gamma\xi = X^* \setminus [O_{X^*}(V')]_{X^*}$ of ξ. Setting $U = \Gamma\xi \cap X$, we have $U \in h\xi \subset \eta$, $V' \in \eta$ and $U \cap V' = \Lambda$, contradicting the fact that η is centralized. Thus we have constructed a one-to-one mapping h of the compactum X^* into the compactum vX. It is easy to see that h is a natural mapping of the compactification X^* of the space X into the compactification vX, i.e., $h(x) = (x) = x$, if $x \in X$.

In order to show now that h is continuous, let $\xi_0 \in X^*$. Consider an arbitrary basic nhood O_U (U is open in X) of the point $h\xi_0$ in vX. Consider also the open set $O(U)$ in X^*, i.e., the maximal open set in X^*, satisfying $O(U) \cap X = U$. By the construction of the mapping h, we surely have $hO(U) \subset O_U$. Furthermore, $U \in h\xi_0$; consequently, $\xi_0 \in O(U)$, i.e., $O(U)$, a nhood of the point ξ_0 in X^*, is mapped by h into the nhood O_U of $h\xi_0$ in vX, establishing that h is a continuous mapping.

Thus, $h: X^* \to vX$ is a natural continuous one-to-one mapping of the compactification X^* of X into the compactification vX of X (consequently, it is a homeomorphism), and hence it is a mapping onto vX. By the same token, it follows that vX coincides with X^*.

156. This follows from Problems 154 and 155.

157. (*a*) Suppose that v coincides with an elementary subordination on X. This implies that $F <_v H$ iff F and $X \setminus H$ are functionally separable. Then, by Problem 150, $[F]_{vX} \subset O_H$, i.e., $[F]_{vX} \cap (vX \setminus O_H) = \Lambda$, but $vX \setminus O_H = [X \setminus H]_{vX}$ (cf. Problem 149); consequently, $[F]_{vX} \cap [X \setminus H]_{vX} = \Lambda$. Then, every functionally separable closed set in X has a disjoint closure in vX (if v is an elementary subordination on X), so that vX coincides with βX (cf. Problem 15).

(*b*) Conversely, if $vX \equiv \beta X$, then (cf. Problem 153) $F <_v H$ iff $[F]_{\beta X} \subset O_{\beta X}(H)$, i.e., iff $[F]_{\beta X} \cap [X \setminus H]_{\beta X} = \Lambda$. But this implies (owing to the fact that $vX \equiv \beta X$) that $F <_v H$ iff F and $X \setminus H$ are functionally seperable, i.e., iff v is an elementary subordination.

158. (*a*) Suppose that A and B are v-distant sets. This implies, by definition, that $[A]_X <_v X \setminus [B]_X$. Put $\bar{A} = [A]_X$, $\bar{B} = [B]_X$, $X \setminus [B]_X = H$. From Problem 153 it follows that $[\bar{A}]_{vX} \subset O_H$, i.e., $[\bar{A}]_{vX} \cap (vX \setminus O_H) = \Lambda$. Since $vX \setminus O_H = [\bar{B}]_{vX} = [X \setminus H]_{vX}$ (cf. Problem 149) and $[A]_{vX} = [\bar{A}]_{vX}$, $[B]_{vX} = [\bar{B}]_{vX}$, so $[A]_{vX} \cap [B]_{vX} = \Lambda$.

(*b*) Conversely, suppose that the sets $A \subset X$, $B \subset X$ are such that $[A]_{vX} \cap [B]_{vX} = \Lambda$. Putting $\bar{A} = [A]_X$, $\bar{B} = [B]_X$, we have $[A]_{vX} = [\bar{A}]_{vX}$, $[B]_{vX} = [\bar{B}]_{vX}$, but since $[A]_{vX} \subset vX \setminus [B]_{vX} = O_H$, where $H = X \setminus \bar{B}$, hence $\bar{A} <_v X \setminus \bar{B}$ (cf. Problem 153); consequently, A and B are v-distant.

159. (*a*) Suppose that $f: X \to Y$ is a (v, w)-uniformly continuous mapping. Clearly, f is continuous, because if $x \in [A]_X$, $A \subset X$, then this signifies that the point x is v-close to the set A, which implies that the point fx is w-close to the set fA, i.e., $fx \in [fA]$; and the inclusion $f[A] \subset [fA]$ for every set $A \subset X$ implies that f is a continuous mapping (cf. Problem 281 of Chap. 2). Suppose now that Φ is closed in Y, O is open in Y and $\Phi \subset_w O$. Then $f^{-1}\Phi$ is closed in X and $f^{-1}O$ is open in X. If the set $f^{-1}\Phi$ is not v-subordinate to the set $f^{-1}O$, then the sets $f^{-1}\Phi$ and $X \setminus f^{-1}O$ are v-close, but then (by hypothesis) the sets $\Phi = ff^{-1}\Phi$ and $O = f(X \setminus f^{-1}O)$ are w-close. This contradicts the fact that $\Phi <_w O$.

(*b*) Conversely, let $A \subset X$, $B \subset X$ and AvB. It is required to show that $fAwfB$. Assume the contrary: let $fA\bar{w}fB$. This implies that $[fA]_Y <_w (Y \setminus [fB])$. Then, by hypothesis, $F = f^{-1}[fA]_Y$ is closed in X, $H = f^{-1}[Y \setminus [fB]_Y)$ is open in X and $F <_v H$. Since $[A]_X \subset F$, $H \subset X \setminus [B]_X$, hence (cf. Axiom K-3) it follows that $[A]_X <_v (X \setminus [B]_X)$, i.e., $A\bar{v}B$, a contradiction.

160. (*a*) The family $\{fU: U \in \xi\}$ (where $\xi \in vX$) is centralized in Y, because the family ξ in X is centralized and the inclusion

$$f \bigcap_{i=1}^{s} U_i \subset \bigcap_{i=1}^{s} fU_i$$

always holds. Hence, since wY is compact, we have

$$\bar{f}\xi = \bigcap \{[fU]_{wY}: U \in \xi\} \neq \Lambda.$$

(*b*) Let $\eta \in wY$ be such that $\eta \in \bar{f}\xi$. Consider the families $\sigma = \{f^{-1}V: V \in \eta\}$ and $\bar{\xi} = \xi \cup \sigma$. We show that $\bar{\xi}$ is v-regular and centralized.

(i) $\bar{\xi}$ is centralized. Indeed, suppose we are given an arbitrary finite subfamily $\{U_1, \ldots, U_k, f^{-1}V_1, \ldots, f^{-1}V_l\}$ of $\bar{\xi}$, where $U_1, \ldots, U_k$ are in ξ and $f^{-1}V_1, \ldots, f^{-1}V_l$ are in σ. Then, by using the fact that ξ is a v-end in (X, v) and η is a w-end in (Y, w) (cf. Problem 146), we conclude that $U_0 = U_1 \cap \ldots \cap U_k \in \xi$, $V_0 = V_1 \cap \ldots \cap V_l \in \eta$, and it remains to show that $U_0 \cap f^{-1}V_0 \neq \Lambda$. However, $\eta \in [fU]_{wY}$ for any $U \in \xi$. Consequently, $\Gamma_V \cap fU \neq \Lambda$ for every $V \in \eta$ and $U \in \xi$ (here $\Gamma_V = \{\eta \in wY: V \in \eta\}$). Hence $V \cap fU \neq \Lambda$ for every $V \in \eta$ and $U \in \xi$. In particular, $V_0 \cap fU_0 \neq \Lambda$. But then $f^{-1}V_0 \cap U_0 \neq \Lambda$, giving the desired result.

(ii) Let $U \in \xi \subset \bar{\xi}$. Since ξ is v-regular, there is $U' \in \xi \subset \bar{\xi}$ such that $[U'] <_v U$. Let $f^{-1}V \in \sigma \subset \bar{\xi}$. Since η is w-regular, there is a $V' \in \eta$ such that $[V'] <_w V$. Then, since the mapping f is (v, w)-uniform, we have $f^{-1}[V'] <_v f^{-1}V$. However, it is easy to see that $[f^{-1}V'] \subset f^{-1}[V']$ (it is just the continuity of the mapping f that is used here). Then $[f^{-1}V'] <_v f^{-1}V$ and $f^{-1}V' \in \sigma \subset \bar{\xi}$.

Thus, the family $\bar{\xi} = \xi \cup \sigma$ is v-regular. Hence, ξ being a v-end we obtain $\bar{\xi} = \xi$; consequently, $\sigma \subset \xi$.

Conversely, let $\eta \in wY$ be such that $\sigma = \{f^{-1}V: V \in \eta\} \subset \xi$. It is required to show that $\eta \in \bar{f}\xi$. Let $V \in \eta$ be arbitrary. By hypothesis, $f^{-1}V \in \xi$. Hence $f^{-1}V \cap U \neq \Lambda$ for any $U \in \xi$ (since ξ is centralized and $f^{-1}V \in \xi$). But, then, $V \cap fU \neq \Lambda$ for any $U \in \xi$, so that $\Gamma_V \cap fU \neq \Lambda$ for any $U \in \xi$, where $\Gamma_V = \{\eta \in wY\colon V \in \eta\}$. Hence, since sets of form Γ_V form a base in wY and $V \in \eta$ is arbitrary, we get $\eta \in [fU]_{wY}$ for any $U \in \xi$. Hence $\eta \in \bar{f}\xi$, as required.

Statements (*c*) and (*d*) of the problem are simple corollaries of (*b*). These are left as an exercise for the reader.

(*e*) Let $x_0 \in X$. Consider the v-end (x_0), where (x_0) is the family of all nhoods of the point $x_0 \in X$ in X. Then $fx_0 = \bar{f}(x_0) = \bigcap\{[fUx_0]_{wY}\colon Ux_0 \in (x_0)\}$.

161. To solve this, it is necessary to make use of Problem 276 of Chap. 2 and Problems 142, 158 and 159 of the present chapter.

162. (*a*) Define (as in Problem 160) $\bar{f}\xi = \bigcap\{[fU]_{wY}\colon U \in \xi\}$ for every point $\xi \in vX$. From Problem 160 it follows that $\bar{f}\xi$ consists of a single point and $\bar{f}\colon vX \to wY$ is a continuous mapping of the extension vX of the space X onto the extension wY of the space Y, which extends the mapping f.

(*b*) Suppose that f is a mapping of X with subordination v onto a space Y with subordination w which can be extended continuously to a mapping $\bar{f}\colon vX \to wY$. It is to be shown that f is a (v, w)-uniformly continuous mapping. This requires, by definition, that every pair of v-close sets A and B in X under the mapping f carry over into w-close sets fA and fB. If A and B are v-close, then $[A]_{vX} \cap [B]_{vX} \neq \Lambda$ (cf. Problem 158). Furthermore, $\bar{f}[A]_{vX} = [\bar{f}A]_{wY} = [fA]_{wY}$ and $\bar{f}[B]_{vX} = [\bar{f}B]_{wY} = [fB]_{wY}$ (here we make use of the continuity and closedness of $\bar{f}$ and the fact that $\bar{f}$ is an extension of f). Hence we obtain $[fA]_{wY} \cap [fB]_{wY} \neq \Lambda$; consequently, the sets fA and fB are w-close (Problem 158).

163. (*a*) Let $\eta \in O_w(U)$. Then $U \in \eta$, whence $U \in \xi$ for every $\xi \in h^{-1}\eta$ (in consequence of Problem 160 and the obvious equality $h^{-1}U = U$). This, in turn, implies that $h^{-1}\eta \subset O_v(U)$, i.e., $\eta \in h^{\#}O_v(U)$.

(*b*) Conversely, let $\eta_0 \in h^{\#}O_v(U)$. Then $h^{-1}\eta_0 \subset O_v(U)$. Since h is a continuous closed mapping, the set $W = h^{\#}O_v(U) = \{\eta \in wX\colon h^{-1}\eta \subset O_v(U)\}$ is open in wX (cf. Problem 321 of Chap. 2) and hence is a nhood of the point η_0 in wX. Put $U' = W \cap X$. It is clear that U' is a nonempty open set in X, where $U' \in \eta_0$ and $U' \subset U$. Hence $U \in \eta_0$, i.e., $\eta_0 \in O_w(U)$. The proof is complete.

164. This is implied by Problems 156 and 162.

165. (i) For establishing that the mapping f is (β, β')-uniform it is necessary to show that for every pair of functionally separable closed sets Φ_0 and Φ_1 in Y, the closed sets $f^{-1}\Phi_0$ and $f^{-1}\Phi_1$ in X are functionally separable (recall the definition of an elementary subordination and that of (β, β')-uniform mapping). In fact, if a continuous function $g(y)$ on Y separates Φ_0 from Φ_1, then the continuous function $h(x) = g(f(x))$ on X separates $f^{-1}\Phi_0$ from $f^{-1}\Phi_1$.

(ii) This follows from (i) and Problem 162 (for another solution, see Problem 31).

166. (*a*) If f is a closed mapping, then $\bar{f}^{-1}y = [f^{-1}y]_{\beta X}$ for every point $y \in Y$ (cf. Problem 73 of Chap. 6). In this connection, it is necessary to note that X and Y are completely regular.

(*b*) Conversely, let $F \subset X$, F being closed. Then $\bar{f}[F]_{\beta X} = [fF]_{\beta Y}$. It is required to to show fF is closed in Y. Let $y_0 \in [fF]_Y$. Assume that $y_0 \notin fF$. Then $f^{-1}y_0 \cap F = \Lambda$, which implies, in view of the normality of space X (cf. Problem 16) that $[f^{-1}y_0]_{\beta X} \cap [F]_{\beta X} = \Lambda$. But $[f^{-1}y_0]_{\beta X} = \bar{f}^{-1}y_0$. Hence $y_0 \in \beta Y \setminus \bar{f}[F]_{\beta X} = \beta Y \setminus [fF]_{\beta Y} = \beta Y \setminus [[fF]_Y]_{\beta Y}$, contradicting the fact that $y_0 \in [fF]_Y$.

167. (*a*) Suppose that the mapping $f: X \to Y$ is (v, w)-uniformly continuous, and (v, w)-open. It is to be shown that the extension $\bar{f}: vX \to wY$ of f on the corresponding compactifications is open. We show first that $\text{Int}[fH]_Y \in \bar{f}\xi$ for every $H \in \xi$. Indeed, $\bar{f}\xi = \bigcap\{[fH]_{wY}: H \in \xi\}$. Take $H' \in \xi$ and $H \in \xi$, so that $[H']_X <_v H$. Since f is (v, w)-open, $[f[H']_X]_Y <_w \text{Int}[fH]_Y$; consequently, $[[f[H']_X]_Y]_{wY} \subset O_w(\text{Int}\,[fH])_Y$, where $O_w(\text{Int}[fH]_Y)$ is a maximal open set in wY, satisfying $O_w(\text{Int}\,[fH]_Y) \cap Y = \text{Int}[fH]_Y$. Furthermore, $[f[H']_X]_Y \supset [fH']_Y$ and $[fH']_{wY} = [[fH']_Y]_{wY}$. Hence $[fH']_{wY} \subset O_w(\text{Int}[fH]_Y)$. Since $\bar{f}\xi \in [fH']_{wY}, \bar{f}\xi \in O_w(\text{Int}[fH]_Y)$, i.e., $\text{Int}[fH]_Y \in \bar{f}\xi$. Thus, $\text{Int}\,[fH]_Y \in \bar{f}\xi$, if $H \in \xi$. In order to show that $\bar{f}: vX \to wY$ is an open mapping, suppose that O is an arbitrary open set in vX. Let $\eta_0 \in \bar{f}O$. Take $\xi_0 \in O$ to be such that $\bar{f}\xi_0 = \eta_0$, and $H_0 \in \xi_0$ such that $[O_v(H_0)]_{vX} \subset O$. Then $\bar{f}[O_v(H_0)] = \bar{f}[H_0]_{vX} = [fH_0]_{wY} = [[fH_0]]_{wY} \subset \bar{f}O$. In addition, by what has been proved above, $\text{Int}[fH_0]_Y \in \bar{f}\xi_0$. We have $\text{Int}[fH_0]_Y \subset [fH_0]_Y$. Hence $O_w(\text{Int}[fH_0]_Y) \subset [[fH_0]_Y]_{wY} \subset \bar{f}O$. Since $\text{Int}[fH_0]_Y \in \bar{f}\xi_0, \bar{f}\xi_0 \in O_w(\text{Int}[fH]_Y)$, i.e., $\eta_0 = \bar{f}\xi_0$ is an interior point of the set $\bar{f}O$. Thus the openness of the mapping $\bar{f}: vX \to wY$ is proved.

(*b*) Conversely, suppose that the mapping $\bar{f}: vX \to wY$, which is an extension of a (v, w)-uniformly continuous mapping $f: X \to Y$, is open. Let $F \subset X$ be closed, $H \subset X$ be open and $F <_v H$. Then $[F]_{vX} \subset O_v(H)$ (cf. Problem 153). Furthermore, $\bar{f}[F]_{vX} = [fF]_{wY} = [[fF]_Y]_{wY}, \bar{f}[F]_{vX} \subset \bar{f}O_v(H)$, and $\Gamma = \bar{f}O_v(H)$ is an open set in wY. Thus $[[fF]_Y]_{wY} \subset \Gamma$. Hence $[fF]_Y <_w (\Gamma \cap Y)$. Since $\bar{f}[O_v(H)] = \bar{f}[H]_{vX} = [fH]_{wY} = [[fH]_Y]_{wY}$, $(\Gamma \cap Y) \subset [fH]_Y$. Consequently, $(\Gamma \cap Y) \subset \text{Int}[fH]_Y$. Hence $[fF]_Y <_w \text{Int}[fH]_Y$, i.e., f is (v, w)-open.

168. It is required to show that the mapping $f: X \to Y$ is (β, β')-open, where β (resp. β') is an elementary subordination on X (resp. Y). Once this has been achieved, we may just make use of Problems 165 and 167. Thus, let $F <_\beta H$ in X, where $F \subset X$ is closed and $H \subset X$ is open. This implies that F is functionally separable from $X \setminus H$. Let $\varphi: X \to [0, 1]$ be a continuous function on X such that $\varphi(F) = \{1\}$ and $\varphi(X \setminus H) = \{0\}$. Define a function ψ on Y thus: $\psi(y) = \sup\{\varphi(x): x \in f^{-1}y\}$. In consequence of Problem 128 of Chap. 6, $\psi(y)$ is a continuous function on Y. It is also clear that $0 \leqslant \varphi(y) \leqslant 1$ for every $y \in Y$. Consider the set $\Phi = fF$ which is closed in Y (since f is closed) and the set $fH = V$ which is open in Y (since f is open). Clearly, $\psi(\Phi) = \{1\}$ and $\psi(Y \setminus V) = \{0\}$, i.e., the sets Φ and $Y \setminus V$ are functionally separable. Hence $fF <_{\beta'} fH \subset \text{Int}[fH]$; consequently, $fF <_{\beta'} \text{Int}[fH]$. The proof is complete.

169. Assume the contrary; let $\bar{f}: \beta X \to \beta Y$ be non-monotonic. Make use of Problem 34 of Chap. 6. For each $\eta \in \beta Y$ consider the partition $\mathcal{L}_\eta$ of the compactum $\bar{f}^{-1}\eta$ into connected components (see p. 154). Consider the partition $\mathcal{L} = \bigcup\{\mathcal{L}_\eta: \eta \in \beta Y\}$ of the compactum βX. This partition $\mathcal{L}$ is continuous (cf. Problem 34 of Chap. 6); the space of this partition, denoted by Z, is a compactum (cf. Problem 92 of Chap. 3). Let $\varphi: \beta X \to Z$ be the natural mapping of the compactum βX onto the space of this partition, and $\psi: Z \to \beta Y$ be a natural 0-demensional mapping of the compactum Z onto the compactum βY. We have $\bar{f} = \psi \cdot \varphi$, φ is a monotonic mapping and ψ a 0-dimensional mapping (Problem 35 of Chap. 6). Since $\bar{f}$ is a non-monotonic mapping, ψ is not a homeomorphism. Since f is closed, $[f^{-1}y]_{\beta X} = \bar{f}^{-1}y$ for every point $y \in Y$ (cf. Problem 166). Then, since f is monotonic, each set $\bar{f}^{-1}y$, as the closure of a connected set $f^{-1}y$, is also connected. Hence $\varphi(\bar{f}^{-1}Y) = Y$; consequently, Z is a compactification of Y and $\psi(y) = y$, if $y \in Y$, i.e., Z is a compactification of Y, following the maximal compactification βY (the Stone-Čech compactification, see Problem 29 or Problems 164, 165), which contradicts the maximality of βY.

170. (a) Suppose that vX is perfectly related to $U \subset X$. Let F be closed in X, $F \subset U$ and $F <_v X \setminus \operatorname{Fr} U$. From Problem 153 it follows that $[F]_{vX} \subset O(X \setminus \operatorname{Fr} U)$. In consequence of the same problem 153, it suffices to show that $[F]_{vX} \subset O(U)$. Assume the contrary, and let $[F]_{vX} \not\subset O(U)$. From $F \subset U$ it follows that $[F]_{vX} \subset [O(U)]$. If $[F]_{vX} \not\subset O(U)$, then $[F]_{vX} \cap \operatorname{Fr}_{vX} O(U) \neq \Lambda$. Hence, since $[\operatorname{Fr}]_{vX} O(U) = [\operatorname{Fr}_X U]_{vX}$ and $O(X \setminus \operatorname{Fr} U) = vX \setminus [\operatorname{Fr} U]_{vX} = vX \setminus \operatorname{Fr}_{vX} O(U)$ (cf. Problem 149), we obtain $[F]_{vX} \not\subset O(X \setminus \operatorname{Fr} U)$. This is a contradiction, because $[F]_{vX} \subset O(X \setminus \operatorname{Fr} U)$.

(*b*) Suppose we are given an open set U in X such that for every closed set F in X satisfying $F \subset U$ and $F <_v (X) \setminus \operatorname{Fr} U$, we have $F <_v (U)$. It is to be shown that $\operatorname{Fr}_{vX} O(U) = [\operatorname{Fr}_X U]_{vX}$. It is clear that we always have $[\operatorname{Fr}_X U]_{vX} \subset \operatorname{Fr}_{vX} O(U)$. To prove the reverse inclusion, let $\xi \in \operatorname{Fr}_{vX} O(U)$, i.e., $\xi \in [O(U)]_{vX} \setminus O(U)$. Hence, we obtain $O(H) \cap O(U) = O(H \cap U) \neq \Lambda$ for every $H \in \xi$. This yields

$$\xi \in [O(U) \cap O(H)_{vX} = [O(H \cap U)]_{vX} = [[H \cap U]_X]_{vX} \text{ for every } H \in \xi; \quad (1)$$

$$\xi \notin O\{U). \quad (2)$$

It is required to show that $\xi \in [\operatorname{Fr}_X U]_{vX}$. Assume the contrary. Then, there is a $H_0 \in \xi$ such that

$$[H_0]_X <_v X \setminus \operatorname{Fr} U. \quad (3)$$

From (1) it follows that

$$\xi \in [[H_0 \cap U]_X]_{vX}. \quad (4)$$

From (3) we obtain, in particular,

$$[H_0]_X \subset U \cup (X \setminus [U]_X), \quad (5)$$

and from (5) it follows that $[H_0]_X \cap U$ is closed in X. Then

$$[H_0 \cap U]_X \subset [H_0]_X \cap U \subset U. \quad (6)$$

Furthermore, $[H_0 \cap U]_X \subset [H_0]_X <_v X \setminus \operatorname{Fr} U$; consequently (subordination axiom K-3),

$$[H_0 \cap U]_X <_v X \setminus \operatorname{Fr} U. \quad (7)$$

By hypothesis, from (7) and (6) we conclude that

$$[H_0 \cap U]_X <_v U. \quad (8)$$

But then (cf. Problem 153),

$$[[H_0 \cap U]_X]_{vX} \subset O(U). \quad (9)$$

Hence from (4) and (9) we obtain

$$\xi \in O(U), \tag{10}$$

which contradicts (2). The proof is complete.

171. It is required to show that for an arbitrary open set $U \subset X$ the elementary subordination β (cf. p. 231), which generates the Stone-Čech compactification βX (cf. Problem 157), satisfies the hypotheses of Problem 170. Suppose that $F \subset X$ is an arbitrary closed set in X such that

$$F \subset U \quad \text{and} \quad F <_\beta X \setminus \operatorname{Fr} U. \tag{1}$$

By the definition of β, there is a continuous function f on X such that

$$f(x) = 0 \text{ if } x \in F, \quad \text{and} \quad f(x) = 1 \text{ if } x \in \operatorname{Fr} U. \tag{2}$$

Define a function $g(x)$ on X as follows:

$$g(x) = \begin{cases} f(x), & \text{if } x \in [U]_X, \\ 1, & \text{if } x \in X \setminus U. \end{cases} \tag{3}$$

It is easily seen that $g(x)$ is a continuous function on X that separates the closed sets F and $X \setminus U$; hence

$$F <_\beta U. \tag{4}$$

It remains to refer to Problem 170.

172. *(a)* Suppose that vX is a perfect compactification of X. Further suppose that U_1, U_2 is a pair of open sets in X such that $U_1 \cap U_2 = \Lambda$. Since $U_1 \cap U_2 = \Lambda$, $\operatorname{Fr}(U_1 \cup U_2) \subset \operatorname{Fr} U_1 \cup \operatorname{Fr} U_2$. Similarly,

$$\operatorname{Fr}_{vX}(O(U_1) \cup O(U_2)) \subset \operatorname{Fr}_{vX} O(U_1) \cup \operatorname{Fr}_{vX} O(U_2), \tag{1}$$

because $O(U_1) \cap O(U_2) = \Lambda$. Since vX is a perfect extension, $\operatorname{Fr}_{vX} O(U_1) = [\operatorname{Fr} U_1]_{vX}$, $\operatorname{Fr}_{vX} O(U_2) = [\operatorname{Fr} U_2]_{vX}$, $\operatorname{Fr}_{vX} O(U_1 \cup U_2) = [\operatorname{Fr}(U_1 \cup U_2)]_{vX}$. This yields

$$\operatorname{Fr}_{vX} O(U_1 \cup U_2) \subset \operatorname{Fr}_{vX} O(U_1) \cup \operatorname{Fr}_{vX} O(U_2). \tag{2}$$

Moreover, we have

$$[O(U_1 \cup U_2)]_{vX} = [U_1 \cup U_2]_{vX} = [U_1]_{vX} \cup [U_2]_{vX} = [O(U_1)]_{vX} \cup [O(U_2)]_{vX}. \tag{3}$$

From formulas (1), (2) and (3) we obtain $O(U_1 \cup U_2) \subset O(U_1) \cup O(U_2)$. It remains to observe that the reverse inclusion always holds.

(b) Conversely, suppose that U is an arbitrary open set in X. Consider the open set $V = X \setminus [U]$ in X. Clearly, $\operatorname{Fr} U = X \setminus (U \cup V)$; consequently,

$$[\operatorname{Fr} U]_{vX} = [X \setminus (U \cup V)]_{vX} = vX \setminus O(U \cup V) = vX \setminus (O(U) \cup O(V)). \tag{4}$$

Furthermore,

$$vX \setminus [O(U)]_{vX} = vX \setminus [[U]_X]_{vX} = O(X \setminus [U]_X) = O(V) \tag{5}$$

(here we have used the equalities proved in Problem 149), which implies that

$$\mathrm{Fr}_{vX}O(U) = vX\setminus(O(U)\cup O(V)). \tag{6}$$

Combine (6) with (4) to obtain $[\mathrm{Fr}U]_{vX} = \mathrm{Fr}_{vX}O(U)$, thus proving that vX is a perfect extension.

173. (*a*) Suppose that X is a normal space, and β is the elementary subordination on X. To prove equality (1) for every pair of open sets in X, it suffices to prove the equality

$$[F_1\cap F_2]_{\beta X} = [F_1]_{\beta X}\cap[F_2]_{\beta X} \tag{1'}$$

for every pair of closed sets F_1 and F_2 in X. Let $\xi\notin[F_1\cap F_2]_{\beta X}$. Then there is a nhood $O\xi$ of the point ξ in βX such that

$$[O\xi]_{\beta X}\cap[F_1\cap F_2]_{\beta X} = \Lambda. \tag{2}$$

Write

$$F_1' = [O\xi]_{\beta X}\cap F_1 \quad\text{and}\quad F_2' = [O\xi]_{\beta X}\cap F_2. \tag{3}$$

Obviously,

$$F_1'\cap F_2' = \Lambda. \tag{4}$$

Then (cf. Problem 16)

$$[F_1']_{\beta X}\cap[F_2']_{\beta X} = \Lambda. \tag{5}$$

Consequently,

$$\xi\notin[F_1]_{\beta X}\cap[F_2]_{\beta X}. \tag{6}$$

This yields the inclusion $[F_1]_{\beta X}\cap[F_2]_{\beta X}\subset[F_1\cap F_2]_{\beta X}$. The reverse inclusion obviously always holds. Thus equality (1′) and, consequently, also equality (1) are proved. This equality also follows from Problems 17 and 5.

(*b*) Suppose that equality (1′) is satisfied for every pair of closed sets F_1 and F_2 in X in an extension vX, in particular, for every pair of disjoint closed sets F_1 and F_2 in X. Thus, every pair of disjoint closed sets F_1 and F_2 in X are v-distant and, consequently, are also functionally separable, which happens only when the space X is normal, v is an elementary subordination, and vX is the Stone-Čech compactification of X.

174. To start with, we formulate an extremely simple statement, whose proof may be left to the reader. The mapping $f\colon X\to Y$ is monotonic iff for every pair of disjoint open sets U_1 and U_2 in X the equality $f^{\#}(U_1\cup U_2) = f^{\#}U_1\cup f^{\#}U_2$ holds, where, as always, $f^{\#}U = \{y\in Y\colon f^{-1}y\subset U\} = Y\setminus f(X\setminus U)$. We now use this proposition to solve our problem.

(*a*) Suppose that the natural mapping $h\colon \beta X\to vX$ is monotonic. Let U_1, U_2 be a pair of arbitrary disjoint open sets in X. Consider the pair of sets $O_\beta(U_1)$ and $O\beta(U_2)$, open in βX. Clearly, $O_\beta(U_1)\cap O_\beta(U_2) = \Lambda$ and hence

$$h^{\#}(O_\beta(U_1)\cup O_\beta(U_2)) = h^{\#}O\beta(U_1)\cup h^{\#}O_\beta(U_2).$$

Furthermore, since βX is a perfect extension, we have

$$O_\beta(U_1 \cup U_2) = O_\beta(U_1) \cup O_\beta(U_2).$$

From Problem 163 we obtain the equalities:

$$h^{\#}(O_\beta(U_1 \cup U_2)) = O_v(U_1 \cup U_2),$$
$$h^{\#}O_\beta(U_1) = O_v(U_1), \quad h^{\#}O_\beta(U_2) = O_v(U_2).$$

The equalities obtained yield

$$O_v(U_1 \cup U_2) = O_v(U_1) \cup O_v(U_2)$$

but this means that vX is a perfect extension (cf. Problem 172).

(*b*) Conversely, it is required to show that the natural mapping $h: \beta X \to vX$ is monotonic, assuming that vX is a perfect extension. Assume that h is not monotonic. Let $\eta_0 \in vX$ be such that $h^{-1}\eta_0$ is disconnected. Then, there exist open sets Γ_1 and Γ_2 in βX such that

$$[\Gamma_1] \cap [\Gamma_2] = \Lambda, \quad h^{-1}\eta_0 \subset \Gamma_1 \cup \Gamma_2, \quad h^{-1}\eta_0 \cap \Gamma_1 \neq \Lambda, \quad h^{-1}\eta_0 \cap \Gamma_2 \neq \Lambda. \qquad (1)$$

Put $U_1 = \Gamma_1 \cap X$, $U_2 = \Gamma_2 \cap X$. It is obvious that U_1 and U_2 are open sets in X and $U_1 \cap U_2 = \Lambda$. Note that we can always assume, without violating condition (1), that $\Gamma_1 = O_\beta(U_1)$, $\Gamma_2 = O_\beta(U_2)$. Using the perfection of extensions βX and vX and Problem 163, we obtain

$$h^{\#}O_\beta(U_1 \cup U_2) = h^{\#}(O_\beta(U_1) \cup O_\beta(U_2)),$$
$$h^{\#}O_\beta(U_1 \cup U_2) = O_v(U_1 \cup U_2) = O_v(U_1) \cup O_v(U_2)$$
$$= h^{\#}O_\beta(U_1) \cup h^{\#}O_\beta(U_2).$$

Since $h^{-1}\eta_0 \subset O_\beta(U_1) \cup O_\beta(U_2)$, $\eta_0 \in O_v(U_1) \cup O_v(U_2)$; on the other hand, $h^{-1}\eta_0 \cap O_\beta(U_1) \neq \Lambda$, $h^{-1}\eta_0 \cap O_\beta(U_2) \neq \Lambda$; hence $\eta_0 \notin h^{\#}O_\beta(U_1) = O_v(U_1)$, $\eta_0 \notin h^{\#}O_\beta(U_2) = O_v(U_2)$. This is a contradiction.

175. First consider the subordination axioms (p. 230). All these subordination axioms are easy to prove for the relation introduced in the hypothesis of the problem. We prove only axiom K-5. Thus let $F \subset U_1 \subset [U_1] \subset H$, $U_1 \in \sigma$. Using the fact that X is a regular space and σ is a base, choose $Ux \in \sigma$ for each point $x \in \mathrm{Fr}\,U_1$ such that $[Ux] \subset H$, $[Ux] \cap F = \Lambda$. As a result, we obtain a cover $\omega = \{Ux: x \in \mathrm{Fr}\,U_1\}$ of the compactum $\mathrm{Fr}\,U_1$, from which we extract a finite subcover $\omega_0 = \{Ux_1, \ldots, Ux_s\}$. Put

$$U_2 = U_1 \setminus \bigcup_{i=1}^{s} [Ux_i], \quad U_3 = U_1 \cup \bigcup_{i=1}^{s} Ux_i.$$

Since σ is an algebraically closed base (cf. the definition on p. 232), $U_2 \in \sigma$, $U_3 \in \sigma$, and since ω_0 is a cover of the set $\mathrm{Fr}\,U_1$ and

$$\bigcup_{i=1}^{s} [Ux_1] \subset H, \quad \bigcup_{i=1}^{s} [Ux_i] \cap F = \Lambda,$$

we have

$$F \subset U_2 \subset [U_2] = [U_1 \setminus \bigcup_{i=1}^{s} [Ux_i]] \subset U_1 \subset [U_1] \subset U_1 \cup \bigcup_{i=1}^{s} Ux_i$$

$$= U_3 \subset [U_3] = [U_1] \cup \bigcup_{i=1}^{s} [Ux_i] \subset H.$$

Thus, we have $U_2, U_3 \in \sigma$ such that $F \subset U_2$, $[U_2] \subset U_1$, $[U_1] \subset U_3$, $[U_3] \subset H$. Since $U_1, U_2, U_3 \in \sigma$, $F <_\sigma U_2$, $[U_2] <_\sigma U_1$, $[U_1] <_\sigma U_3$. Incidentally, note that $F <_\sigma U$ if $F \subset U$ and $U \in \sigma$, and $[U_1] <_\sigma U_2$ if $U_1 \in \sigma$ and $[U_1] \subset U_2$.

176. (*a*) Regularity is easy to establish via Problem 55 of Chap. 3.

(*b*) In order to show that X is completely regular note first that the family σ of all open sets with compact boundary in a π-compact space X forms an algebraically closed base. Now impose on X a subordination relation exactly as in Problem 175 and refer to Problem 143.

177. Let $U \in \sigma$. If the closed set $F \subset X$ is such that $F \subset U$, then surely $F <_\sigma U$ in the sense of the subordination relation is definable from the algebraically closed π-compact base for X (cf. Problem 175). Then, in view of Problem 170, the compactification σX generated by this subordination (cf. Problem 175) is perfectly related to the open set U.

178. The fact that σ is an algebraically closed base is trivial, σ generates a subordination we denote also by σ (cf. Problem 175), and, consequently, a compactification σX. Suppose that U is an arbitrary open set in X, and $F \subset X$ is a closed set such that $F \subset U$ and $F <_\sigma X \setminus \mathrm{Fr}U$. By the definition of σ there is a $U' \in \sigma$ such that $F \subset U' \subset [U'] \subset X \setminus \mathrm{Fr}U$. Then $F <_\sigma U'$, $[U'] <_\sigma X \setminus \mathrm{Fr}U$. Moreover, it is clear that $\mathrm{Fr}(U \setminus [U']) \subset \mathrm{Fr}U \cup \mathrm{Fr}U'$. Hence we conclude that the subspace $\mathrm{Fr}(U \setminus [U'])$ is compact. Since σ consists of all open sets with compact boundary, $U \setminus [U'] \in \sigma$. Using the fact that $F \subset U \setminus [U'] \subset U$ and the definition of σ, we immediately obtain $F <_\sigma U$. Hence σX is perfect (cf. Problem 170).

179. By Problems 164 and 162, the identity mapping $h: X \to X$ is (σ, v)-uniformly continuous, but not (v, σ)-uniformly continuous, i.e., there exists a closed set $F \subset X$ and an open set $H \subset X$ such that $F <_\sigma H$, but $F \not<_v H$. Since $F <_\sigma H$, there is a $H' \in \sigma$ such that $F <_\sigma H'$, $[H'] <_\sigma H$. Then $F \not<_v H'$, because $H' \subset H$ and $F \not<_v H$. Since $H' \in \sigma$, the subspace $\mathrm{Fr}H'$ is compact. Hence we conclude that $F <_v X \setminus \mathrm{Fr}H'$. Noticing that $F \subset H'$ and $F \not<_v H'$, we see from Problem 170 that the extension vX is not perfectly related to the open set H'.

180. (*a*) Let $\xi \in \sigma X$, and let $\Gamma\xi$ be an arbitrary nhood of the point ξ at σX. Take $H_1, H_2 \in \xi$ such that $[H_1] <_\sigma H_2$ and $O_\sigma(H_2) \subset \Gamma\xi$. By the definition of σ (cf. Problem 175), there is a $U \in \sigma$ such that $[H_1] <_\sigma U \subset [U] <_\sigma H_2$. Clearly, $U \in \xi$ and $O_\sigma[U] \subset O_\sigma(H_2) \subset \Gamma\xi$. Thus, the family $\Sigma = \{O_\sigma(U) : U \in \sigma\}$ is a base for the compactum σX.

(*b*) Use Problem 177 from which it follows that σX is perfectly related to $U \in \sigma$. Then, using the compactness of the subspace $\mathrm{Fr}U$, if $U \in \sigma$, we get $\mathrm{Fr}O_\sigma(U) = [\mathrm{Fr}U]_{\sigma X} = \mathrm{Fr}U$ for every $U \in \sigma$, whence $\mathrm{Fr}O_\sigma(U) \cap (\sigma X \setminus X) = \Lambda$. Hence, for every $U \in \sigma$, $O_\sigma(U) \cap (\sigma X \setminus X)$ is a clopen set in the subspace $\sigma X \setminus X$.

(*c*) The last statement, which is a simple consequence of the regularity of the subspace $\sigma X \setminus X$, has just been proved in (*b*). Recall the definition of connectedness (p. 42).

181. Consider the natural mapping $h: \beta X \to \sigma X$ of the Stone-Čech compactification βX of X onto the extension σX. The extension σX is perfect (cf. Problem 178), so that h is a monotonic mapping (cf. Problem 174). We show that, if $\eta \in \sigma X \setminus X$, then $h^{-1}\eta$ is a maximal compact subset of subspace $\beta X \setminus X$ (maximal in the sense that $h^{-1}\eta$ is not contained in any connected compact subset of the subspace $\beta X \setminus X$ other than $h^{-1}\eta$). Let $C \subset \beta X \setminus X$, C connected and compact. Then hC is connected and $hC \subset \sigma X \setminus X$. But the only connected sets belonging to $\sigma X \setminus X$ are singletons (cf. Problem 180); hence there is a (unique) point $\eta \in \sigma X \setminus X$ such that $C \subset h^{-1}\eta$. Suppose now that vX is an arbitrary perfect compactification of X, and $f: \beta X \to vX$ is the natural (monotonic in consequence of Problem 174) mapping of βX onto vX. As a corrollary to what has been just proved, for each point $\zeta \in vX \setminus X$ there exists a unique point $\eta \in \sigma X \setminus X$ such that $f^{-1}\zeta \in h^{-1}\eta$. Since, in addition, $f^{-1}x = h^{-1}x = x$ for any $x \in X$, we get a natural mapping $g: vX \to \sigma X$ of vX onto σX, where $h = gf$. Clearly, g is a continuous natural mapping.

182. Clearly, such a compactification is the compactification σX, constructed on the base σ consisting of all open sets with compact boundary (cf. Problems 178, 179 and 181). It is also obvious that conditions (i), (ii) and (iii) ensure the existence and uniqueness of such a compactification.

183. This is false. Suppose that X is an arbitrary completely regular space and θX is any compactification other than the Stone-Čech compactification. Then there always exists a non-perfect compactification, following θX. In fact, let $f: \beta X \to \theta X$ be the natural mapping. Since θX is different from $\beta X, f^{-1}\eta$ is not a singleton for some point $\eta \in \theta X \setminus X$. Select a pair of points $\xi_1, \xi_2 \in f^{-1}\eta$, $\xi_1 \neq \xi_2$. Define a partition $\mathcal{L}$ of the compactum βX as follows: all elements of the partition are singletons of points in βX, except one element which consists of two points ξ_1 and ξ_2. Clearly, the partition $\mathcal{L}$ is continuous, and the space of this partition (cf. Problem 92 of Chap. 3) is a non-perfect Hausdorff compactification of X, following θX.

184. Suppose that bR, a compactification of the real line, is homeomorphic to a segment of the real line. It is easy to see that bR is a perfect extension of R. On the other hand, the unique compactification preceding bR, a circle, is not a perfect extension. The solution of the problem now follows from these observations and Problems 182 and 174.

Chapter 5

Metrization and Paracompactness

5.1. Definitions

Let X be a set and let γ be a family of subsets of X.

An element U of γ is called a *maximal* element if it is not contained in any element of γ other than U. We denote by $m(\gamma)$ the set of all maximal elements of γ. We call $m(\gamma)$ the *surface* of γ.

We say that γ is a *Noetherian family* of sets, if every element is included in at most finitely many elements of γ.

Let $x \in X$ and let n be a natural number. We write $\mathrm{Kp}(x, \gamma) \leqslant n$, if not more than n elements of γ contain the point x; otherwise, we write $\mathrm{Kp}(x, \gamma) > n$. We call γ *point-finite* if for each point $x \in X$, there is a natural number n such $\mathrm{Kp}(x, \gamma) \leqslant n$. Recall that a family of sets is called *disjoint*, if its elements are mutually disjoint. We say that γ is *σ-point-finite* (resp. *σ-disjoint*) if it is the union of a countable collection of point-finite (resp. disjoint) families of sets. If each element of γ intersects at most a finite (resp. countable) set of elements of γ, then we say that γ is *star-finite* (resp. *star-countable*).

Let $A \subset X$. We put $\gamma(A) = \cup\{V \in \gamma \colon V \cap A \neq \Lambda\}$ and call $\gamma(A)$ *the star of the set A* with respect to γ, or the *star of* γ with respect to A. If $x \in X$ and $A = \{x\}$, then we write $\gamma(x)$ instead of $\gamma(\{x\})$. Also, we put $\gamma^+(x) = \cap \{U \colon U \in \gamma \text{ and } U \ni x\}$, $\gamma^-(A) = X \setminus \cup \{[P] \colon P \in \gamma \text{ and } [P] \cap A = \Lambda\}$, $\gamma^-(x) = \gamma^-(\{x\})$, $\gamma^0(x) = \gamma^-(x) \cap \gamma^+(x)$ and $\langle\gamma\rangle = \{\gamma^0(x) \colon x \in X\}$. The set $\gamma^-(A)$ is called the *anti-star* of γ with respect to A.

Suppose that $\gamma = \{U_\alpha \colon \alpha \in A\}$ and $\lambda = \{V_\alpha \colon \alpha \in A\}$ are two indexed families of subsets of X, with the same index set A. We say that λ is a *combinatorial refinement* of γ if $V_\alpha \subset U_\alpha$ for each $\alpha \in A$. Note that for an indexed family $\gamma = \{U_\alpha \colon \alpha \in A\}$ we may have $U_{\alpha'} = U_{\alpha''}$ for some $\alpha', \alpha'' \in A$ such that $\alpha' \neq \alpha''$. This property is extremely useful for certain arguments. For indexed families of sets the definitions of point-finiteness, disjointness, σ-point-finiteness and so on are obtained by the obvious modification of the corresponding definitions presented above. For example, in this case

$\mathrm{Kp}(x, \gamma) \leqslant n$ if $| \{\alpha \in A: U_\alpha \ni x\} | \leqslant n$, and γ is disjoint if for $\alpha', \alpha'' \in A$ and $\alpha' \neq \alpha''$ we have $U_{\alpha'} \cap U_{\alpha''} = \Lambda$. This convention also carries over to some of the following definitions.

Recall that λ is a *refinement* of γ (or λ *refines* γ) if for each $V \in \lambda$ there is a $U \in \gamma$ such that $V \subset U$. We say that λ is a *star-refinement* of γ if the family $\{\lambda(x): x \in X\}$ is a refinement of γ, i.e., if for each point $x \in X$ there is a $U \in \gamma$ such that $\lambda(x) \subset U$. If for each $V \in \lambda$ there is an element U of γ such that $\lambda(V) \subset U$, then we say that λ is a *strong star-refinement af* γ.

All the above definitions obviously carry over to the special case when the family of sets under consideration is a cover of a given set X. Thus, for example, we can speak of σ-point-finite covers, one cover being a star refinement of another and so on. By a *refinement* of a cover γ we always mean a *cover* which refines γ.

We now introduce three properties of covers.

A cover γ of X is called *binary* if it consists of exactly two elements.

A cover γ of X is called *minimal* if no subfamily λ of family γ other than γ covers the set X.

Suppose that $\mathcal{F}$ is a family of covers of X. For each $\gamma \in \mathcal{F}$ select $U_\gamma \in \gamma$ and put $P = \bigcap \{U_\gamma: \gamma \in \mathcal{F}\}$. The family of all sets P, which can be thus obtained obviously covers X; it is called the *derivative* of $\mathcal{F}$.

In what follows, we assume that X is a topological space.

We say that a cover λ of X is a *closure-refinement* of the cover γ of X if the cover $\{[V]: V \in \lambda\}$ is a refinement of γ.

A family γ of subsets of X is called *locally finite in* X, if every point $x \in X$ has a nhood Ox which intersects at most finitely many elements of γ. If every point $x \in X$ has a nhood Ox, which intersects at most one element of γ, then γ is called *discrete in* X.

A family of sets is σ-*locally finite* (resp. σ-*discrete*) iff it is the union of countably many locally finite (resp. discrete) families.

A family of sets γ is said to be *weakly discrete*, if for any choice of a point $x(U)$ from each element U of γ, the set $\{x(U): U \in \gamma\}$ is a closed discrete subspace of X (such a subspace of X is called *discrete* in X).

A family γ is called *conservative* if for every subfamily γ^* of γ we have $[\bigcup\{P: P \in \gamma^*\}] = \bigcup\{[P]: P \in \gamma^*\}$.

A space X is called *collectionwise normal*, if for every discrete family of its subsets $\lambda = \{F_\alpha: \alpha \in A\}$ there is a discrete family $\gamma = \{U_\alpha: \alpha \in A\}$ of open sets in X such that $F_\alpha \subset U_\alpha$ for each $\alpha \in A$.

A space X is called *paracompact* (resp. *strongly paracompact*) iff each open cover of X has an open locally finite (resp. star-finite) cover which refines it.

A space X is called *weakly paracompact* (resp. *σ-metacompact*) if every open cover of X has an open point-finite (resp. σ-point finite) refinement. Weakly paracompact spaces are also called *pointwise paracompact* or *metacompact*.

A paracompact Hausdorff space is called a *paracompactum*.

A space X is *countably paracompact* if every countable open cover of X has a locally finite open refinement.

A space X is called *screened* if every open cover of X has a σ-disjoint open refinement.

Naturally, a space is called *hereditarily paracompact*, if every subspace is paracompact.

The concept of a *partition of unity* subordinate to a given open cover γ is closely related to the concept of paracompactness. A partition of unity on X is a family $\mathcal{F}$ of continuous non-negative real-valued functions on X such that the following conditions are satisfied:

(i) for each function f, the set $\{x \in X: f(x) > 0\}$ is contained in some element of γ;

(ii) each point x has a nhood Ox such that the set $\{f \in \mathcal{F}: f(Ox) \neq \{0\}\}$ is finite;

(iii) $\Sigma\{f(x): f \in \mathcal{F}\} = 1$ for each point $x \in X$.

A space X is called *star-normal* (sometimes *star-paracompact*) if every open cover of X has a star refinement.

The following definitions relate to bases and families of covers.

A base $\mathcal{B}$ for a space X is called *uniform* if for each point $x \in X$ and every nhood Ox of x the family $\{U \in \mathcal{B}: x \in U \text{ and } U \cap (X \setminus Ox \neq \Lambda\}$ is finite.

A base $\mathcal{B}$ for a space X is called *regular* if for each point $x \in X$ and every nhood Ox of x there is a nhood O_1x of x such that the family $\{U \in \mathcal{B}: U \cap O_1x \neq \Lambda \text{ and } U \cap (X \setminus Ox) \neq \Lambda\}$ is finite.

A family γ of sets is called *regular* if for each $U \in \gamma$ and every point $x \in U$ there is a nhood Ox of x such that the set $\{V \in \gamma: V \cap Ox \neq \Lambda$ and $V \cap (X \setminus U) \neq \Lambda\}$ is finite.

A base for a sapce X is called an *NS-base* if it is a σ-locally finite family.

Suppose that $\mathcal{B}$ is a base for a space X satisfying the condition: for any point $x \in X$ and any sequence $\{U_n: n \in N^+\}$ of elements of this base such that $x \in U_n$ and $U_{n+1} \subset U_n$, $U_{n+1} \neq U_n$ for all $n \in N^+$, the family $\{U_n: n \in N^+\}$ is a base for X at the point x. Then we say that $\mathscr{B}$ is a base of *countable order*.

Suppose that $\mathcal{E}$ is a family of covers of X. We call $\mathcal{E}$ *refined* if for any point $x \in X$ and any nhood Ox of x, there is a $\gamma \in \mathcal{E}$ such that $\gamma(x) \subset Ox$. If all members of $\mathcal{E}$ are open covers, then we say that $\mathcal{E}$ is a *refinement*. A refinement $\mathcal{E}$ is called *countable* if $|\mathcal{E}| \leqslant \aleph_0$.

A family $\mathcal{E}$ of open covers of X is called a T_1-*refinement* if for each point $x \in X$ we have $\bigcap \{\gamma(x): \gamma \in \mathcal{E}\} = \{x\}$.

A refinement $\mathcal{E}$ is called a *k-refinement* if for each compact subset Φ of X and every nhood U of Φ there is a $\gamma \in \mathcal{E}$ such that $\gamma(\Phi) \subset U$.

The concepts of filter and T_1-filter are also fruitful.

A set $\mathcal{F}$ of families of open sets in X is called a *filtration* (resp. T_1-*filtration*) if for each point $x \in X$ and every nhood Ox of x (resp. also for each point y distinct from x) there is a $\gamma \in \mathcal{F}$ such that the set $\gamma(x)$ is nonempty and $\gamma(x) \subset Ox$ (resp. also $\gamma(x) \not\ni y$).

Naturally, a filtration (T_1-filtration) $\mathcal{F}$ is *countable* if $|\mathcal{F}| \leqslant \aleph_0$.

Let ξ be a set of families of open sets in X.

The set of all points x of a set U such that for some $\gamma \in \xi$ we have $\gamma(x) \neq \Lambda$ and $\gamma(x) \subset U$, is called the ξ-*kernel* of $U \subset X$. We agree to denote the ξ-kernel of U by $\langle U \rangle_\xi$, or simply by $\langle U \rangle$ if there is no ambiguity about the family ξ. U is called ξ-*perfect* if $\langle U \rangle = U$.

We say that a collection ξ of families of open sets of X is *majorized by a cover* λ of X if for each point $x \in X$ there is a nhood Ox of x, an element U of λ and an element γ of ξ such that $\gamma(Ox) \neq \Lambda$ and $\gamma(Ox) \subset U$.

A set of open covers of a space X is called *fundamental* if it majorizes every open cover of this space.

Countable fundamental sets of covers are of great importance. Star-monotonic sequences of open covers also play a central role. A sequence $\xi = \{\gamma_n: n \in N^+\}$ of open covers of X is called *star-monotonic* (resp. *strongly star-monotonic*) if for each $n \in N^+$ the cover γ_{n+1} is a star-refinement of γ_n (resp. if γ_{n+1} is a strong star-refinement of γ_n).

X is called a *q-space* if for each point x there is a sequence $\{O_n(x): n \in N^+\}$ of nhoods of x such that if $x_n \in O_n x$ for each $n \in N^+$, then the set $\{x_n: n \in N^+\}$ has a limit point in X.

In various problems we also come across the following variants of the concepts introduced above.

A collection $\mathcal{F}$ of families of open sets in X is *weakly majorized* by a cover γ of X if for each point $x \in X$ there is an element U of γ and $\lambda \in \mathcal{F}$ such that $\lambda(x) \neq \Lambda$ and $\lambda(x) \subset U$. If for each open cover γ of X there is a countable family $\mathcal{F}$ of open covers of X which weakly majorizes γ, then X is called a *σ-paracompact* space. Finally, a space X is called *weakly screened* if for each open cover γ of X there is a countable set $\mathcal{F}$ of families of open sets in X which weakly majorizes γ.

We also introduce the following especial concept. A completely regular space X is called a *feathered space* (*p-space*), if there is a countable family $\mathcal{P}$ of covers of X by sets open in its Stone-Čech compactification βX such that $\cap \{\gamma(x): \gamma \in \mathcal{P}\} \subset X$ for each point $x \in X$.

$\mathcal{P}$ is called a *feathering* of the space X in βX.

5.2. General Problems on Covers and Bases

1. Show that every open cover of a regular space has an open closure refinement.

2. Show that a locally finite family of sets in a space X is conservative.

3. Show that the derivative of a finite set of open covers of a space is an open cover of the space.

4. Show that for every finite set θ of open covers of a space X, there is an open cover of X, which refines all the elements of θ.

5. Show that in a separable space X, every pointwise countable family γ of nonempty open sets is countable.

6. Suppose that X is a space, Y a separable subspace of X and γ a family of open sets in X such that $|\{U \in \gamma: U \ni y\}| \leqslant \aleph_0$ for all $y \in Y$. Show that $|\{U \in \gamma: U \cap Y \neq \Lambda\}| \leqslant \aleph_0$.

7. Suppose that γ is a star-countable family of open sets in a space X. Show that γ can be represented as a union of subfamilies γ_α (where α runs through some set A) in such a way that (i) $|\gamma_\alpha| \leqslant \aleph_0$ for all $\alpha \in A$, and (ii) if $U \in \gamma_{\alpha'}$ and $V \in \gamma_{\alpha''}$, where $\alpha' \neq \alpha''$, then $U \cap V = \Lambda$.

8. If γ is a star-countable open cover of a space X, show that there exist a set A and, for each $\alpha \in A$, a subfamily γ_α of γ and a subspace X_α of X such that (i) $|\gamma_\alpha| \leqslant \aleph_0$; (ii) if $\alpha' \neq \alpha''$, then $\gamma_{\alpha'} \cap \gamma_{\alpha''} = \Lambda$; (iii) $\cup \{U: U \in \gamma_\alpha\} = X_\alpha$; (iv) X_α is clopen in X; (v) $\cup \{\gamma_\alpha: \alpha \in A\} = \gamma$; (vi) $\cup \{X_\alpha: \alpha \in A\} = X$; (vii) if $\alpha' \neq \alpha''$, then $X_{\alpha'} \cap X_{\alpha''} = \Lambda$ (where α, α' and α'' are arbitrary elements of A).

9. Show that every star-countable open cover γ of a topological space X is σ-discrete.

10. Show that every pointwise countable open cover of a space X by separable subspaces is star-countable.

11. If there is a pointwise countable open cover γ of a space X by subspaces, each of which has a countable base, show that X is the union of disjoint clopen subspaces, each of which has a countable base.

12. Show that every star-countable open cover γ of a nonempty connected space is countable.

13. If an open cover γ of a space X can be refined to a locally finite (point-finite, star-finite) open cover, show that γ can be combinatorially refined to a locally finite (point-finite, star-finite) open cover.

14. Suppose that γ is a point-finite cover of a topological space X. Show that there is a minimal cover λ of X included in γ.

15. Suppose that γ is a minimal open cover of a T_1-space X. Show that there is a closed discrete set in X which is equipotent to γ.

16. Let γ^* be a point-finite family of subsets of X, $X^* \subset X$ and let each $A \in \gamma^*$ be placed in correspondence with some set $p(A) \subset A$. Suppose further that a well-ordering $<$ is defined on γ^* such that $(\cup \{p(A): A \leqslant B\}) \cup (\cup \{A: B <_c A\}) \supset X^*$ is satisfied for each $B \in \gamma^*$. Show that

$\cup \{p(A): A \in \gamma^*\} \supset X^*$.

17. Suppose that γ is an open point-finite cover of a normal space X. Show that for each $U \in \gamma$ an open set $p(U)$ can be chosen in such a way that the following conditions are satisfied: (i) $[p(U)] \subset U$ and (ii) $\cup \{[p(U)]: U \in \gamma\} = X$.

18. Show that every point-finite open cover of a normal space X can be combinatorially refined to a cover by closed sets.

19. Suppose that γ is a point-finite open cover of a space X and $X_n = \{x \in X: |\{U \in \gamma: U \ni x\}| = n\}$ for each $n \in N^+$. Let V_{n-1} be some open set containing X_{n-1} with $X_0 = \Lambda$ and $V_0 = \Lambda$. Show that: (i) every X_n is closed in X, (ii) $X_n \setminus V_{n-1}$ is covered by a discrete family of closed sets in X which refines γ.

20. If a family of sets γ is a star-refinement of a family of sets λ, show that $\gamma(\gamma(A)) \subset \lambda(A)$ for every set A.

21. Suppose that γ, γ' and γ'' are families of subsets of X and γ' (resp. γ'') is a star-refinement of γ (resp. γ'). Show that γ'' is a strong star-refinement of γ.

22. If every open cover of X has an open star-refinement, show that every open cover of X also has a strong star refinement.

23. If every binary open cover of X has an open star-refinement, show that X is a normal space.

24. If every open cover of a space X has an open star-refinement, show that X is collectionwise normal.

25. If ξ is an open cover of a space X and if there is a countable set γ of open covers of X that majorizes ξ, show that ξ has an open star-refinement.

26. Show that every discrete (σ-discrete) family of sets of a space is locally finite (σ-locally finite).

27. If a topological space has a countable fundamental set of covers, show that every open cover of this space has an open star-refinement.

28. Show that every open cover of a metric space has an open star-refinement.

29. Show that every metric space is collectionwise normal.

30. Suppose that ξ is a set of collections of open sets in a space X. Show that every ξ-perfect set is open.

31. Suppose that $\xi = \{\gamma_n : n \in N\}$ is a star-monotonic sequence of open covers of X. Show that the ξ-kernel $\langle U \rangle$ of an arbitrary set $U \subset X$ is an open ξ-perfect set.

32. If λ is a cover of X and ξ is a star-monotonic sequence of open covers of X that majorizes λ, show that there is an open refinement γ of λ which consists of ξ-perfect sets.

33. Suppose that $\xi = \{\gamma_n : n \in N\}$ is a strong star-monotonic sequence of open covers of a space X and that γ is a cover of X by ξ-perfect sets, well-ordered by some relation $<$. Put $t(U) = U \setminus \bigcup \{V \in \gamma : V < U \text{ and } V \neq U\}$, $t_n(U) = \{x \in t(U) : \gamma_n(x) \subset U\}$, $O_n(U) = \gamma_{n+1}(t_n(U))$. Show that

(*a*) $\{t(U) : U \in \gamma\}$ is a cover of X;

(*b*) $\{t_n(U) : U \in \gamma, n \in N\}$ is a cover of X;

(*c*) no element of the family γ_n intersects two distinct elements of the family $\{t_n(U) : U \in \gamma\}$, where $n \in N$;

(*d*) no element of the family γ_{n+1} intersects two distinct elements of the family $\{O_n(U) : U \in \gamma\}$, where $n \in N$.

(*e*) $\{O_n(U) : U \in \gamma, n \in N\}$ is a σ-discrete open cover of X which refines γ.

34. If a star-monotonic sequence $\xi = \{\gamma_n : n \in N\}$ of open covers of a space X majorizes a cover λ of this space, show that λ has a σ-discrete open refinement.

35. Show that the union of a finite family ξ of locally finite collections of sets in a space X is a locally finite family of sets.

36. Show that the derivative η of a finite family ξ of locally finite collections of sets in a space X is a locally finite family of sets.

37. If ξ is a locally finite family of sets in a space X and $\lambda = \{[Q]: Q \in \xi\}$, show that λ also is locally finite (in X).

38. If γ is a locally finite family of sets in a space X and F is a compact subspace of X, show that F intersects at most finitely many elements of γ.

39. If $c(X) \leqslant \aleph_0$, show that every family γ of nonempty open subsets locally finite in X is countable.

40. Show that every σ-locally finite open cover ξ of X has a locally finite refinement.

41. Show that each weakly discrete disjoint cover γ of any q-space X is locally finite.

42. If $\gamma = \{U_\alpha: \alpha \in A\}$ and $\lambda = \{V_\alpha: \alpha \in A\}$ are families of sets in a space X such that $V_\alpha \subset U_\alpha$ for all $\alpha \in A$ and γ is weakly discrete, show that λ is also weakly discrete.

43. Show that every weakly discrete cover of a topological space has a disjoint weakly discrete refinement.

44. If X is a regular space and every open cover of X has a locally finite refinement (whose elements are not necessarily either open or closed), show that X is paracompact.

45. Show that every metric space is paracompact.

46 Show that a regular q-space is paracompact iff every open cover has a weakly discrete refinement.

47. Show that a regular pointwise countable space is paracompact iff every open cover has a weakly discrete refinement.

48. If X is a normal space, $\xi = \{F_\alpha: \alpha \in A\}$ is a discrete family of closed sets in X and if $\gamma = \{U_\alpha: \alpha \in A\}$ is a family of mutually disjoint nhoods such that $U_\alpha \supset F_\alpha$ for each $\alpha \in A$, show that there exist open sets $V_\alpha \supset F_\alpha$ such that $\{V_\alpha: \alpha \in A\}$ is a discrete family.

49. Suppose that γ is some family of subsets of a space X, and λ is a locally finite cover of closed sets, each of which intersects at most a finite number of elements of γ. Show that $\{\lambda^-(A): A \in \gamma\}$ is a locally finite family of open sets, and $\lambda^-(A) \supset A$ for all $A \in \gamma$.

50. If λ is a locally finite open cover of X, show that $\langle\lambda\rangle = \{\lambda^0(x): x \in X\}$ is an open refinement of λ.

51. If λ and γ are open locally finite covers of a space X and λ is a closure-refinement of γ, show that $\langle \lambda \cup \gamma \rangle$ is a star-refinement of γ; more precisely, the family $\{\langle \lambda \cup \gamma \rangle(V): V \in \lambda\}$ is a refinement of γ.

52. Suppose that X is a topological space, γ is a countable open cover of X and for each $U \in \gamma$ the open set $t(U)$ in X is so defined that (i) $[t(U)] \subset U$, and (ii) $\cup \{t(U): U \in \gamma\} = X$. Show that γ has an open locally finite refinement (which is also countable).

53. Show that every locally finite family of compact subspaces of a space X is star-finite.

54. Suppose that μ is a star-finite open cover of X and for each $U \in \mu$ λ_U is a finite family of open sets in X which covers U. Show that $\{U \cap V: V \in \lambda_U, U \in \mu\}$ is a star-finite open cover of X.

55. Suppose that $\xi = \{F_k: k \in N^+\}$ (resp. $\eta = \{V_k: k \in N^+\}$ is a sequence of closed (resp. open) sets in X and also let $\cup \{F_k: k \in N^+\} = X$ and $F_k \subset V_k \subset F_{k+1}$ for all $k \in N^+$. Suppose further that, for every k, λ_k is a finite family of open sets in X which covers F_k, and let $\lambda = \cup \{\lambda_k: k \in N^+\}$. Show that λ has a star-finite open refinement.

56. Suppose that X is a topological space, $\lambda = \{U_{i,j}: i \in N, j \in N\}$ is an open cover of X such that $[U_{i,j'}] \subset U_{i,j''}$ when $j' < j''$. Put $U_i = \cup \{U_{i,j}: j \in N\}$ and $\gamma = \{U_i: i \in N\}$. Construct an open star-finite refinement of γ.

57. Suppose that X is a normal space and for each $k \in N$ there is an open set V_k and a closed set F_k such that (i) $V_k \supset F_k$ and (ii) $\cup \{F_k: k \in N\} = X$. Show that the cover $\mu = \{V_k: k \in N\}$ has an open star-finite refinement.

58. If X is a normal space and γ is a countable open cover of X, with a countable closed refinement, show that γ also has a star-finite open refinement.

59. Show that every finite family of sets has a maximal element.

60. If a cover γ of X is Noetherian, show that the surface $m(\gamma)$ of γ covers X.

61. Show that every regular family of sets is Noetherian.

62. Show that a subfamily of a regular family of sets is regular.

63. Show that a regular family γ of sets, every member of which is maximal, is locally finite.

64. Show that the surface $m(\gamma)$ of a regular family of sets γ is locally finite.

65. Show that the surface $m(\mathscr{B})$ of a uniform base $\mathscr{B}$ for a space X is a point-finite open cover of X.

66. Show that every T_1-space X with a uniform base $\mathcal{B}$ has a countable refined family of point-finite open covers.

67. If $\xi = \{\gamma_n: n \in N\}$ is a refined sequence of point-finite open covers of a space X and $\gamma_{n'}$ is a refinement of $\gamma_{n''}$ for $n' > n''$, show that $\gamma = \bigcup \{\gamma_n: n \in N\}$ is a uniform base for X.

68. If a regular space X has a σ-discrete base, show that X also has a regular base.

69. Show that a T_1-space X, possessing a regular base $\mathcal{B}$, is regular.

70. Show that a space X, possessing a regular base $\mathcal{B}$, is paracompact.

71. Show that a regular space X with a σ-discrete base is paracompact.

72. If every open cover of a regular space X has a σ-locally finite open refinement, show that X is paracompact.

73. Show that every regular space X with an NS-base is normal.

74. Show that a regular space X with an NS-base is perfectly normal.

5.3. Fundamental Metrization Theorems

75. Suppose that X is a T_0-space and $\xi = \{\gamma_n: n \in N\}$ is a strongly star-monotonic and refined sequence of open covers of X. For $x, y \in X$, put $n(x, y) = \min\{n \in N: \gamma_n(x) \not\ni y\}$ and $f(x, y) = 1/2^{n(x,y)}$ if $x \neq y$, and $f(x, y) = 0$ if $x = y$. A finite sequence $\theta = \{x_r, \ldots, x_{r+s}\}$ of points of X, indexed by the natural numbers from r to $r + s$ is called a *chain* of ranks s joining x_r to x_{r+s}; the number

$$\sum_{i=0}^{s-1} f(x_{r+i}, x_{r+i+1})$$

is called the length of this chain and is denoted by $L(\theta)$. The rank of θ we denote by $R(\theta)$. Let us define $d(x, y)$ to be the greatest lower bound of the length of a chain (of all possible ranks), joining x to y. Show that:

(i) $f(x, y) = f(y, x)$ for every $x, y \in X$, where $x \neq y$;

(ii) $f(x_0, x_{l+1}) \leqslant 2 \sum_{i=0}^{l} f(x_i, x_{i+1}) = 2L(\theta)$ for every chain $\theta = \{x_0, \ldots, x_{l+1}\}$, joining x_0 to x_{l+1};

(ii′) $d(x, y) \leqslant f(x, y) \leqslant 2d(x, y)$, $x, y \in X$;

(iii) $d(x, y)$ is a metric on the set X;

(iv) $\{y: d(x, y) < 1/2^n\} \supset \gamma_n(x)$;

(v) $\{y: d(x\, y) < 1/2^n\} \subset \gamma_{n-1}(x)$;

(vi) d generates a topology on X.

76. If a paracompactum X has a countable refinement $\{\gamma_i: i \in N^+\}$, then it is metrizable (and conversely).

77. For a T_0-space X to be metrizable it is necessary and sufficient that there exist a countable fundamental set of covers of X.

78. Show that a T_0-space X is metrizable iff there is a countable k-refinement on X.

79. A regular space with a countable base is metrizable.

80. A regular space with a σ-discrete base is metrizable. Show whether the converse statement also holds.

81. True or false? Every paracompactum with a σ-disjoint base is metrizable.

82. True or false? Every perfectly normal space with a σ-disjoint base is metrizable.

83. If a regular space X has an NS-base, then it is metrizable.

84. Show that a regular space X with an NS-base is metrizable, thereby generalizing the statement of Problem 39 of Chap. 3.

85. A paracompactum with a uniform base is metrizable (and conversely).

86. True or false? Every regular Lindelöf space with a pointwise countable base is metrizable.

87. Show that (*a*) if a regular space X has a pointwise countable base and each point $x \in X$ has a nhood which is a space with a countable base, then X is metrizable; (*b*) a locally compact Hausdorff space with a pointwise countable base is metrizable.

88. Every T_1-space with a regular base is metrizable (and conversely).

89. If $\gamma = \{U_\alpha: \alpha \in A\}$ is a locally finite open cover of a regular space X by metrizable subspaces U_α of X, then X is metrizable.

90. If a paracompactum is locally metrizable, then it is metrizable.

91. A paracompactum, locally metrizable by a complete metric, is metrizable by a complete metric.

92. A metric space X is metrizable by a complete metric iff it is Čech complete, i.e., it is a G_δ-set in some (and then even in every) Hausdorff compactification of X.

93. A locally metrizable, locally compact, Hausdorff, weakly paracompact space X is metrizable.

94. Every metrizable locally compact space X is representable as the union of a disjoint family of clopen subspaces of X, each of which is the union of a countable set of compacta.

95. Every semi-metrizable compactum is metrizable.

96. Suppose that the topology of the Hausdorff space X is induced by a semi-metric d, which satisfies the following condition: if A is closed in X and Φ is a compact subset of X not intersecting A, then $d(\Phi, A) > 0$. Show that: (a) X is first countable; (b) X is metrizable.

97. Suppose that X is a metrizable non-compact space. Show that there exists a metrizable (generally speaking, non-compact) extension $\tilde{X}$ of X (distinct from X).

98. Show that a metrizable space X is compact iff every metric on X compatible with its topology is complete.

99. If X is a metrizable space, A is a closed subspace of X and ρ is a metric on A, which generates the topology on A inherited from X, then there is a metric $\tilde{\rho}$ on X, compatible with its topology, whose restriction to A is ρ.

100. Construct a regular space $\mathscr{P}$ with the following properties:

(i) $\mathscr{P} = \mathscr{P}_0 \cup \mathscr{P}_1$, where $\mathscr{P}_0$ is an open discrete subspace of $\mathscr{P}$ and $\mathscr{P}_1$ is a closed discrete subspace of $\mathscr{P}$, with $\mathscr{P}_0 \cap \mathscr{P}_1 = \Lambda$, $|\mathscr{P}_0| = \aleph_0$; $|\mathscr{P}_1| = 2^{\aleph_0}$;

(ii) every infinite subset C of $\mathscr{P}_0$ has a limit point in $\mathscr{P}$ (more precisely, in $\mathscr{P}_1$).

101. Show that every regular space $\mathscr{P}$, satisfying conditions (i) and (ii) of the preceding problem is: (a) locally compact, (b) completely regular, (c) locally metrizable, (d) pseudocompact, (e) separable, (f) non-metrizable, (g) not countably compact, (h) not Lindelöf, (i) not normal, (j) not weakly paracompact, and that it (k) has a refined sequence of covers, (l) does not have a pointwise countable base, (m) admits a continuous closed mapping onto an infinite compactum such that the frontier of the pre-image of a non-isolated point is non-compact, (n) is not a Q-space.

5.4. Almost Metrizable Spaces. Special Theorems on Metrization and Metric Spaces

102. If a space X has a countable refinement, then every closed set in X is a G_δ.

103. If the space X has a countable T_1-filtration (resp. a filtration) and every closed set in X is a G_δ, then X also has a countable T_1-refinement (resp. a refinement).

104. If a space X can be contracted to a space with a countable T_1-refinement (resp. with a countable T_1-filtration), then X has a countable T_1-refinement (resp. a countable T_1-filtration).

105. A space X has a countable T_1-refinement iff the diagonal Δ is a G_δ-set in the Tychonoff product $X \times X$.

106. If $\varphi = \{\gamma_\alpha : \alpha \in A\}$ is a T_1-refinement on a space X, then $U\{\gamma_\alpha : \alpha \in A\}$ is a pseudobase for X.

107. If a Lindelöf space X has a countable T_1-refinement, then X has also a countable pseudobase.

108. A compactum with a countable T_1-filtration is perfectly normal.

109. Show that every compactum X, having a countable T_1-filtration, has a countable base and is hence metrizable.

110. True or false? Every completely regular, pseudocompact, seperable space with a countable refinement is metrizable.

111. If $2^{\aleph_1} > 2^{\aleph_0}$ and X is a normal separable space with a countable refinement, then every uncountable set $A \subset X$ has a limit point in X.

112. If a space X has a countable refinement ξ and every uncountable set $A \subset X$ has a limit point in X, then X is Lindelöf.

113. If X is a regular space with a countable refinement and every uncountable set $A \subset X$ has a limit point in X, then X is a metrizable space with a countable base.

114. If $2^{\aleph_1} > 2^{\aleph_0}$, then every separable normal space X with a countable refinement is metrizable and has a countable base.

115. Every separable normal space with a countable refinement is metrizable iff every uncountable subspace X of the real line contains a subset which is not an F_σ in X.

116. If every normal space with a uniform base is metrizable, then every separable normal space with a countable refinement is also metrizable.

117. Every metric space X of weight τ is homeomorphic to some subspace of the space $(Kw(\tau))^{\aleph_0}$, a countable Tychonoff hedgehog space of weight τ (cf. Problem 251 of Chap. 2).

118. The cardinality of the space $(Kw(\tau))^{\aleph_0}$ is $\tau^{\aleph_0}$.

119. If τ is a non-measurable cardinal number, then $Kw(\tau)$ and $\mathcal{E}(\tau) = (Kw(\tau))^{\aleph_0}$ are Q-spaces.

120*. The Q-weight (see p. 225) of $Kw(\tau)$, a hedgehog space of weight τ, is $Q(\tau)$, the Q-weight of a discrete space of cardinality τ.

121*. The Q-weight of the space $(Kw(\tau))^{\aleph_0}$ is equal to $Q(\tau)$ for any non-measurable cardinal number τ.

122*. The Q-weight of any subspace Y of the space $\mathcal{E}(\tau) = (Kw(\tau))^{\aleph_0}$ does not exceed max $\{Q(\tau), \tau^{\aleph_0}\}$, where τ is an arbitrary non-measurable cardinal number.

123*. If τ is a non-measurable cardinal number and X a metric space of weight τ, then X is a Q-space and its Q-weight does not exceed $Q(\tau) \cdot \tau^{\aleph_0}$.

124*. If the generalized continuum hypothesis holds and τ is a cardinal number less than the first strongly inaccessible cardinal number, then every metric space of weight τ is homeomorphic to a closed subspace of the space $R^{\tau^{\aleph_0}}$ of the Tychonoff product of $\tau^{\aleph_0}$ copies of the usual real line.

125*. Every metric space of weight τ, where τ is a non-measurable cardinal number, has a Q-extension whose weight does not exceed $Q(\tau) \cdot \tau^{\aleph_0}$.

5.5. PARACOMPACTA

126. Give an example of a normal, seperable, non-paracompact space.

127. If each open subspace of a space X is paracompact, then every subspace of X is paracompact.

128. Suppose that X is a paracompact space, B is a subset of X, $\mathcal{A}$ is a family of subsets of X, and the following conditions are satisfied:

(a) $P = \bigcup\{A\colon A \in \mathcal{A}\}$ is a closed set in X;
(b) for each $A \in \mathcal{A}$ there is an open set U_A in X such that $U_A \supset A$ and $[U_A] \cap B = \Lambda$.

Show that P has a nhood in X whose closure does not intersect B.

129. Show that a paracompact Hausdorff space X is regular.

130. Show that a paracompact Hausdorff space is normal.

131. If a regular space X has a locally finite cover by paracompact subspaces, show that X is itself paracompact.

132. Show that every regular Lindelöf space is paracompact.

133. If X is a paracompact space and $c(X) \leqslant \aleph_0$, show that X is Lindelöf.

134. Suppose that X is a paracompactum and γ is an open cover of X.

Show that there exists a family $\mathcal{Y}$ of continuous real-valued functions on X such that:

(i) for each $f \in \mathcal{Y}$, the set $\{x \in X: f(x) \neq 0\}$ (called the *support* of the function f) is contained in some element of the cover γ;

(ii) each point $x \in X$ has a nhood Ox such that the set $\{f \in \mathcal{Y}: Ox \setminus f^{-1}(0) \neq \Lambda\}$ is finite;

(iii) $\Sigma\{f(x): f \in \mathcal{Y}$ and $f(x) \neq 0\} = 1$ for all $x \in X$ (note that by virtue of (ii) the set $\{f \in \mathcal{Y}: f(x) \neq 0\}$ is finite for all $x \in X$).

135. Let X be the set of all points of the plane, with integral coordinates. Every point $(m, n) \in X$ different from $(0, 0)$ is taken as isolated. A set $A \subset X$, $A \ni (0, 0)$, is said to be *open* if $X \setminus A$ intersects every straight line, parallel to the y-axis—apart from at most finitely many such straight lines—in a finite set.

Show that X with such a topology is a non-discrete paracompactum without a countable base, every compact subspace of which is finite (consequently, X is not a k-space).

136. Does there exist a countable paracompactum which is not first countable at any point?

137. Suppose that X is a paracompactum and $X_0 \subset X$ is a F_σ-subset of X. Show that every cover ω of X_0 by sets open in X can be refined to a σ-locally finite (in X) cover of X_0 by sets open in X.

138. Show that every F_σ-subspace X_0 of a paracompactum X is paracompact.

139. Show that every paracompact perfectly normal space is hereditarily paracompact.

140. If a space X is hereditarily paracompact and $c(X) \leqslant \aleph_0$, show that X is Lindelöf.

141. Is every compactum, each of whose subspaces is paracompact, perfectly normal ?

142. If every subspace of a compactum is paracompact, then is it necessarily a Fréchet-Urysohn space (or a sequential space)?

143. If a space X is hereditarily paracompact and $A(X) = X \cup d(X)$ is the space of Problem 81 of Chap. 3, show that $A(X)$ is also hereditarily paracompact.

144. Is every hereditarily paracompact, first countable compactum perfectly normal?

145. If a space X is perfectly normal and every open cover of X has a σ-disjoint open refinement, show that X is paracompact.

146. Suppose that X is a collectionwise normal T_1-space, $X \supset Y$ and $Y = \bigcup\{F_i: i \in N^+\}$, where each F_i is closed in Y. Suppose further that (i) each F_i, as a subspace of X, is paracompact, and (ii) for any open set U in X containing Y, the subspace $X \setminus U$ is paracompact. Show that X is also paracompact.

147. Show that every open cover ξ of a paracompact Hausdorff space X has an open star-refinement.

148. Show that a space X is paracompact iff it is a T_1-space and every open cover of X has an open star-refinement.

149. Show that every open cover of a paracompactum has a strong open star-refinement.

150. Show that the following statements are equivalent:

(*a*) X is paracompact;

(*b*) X is a T_1-space, every open cover of which has an open star-refinement;

(*c*) X is a regular space, every open cover of which has a σ-locally finite open refinement;

(*d*) X is a regular space, every open cover of which has a locally finite refinement;

(*e*) X is a regular space, every open cover of which has a σ-discrete open refinement;

(*f*) X is a regular space, every open cover of which has a conservative refinement.

151. Suppose a completely regular compact space X has a conservative cover. Is X necessarily paracompact?

152. Suppose that X is a completely regular space and bX a Hausdorff compactification of X. Show that X is paracompact iff for each compactum F contained in $bX \setminus X$, there is a locally finite (in X) open cover of X by subsets whose closures (in bX) do not intersect F.

153. Suppose that X is paracompact. Show that the following conditions are equivalent:

(i) $\dim X = 0$, i.e., every finite open cover of X has a (finite) disjoint open refinement;

(ii) $\operatorname{Ind} X = 0$, i.e., for every closed set F in X and an arbitrary nhood UF of F, there is a clopen nhood $U'F$ of F such that $U'F \subset UF$;

(iii) every open cover ω of X has a disjoint open refinement.

154. Suppose that X is a metrizable space. Show that the following conditions are equivalent:

(i) $\dim X = 0$;

(ii) there is a base σ for X, which is the union of a countable family of disjoint open covers of X: $\sigma = \bigcup\{\omega_i : i \in N^+\}$.

155. Suppose that $B(\tau)$ is a Baire space of weight τ (cf. Problem 246 of Chap. 2). Show that $\dim B(\tau) = 0$.

156. Suppose that X is a metrizable space and let $\dim X = 0$. Show that $\dim X_0 = 0$ for every subspace $X_0 \subset X$.

5.6. Paracompact-like Properties: Countable Paracompactness, Strong Paracompactness, Weak Paracompactness . . .

157. If X is a normal space, show that the following statements are equivalent:

(*a*) every countable open cover of X has a point-finite open refinement;

(*b*) every countable open cover of X has a closed combinatorial refinement;

(*b'*) every countable open cover of X has a countable closed refinement;

(*c*) every countable open cover of X has a locally finite open refinement;

(*d*) every countable open cover of X has a finite open star-refinement.

158. Show that every perfectly normal space is countably paracompact.

159. Show that a normal space X is countably paracompact iff it satisfies the condition: if ξ is any countable centralized family of closed sets in X, the intersection of whose elements is empty, then for each $F \in \xi$ there is an open set $U(F)$ containing F such that $\bigcap\{U(F): F \in \xi\} = \Lambda$.

160. If X is normal, show that it is countably paracompact iff every countable open cover $\gamma = \{U_i: i \in N^+\}$ of X has a countable open star-refinement.

161. Show that every regular Lindelöf space X is strongly paracompact.

162. Show that every regular space with a countable net is strongly paracompact.

163. Show that the metric space $K\omega(\tau)$ (a hedgehog space of weight τ), constructed in Problem 251 of Chap. 2, is not strongly paracompact for $\tau > \aleph_0$.

164. If a space X admits a disjoint open cover γ, every element of which is a strongly paracompact subspace of X, show that X is strongly paracompact.

165. Show that a regular space X is strongly paracompact iff every open cover of X has a countable open star-refinement.

166. If a regular space X has a star-finite open cover, each member of which is the union of a countable set of compacta, show that X is strongly paracompact.

167*. Show that every locally compact, paracompact Hausdorff space X has an open cover γ such that: (i) the closure of each $U \in \gamma$ is compact, and (ii) each point $x \in X$ has a nhood Ox which intersects at most two elements of γ.

168. Show that every locally compact, paracompact space X can be covered by a family of mutually disjoint clopen Lindelöf subspaces.

169. Show that every paracompact locally compact Hausdorff space X is strongly paracompact.

170. Let X be a completely regular space and let βX be its Stone-Čech compactification. Show that the following statements are equivalent:

(i) X is strongly paracompact;

(ii) for each open set G in βX, containing X, there is a set $\tilde{X} \subset G$, containing X, such that $\tilde{X}$, as a subspace of βX, is strongly paracompact;

(iii) if G is open in βX and $G \supset X$, then there is an open strongly paracompact subspace $\tilde{X} \subset G$ of βX containing X.

171. Let X be a completely regular space and bX a Hausdorff compactification of X. Show that X is strongly paracompact iff for each compactum F contained in $bX \setminus X$, there is a star-finite cover of X by open sets in bX, whose closures do not intersect F.

172. Show that any paracompactum is collectionwise normal.

173. Show that every connected strongly paracompact space X is Lindelöf.

174. Show that every paracompact space is weakly paracompact.

175. Show that every space X with a uniform base $\mathcal{B}$ is weakly paracompact.

176. Is every locally compact weakly paracompact space paracompact?

177. Is every normal, locally compact, weakly paracompact space paracompact?

178. Does there exist a non-paracompact, normal, locally compact, weakly paracompact and first countable space?

179. Show that a normal weakly paracompact space is countably paracompact.

180. If X is a non-compact, weakly paracompact T_1-space, show that X has an infinite closed discrete subset.

181. Show that a countably compact, weakly paracompact T_1-space is compact.

182. Is the space $T(\omega_1)$ (cf. p. 136) weakly paracompact ?

183. Show that every open point-finite cover γ of a collectionwise normal space X has a σ-discrete open closure refinement.

184. Show that a collectionwise normal space with a countable refinement is metrizable.

185. Show that a Hausdorff space is paracompact iff it is collectionwise normal and weakly paracompact.

186. If a space X has a σ-point-finite base, show that X also has a countable filtration, which consists of point-finite families.

187. If every closed set in X is a G_δ and X has a countable filtration consisting of point-finite families, show that X has a uniform base.

188. If every closed set in X is a G_δ and X has a σ-point-finite base, show that X also has a uniform base.

189. Show that every collectionwise normal, perfectly normal space with a σ-point-finite base is metrizable.

190. If a space X is σ-metacompact, show that it is also screened.

191. If X is a σ-metacompact space and every closed set in X is a G_δ, show that X is weakly paracompact.

192. If a space X has the Souslin property and λ is a family of non-empty open sets in X such that $\mathrm{Kp}(x, \lambda) \leqslant k$ for some $k \in N^+$ and all $x \in X$, show that λ is countable.

193. If X has the Souslin property, $\mathscr{B}$ is a base for X and λ is an uncountable family of nonempty open sets in X of finite multiplicity, show that for each $k \in N^+$ there is a $V \in \mathscr{B}$ such that $\mathrm{Kp}(x, \lambda) \geqslant k$ for all $x \in V$ and $|\{U \in \lambda: U \cap V \neq \Lambda\}| > \aleph_0$.

194. Show that in every Čech-complete space X, there is a sequence $\{\mathscr{B}_n: n \in N^+\}$ of bases such that if $U_n \in \mathscr{B}_n$ and $U_{n+1} \subset U_n$ for each $n \in N^+$, then $\bigcap\{[U_n]: n \in N^+\} \neq \Lambda$.

195. If X is a Čech-complete space and $c(X) \leqslant \aleph_0$, show that every point-finite family λ of nonempty open sets of X is countable.

196. If X is a Čech-complete space and $c(X) \leqslant \aleph_0$, show that the following statements are equivalent:

(i) X is σ-metacompact;
(ii) X is weakly paracompact;
(iii) X is paracompact;
(iv) X is strongly paracompact;
(v) X is Lindelöf.

197. If a regular space X is locally Čech-complete and locally possesses the Souslin property, show that the following statements are equivalent:

(i) X is weakly paracompact;
(ii) X is paracompact;
(iii) X is strongly paracompact;
(iv) X is a free sum of Lindelöf spaces.

198. If a σ-metacompact space X is locally Čech-complete and locally possesses the Souslin property, show that X is screened.

199. If X is a locally Čech-complete space, locally possesses the Souslin property, is perfectly normal and σ-metacompact, show that X is paracompact.

200. If X is a perfectly normal locally compact space, show that the following statements are equivalent: (i) X is σ-metacompact; (ii) X is weakly paracompact; (iii) X is paracompact; (iv) X is strongly paracompact.

201. If a compactum X has the Souslin property and F is a closed set in X, show that the following statements are equivalent; (i) the character $\chi(F, X)$ of $F \in X$ is countable in X; (ii) the space $X \setminus F$ is σ-metacompact.

202. Show that a compactum X possessing the Souslin property is perfectly normal iff every (open) subspace of X is σ-metacompact.

203. Show that every Čech-complete space with a σ-point-finite base, and which has the Souslin property, is a metrizable space with a countable base.

204. Show that a normal space with a countable refinement has a σ-discrete net.

205. Can we assert (without any set-theoretic considerations) that every normal space with a countable refinement is metrizable?

206. Show that a space X is σ-paracompact iff every open cover of X has a closed σ-discrete refinement.

5.7. More Problems

207. Show that for every locally finite open cover $\omega = \{U_\alpha: \alpha \in A\}$ of a normal space X, there is an ω-mapping f of X onto some metric space Y of weight $\tau = |\omega|$. A continuous mapping $f: X \to Y$ is called an *ω-mapping*, where ω is an open cover of X, if there is an open cover λ of Y such that the family $\{f^{-1}V: V \in \lambda\}$ is a refinement of ω.

208. Suppose that X is a set and ξ is a star-monotonic sequence of covers by ξ-perfect sets. Suppose that $\bigcap\{\gamma(x): \gamma \in \xi\} = \{x\}$ for all $x \in X$. Show that the collection $\mathcal{T}$ of all ξ-perfect subsets of X forms a metrizable topology on X.

209. Show that a Hausdorff space X is paracompact iff for every open cover ω of X there is an ω-mapping of X onto some metric space of weight $|\omega|$.

210. Suppose that X is a paracompact Hausdorff space. Show that the following statements are equivalent:

(*a*) the diagonal Δ in a Tychonoff product $X \times X$ is a G_δ-set;
(*b*) X has a T_1-refinement;
(*c*) X contracts to a metric space (cf. p. 41).

211. Show that every open cover ω of a paracompact space X has an open star-refinement ω' such that $|\omega'| \leqslant |\omega|$.

212*. Suppose that X is a paracompactum, τ is a non-measurable cardinal number, $\tau \geqslant \aleph_0$, bX is a Hausdorff compactification of X, and also that the following conditions are satisfied:

(i) the weight of X does not exceed τ;
(ii) $X = \bigcap\{U: U \in \mathcal{E}\}$, where $\mathcal{E}$ is some family of open sets in bX and $|\mathcal{E}| \leqslant \tau$.

Show that the Q-weight of X does not exceed $Q(\tau) \cdot \tau^{\aleph_0}$.

213*. Show that every paracompactum of weight τ is homeomorphic to a closed subspace of the Tychonoff product of some collection of metric spaces, the weight of each of which does not exceed τ.

214*. If the weight of a paracompactum X is a non-measurable cardinal number, show that X is a Q-space.

215*. Suppose that X is a paracompactum, τ is an infinite cardinal number less than the first strongly inaccessible cardinal number, $\tau > \aleph_0$, and the generalized continuum-hypothesis holds. Show that the following statements are equivalent:

(i) the Q-weight of X does not exceed $\tau^{\aleph_0}$.

(ii) X is homeomorphic to a closed subspace of the space $R^{\tau^{\aleph_0}}$ (where R is the real line);

(iii) the weight of X does not exceed $\tau^{\aleph_0}$ and there is a Hausdorff compactification bX of X in which the pseudo-character of X does not exceed $\tau^{\aleph_0}$.

216. Is the product of two paracompacta always a paracompactum?

217. Show that (*a*) if X is a sequential paracompactum and Y is normal and countably compact, then the product $X \times Y$ is normal; (*b*) if X is a paracompactum, $t(X) \leqslant \aleph_6$, Y is normal, and for each countable $B \subset Y$ the set $[B]$ is compact, then $X \times Y$ is normal.

218. Show that the product of a paracompactum X and a locally compact paracompactum Y is a paracompactum.

219. Show that the product $Z = X \times Y$ of a perfectly normal paracompactum X and a metrizable space Y is a perfectly normal paracompactum.

220. Is it always true that the product of a hereditarily paracompact space X and a metrizable space Y is normal?

221. Let X be the segment $[0, 1]$ of the real line, and let X_0 be the set of all irrational points in X. Let $\mathcal{T}$ be the usual topology on X. The topology $\mathcal{T}^*$ on X consists of all sets of the form $U \cup P$, where U is open in the usual topology of the segment and $P \subset X_0$ is an arbitrary subset of X_0. Show that

(*a*) $(X, \mathcal{T}^*)$ is a topological space;

(*b*) $(X, \mathcal{T}^*)$ is hereditarily paracompact;

(*c*) the space $Z = (X, \mathcal{T}^*) \times X_0$ is not normal (where X_0 has the usual topology).

222. Show that the product of a countably paracompact normal space X and a metrizable compactum Y is normal.

223. Show that the product $X \times I$ of a space X and the interval $I = [0, 1]$ of the real line is normal iff X is normal and countably paracompact.

224. True or false? The product of every normal space with an interval is normal.

225. True or false? Every normal space is countably paracompact.

226. Show that every completely regular space with a countable refinement is a p-space.

227. If a p-space has a net of cardinality τ, show that it also has a base of cardinality τ.

228. Show that the paracompact p-spaces and only they are mapped perfectly onto metric spaces.

229. Show that a paracompact p-space with a pointwise countable base is metrizable.

230. Is the image of a p-space under a perfect mapping always a p-space?

231. Show that the image of a paracompact p-space under a perfect mapping is a paracompact p-space.

5.8. SOLUTIONS

5. Fix a countable dense set $A \subset X$. Then $\gamma = \bigcup\{\gamma_a : a \in A\}$, where $\gamma_a = \{U \in \gamma : U \ni a\}$, with $|\gamma_a| \leqslant \aleph_0$ for all $a \in A$. Hence (Problem 19 of Chap. 1) we also have $|\gamma| < \aleph_0$.

6. The solution is similar to that of the preceding problem.

7. For each $U \in \gamma$ put $\lambda_U^1 = \{V \in \gamma : V \cap U \neq \Delta\}$ and if λ_U^k is already defined for some $k \in N^+$, put $\lambda_U^{k+1} = \bigcup\{\lambda_V^1 : V \in \lambda_U^k\}$. The set λ_U^1 is countable; if λ_U^k is countable, then λ_U^{k+1} is also countable. Hence λ_U^n is countable for all $n \in N^+$. Put $\lambda_U = \bigcup\{\lambda_U^n : n \in N^+\}$. Then, obviously, the following hold:

(i) $|\lambda_U| \leqslant \aleph_0$;
(ii) if $V \in \lambda_U$, then $\lambda_V \subset \lambda_U$;
(iii) for every $U \in \gamma$, $V \in \gamma$, either $\lambda_U = \lambda_V$, or $\lambda_U \cap \lambda_V = \Delta$;
(iv) $U \in \lambda_U$; and hence
(v) $\bigcup\{\lambda_U : U \in \gamma\} = \gamma$.

Note that (ii) implies (iii). It is now easy to complete the solution.

8. The family $\{\gamma_\alpha : \alpha \in A\}$, satisfying conditions (i), (ii) and (v), exists by virtue of the preceding problem. Put $X_\alpha = \bigcup\{U : U \in \gamma_\alpha\}$. Then condition (iii) is satisfied; it follows from (iii) and (v) that (vi) holds and (vii) follows from (ii) and (iii). In other words, $\{X_\alpha : \alpha \in A\}$ is a disjoint open cover of X. Consequently, the X_α are closed in X and condition (iv) is satisfied.

9. Suppose that $\{\gamma_\alpha : \alpha \in A\}$ is a family of subfamilies of the covering, possessing properties (i)—(vii) of the preceding problem. By (i) the elements of every γ_α can be enumerated (possibly with repetitions): $\gamma_\alpha = \{U_i^\alpha : i \in N\}$. Put $\theta_i = \{U_i^\alpha : \alpha \in A\}, i \in N$. From properties (iii), (iv), (vi) and (vii) it follows that the θ_i are discrete in X. In addition, by (v), $\bigcup\{\theta_i : i \in N\} = \gamma$. Thus γ is a σ-discrete cover.

10. This follows from Problem 6.

11. The cover γ is certainly star-countable (cf. Problem 10). Consider the families $\{\gamma_\alpha: \alpha \in A\}$ and $\{X_\alpha: \alpha \in A\}$ defined with respect to γ as in Problem 8. Then every X_α has a countable base—to see this it suffices to select a countable base from every element of γ_α and take the union of all such bases.

12. Consider a family $\{\gamma_\alpha: \alpha \in A\}$ of subfamilies of γ, satisfying conditions (i)—(vii) of Problem 8. From (ii), (iv) and (iii) it follows that $|A| = 1$, i.e., that $\gamma = \gamma_\alpha$, where $A = \{\alpha\}$. By (i), $|\gamma| = |\gamma_\alpha| < \aleph_0$ and the problem is solved.

13. Let λ be a locally finite (resp. a point-finite, or star-finite) open refinement of γ. Fix for each $V \in \lambda$ an element of γ containing V and denote it by $\varphi(V)$. Furthermore, put $g(U) = \bigcup\{V \in \lambda: \varphi(V) = U\}$ for each $U \in \gamma$. Then, as is easy to see, $\{g(U): U \in \gamma\}$ is a locally finite (resp. point-finite, or star-finite) open combinatorial refinement of γ.

14. Consider the set $\mathcal{E}$ of all subcovers contained in γ. Impose on $\mathcal{E}$ the partial order relation: $\lambda' < \lambda''$, if $\lambda'' \subset \lambda'$. From $\lambda' < \lambda''$ and $\lambda'' < \lambda'$ it follows that $\lambda' = \lambda''$. To show that the relation $<$ is inductive, suppose that ξ is an arbitrary chain in $\mathcal{E}$. Put $\lambda^* = \bigcap\{\lambda: \lambda \in \xi\}$. λ^* covers X. To see this, let $x \in X$. Consider the set θ of all elements of the cover γ containing x. The set θ is finite: $\theta = \{U_1, \ldots, U_k\}$. If $\lambda^* \cap \theta = \Lambda$, then there are $\lambda_1, \ldots, \lambda_k \in \xi$ such that $U_i \notin \lambda_i$. But among $\lambda_1, \ldots, \lambda_k$ of ξ there is a largest element, let it be λ_{i^*}. Then $\theta \cap \lambda_{i^*} = \Lambda$ but this contradicts the fact that λ_{i^*} covers X. Consequently $\lambda^* \cap \theta \neq \Lambda$, i.e., λ^* is a cover of X. Obviously, $\lambda^* \subset \gamma$, hence $\lambda^* \in \mathcal{E}$. Also $\lambda < \lambda^*$ for all $\lambda \in \xi$. This shows that the set ξ has been ordered inductively, and hence has a maximal element (cf. p. 6). Obviously, the latter is also a maximal cover of X contained in γ.

15. For each $U \in \gamma$, put $1(U) = X \setminus \bigcup\{V \in \gamma: V \neq U\}$. The minimality of γ implies that $1(U) \neq \Lambda$ for each $U \in \gamma$. Moreover, it is clear that $1(U)$ is closed. To show that the family $\eta = \{1(U): U \in \gamma\}$ is discrete in X, let $x \in X$, fix $U^* \in \gamma$ for which $x \in U^*$, and let $V \in \gamma$, $V \neq U^*$. Then $x \in U^* \cap V$ implies that $x \notin 1(V)$, i.e., $U^* \cap 1(V) = \Lambda$. Hence U^* is a nhood of x which intersects at most one element of η (namely, $1(U^*)$) and η is a discrete family. Choose $x(U) \in 1(U)$. Then $\{\{x(U)\}: U \in \gamma\}$ is a discrete family and since each $\{x(U)\}$ is closed, $\{x(U): (U) \in \gamma\}$, a closed discrete set in X, is obviously equipotent to γ.

16. Consider an arbitrary point $x^* \in X^*$. Put $\lambda = \{A \in \gamma: A \ni x^*\}$. From the hypothesis it follows that λ is a nonempty finite set. Let B^* be the largest element of λ (with respect to the given well-ordering $<$). Then $(\bigcup\{p(A): A < B^*\}) \cup (\bigcup\{A: B^* <_c A\}) \supset X^* \ni x^*$. But $\bigcup\{A: B^* <_c A\} \not\ni x^*$. Hence $x^* \in \bigcup\{p(A): A < B^*\} \subset \bigcup\{p(A): A \in \gamma^*\}$ and, consequently, $X^* \subset \bigcup\{p(A): A \in \gamma^*\}$.†

17. Fix some well-ordering $<$ on γ. Construct the set $p(U)$ by induction on $(\gamma, <)$. Let $U^* \in \gamma$ and suppose that for all $U < U^*$, where $U \neq U^*$, an open set $p(U)$ is defined such that the following conditions are satisfied: (i) $[p(U)] \subset U$ and (ii′) if $V < U^*$ and $V \neq U^*$, then $(\bigcup\{p(U): U < V\}) \cup (\bigcup\{U: V <_c U\}) = X$. Put

$$X^* = X \setminus \bigcup\{U \in \gamma: U^* < U\} \quad \text{and} \quad \gamma^* = \{U \in \gamma: U <_c U^*\}.$$

†Recall that $B^* <_c A$ implies that $B^* < A$ and $B \neq A$ (see footnote on p. 7).

Then the conditions of the preceding problem are satisfied. Consequently, $\bigcup\{p(U): U <_c U^*\} \supset X^*$ and we have

$$(\bigcup\{p(U): U <_c U^*\}) \cup (\bigcup\{U: U^* \leqslant U\}) = X. \qquad (*)$$

Put $F^* = X \setminus (\bigcup\{U: U^* <_c U\}) \cup (\bigcup\{p(U): U <_c U^*\})$. Then F^* is closed in X and by (*) $F^* \subset U^*$. Let $p(U^*)$ be any open set containing F^* and whose closure is contained in U^*. Then $(\bigcup\{p(U): U \leqslant U^*\}) \cup (\bigcup\{U: U^* <_c U\}) = X$ and the construction may be continued.

Suppose that $p(U)$ is defined in accordance with the above procedure for all $U \in \gamma$. Then $\{p(U): U \in \gamma\}$ is a cover of X by the preceding problem. Hence conditions (i) and (ii) are fulfilled.

18. Cf. the preceding problem.

19. Let us prove (i). If $x \notin X_n$, there exist mutually distinct elements $U_1, \ldots, U_{n+1}$ of the family γ containing x. Then the set $U = \bigcap\{U_i: i = 1, \ldots, n+1\}$ is open, contains x and does not intersect X_n.

Consider now the closed (by virtue of (i)) set $P_n = X_n \setminus V_{n-1}$ and prove (ii). For each $x \in P_n$, put $F_n(x) = \bigcap\{U \cap P_n: U \in \gamma \text{ and } U \ni x\}$. The family $\xi = \{A \subset X$: there exists $x \in P_n$ such that $A = F_n(x)\}$ obviously refines γ and covers P_n. To show that ξ is discrete in P_n and, consequently, in X, it suffices to show that each $F_n(x)$ is open in P_n and that, for each $x', x'' \in P_n$, either $F_n(x') = F_n(x'')$, or $F_n(x') \cap F_n(x'') = \Lambda$. Since γ is point-finite, $F_n(x)$ is open in P_n (cf. the definition of $F_n(x)$). Assume that $F_n(x') \cap F_n(x'') \neq \Lambda$. Fix $x^* \in F_n(x') \cap F_n(x'')$. Put $\gamma' = \{U \in \gamma: U \ni x'\}$, $\gamma'' = \{U \in \gamma: U \ni x''\}$ and $\gamma^* = \{U \in \gamma: U \ni x^*\}$. From $x' \in X_n$, $x'' \in X_n$, $x^* \in X_n$ it follows that $|\gamma'| = |\gamma''| = |\gamma^*| = n < \aleph_0$. But since $x^* \in F_n(x')$, $\gamma' \subset \gamma^*$, and hence $\gamma' = \gamma^*$. Since $x^* \in F_n(x'')$, $\gamma'' \subset \gamma^*$, and hence $\gamma'' = \gamma^*$. Therefore $\gamma' = \gamma''$ and $F_n(x') = F_n(x'')$. Thus ξ is a discrete (in X) cover of space P_n by closed sets in X and (ii) is proved.

21. It is required to show that for each $H \in \gamma''$ there is a $G \in \gamma$ such that every element H^* of γ'', which intersects H, is contained in G. Fix any $x_0 \in H$ and select $G \in \gamma$ such that $\gamma'(x_0) \subset G$. G is the desired set. Indeed, select $x^* \in H \cap H^*$. Then $x_0 \in H \cup H^* \subset \gamma''(x^*) \subset E$ for some $E \in \gamma'$ (because γ'' is a star-refinement of γ'). Hence $H^* \subset E \subset \gamma'(x_0) \subset G$, so that $\gamma''(H) \subset G$.

24. Suppose that $\xi = \{F_\alpha: \alpha \in A\}$ is a discrete family of closed sets in X. Then the family η of all open sets in X, each of which intersects at most one element of ξ, covers X. Let γ be a star-refinement of η. Then $\gamma(F_{\alpha'}) \cap \gamma(F_{\alpha''}) = \Lambda$ for $\alpha' \neq \alpha''$. Indeed, if $x \in \gamma(F_{\alpha'}) \cap \gamma(F_{\alpha''})$, then $\gamma(x) \cap F_{\alpha'} \neq \Lambda$ and $\gamma(x) \cap F_{\alpha''} \neq \Lambda$. But $\gamma(x) \subset G$ for some $G \in \eta$. Then $G \cap F_{\alpha'} \neq \Lambda$ and $G \cap F_{\alpha''} \neq \Lambda$, contradicting the definition of η. Hence $\gamma(F_\alpha)$, $\alpha \in A$, is a family of mutually disjoint nhoods of F_α, $\alpha \in A$, and X is collectionwise normal.

25. Enumerate the elements of γ: $\gamma = \{\lambda_1, \lambda_2, \ldots\}$ (possibly with repetitions). Without loss of generality, it may be assumed that λ_m is a refinement of λ_n when $m > n$ (because a finite family θ of open covers can be simultanously refined).

We call an open set $H \subset X$ *distinguished* if there exists a natural number n, $\Gamma \in \lambda_n$ and $G \in \xi$ such that $H \subset \Gamma$ and $\lambda_n(H) \subset G$. Set in correspondence to each H some $n = n(H)$ satisfying the above condition. Since γ majorizes ξ, the family η of all distinguished sets covers X. For any point $x \in X$, we claim that the star $\eta(x)$ of x with respect to the cover η is contained in some element of the cover ξ. Put $n^* = \min\{n(H): H \in \eta \text{ and } H \ni x\}$ and select $H^* \in \eta$, $H^* \ni x$ such that $n(H^*) = n^*$. By the definition of $n(H^*)$, there is a $G^* \in \xi$ such that $\lambda_{n^*}(H^*) \subset G^*$. Let us show that $\eta(x) \subset \lambda_{n^*}(H^*)$. If $H' \in \eta$ and $H' \ni x$, then by the definition of n^*, $n' = n(H') \geqslant n^*$. There exists $\Gamma' \in \lambda_{n'}$ such that $H' \subset \Gamma'$. But $\lambda_{n'}$ is a refinement of λ_{n^*} (because $n' \geqslant n^*$); hence there is a $\Gamma^* \in \lambda_{n^*}$ such that $\Gamma' \subset \Gamma^*$. Then $x \in H' \subset \Gamma^*$ and hence $H' \subset \Gamma^* \subset \lambda_{n^*}(x) \subset \lambda_{n^*}(H^*) \subset G^*$. This shows, by the choice of H', that $\eta(x) \subset G^*$. Thus the open cover η is a star-refinement of ξ.

27. This follows directly from Problem 25.

28. For each $i \in N^+$ let γ_i be the cover of the metric space (X, ρ) by all open sets, whose diameter does not exceed $1/i$. Then $\xi = \{\gamma_i: i \in N^+\}$ is obviously a fundamental set of covers of X. It remains to refer to the preceding problem.

29. This is implied by Problems 24 and 28.

30. $\gamma(x)$ is an open set for every $x \in X$ and each ξ-perfect set is the union of such sets.

31. Let $x \in \langle U \rangle$. Select $n \in N$ such that $\gamma_n(x) \subset U$. Then $\gamma_{n+1}(\gamma_{n+1}(x)) \subset U$ (see Problem 20), whence $\gamma_{n+1}(x) \subset \langle U \rangle$. This signifies that whenever $\langle U \rangle$ is open, $\langle U \rangle$ is also ξ-complete.

32. We take as γ the collection of ξ-kernels of all elements of λ. Then γ covers X (because ξ majorizes λ). Furthermore, every element of γ is open and ξ-complete (see Problem 31).

33. (*a*) is obvious and (*b*) follows from (*a*). Let us prove (*c*): if $V \in \gamma_n$, $U' \in \gamma$, $U'' \in \gamma$, $U' < U''$, $U' \neq U''$, $V \cap t_n(U') \neq \Lambda$, then $V \subset U'$ and hence $V \cap t(U'') = \Lambda$; moreover, $V \cap t_n(U'') = \Lambda$. We shall prove (*d*) by contradiction: if $W \in \gamma_{n+1}$, $U' \in \gamma$, $U'' \in \gamma$, $U' \neq U''$ and $W \cap (O_n(U')) \neq \Lambda$, $W \cap (O_n(U'')) \neq \Lambda$, then $\gamma_{n+1}(W) \cap t_n(U') \neq \Lambda$ and $\gamma_{n+1}(W) \cap t_n(U'') \neq \Lambda$. But $\gamma_{n+1}(W)$ is contained in some element V of γ_n which, consequently, satisfies $V \cap t_n(U') \neq \Lambda$ and $V \cap t_n(U'') \neq \Lambda$, but this contradicts (*c*). From (*b*) and (*d*) it obviously follows that $\{O_n(U): U \in \gamma, n \in N\}$ is a σ-discrete open cover of X. It is a refinement of γ, because $O_n(U) \subset U$ for each $U \in \gamma$.

34. We may assume that the sequence ξ is strongly star-monotonic, otherwise we pass to the subsequence consisting of members with even indices. Furthermore, the ξ-kernels of elements of λ form a refinement γ of λ, consisting of ξ-complete (open) sets (Problem 32). Finally, γ can be refined to a σ-discrete open cover by virtue of Problem 33. The latter is obviously also a refinement of λ.

38. Fix for each point $x \in F$ a nhood Ox which intersects at most finitely many elements of γ. From the cover $\lambda = \{Ox: x \in F\}$ of F we select a finite cover of F: $\lambda^* = \{Ox_1, \ldots, Ox_k\}$. Then $F \subset \bigcup\{Ox_i: i = 1, \ldots, k\}$ and $\bigcup\{Ox_i: i = 1, \ldots, k\}$ intersects at most finitely many elements of γ. Moreover, this holds true for F as well.

39. Put $\mathscr{B} = \{U \subset X: U \text{ is open in } X \text{ and } |\{V \in \gamma: V \cap U \neq \Lambda\}| < \aleph_0\}$. Our assumption on γ implies that $\mathscr{B}$ is a base for X. Hence (see Problem 102 of Chap. 2) there is a disjoint family λ of elements of $\mathscr{B}$ such that $[\bigcup\{V: V \in \lambda\}] = X$. Since $c(X) \leqslant \aleph_0$, $|\lambda| \leqslant \aleph_0$. But every element of γ intersects some element of λ, because $\bigcup\{V: V \in \lambda\}$ is dense in X. From $\lambda \subset \mathscr{B}$ and the definition of $\mathscr{B}$ it now follows that $|\gamma| < \aleph_0$.

40. Let $\xi = \bigcup\{\xi_i: i \in N\}$, where every ξ_i is locally finite. Put $W_i = \bigcup\{U: U \in \xi_i\}$, $G_i = \bigcup\{W_j: j \in N \text{ and } j < i\}$ and $\eta_i = \{U \setminus G_i: U \in \xi\}$, $\eta = \bigcup\{\eta_i: i \in N\}$. Then

(i) η is a refinement of ξ;
(ii) η covers X, and
(iii) η is locally finite.

Statement (i) is obvious. To prove (ii) and (iii), fix $x \in X$ and put $i_0 = \min\{i \in N: W_i \ni x\}$. Since ξ covers X, $\bigcup\{W_i: i \in N\} = X$ and i_0 is defined correctly. From $W_{i_0} \ni x$, $W_j \not\ni x$ for $j < i_0$ it follows that ξ_{i_0} covers both x and G_{i_0}. Hence η_{i_0} covers x, i.e., η is a covering. Furthermore, since every ξ_i is locally finite, there is a nhood Ox of x such that $|\{U \in \bigcup\{\xi_i: i \leqslant i_0\}: U \cap Ox \neq \Lambda\}| < \aleph_0$. In addition, the set $\{V \in \bigcup\{\eta_i: i \leqslant i_0\}: V \cap Ox \neq \Lambda\}$ is finite. But W_{i_0} is a nhood of the point x_{i_0}, which does not intersect any element of the family $\bigcup\{\eta_i: i \geqslant i_0\}$. Hence $O^*x = W_{i_0} \cap Ox$ is a nhood of x, which intersects at most finitely many elements of η, i.e., η is locally finite.

41. Consider an arbitrary point $x \in X$ and a sequence $\xi = \{O_i(x): i \in N^+\}$ of nhoods of x, mentioned in the definition of a q-space (see p. 273). Suppose that for every $i \in N^+$ the set $O_i(x)$ intersects infinitely many elements of γ. Then we can select $x_i \in O_i(x)$ and $U_i \in \gamma$ in such a way that $U_i \neq U_j$ for $i \neq j$ and $x_i \in U_i \cap O_i(x)$ for all $i \in N^+$. Since γ is weakly discrete, the set $\{x_i: i \in N^+\}$ must be discrete in X, but it has a limit point because $x_i \in O_i(x)$, a contradiction.

43. It is clear that every cover of an arbitrary set can be combinatorially refined to a disjoint cover (by well-ordering the given cover and taking the difference of each element with the union of all preceding ones). By Problem 42, the latter covering meets the requirements.

44. Let ξ be an open cover of X and γ a locally finite refinement of ξ. Let η be the family of all open sets, which intersect at most finitely many elements of γ. Then η covers X. Let λ be an open closure-refinement of η. Let θ be a locally finite refinement of λ and put $\mu = \{[P]: P \in \theta\}$. Then μ is locally finite, covers X, consists of closed sets and refines η (see Problem 37). Hence every element of μ intersects at most finitely many elements of γ. For each $A \in \gamma$, fix a set $U(A) \in \xi$ such that $A \subset U(A)$. Then $\delta = \{\mu^-(A): A \in \gamma\}$ is a locally finite open cover of X, where $\mu^-(A) \supset A$ for all $A \in \gamma$ (see Problem 49). Then $U(A) \cap \mu^-(A) \supset A$, and hence the family $\{U(A) \cap \mu^-(A): A \in \gamma\}$ covers X. Obviously, it is a refinement of ξ, consists of open sets and is locally finite, because δ is locally finite. Hence X is paracompact.

45. The solution is immediate from Problems 28, 34, 40 and 44.

46. This follows from Problems 41, 43 and 44.

47. It is easy to see that every pointwise countable space is a q-space. The assertion is now immediate from Problem 46.

48. Put $F = \bigcup\{F_\alpha: \alpha \in A\}$, $U = \bigcup\{U_\alpha: \alpha \in A\}$ and $P = X \setminus U$. The set F is closed, since ξ is discrete. Moreover, U is open but P is closed and $U \supset F$, $P \cap F = \Lambda$. The space X is normal, hence there exist open disjoint sets $W \supset P$ and $V \supset F$. Put $V_\alpha = V \cap U_\alpha \supset F \cap U_\alpha = F_\alpha$; we verify that the family $\eta = \{V_\alpha: \alpha \in A\}$ is discrete. Let $x \in X$. If $x \notin U$, then W is a nhood of the point x, which intersects no element of η. If $x \in U$, then $x \in U_{\alpha^*}$ for some $\alpha^* \in A$. Thereupon $U_\alpha \supset V_\alpha$ and the disjointness of

γ imply that U_{α^*} is a nhood of x which intersects at most one element of η. Hence η is discrete.

49. We have $\lambda^-(A) = X \setminus \bigcup \{P: P \in \lambda \text{ and } P \cap A = \Lambda\}$. Obviously, $\lambda^-(A) \supset A$ and $\lambda^-(A)$ is open, since λ is a locally finite closed cover (see Problem 2). Fix a point $x \in X$. Select a nhood Ox of x and a finite family $\mu \subset \lambda$ such that $Ox \subset \bigcup \{P: P \in \mu\}$. Put $\theta = \{A: A \in \gamma \text{ and } Ox \cap \lambda^-(A) \neq \Lambda\}$ and $\theta_P = \{A: A \in \gamma \text{ and } P \cap \lambda^-(A) \neq \Lambda\}$. Then $\theta \subset \bigcup \{\theta_P: P \in \mu\}$. But $P \cap \lambda^-(A) \neq \Lambda$ implies that $P \cap A \neq \Lambda$, hence $\theta_P \subset \{A \in \gamma: P \cap A \neq \Lambda\}$ and, consequently, θ_P is finite. Hence θ also is finite (because μ is finite).

50. Since λ is locally finite, for each point $x \in X$ the set $P(x) = \bigcup \{[G]: G \in \lambda \text{ and } x \notin [G]\}$ is closed in X. Hence $\lambda^-(x) = X \setminus P(x)$ is open. Obviously, $P(x) \not\ni x$ and, consequently, $\lambda^-(x) \ni x$. Put $\xi = \{U \in \lambda: U \ni x\}$. Then $|\xi| < \aleph_0$ and $\bigcap\{U: U \in \xi\} = \lambda^+(x)$ is a nonempty open set containing x (since the family λ is point-finite). Hence $x \in \lambda^0(x) = \lambda^+(x) \cap \lambda^-(x)$ and $\lambda^0(x)$ is an open set, $\lambda^0(x) \in \langle \lambda \rangle$. This shows that $\langle \lambda \rangle$ is an open cover of X. Let $U \in \langle \lambda \rangle$; then $U = \lambda^0(x)$ for some $x \in X$. Fix $V \in \lambda$, $V \ni x$. Then $U = \lambda^0(x) \subset \lambda^+(x) \subset V$, i.e., $U \subset V$, whence $\langle \lambda \rangle$ is a refinement of λ.

51. Put $\lambda \cup \gamma = \mu$. Let $V \in \lambda$. Fix $U \in \gamma$ such that $[V] \subset U$. We verify that $\langle \mu \rangle (V) \subset U$. Let $G \in \langle \mu \rangle$ and $G \cap V \neq \Lambda$. Then $G = \mu^0(x)$ for some $x \in X$. Surely, $x \in [V]$. Otherwise, $\mu^0(x) \subset \mu^-(x) \subset \lambda^-(x) \subset X \setminus [V]$ which contradicts $\mu^0(x) \cap V \neq \Lambda$. We have $\mu^0(x) \subset \gamma^+(x) \subset U$, i.e., $G \subset U$. Thus $\langle \mu \rangle (V) \subset U$. But λ covers X, hence $\{\langle \mu \rangle (x): x \in X\}$ is a refinement of γ. The family $\langle \mu \rangle$ is an open cover of X, since $\lambda \cup \gamma$ is a locally finite open cover (see Problem 50).

52. Enumerate the elements of γ as $\gamma = \{U_i: i \in N\}$, and put $V_i = t(U_i)$. Set $G_i = U_i \setminus \bigcup \{[V_j]: j < i, j \in N\}$, $\lambda = \{G_i: i \in N\}$. Obviously, λ consists of open sets and is a refinement of γ. To show that λ covers X, let $x \in X$. Put $i(x) = \min \{i \in N: U_i \ni x\}$. Then $x \in U_{i(x)}$ and $x \notin U_j$ for $j < i(x)$; moreover, $x \notin [V_j]$ for $j < i(x)$. Hence $x \in G_{i(x)}$ and λ covers X. It is to be shown that λ is locally finite in X. Since $\bigcup \{V_j: j \in N\} = X$, we can select $j(x) \in N$ to satisfy $V_{j(x)} \ni x$. Then $G_i \cap V_{j(x)} = \Lambda$ for all $i > j(x)$ and, consequently, $V_{j(x)}$ is a nhood of x, intersecting at most finitely many elements of λ. Thus λ is locally finite, countable, and refines γ.

53. Refer to Problem 38.

55. Put $W_{k-1} = V_k \setminus F_{k-1}$, where $k \in N^+$, $F_0 = \Lambda$ and $\lambda_0 = \{\Lambda\}$. Obviously, W_{k-1} is open and contains the set $F_k \setminus F_{k-1}$. From $\bigcup \{F_k \setminus F_{k-1}: k \in N^+\} = F_1 \cup (F_2 \setminus F_1) \cup (F_3 \setminus F_2) \cup \ldots = \bigcup \{F_k: k \in N^+\} = X$ it follows that the family $\mu = \{W_k: k \in N\}$ covers X. Since $F_k \subset V_k \subset F_{k+1}$ for all k, $W_{k_1} \supset W_{k_2} = \Lambda$ if $|k_1 - k_2| \geqslant 2$ and μ is star-finite (every element of μ has at most two nhoods). Moreover, λ_k is a finite open cover of W_k. Then $\theta = \bigcup \{\theta_k: k \in N\}$, where $\theta_k = \{W_k \cap U: U \in \lambda_k\}$, is a star-finite open cover of X (see Problem 54), which clearly refines λ.

56. Put $F_0 = \Lambda$, $F_k = \bigcup \{[U_{i,k}]: i \in N \text{ and } i \leqslant k\}$, $V_k = \bigcup \{U_{i,k+1}: i \in N \text{ and } i \leqslant k\}$. V_k is open, F_k is closed, $F_k \subset V_k \subset F_{k+1}$ and $\bigcup \{F_k: k \in N\} = X$. Moreover, $\bigcup \{U_j: j \leqslant k\} \supset F_k$. Then the conditions of Problem 55 are satisfied (the role of λ_k is played by the family $\{U_j: j \leqslant k\}$). Put $W_{k,j} = (V_k \setminus F_{k-1}) \cap U_j$, and obtain an open star-finite refinement $\{W_{k,j}: k \in N, j \in N \text{ and } j \leqslant k\}$ of γ (see solution of the preceding problem).

57. Since X is normal, for each $k \in N$ we can define a sequence $U_{k,j}$ of open sets so that $F_k \subset U_{k,j} \subset [U_{k,j}] \subset U_{k,j+1} \subset V_k$ for all $j \in N$. Put $U_k = \bigcup \{U_{k,j}: j \in N\}$ and $\gamma = \{U_k: k \in N\}$. Then γ covers X, refines μ and the family $\{U_{k,j}: j \in N, k \in N\}$ satisfies the conditions of Problem 56. Hence, γ has a star-finite open refinement and, therefore, so does μ.

58. Suppose that λ is a countable closed refinement of γ. Enumerate λ as $\lambda = \{F_i: i \in N\}$. For each $i \in N$, let U_i be an element of γ containing F_i. Then the families $\{U_i: i \in N\}$ and $\{F_i: i \in N\}$ satisfy the conditions of Problem 57. Hence γ has a star-finite open refinement.

60. Fix $x \in X$ and $U \in \gamma$, $U \ni x$. Then the set $\lambda_U = \{V \in \gamma: V \supset U\}$ is finite; consequently, it has a maximal element, V^* say. Obviously, $V^* \ni x$ and $V^* \in m(\gamma)$, i.e., $m(\gamma)$ is a cover of X.

63. Fix $x \in X$ and $U \in \gamma$, $U \ni x$. Since γ is regular, there is a nhood Ox of x in X, for which $\lambda = \{V \in \gamma: V \cap Ox \neq \Lambda$ and $V \setminus U \neq \Lambda\}$ is finite. But $V \setminus U \neq \Lambda$ for all $V \in \lambda$ distinct from U (because every element of V in γ is maximal). Hence $\{V \in \gamma: V \cap Ox \neq \Lambda\} = \lambda \cup \{U\}$ is a finite set and γ is a locally finite family.

64. In fact, $\lambda = m(\gamma)$ itself is a regular family, as a subfamily of a regular family (Problem 62). Obviously, every element of λ is maximal in λ (because it is maximal in γ and $\gamma \supset \lambda$). But then λ is locally finite, see the preceding problem.

65. It is easy to see that a uniform base is Noetherian. Consequently (Problem 60), $m(\mathcal{B})$ is an open cover of X. It is now required just to show that $m(\mathcal{B})$ is point-finite. However, if a point $x \in X$ belongs to an infinite set ξ of elements of $m(\mathcal{B})$, then ξ is a base for X at x (because $\xi \subset \mathcal{B}$ and $\mathcal{B}$ is a uniform base). But this guarantees that there exist two elements in ξ, one of which contains the other, contradicting that $\xi \subset m(\mathcal{B})$ and that all elements of $m(\mathcal{B})$ are maximal.

66. Define $\mathcal{B}_n$ by induction for all $n \in N^+$ thus: put $\mathcal{B}_1 = \mathcal{B}$ and if $\mathcal{B}_k$ is already defined, then $\mathcal{B}_{k+1} = \mathcal{B}_k \setminus m^*(\mathcal{B}_k)$, where $m^*(\mathcal{B}_k)$ is the collection of all maximal elements of $\mathcal{B}_k$, each of which contains more than one point. From $\mathcal{B}_k \subset \mathcal{B}$ and the uniformity of $\mathcal{B}$ it follows that $m^*(\mathcal{B}_k)$ is a point-finite family of open sets each of which contains more than one point (Problem 65), and it is now easy to see that $\mathcal{B}_{k+1}$ is a uniform base for X. Hence $m(\mathcal{B}_{k+1})$ is an open point-finite cover of X (see Problem 65) and hence $m(\mathcal{B}_n)$, $n \in N^+$, is also. To show that the family $\{m(\mathcal{B}_n): n \in N^+\}$ is refined, fix $x \in X$ and choose $U_n \in m(\mathcal{B}_n)$ such that $x \in U_n$ for each $n \in N^+$. If there is $n \in N^+$ satisfying $|U_n| = 1$, then $\{U_n: n \in N^+\}$ is a base at x. If $|U_n| \geqslant 2$ for all $n \in N$, then from the definition of $\mathcal{B}_n$ it follows that $U_{n'} \neq U_{n''}$ for $n' \neq n''$. Hence $\{U_n: n \in N^+\}$ is an infinite set of elements of the uniform base $\mathcal{B}$ containing x; consequently, it is a base for X. This implies that $\{m(\mathcal{B}_n): N \in N^+\}$ is refined.

68. Suppose that $\mathcal{B}$ is a base for X and $\mathcal{B} = \bigcup \{\gamma_i: i \in N^+\}$, where each γ_i is discrete. For each pair $(i, j) \in N^+ \times N^+$, put $\chi_{i,j} = \{U \in \gamma_i$: there is $V \in \gamma_j$ such that $[V] \subset U\}$, $\lambda_{i,j} = \{V \in \gamma_j$: there is $U \in \gamma_i$ such that $[V] \subset U\}$, $P_{i,j} = \bigcup \{[V]: V \in \lambda_{i,j}\}$, $G_{i,j} = X \setminus P_{i,j}$, $\theta_{i,j} = \chi_{i,j} \cup \{G_{i,j}\}$.

The set $P_{i,j}$ is closed, $G_{i,j}$ is open, $\theta_{i,j}$ is locally finite and covers X and every point of $P_{i,j}$ is uniformly covered by an element of $\theta_{i,j}$, which belongs to $\chi_{i,j}$. Enumerate the elements of the countable family $\xi = \{\theta_{i,j}: (i, j) \in N^+ \times N^+\}$ in a sequence: $\xi = \{\mu_1, \ldots, \mu_k, \ldots\}$ and let ν_k be the derivative (see p. 271) of $\{\mu_1, \ldots, \mu_k\}$. Since all $\mu_1, \ldots, \mu_k$ are locally finite open covers of X, ν_k is also a locally finite open cover of X (Problems 3 and 36). To show that $\nu = \bigcup\{\nu_k: k \in N^+\}$ is a regular base for X, choose any point $x \in X$ and any nhood Ox of x. Select $U \in \mathcal{B}$ and $V \in \mathcal{B}$ such that $x \in V \subset [V] \subset U \subset Ox$. Fix $i^* \in N^+$ and $j^* \in N^+$ such that $U \in \gamma_{i^*}$ and $V \in \gamma_{j^*}$. Then $U \in \chi_{i^*,j^*}$, $V \in \lambda_{i^*,j^*}$ and consequently $[V] \subset P_{i^*,j^*}$. Hence it follows that at most one element of θ_{i^*,j^*} intersects $[V]$, namely U. Let $\theta_{i^*,j^*} = \mu_{k^*}$. Then for each element W of ν_k, where $k \geqslant k^*$, such that $[V] \cap W \neq \Lambda$, the element of μ_{k^*} associated with W must coincide with U. Consequently, $W \subset U$, i.e., $\bigcup\{\nu_k([V]): k \in N^+$ and $k > k^*\} \subset U$. Since ν_{k^*} is a cover, there is a $W \in \nu_{k^*}$ such that $x \in W$. Then $W \cap [V] \neq \Lambda$ and hence $x \in W \subset U \subset Ox$, i.e., ν is a base. But $\beta_{k^*} = \bigcup\{\nu_k: k \in N^+$ and $k < k^*\}$ is a locally finite family of sets. Consequently, there is a nhood O_1x of x in X such that $|\{G \in \beta_{k^*}: G \cap O_1x \neq \Lambda\}| < \aleph_0$. Put $O_2x = V \cap O_1x$. Then, obviously, $\{G \in \nu: G \cap O_2x \neq \Lambda$ and $G \cap (X \setminus U) \neq \Lambda\} \subset \{G \in \beta_k: G \cap O_1x \neq \Lambda\}$ is finite. Hence ν is a regular base.

69. Suppose that P is closed in X and $x \in X \setminus P$. There is a nhood Ox of x such that $Ox \cap P = \Lambda$ and $\lambda = \{U \in \mathcal{B}: U \cap Ox \neq \Lambda$ and $U \cap P \neq \Lambda\}$ is finite. For each $U \in \lambda$, we have $|U| \geqslant 2$. Since X is a T_1-space, we conclude that $\mathcal{B}' = \mathcal{B} \setminus \lambda$ is a base for X. Then $\gamma = \{U \in \mathcal{B}': U \cap P \neq \Lambda\}$ covers P and $(\bigcup\{U: U \in \gamma\}) \cap Ox = \Lambda$, i.e., $G = \bigcup\{U: U \in \gamma\}$ and Ox are disjoint nhoods of P and x. Hence X is regular.

70. Let γ be an open cover of X. Put $\gamma_{\mathcal{B}} = \{U \in \mathcal{B}$: there exists $G \in \gamma$ such that $U \subset G\}$. Then, obviously, $\gamma_{\mathcal{B}}$ is a base for X and the family $\gamma\mathcal{B}$ is a refinement of γ. By Problem 62, $\gamma_{\mathcal{B}}$ is regular. Then $m(\gamma_{\mathcal{B}})$, the surface of $\gamma_{\mathcal{B}}$, is a locally finite refinement of γ (see Problem 64).

71. X has a regular base (Problem 68), and every space with a regular base is paracompact (preceding problem). Other solutions are also possible.

72. From Problem 40 it follows that every open cover of X has a locally finite refinement. But X is regular, hence Problem 44 can be invoked to conclude that X is paracompact.

73. Suppose that $\sigma = \bigcup\{\omega_i: i \in N^+\}$ is a base for X, where every ω_i is locally finite in X. Consider arbitrary disjoint closed sets F_1 and F_2 in X. Put $\sigma_1 = \{U \in \sigma: U \cap F_1 \neq \Lambda$ and $[U] \cap F_2 = \Lambda\}$, $\sigma_2 = \{U \in \sigma: U \cap F_2 \neq \Lambda$ and $[U] \cap F_1 = \Lambda\}$, $\omega_i' = \sigma_1 \cap \omega_i$, $\omega_i'' = \sigma_2 \cap \omega_i$. It is clear that $\sigma_1 = \bigcup\{\omega_i': i \in N^+\}$ and $\sigma_2 = \bigcup\{\omega_i'': i \in N^+\}$. Moreover, $F_1 \subset \bigcup\{U: U \in \sigma_1\}$ and $F_2 \subset \bigcup\{U: U \in \sigma_2\}$. Put $\Gamma_i' = \bigcup\{U: U \in \omega_i'\}$ and $\Gamma_i'' = \bigcup\{U \in \omega_i''\}$. We have $F_1 \subset \bigcup\{\Gamma_i': i \in N^+\}$ and $F_2 \subset \bigcup\{\Gamma_i'': i \in N^+\}$. By virtue of the local finiteness of ω_i' and ω_i'', we have $[\Gamma_i'] = \bigcup\{[U]: U \in \omega_i'\}$, $[\Gamma_i''] = \bigcup\{[U]: U \in \omega_i''\}$ (see Problem 2). The definition of σ_1 and σ_2 yields $[\Gamma_i'] \cap F_2 = \Lambda$ and $[\Gamma_i''] \cap F_1 = \Lambda$ for all $i \in N^+$. Put $V_i' = \Gamma_i' \setminus \bigcup\{[\Gamma_j'']: j \in N^+$ and $j \leqslant i\}$, $V_i'' = \Gamma_i'' \setminus \bigcup\{[\Gamma_j']: j \in N^+$ and $j \leqslant i\}$. Clearly, $F_1 \subset \bigcup\{V_i': i \in N^+\}$, $F_2 \subset \bigcup\{V_i'': i \in N^+\}$ and $(\bigcup\{V_i': i \in N^+\}) \cap \bigcup\{V_i'': i \in N^+\}) = \Lambda$.

74. The normality of such a space has already been established in the preceding problem. It remains to show that every open set V in X is an F_σ. Suppose that $\sigma = \bigcup\{\omega_i: i \in N^+\}$ is a base for X, and each ω_i is locally finite. Put $\sigma(V) = \{U \in \sigma: [U] \subset V\}$ and $\omega_i' = \sigma(V) \cap \omega_i$. It is obvious that $\sigma(V) = \bigcup\{\omega_i': i \in N^+\}$ and $V = \bigcup\{[U]: U \in \sigma(V)\}$.

Since ω_i' is locally finite in X, $F_i = \bigcup\{[U]: U \in \omega_i'\}$ is closed in X (see Problem 2). Moreover, $F_i \subset V$ and $V = \bigcup\{F_i: i \in N^+\}$.

75. If $\gamma_n(x) \ni y$, then $\gamma_n(y) \ni x$ and conversely; hence $n(x, y) = n(y, x)$ and $f(x, y) = f(y, x)$, proving (i). If $x \neq y$, then there exists $n \in N$ such that $\gamma_n(x) \not\ni y$ (or $\gamma_n(y) \not\ni x$); hence $n(x, y)$ is defined, $n(x, y) \geqslant 1$ and $f(x, y) \neq 0$. From (i) it follows that $d(x, y) = d(y, x)$. Let $x, y, z \in X$ and let a, b be the integers, $a > 0$ and $b > 0$, and suppose that there exist chains θ_1 from x to y of length less than a and θ_2 from y to z of length less than b. Then there is a chain θ from x to z, whose length does not exceed $a + b$; in order to obtain such a chain, it suffices to extend θ_1 to θ_2. Consequently, $d(x, z) \leqslant d(x, y) + d(y, z)$. From (ii) follows (ii′) (the left-hand side of (ii′) is generally trivial), and from (ii′) it follows that if $x \neq y$, then $d(x, y) \neq 0$. Thus, in order to prove (iii) it remains to prove (ii).

The proof of (ii) will be carried out by induction with respect to the rank of a chain. If $R(\theta) = 1$, then (ii) holds. We assume that for the chain of rank not exceeding $l + 1$ relation (ii) holds, and consider the chain $\theta = \{x_0, \ldots, x_{l+1}\}$ of rank $l + 2$. Put $a = L(\theta)$ and suppose that k is a greatest integer, for which $L\{x_0, \ldots, x_k\} \leqslant a/2$. Then $L\{x_0, \ldots, x_k, x_{k+1}\} > a/2$ and hence $L\{x_{k+1}, \ldots, x_{l+1}\} = a - L\{x_0, \ldots, x_k, x_{k+1}\} < a/2$. By the induction hypothesis $f(x_0, x_k) \leqslant 2L\{x_0, \ldots, x_k\} \leqslant 2\cdot(a/2) = a$ and $f(x_{k+1}, x_{l+1}) \leqslant 2L\{x_{k+1}, \ldots, x_{l+1}\} < 2a/2 = a$. Moreover, $f(x_k, x_{k+1}) \leqslant L(\theta) = a$. Put $m = \min\{n \in N\colon (1/2^n) \leqslant a\}$. Then, $f(x_0, x_k) \leqslant 1/2^m, f(x_k, x_{k+1}) \leqslant 1/2^m$ and $f(x_{k+1}, x_{l+1}) \leqslant 1/2^m$. Hence $\gamma_{m-1}(x_0) \ni x_k, \gamma_{m-1}(x_k) \ni x_{k+1}$ and $\gamma_{m-1}(x_{k+1}) \ni x_{l+1}$, i.e., U, V and $W \in \gamma_{m-1}$ can be found such that $x_0, x_k \in U$, $x_k, x_{k+1} \in V$, $x_{k+1}, x_{l+1} \in W$. Then $x_0, x_{l+1} \in U \cup V \cup W \subset \gamma_{m-1}(V) \subset G$ for some $G \in \gamma_{m-2}$. Hence $\gamma_{m-2}(x_0) \ni x_{l+1}$ and $n(x_0, x_{l+1}) > m - 2$, $f(x_0, x_{l+1}) \leqslant 1/2^{m-1} \leqslant 2a$, proving relation (ii) (and together with this the relations (ii′) and (iii) as well).

Let us prove (v): if $d(x, y) < 1/2^n$, then $f(x, y) < 1/2^{n-1}$ by (ii′), and hence $\gamma_{n-1}(x) \ni y$. To prove (iv): if $y \in \gamma_n(x)$, then $n(x, y) > n$ and $f(x, y) < 1/2^n$. Moreover (by the trivial side of (ii′)), $d(x, y) < 1/2^n$.

Since $\{\gamma_n\colon n \in N\}$ is a refinement, the sets $\{\gamma_n(x)\colon n \in N\}$ form a base for the given topology on X at the point x. Hence, from the relations (iv) and (v) it follows immediately that the metric d generates the same topology on X.

76. Every γ_i has a locally finite open refinement λ_i. $\lambda = \bigcup\{\lambda_i\colon i \in N\}$ is then a base for X, because $\{\lambda_i\colon i \in N\}$ is a refinement. This base is σ-locally finite. Moreover, X is regular (Problem 129). Hence (Problem 83) X is metrizable. Another solution can also be obtained from Problems 51, 21 and 75.

77. Suppose that $\xi = \{\lambda_1, \lambda_2, \ldots\}$ is a countable fundamental set of covers of X. Define by induction the sequence $\{\gamma_1, \gamma_2, \ldots\}$ of open covers of X thus: put $\gamma_1 = \lambda_1$ and if γ_k is already defined, then take γ_{k+1} as any open cover of X which strongly star-refines both λ_{k+1} and γ_k (such γ_{k+1} exists, see Problems 25 and 21). Then $\gamma_1, \gamma_2, \ldots, \gamma_n, \ldots$ is a strongly star-monotonic and refined sequence of open covers of the T_0-space X, and from Problem 75 it follows that X is metrizable. The proof of the converse statement is trivial (see solution of Problem 28).

78. (*a*) Suppose that there exists in X a k-refined sequence $\xi = \{\omega_i\colon i \in N^+\}$ of open covers. We may assume without loss of generality that ω_{i+1} is a refinement of ω_i. To prove the metrizability of X (by Problem 77), it suffices to show that ξ forms a fundamental set of open covers of X. Let $x_0 \in X$ and let Ox_0 be any nhood of x_0. It is required to show that there is a smaller nhood O_1x_0, satisfying $\omega_n(O_1x_0) = \bigcup\{U\colon U \in \omega_n,$

$U \cap O_1x_0 \neq \Delta\} \subset Ox_0$ for some n. Assume the contrary. Consider the nhoods $O_nx_0 = \omega_n(x_0) = \bigcup\{U\colon U \in \omega_n, x_0 \in U\}$ of x_0. Since ξ is a refinement (by assumption ξ is even a k-refinement), the family of nhoods $\{O_nx_0\colon n \in N^+\}$ forms a base at x_0. Moreover, $O_{n+1}x_0 \subset O_nx_0$ (because ω_{n+1} is a refinement of ω_n). By virtue of our assumption we have $\omega_n(O_nx_0) \cap (X \setminus Ox_0) \neq \Delta$, $n \in N^+$. Hence $x_n \in O_nx_0$ can be chosen such that $\omega_n(x_n) \cap (X \setminus Ox_0) \neq \Delta$. It is obvious that $\lim_{n\to\infty} x_n = x_0$. Hence $F = \{x_n\colon n \in N+$ and $x_n \in Ox_0\} \cup \{x_0\}$ is a compact subspace of X, of which Ox_0 is a nhood. From the definition of x_n we obtain $\omega_n(F) \cap (X \setminus Ox_0) \supset \omega_n(x_n) \cap (X \setminus Ox_0) \neq \Delta$ for all n. But this contradicts that the sequence ξ is a k-refinement.

(*b*) Conversely, suppose that X is a metrizable space and ρ is a metric on X, compatible with the topology on X. Consider a sequence $\xi = \{\omega_n\colon n \in N+\}$ of open covers of X such that for each $U \in \omega_n$ we have diam $U < 1/n$. To show that ξ is a k-refinement, suppose that $F \subset X$ is an arbitrary compact subspace and UF is any nhood of F. It is clear that $\rho(F, X \setminus UF) = \rho_0 > 0$ (see Problem 288 of Chap. 3). Then, if $n \in N+$ and $(1/n) < \rho_0$, we have $\omega_n(F) \subset UF$.

79. See Problem 39 of Chap. 3. To obtain another solution, suppose that $\mathcal{B}$ is a countable base for the regular space X. For each pair U, V of elements of $\mathcal{B}$ such that $[V] \subset U$, take an open cover $\gamma_{U,V}$ of X, consisting of two elements U and $X \setminus [V]$. The set of all $\gamma_{U,V}$ so defined is obviously a countable fundamental set of covers of X. Hence (Problem 77) X is metrizable.

80. Suppose that X is a regular space and $\mathcal{B} = \bigcup\{\lambda_n\colon n \in N^+\}$ is a base for X, where every λ_n is a discrete (in X) family of sets, i.e., every point of X has a nhood which intersects at most one element of λ_n. Fix an arbitrary ordered pair (i, j) of natural numbers. Put $\theta_{i,j} = \{U \in \lambda_i\colon$ there exists $V \in \lambda_j$ such that $[V] \subset U\}$, $\chi_{i,j} = \{V \in \lambda_j\colon$ there exists $U \in \lambda_i$ such that $[V] \subset U\}$ and $P_{i,j} = \bigcup\{[V]\colon V \in \chi_{i,j}\}$, $G_{i,j} = X \setminus P_{i,j}$, $\gamma_{i,j} = \theta_{i,j} \cup \{G_{i,j}\}$. Since $\chi_{i,j} \subset \lambda_j$ and λ_j is a discrete (consequently conservative, see Problem 2) family of sets, $P_{i,j}$ (resp. $G_{i,j}$) is closed (resp. open) in X. It is clear that $\theta_{i,j}$ covers $P_{i,j}$, hence $\gamma_{i,j}$ is an open cover of X. It is to be shown that $\xi = \{\gamma_{i,j}\colon (i, j) \in N^+ \times N^+\}$ is a fundamental set of covers of X.

Let x be any point of X and Ox any nhood of x in X. There exists $U^* \in \mathcal{B}$ such $x \in U^* \subset Ox$. Fix a number i^* such that $U^* \in \lambda_{i^*}$. By virtue of the regularity of X, there is a $V^* \in \mathcal{B}$ such that $x \in [V^*] \subset U^*$. Let $V^* \in \lambda_{j^*}$. Then $U^* \in \theta_{i^*,j^*} \subset \gamma_{i^*,j^*}$, $V^* \in \chi_{i^*,j^*}$ and hence $V^* \subset P_{i^*,j^*}$. Consequently, $G_{i^*,j^*} \not\ni x$. Hence even from the discreteness of θ_{i^*,j^*} it follows that no element of γ_{i^*,j^*} other than U^*, contains the point x. Therefore $\gamma_{i^*,j^*}(x) = U^* \subset Ox$ and ξ is fundamental. ξ is obviously countable. By Problem 77, X is metrizable. There are other possible solutions, find them. Prove also the converse statement.

81. False. The space I^* given in Problem 209 of Chap. 3 is a non-metrizable paracompactum with a σ-disjoint base.

82. True. Suppose that $\gamma = \{U_\alpha\colon \alpha \in A\}$ is a family of disjoint open sets in X. Put $F = X \setminus \bigcup\{U_\alpha\colon \alpha \in A\}$. The set F is closed. If $\{G_i\colon i \in N^+\}$ is a (countable) family of open sets in X such that $F = \bigcap\{[G_i]\colon i \in N^+\}$, put $V_\alpha^i = U_\alpha \cap (X \setminus [G_i])$ and $\lambda_i = \{V_\alpha^i\colon \alpha \in A\}$, $i \in N^+$. Obviously, every λ_i is not only a disjoint but also a discrete family of open sets in X. Also, $\bigcup\{\lambda_i\colon i \in N+\}$ covers exactly the same points of X as γ does.

From this it follows that every perfectly normal space with a σ-disjoint base has a σ-discrete base, and hence (Problem 80) is metrizable.

83. Suppose that $\{\gamma_i: i \in N^+\}$ is a sequence of locally finite families of open sets in X, where $\mathcal{B} = \bigcup\{\gamma_i: i \in N^+\}$ is a base for X. Without loss of generality, we may assume that every γ_i is a cover of X. Put $\lambda_k = \bigcup\{\gamma_i: i \leqslant k\}$. Then λ_k is a locally finite family of sets (Problem 35), $\lambda_i \subset \lambda_j$ for $j > i$ and $\bigcup\{\lambda_i: i \in N^+\} = \mathcal{B}$. Put $\theta_k = \langle \lambda_k \rangle = \{\lambda_k^0(x): x \in X\}$; we verify that $\xi = \{\theta_k: k \in N^+\}$ is a fundamental set of covers of X. By virtue of Problem 50, θ_k is an open cover of X. Fix a point $x \in X$ and a nhood Ox of x in X. Choose numbers j^* and i^* and sets $V \in \lambda_{j^*}$, $U \in \lambda_{i^*}$ such that $j^* > i^*$ and $x \in V \subset [V] \subset U \subset Ox$: this is possible (why?). We shall show that $\theta_{j^*}(V) = U \subset Ox$. An arbitrary element of θ_{j^*} has the form $\lambda_{j^*}^0(x^*)$ for some $x^* \in X$. Assume that $\lambda_{j^*}^0(x^*) \cap V \neq \Lambda$. Then $x^* \in [V]$: otherwise we would have $\lambda_{j^*}^0(x^*) \subset \lambda_{j^*}^-(x^*) \subset X \setminus [V]$, since $V \in \lambda_{j^*}$. But $[V] \subset U$, hence $x^* \in U$ and $\lambda_{j^*}^0(x^*) \subset \lambda_{j^*}^+(x^*) \subset U$, because $U \in \lambda_{i^*} \subset \lambda_{j^*}$. This shows that $\theta_{j^*}(V) \subset U \subset Ox$. Hence ξ is a fundamental (countable) family of covers of X and X is a metrizable space (Problem 77).

84. Suppose that $\sigma = \bigcup\{\omega_i: i \in N^+\}$ is an NS-base for X. X is perfectly normal. Without loss of generality, we may assume that every ω_i is a locally finite open cover of X. By modifying slightly the reasoning in the solution of Problem 207, we find for each ω_i a metrizable space Y_i and a continuous mapping $g_i: X \to Y_i$ such that for every $U \in \omega_i$ the set g_iU is open in Y_i and $g_i^{-1}g_iU = U$. Consider the diagonal product g of the mappings g_i, $g: X \to Y$, where $Y \subset \Pi\{Y_i: i \in N^+\}$ (and $g(X) = Y$). As a subspace of the metrizable space $\Pi\{Y_i: i \in N^+\}$ Y is metrizable (Problem 194 of Chap. 2). The mapping g is continuous (Problem 290 of Chap. 2). Moreover, for every $U \in \sigma$, gU is certainly open in $g(X)$ and $g^{-1}gU = U$. Since σ is a base and gU is open in $g(X)$ for every $U \in \sigma$, g is an open mapping. Furthermore, if $x_1 \neq x_2$, $x_1 \in X$, $x_2 \in X$, we take $U_1, U_2 \in \sigma$ such that $x_1 \in U_1$, $x_2 \in U_2$, $U_1 \cap U_2 = \Lambda$. Then $U_1 = g^{-1}gU_1$, $U_2 = g^{-1}gU_2$, $gU_1 \cap gU_2 = \Lambda$ and, consequently, $gx_1 \neq gx_2$. Thus g is one-to-one, continuous and open, i.e., it is a topological mapping of X onto the metrizable space $g(X)$.

85. A T_1-space with a uniform base has a countable refinement (Problem 66). Also every paracompactum with a refinement is metrizable (Problem 76).

86. False. The space I^* in Problem 209 of Chap. 3 is an appropriate example in this respect.

87. (*a*) Suppose that σ is a pointwise countable base for X. Without loss of generality, we may assume that every subspace $U \in \sigma$ has a countable base (removing, if necessary, all those U from σ which do not have a countable base). Then (Problem 11) X is representable as a union of disjoint clopen subspaces, each of which has a countable base.

From Problem 89 it now follows that X is metrizable. The statement (*b*) follows from (*a*) and Problem 121 of Chap. 3.

88. X is regular (Problem 69) and paracompact (Problem 70). Consequently, X is a paracompactum. Since a regular base is a special case of a uniform base, it remains to refer to Problem 85. The converse is left as an exercise to the reader.

Another solution is possible, which is based on the successive separation of the surfaces of a regular base; see Problem 64 and the solution of Problem 66, followed by reference to Problem 83.

89. In every U_α, fix a σ-locally finite base $\mathscr{B}_\alpha = \bigcup\{\lambda^i_\alpha : i \in N\}$. Put $\lambda^i = \bigcup\{\lambda^i_\alpha : \alpha \in A\}$. Since γ and each λ^i_α is locally finite, λ^i is as well. Hence it is obvious that $\mathscr{B} = \bigcup\{\lambda^i : i \in N\}$ is a σ-locally finite base for X. Consequently, X is metrizable (Problem 83).

90. See Problem 89.

91. See Arkhangel'skii [11].

92. If X is a complete metrizable metric space, then it is a G_δ-set in every Hausdorff extension of X (Problem 235 of Chap. 2). We prove the converse. Suppose that X is a metric space, $\tilde{X}$ is a completion of X and F is any Hausdorff compactification of $\tilde{X}$. Since X is dense in $\tilde{X}$, F is also a Hausdorff compactification of X. But then it follows that X is a G_δ-set in F (see Problem 28 of Chap. 4). Moreover, X is a G_δ-set in completion $\tilde{X}$. Then X itself is also a complete metrizable metric space (Problem 234 of Chap. 2).

93. Locally X has a countable base (because a metrizable compactum has always a countable base). Since X is also weakly paracompact, X has a pointwise countable base. Hence (see Problem 87) X is metrizable.

94. Make use of the constructions described in the solutions of Problems 7, 8 and 11 (see also Problem 168).

95. See Problem 139 of Chap. 3.

96. We first prove (*a*) and then, with its aid the stronger assertion (*b*). Let $A \subset X$, $\tilde{A} = \{y \in X, d(y, A) = 0\}$. If we show that $\tilde{A}$ is closed in X, then from Problem 114 of Chap. 2 it follows that (*a*) holds. Assume that $z \in X \setminus \tilde{A}$ and $d(z, \tilde{A}) = 0$. Choose points $y_n \in \tilde{A}$ for all $n \in N+$ such that $d(y_n, z) < 1/n$. The set S consisting of all the chosen y_n and the point z is a compactum (since every subsequence of the sequence $\{y_n : n \in N^+\}$ converges to z). Since $y_n \in \tilde{A}$, for each $n \in N^+$ there is $x_n \in A$ such that $d(x_n, y_n) < 1/n$.

Denote by P the set of all points of the sequence $\eta = \{x_n : n \in N^+\}$. Surely, $y_n \notin A$ for all sufficiently large n; otherwise, we would have $z \in \tilde{A}$. Hence we may assume that $y_n \notin A$ for all $n \in N+$. Then $P \cap S = \Lambda$ and $d(P, S) \leqslant d(x_n, y_n) < 1/n$ for all $n \in N+$, i.e., $d(P, S) = 0$. But then P is not closed. Hence there is $z^* \in \tilde{A}$ and a subsequence $\eta^* = \{x_{n_k} : k \in N^+\}$ of the sequence η such that $\lim_{k\to\infty} d(z^*, x_{n_k}) = 0$ and consequently $\lim_{k\to\infty} x_{n_k} = z^*$. From $z^* \in \tilde{A}$ it follows that $z^* \neq z$. The set P^*, consisting of all points in η^* and the point z^*, is a compactum and hence is closed in X (since X is a Hausdorff space; see Problem 56 of Chap. 3). From $P^* \cap S = \Lambda$ it now follows that $d(P^*, S) \neq 0$. But $d(P^*, S) \leqslant d(x_{n_k}, y_{n_k}) < 1/n_k$ for all $k \in N+$ and hence $d(P^*, S) = 0$, a contradiction. The statement (*a*) is proved.

Let us prove (*b*). For each $x \in X$, put $U_n x = \operatorname{Int}\{y \in X : d(x, y) < 1/n\}$ and $\gamma_n = \{U_n x : x \in X\}$ for all $n \in N^+$. From (*a*) and Problem 114 of Chap. 2 it follows that $U_n x \ni x$ for all $x \in X$. Hence every γ_n is an open cover of X. We show that the family $\{\gamma_n : n \in N+\}$ of covers is k-refined. Then the metrizability of X follows from Problem 78.

Suppose that F is any compactum contained in X, and V is an arbitrary open set in X containing F. Assume that for each $n \in N+$ we are able to choose $U_n x_n \in \gamma_n$

such that we have $U_n x_n \cap F \neq \Lambda$ and $U_n x_n \cap (X \setminus V) \neq \Lambda$. Fix any $z_n \in U_n x_n \cap (X \setminus V)$ and $y_n \in U_n x_n \cap F$. Since F is a metrizable compactum (see Problem 95), there is a subsequence $\{y_{n_k}: k \in N^+\}$ which converges in F. But $d(x_{n_k}, y_{n_k}) < 1/n_k$, $k \in N^+$. Arguing as above (in the proof of (*a*)) we conclude that $\lim_{k\to\infty} x_{n_k} = \lim_{k\to\infty} y_{n_k} \in F$. Reiterating this procedure we deduce from $d(x_{n_k}, z_{n_k}) < 1/n_k$ that $\lim_{k\to\infty} z_{n_k} = \lim_{k\to\infty} x_{n_k} \in F$. But $z_{n_k} \in X \setminus V$, and since $X \setminus V$ is closed, $\lim_{k\to\infty} z_{n_k} \in X \setminus V$. Thus $(X \setminus V) \cap F \neq \Lambda$, a contradiction.

97. Suppose that ρ is a metric compatible with the topology of X. Since X is not compact, there is an infinite countable set $A = \{a_i: i \in N^+\}$ without a limit point in X. Consider the family $\gamma = \{O_i^j: i \in N^+, j \in N^+\}$ of open sets in X, having the following properties:

(1) $a_i \in O_i^j$, $j \in N^+$, $i \in N^+$;
(2) $O_i^j \cap O_{i'}^j = \Lambda$, if $i \neq i'$ for every $j \in N^+$;
(3) $\operatorname{diam} O_i^j < 1/i$, $i \in N^+$;
(4) $[O_i^{j+1}] \subset O_i^j$.

It is obvious that such a family exists. Note also that the family $\gamma_j = \{O_i^j: i \in N^+\}$ is discrete in X.

Now define the open sets U_k by $U_k = \bigcup \{O_i^k: i \geqslant k\}$ in X. Obviously, we have $\bigcap \{U_k: k \in N^+\} = \Lambda$ and $[U_{k+1}] \subset U_k, k \in N^+$. Consider the space $\tilde{X} = X \cup \{\xi\}$, obtained from X by adding the point ξ, the basic nhoods of which are defined to be the sets $\{\xi\} \cup U_k = V_k$. Obviously $\tilde{X}$ with this topology is a T_1-space. To show that $\tilde{X}$ is regular, let $x \in \tilde{X}$ and Ox be any nhood of x. If $x = \xi$, we may assume that $Ox = V_k$. Then $[V_{k+1}]_{\tilde{X}} \subset V_k$, because $[U_{k+1}] \subset U_k$. If $x \neq \xi$, we take k_0 to be such that $x \notin U_{k_0}$. Putting $O'x = X \setminus [U_{k_0+1}]_X$, obviously $x \in O'x$, $O'x \cap V_{k_0} = \Lambda$, $\xi \in V_{k_0}$. Take a nhood $O_1 x$, satisfying $[O_1 x]_X \subset O'x$. Then $[O_1 x]_X = [O_1 x]_{\tilde{X}} \subset O'x \subset Ox$. Thus, $\tilde{X}$ is regular. It is to be shown that $\tilde{X}$ has a σ-locally finite base.

Consider a σ-locally finite base of X: $\sigma = \bigcup \{\sigma_n: n \in N^+\} = \{G_{nt}: t \in T_n, n \in N^+\}$, where $\sigma_n = \{G_{nt}: t \in T_n\}$ is locally finite in X. Without loss of generality, we may assume that every $O_i^j \in \gamma$ intersects at most finitely many elements of the system σ_j. Then the set U_j intersects at most a countable number of elements of σ. Suppose that U_j has nonempty intersection at most with the sets $G_{jt_1}, G_{jt_2}, \ldots, G_{jt_n}, \ldots$

Consider the family $\theta^j = \{G_{st}^j: t \in T_s, s = 0, 1, 2, \ldots\}$, where, for $s > 0$, $\theta_s^j = \{G_{st}^j: t \in T_s\}$ consists of just one element of G_{jt_s} but for $s = 0$ it consists of all elements of $G_{jt} \in \sigma_j$ disjoint from U_j. The family $\theta = \bigcup \{\theta^j: j \in N^+\}$ is obviously a σ-locally finite base for X, and the family $\Sigma = \theta \cup \xi$, where $\xi = \{V_k: k \in N^+\}$, is a σ-locally finite base for $\tilde{X}$ (here $V_k = \{\xi\} \cup U_k$, $k \in N^+$).

98. (*a*) If a metrizable space X is compact, then every metric ρ on X, compatible with its topology, is complete (see Problem 255 of Chap. 3).

(*b*) Conversely, if a metrizable space X is non-compact, then there is (see Problem 97) a metrizable extension (generally speaking, non-compact) $\tilde{X}$, distinct from X. Consider a metric $\tilde{\rho}$ on $\tilde{X}$, compatible with its topology. Then $\tilde{\rho}$ induces on X a metric $\rho = \tilde{\rho} \mid X$ such that (X, ρ) is a non-complete metric space, though ρ generates the same topology on X. This is a contradiction.

99. See Hausdorff [2].

100. Put $\mathcal{P}_0 = N$, where N is the set of natural numbers, $\mathcal{P}_1 = \mathcal{E}^*$, where $\mathcal{E}^*$ is some family of subsets of N such that: (*a*) if $A \in \mathcal{E}^*$, $B \in \mathcal{E}^*$ and $A \neq B$, then $A \cap B$ is finite; (*b*) $|A| = \aleph_0$ for all $A \in \mathcal{E}^*$; (*c*) $|\mathcal{E}^*| = 2^{\aleph_0}$ and (*d*) if $C \subset N$ and $|C| = \aleph_0$, then $A \cap C$ is infinite for some $A \in \mathcal{E}^*$. Such $\mathcal{E}^*$ exists by Problem 50 of Chap. 4 and Problem 118 of Chap. 1. Now define a topology on the set $\mathcal{P} = \mathcal{P}_0 \cup \mathcal{P}_1$ as follows. All points of $\mathcal{P}_0$ are isolated in $\mathcal{P}$. If $\xi \in \mathcal{P}_1$, then a base for ξ in $\mathcal{P}$ is made up of sets of the form $O_k\xi = \{\xi\} \cup (\xi \setminus K)$, where K is some finite subset of N.

From (*a*) it follows that $\mathcal{P}$ is a Hausdorff space. Condition (i) is obviously satisfied. Condition (ii) is also fulfilled by virtue of (*d*) and the choice of the base for topology at the points of $\mathcal{P}_1$. The space $\mathcal{P}$ is regular, because the sets $O_k\xi$ are not only open but also (obviously) closed.

101. (*a*) If $x \in \mathcal{P}_0$, then $\{x\}$ is a compact nhood of x by (i) of the preceding problem. If $x \in \mathcal{P}_1$, then from condition (i) of the preceding problem it follows that there is a nhood Ox of x in $\mathcal{P}$ such that $Ox \cap \mathcal{P}_1 = \{x\}$. But $\mathcal{P}$ is a regular space, so that there is a nhood O_1x of x in $\mathcal{P}$ such that $[O_1x] \subset Ox$. Then $[O_1x]$ is a compactum. Indeed, if $C \subset [O_1x]$ and $|C| = \aleph_0$, then $C' = C \setminus \{x\} \subset \mathcal{P}_0$ and $|C'| = \aleph_0$. By (ii) of the preceding problem $x' \in \mathcal{P}$ is a limit point of C'. However, (i) of the preceding problem implies that $x' \in \mathcal{P}_1$, but then $x' \in [O_1x]$ and $x' = x$ by the definitions of Ox and O_1x. Hence $[O_1x] = O_1x$ is compact, proving (*a*).

(*b*) A locally compact Hausdorff space is always completely regular (see Problem 177 of Chap. 3).

(*c*) The set $O_1(x)$ (see (*a*)) is countable (because $O_1x \setminus \{x\} \subset \mathcal{P}_0$ and $|\mathcal{P}_0| = \aleph_0$). But a countable compactum is metrizable (Problems 105, 39 of Chap. 3). Hence $\mathcal{P}$ is locally metrizable.

(*d*) Suppose that f is any continuous real function on $\mathcal{P}$, $f: \mathcal{P} \to R$. Then $f\mathcal{P} = f[\mathcal{P}_0] \subset [f\mathcal{P}_0]_R$ by the continuity of f (see Problem 281 of Chap. 2). Thus it suffices to show that f is bounded on $\mathcal{P}_0$. If this were not so, then we would have a sequence $\{x_n: n \in N^+\}$ of mutually distinct elements of $\mathcal{P}_0$ such that $\lim |f(x_n)| = \infty$. But $C = \{x_n: n \in N^+\} \subset \mathcal{P}_0$ and $|C| = \aleph_0$. Hence C would have a limit point $x^* \in \mathcal{P}$ (see (i) of the preceding problem). But then f would not be continuous at x^*, a contradiction. The statement (*d*) is proved. From $[\mathcal{P}_0] = \mathcal{P}$ and $|\mathcal{P}_0| = \aleph_0$ (*e*) follows.

The statements (*f*), (*g*) *and* (*h*). If $\mathcal{P}$ were metrizable, then, since $\mathcal{P}$ is seperable, $\mathcal{P}$ would have a countable base. But then every subspace of $\mathcal{P}$ would be separable, and the same would be true of $\mathcal{P}_1$ too, contradicting the fact that $\mathcal{P}_1$ is discrete and $|\mathcal{P}_1| = 2^{\aleph_0}$. $\mathcal{P}_1$ is thus an uncountable set which has no limit point in $\mathcal{P}$. Hence $\mathcal{P}$ is neither countably compact (Problem 189 of Chap. 3), nor Lindelöf (Problem 200 of Chap. 3). It should also be noted that from the open uncountable cover $\gamma = \{Ox: x \in \mathcal{P}\}$ of $\mathcal{P}$, where $Ox = \{x\}$ if $x \in \mathcal{P}_0$ and $Ox \cap \mathcal{P}_1 = \{x\}$ if $x \in \mathcal{P}_1$, it is impossible to extract a cover of $\mathcal{P}$ other than γ (i.e., γ is minimal).

(*i*) If $\mathcal{P}$ were normal then, being pseudocompact, it would also be countably compact (see Problem 192 of Chap. 3), contradicting (*g*). Hence $\mathcal{P}$ is not normal.

(j) Every point-finite open cover of a separable space is obviously countable. Consequently, if $\mathscr{P}$ were weakly paracompact, then it would also be Lindelöf, contradicting **(h)**.

The statements (*k*) *and* (*l*). Fix for each $x \in \mathscr{P}$ a compact nhood Ox such that $Ox \cap \mathscr{P}_1 = \{x\}$ if $x \in \mathscr{P}_1$ and $Ox = \{x\}$ if $x \in \mathscr{P}_0$, which is possible (see proof of (*a*)). Put $\lambda = \{Ox: x \in \mathscr{P}\}$ and $\lambda_K = \{(Ox \setminus K) \cup \{x\}: x \in \mathscr{P}\}$ for an arbitrary finite set $K \subset \mathscr{P}_0$. Every λ_K is obviously an open cover of $\mathscr{P}$. If $x_0 \in \mathscr{P}_0$ and $K \ni x_0$, then $\lambda_K(x_0)$ is the star of x_0 with respect to λ_K, coinciding with $\{x_0\}$. If $x_1 \in \mathscr{P}_1$, then $\lambda_K(x_1)$ is $Ox_1 \setminus K$.

Now consider an arbitrary nhood Ux_1 of x_1 in $\mathscr{P}$; we show that we can select a finite set $K_1 \subset \mathscr{P}_0$ such that $Ox_1 \setminus K_1 \subset Ux_1$. In other words, it is to be shown that $|Ox_1 \setminus Ux_1| < \aleph_0$. Put $C = Ox_1 \setminus Ux_1$. Since x_1 is the unique non-isolated point of the compactum Ox_1 and Ux_1 is a nhood of x_1, the set C does not have a limit point in Ox_1 and consequently C is finite. This shows that $\{\lambda_K: K \subset \mathscr{P}_0 \text{ and } |K| < \aleph_0\}$ is a (countable) refinement of $\mathscr{P}$. $\mathscr{P}$ has no pointwise countable base because otherwise, since $\mathscr{P}$ is separable, $\mathscr{P}$ would have a countable base contradicting (*f*) and (*h*).

(m). To prove **(m)**, consider the natural mapping π of $\mathscr{P}$ onto the decomposition space Z of $\mathscr{P}$, whose only non-trivial element is the set $\mathscr{P}_1$. This mapping is closed (Problems 318 and 322 of Chap. 2). We have $|Z| = \aleph_0$. It is to be shown that Z is a compactum whose only non-isolated point is $z_1 = \pi(\mathscr{P}_1)$. If $z \in Z \setminus \{z_1\}$, then $\pi^{-1}(z) \subset \mathscr{P}_0$ and hence $\pi^{-1}(z)$ is open in $\mathscr{P}$ (see (i) of the preceding problem). Hence $\{z\}$ is open in Z, i.e., z is an isolated point in Z. Furthermore, if $A \subset Z \setminus \{z_0\}$ and $|A| = \aleph_0$, then $C = \pi^{-1}(A) \subset \mathscr{P}_0$ and $|C| = \aleph_0$, but by virtue of (ii) of the preceding problem, $[C] \cap \mathscr{P}_1 \neq \Delta$. Then, in view of the continuity of π, we have $z_1 = \pi(\mathscr{P}_1) \in [\pi(C)] = [A]$. Thus z_1 is a limit point of any infinite subset A of Z. Consequently, Z is a compactum (Z is a Hausdorff space, being a T_1-space with a unique non-isolated point; see Problem 15 of Chap. 2). Thus statement (*m*) is proved. Since every pseudocompact Q-space is compact (Problem 80 of Chap. 4), $\mathscr{P}$ is not a Q-space.

102. Suppose that A is a closed set in X and $\xi = \{\gamma_n: n \in N^+\}$ is a refinement. Then $P = \bigcap\{\gamma_n(A): n \in N^+\} \setminus A = \Delta$; otherwise, the refinement condition would break down at every point of P.

103. Suppose that $\varphi = \{\gamma_i: i \in N^+\}$ is a T_1-filtration on X. Put $F_i = X \setminus \bigcup\{U: U \in \gamma_i\}$. The set F_i is closed, hence there exist open sets $G_{i,j}$ in X, where $j \in N^+$, such that $\bigcap\{G_{i,j}: j \in N^+\} = F_i$ for all $i \in N^+$. Set $\lambda_{i,j} = \gamma_i \cup \{G_{i,j}\}$. It is then routine to verify that $\{\lambda_{i,j}: i, j \in N^+\}$ is a (countable) T_1-refinement on X.

105. Suppose that $\mathscr{E} = \{\gamma_i: i \in N^+\}$ is a T_1-refinement on X. Put $G_i = \bigcup\{U \times U: U \in \gamma_i\}$; we show that $\bigcap\{G_i: i \in N^+\} = \Delta$. Since γ_i covers X, for each $x \in X$ there is a $U \in \gamma_i$ such that $x \in U$. Then $(x, x) \in U \times U \subset G_i$ and hence $\Delta \subset G$ for all $i \in N^+$. Consider any element (x, y) of the set $(X \times X) \setminus \Delta$. Then $x \neq y$ and, since $\mathscr{E}$ is a T_1-refinement, there is $i' \in N^+$ such that $\gamma_{i'}(x) \not\ni y$. Hence from $U \in \gamma_{i'}$ and $U \ni x$ it follows that $U \times U \not\ni (x, y)$. Obviously, if $U \in \gamma_{i'}$ and $U \not\ni x$, then we have also $U \times U \not\ni (x, y)$. Thus $(x, y) \notin U \times U$ for all $U \in \gamma_{i'}$, i.e., $(x, y) \notin G_{i'}$. Hence $\bigcap\{G_i: i \in N^+\} = \Delta$. The proof of sufficiency is left to the reader.

107. Suppose that $\varphi = \{\gamma_i: i \in N^+\}$ is a T_1-refinement on X. Choose in each γ_i a countable cover λ_i of X. Then $\{\lambda_i: i \in N+\}$ is a T_1-refinement. Consequently (see the preceding problem), $\mathscr{B} = \bigcup \{\lambda_i: i \in N+\}$, a pseudobase for X, is obviously countable.

108. For each $n \in N+$ and every $x \in X$, put $n(x) = X$, if γ_n does not cover x. Otherwise, put $n(x) = U$, where U is any (later fixed) element of γ_n containing x. Furthermore, let $O_k x = \bigcap \{n(x): n \in N+, n \leqslant k\}$ for all $k \in N+$ and $x \in X$. Obviously, every $O_k x$ is open and contains x. We say that a point $x \in A$ is the n-center of a set $A \subset X$, if $O_n x \supset A$. If A has even one n-center, then A is called *n-small*. Suppose that F is a closed set in the compactum X; then for each finite cover γ of F by open sets in X and for each $n \in N+$ there is a finite family λ of n-small open sets in X, each of which has a n-center in F, which covers F and is a closure-refinement of γ. Applying this statement at each step, there is no difficulty in constructing by induction a sequence $\xi = \{\gamma_i: i \in N+\}$ of finite families of open sets in X such that

(i) $F \subset \bigcup \{V: V \in \gamma_i\}$;

(ii) every element V of γ_n is n-small, has n-center in F and intersects F;

(iii) the closure of every element of γ_{n+1} is contained in some element of γ_n, for all $n \in N+$.

Put $G_n = \bigcup \{V: V \in \gamma_n\}$. Then G_n is open in X and $G_n \supset F$ by (i). We show that $\bigcap \{G_n: n \in N+\} = F$. Assume the contrary. Let $z \in \bigcap \{G_n: n \in N+\} \setminus F$. Set $\lambda_n = \{V \in \gamma_n: V \ni z\}$. Then $\lambda_n \neq \Lambda$ and the family $\eta = \{\lambda_n: n \in N+\}$ obviously satisfies conditions (ii) and (iii) and the following condition as well:

(i′) every element of every $\lambda_n \in \eta$ contains z.

Let x_n be any n-center of the set V_n lying in F (see (ii)). The sequence $\{x_n: n \in N+\}$ has a limit point in F; let $x^* \in F$ be such a point. Since $x_i \in V_i \subset V_n$ for all $i > n$, $x^* \in [V_{n+1}] \subset V_n$ for all $n \in N+$, i.e., $x^* \in \bigcap \{V_n: n \in N+\}$. Since φ is a T_1-filtration, from $x^* \in F \subset X \setminus \{z\}$ it follows that there is a number $n^* \in N^*$, for which: *(a)* γ_{n^*} covers x^*, and *(b)* no element of γ_{n^*} contains both x^* and z. Briefly, *(a)*: $\gamma_{n^*}(x^*) \neq \Lambda$ and *(b)*: $\gamma_{n^*}(x^*) \not\ni z$. By *(a)* $\gamma_{n^*}(x^*)$ is a nhood of x^*. Hence there is a number $i^* \in N+$ such that $i^* > n^*$ and $x_{i^*} \in \gamma_{n^*}(x^*)$. Since x_{i^*} is the i^*-center of V_{i^*}, we have $x_{i^*} \in V_{i^*} \subset O_{i^*}(x_{i^*}) = \bigcap \{n(x_{i^*}): n \in N+ \text{ and } n \leqslant i^*\}$. Hence from $i^* > n^*$ it also follows that $V_{i^*} \subset n^*(x_{i^*})$. But by the choice of i^*, $\gamma_{n^*}(x^*) \ni x_{i^*}$, i.e., γ_{n^*} covers x_{i^*}. Hence, by the definition of $n(x)$, we have $n^*(x_{i^*}) \in \gamma_{n^*}$. From $V_{i^*} \subset n^*(x_{i^*})$ it follows that $z \in n^*(x_{i^*})$ and $x^* \in n^*(x_{i^*})$, contradicting *(b)*.

109. By Problem 108, every closed set in X is a G_δ. Hence X has a countable T_1-refinement (Problem 103). Therefore, X has a countable pseudobase (Problem 107). But then X also has a countable base {Problem 122 of Chap. 3)

110. False. An example is the space $\mathscr{P}$ in Problems 100 and 101.

111. This theorem is due to Jones [1]. Recently, Shapirovskii [1] has proved a stronger statement (replacing separability by the Souslin property). However, see the solution of Problem 121 of Chap. 2.

112. Suppose that γ is an arbitrary open cover of X. For each $\lambda \in \xi$, put $X_\lambda = \{x \in X$: there exists a $G \in \gamma$ such that $\lambda(x) \subset G\}$. Since γ covers X and ξ is refined, $\bigcup \{X_\lambda: \lambda \in \xi\} = X$. It suffices to show that the cover $\eta_\lambda = \{\lambda(x): x \in X_\lambda\}$ of X_λ has a countable subcover (then, even from γ we can straightaway extract a countable covering of X_λ, because η_λ is a refinement of γ).

The set $A \subset X_\lambda$ is called λ-*discrete*, if no element of λ contains more than one element of A. From the Kuratowski-Zorn principle it follows that X_λ contains a maximal λ-discrete set A_λ. The maximality of A_λ implies that $\bigcup \{\lambda(x): x \in A_\lambda\} \supset X_\lambda$. But $|A_\lambda| \leqslant \aleph_0$. Indeed, otherwise, A_λ would have a limit point x^* in X. Choose $U^* \in \lambda$, $U^* \ni x^*$. We have $|A_\lambda \cap U^*| \geqslant \aleph_0 > 1$, which contradicts the λ-discreteness of A_λ. Thus η_λ contains the countable cover $\{\lambda(x): x \in A_\lambda\}$ of X_λ.

113. By Problem 112, X is Lindelöf; hence it is metrizable (Problem 76) and has a countable base (Problem 213 of Chap. 2).

114. Every uncountable set $A \subset X$ has a limit point in X (Problem 111). Hence X is metrizable and has a countable base (Problem 113).

115. See, Heath [1] and Bing [2].

116. See Heath [1].

117. Select in X a σ-discrete base $\mathscr{B}$: $\mathscr{B} = \bigcup \{\gamma_i : i \in N^+\}$, where every γ_i is discrete in X and consists of open sets. Put $G_i = \bigcup \{V: V \in \gamma_i\}$ and $F_i = X \setminus G_i$. The set F_i is closed in X and hence is G_δ in X. Consequently, there is a continuous real function f_i on X such that $f_i^{-1}(0) = F_i$ and $f(X) \subset [0, 1]$. Fix any one-to one mapping φ_i of γ_i into the set A, in accordance with Problem 251 of Chap. 2 (furthermore, we utilize also some notations from Problem 251 of Chap. 2). This is possible, since $|\gamma_i|$ does not exceed the weight of X, namely τ. Define a mapping $\psi_i: X \to Kw(\tau)$ as follows: if $x \in F_i$, then $\psi_i(x) = 0$; if $x \in U \in \gamma_i$, then $\psi_i(x) = (f_i(x), \varphi_i(U)) \in (0, 1] \times A$, the latter convention is correct, because from $x \in U \in \gamma_i$ it follows that $f_i(x) \in (0, 1]$. The family $\mathscr{F}$ of ψ_i, where $i \in N+$, is a partitioning family of mappings: prove this. Let Q be any closed set in X and x a point of $X \setminus Q$. It is required to find $i \in N+$ for which $\psi_i(x) \notin [\psi_i(Q)]_{Kw(\tau)}$. Since $\mathscr{B}$ is a base for X, there exist $i^* \in N+$ and $U^* \in \gamma_{i^*}$ such that $x \in U^* \subset X \setminus Q$. Then $\psi_{i^*}(x) \in (0, 1] \times \varphi_{i^*}(U^*) \subset Kw(\tau)$ and $\psi_{i^*}(Q) \cap ((0, 1] \times \varphi_{i^*}(U^*)) = \Lambda$. Note that the set $(0, 1] \times \varphi_i(U^*)$ is open in $Kw(\tau)$. Consequently, $[\psi_{i^*}(Q)]_{Kw(\tau)} \not\ni \psi_{i^*}(x)$ and $\mathscr{F}$ is a partitioning family. Hence (see Problem 37 of Chap. 3) the diagonal product of the mappings ψ_i is a homeomorphism of X into $(Kw(\tau))^{\aleph_0}$.

118. Since $|Kw(\tau)| = \tau \cdot 2^{\aleph_0}$, $|Kw(\tau)^{\aleph_0}| = (\tau \cdot 2^{\aleph_0})^{\aleph_0} = \tau^{\aleph_0} \cdot 2^{\aleph_0 \cdot \aleph_0} = \tau^{\aleph_0}$ (because $\tau \geqslant \aleph_0 > 2$).

119. Use Problem 86 of Chap. 4.

126. Suppose that X is any non-paracompact, completely regular space of weight not exceeding $2^{\aleph_0}$ such that, of any two closed disjoint sets, one is certainly compact ($T(\omega_1)$ is an example of such space; see Problem 186 of Chap. 3). We may assume that X is a subspace of the Tychonoff product $I^{\mathfrak{c}}$ of many continuum segments (see Problems 36 and 37 of Chap. 3).

Fix a countable set S, dense in $I^{\mathfrak{c}}$ (such a set S exists by Problem 381 of Chap. 2). Consider the result of doubling the space $I^{\mathfrak{c}}$ in the Alexandroff sense: $A(I^{\mathfrak{c}}) = I^{\mathfrak{c}} \cup d(I^{\mathfrak{c}})$, in which $I^{\mathfrak{c}}$ inherits the usual topology, and $d(I^{\mathfrak{c}})$ (considered a copy of the set $I^{\mathfrak{c}}$) is discrete in $A(I^{\mathfrak{c}})$ (see Problem 81 of Chap. 3). Let $d(S)$ be a copy of S, in $d(I^{\mathfrak{c}})$. From the definition of the topology in $A(I^{\mathfrak{c}})$ and the density of S in $I^{\mathfrak{c}}$ it follows that $d(S)$ is dense in $A(I^{\mathfrak{c}})$. Put $Y = d(S) \cup X$. With respect to the topology induced in $A(I^{\mathfrak{c}})$, Y is a normal, separable, but a non-paracompact space. The set $d(S)$ is countable and dense in Y, so that Y is seperable. The points of $d(S)$ are isolated in $A(I^{\mathfrak{c}})$, and so they are isolated in Y, too. Consequently (since $X = Y \setminus d(S)$), X is closed in Y. Hence Y is not paracompact (otherwise, X also would be paracompact, contradicting assumption). To show that Y is normal, let P and Q be disjoint and closed in Y. The set $P' = P \cap X$ and $Q' = Q \cap X$ are disjoint and closed in X. Hence one of them is compact. Assume for definiteness that P' is compact. Y, as a subspace of $A(I^{\mathfrak{c}})$ ($A(I^{\mathfrak{c}})$ is compact by Problem 81 of Chap. 3), is completely regular. Also, in a completely regular space, a compactum and a disjoint closed set can be separated by disjoint nhoods. Hence there are open sets U and V in Y such that $U \supset P'$, $V \supset Q$ and $U \cap V = \Lambda$. Put $W = V \cap (X \setminus P)$ and $G = U \cup P = U \cup (P \setminus U)$. Since $P \setminus U \subset d(S)$, the set $P \setminus U$ is open in Y. Thus $W \supset Q$, $G \supset P$, $W \cap G = \Lambda$ and W, G are open sets in Y. Hence Y is normal.

128. Fix for each $A \in \mathcal{A}$ an open set U_A such that $[U_A] \cap B = \Lambda$, $A \subset U_A$, and consider the family $\gamma = \{U_A : A \in \mathcal{A}\} \cup \{X \setminus P\}$; the latter is obviously an open cover of X. Since X is paracompact, γ has a locally finite open refinement λ. Let $V \in \lambda$ and $V \cap P \neq \Lambda$. Then V is contained in some element G of γ. Obviously, $G \neq X \setminus P$. Consequently, $G = U_{A^*}$ for some $A^* \in \mathcal{A}$. But then $[V] \subset [G] = [U_{A^*}] \subset X \setminus B$. Put $U = \{V : V \in \lambda \text{ and } V \cap P \neq \Lambda\}$. Then $U \supset P$ (because λ is a cover) and U is an open set. Since λ is locally finite, $[U] = \bigcup \{[V] : V \in \lambda \text{ and } V \cap P \neq \Lambda\}$ (see Problem 2), and from the preceding remark we conclude that $[U] \subset X \setminus B$.

129. Let A be a closed set in X, $x \in X \setminus A$. For each point $y \in A$ fix a nhood Oy whose closure does not contain the point x, Refine the open cover $\gamma = \{Oy : y \in A\} \cup \{X \setminus A\}$ to a locally finite open cover λ and put $\lambda^* = \{U \in \lambda : U \cap A \neq \Lambda\}$. Then $[U] \not\ni x$ for each $U \in \lambda^*$ (of course, U is contained in some element of γ distinct from $X \setminus A$; hence in some Oy). Since λ is locally finite, $[\bigcup \{U : U \in \lambda^*\}] = \bigcup \{[U] : U \in \lambda^*\} \subset X \setminus \{x\}$, i.e., $W = \bigcup \{U : U \in \lambda^*\}$ is a nhood of A whose closure does not contain x.

130. This is immediate from the statements of Problems 128 and 129.

131. Use Problem 44.

132. See Problem 72.

133. This is immediate from Problem 39.

134. Since X is a paracompactum, we may assume that γ is a locally finite open cover: $\gamma = \{U_\alpha : \alpha \in A\}$. There exists (see Problems 18 and 13) a closed locally finite cover $\lambda = \{F_\alpha : \alpha \in A\}$ of X which combinatorially refines γ, i.e., such that $F_\alpha \subset U_\alpha$ for all $\alpha \in A$. X is a normal space (Problem 130). Thus for each $\alpha \in A$ there is a non-negative continuous real function f_α on X such that $f_\alpha(F_\alpha) = \{1\}$ and $f_\alpha(X \setminus U_\alpha) = \{0\}$. The family $\mathcal{P} = \{f_\alpha : \alpha \in A\}$ obviously satisfies conditions (i), (ii) and the following condition (iii′) (weaker than condition (iii)): $\varphi(x) = \Sigma \{f_\alpha(x) : \alpha \in A \text{ and } f_\alpha(x) \neq 0\} > 0$ for all $x \in X$. From (i) and (ii) it easily follows that $\varphi(x)$ is continuous on X (indeed at all points of the nhood Ox appearing in (ii), $\varphi(x) = \Sigma \{f_\alpha(x) : \alpha \in A'\}$, where $A' = \{\alpha \in A : f_\alpha(Ox) \neq \{0\}\}$ is a finite set). Also, function ψ_α defined by the formula: $\psi_\alpha(x) = f_\alpha(x)/\varphi(x)$ for all $x \in X$ (note that we always have $\varphi(x) \neq 0$ by (iii′)) is continuous on X. Obviously, its carrier coincides with the carrier of f_α. Hence it follows that conditions (i) and (ii) are satisfied by the family $\mathcal{Y} = \{\psi_\alpha : \alpha \in A\}$. But $\Sigma \{\psi_\alpha(x) : \alpha \in A \text{ and } \psi_\alpha(x) \neq 0\} = \Sigma \{(f_\alpha(x)/(\varphi(x)) : \alpha \in A \text{ and } f_\alpha(x) \neq 0\} = \varphi(x)/\varphi(x) = 1$. Hence condition (iii) is also satisfied by the family $\mathcal{Y}$.

135. Clearly, (0, 0) is not an isolated point in X. Consequently, X is not discrete,

As a space with exactly one non-isolated point, X is normal (see Problem 15 of Chap. 3). But $|X| = \aleph_0$, hence X is Lindelöf. Since a regular Lindelöf space is always paracompact, X is a paracompactum. If X were first countable at $(0, 0)$, then we would have in X an infinite compactum containing $(0, 0)$. Thus it remains to show that every compact subspace Φ of X is finite.

Suppose that l is an arbitrary straight line parallel to the y-axis. Since $X \cap l$ is a closed discrete subspace of X, $\Phi \cap l$ is finite. But then $U = \{(0, 0)\} \cup (X \setminus \Phi)$ is an open set in X and from $U \cap \Phi = \{(0, 0)\}$ we conclude that all points of the compactum Φ are isolated in Φ. Consequently, $|\Phi| < \aleph_0$.

136. The answer is affirmative. Take a countable dense subspace of $I^{\mathfrak{c}}$ (see Problem 381 of Chap. 2).

137. Let $X_0 = \bigcup \{F_i: i \in N+\}$, where F_i is closed in X. Consider for each i the open cover Ω_i of the space X, which consists of the elements of the family ω together with the open set $X \setminus F_i$. Refine Ω_i to a locally finite open cover Ω_i' of X. Let ω_i' be the family of those elements $\Gamma \in \Omega_i'$ which satisfy $\Gamma \cap Fi \neq \Lambda$. Clearly, ω_i' is locally finite in X, consists of open sets in X, is a refinement of ω and covers at least F_i. Then the family $\omega' = \bigcup \{\omega_i': i \in n^+\}$ is locally finite in X, consists of open sets in X, is a refinement of ω and covers (at least) the subspace X_0.

138, This follows from the preceding problem.

139. This follows from Problems 127 and 138.

140. It suffices to show (see Problem 206 of Chap. 3) that every open subspace Y of X is Lindelöf. Since $c(X) \leqslant \aleph_0$ and Y is open in X, $c(Y) \leqslant \aleph_0$. Moreover, Y is paracompact by hypothesis. Hence (see Problem 133) Y is Lindelöf.

141. The answer is negative. If X is the Alexandroff compactification (see Problem 176 of Chap. 3) of an uncountable discrete space, then every subspace of X is either discrete or compact and in both cases it is paracompact (even strongly paracompact).

142. This is still an open question.

143. It suffices to show that an arbitrary open subspace Y of $A(X)$ is paracompact (see Problem 127). Suppose that γ is an arbitrary open cover of Y. Put $Z = Y \cap X$ and $\lambda = \{U \cap X: U \in \gamma\}$. Then Z is open in X and λ is an open cover of Z. Refine λ to a locally finite (in Z) cover μ of Z by sets open in Z (and hence also in X). For each $V \in \mu$, fix some element of γ, containing it, and denote it by $t(V)$. Furthermore, set $r(V) = V \cup d(V)$, where $d(V)$ is the natural projection of V into $d(X)$, i.e., $d(V)$ is a copy of V lying in $d(X)$. Since V is open in X, $r(V)$ is open in $A(X)$. From the local finiteness of μ at the points of Z it follows that the family $\{r(V): V \in \mu\}$ is also locally finite at the points of Z. Put $q(V) = r(V) \cap t(V)$. Then $\xi = \{q(V): V \in \mu\}$ is also locally finite at the points of Z. But $q(V) \supset V$, $q(V)$ is open in $A(X)$ and $q(V) \subset t(V) \in \gamma$. Hence ξ is a refinement of γ, consists of open sets in Y, is locally finite at the points of Z and covers Z. The set $Y \setminus \bigcup \{q(V): V \in \mu\}$ is closed and discrete in Y. Hence $\eta = \xi \cup \{\{y\}: y \notin \bigcup \{q(V): V \in \mu\}\}$, a locally finite open cover of Y, obviously refines γ.

144. The answer is negative. Consider the compact Alexandroff 'double circle' (cf. Problem 80 of Chap. 3). It is obtained by doubling the usual circle, and the latter, like every metric space, is hereditarily paracompact. But a space obtained by doubling a hereditarily paracompact space, is also hereditarily paracompact (Problem 143). Hence the compact double circle is a hereditarily paracompact space. This compactum is first countable and contains a discrete subspace of cardinality $2^{\aleph_0}$; the latter implies that it is not perfectly normal (Problem 206 of Chap. 3).

145. If γ is a disjoint family of open sets in X and $G = \bigcup \{U: U \in \gamma\}$, $P = X \setminus G$, then P is closed in X and hence $P = \bigcap \{W_i: i \in N^+\}$, where all W_i are open in X. Put $F_i = X \setminus W_i$. Obviously, $F_i \subset G$, where F_i is closed but G is open in X. Using the normality of X choose an open set V_i, satisfying $F_i \subset V_i \subset [V_i] \subset G$. Put $\lambda_i = \{V_i \cap U: U \in \gamma\}$. The family λ_i consists of open sets and is obviously discrete in X. Hence $\lambda = \bigcup \{\lambda_i: i \in N^+\}$, a σ-discrete family of open sets, refines γ. From $\bigcup \{F_i: i \in N^+\} = G$ and the definition of V_i it follows that $\bigcup \{A: A \in \lambda\} = G = \bigcup \{U: U \in \gamma\}$. This reasoning shows that every σ-disjoint open cover of X has σ-discrete open refinement. Consequently (Problems 72 and 26), X is paracompact.

146. Consider an arbitrary open cover γ of X. For each $i \in N^+$, fix a σ-discrete cover λ_i of F_i by closed sets in F_i, where each λ_i is a refinement of γ. Such covers obviously exist (see Problem 150). Since X is collectionwise normal, there is a σ-discrete family μ_i of open (in X) nhoods of the elements of family λ_i refining γ (see Problem 48). Put $\mu = \bigcup \{\mu_i: i \in N^+\}$ and $U = \bigcup \{G: G \in \mu\}$. Then μ is a σ-discrete family of open sets in X, refines γ, U is open in X and $U \supset Y$. The subspace $P = X \setminus U$ of X is paracompact, and hence there is a σ-discrete cover θ of P by closed sets in P and, consequently, in X, refining γ. From θ we obtain a σ-discrete (in X) family $\tilde{\theta}$ of nhoods (in X) of elements of θ; in addition, $\tilde{\theta}$ may be regarded as being a refinement of γ. Here we have again made use of the collectionwise normality of X. Then $\tilde{\theta} \cup \mu$ is a σ-discrete open cover of X refining γ. But X is a regular space, hence, since γ was chosen arbitrarily, we conclude that X is a paracompactum (see Problem 150).

147. Refine ξ to a locally finite open cover γ. X is a regular space (Problem 129). Hence the family $\eta = \{U \subset X: U$ is open in X and there is a $G \in \gamma$ such that $[U] \subset G\}$ covers X. Refine η to a locally finite open cover λ. Then γ and λ satisfy the conditions of Problem 51, and hence there is an open cover of X which star-refines γ; for example, $\langle \gamma \cup \lambda \rangle$.

148. Necessity is proved in the preceding problem. Let us establish the sufficiency. The space X is regular and also collectionwise normal (see Problem 24). For every open cover λ of X we can construct a star-monotonic sequence $\xi = \{\gamma_n: n \in N^+\}$ of open covers of X, where γ_1 is a refinement of λ. Then ξ majorizes the cover λ, and hence λ can be refined to a σ-discrete open cover (Problem 34). But λ is an arbitrary open cover of X. Consequently, X is a paracompactum (see Problem 72), because a σ-discrete cover is obviously σ-locally finite.

149. See Problems 147 and 22.

150. The equivalence of *(a)*, *(b)*, *(c)*, *(d)* and *(e)* is established in Problems 148, 72, 44, and 34. For the proof of equivalence of *(f)* to them, see Michael [1], [2].

151. No, not always. See Potoczny [1]. A stronger example has recently been constructed by Peregud.

152. *Necessity.* Since $X \cap F = \Lambda$ and F is closed in bX and bX is regular, for each point $x \in X$ there is a nhood Ox of x in bX such that $[Ox] \cap F = \Lambda$. The open cover $\gamma = \{Ox \cap X: x \in X\}$ of X can be refined to a locally finite open cover λ, which obviously gives the desired result.

Sufficiency. Let ξ be an arbitrary open cover of X. For each $U \in \xi$ fix an open set $t(U)$ in bX such that $t(U) \cap X = U$. Put $G = \bigcup \{t(U): U \in \xi\}$ and $F = bX \setminus G$. Obviously G is open in bX and $G \supset X$. Hence F is a compactum contained in $bX \setminus X$. By hypothesis, there is a locally finite open cover η of X such that $[V]_{bX} \cap F = \Lambda$ for all $V \in \eta$. From $[V]_{bX} \cap F = \Lambda$ it follows that $[V]_{bX} \subset G$. But $t(\xi) = \{t(U): U \in \xi\}$ is an open cover of G and $[V]_{bX}$ is compact. Consequently, from $t(\xi)$ we can extract a finite cover of $[V]_{bX}$. Hence for each $V \in \eta$ there is a finite subfamily ξ_V of ξ such that $\bigcup \{U: U \in \xi_V\} \supset V$. Put $\eta_V = \{V \cap U: U \in \xi_V\}$. Then η_V, a finite family of open sets in X, is a refinement of ξ, the union of whose elements is exactly V. Consequently, $\lambda = \bigcup \{\eta_V: V \in \eta\}$ is a locally finite open cover of X, refines ξ, and so X is paracompact.

153. (i) $\Rightarrow$ (ii). Suppose that $F \subset X$ is an arbitrary closed set in X and UF is any nhood of F. Consider the open cover $\omega = \{UF, X \setminus [U'F]\}$, consisting of two elements: UF and $X \setminus [U'F]$, where $U'F$ is a nhood of F satisfying $[U'F] \subset UF$ (we have used the normality of the paracompactum X, see Problem 130). Refine ω to a disjoint open cover ω'. Clearly, each $U \in \omega'$ is clopen in X. Consider a nhood $U''F = \bigcup \{U \in \omega': U \cap F \neq \Lambda\}$ of F. Clearly, $U''F \supset F$ and $U''F$ is clopen. In addition, $U''F \subset UF$. The implication (i) $\Rightarrow$ (ii) is thus proved (note also that it holds for every normal, not necessarily paracompact, space).

(ii) $\Rightarrow$ (iii). Suppose that ω is an arbitrary open cover of X. It is refinable to a locally finite open cover $\omega' = \{U_t: t \in T\}$. The cover ω' (using Problem 17) is combinatorially refinable to an open cover $\omega'' = \{V_t: t \in T\}$ such that $[V_t] \subset U_t$ for all $t \in T$. Using (ii), choose clopen sets Γ_t such that $[V_t] \subset \Gamma_t \subset U_t$. Clearly, $\gamma = \{\Gamma_t: t \in T\}$ is a locally finite (in X) cover by clopen sets, refines ω' and hence also refines ω. Assuming T to be well-ordered, put

$$W_t = \Gamma_t \setminus \bigcup \{\Gamma_{t'}: t' <_c t\}.$$

Noticing that $[\Gamma_t] = \Gamma_t$ for every t and that the entire family γ is locally finite (hence conservative), we conclude that W_t is clopen in X for every $t \in T$. Obviously, the family $\gamma' = \{W_t: t \in T\}$ is disjoint, covers X, is a refinement of ω' and hence also of ω. The implication (ii) $\Rightarrow$(iii) is thus proved.

(iii) $\Rightarrow$ (i). Suppose that $\omega = \{U_1, \ldots, U_s\}$ is an arbitrary finite open cover of X. By (iii), it is refinable to a disjoint open cover γ. Clearly, γ is locally finite in X and consists of clopen sets in X. Put $\gamma_1 = \{V \in \gamma: V \subset U_1\}$, $\gamma_2 = \{V \in \gamma \setminus \gamma_1: V \subset U_2\}, \ldots,$

$\gamma_s = \{V \in \gamma \setminus \bigcup_{i=1}^{s-1} \gamma_i: V \subset U_s\}$. Consider the sets $\Gamma_1 = \bigcup \{V: V \in \gamma_1\}$, $\Gamma_2 = \bigcup \{V: V \in \gamma_2\}, \ldots, \Gamma_s = \bigcup \{V: V \in \gamma_s\}$. The sets Γ_i, $i = 1, 2, \ldots, s$, are open. In addition, $\Gamma_i \cap \Gamma_j = \Lambda$, if $i \neq j$. It is also obvious that $\gamma' = \{\Gamma_1, \Gamma_2, \ldots, \Gamma_s\}$ is a cover of X which refines ω. The implication (iii) $\Rightarrow$(i) is therefore proved.

154. (i) $\Rightarrow$ (ii). Note that X is a paracompactum (Problem 45). Every open cover ω of X can be refined to a disjoint open cover (Problem 153). For each $i \in N^+$ the open (in X) cover by sets of diameter not exceeding $1/2^i$ is refinable to a disjoint open cover ω_i. The family $\sigma = \bigcup \{\omega_i: i \in N^+\}$ is the required base.

(ii) $\Rightarrow$ (i). Without loss of generality, we may assume that ω_{i+1} is a refinement of ω_i. Each element of every cover ω_i is a clopen set in X. It suffices (in consequence of

Problem 153) to show that Ind $X = 0$. Suppose that F is a closed set in X and OF is any nhood of F. For each i consider the subfamily $\omega_i(F, OF) \subset \omega_i$, defined by

$$\omega_i(F, OF) = \{U \in \omega_i\colon U \cap F \neq \Lambda \text{ and } U \subset OF\}. \tag{1}$$

Since every ω_i is locally finite in X, and all $U \in \omega_i$ are clopen in X,

$$\Gamma_i = \bigcup\{U\colon U \in \omega_i(F, OF)\}$$

are clopen in X for all $i \in N^+$ (see Problem 2). Since σ is a base for X, $F \subset \bigcup\{\Gamma_i\colon i \in N^+\} \subset OF$. Put

$$H_1 = \Gamma_1,\ H_2 = \Gamma_2 \setminus \Gamma_1, \dots,\ H_n = \Gamma_n \setminus \bigcup\{\Gamma_i\colon i < n\}.$$

Clearly, $H_1, H_2, \dots, H_n, \dots$ are clopen sets in X. It is also obvious that $H_{n_1} \cap H_{n_2} = \Lambda$ if $n_1 \neq n_2$, and $F \subset \bigcup\{H_n\colon n \in N^+\} = \bigcup\{\Gamma_i\colon i \in N^+\} \subset OF$.

It remains to show that the set $\bigcup\{H_n\colon n \in N^+\}$ is clopen. For this, it suffices to show that the family $\xi = \{H_1, H_2, \dots, H_n, \dots\}$ is locally finite in X. The family ξ is locally finite at the points of F, since it is disjoint and covers F. Let $x_0 \notin F$. Let i_0 be the first number such that $x_0 \in U^* \in \omega_{i_0}$, $F \cap U^* = \Lambda$. Then $U^* \cap \Gamma_i = \Lambda$ for all $i \geqslant i_0$. Hence $U^* \cap H_i = \Lambda$ for all $i \geqslant i_0$. Since $x_0 \in U^*$, it follows that ξ is a locally finite family at x_0.

155. This follows from Problem 246 of Chap. 2 and Problem 154 of the present chapter.

156. Since X is metrizable and dim $X = 0$, there is (see Problem 154) a base σ for X, which is the union of a countable family of disjoint open covers of X. But X_0 also has such a base — for example, $\sigma_0 = \{U \cap X_0\colon U \in \sigma\}$. Now, refer again to Problem 154.

157. (*b*) implies (*d*) by Problem 58. Obviously, (*d*) implies (*c*) and (*c*) implies (*a*). The condition (*b'*) is merely a weaker form of (*b*): the elements of this cover can be so enumerated as to yield a combinatorial refinement. It remains to deduce (*b*) from (*a*), and the equivalence of all the four conditions will be established. Fix a countable open cover $\gamma = \{U_i\colon i \in N^+\}$ of X and a point-finite open cover λ combinatorially refining it (this is easily achieved, see Problem 13). Since X is normal and λ is point-finite, λ can be combinatorially refined to a closed cover μ of X (Problem 17). Also, μ is a combinatorial refinement of γ.

158. See Problem 157.

159. *Necessity.* Put $\gamma = \{X \setminus F\colon F \in \xi\}$. Then γ is a countable open cover of X. γ can be combinatorially refined to a closed cover λ of X (Problem 157). For each element $U \in \gamma$ let $\varphi(U)$ be the corresponding element of λ. Put $U(F) = X \setminus \varphi(X \setminus F)$. Then $U(F)$ is open, $U(F) \supset F$, since $\varphi(X \setminus F) \subset X \setminus F$ and $\bigcap\{U(F)\colon F \in \xi\} = \Lambda$, because λ is a covering.

Sufficiency. Since X is normal, it suffices to show (see Problem 157) that every countable open cover γ of X, containing no finite cover of X, can be combinatorially refined to a closed cover. But then $\xi = \{X \setminus U\colon U \in \gamma\}$ is a centralized family of closed sets, the intersection of whose elements is empty. Fix for each $U \in \gamma$ an open set $W(U)$ containing $X \setminus U$ such that $\bigcap\{W(U)\colon U \in \gamma\} = \Lambda$. Put $F(U) = X \setminus W(U)$. Then $F(U)$ is closed, $F(U) \subset U$ and $\bigcup\{F(U)\colon U \in \gamma\} = X$, i.e., $\{F(U)\colon U \in \gamma\}$ is the required cover.

160. Solve this along the same lines as Problem 148.

161. X is a paracompact space (see Problem 132). From an arbitrary open cover γ of X, we can extract a countable cover λ. But every countable open cover of a paracompactum (and, furthermore, of a normal, countable paracompact space) can be star-refined to a finite open cover (Problem 157).

162. A space with a countable net is Lindelöf (Problem 143 of Chap. 2), and every regular Lindelöf space is strongly paracompact (Problem 161).

163. Use the notation in Problem 251 of Chap. 2. Note that $T(1) = \{(x, a) \in Kw(\tau): x = 1\}$ is a discrete subspace of $Kw(\tau)$. If $Kw(\tau)$ were strongly paracompact, then it would also be Lindelöf (see Problem 173; the important feature here is that $Kw(\tau)$ is connected). Then $Kw(\tau)$ would have a countable base (Problem 213 of Chap. 2) and $T(1)$ would be separable, contradicting $|T(1)| = \tau > \aleph_0$.

165. We need only prove sufficiency. The space X is paracompact; in fact, it is regular and every open cover of X can be star-refined to a countable open cover, but the latter (see Problem 9) is automatically a σ-discrete cover; now we can refer to Problem 150.

Suppose now that λ is an arbitrary open cover of X, γ is a star-countable open refinement of λ, and let $\{\gamma_\alpha: \alpha \in A\}$, $\{X_\alpha: \alpha \in A\}$ be the families, satisfying with respect to γ and X the properties (i) through (vii) formulated in Problem 8. Since X_α is closed in X, X_α is also a paracompactum. The family γ_α is a countable open cover of X_α, hence γ_α can be refined to a star-finite open cover μ_α (see Problem 157). If $W' \in \mu_{\alpha'}$, $W'' \in \mu_{\alpha''}$ and $\alpha' \neq \alpha''$, then $W' \cap W'' = \Lambda$ (see (vii) of Problem 8). Hence (see Problem 8), $\mu = \bigcup \{\mu_\alpha: \alpha \in A\}$ is a star-finite open refinement of γ.

166. Use Problems 8, 161 and 164 of the present chapter and 208 of Chap. 3.

168. Take a locally finite open cover γ of X such that $[U]$ is a compactum for every $U \in \gamma$. By Problem 53, this cover is star-finite. Now use Problem 8 and refer to Problem 208 of Chap. 3.

169. This follows from Problems 168, 161 and 164.

170. (i) $\Rightarrow$ (iii). Fix for each point $x \in X$ a nhood Ox of x in βX such that $[Ox]_{\beta X} \subset G$. The cover $\lambda = \{Ox: x \in X\}$ of X is refinable to a star-finite open cover γ. There is an open cover $\mu = \{V(U): U \in \gamma\}$ of X such that $[V(U)]_X \subset U$ for all $U \in \gamma$. Put $F(U) = [V(U)]_{\beta X}$ and $\hat{U} = \operatorname{Int} [U]_{\beta X}$. Then (since X is a normal space and βX is the Stone-Čech compactification of X) $F(U) \subset \hat{U}$ and it is straighforward to construct in βX an open set $W(U)$ which is the union of a countable set of compacta, such that $F(U) \subset W(U) \subset \tilde{U}$. Put $\tilde{X} = \bigcup \{W(U): U \in \gamma\}$. From the density of X in βX it follows that $\tilde{U} \cap \tilde{U}' \neq \Lambda$ iff $U \cap U' \neq \Lambda$. Hence $\{W(U): U \in \gamma\}$ is a star-finite open cover of $\tilde{X}$ by sets each of which is the union of countably many compacta. Hence $\tilde{X}$ is strongly paracompact (see Problem 166). Moreover, obviously, $W(U) \subset G$ for each $U \in \gamma$. Then $\tilde{X} \subset G$.

Clearly, (iii) implies (ii). Let us now deduce (i) from (ii). Consider an arbitrary open cover λ of X. For each $U \in \lambda$, fix an open set $\tilde{U}$ in βX such that $\tilde{U} \cap X = U$. Put $G = \bigcup \{\tilde{U}: U \in \gamma\}$. Then G is open in βX and $X \subset G$. Using (ii) choose a strongly paracompact subspace $\tilde{X}$ of βX such that $X \subset \tilde{X} \subset G$. The family $\xi = \{\tilde{U}: U \in \gamma\}$ covers G. Moreover, ξ covers $\tilde{X}$, and ξ has a star-finite open refinement η. Then, obviously, $\theta =$

$\{V \cap X: V \in \eta\}$ is a star-finite open refinement of λ. This completes the proof of the equivalence of (i), (ii) and (iii).

171. Sufficiency is proved in the same way as in Problem 152; it is necessary merely to note that if each element U of a star-finite family η of sets is replaced by some finite family λ_U of elements whose union equals U, then the resulting family $\lambda = \bigcup \{\lambda_U: U \in \eta\}$ is again star-finite.

Necessity is also proved in the same way as in Problem 152; the requirement is to note just that if ξ is a family of open sets in X and, for each $U \in \xi$, $t(U)$ is an open set in bX such that $t(U) \cap X = U$, then the star-finiteness of ξ implies the star-finiteness of $\{t(U): U \in \xi\}$ (in fact, $U \cap U' \neq \Delta$ iff $t(U) \cap t(U') \neq \Delta$).

172. See Problems 148 and 24.

173. An arbitrary open cover γ of X can be refined to a star-finite open cover λ. But λ is countable by Problem 12. Consequently, a countable cover can be extracted from γ (see Problem 75 of Chap. 2).

175. Consider an open cover ξ of X and put $\mathscr{B}_\xi = \{U \in \mathscr{B}$: there exists $G \in \xi$ such that $U \subset G\}$. Then $\mathscr{B}_\xi$ is a base of X and is also uniform. The surface $m(\mathscr{B}_\xi)$ of $\mathscr{B}_\xi$ is then a point-finite open cover of X refining ξ (see Problem 65).

176. The answer is negative. In a three-dimensional Euclidean space consider the usual cube K^3, defined with respect to some orthogonal frame of reference as follows:

$$K^3 = \{(x, y, z): 0 < x \leqslant 1, \quad 0 \leqslant y \leqslant 1, \quad 0 < z < 1\}.$$

The sets $\alpha = \{(x, y, z): 0 \leqslant x < 1, y = 0, 0 < z \leqslant 1\}$ and $\beta = \{(x, y, z): 0 < x \leqslant 1, y = 1, 0 < z \leqslant 1\}$ are opposite faces of K^3. The set $\Phi = \{(x, y, z): 0 \leqslant x \leqslant 1, y = 0, z = 0\}$ is the lower bound of basic (edge) face α. Put $\alpha^* = \alpha \setminus \Phi$. On the set $X = \alpha^* \cup \beta$ we impose a non-standard topology. Set $F_\lambda = \{(x, y, z): x = \lambda, y = 1, 0 < z \leqslant 1\}$, where $\lambda \in [0, 1]$. Then $\beta = \bigcup \{F_\lambda: \lambda \in [0, 1]\}$, and $F_{\lambda'} \cap F_{\lambda''} = \Delta$ for $\lambda' \neq \lambda''$. Each F_λ is endowed with the usual topology of the interval. It is further required that each F_λ be an open set in X. This defines the topology of X at all points of β. A basic nhood Op of each point $p \in \alpha^*$ in X has the form $Op = Up \cup (\pi(Up) \setminus B)$, where Up is any nhood of p in α^*, equipped with the usual topology of a subspace of an Euclidean space, $\pi(Up)$ is an (orthogonal) projection of Up on β and $B = \bigcup \{F_\lambda: \lambda \in E\}$, where E is an arbitrary finite subset of the interval $[0, 1]$. This completes the definition of X. Clearly, X is Hausdorff and obviously locally compact at the points of β. The topology induced by X on α^* is identical with the usual topology. If an infinite set $A \subset B$ intersects every F_λ, where $\lambda \in [0, 1]$, in a finite (possibly empty) set, then the projection of A on α has a complete accumulation point in α (with respect to the usual topology), and if this point belongs to α^*, then, it is obviously a complete accumulation point for the set A in X itself. It is easy to deduce from this that X is locally compact at the points of α^*. Let Φ' be the projection of the edge Φ on the face β, i.e., $\Phi' = \{(x, y, z): 0 < x \leqslant 1, y = 1, z = 0\}$. It is obvious that Φ' is a closed discrete subspace of X. The points of Φ' can be enclosed in a disjoint family of nhoods. If X were normal, then from this family we could obtain a discrete family of nhoods. But each of these nhoods would be assigned to the set $\beta \setminus \Phi'$. This would enable us to obtain an uncountable discrete closed set in X, contained in $\beta \setminus \Phi'$. But this set would have, by assumption, a complete accumulation point in α^*, a contradiction.

Consequently, X is not normal, nor even paracompact. But X is weakly paracompact. Indeed, let γ be an arbitrary open cover of X. γ can be refined to a disjoint family μ of mutually disjoint open sets, covering Φ' (see above). But $X\setminus\Phi'$, as is easy to see, is an open, Lindelöf subspace of X. Consequently, γ can be refined to an open point-finite cover λ of $X\setminus\Phi'$. Obviously, $\lambda \cup \mu$ is an open point-finite cover of X, refining γ. Hence X is weakly paracompact.

177. This is still an open question.

178. This is still an open question.

179. This problem follows from Problem 157.

180. Let γ be an open cover of X, no finite subfamily of which covers X. Refine γ to a point-finite open cover λ and from λ select a minimal subcover μ (see Problem 14). Then μ is open and point-finite, with $|\mu| \geqslant \aleph_0$. By Problem 15, X has a closed discrete subset equipotent to μ, which is infinite.

181. This problem follows from the preceding problem and Problem 189 of Chap. 3.

182. No. The space $T(\omega_1)$ is normal and countably compact (Problems 187 and 193 of Chap. 3). If $T(\omega_1)$ were weakly paracompact, then $T(\omega_1)$ would be compact (see Problem 181) but the latter is false (Problem 185 of Chap. 3).

183. Put $X_n = \{x \in X: |\{U \in \gamma: U \ni x\}| = n\}, n \in N^+$. Every X_n is closed in X (see Problem 19) and $\bigcup \{X_n: n \in N^+\} = X$. We now define by induction a sequence $\{\eta_n: n \in N^+\}$ of discrete families η_n of open sets, closure-refining γ such that $\bigcup \{\eta_n: n \leqslant k\}$ covers X_k for each $k \in N^+$. By Problem 19, there is a discrete cover λ_1 of X_1 by closed sets in X, refining γ. As η_1 take any discrete family of nhoods of λ_1, the closure of each of which is contained in an element of γ; such an element exists, since X is also collectionwise normal by Problem 48. Assume that the families η_n are already defined for $n < k$, in such a way that $V_k = \bigcup \{U: U \in \bigcup \{\eta_n: n \leqslant k\}\}$ is an open set containing X_k. Then, by Problem 19 there is a discrete cover λ_{k+1} of X_{k+1} by closed sets in X, refining γ. As η_{k+1} take any discrete family of nhoods of the elements of λ_{k+1}; such a family exists by Problem 48. Since X is normal, we may assume in this connection that the closure of every element of η_{l+1} is contained in some element of the cover γ. Continuing in this way, we obtain the sequence $\{\eta_n: n \in N^+\}$. Obviously, $\eta = \bigcup \{\eta_n: n \in N^+\}$ is a σ-discrete open cover of X, closure-refining γ.

184. Using Problems 204 and 48, we can construct not only a σ-discrete net in X, but also a σ-discrete base for X. Now use Problem 80.

185. See Problems 183, 150, 172 and 174.

186. Suppose that $\mathcal{F}$ is a countable set of point-finite families of open sets in X and $\mathcal{B} = \bigcup \{\gamma: \gamma \in \mathcal{F}\}$ is a base for X. For each $\gamma \in \mathcal{F}$ and each $x \in X$ put $\gamma_x = \{U \in \gamma: U \ni x\}$. Let $k \in N^+$. Let $X_{\gamma,k}$ be the set of all points $x \in X$ such that $|\gamma_x| = k$. Also put $V_\gamma(x) = \bigcap \{U: U \in \gamma_x\}$ if $\gamma_x \neq \Lambda$. Note that if $x, y \in X_{\gamma,k}$, then either $V_\gamma(x) = V_\gamma(y)$, or $x \notin V_\gamma(y)$ and $y \notin V_\gamma(x)$. Indeed, if $x \in V_\gamma(y)$, then $\gamma_y \subset \gamma_x$. But from $|\gamma_y| = |\gamma_x|$ it then follows that $\gamma_y = \gamma_x$. Now put $\lambda_{\gamma,k} = \{V_\gamma(x): x \in X_{\gamma,k}\}$ and $\mathcal{P} = \{\lambda_{\gamma,k}: k \in N^+, \gamma \in \mathcal{F}\}$. We show that the family $\mathcal{P}$ is a filtration on X. Let $x \in X$ and let the nhood Ox of x be selected arbitrarily. Take $V \in \mathcal{B}$ such that $x \in V \subset Ox$. For some $\gamma^* \in \mathcal{F}$ we have $V \in \gamma^*$. Put $k = |\gamma^*_x|$. Then $k \in N^+$, $x \in V_{\gamma_*}(x) \subset Ox$, $x \in X_{\gamma_*,k}$ and $V_{\gamma_*}(x) \in \lambda_{\gamma_*,k}$. Furthermore, no element of $\lambda_{\gamma_*,k}$ other than $V_{\gamma_*}(x)$

contains x: this has been established several times previously. Hence $\lambda_{\gamma *, k}(x) = V_{\gamma *}(x)$ is a nonempty set contained in Ox. This shows that $\mathcal{P}$ is a filtration.

Let $z \in X$, $x \in X_{\gamma, k}$ and $V_\gamma(x) \ni z$. Then $\gamma_z \supset \gamma_x$. But $|\gamma_z| < \aleph_0$ and $V_\gamma(x) = \bigcap \{U: U \in \gamma_x\}$. Hence every element of $\lambda_{\gamma, k}$ containing z is the intersection of finitely many elements of the fixed finite family γ_z. Consequently, $\lambda_{\gamma, k}$ is a point-finite family of sets.

187. We argue along the same lines as we did in the solution of Problem 103 for constructing a refinement. The refinement obtained in the given case (automatically) consists of point-finite families of sets. Now follow Problem 67.

188. For the proof it suffices to compare the statements of Problems 186 and 187.

189. This is immediate from Problems 188, 175, 185 and 85.

190. It is necessary to argue along the same lines as in the solution of Problem 186; however, we start not with a σ-point-finite base but with a σ-point-finite open refinement of the given cover.

191. Let γ be an open cover of X. First 'screen' this cover by a countable family of point-finite families of open sets in X in accordance with the preceding problem. Then, as in the solution of Problem 187, construct a family of open sets refining γ by recalling the properties of a uniform base (in order to obtain the refinement of γ, we simply discard all sets violating this refinement). The collection of maximal elements of this family is also a point-finite open cover, refining γ.

192. If $k = 0$, then the statement is obvious. Suppose that for $k = n$ our statement is proved. Assume that $\mathrm{Kp}(x, \lambda) \leqslant k + 1$. There exists (since $c(X) \leqslant \aleph_0$) a countable $\gamma \subset \lambda$ such that $[\bigcup\{U \in \gamma\}] \supset \bigcup\{V: V \in \lambda\}$ (see solution of Problem 39). Put $G = \bigcup\{U: U \in \gamma\}$, $\lambda^* = \lambda \setminus \gamma$ and $\mu = \{U \cap G: U \in \lambda^*\}$. μ may be regarded as a family of open sets in the space G. Also $\mathrm{Kp}(x, \mu) \leqslant k$ for all $x \in G$ and, since G is open in X, $c(G) \leqslant c(X) \leqslant \aleph_0$. Moreover, from the definition of G it follows that $V \cap G \neq \Delta$ for each $V \in \lambda$. The inductive assumption is thus admissible. We conclude: $|\mu| \leqslant \aleph_0$. Hence, even $|\lambda^*| \leqslant \aleph_0$. Finally, $|\lambda| \leqslant |\lambda^*| + |\gamma| \leqslant \aleph_0 + \aleph_0 = \aleph_0$.

193. Put $E = \{x \in X: \mathrm{Kp}(x, \lambda) \leqslant k - 1\}$, $Y = \{x \in X: \mathrm{Kp}(x, \lambda) \geqslant k\}$. Then $X = E \cup Y$. Obviously, Y is open in X. Hence $c(Y) \leqslant \aleph_0$. By Problem 102 of Chap. 2, there exists a countable family $\gamma \subset \mathcal{B}$ such that $\bigcup\{U: U \in \gamma\} \subset Y \subset [\bigcup\{U: U \in \gamma\}]$. Put $G = \bigcup\{U: U \in \gamma\}$, $\mu = \{U: U \in \lambda$ and $U \cap G = \Delta\}$, $F = [G]$ and $W = X \setminus F$. W is an open set. Since $\mathrm{Kp}(x, \lambda) \leqslant k - 1$ for all $x \in W$ and $c(W) \leqslant \aleph_0$, it follows from Problem 192 that $|\mu| \leqslant \aleph_0$. Hence $|\lambda \setminus \mu| > \aleph_0$. Clearly, $\lambda \setminus \mu \subset \lambda^* = \{U \in \lambda: U \cap G \neq \Delta\}$. Consequently, $|\lambda^*| > \aleph_0$. But $\lambda^* = \bigcup\{\theta(U): U \in \gamma\}$, where $\theta(U) = \{V \in \lambda: V \cap U \neq \Delta\}$. Since $|\gamma| \leqslant \aleph_0$, there is $U^* \in \gamma$, satisfying $|\theta(U^*)| > \aleph_0$. U^* is obviously the required set.

194. Fix a Hausdorff compactification bX of X and a family $\{G_i: i \in N^+\}$ of open sets in bX such that $\bigcap\{G_i: i \in N^+\} = X$. Put $\mathcal{B}_n = \{U \subset X: U$ is open in X, $U \neq \Delta$ and $[U]_{bX} \subset G_n\}$. It is routine to verify that $\{\mathcal{B}_n: n \in N^+\}$ is the desired family.

195. Fix a family $\{\mathcal{B}_n: n \in N^+\}$ of bases for X, as in the statement of the preceding problem. Suppose that $|\lambda| > \aleph_0$. Since X is a regular space and for every open set U in X we have $c(U) \leqslant c(X) \leqslant \aleph_0$, by successively applying the statements of Problem 193, we can select a sequence $\{U_n: n \in N^+\}$ such that for each $n \in N^+$ we have: (i_1) $[U_{n+1}] \subset U_n$; (i_2) $U_n \in \mathcal{B}_n$; (i_3) $|\{V \in \lambda: V \cap U_n\}| > \aleph_0$ and (i_4) $\mathrm{Kp}(x, \lambda) \geqslant$

n for all $x \in U_n$. The satisfaction of condition (i_1) is a somewhat subtle affair. For this, while defining U_{n+1}, passage has to be effected from $\mathcal{B}_n$ to the base $\mathcal{B}_n^*$ for U_n, consisting of all elements of $\mathcal{B}_n$ whose closure is included in U_n. In other respects the sequence $\{U_n: n \in N^+\}$ is constructed without difficulty. Then $\bigcap\{U_n: n \in N^+\} = \bigcap\{[U_n]: n \in N^+\} \neq \Lambda$ and obviously $\mathrm{Kp}(x, \lambda) > k$ for every $x \in \bigcap\{U_n: n \in N^+\}$ and each $k \in N^+$, i.e., every point of $\bigcap\{U_n: n \in N^+\}$ belongs to an infinite set of elements of λ. This contradicts that the latter is point-finite.

196. Since (i) is the weakest and (v) is the strongest of our statements, it suffices to deduce (v) from (i). By the preceding problem, any σ-point-finite open cover of X is countable. Hence if X is σ-metacompact, it is also Lindelöf.

197. It suffices to derive (iv) from (i). If X is weakly paracompact, then there is an open point-finite cover γ of X by Čech-complete subspaces, satisfying the Souslin property. From Problem 195 it follows immediately that γ is star-countable and that every element of γ is a Lindelöf space. It is now routine to deduce the statement (iv) from Problem 8 (see also the solution of Problem 166).

198. Suppose that $\gamma = \bigcup\{\gamma_i: i \in N^+\}$ is an open cover of X, where every γ_i is point-finite and $c(U) \leqslant \aleph_0$, and that U is Čech-complete for each $U \in \gamma$. By Problem 195, γ_i is star-finite. From Problem 9 it follows directly that γ_i is the union of a countable family of disjoint families of sets, i.e., that γ_i is σ-disjoint. But then γ is also σ-disjoint. Since such a cover as γ refines every open cover of X, we conclude that X is screened.

199. By the preceding problem, the space X is screened. But a screened perfectly normal space X is paracompact (see Problem 145).

200. The statements (iii) and (iv) are equivalent for all locally compact Hausdorff spaces (Problem 169), and (iii) is stronger than (ii) and (i). Hence it suffices to deduce (iii) from (i). Every perfectly normal compactum has the Souslin property (Problem 207 of Chap. 3), and is obviously a Čech-complete space. Hence, the required conclusion is immediate from the statement of the preceding problem.

201. Since the set $X \setminus F$ is open in X, $c(X \setminus F) \leqslant c(X) \leqslant \aleph_0$. In addition, $X \setminus F$ is locally compact and is also Čech-complete. Hence, by Problem 196 it follows from (ii) that $X \setminus F$ is Lindelöf. This enables us to conclude that F is a G_δ-set in X. But then $\chi(F, X) \leqslant \aleph_0$ (Problem 68 of Chap. 3).

Conversely, if $\chi(F, X) \leqslant \aleph_0$, then $X \setminus F$ is the union of a countable family of compacta and, consequently, is a Lindelöf space. Furthermore, $X \setminus F$ is a σ-metacompactum.

202. Clearly, if every open subspace of X is σ-metacompact, then each subspace of X is also. It now suffices to follow Problem 201.

203. By problem 195, every σ-point-finite base of such a space is countable. The space under consideration is metrizable, being a completely regular space with a countable base.

204. It suffices to show that every open cover ω has a σ-discrete closed refinement. Suppose the elements of ω are indexed by the elements of some well-ordered set. Suppose that $\gamma = \{\gamma_i: i \in N^+\}$ is a countable refinement and γ_{i+1} is a refinement of γ_i. For each $U \in \omega$, let $t(U, i)$ be the set of all points $x \in X$ such that no element of ω containing x precedes $U \in \omega$ in the given ordering of ω and every element in γ_i containing x is contained in U. The family $\varphi_i = \{t(U, i): U \in \omega\}$ is discrete and closed, and $\varphi = \bigcup\{\varphi_i: i \in N^+\}$ is a σ-discrete closed cover refining ω.

205. This is still an open question.

206. Necessity is proved basically by the same arguments as in the solution of Problem 204. The sufficiency has been estabiished by Choban [3] and Burke [1].

207. For each $\alpha \in A$ take a closed set F_α in X such that $F_\alpha \subset U_\alpha$ and $\bigcup\{F_\alpha: \alpha \in A\} = X$. Let E be the set of all real functions, defined on A, each of which vanishes outside some finite set. For each $\alpha \in A$, fix a continuous real function f on X such that $f(X) \subset [0, 1], f(X \setminus U_\alpha) = \{0\}$ and $f(F_\alpha) = \{1\}$. Let φ be the mapping of X into E, defined as follows: $(\varphi(x))(\alpha) = f_\alpha(x)$ for all $\alpha \in A$. Since ω is a point-finite cover, $\varphi(x) \in E$. The distance between any two elements g' and g'' of E is defined as usual by $\rho(g', g'') = \max\{|g'(\alpha) - g''(\alpha)|: \alpha \in A\}$. Since g' and g'' are different from zero only on a finite set of elements of A, this definition is coherent. We now see that $\varphi: X \to E$ is an ω-mapping of X into the space (E, ρ).

Consider an arbitrary element e of $\varphi(X) \subset E$. Fix $x^* \in X$ satisfying $\varphi(x^*) = e$. There is a $\alpha^* \in A$ such that $F_{\alpha^*} \ni x^*$. Then $e(\alpha^*) = f_{\alpha^*}(x^*) = 1$. It is to be shown that if $\rho(e, \varphi(x)) < 1$, then $x \in U_{\alpha^*}$; in other words, that the pre-image under φ of the sphere of radius 1 in $\varphi(X)$ with center at e is contained in U_{α^*}. Put $\varphi(x) = e'$. Since $\rho(e, e') < 1$, $|e(\alpha^*) - e'(\alpha^*)| < 1$. But $e'(\alpha^*) = f_{\alpha^*}(x)$, $e(\alpha^*) = 1$. Hence $f_{\alpha^*}(x) > 0$, i.e., $x \in U_{\alpha^*}$. It remains to verify that φ is a continuous mapping. Let $\varphi(x) = e$. Since ω is locally finite, there is a nhood Ox of x in X, outside which all $\{f_\alpha: \alpha \in A\}$ vanish, except for some finite family $\{f_\alpha: \alpha \in A'\}$. Now, take, with respect to $\varepsilon > 0$, a nhood $O'x \ni x$ so that $|f_\alpha(x) - f_\alpha(x')| < \varepsilon$ for all $\alpha \in A'$ and $x \in O'x$. Then $\rho(\varphi(x'), \varphi(x)) < \varepsilon$ for all $x \in O'x \cap Ox$. Since ε can be taken arbitrarily small, φ is a continuous mapping. Clearly, the weight of (E, ρ) is τ.

208. It is routine to observe that $(X, \mathcal{T})$ is a T_1-space and ξ is a star-monotonic refined sequence of open covers of X. Consequently (Problem 75), $(X, \mathcal{T})$ is metrizable.

209. Every paracompactum is normal (Problem 130). To establish necessity, it remains to refer to Problem 207. Sufficiency follows from the definition of ω-mapping and the paracompactness of metric spaces.

210. Conditions (*a*) and (*b*) are equivalent (Problem 105). Also (*a*) follows from (*c*): if $f: X \to Y$ is a continuous one-to-one mapping, where Y is a metric space, then its square $\varphi = f \times f: X \times X \to Y \times Y$ is also continuous and the pre-image of the diagonal $\Delta' \subset Y \times Y$ under φ is the diagonal $\Delta \subset X \times X$. But the space $Y \times Y$ is metrizable and Δ' is a closed set in it. Hence Δ' is a G_δ-set in $Y \times Y$. Thus Δ, as the pre-image of Δ' under a continuous mapping, is a G_δ-set in $X \times X$.

We now deduce (*c*) from (*b*) under the assumption that X is a paracompactum. Suppose that $\mathcal{E} = \{\gamma_i: i \in N^+\}$ is some T_1-refinement on X. Using the paracompactness of X, it is easy to construct by induction a strongly star-monotonic (countable) sequence ξ of open covers of X, which is a T_1-refinement on X. By Problem 32, all the elements of these covers may be regarded as ξ-complete sets. Then the family $\tilde{\mathcal{T}}$ of all ξ-complete subsets of X forms a topology, included in the original topology on X and (with respect to $\tilde{\mathcal{T}}$) ξ is a star-monotonic refined set of open covers: the latter is immediate from the definition of $\tilde{\mathcal{T}}$. Also $(X, \tilde{\mathcal{T}})$ is a T_1-space. Consequently, $(X, \tilde{\mathcal{T}})$ is metrizable (Problem 75, see also Problem 208). The identity mapping of $(X, \mathcal{T})$ onto $(X, \tilde{\mathcal{T}})$ is the desired contraction.

211. There is an ω-mapping f of X onto an metric space Y whose weight does not exceed $|\omega|$ (see Problem 207). The covering of Y by all open sets, the pre-image of each of which is contained in some element of ω, can be star-refined to an open cover (see Problem 28), and from the latter we can extract a cover λ of Y, for which $|\lambda|$ does not exceed the weight of Y. Obviously, pre-image of λ under f is the required cover of X.

216. In Problem 11 of Chap. 3, we described a regular Lindelöf space (which is therefore a paracompactum, see Problem 132), whose product with itself is neither normal, nor paracompact (Problem 130, see also Problem 221).

217. See Kombarov [2].

218. The space Y is the union of a locally finite family $\gamma = \{\Phi_t : t \in T\}$ of compact subspaces. For each $t \in T$, the product $X \times \Phi_t$ is a paracompactum (see Problem 43 of Chap. 6). Clearly, $\gamma^* = \{(X \times \Phi_t) : t \in T\}$ is a locally finite cover of the space $X \times Y$ by closed paracompact sets. Hence $X \times Y$ is a paracompactum (see Problem 131).

219. The space Z is regular, as the product of regular spaces. Fix a σ-locally finite base $\theta = \{V_\alpha : \alpha \in A_i, i \in N^+\}$ of Y (each $\theta_i = \{V_\alpha : \alpha \in A_i\}$ is locally finite). Put $A = \bigcup \{A_i : i \in N^+\}$. First we show that any open set O in Z is an F_σ. Obviously, there is a family $\{R_\alpha : \alpha \in A\}$ of open sets in X such that $\bigcup \{R_\alpha \times V_\alpha : \alpha \in A\} = O = \bigcup \{R_\alpha \times [V_\alpha] : \alpha \in A\}$ (for some $a \in A$, we may have $R_\alpha = \Lambda$). In X every open set is an F_σ, hence for all $\alpha \in A$ we have $R_\alpha = \bigcup \{B_{\alpha,j} : j \in N^+\}$, where for each $j \in N^+$, $B_{\alpha,j}$ is a closed set in X. Put $\gamma_{i,j} = \{B_{\alpha,j} \times [V_\alpha] : V_\alpha \in \theta_i\}$ and $P_{i,j} = \bigcup \{C : C \in \gamma_{i,j}\}$. Clearly $\gamma_{i,j}$ is a locally finite family of closed sets in Z. Obviously, $O = \bigcup \{P_{i,j} : i, j \in N^+\}$. Thus O is an F_σ-set.

Now consider an arbitrary open cover ω of Z. For each $\alpha \in A$, there is a family $\eta_\alpha = \{R_t : t \in T(\alpha)\}$ of open sets in X such that $\{R_t \times V_\alpha : t \in T(\alpha), \alpha \in A\}$ is a cover of Z, refining ω. Put $W_\alpha = \bigcup \{R_t : t \in T_\alpha\}$.

By Problem 137, we can refine the cover η_α of the space W_α to an open cover ξ_α, which is the union of a countable family $\{\xi_{\alpha,i} : i \in N^+\}$ of locally finite collections $\xi_{\alpha,i}$ of subsets of X. It is then easy to see that $\mu_{i,j} = \{G \times V_\alpha : G \in \xi_{\alpha,j}, \alpha \in A_i\}$ is a locally finite family of open sets in Z. Moreover $\mu = \bigcup \{\mu_{i,j} : i, j \in N^+\}$ covers Z and is a refinement of ω. Hence each open cover of Z has a σ-locally finite open refinement. But Z is regular. Hence (Problem 72), Z is paracompact.

220. No, not necessarily. See Problem 221.

221. Statement (*a*) is obvious. It is easy to see that $(X, \mathcal{T}^*)$ is regular. To establish (*b*), it suffices (Problem 127) to show that every open subspace V of $(X, \mathcal{T}^*)$ is paracompact. We have $V = U \cup P$, where $U \in \mathcal{T}$, $P \subset X_0$. Put $\mathcal{B} = \{U \in \mathcal{T} : U \subset V\}$ and $\xi = \{\{y\} : y \in P\}$. Clearly, $\mathcal{B} \cup \xi$ is a base for V (because $\mathcal{T} \cup \{\{y\} : y \in X_0\}$ is obviously a base for $(X, \mathcal{T}^*)$). For an arbitrary open cover ω of V we must find a locally finite open refinement. We may assume that $\omega \subset \mathcal{B} \cup \xi$. Then $\omega = \omega_1 \cup \omega_2$, where $\omega_1 = \mathcal{B} \cap \omega$, and $\omega_2 = \xi \cap \omega$. Obviously, $\omega_1 \subset \mathcal{T}$ and ω_1 covers U. Since U is a paracompact subspace of $(X, \mathcal{T})$, ω_1 has locally finite open refinement γ_1 which covers U. This γ_1 is a locally finite family of open sets in the space V. Obviously, ω_2 is discrete in V. Hence $\gamma_1 \cup \omega_2$ is a locally finite open cover of V, refining ω. It remains to prove (*c*).

Put $Q = X \setminus X_0$, $A = Q \times X_0$, $B = \{(x, x): x \in X_0\}$. The sets A and B are closed in Z and $A \cap B = \Lambda$. We must show that A and B do not have disjoint nhoods. Let V be an arbitrary nhood of B in Z. For each $n \in N^+$, put

$$S_{1/n}(x) = \{y \in X_0: |x - y| < 1/n\} \text{ for } x \in X_0,$$
$$U_n = \{x \in X_0: (\{x\} \times S_{1/n}(x)) \subset V\}.$$

Clearly, $X_0 = \bigcup \{U_n: n \in N^+\}$. Since X_0 is not F_σ in X, there is a $k \in N^+$ such that $[U_k]_X \cap Q \neq \Lambda$. Fix $x^* \in [U_k]_X \cap Q$ and $y^* \in X_0$, so that $|x^* - y^*| < 1/2k$. Every nhood W of (x^*, y^*) in Z contains a point (x', y') such that $x' \in U_k$, $|x^* - x'| < 1/2k$ and $y' = y^*$. But $(\{x'\} \times S_{1/k}(x')) \subset V$ and since $|y^* - x'| \leqslant |y^* - x^*| + |x^* - x'| < (1/2k) + (1/2k) = 1/k$, $(x', y^*) \in (\{x'\} \times S_{1/k}(x')) \subset V$. Thus $W \cap V \neq \Lambda$, which implies that $(x^*, y^*) \in [V]$. But $(x^*, y^*) \in A$. This establishes the non-separability of A and B.

222. Consider the projections $p: X \times Y \to X$ and $\pi: X \times Y \to Y$. The mapping p is perfect (see Problem 33 of Chap. 6). Suppose that $\sigma = \{V_i: i \in N^+\}$ is a countable base for Y. For each finite set $S \subset N^+$, put $V_S = \bigcup \{V_i: i \in S\}$. Clearly, $\Sigma = \{V_S: S \subset N^+, |S| < \aleph_0\}$ is also a countable base for Y. For every $M \subset X \times Y$, let M_x be the set $\pi((\{x\} \times Y) \cap M)$. Suppose that A and B are two closed disjoint sets in $X \times Y$. For each $V_S \in \Sigma$, put

$$U_S = \{x \in X: A_x \subset V_S \subset [V_S] \subset Y \setminus B_x\}.$$

Since

$$X \setminus U_S = p(A \cap (X \times (Y \setminus V_S))) \cup p(B \cap (X \times [V_S]))$$

and p is a closed mapping, U_S is open in X. It is easily seen that

$$\omega = \{U_S: S \subset N^+, |S| < \aleph_0\}$$

is a countable open cover of X. Since X is normal and countably paracompact, there is a locally finite countable open cover $\omega' = \{U'_S: S \subset N^+, |S| < \aleph_0\}$ such that $U'_S \subset [U'_S] \subset U_S$ (see Problem 157). Consider the set $\Gamma = \bigcup \{(U'_S \times V_S): S \subset N^+, |S| < \aleph_0\}$, open in $X \times Y$. We show that

$$A \subset \Gamma \text{ and } B \cap [\Gamma] = \Lambda.$$

For each point $(x, y) \in A$, we have $x \in U'_S \subset U_S$ for a certain $S \subset N^+$, $|S| < \aleph_0$. Then $(x, y) \in U'_S \times V_S \subset \Gamma$. Now, let $(x, y) \in B \cap [\Gamma]$.

Consider a nhood Ox of $x \in X$ intersecting at most finitely many elements of ω'. Then the nhood $Ox \times Y$ of (x, y) in the space $X \times Y$ has nonempty intersection with at most finitely many elements of the family $\{U'_S \times V_S: S \subset N^+, |S| < \aleph_0\}$. Let these be $U'_{S_1} \times V_{S_1}, \ldots, U'_{S_k} \times V_{S_k}$. Since $(x, y) \in [\Gamma]$, $(x, y) \in \bigcup_{i=1}^{k} [(U'_{S_i} \times V_{S_i})]$ and $(x, y) \in B \cap [U'_{S_i} \times V_{S_i}] = B \cap ([U'_{S_i}] \times [V_{S_i}]) \subset B \cap (U_{S_i} \times [V_{S_i}])$ for a certain $i \leqslant k$. Then $x \in U_{S_i}$, $y \in B_x \cap [V_{S_i}]$. But, by the definition of U_{S_i}, $x \in U_{S_i}$ implies

$B_\alpha \cap [Vs_i] = \Delta$. The contradiction thus engendered implies that $B \cap [\Gamma] = \Delta$. Consequently, $X \times Y$ is normal.

223. (*a*) If X is normal and countably paracompact, then $X \times I$ is normal by the preceding problem.

(*b*) X is homeomorphic to a closed subspace of $X \times I$ and hence is normal. Suppose that $\{F_i: i \in N^+\}$ is a family of closed sets in X, where $F_{i+1} \subset F_i$ for each $i \in N^+$ and $\bigcap\{F_i: i \in N^+\} = \Delta$. Consider the set $U_j = (X \setminus F_j) \times [0, (1/j)) \subset X \times I$, open in $X \times I$, where $j \in N^+$ and the disjoint closed sets $A = (X \times I) \setminus \bigcup \{U_j: j \in N^+\}$, $B = X \times \{0\}$. Since $X \times I$ is normal, there are open sets U and V in $X \times I$, for which $A \subset U$, $B \subset V$, $U \cap V = \Delta$.

Consider further the set $W_i = \{x \in X: (x, (1/i)) \in U\}$, $i \in N^+$, open in X. Since $B \cap [U] = \Delta$, $\bigcap\{W_i: i \in N^+\} = \Delta$. Obviously, $F_i \times \{(1/i)\} \subset A \subset U$. Hence $F_i \subset W_i$. The conclusion that X is a countable paracompact space is now immediate from Problem 159.

224. False. For an example, see M. E. Rudin [2].

225. False. This follows from Problems 222 and 224.

227. See Arkhangel'skii [2].

228. See Arkhangel'skii [2].

229. See Fillipov [4], or Ponomarev [3].

230. The answer is, in general, negative. But this is not always the case. An appropriate example has been quite recently constructed by Borrel.

231. See Fillipov [1].

Chapter 6

Spaces and Continuous Mappings

6.1. Theoretical Introduction

In Section 6.1, a detailed study will be made of the following two classes of mappings: *pseudo-open* and *biquotient* mappings.

Pseudo-open mappings were defined on p. 78. Every pseudo-open mapping is a quotient mapping (for the definition of quotient mapping see p. 45). The converse is not true. On the other hand, every continuous open mapping and every continuous closed mapping is pseudo-open.

The class of biquotient mappings, introduced quite recently, is closely related to the class of pseudo-open mappings. The latter occupies an intermediate position between quotient and biquotient mappings, i.e., every biquotient mapping is pseudo-open and every pseudo-open mapping is a quotient mapping. On the other hand, of course, not every quotient mapping is pseudo-open, nor is every pseudo-open mapping a biquotient mapping. However, if the continuous mapping $f: X \to Y$ is compact (see p. 45), then, if it is pseudo-open, it is a biquotient mapping (Problem 17).

$f: X \to Y$ is called a *biquotient mapping at the point* $y \in Y$ if every cover γ of $f^{-1}y$ by open sets in X has a finite subfamily $\lambda \subset \gamma$, $|\lambda| < \aleph_0$ such that $y \in \operatorname{Int} \bigcup \{fU: U \in \lambda\}$. If $f: X \to Y$ is biquotient at each point $y \in Y$, then it is simply called a *biquotient mapping*.

If X is a topological space and $A \subset X$, then Int A denotes, as always, the set of all interior points of A, i.e., the interior of A in X.

In this chapter we shall also use various sorts of mappings investigated earlier, e.g., closed mappings (see p. 45), perfect mappings (see p. 45), quotient mappings (see p. 45), open mappings (see p. 45), irreducible mappings (see p. 144) and so forth.

We shall deal here in depth with the variety of mappings mentioned above, particularly in the classification of topological spaces. Alexandroff [2] posed the following general questions in 1961:

(*a*) which spaces can be represented as images of 'nice' spaces (metric, paracompact, zero-dimensional and so on) under 'nice' continuous mappings (closed, perfect, open and so on)?; and
(*b*) which spaces can be mapped onto 'nice' spaces by means of 'nice' mappings?

Intimately related to these questions is the problem of the conservation of various topological properties under mappings, both under passage to the image and the pre-image. The following question is of primary interest: under what mappings is the image of a metrizable space metrizable?

A good portion of Sections 6.1—6.4 is devoted to these problems.

We start with the definitions of some classes of mappings which we shall meet occasionally in this chapter.

M-1. A continuous mapping $f: X \to Y$ is called an *s-mapping* if for each $y \in Y$ the subspace $f^{-1}y \subset X$ is separable. If X is metrizable, then $f: X \to Y$ is an *s-mapping* iff, for each $y \in Y, f^{-1}y \subset X$ has a countable base.

M-2. A continuous mapping $f: X \to Y$ is called *monotonic* if the subspace $f^{-1}y \subset X$ is connected for each $y \in Y$.

M-3. *A continuous mapping* $f: X \to Y$ is 0-*dimensional* if, for each $y \in Y$, $f^{-1}y$ is 0-dimensional (precisely, 0-dimensional in the sense of p. 42).

M-4. A mapping $f: X \to Y$ is of *finite multiplicity* if $|f^{-1}y| < \aleph_0$ for every point $y \in Y$; *of multiplicity n*, where n is a positive natural number, if $|f^{-1}y| = n$ for each $y \in Y$, and of *countable multiplicity* if $|f^{-1}| \leqslant \aleph_0$ for each $y \in Y$.

M-5. A point $x_0 \in X$ is a *point of local homeomorphism for the mapping* $f: X \to Y$ if there is a nhood Ux_0 of x_0 such that the restriction $f \mid Ux_0$ is a homeomorphism on some nhood of fx_0 into Y.

If each point $x \in X$ is a point of local homeomorphism for the mapping $f: X \to Y$, then f is called a *local homeomorphism* or a *local topological mapping*.

M-6. A mapping $f: X \to Y$ is called a c-*covering* (here c is the starting letter of the word 'compactness'), if for every compact subspace $\Phi \subset Y$ there is a compact subspace $F \subset X$ such that $\Phi \subset fF$.

We now recall the classical concepts of retract and retraction. A subspace F of X is called a *retract* of X, if there is a continuous mapping $f: X \to F, f(X) = F$ such that $fx = x$ for each $x \in F$. The mapping f in this case is called a *retraction of X onto F*. If f is a closed mapping, it is called a *closed retraction of X* onto F.

What kinds of spaces will arise most frequently in this chapter? First and foremost are of course compact and metrizable spaces, followed by paracompacta, τ-paracompacta†, where τ is a cardinal number, countably compact spaces, τ-compact spaces†, countably paracompact spaces, Lindelöf and strongly paracompact spaces. These spaces have already been defined. Spaces satisfying the first axiom of countability (see pp. 43-44), with a point-countable base (see p. 56), with a uniform base (see p. 272), with a countable refinement or, equivalently, spaces with a refined countable family of open coverings (see p. 273), and spaces of countable type or pointwise countable spaces (see p. 137), occupy an important place in the present chapter. All these sorts of spaces have been encountered frequently in previous chapters,

A new class of spaces, the so-called *extremally disconnected* spaces, is studied in detail in Sections 6.5 and 6.6. Extremally disconnected spaces play a prominent role in set-theoretic mathematics. In their own right, if we start just from their formal definition, the extremally disconnected spaces form a fascinating class possessing a whole series of unusual properties. However, to emphasize their versatility, it suffices to point out that extremally disconnected spaces play a crucial role in the theory of Boolean algebras, in axiomatic set theory and in some branches of functional analysis as well. The importance of extremally disconnected spaces, particularly in general topology, becomes transparent in the context of the theory of absolute topological spaces, which we consider at the end of Section 6.6. We shall further dwell upon it later.

We now give the definition: a space X is called *extremally disconnected* if for every open set $U \subseteq X$ in X the set $[U]$ is not only closed but also open in X.

As a rule, we shall consider regular extremally disconnected spaces (note that they are then automatically completely regular). We now set up an important concept, which will be investigated in Section 6.4. This relates to the conventional concept of A-operation in descriptive set theory; in fact, it plays a subsidiary role in Section 6.4.

Let W be a set of cardinality τ, $\tau \geqslant \aleph_0$ and let X be a topological space. Assume that open sets $U_{\lambda_1 \ldots \lambda_n}$ of X are associated with each finite

†A space X is called *τ-paracompact*, if every open cover ω of X, of cardinality $|\omega| < \tau$, has a locally finite open refinement. If every covering ω, $|\omega| \leqslant \tau$ of X, has a finite subcovering, then X is called a τ-compact space.

sequence $\lambda_1, \ldots, \lambda_n$ of elements of W (where $n \in N^+$ is arbitrary). Moreover, this correspondence must be such that we have the equality

$$U_{\lambda_1 \ldots \lambda_n} = \bigcup\{U_{\lambda_1 \ldots \lambda_n \lambda_{n+1}}: \lambda_{n+1} \in W\} \tag{1}$$

for each $n \in N^+$ and any finite sequence $\lambda_1, \ldots, \lambda_n$, $\lambda_1 \in W$, $\lambda_2 \in W$, $\ldots, \lambda_n \in W$.

In this case, the family

$$\varphi = \{U_{\lambda_1 \ldots \lambda_n}: \lambda_1 \in W, \ldots, \lambda_n \in W, n \in N^+\} \tag{2}$$

is called an *A_τ-family of open sets of X*. Note firstly that the finite sequence $(\lambda_1, \ldots, \lambda_n)$, $\lambda_1 \in W, \ldots, \lambda_n \in W$ does not necessarily consist of mutually distinct elements of W and secondly, that distinct sequences $(\lambda_1, \ldots, \lambda_n)$ and $(\lambda_1', \ldots, \lambda_n')$ (i.e., such that $\lambda_i \neq \lambda_i'$ for all $i \leqslant n$) are not necessarily associated with distinct sets $U_{\lambda_1 \ldots \lambda_n}$ and $U_{\lambda_1' \ldots \lambda_n'}$.

For each fixed $n \in N^+$, put

$$\varphi_n = \{U_{\lambda_1 \ldots \lambda_n}: \lambda_1 \in W, \ldots, \lambda_n \in W\}. \tag{3}$$

In this case, when φ_n is a cover of X for each $n \in N^+$ (incidentally, for this it suffices to prescribe that φ_1 be a cover of X), an A_τ-family φ is called an *A_τ-sequence of open covers* of X. If the covers φ_n are refined in X (see p. 273), then the family φ is called a *refined A_τ-sequence of open covers of X*.

Furthermore, an A_τ-sequence $\varphi = \{U_{\lambda_1 \ldots \lambda_n}: \lambda_1 \in W, \ldots, \lambda_n \in W, n \in N^+\}$ of open covers in X is called *complete* if for every countably infinite sequence $\lambda_1, \ldots, \lambda_n, \ldots$ of elements of W we have $\bigcap\{U_{\lambda_1 \ldots \lambda_n}: n \in N^+\} \neq \Lambda$.

Finally, an A_τ-sequence φ is *disjoint* if for every $n \in N^+$ and distinct sequences $(\lambda_1, \ldots, \lambda_n)$ and $(\lambda_1', \ldots, \lambda_n')$, $\lambda_1 \in W, \ldots, \lambda_n \in W, \lambda_1' \in W, \ldots, \lambda_n' \in W$, we have $U_{\lambda_1 \ldots \lambda_n} \cap U_{\lambda_1' \ldots \lambda_n'} = \Lambda$.

We now arrive at the concluding stage of our discussions, in which the concept of absolute of regular spaces plays a central role. We say that Y is an *irreducible perfect pre-image* of a regular space X if there is an irreducible perfect mapping f of Y onto X.

A space $\dot{X}$ is called a (*regular*) *absolute of the space* X if, firstly $\dot{X}$ is an irreducible perfect pre-image of X and secondly, every irreducible perfect pre-image of $\dot{X}$ is homeomorphic to $\dot{X}$.

Every regular space X has a (unique, to within homeomorphism) absolute $\dot{X}$. This absolute is completely regular and extremally disconnected.

We say that two regular spaces X and Y are *co-absolute* if their absolutes $\dot{X}$ and $\dot{Y}$ are homeomorphic. Clearly, the co-absolute relationship is symmetric and transitive. Thus all regular spaces are classified into co-absolute spaces. Two co-absolute spaces have many common properties. In particular, the focus of our interest are those spaces which are co-absolute with some metrizable space.

The first series of problems is devoted to the construction of an absolute of a given regular space, the proof of its uniqueness and also its basic properties (see Problem 218). These will reveal the basic construction and introduce the fundamental concepts needed for constructing absolutes.

Let X be a regular space, and let $\mathfrak{B}(X)$ be the family of all nonempty regular closed sets in X.

Recall that a closed set A in X is called regular if it is the closure of an open set in X, hence if it is the closure of its interior. A family $\xi \subset \mathfrak{B}(X)$ of regular closed sets in X is $\varkappa$-centralized if $\cap\{\text{Int } A: A \in \xi'\} \neq \Lambda$ for every finite subfamily $\xi' \subset \xi$.

Every $\varkappa$-centralized family ξ of regular closed sets of X is called a $\varkappa$-*family*. If a $\varkappa$-family ξ is not contained in any other (different from ξ) $\varkappa$-family, then ξ is called a $\varkappa$-*ultrafilter*. If $\cap\{A: A \in \xi\} \neq \Lambda$, where ξ is a $\varkappa$-family, then ξ is called *distinguished*. Finally, if $x \in \cap\{A: A \in \xi\}$, where $x \in X$ is some point and ξ is a $\varkappa$-family, then ξ is said to be *distinguished at the point* x.

A finite closed cover α of X is called *regular* if all elements of α are regular closed sets and the interiors of distinct elements of α do not intersect. Sometimes it is useful to admit into consideration also infinite closed covers with these properties; it is of course required that they be locally finite in X. Such coverings are called *locally finite regular covers of X*.

If α is a regular covering, then we put $\alpha^0 = \{\text{Int } A: A \in \alpha\}$; let $\hat{\alpha}$ be the set α equipped with the discrete topology, and let $\alpha(x)$, where $x \in X$, be the star of the point x with respect to α, i.e., $\alpha(x) = \cup\{A \in \alpha: x \in A\}$.

If a regular cover α_2 is a refinement of a regular cover α_1, then we say that α_2 follows α_1, symbolically $\alpha_2 > \alpha_1$.

We denote by $\mathfrak{U}(X)$ the family of all regular covers of a given regular space X. This family is refined and *directed*. The latter means that for every pair of covers $\alpha_1, \alpha_2 \in \mathfrak{U}(X)$ there is a cover $\alpha_3 \in \mathfrak{U}(X)$ such that $\alpha_3 > \alpha_1, \alpha_3 > \alpha_2$. If $\alpha \in \mathfrak{U}(X)$ and $\beta \in \mathfrak{U}(X)$, then $\gamma = \alpha \wedge \beta = \{[\text{Int } A \cap \text{Int } B]: A \in \alpha, B \in \beta \text{ and Int } A \cap \text{Int } B \neq \Lambda\}$.

Suppose now that $\mathfrak{U}_0(X)$ is some refined directed family of regular covers of X. Consider the Tychonoff product $P_0 = \Pi\{\hat{\alpha}: \alpha \in \mathfrak{A}_0(X)\}$. If for each $\alpha \in \mathfrak{A}_0(X)$ we fix an element $A^\alpha \in \alpha$, then the family $\xi = \{A^\alpha: \alpha \in \mathfrak{A}_0(X)\}$ may be regarded as a point in P_0. The point $\xi = \{A^\alpha: \alpha \in \mathfrak{A}_0(X)\}$ is called a *thread in the family* $\mathfrak{A}_0(X)$ if $A^{\alpha_2} \subset A^{\alpha_1}$ whenever $\alpha_2 > \alpha_1$, where α_2, $\alpha_1 \in \mathfrak{A}_0(X)$. Denote by $\Xi_0(X)$ *the set of all threads in* $\mathfrak{A}_0(X)$ and endow it with the topology induced by the Tychonoff product $P_0 = \Pi\{\hat{\alpha}: \alpha \in \mathfrak{A}_0(X)\}$. A *thread* $\xi \in \Xi_0(X)$ is said to be *distinguished* if $\cap \{A: A \in \xi\} \neq \Lambda$. Moreover, if $x \in \cap\{A: A \in \xi\}$, then ξ is *distinguished at the point* x. We denote by $\dot{X}_0$ *the set of all distinguished threads in* $\mathfrak{A}_0(X)$, which also is endowed with the topology induced by P_0. If $\xi \in \dot{X}_0$, we put $\pi_X^0(\xi) = \cap\{A: A \in \xi\}$. From the fact that $\mathfrak{A}_0(X)$ is refined it follows that $\pi_X^0\xi$ consists of exactly one point. Thus we have defined a mapping $\pi_X^0: \dot{X}_0 \to X$ of the space $\dot{X}_0$ of all distinguished threads in $\mathfrak{A}_0(X)$ into the space X. This mapping is called a *natural mapping generated by the family* $\mathfrak{A}_0(X)$. If we apply this construction to the family $\mathfrak{A}(X)$ of all regular covers of a given regular space X, then we get the space of all threads in $\mathfrak{A}(X)$, denoted by $\Xi(X)$; the space of all distinguished threads in $\mathfrak{A}(X)$, denoted by $\dot{X}$, and a natural mapping $\pi_X: \dot{X} \to X$. As a result, $\dot{X}$ is an absolute of X, and $\Xi(X)$ is the Stone-Čech compactification of this absolute (see Problem 204).

Let us introduce some notations for use in Section 6.6:

1°. If $A_1 \in \mathfrak{B}(X), \ldots, A_s \in \mathfrak{B}(X)$, then
$\overline{\langle A_1, \ldots, A_s \rangle} = \{\xi \in \Xi(X): A_1 \in \xi, \ldots, A_s \in \xi\}$,
$\langle A_1, \ldots, A_s \rangle = \{\xi \in \dot{X}: A_1 \in \xi, \ldots, A_s \in \xi\}$.

2°. If $\alpha \in \mathfrak{A}(X)$ is a regular cover, then
$\overline{\langle \alpha \rangle} = \{\overline{\langle A \rangle}: A \in \alpha\}$ and $\langle \alpha \rangle = \{\langle A \rangle: A \in \alpha\}$.

3°. $\langle \alpha \rangle(\xi) = \cup\{\langle A \rangle \in \langle \alpha \rangle: \xi \in \langle A \rangle\}$, where $\xi \in \dot{X}$.

6.2. Quotient, Biquotient and Pseudo-open Mappings

1. Show that a continuous mapping $f: X \to Y$ of X onto Y is a quotient mapping iff for every mapping $g: Y \to Z$ the continuity of gf implies the continuity of g.

2. Show that every T_1-space X can be mapped one-to-one and continuously onto a compact T_1-space.

3. Show that every locally compact Hausdorff space can be mapped one-to-one and continuously onto some compactum.

4. Show that the product (p. 5) of every set of compact mappings (p. 45) is a compact mapping.

5. Show that if X is a Hausdorff space and f is a retraction onto $f(X) \subset X$, then $f(X)$ is closed in X.

6. Suppose that (X, ρ) is a metric space, $f: X \to Y$ is a compact quotient mapping of X onto Y and $d(y', y'') = \rho(f^{-1}y', f^{-1}y'')$ for all $y', y'' \in Y$. Show that d is a semi-metric on Y, generating the topology of Y.

7. If a compactum is the image of a metric space by a quotient compact mapping, then it is metrizable.

8. Suppose that f is a perfect mapping of a metric space (X, ρ) onto a T_1-space Y and $d(y', y'') = \rho(f^{-1}y', f^{-1}y'')$ for all $y', y'' \in Y$. Show that

(i) d is a semi-metric on Y;

(ii) d generates the topology of Y;

(iii) d satisfies the condition (j) of Problem 96, Chap. 5.

Finally, deduce from Problem 96 of Chap. 5 that Y is metrizable.

9. If a dyadic compactum X is a quotient space of a metric space, then X has a countable base.

10. True or false? The product of every set of pseudo-open compact mappings on a Hasudorff space is a pseudo-open compact mapping.

11. If X is a completely normal space (for the definition, see p. 44), then for every pseudo-open compact mapping $f: X \to Y$ such that $f(X) = Y$, every closed set in Y is a G_δ.

12. If a space Y is the image of X under a compact pseudo-open mapping, then the weight of Y does not exceed the weight of X.

13. Every biquotient mapping is quotient.

14. Every biquotient mapping is pseudo-open.

15. If $f: X \to Y$ is a biquotient mapping and $Y' \subset Y$, $X' = f^{-1}Y'$, then $f' = f \mid X'$, $f': X' \to Y'$ is a biquotient mapping.

16. A mapping $f: X \to Y$, where $f(X) = Y$, is a biquotient mapping iff for every prefilter ξ on Y and for any adherent point y of ξ (see p. 39) there is a point $x \in f^{-1}y$, which is an adherent point of $f^{-1}\xi = \{f^{-1}B, B \in \xi\}$.

17. A compact mapping is a biquotient mapping iff it is pseudo-open.

18. The image of a locally compact space under a biquotient mapping is a locally compact space.

19. The product of every set of biquotient mappings is a biquotient mapping.

20. Suppose that $f: X \to Y$, $f(X) = Y$, is a biquotient mapping. Then the weight of Y does not exceed the weight of X.

21. Suppose that $f: X \to Y$ is a quotient mapping, Y is a Hausdorff space and $f(X) = Y$. Suppose further that y is a point of Y, for which $f^{-1}y$ is a Lindelöf subspace of X. Assume also that Y is first countable at y. Then f is a biquotient mapping at y.

22. If $f: X \to Y$ is a quotient mapping of a space X with a countable base onto a Hausdorff first countable space Y, then Y has a countable base.

23. If a Hausdorff first countable space Y is a quotient space of a locally compact space with a countable base, then Y is a locally compact metrizable space with a countable base.

24. Does there exist a quotient not pseudo-open mapping of a locally compact Hausdorff space onto a compactum?

25. Suppose that $f: X \to Y$ is a quotient mapping, $f(X) = Y$ and $i: F \to F$ is the identity mapping of a compactum F onto itself. Then, the product $\varphi = f \times i: X \times F \to Y \times F$ of f and i is a quotient mapping (cf. Problem 307 of Chap. 2).

26. If $f: X \to Y$ is a quotient mapping, $f(X) = Y$, Z is a locally compact Hausdorff space and $j: Z \to Z$ is the identity mapping, then $f \times j: X \times Z \to Y \times Z$ is also a quotient mapping.

27. If $f: X \to Y$ is a monotonic quotient mapping of a strongly paracompact metrizable space X onto a first countable space Y, then Y is metrizable and strongly paracompact.

28. In the class of Hausdorff spaces, the Fréchet-Urysohn spaces, and only they, are the images of metric spaces under continuous pseudo-open mappings.

6.3. Perfect Mappings

29. The composition of perfect mappings is a perfect mapping.

30. The restriction of a perfect mapping to a closed subspace is again a perfect mapping.

31. The product (of any set) of perfect mappings on completely regular spaces is a perfect mapping.

32. For every perfect mapping $f: X \to Y$, $f(X) = Y$, there is a closed set $F^* \subset X$ such that $f(F^*) = Y$ and $f \mid F^*$ is an irreducible perfect mapping.

33. Suppose that X is an arbitrary topological space, F a compactum, $X \times F$ a topological product and π: $X \times F \rightarrow X$ is the natural projection onto the first factor, i.e., $\pi(x, a) = x$ for all $x \in X$, $a \in F$. Show that π is a perfect mapping.

34. Suppose that X is a Hausdorff space and $\mathcal{Z} = \{F_t: t \in T)$ a continuous decomposition of X into compacta (see p. 77). Every compactum F_t is decomposed into its connected components (p. 154); $F_t = \bigcup\{F_{ts}: s \in S_t\}$ (here F_{ts} is a connected component of F_t). Consider the decomposition $\mathcal{K}$ of X: $X = \bigcup\{F_{ts}: t \in T, s \in S_t\}$. Show that $\mathcal{K}$ is also a continuous decomposition of X.

35. Suppose that f: $X \rightarrow Y$ is a perfect mapping of X onto Y, both Hausdorff spaces. Show that there is a Hausdorff space Z and perfect mappings φ: $X \rightarrow Z$ and ψ: $Z \rightarrow Y$ such that $f = \psi\varphi$, φ is monotonic and ψ is 0-dimensional.

36. If f: $X \rightarrow Y$ is a perfect mapping of X onto a compact space Y, then X is compact.

37. If f: $X \rightarrow Y$ is a closed continuous mapping of X onto a Lindelöf space Y such that the subspace $f^{-1}y$ is Lindelöf for each point $y \in Y$, then X is Lindelöf.

38. If f is a closed continuous mapping of X onto a τ-paracompact (p. 325) (where τ is a fixed cardinal number, $\tau \geqslant \aleph_0$) space Y such that $f^{-1}(y)$ is a τ-compact subspace (p. 325) of X for each $y \in Y$, then X is τ-paracompact.

39. Suppose that f is a closed continuous mapping of X onto a τ-compact (where τ is a fixed cardinal number, $\tau \geqslant \aleph_0$) space Y, where $f^{-1}y$ is τ-compact for each point $y \in Y$. Then X is τ-compact.

40. If f is a closed continuous mapping of X onto a countably paracompact space Y and the pre-images of all points are countably compact, then X is countably paracompact.

41. If Y is the image of X under a perfect mapping, then the weight of Y does not exceed the weight of X.

42. If X is perfectly mapped onto Y of weight $\leqslant \tau$, then every open cover of X has a subcover of cardinality $\leqslant \tau$.

43. The product of a paracompactum X with a compactum Y is a paracompactum.

44. A continuous mapping f: $X \rightarrow Y$ is perfect iff for every prefilter ξ

in X and for each adherent point $y \in Y$ of $f\xi = \{fP\colon P \in \xi\}$, there is an adherent point $x \in f^{-1}y$ of ξ.

45. A continuous mapping f of a locally compact Hausdorff space X onto a (non-compact) locally compact Hausdorff space Y is a perfect mapping iff f can be extended to a continuous mapping $\tilde{f}\colon \alpha X \to \alpha Y$, where αX and αY are Alexandroff compactifications of X and Y (i.e., $|\alpha X \setminus X| \leqslant 1$ and $|\alpha Y \setminus Y| \leqslant 1$; see Problem 176 of Chap. 3).

46. Suppose that bX is a Hausdorff compactification of X, bY a Hausdorff compactification of Y and $\tilde{f}\colon bX \to bY$ is a continuous mapping, where $\tilde{f}(bX) = bY$ and $\tilde{f}(bX \setminus X) \subset bY \setminus Y$. Then $f\colon X \to Y$ is a perfect mapping of X onto Y, where $f = \tilde{f} \mid X$.

47. Suppose that bX, bY are Hausdorff compactifications of X and Y respectively, and $f\colon bX \to bY$ is a continuous mapping, whose restriction $f_0 = f \mid X$ to X is a perfect mapping of X into Y. Then $f(bX \setminus X) \subset bY \setminus Y$.

48. Suppose that $f\colon X \to Y$ is a perfect mapping of X onto Y, where X and Y are completely regular spaces. Show that

(*a*) when either of these spaces is locally compact, then the other is as well;

(*b*) the Čech-completeness of one implies the Čech-completeness of the other.

49. If X is a metrizable space, and f a perfect mapping of X onto a T_1-space Y, then Y is metrizable.

50. A completely regular space Y, which is the image of a completely regular pointwise countable space X under a perfect mapping, is not necessarily a pointwise countable space.

51. If X has a base of multiplicity τ, $\tau > \aleph_0$ and $f\colon X \to Y$, $fX = Y$ is a perfect mapping, then Y also has a base of multiplicity τ.

52. Give an example of a regular Lindelöf space, which cannot be mapped perfectly onto any seperable space.

53. Suppose that $f\colon X \to Y$ is a continuous mapping of X onto Y and bX, bY are Hausdorff compactifications of X and Y, and $\tilde{f}\colon bX \to bY$ is a continuous extension of f. Assume that: (*a*) the restriction $\tilde{f} \mid (bX \setminus X)$ is a homeomorphism of $bX \setminus X$ onto $bY \setminus Y$, and (*b*) there is a continuous mapping $\varphi\colon Y \to X$ such that $f\varphi(y) = y$ for each $y \in Y$.

Show that the restriction $f \mid R(X)$ is a homeomorphism of $R(X)$ onto

$R(Y)$ (where $R(X)$ consists of all points of X, at which X is not locally compact; and similarly for $R(Y)$).

54. Suppose that X is a compactum, containing more than one point, Y is a completely regular but not locally compact space, π: $X \times Y \to Y$ is a projection and $b(X \times Y)$, bY are Hausdorff compactifications of $X \times Y$ and Y respectively such that π extends to a continuous mapping π^*: $b(X \times Y) \to bY$. Then the mapping $\pi^* \mid (b(X \times Y) \setminus (X \times Y))$ cannot be a homeomorphism.

55. If f: $X \to Y$ is a perfect mapping and g: $X \to Z$ is a continuous mapping, where Z is a Hausdorff space, then the diagonal (p. 5) product $f \Delta g$: $X \to Y \times Z$ is also a perfect mapping (without assuming that $f(X) = Y$ and $g(X) = Z$).

56. Suppose that f: $X \to Y$ and g: $Y \to Z$ are continuous mappings, gf: $X \to Z$ being a perfect mapping, and Y is a Hausdorff space. Then f is also perfect. Also, if $f(X) = Y$, then g is also perfect.

57. If f: $X \to Y$ is a continuous mapping, X a Hausdorff space, $X_1 \subset X$ and if f: $X_1 \to Y$ is a perfect mapping (the last implies that f: $X_1 \to Y$ is a perfect mapping of X_1 onto a closed subspace of Y), then X_1 is closed in X.

58. The diagonal product of any set of mappings on a Hausdorff space, of which one is perfect, is a perfect mapping.

59. If X admits a perfect mapping into Y and a continuous one-to-one mapping (contraction) into a Hausdorff space Z, then X is homeomorphic to some closed subspace of $Y \times Z$.

60. Suppose that Y is an arbitrary space and X is a completely regular space. Then the following statements are equivalent:

(*a*) there is a perfect mapping f: $X \to Y$;

(*b*) there is a compactum Z such that X is homeomorphic to some closed subspace of $Y \times Z$.

61. If a Hausdorff space X, having a net of cardinality not greater than τ, can be mapped perfectly onto a space of weight not exceeding τ, then X also has weight not exceeding τ.

62. If X can be perfectly mapped onto a metric space and is contractible to another metric space, then it is metrizable.

63. If X can be perfectly mapped onto a space of weight not greater than τ and is contractible to a Hausdorff space of weight not greater than τ, then the weight of X itself does not exceed τ.

64. If X can be perfectly mapped onto a first countable space and is contractible to a first countable space, then X is also first countable.

65. If X can be perfectly mapped onto a space with a uniform base and is contractible to a Hausdorff space with a uniform base, then X itself has a uniform base.

66. If X can be perfectly mapped onto a space with a countable refinement and is contractible to a Hausdorff space with a countable refinement, then X itself has a countable refinement.

67. A Hausdorff space which is the pre-image of a k-space under a perfect mapping is also a k-space.

68. Suppose that $f: X \to Y$ is a 0-dimensional closed mapping of a normal locally connected space X onto a space Y. Let $\theta = \{V_t: t \in T\}$ be a base for Y. For each $t \in T$, consider the family $\sigma_t = \{U_{t\alpha}: \alpha \in A_t\}$ of all connected components of $f^{-1}V_t$. Show that $\sigma = \bigcup\{\sigma_t: t \in T\} = \{U_{t\alpha}: \alpha \in A_t, t \in T\}$ is a base for X.

69. Suppose that $f: X \to Y$ is a perfect mapping of a locally connected space X onto a space Y of weight not exceeding τ, where τ is an infinite cardinal number. Then, for every open set $V \subset Y$ in Y the family σ_V of connected components of $f^{-1}V$ has cardinality not exceeding τ.

70. Suppose that $f: X \to Y$ is a closed 0-dimensional mapping of a normal locally connected space X onto a metrizable space Y. Then X is also metrizable.

71. Suppose that $f: X \to Y$ is a perfect 0-dimensional mapping of a normal locally connected space X onto Y. Then the weight of X coincides with the weight of Y.

72. Suppose that $f: X \to Y$ is a perfect 0-dimensional mapping of a normal locally connected space X onto a space Y of countable weight. Show that X is metrizable.

6.4. Closed Mappings

73. Suppose that bX and bY are Hausdorff compactifications of X and Y respectively and that $f: bX \to bY$ is a continuous mapping, whose restriction $f_0 = f \mid X$ on X is a closed mapping of X into Y. Then $[f_0^{-1}y]_{bX} = f^{-1}y$ for each $y \in Y$.

74. If $f: X \to Y$ is a closed continuous mapping, $f(X) = Y$ and every closed set in X is a G_δ, then every closed set in Y is also a G_δ. Does a similar statement also hold for other cardinalities?

75. The image of a collectionwise normal space under a continuous closed mapping is collectionwise normal.

76. A regular space Y, which is the image of a paracompactum X under a continuous closed mapping, is also paracompact.

77. Suppose that X is a metrizable space: $\dim X = 0$ (see Problem 153 of Chap. 5), and that F is a closed subset of X. Then there is a closed continuous mapping $f: X \to F$ such that $fx = x$ if $x \in F$ (F is a retract of X under a closed retraction).

78. Suppose that $B(\tau)$ is a Baire space of weight τ (Problem 246 of Chap. 2) and $X \subset B(\tau)$ is a closed subspace. Show that there is a closed retraction of $B(\tau)$ onto X.

79. Suppose that X is a Hausdorff space of countable tightness (see p. 43) and $f: T(\omega_1) \to X$ (see p. 136) is a continuous mapping. Then f is closed.

80. Suppose that $f: X \to Y$ is a continuous closed mapping of X onto a countably compact T_1-space Y. Then, for any point $y \in Y$ and continuous real function φ on X, the restriction φ to the set $\mathrm{Fr}\,(f^{-1}y)$ is a bounded function (here $\mathrm{Fr}\,(f^{-1}y)$ is the frontier of $f^{-1}y$ in X).

81 If $f: X \to Y$ is a closed continuous mapping of a normal space X into a first countable T_1-space Y, then for each $y \in Y$ the frontier of $f^{-1}y$ is countably sequentially compact.

82. If $f: X \to Y$ is a closed continuous mapping of a weakly paracompact normal space X into a first countable T_1-space Y, then for each point $y \in Y$ the frontier of $f^{-1}y$ is a compactum.

83. Under a closed continuous mapping of a normal space X onto a countably sequentially compact T_1-space Y, the frontier of the pre-image of each point is countably compact.

84. Suppose that $f: X \to Y$ is a closed continuous mapping of a normal space X onto a T_1-space Y and that y is some point of Y, with a countable family γ of nhoods in Y such that for any choice of a point $y(U)$ from each $U \in \gamma$, where $y(U) \neq y(V)$ for $U \neq V$, the set $\{y(U): U \in \gamma\}$ has a limit point in Y. Then the frontier of $f^{-1}y$ is a countably compact space.

85. If X is a paracompactum, Y is a pointwise countable T_1-space and $f: X \to Y$ is a closed continuous mapping, $f(X) = Y$, then the frontier of $f^{-1}y$ in X is compact for each $y \in Y$.

86. Suppose that Y is a countable compactum, $Z = T(\omega_1)$ (see p. 136), $X = Y \times Z$ (a topological product) and that $f: X \to Y$ is the natural projection of X onto the first factor. Then: (*a*) X is normal; (*b*) X is countably compact; (*c*) f is closed; (*d*) Fr $(f^{-1}y^*) = f^{-1}(y^*)$ is homeomorphic to Z and is not a compactum, where y^* is the unique non-isolated point in Y.

87. If $f: X \to Y$ is a closed continuous mapping of a paracompactum X onto a first countable T_1-space Y, then there is a closed subspace X^* of X such that $fX^* = Y$ and $f_0 = f \mid X^*$, $f_0: X^* \to Y$ is a perfect mapping.

88. Suppose that $f: X \to Y$ is a closed continuous mapping of a T_1-space X onto a T_1-space Y, under which the frontier of the pre-image of every point is compact. Then there is a closed subspace X^* of X such that $f(X^*) = Y$ and the restriction $f_* = f \mid X^*$; $f_*: X^* \to Y$ is a perfect mapping.

89. If a pointwise-countable T_1-space Y is the image of a metric space X under a continuous closed mapping, then Y is metrizable.

90. Suppose that (X, ρ) is a metric space and $\mathcal{Z}$ is a decomposition of X into mutually disjoint closed sets such that for every $\varepsilon > 0$ the set of elements of $\mathcal{Z}$ of diameter $> \varepsilon$ is finite. Then $\mathcal{Z}$ is a continuous decomposition.

91. Give an example of a continuous partition $\mathcal{Z}$ of a separable metric space X into mutually disjoint closed non-compact sets with empty interior.

92. Give an example of a normal space, not first countable at any point, which is the image of a separable metrizable space under a closed continuous mapping.

93. Suppose that $\gamma = \{G_\alpha: \alpha \in A\}$ is a point-finite open cover of a T_1-space X and $f: X \to Y$ is a closed continuous mapping of X onto a k-space Y (see p. 137). Then the set $D = \{y \in Y: f^{-1}y \cap (X \setminus \bigcup \{G_\alpha: \alpha \in A'\}) \neq \Lambda$ for any finite set $A' \subset A\}$ is discrete in Y (and hence also closed, see p. 271).

94. If $f: X \to Y$ is a closed continuous mapping of a weakly paracompact, locally compact T_1-space X onto a space Y, then the subspace $f^{-1}y$ is compact for all $y \in Y$, contained in the complement of some discrete set A in Y.

95. The image of a locally compact weakly paracompact T_1-space under a closed continuous mapping is representable as the union of two locally compact subspaces, of which one is open and the other is closed.

96. If X is a regular space with a countable net $\mathcal{S}$, $f: X \to Y$ is a closed continuous mapping and $f(X) = Y$, then the set of all points $y \in Y$, for which $f^{-1}y$ is non-compact, has cardinality not exceeding $\aleph_0$.

97. If $f: X \to Y$ is a closed continuous mapping of countable multiplicity from a regular space X with a countable net $\mathcal{S}$ onto a T_1-space Y, then

Y can be represented as the union of a countable family of subspaces each of which is homeomorphic to a subspace of X.

98. Suppose that $\{F_\alpha: \alpha \in M\}$ is the family of all elements of some continuous partition of a topological space X and $\xi = \{U_\alpha: \alpha \in M\}$ is a family of open sets in X such that $U_\alpha \supset F_\alpha$ for each $\alpha \in M$. Put $M_\xi = \{\alpha \in M$: there is an $\alpha' \in M\setminus\{\alpha\}$ such that $U_{\alpha'} \supset F_\alpha\}$. Then for every $\beta \in M$, $|\{\alpha \in M\setminus M_\xi: \langle U_\beta \rangle \cap F_\alpha \neq \Lambda\}| \leqslant 1$, where $\langle U_\beta \rangle$ is the largest distinguished open set contained in U_β, i.e., the union of all elements of the decomposition contained in U_β (see p. 77). Furthermore, $\{\{F_\alpha\}: \alpha \in M\setminus M_\xi\} = Z\setminus Z_\xi$, where Z is the space of the decomposition under consideration, $Z_\xi = \{\{F_\alpha\}: \alpha \in M_\xi\}$ and $Z\setminus Z_\xi$ is discrete in Z (see p. 271).

99. Let (X, ρ) be a metric space, and let f be a closed continuous mapping of (X, ρ) onto some T_1-space Z. Then $Z = \bigcup\{Y_i: i = 0, 1, 2, \ldots\}$, where, for every $i \geqslant 1$, Y_i is a discrete closed set in Z and $f^{-1}z$ is a compactum for every $z \in Y_0$.

100. A T_1-space Y, which is the image of a metric space X under a closed continuous mapping f, is representable as the union of a metrizable subspace and a countable family of closed discrete (hence, also metrizable) subspaces.

101. If a T_1-space Y is the image of some metrizable space X with a countable base under a closed continuous mapping, then Y has a subspace Y^* with a countable base such that $|Y\setminus Y^*| \leqslant \aleph_0$.

102. Suppose that $f: X\to Y$ is a closed continuous mapping, X is a paracompactum, $fX = Y$, and that Y is a T_1-space, every non-discrete subset of which contains a countable but not closed (in Y) subset. Then, the cardinality of the set of all those points of Y whose pre-images are not compact, does not exceed $\tau^{\aleph_0}$, where τ is the weight of X.

103. Let τ be a cardinal number, X a first countable paracompactum having a net of cardinality not exceeding τ. Then, under any closed continuous mapping of X, the pre-images of all points—with the possible exception of a set of cardinality not exceeding τ—are compact.

104. Suppose that f is a closed continuous mapping of a paracompactum X of weight τ onto a T_1-space Y, satisfying the condition that every non-discrete subspace of Y contains a countable non-closed set (in Y); equivalently, Y is a space of countable tightness. Show that Y has a subspace Y^* of weight not greater than τ such that $[Y\setminus Y^*] \leqslant \tau$.

105*. Give an example of a paracompactum X, $|X| > wX$, and a closed continuous mapping f of X onto a paracompactum Y such that $f^{-1}y$ is not compact for any point $y \in Y$.

106. If a paracompactum X is a k-space and if $f: X \to Y$ is a closed continuous mapping of X onto a T_1-space Y, then $Y = Y^* \cup Y'$, where the weight of Y^* and the cardinality of Y' do not exceed the weight of X.

107. If a homogeneous (see p. 125) T_1-space Y, which is not the union of any countable family of discrete subspaces, is the image of a metrizable space under a closed continuous mapping, then Y is itself metrizable.

108. If an uncountable homogeneous T_1-space Y is the image of a regular space X with a countable base under a closed continuous mapping f, then Y itself has a countable base.

109. Suppose that $f: X \to Y$ is an irreducible closed mapping, $f(X) = Y$. Then, for every open set U in X, the set $f^{-1}f^{\#}U$ is open, nonempty, is contained in U and is dense in U. Here $f^{\#}U = \{y \in Y: f^{-1}y \subset U\}$.

110. Suppose that $f: X \to Y$ is an irreducible closed mapping. Then

$$f[U] = [f^{\#}U]$$

for every open set U in X and the image of every regular closed set in X is a regular closed set in Y (for the definition, see p. 48).

111. Suppose that $f: X \to Y$ is an irreducible closed mapping of a T_1-space X onto a T_1-space Y. Then the point $x_0 \in X$ is isolated in X iff the point fx_0 is isolated in Y.

112. If $f: X \to Y$ is an irreducible mapping of a T_1-space X onto a space Y, then for each point $y \in Y$ either the frontier of $f^{-1}y$ coincides with $f^{-1}y$ in X, or $|f^{-1}y| = 1$.

113. Suppose that $f: X \to Y$ is an irreducible closed continuous mapping of a completely regular space X onto a completely regular space Y. Show that a continuous extension $f: \beta X \to \beta Y$ (to the Stone-Čech compactifications of these spaces) is also irreducible (see Problem 31 of Chap. 4).

114. Every irreducible closed continuous mapping of a paracompactum onto a first countable T_1-space is perfect.

6.5. Open Mappings

115. Every first countable T_1-space X is the image of a metric space S (it may be assumed that $S \subset B(\tau)$, where $\tau = wX$ and $B(\tau)$ is a Baire space) under an open continuous mapping (the converse statement is also true).

116. A compactum which is the image of a metric space under an open continuous mapping, is always metrizable. True or false?

117. A T_1-space X has a point-countable base iff it is the image of some metric space S under an open continuous mapping such that the pre-image of each point has a countable base (here it may be assumed that $S \subset B(\tau)$, where $\tau = wX$ and $B(\tau)$ is a Baire space of weight τ; see Problems 246-248 of Chap. 2).

118. A T_1-space Y has a uniform base iff it is the image of some metric space S under an open continuous compact mapping (it is possible to assume that $S \subset B(\tau)$, $\tau = wY$).

119. If a locally compact Hausdorff space X is the image of a metric space under an open continuous s-mapping, then it is metrizable.

120. Suppose that $f: M \to X$ and $g: X \to Y$ are open continuous compact 'onto' mappings, where M is a metric space and X and Y are completely regular spaces. Then Y has a uniform base and, consequently (see Problem 118), is the image of some metric space under an open continuous compact mapping. True or false?

121. The pre-image of a compactum under an open continuous compact mapping is a compactum. True or false?

122. Suppose that X is a completely regular space, in which every closed set is a G_δ and that f is an open continuous compact mapping of X onto a completeley regular space Y. Then in Y every closed set is a G_δ. True or false?

123. Show that every irreducible open continuous mapping $f: X \to Y$ of a Hausdorff space is a homeomorphism.

124. Every locally homeomorphic mapping of multiplicity k from a Hausdorff space is closed.

125. Every open mapping of multiplicity k from a Hausdorff space X onto a space Y is a local homeomorphism (i.e., for each point there is a nhood which is mapped homeomorphically onto an open set in the image) and closed.

126. Suppose that $f: X \to Y$ is an open continuous mapping of finite multiplicity, the weight of Y does not exceed τ, X is a Hausdorff space and that $\tau \geqslant \aleph_0$. Then, X has a net of cardinality not greater than τ.

127. (*a*) If a Hausdorff space X can be mapped onto a space Y by an open and closed mapping of finite multiplicity, and the weight of Y is not less than $\aleph_0$, then the weight of X equals the weight of Y.

(*b*) If X is a p-space (in particular, X is Čech-complete) and if $f: X \to Y$

is a continuous open (not necessarily closed) mapping of finite multiplicity onto Y, then $wX = wY$ (for definition, see pp. 125 and 274).

128. If $f: X \to Y$ is an open and closed mapping of X onto Y and $\varphi: X \to I$ is a continuous bounded real-valued function, then the mapping ψ defined on Y by $\psi(y) = \sup \{\varphi(x): x \in f^{-1}(y)\}$ for each $y \in Y$, is also continuous on Y.

129. Is every open continuous mapping of one metric space onto another a c-cover?

130. Suppose that S_n is a finite set for each $n \in N^+$, where $S = \bigcup \{S_n: n \in N^+\}$ is infinite. Let $f: S \to S$ be a many-valued mapping, satisfying the conditions: (α) for every $U \in S_n$ and each $n \in N^+$, we have the equality $f^{-1}(f^{-1}(U) \cap S_{n+1}) = f^{-1}U \cap P_{n+1}$, where $P_n = S \setminus \bigcup \{S_i: i \leqslant n\}$ and $f^{-1}(U) = \{U' \in S: fU' \ni U\}$; ($\beta$) $f^{-1}(S_n) \supset P_n$, $n \in N^+$. Then $U_n \in S_n$ can be so chosen that we have $f(U_{n+1}) \ni U_n$ for all $n \in N^+$.

131. Suppose that we are given for each $n \in N^+$ a finite set S_n and single-valued mapping $f_n: S_{n+1} \to S_n$. Then $U_n \in S_n$ can be chosen such that $f_n(U_{n+1}) = U_n$, $n \in N^+$.

132. Suppose that S_n (for each $n \in N^+$) is a finite family of sets (in a topological space X), where each element of S_{n+1} is contained in some element of S_n. Then $U_n \in S_n$ can be chosen such that $U_{n+1} \subset U_n$ for every $n \in N^+$.

133. If f is an open mapping of a metric space (X, ρ) onto a compact space Y, then, in (X, ρ), there is a sequence $\{\gamma_i: i \in N^+\}$ of finite families of open sets in (X, ρ) such that for all $i \in N^+$:

(i) if $U \in \gamma_i$, then diam $(U) < 1/i$;
(ii) for each $U \in \gamma_{i+1}$, there is a $V \in \gamma_i$ such that $U \subset V$;
(iii) $f\gamma_i = \{fU: U \in \gamma_i\}$ covers Y.

134. Suppose that f is an open (not necessarily continuous) mapping of a metric space (X, ρ) onto a compact Hausdorff space Y, where $(f^{-1}y, \rho)$ is complete for all $y \in Y$. Then there is a closed totally bounded subspace (Φ, ρ) in (X, ρ) such that $f\Phi = Y$. If, in addition, (X, ρ) is complete, then by Problem 255 of Chap. 3, the subspace Φ is compact.

135. If f is an open (not necessarily continuous) mapping of X onto a space Y such that the pre-image $f^{-1}y$ of each point $y \in Y$ is closed in X, and $\mathcal{B}_x$ is a base for X at $x \in X$, then $f\mathcal{B}_x = \{fU: U \in \mathcal{B}_x\}$ is a pseudo-base for Y at $fx \in Y$.

136. Suppose that f is an open mapping of a metric space (X, ρ) onto a compactum Y, where $(f^{-1}y, \rho)$ is complete for each $y \in Y$. Then Y has a countable base.

137. If a compactum Y is the image of a complete metric space X under an open continuous mapping f, then Y is metrizable.

138. Every open mapping of a metric space X onto a space Y, under which the pre-images of all points are complete with respect to the metric on X, is a c-covering.

139. Suppose that f is an open and closed continuous mapping of a compactum F onto a compactum Φ and $\{y_n: n \in N^+\}$ is a sequence of points of Φ converging in Φ to some point y. Then there is a sequence $\{x_n: n \in N^+\}$ of points of F converging to some point x such that $fx_n = y_n$ for all $n \in N^+$. True or false?

140. Give an example of an open continuous mapping of finite multiplicity from a normal space of countable type without non-trivial convergent sequences onto an infinite compactum.

141. Suppose that $f: X \to Y$ is an open mapping of a Čech-complete space X onto a paracompactum Y. Then there is a closed set $F \subset X$ such that $f(F) = Y$ and the restriction $f_0 = f \mid F$ of f to F is a perfect mapping.

142. Show that a paracompactum Y, which is the image of a Čech-complete space X under an open mapping, is also Čech-complete.

143. Suppose that $f: X \to Y$ is an open mapping of a metrizable complete metric space X onto a metrizable space Y. Show that Y is also metrizable by a complete metric.

144. Show that a paracompactum Y, which is the image of a metrizable complete metric space X under an open mapping f, is also metrizable by a complete metric.

145. Suppose that $f: X \to Y$ is an open mapping of a Čech-complete space X onto a Hausdorff space Y. Then f is a c-covering.

146.* If $f: X \to Y$ is an open mapping, $f(X) = Y$, X is a complete metric space with metric ρ, Y is a paracompactum of dimension 0 (see Problem 153 of Chap. 5), then there is a closed set F in X such that $f(F) = Y$ and the restriction $f_0 = f \mid F$ is a homeomorphism.

147. X is the image of a Baire space $B(\tau)$ (see Problem 246 of Chap. 2) under an open mapping iff X has a complete refined A_τ-family of open covers (see p. 326).

148. X is homeomorphic to the Baire space $B(\tau)$ of weight τ iff there is, in X, a disjoint complete refined A_τ-family of open covers.

149. If (X, ρ) is a complete metric space of dimension 0, homogeneous with respect to weight (i.e., $wU = wX$ for every open subspace $U \subset X$) and having no points of local compactness, then X is homeomorphic to the Baire space $B(\tau)$, where $\tau = wX$.

150. Every complete metric space (X, ρ) of weight τ is an image of the Baire space $B(\tau)$ of weight τ under an open mapping.

151. If X is a paracompactum, then X is metrizable by a complete metric iff it is an image of the Baire space $B(\tau)$ under an open mapping.

152. Every complete metric space (X, ρ) is an image of the Baire space under a closed continuous mapping.

6.6. Extremally Disconnected Spaces

153. Show that every discrete T_1-space X is extremally disconnected.

154. A regular extremally disconnected space X is completely regular.

155. X is extremally disconnected iff disjoint open sets in X have disjoint closures.

156. Show that every open subspace U of an extremally disconnected space X is an extremally disconnected space.

157. Show that a dense subspace X_0 of an extremally disconnected space X is also extremally disconnected.

158. Show that the Stone-Čech compactification βX of an extremally disconnected completely regular space X is also extremally disconnected.

159. Show that the Stone-Čech compactification βD of a discrete space D is extremally disconnected.

160. Suppose that $f\colon X \to Y$ is an irreducible closed mapping of a Hausdorff space X onto an extremally disconnected space Y. Show that f is a homeomorphism and that X is also extremally disconnected.

161. Suppose that X is an extremally disconnected normal space, Y is a closed subspace of X and $P \subset Y$ is a clopen set in Y. Then There is a clopen set G in X such that $G \cap Y = P$.

162. Suppose that X is an extremally disconnected compactum and X_0 is a dense subspace of X. Show that the Stone-Čech compactification βX_0 of X_0 coincides with the compactum X.

163. Suppose that X is an extremally disconnected regular space and $D_1, D_2 \subset X$ are two Lindelöf subspaces of X, satisfying $D_1 \cap [D_2] = \Lambda$ and $D_2 \cap [D_1] = \Lambda$. Then $[D_1] \cap [D_2] = \Lambda$.

164. Suppose that X is a regular extremally disconnected space and $D_1, D_2 \subset X$ are two countable discrete subspaces of X, satisfying $[D_1] \cap D_2 = \Lambda$ and $D_1 \cap [D_2] = \Lambda$. Then $[D_1] \cap [D_2] = \Lambda$.

165. Suppose that X is an extremally disconnected compactum and D is a countable discrete subspace of X. Show that $[D]_X$ coincides with the Stone-Čech compactification βD of D and hence, since D is homeomorphic to the space N of natural numbers with the discrete topology, it coincides with Stone-Čech compactification βN of N.

166. Show that an infinite extremally disconnected compactum contains a topological image of the Stone-Čech compactification βN of N.

167. Show that in an infinite extremally disconnected compactum there are no non-trivial countable convergent sequences.

168. Show that in an extremally disconnected completely regular space there are no non-trivial convergent sequences.

169. Suppose that X is an extremally disconnected compactum, and $D \subset X$ is a subspace of X satisfying the following condition: (k) for each point $d \in D$ there is an nhood Ud of d in X such that the family $\gamma = \{Ud: d \in D\}$ is disjoint. Then $[D]_X$ coincides with the Stone-Čech compactification βD of D.

170. Show that an extremally disconnected regular space is 0-dimensional (in the sense of p. 42). Give an example of a compactum that is 0-dimensional but not extremally disconnected.

171. Show that the weight of every infinite extremally disconnected compactum X is not less than $2^{\aleph_0}$ and its cardinality is not less than $2^{\mathfrak{c}}$.

172. Is every closed subspace of an extremally disconnected compactum extremally disconnected ?

173. In order for a compactum X to be extremally disconnected, it is necessary that X be the Stone-Čech compactification of each of its dense subspaces and sufficient that X be the Stone-Čech compactification of each of its dense open subspaces.

174. Suppose that X is an infinite extremally disconnected compactum and $D \subset X$ is a countably infinite discrete subspace of X, none of whose points are isolated in X. Then the subspace $X_0 = X \setminus D$ is not contractible to any compactum, i.e., there exists no one-to-one continuous mapping of X_0 onto any compactum.

175. Every infinite extremally disconnected compactum X can be mapped continuously onto the Stone-Čech compactification βN of the space of natural numbers.

176. Show that no infinite extremally disconnected compactum is hereditarily normal.

177. Show that the image Y of an extremally disconnected space X

under a continuous open mapping f is also an extremally disconnected space.

178. The growth $X = \beta N \setminus N$ of the Stone-Čech compactification of the discrete space of natural numbers is not extremally disconnected.

179. Is the property of extremal disconnectedness preserved under perfect mappings?

180. Suppose that D is a discrete space of cardinality $|D| = \tau$. Show that $|\beta D| = 2^{2^\tau}$, where βD is the Stone-Čech compactification of D.

181. Suppose that f is a mapping of X into itself and k is an integer, $k \geqslant 2$. A family γ of mutually disjoint clopen sets in X is called an (f, k)-*family*, if (i) $fU \cap U = \Lambda$ for all $U \in \gamma$, (ii) $f(\cup \{U: U \in \gamma\}) \subset \cup \{U: U \in \gamma\}$ and (iii) $|\gamma| = k$. In addition, we agree to call $\cup \{U: U \in \gamma\}$ an (f, k)-*separable* set and denote it by $\langle \gamma \rangle$. Assuming that f is a continuous mapping, show that:

(*a*) if γ is an (f, k)-family, then there is an (f, k)-family λ such that (i) γ is a refinement of λ, and (ii) $f^{-1}\langle \gamma \rangle = \langle \lambda \rangle$;

(*b*) if γ is an (f, k)-family, X is extremally disconnected and f is a homeomorphism onto a closed subspace of X, then there is an (f, k)-family μ such that: (i) γ is a refinement of μ, and (ii) $f^{-1}\langle \mu \rangle = \langle \mu \rangle$;

(*c*) if X is extremally disconnected and $\mathcal{S}$ is a disjoint family of (f, k)-separable sets (possibly infinite), then $[\cup \{A: A \in \mathcal{S}\}]$ is also an (f, k)-separable set.

182. If f is a homeomorphism of an extremally disconnected normal space X into itself, not the identity, and $f(X)$ is closed in X, then

(i) X contains a nonempty $(f, 3)$-separable set;

(ii) X contains a nonempty $(f, 3)$-separable set A such that (II′) $f^{-1}A = A$, and (II″) if $A' \neq \Lambda$ and A' is a $(f, 3)$-separable set, then $A' \cap A \neq \Lambda$;

(iii) X contains a nonempty $(f, 3)$-separable set A such that $f(x) = x$ for all $x \in X \setminus A$.

183. If f is a homeomorphism of an extremally disconnected normal space X into itself such that $f(X)$ is closed in X, then $X = B_0 \cup B_1 \cup B_2 \cup B_3$, where B_0, B_1, B_2, B_3 are mutually disjoint clopen sets in X, with $f(x) = x$ for all $x \in B_0$ and $f(B_i) \cap B_i = \Lambda$ for $i = 1, 2, 3$. In particular, the set of all fixed points of f is clopen in X.

184. For every homeomorphism f of an extremally disconnected compactum onto itself (not the identity), does there exist a nonempty $(f, 2)$-separable set?

185. Suppose that X is an extremally disconnected normal space and f is a homeomorphism of X into itself such that $f(X)$ is a closed nowhere dense set in X. Then f has no fixed point.

186. Suppose that X is an extremally disconnected compactum, $F \subset X$ is a closed nowhere dense set in X. Assume that there is a subspace $Y \subset F$, homeomorphic to X. Then F is not homogeneous.

187. Show that the compactum $X = \beta N \setminus N$, where βN is the Stone-Čech compactification of the discrete space of natural numbers, is not homogeneous.

6.7. Absolute Regular Spaces and Irreducible Perfect Mappings. Co-absolute Spaces

188. Suppose that α is a regular cover of X and x_0 is any point in X. Then $x_o \in \operatorname{Int} \alpha(x_0)$, where $\alpha(x_0)$ is, as usual, the star of x_0 with respect to α.

189. Suppose that α_1 and α_2 are two regular covers of X and that $\alpha_2 > \alpha_1$, i.e., α_2 is a refinement of α_1. Then:

(*a*) each element of α_2 is contained in a unique element of α_1;

(*b*) each element of α_1 is the union of all elements of α_2, contained in the given element of α_1.

190. Suppose that α and β are two regular covers of X. Show that $\gamma = \alpha \wedge \beta$ (for definition, see p. 327) is a regular cover of X, refining both α and β. This also holds in the case in which α and β are locally finite regular covers. In this case $\alpha \wedge \beta$ is also a locally finite regular cover.

191. Let X be a space, $\mathfrak{A}(X)$ the family of all regular covers of the space with the natural ordering (for the definition, see p. 327). Then $\mathfrak{A}(X)$ is partially ordered and directed in the sense that for every $\alpha \in \mathfrak{A}(X)$, $\beta \in \mathfrak{A}(X)$, there is a $\gamma \in \mathfrak{A}(X)$ such that $\gamma > \alpha$, $\gamma > \beta$.

192. Show that the family $\mathfrak{A}(X)$ of all regular covers of a T_1-space X is refined iff X is regular.

193. Suppose that X is a Hausdorff space and $\mathfrak{A}(X)$ is the family of all regular covers of X. If ξ is a distinguished thread in $\mathfrak{A}(X)$, then $\bigcap \{A: A \in \xi\}$ consists of exactly one point.

194. Suppose that X is Hausdorff and $\mathfrak{P}(X)$ is the family of all regular closed sets in X. Show that:

(*a*) every $\varkappa$-family η (distinguished at the point $x_0 \in X$) belongs to some $\varkappa$-ultrafilter (distinguished at the point $x_0 \in X$);

(*b*) the $\varkappa$-family η is a $\varkappa$-ultrafilter iff the following conditions are satisfied:

(i) if $A_1, A_2 \in \eta$, then $[\operatorname{Int} A_1 \cap \operatorname{Int} A_2] \in \eta$;
(ii) if $A_1 \in \eta$ and $A_1 \subset A_2$, $A_2 \in \mathfrak{P}(X)$, then $A_2 \in \eta$;
(iii) if $B \in \mathfrak{P}(X)$ and $\operatorname{Int} B \cap \operatorname{Int} A \neq \Lambda$ for every $A \in \eta$, then $\mathfrak{P} \in \eta$;

(*c*) if η is a $\varkappa$-ultrafilter and $A \in \mathfrak{P}(X)$, then either $A \in \eta$ or $[X \setminus A] \in \eta$.

195. Suppose that X is Hausdorff. Then a $\varkappa$-family ξ is a $\varkappa$-ultrafilter in $\mathfrak{P}(X)$ (resp. a distinguished $\varkappa$-ultrafilter) iff ξ is a thread (resp. a distinguished thread) in the family $\mathfrak{A}(X)$ of all regular covers of X.

196. Suppose that X is a Hausdorff space, $x_0 \in X$ is an arbitrary point and A_0 is a regular closed set of X containing x_0. Then there is a distinguished thread ξ in $\mathfrak{A}(X)$, for which $\{x_0\} = \cap\{A: A \in \xi\}$ and $A_0 \in \xi$.

197. Show that the natural mapping $\pi_X: \dot{X} \to X$ of the set $\dot{X}$ of all distinguished threads into X (see p. 328) is a mapping onto X.

198. Suppose that X is a Hausdorff space and $\mathfrak{A}(X)$ is the family of all regular covers of X. Consider the Tychonoff product $\Pi\{\hat{\alpha}: \alpha \in \mathfrak{A}(X)\} = P$. Then $\dot{X} \subset \Xi(X) \subset P$. Assume that $\dot{X}$ and $\Xi(X)$ are assigned the topologies induced by the Tychonoff product $\Pi\{\hat{\alpha}: \alpha \in \mathfrak{A}(X)\} = P$ (see p. 328). Show that:

(*a*) the sets of the form $\langle \bar{A}_0 \rangle = \{\xi \in \Xi(X): A_0 \in \xi\}$ (resp. $\langle \dot{A}_0 \rangle = \{\xi \in \dot{X}: A_0 \in \xi\}$) form a base for a topology for $\Xi(X)$ (resp. $\dot{X}$);
(*b*) $\Xi(X)$ is closed in P;
(*c*) $\Xi(X)$ is a compactum;
(*d*) the subspace $\dot{X} \subset \Xi(X)$ is dense in $\Xi(X)$.

199. Suppose that X is a Hausdorff space, $\mathfrak{A}(X)$ is the family of all regular covers of X, $\dot{X}$ is the space of all distinguished threads in $\mathfrak{A}(X)$ and $\pi_X: \dot{X} \to X$ is the natural mapping. Then, for every regular closed set $A_0 \in \mathfrak{P}(X)$, we have the equalities:

(*a*) $\pi_X(\langle A_0 \rangle) = A_0$;

(*b*) $\pi^{\#}_{\dot{X}}(\langle A_0 \rangle) = \text{Int } A_0$, where, as usual, $\pi^{\#}_{X}(\langle A_0 \rangle) = \{x \in X: \pi_X^{-1}x \subset \langle A_0 \rangle\}$.

200. Suppose that X is a Hausdorff space. Then the natural mapping $\pi_X: \dot{X} \to X$ is irreducible and compact. If X is regular, then π_X is also continuous.

201. Suppose that X is regular, $\mathfrak{A}(X)$ is the family of all regular covers of X. For each $\alpha \in \mathfrak{A}(X)$, put $\langle \bar{\alpha} \rangle = \{\langle \bar{A} \rangle: A \in \alpha\}$ and $\langle \alpha \rangle = \{\langle A \rangle: A \in \alpha\}$. Then:

- (i) $\langle \bar{A} \rangle = [\langle A \rangle]_{\Xi(X)}$ for each regular closed set A in X;
- (ii) for each $\alpha \in \mathfrak{A}(X)$ the family $\langle \bar{\alpha} \rangle$ (resp. $\langle \alpha \rangle$) is a disjoint clopen cover of $\Xi(X)$ (resp. $\dot{X}$); in particular, $\langle \bar{\alpha} \rangle$ (resp. $\langle \alpha \rangle$) is a regular cover of $\Xi(X)$ (resp. $\dot{X}$);
- (iii) the family $\{\langle \bar{\alpha} \rangle: \alpha \in \mathfrak{A}(X)\}$ (resp. $\{\langle \alpha \rangle: \alpha \in \mathfrak{A}(X)\}$ is refined in $\Xi(X)$ (resp. $\dot{X}$);
- (iv) $\langle \bar{\alpha}_2 \rangle$ (resp. $\langle \alpha_2 \rangle$) is a refinement of $\langle \bar{\alpha}_1 \rangle$ (resp. $\langle \alpha_1 \rangle$) iff α_2 is a refinement of α_1.

202. If X is regular, then the natural mapping $\pi_X: \dot{X} \to X$ is an irreducible perfect mapping onto X.

203. If $f: X \to Y$ is an irreducible closed mapping, $f(X) = Y$, A_1, $A_2 \subset X$ are regular closed sets in X for which $fA_1 = fA_2$, then $A_1 = A_2$.

204. Suppose that X is regular and $\mathfrak{A}(X)$ is the family of all regular covers of X. Then:

(*a*) $\Xi(X)$ is an extremally disconnected compactum;

(*b*) $\dot{X}$ is an extremally disconnected completely regular space;

(*c*) $\dot{X}$ is an absolute of X;

(*d*) the Stone-Čech compactification $\beta\dot{X}$ of $\dot{X}$ coincides with $\Xi(X)$.

205. Show that every regular space X has an absolute, and that it is an extremally disconnected completely regular space.

206. Every regular space X is the image of a 0-dimensional completely regular space under an irreducible perfect mapping.

207. Show that every regular space of weight τ has a refined directed family of regular covers of cardinality τ.

208. Show that every regular space X of weight τ is the image of a 0-dimensional completely regular space Y of weight τ under an irreducible perfect mapping. Y may be taken to be a subspace of the generalized discontinuum D^τ (see p. 86).

209. Every compactum X of weight τ is the image of a 0-dimensional compactum of weight τ under an irreducible continuous mapping.

210. Every compactum of weight τ is the image of a closed set in D^τ under an irreducible perfect mapping.

211. Every 0-dimensional regular space X, whose weight does not exceed τ, is homeomorphic to a subset of D^τ.

212. Show that every 0-dimensional compactum of weight not greater than τ is homeomorphic to a closed subset of D^τ.

213. Suppose that $f\colon X \to Y$ is an irreducible closed mapping of a regular space X onto a regular space Y and $\mathfrak{U}(X)$ (resp. $\mathfrak{U}(Y)$) is the family of all regular covers of X (resp. Y). Show that:

(*a*) if ξ is a thread in $\mathfrak{U}(X)$, then $\tilde{f}\xi = \{fA\colon A \in \xi\}$ is a thread in $\mathfrak{U}(Y)$;

(*b*) if ξ is a distinguished thread in $\mathfrak{U}(X)$, then $\dot{f}\xi = \{fA\colon A \in \xi\}$ is a distinguished thread in $\mathfrak{U}(Y)$;

(*c*) if ξ_1 and ξ_2 are distinct threads in $\mathfrak{U}(X)$, then $\tilde{f}\xi_1 = \{fA\colon A \in \xi_1\}$ and $\tilde{f}\xi_2 = \{fA\colon A \in \xi_2\}$ are distinct threads in $\mathfrak{U}(Y)$.

214. Suppose that $f\colon X \to Y$ is an irreducible closed 'onto' mapping, and X and Y are regular. If η is a thread in $\mathfrak{U}(Y)$, then the family $\xi = \{[f^{-1}(\operatorname{Int} B)]\colon B \in \eta\}$ is a thread in $\mathfrak{U}(X)$.

215. If $f\colon X \to Y$ is an irreducible perfect 'onto' mapping, X and Y are regular and η is an arbitrary distinguished thread in the set $\mathfrak{U}(Y)$ of all regular covers of Y, then $\xi = \{[f^{-1}(\operatorname{Int} B)]\colon B \in \eta\}$ is a distinguished thread in the set $\mathfrak{U}(X)$ of all regular covers of X.

216. Suppose that $f\colon X \to Y$ is an irreducible closed 'onto' mapping, and X and Y are regular. If ξ is a thread in $\mathfrak{U}(X)$, put $\tilde{f}\xi = \{fA\colon A \in \xi\}$. By Problem 213, $\tilde{f}\xi = \{fA\colon A \in \xi\}$ is a thread in $\mathfrak{U}(Y)$. Show that $\tilde{f}$ is a one-to-one mapping of $\Xi(X)$ onto $\Xi(Y)$.

We say that the mapping $\tilde{f}\colon \Xi(X) \to \Xi(Y)$ is generated by the irreducible closed mapping $f\colon X \to Y$.

217. Suppose that $f\colon X \to Y$ is an irreducible closed 'onto' mapping, and X and Y are regular. Put $\dot{f}\xi = \{fA\colon A \in \xi\}$, where $\xi \in \dot{X}$. Then:

(*a*) show that $\dot{f}$ is a one-to-one mapping of $\dot{X}$ into $\dot{Y}$;

(*b*) if f is perfect and irreducible, then $\dot{f}$ is a one-to-one mapping onto $\dot{Y}$;

(*c*) $\tilde{f}$: $\Xi(X) \to \Xi(Y)$ is an extension of $\dot{f}$: $\dot{X} \to \dot{Y}$.

We say that $\dot{f}$: $\dot{X} \to \dot{Y}$ is generated by the irreducible closed mapping f: $X \to Y$.

218. Suppose that X and Y are regular, f: $X \to Y$ is perfect and irreducible, $f(X) = Y$.

Consider $\dot{f}$: $\dot{X} \to \dot{Y}$, generated by f (see the preceding problem). Then:

(*a*) $\dot{f}$ is a homeomorphism onto $\dot{Y}$;

(*b*) the diagram in Figure 1 is commutative, i.e., $f(\pi_X(\xi)) = \pi_Y(\dot{f}(\xi))$ for every $\xi \in \dot{X}$.

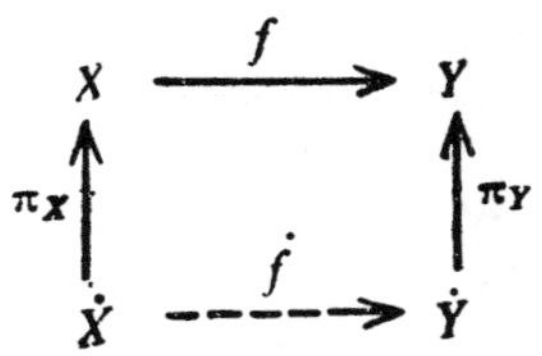

Figure 1.

219. Suppose that f: $X \to Y$ is an irreducible closed 'onto' mapping, X and Y are regular, π_X: $\dot{X} \to X$ and π_Y: $\dot{Y} \to Y$ are the corresponding natural mappings, $\dot{f}$: $\dot{X} \to \dot{Y}$ and $\tilde{f}$: $\Xi(X) \to \Xi(Y)$ are the mappings generated by the given f: $X \to Y$. Then:

(*a*) the diagram in Figure 2 is commutative;

(*b*) $\dot{f}$ is a homeomorphism into $\dot{Y}$;

(*c*) $\tilde{f}$ is a homeomorphism of $\Xi(X)$ onto $\Xi(Y)$, which is an extension of $\dot{f}$.

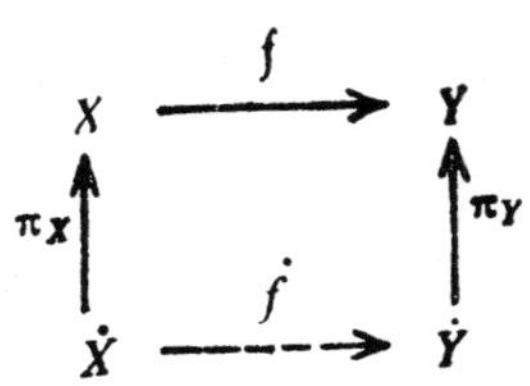

Figure 2.

220. Suppose that $f: X \to Y$ is an irreducible closed onto mapping between regular spaces X and Y. Then the Stone-Čech compactifications $\beta\dot{X}$ and $\beta\dot{Y}$ of $\dot{X}$ and $\dot{Y}$, respectively, are homeomorphic.

221. Each regular space X has an absolute, unique up to homeomorphism. In addition, the absolute of each regular space is a completely regular extremally disconnected space.

222. Suppose that X is regular. Show that a space pX is an absolute of X iff pX is extremally disconnected and completely regular and there is an irreducible perfect mapping of pX onto X.

223. If for two regular spaces X and Y there is an irreducible perfect mapping of one of them onto the other, then X and Y are co-absolute.

224. Two regular spaces X and Y are co-absolute iff there exist a regular space Z and irreducible perfect mappings $p_X: Z \to X$ and $p_Y: Z \to Y$, such that $p_X(Z) = X$, $P_Y(Z) = Y$.

225. Show that a regular space X is its own absolute iff it is extremally disconnected.

226. Suppose that $f: X \to Y$ is an irreducible perfect 'onto' mapping, where X and Y are regular. Suppose that Z is an extremally disconnected completely regular space and $g: Z \to Y$ is an irreducible perfect mapping onto Y. Then there is an irreducible perfect mapping $h: Z \to X$ onto X such that the diagram in Figure 3 commutes.

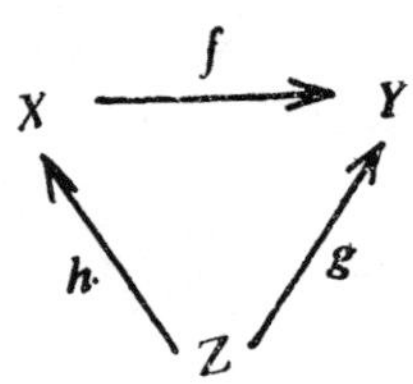

Figure 3.

227. Suppose that X is completely regular. Show that every pair of Hausdorff compactifications b_1X and b_2X of X are co-absolute.

228. Suppose that X and Y are completely regular spaces and $f: X \to Y$ is an irreducible closed mapping such that $f(X) = Y$. Show that any compactifications bX and cY of these spaces are co-absolute.

229. Suppose that X is completely regular. Prove the equality $(\beta X)^{\cdot} = \beta\dot{X}$ (here $\dot{X}$ is the absolute of X; $\beta\dot{X}$ is the Stone-Čech compactification of $\dot{X}$, $(\beta X)^{\cdot}$ is the absolute of the Stone-Čech compactification βX of X).

230. Show that all countable compacta are co-absolute, their common absolute being homeomorphic to the Stone-Čech compactification βN of the space of the natural numbers N.

231. Is it always true that an absolute $\dot{X}$ of a normal space is also a normal space ?

232. Suppose that X is completely regular and that the regular space Y is co-absolute with X. Does this imply that Y is completely regular ?

233. Suppose that the regular spaces X and Y are co-absolute. Show that in the case in which either of these spaces is compact, Lindelöf, paracompact, locally compact, or Čech-complete, the other space is as well respectively.

234. (*a*) Show that an absolute $\dot{X}$ of every infinite compactum X has cardinality $|\dot{X}| \geqslant 2^{\mathfrak{c}}$ and weight $w\dot{X} \geqslant \mathfrak{c}$.

(*b*) Give an example of an infinite compactum X for which $|\dot{X}| = 2^{\mathfrak{c}}$ and $w\dot{X} = \mathfrak{c}$.

235. Suppose that $f\colon X \to Y$ is an irreducible closed mapping of a regular space X onto a regular space Y. Then the π-weight of X coincides with the π-weight of Y.

236. Suppose that the regular spaces X and X are co-absolute. Show that in this case the π-weight of X coincides with the π-weight of Y, i.e., $\pi wX = \pi wY$ (see p. 43).

237. Show that a compactum X is co-absolute with some metrizable compactum iff $\pi wX \leqslant \aleph_0$.

238. Show that every pair of compacta X and Y without isolated points are co-absolute.

239. Show that a compactum X is co-absolute with the Cantor perfect set C iff $\pi wX \leqslant \aleph_0$ and X has no isolated points. Show that every pair of countable compacta with countable π-bases and without isolated points are co-absolute.

240. If two regular spaces are co-absolute, then their density and also their Souslin numbers coincide.

241. A first countable compactum X is co-absolute with a metrizable compactum iff X is seperable.

242. A perfectly normal compactum X is co-absolute with a metrizable compactum iff X is separable.

243. Every dyadic compactum X, co-absolute with some metrizable compactum, is metrizable and, consequently, is a metrizable compactum. If a completely regular space X is co-absolute with some metrizable space

Y, and some compactification bX of X is dyadic, then X and even bX and Y have countable bases so that X and bX are metrizable. If a dyadic compactum is first countable, then it has a countable base and is hence metrizable.

244. A compactum of π-weight $\tau \geqslant \aleph_0$ is co-absolute with a compactum of weight τ.

245. A regular space X is paracompact iff every open cover of X can be refined to a locally finite regular cover.

246. Every extremally disconnected paracompactum is perfectly 0-dimensional, i.e., every open cover has a disjoint open refinement.

247. Show that an absolute $\dot{X}$ of a paracompactum X is perfectly 0-dimensional and, consequently, is a strongly paracompact space.

248. Suppose that $f: X \to Y$ is a perfect mapping of a strongly paracompact space X onto Y. Is Y necessarily strongly paracompact?

249. Suppose that α_1 and α_2 are two locally finite regular covers of a space X. Let $\alpha_2 > \alpha_1$, i.e., α_2 is a refinement of α_1. Then: (*a*) every element of α_2 is contained in exactly one element of α_1; (*b*) every element of α_1 is the union of the elements of α_2 contained in the given element of α_1.

250. Suppose that X is a metrizable space. Then, X has a refined countable sequence† $\{\alpha_i: i \in N^+\}$ of locally finite regular covers. If, in addition, X is homeomorphic to a complete metric space, then $\{\alpha_i: i \in N^+\}$ can be taken as complete †† and refined.

251. Suppose that X is regular, $\mathfrak{A}_0 = \{\alpha_i: i \in N^+\}$ is a refined sequence of locally finite regular covers of X. Let $|\alpha_i| = |\alpha_{i+1}| = \tau$ for every $i \in N^+$, where $\tau \geqslant \aleph_0$. Then there is a subspace Ξ of the Baire space $B(\tau)$ (see Problem 246 of Chap. 2) and an irreducible perfect mapping $f: \Xi \to X$ such that $f\Xi = X$. If, in addition, the sequence $\mathfrak{A}_0$ is complete, then Ξ may be taken as closed in $B(\tau)$. If α_i is disjoint for each $i \in N^+$, and $\{\alpha_i: i \in N^+\}$ is refined, then X is homeomorphic to the subspace Ξ of $B(\tau)$. Finally, if α_i is disjoint for every $i \in N^+$, and $\{\alpha_i: i \in N^+\}$ is complete and refined, then X is homeomorphic to a closed subspace Ξ of $B(\tau)$.

252. Every metric space X of weight τ is the image of a subspace of the Baire space $B(\tau)$ under an irreducible perfect mapping. If, in addition, X is a complete space, then X is the image of a closed subspace of $B(\tau)$ under an irreducible perfect mapping.

† That is, $\alpha_{i+1} > \alpha_i$ for every $i \in N^+$.

†† That is, every thread in $\{\alpha_i : i \in N^+\}$ is distinguished.

6.8. SOLUTIONS

1. *Necessity.* If U is open in Z, then $(gf)^{-1}(U) = f^{-1}(g^{-1}(U))$ is open in X, and hence (f of course is a quotient mapping) $g^{-1}U$ is open in Y. This implies that g is continuous.

Sufficiency. Give Y the quotient topology $\mathcal{T}(f)$ induced by f. Take Z to be the space $(Y, \mathcal{T}(f))$ and for g take the identity mapping $i: Y \to Z$. The mapping if is continuous, because it is just a quotient mapping of X onto $(Y, \mathcal{T}(f))$. Hence, i is also continuous. Therefore, $\mathcal{T}(f) \subset \mathcal{T}$, where $\mathcal{T}$ is the original topology on Y. But since f is continuous and $\mathcal{T}(f)$ is the quotient topology on Y, $\mathcal{T} \subset \mathcal{T}(f)$ (Problem 297 of Chap. 2). Hence $\mathcal{T} = \mathcal{T}(f)$ and f is a quotient mapping.

2. This can be solved along the same lines as the next problem, further relying on the conclusion of Problem 1 of Chap. 4.

Another solution. Let $\tilde{\mathcal{T}}$ be the topology on X consisting of the empty set and all those $U \subset X$ for which $X \setminus U$ is finite. Thereupon, if $\mathcal{T}$ is a topology on X, then the identity mapping $i: (X, \mathcal{T}) \to (X, \tilde{\mathcal{T}})$ is a contraction (because $\mathcal{T}$ is a T_1-topology and hence $\mathcal{T} \supset \tilde{\mathcal{T}}$), whence $(X, \tilde{\mathcal{T}})$ is a compact T_1-space.

3. Suppose that X is a locally compact Hausdorff space. Consider its Alexandroff compactification $\bar{X}$ (see Problem 176 of Chap. 3): $\bar{X} \setminus X$ consists of a point ξ. Consider also any point $x_0 \in X$. Define a decomposition $\mathcal{A}$ (see p. 45) of $\bar{X}$ as follows: for each point $x \in \bar{X}$ other than x_0 and ξ, $\{x\}$ is an element of $\mathcal{A}$; in addition, the pair $\{x_0, \xi\}$ is an element of $\mathcal{A}$. Let Y be the set of all elements of $\mathcal{A}$ and let $f: \bar{X} \to Y$ be the natural mapping of $\bar{X}$ onto the set Y of all elements of $\mathcal{A}$, which assigns to each point $x \in \bar{X}$ the unique element of $\mathcal{A}$ which it contains. Give Y The quotient topology, i.e., the topology such that $f: \bar{X} \to Y$ is a quotient mapping. Y is thus a compact T_1-space, as a continuous image of the compactum $\bar{X}$.

The decomposition $\mathcal{A}$ is obviously continuous (see p. 77) and the mapping f is closed (Problem 322 of Chap. 2). Hence Y is a compactum (Problem 92 of Chap. 3) and the restriction $f|X$ is a one-to-one mapping of X onto Y.

5. Let $y \in [fx] \setminus f(X)$. Then $y' = f(y) \in f(X)$ and hence $y \neq y'$. Fix non-intersecting nhoods Oy and Oy' of y and y' in X such that $f(Oy) \subset Oy'$. Since f is continuous and X is Hausdorff, this can be done. Put $A = Oy \cap f(X)$. Then $A \neq \Lambda$ but, on the one hand, $fA = A \subset Oy$ and, on the other, $fA \subset fOy \subset Oy' \subset X \setminus Oy$. This is a contradiction, whence fX is closed in X.

6. Since (X, ρ) is a metric space and $f^{-1}y', f^{-1}y''$ are disjoint compacta, when $y' \neq y''$, $d(y', y'') = 0$ iff $y' = y''$. Obviously, $d(y', y'') = d(y'', y')$. Consequently, d is a semi-metric on Y (note also that $f(X) = Y$). Suppose that A is closed in Y and $y_0 \notin A$. Then $f^{-1}A$ is closed and $f^{-1}y_0$ is a compactum, with $f^{-1}A \cap f^{-1}y_0 = \Lambda$. Hence $\rho(f^{-1}y_0, f^{-1}A) > 0$. But obviously $d(y_0, A) = \rho(f^{-1}y_0, f^{-1}A)$; hence $d(y_0, A) > 0$. Now suppose that A is not closed in Y. Since f is a quotient mapping, $f^{-1}A$ is not closed in X. Fix $x^* \in [f^{-1}A] \setminus f^{-1}A$, put $y^* = fx^*$. Then $\rho(x^*, f^{-1}A) = 0$ obviously implies that $d(y^*, A) = 0$ and $x^* \notin f^{-1}A$ that $y^* \notin A$. Thus we have shown that d generates the topology of Y.

7. From the preceding problem it follows that the compactum under consideration is semi-metrizable. Hence by Problem 95 of Chap. 5 it follows that it is metrizable.

8. Statements (i) and (ii) are immediate from Problem 6, since every perfect mapping is a compact quotient mapping. Let $\Phi \subset Y$, $P \subset Y$, Φ is a compactum, P is closed in Y and $\Phi \cap P = \Lambda$. Put $F = f^{-1}\Phi$, $Q = f^{-1}P$. Then $F \cap Q = \Lambda$, F is a compactum (since f is a perfect mapping; see Problem 36), but Q is a closed set in X. Hence $\rho(F, Q) > 0$ (Problem 288 of Chap. 3). But obviously $d(\Phi, P) = \rho(F, Q)$. Consequently, $d(\Phi, P) > 0$. For the rest, refer to Problem 96 of Chap. 5.

9. See Problem 312 of Chap. 2 and then refer to Arkhangel'skii and Ponomarev [1].

10. True. The solution is based on Wallace's theorem (see Problems 76 and 77 of Chap. 3).

11. Suppose that Φ is closed in Y, $F = f^{-1}\Phi$, $F = \bigcap\{G_i: i \in N^+\}$, where every G_i is open in X and $[G_{i+1}] \subset G_i$. For every compactum $P \subset X \setminus F$ there is a G_i such that $[G_i] \cap P = \Lambda$ (otherwise, the family $\{[G_i] \cap P: i \in N^+\}$ would be centralized). Taking now for P the pre-image of each point $y \in Y$ we conclude that $\bigcap\{fG_i: i \in N^+\} = \Phi$. Since f is pseudo-open, $\Phi \subset \text{Int}\, fG_i$ for each $i \in N^+$. Hence $\Phi = \bigcap\{\text{Int}\, fG_i: i \in N^+\}$.

12. Since every pseudo-open compact mapping is a biquotient mapping (Problem 17), the given statement is a particular case of Problem 20.

16. *Necessity.* Assume the contrary; suppose that there is a prefilter ξ on Y and an adherent point $y \in Y$ for ξ such that each point $x \in f^{-1}y$ has a nhood Ux in X which does not intersect some set of the form $f^{-1}(Bx)$, where $Bx \in \xi$. Then, $\gamma = \{Ux: x \in f^{-1}y\}$ is a cover of $f^{-1}y$ by open sets in X. There is a finite family $\lambda = \{Ux_i: i = 1, \ldots, k\}$ of elements of γ such that $V = \text{Int}\,(\bigcup\{fUx_i: i = 1, \ldots, k\}) \ni y$ (because f is biquotient). But $fUx_i \cap Bx_i = \Lambda$ and $Bx_i \in \xi$, $i = 1, \ldots, k$. Since ξ is a prefilter, there is a $B^* \in \xi$ such that $B^* \subset \bigcap\{Bx_i, i = 1, \ldots, k\}$. Then $B^* \cap (\bigcup\{fUx_i: i = 1, \ldots, k\}) = \Lambda$; furthermore, $B^* \cap V = \Lambda$. But V is a nhood of y. Hence $y \notin [B^*]$, contradicting that y is an adherent point of ξ.

Sufficiency. Suppose that f is not a biquotient mapping. Then there is a point y and a cover γ of $f^{-1}y$ by open sets in X such that no nhood of y is covered by finitely many sets of the form $f(U)$, where $U \in \gamma$. Put $\xi = \{Y \setminus \bigcup\{fU: U \in \lambda\}: \lambda$ is an arbitrary finite family of elements of $\gamma\}$. Then ξ is a prefilter and y is an adherent point of ξ, but no point of $f^{-1}y$ is an adherent point of $f^{-1}\xi$. To show the latter, let $x^* \in f^{-1}y$. Pick $U^* \in \gamma$ such that $U^* \ni x^*$. Then $P^* = Y \setminus fU^* \in \xi$, $f^{-1}P^* \cap U^* = \Lambda$; hence $x^* \notin [f^{-1}P^*]$.

18. Let $f: X \to Y$ be the mapping in question. We are given that X is locally compact. Consider an arbitrary point $y \in Y$. For each $x \in f^{-1}y$ fix an open set Ux in X and a compact subspace Φx of X satisfying $x \in Ux \subset \Phi x$. Since f is a biquotient mapping and $\gamma = \{Ux: x \in f^{-1}y\}$ is a cover of $f^{-1}y$ by open sets in X, there is a finite family $\lambda = \{Ux_1, \ldots, Ux_k\} \subset \gamma$ such that $\text{Int}\,(\bigcup\{fUx_i: i = 1, \ldots, k\}) \ni y$. But $F = f(\bigcup\{f\Phi x_i: i = 1, \ldots, k\})$ is a compact subspace of Y (because f is continuous) and $F \supset \text{Int}\,(\bigcup\{fUx_i: i = 1, \ldots, k\})$. Consequently, Y is locally compact.

19. Given a family of biquotient mappings $\{f_\alpha: X_\alpha \to Y_\alpha: \alpha \in A\}$, where $f(X_\alpha) = Y_\alpha$ for each $\alpha \in A$, and $f = \Pi\{f_\alpha: \alpha \in A\}$, $f: X \to Y$, where $X = \Pi\{X_\alpha: \alpha \in A\}$ and $Y = \Pi\{Y_\alpha: \alpha \in A\}$ are topological products (i.e., f is the product of mappings f_α with respect to $\alpha \in A$).

Consider any prefilter ξ in Y, and an adherent point $y \in Y$. Put $\eta = f^{-1}\xi = \{f^{-1}P:$

$P \in \xi\}$. We must show that η has an adherent point $x \in f^{-1}y$, whence, by Problem 16, f is a biquotient mapping. There is an ultrafilter $\bar{\xi}$ (Problem 4 of Chap. 2) on Y, containing ξ and converging to y. Then the prefilter $\pi_\alpha\bar{\xi} = \{\pi_\alpha Q: Q \in \bar{\xi}\}$ (where π_α is a projection, $\pi_\alpha: Y \to Y_\alpha$) obviously converges to $y_\alpha = \pi_\alpha y$ for all $\alpha \in A$. Since f_α is a biquotient mapping, there is $x_\alpha \in f_\alpha^{-1} y_\alpha$ which is an adherent point of the prefilter $f_\alpha^{-1}(\pi_\alpha \bar{\xi})$ (see Problem 16). Suppose that x is a point of the product X whose αth coordinate is x_α for all $\alpha \in A$. Then, by the definition of f, we have $x \in f^{-1}y$ and thus it remains to show that x is an adherent point of $f^{-1}\xi$.

Suppose that U_α is a nhood of x_α in X_α and $P \in \bar{\xi}$. Then $U_\alpha \cap f_\alpha^{-1} \pi_\alpha P \neq \Lambda$. But $f_\alpha^{-1}\pi_\alpha P = q_\alpha f^{-1}P$ (where q_α is a projection, $q_\alpha: X \to X_\alpha$). Hence $P \cap f(q_\alpha^{-1}(U_\alpha)) \neq \Lambda$. Since this holds for every $P \in \bar{\xi}$ and $\bar{\xi}$ is an ultrafilter, $fq_\alpha^{-1}(U_\alpha) \in \bar{\xi}$. Hence every set of the form $\bigcap\{fq_{\alpha_i}^{-1}(U_{\alpha_i}): i = 1, \ldots, n\}$, where U_{α_i} is a nhood of x_{α_i} in X_{α_i}, also belongs to $\bar{\xi}$ and, consequently, intersects every $P \in \xi$. But then we surely have $\bigcap\{q_{\alpha_i}^{-1} U_{\alpha_i}: i = 1, \ldots, n\} \cap f^{-1}P \neq \Lambda$ for each $P \in \xi$, i.e., $x \in [f^{-1}P]$ for all $P \in \xi$, implying that x is an adherent point of $f^{-1}\xi$.

20. In X fix a base $\mathcal{B}$, whose cardinality equals the weight of X. We may assume that the union of any finite set of elements of $\mathcal{B}$ belongs to X. Put $C = \{\operatorname{Int} fU: U \in \mathcal{B}\}$. It is routine to show that C is a base for Y. Obviously, $|C| \leqslant |\mathcal{B}|$; hence the weight of Y does not exceed the weight of X.

21. Consider an arbitrary cover γ of $f^{-1}y$ by open sets in X. Since $f^{-1}y$ is Lindelöf, we may assume that γ is countable: $\gamma = \{U_n: n \in N^+\}$, $U_n \subset U_{n+1}$ and $U_n \cap f^{-1}y \neq \Lambda$ for all values of $n \in N^+$. Fix a countable base $\{W_i: i \in N^+\}$ for Y at y such that $W_i \supset W_{i+1}$ for all $i \in N^+$. Then there is an $n_0 \in N^+$ such that $f(U_{n_0}) \supset W_{n_0}$ and, consequently, f is a biquotient mapping at y. We shall prove this by contradiction.

Let $W_n \setminus f(U_n) \neq \Lambda$ for all $n \in N^+$. Pick a point $y_n \in W_n \setminus f(U_n)$ for each $n \in N^+$. Put $Q = \{y_n: n \in N^+\}$. Obviously, Q is not closed in Y, where y is the only point in Y which is a limit point of Q. Put $P = f^{-1}Q$. P is not closed in X (because Q is not closed in Y and f is a quotient mapping). Select $x^* \in [P]\setminus P$. Then $f(x^*) \notin Q$ but $f(x^*) \in [fP]_Y = [Q]_Y$, i.e., $f(x^*)$ is a limit point of Q. Consequently, $y = f(x^*)$ and $x^* \in f^{-1}y$.

Fix a number n' such that $x^* \in U_{n'}$. Put $P_{n'} = \bigcup\{f^{-1}(y_n): n = n', n' + 1, \ldots\}$. From $x^* \in [P]\setminus P$ it follows that $x^* \in [P_{n'}]$ and $U_{n'} \cap P_{n'} \neq \Lambda$. Consequently, there is a number n^*, for which $U_{n'} \cap f^{-1}(y_{n^*}) \neq \Lambda$, and $n^* > n'$. But then we also have $U_{n^*} \cap f^{-1}(y_{n^*}) \neq \Lambda$, i.e., $f(U_{n^*}) \ni y_{n^*}$, contradicting the choice of y_{n^*} (recall that we have $y_n \in W_n \setminus f(U_n)$ for each $n \in N^+$).

22. This follows from Problems 20 and 21.

23. This follows from Problems 18, 21 and 22.

24. The answer is affirmative. Suppose that Y is an uncountable set well-ordered by the relation $<$, having a largest element Ω, and the ordering $<$ is minimal on $Y\setminus \Omega$, with $|\{x < x': x \in Y\}| < \aleph_0$ for every $x' \in Y\setminus\{\Omega\}$. Give Y the usual topology generated by the ordering (see p. 49) and regard $X_1 = Y\setminus\{\Omega\}$, $X_2 = \{x \in Y$: the point x is not isolated in $Y\}$ as subspaces of Y. Let X be the disjoint union of X_1 and X_2 (see p. 44) and let $\pi: X \to Y$ be the natural mapping (i.e., the mapping, for which πi_1: $X_1 \to Y$, πi_2: $X_2 \to Y$, where i_1 and i_2 are natural embeddings in the disjoint union (see

p. 44), are identity mappings). One can now verify that π is the required mapping.

25. Consider in $X \times F$ an arbitrary closed set P, distinguished with respect to the mapping $\varphi = f \times i$, i.e., such that $\varphi^{-1}(\varphi P) = P$. Fix any $z = (y, a) \in (Y \times F) \setminus \varphi P$, where $y \in Y$, $a \in F$. Put $\Phi = f^{-1}y$. Then $\varphi^{-1}(z) = \Phi \times \{a\}$ is a closed set in $X \times F$, disjoint from P. If for every nhood Oa of a in F we have $(\Phi \times Oa) \cap P \neq \Lambda$, then $(\{x\} \times Oa) \cap P \neq \Lambda$ for any point $x \in \Phi$, beeause P is a distinguished set (and, while intersecting the set $\Phi \times \{a\}$, P surely contains the whole of it). Then $(x, a) \in [P]$, but $(x, a) \in X \setminus P$, which contradicts that P is closed. Consequently, there is a nhood O^*a of a in F, for which $(\Phi \times [O^*a]) \cap P = \Lambda$. Put $P^* = \{(x, a') \in P: a' \in [O^*a]\}$ and $P^{**} = \{(x, a') \in X \times F: a' \in [O^*a]$ and there exists $(x', a') \in P^*$, for which $x = x'\}$. The set P^{**} is obtained if the second coordinate of every element of P^* varies along $[O^*a]$. Put $Q = \{(x, a') \in X \times F: a' \in [O^*a]\}$. Then, obviously, $P^* = P \cap Q$ is a closed set and $P^{**} = \pi^{-1}(\pi P^*) \cap Q$, where π is the projection of $X \times F$ onto X. The mapping π is perfect (Problem 33). Hence πP^* is a closed set and therefore P^{**} is also closed. Obviously, from the definition of φ it follows that both P^* and P^{**} are distinguished sets. From $(\Phi \times [O^*a]) \cap P = \Lambda$ it also follows that $P^{**} \cap (\Phi \times [O^*a]) = \Lambda$. Put $V = X \setminus \pi P^*$. Then V is open and distinguished with respect to f (because πP^* is closed and distinguished with respect to f), $\Phi \subset V$ and $(V \times O^*a) \cap P = \Lambda$. Hence fV is open in Y and $fV \times O^*a$ is a nhood of the point $z = (y, a)$ in $Y \times F$, not intersecting φP. Consequently, φP is closed in $Y \times F$ and φ is a quotient mapping.

26. Represent Z as an open subspace of some compactum F (see Problem 179 of Chap. 3). Let i be the identity mapping of F onto itself. Then, $f \times i: X \times F \to Y \times F$ is a quotient mapping (Problem 25), $i \mid Z = j$ and $(f \times i) \mid X \times Z = f \times j$. Last but not least, $(f \times i)^{-1}(Y \times Z) = X \times Z$ (this is obvious). Hence, since $X \times Z$ is open in $X \times F$, we conclude that $(f \times i) \mid X \times Z = f \times j$ is a quotient mapping.

27. See Ponomarev [4] and Choban [1].

28. Sufficiency is proved in the obvious way. The proof of necessity is similar to the relevant portion of the proof of theorem on k-spaces (see Problem 247 of Chap. 3); the only requirement is to take the disjoint union not of all compact subspaces of the Fréchet-Urysohn space under consideration, but of all its compact metrizable subspaces. It is also necessary to observe that a quotient mapping on a Fréchet-Urysohn space is always pseudo-open.

31. Suppose that $f_\alpha: X_\alpha \to Y_\alpha$ is a perfect mapping for each $\alpha \in A$. We may assume that $f(X_\alpha) = Y_\alpha$ for all $\alpha \in A$. Let $\tilde{f}_\alpha$ be the natural extension of f_α to a mapping of the Stone-Čechcompactifications: $\tilde{f}_\alpha: \beta X_\alpha \to \beta Y_\alpha$. $\tilde{f}_\alpha$ is continuous and $\tilde{f}_\alpha(\beta X_\alpha \setminus X_\alpha) = \beta Y_\alpha \setminus Y_\alpha$ (Problem 47), since f_α is perfect. Hence $\tilde{f} = \Pi\{\tilde{f}_\alpha: \alpha \in A\}$ is a continuous mapping of $\tilde{X} = \Pi\{\beta X_\alpha: \alpha \in A\}$ onto $\tilde{Y} = \Pi\{\beta Y_\alpha: \alpha \in A\}$; in addition, $\tilde{f}(\tilde{X} \setminus X) = \tilde{Y} \setminus Y$, where $X = \Pi\{X_\alpha: \alpha \in A\}$ and $Y = \Pi\{Y_\alpha: \alpha \in A\}$. But $\tilde{X}$ and $\tilde{Y}$ are compact and hence are compactifications of X and Y. Furthermore, $\tilde{f} \mid X = f = \Pi\{f_\alpha: \alpha \in A\}$. Since $\tilde{f}$ maps the growth $\tilde{X} \setminus X$ into the growth $\tilde{Y} \setminus Y$, $\tilde{f} \mid X$ is perfect (see Problem 46), i.e., f is perfect.

32. Consider the family $\mathcal{P} = \{F_\alpha: \alpha \in A\}$ of all closed sets F_α in X such that $f(F_\alpha) = Y$. $\mathcal{P}$ with a natural image is naturally partially ordered: put $F_{\alpha'} > F_\alpha$, where $F_{\alpha'}, F_\alpha \in \mathcal{P}$ iff $F_{\alpha'} \subset F_\alpha$.

Consider a nest (see p. 6) $\xi \subset \mathcal{P}$. It is to be shown that ξ has a maximal element in

$\mathcal{P}$. Consider the closed set $F_0 = \bigcap\{F: F \in \xi\}$ in X. We show that F_0 is a maximal element of ξ in $\mathcal{P}$. Clearly, this requires that $F_0 \in \mathcal{P}$, i.e., $f(F_0) = Y$. Let $x \in X$ be arbitrary. Since $f(F) = Y$ for every $F \in \xi$, $\Phi(F) = f^{-1}fx \cap F$ is nonempty for every $F \in \xi$ and is obviously compact. Consider the family $\eta = \{\Phi(F): F \in \xi\}$ of these nonempty compact sets. Clearly, η is linearly ordered by inclusion. Hence $\bigcap\{\Phi(F): F \in \xi\} = \Phi_0 \neq \Lambda$; obviously, $\Phi_0 \subset F_0$. Therefore, $f^{-1}fx \cap F_0 \neq \Lambda$, i.e., $f(F_0) = Y$. Thus, every nest $\xi \in \mathcal{P}$ has a maximal element in $\mathcal{P}$. Hence, by the maximum principle (see p. 6), $\mathcal{P}$ itself has a maximal element. This implies that there is a closed set $F^* \subset X$ in X, for which $f(F^*) = Y$ and $f \mid F^*$ is irreducible. It is also obvious that $f \mid F^*$ is perfect, because F^* is closed in X.

33. To show that π is a closed mapping, let P be closed in $X \times F$ and $y \in X \setminus \pi P$. Then $(X \times F) \setminus P = U$ is an open set in $X \times F$, containing the set $F_y = \pi^{-1}\{y\}$. But F_y can be represented as a product of compact spaces; explicitly, $F_y = \{y\} \times F$. By Problem 76 of Chap. 3, there is an open set W in X, $W \ni y$, for which $W \times F \subset U$. But then $W \cap \pi P = \Lambda$. Consequently, πP is closed in X and π is a closed mapping. Obviously, π is continuous and compact, and hence also perfect.

34. Suppose that $F_{t_0 s_0}$ is an arbitrary connected component of some (also arbitrary) element F_{t_0} of the decomposition $\mathcal{Z}$, and $UF_{t_0 s_0}$ is a nhood of $F_{t_0 s_0}$ in X. Using Problems 365 and 366 of Chap. 3, pick a clopen set $H_{t_0} \subset F_{t_0}$ in the subspace F_{t_0}, for which $F_{t_0 s_0} \subset H_{t_0} \subset UF_{t_0 s_0}$. Write $H'_{t_0} = F_{t_0} \setminus H_{t_0}$. H'_{t_0} is also clopen in F_{t_0}. Since F_{t_0} is a compactum, H_{t_0} and H'_{t_0} are as well and, in addition (by construction), $H_{t_0} \cap H'_{t_0} = \Lambda$. Hence there are open sets V_{t_0} and V'_{t_0} in X for which

$$H_{t_0} \subset V_{t_0},\ H'_{t_0} \subset V'_{t_0}; \tag{1}$$

$$V_{t_0} \cap V'_{t_0} = \Lambda; \tag{2}$$

$$V_{t_0} \subset UF_{t_0 s_0} \tag{3}$$

(here we have used the fact that disjoint compact sets in a Hausdorff space X can be separated by open sets in X).

Consider the nhood $WF_{t_0} = V_{t_0} \cup V'_{t_0}$ of F_{t_0} in X. Since the decomposition $\mathcal{Z}$ is continuous, there is a nhood $W'F_{t_0} \subset WF_{t_0}$ of F_{t_0} for which

$$F_t \subset W'F_{t_0}, \text{ if only } F_t \cap W'F_{t_0} \neq \Lambda, \tag{4}$$

i.e., there is a distinguished nhood for the decomposition $\mathcal{Z}$ (see p. 77). Finally, we write

$$U'F_{t_0 s_0} = W'F_{t_0} \cap V_{t_0}. \tag{5}$$

Obviously, $U'F_{t_0 s_0}$ is open in X and $U'F_{t_0 s_0} \subset UF_{t_0 s_0}$.

Now let the element F_{ts} of the decomposition $\mathcal{K}$ satisfy

$$F_{ts} \cap U'F_{t_0 s_0} \neq \Lambda. \tag{6}$$

By the definition of $\mathcal{K}$, F_{ts} is a connected component of F_t.

By (6), we have

$$F_t \cap W'F_{t_0} \neq \Lambda, \tag{7}$$

which, by virtue of (4), implies

$$F_t \subset W'F_{t_0} \subset V_{t_0} \cup V'_{t_0}. \tag{8}$$

But from (5) it also follows that

$$F_{ts} \cap V_{t_0} \neq \Lambda. \tag{9}$$

Since V_{t_0} and V'_{t_0} are open in X and F_{ts} is a connected set, F_{ts} (by virtue of (8) and (2)) belongs to either V_{t_0} or V'_{t_0} whence, because of (9), we obtain $F_{ts} \subset V_{t_0}$. Combining the last conclusion with (8), we get

$$F_{ts} \subset W'F_{t_0} \cap V_{t_0} = U'F_{t_0s_0}. \tag{*}$$

Thus $U'F_{t_0s_0}$ is a distinguished nhood for $\mathcal{K}$ of the element $F_{t_0s_0}$ contained in $UF_{t_0s_0}$, which implies that $\mathcal{K}$ is continuous (see p. 77). The proof is complete.

35. Since f is closed and X and Y are first countable T_1-spaces, the decomposition $\mathcal{K} = \{f^{-1}y: y \in Y\}$ of X is continuous (see Problem 322 of Chap. 2). Partition each compactum $f^{-1}y$ into its components: $f^{-1}y = \bigcup\{F_{ys}: s \in S_y\}$ to obtain the decomposition $\mathcal{Z}$: $X = \bigcup\{F_{ys}: s \in S_y, y \in Y\}$ into connected compact sets F_{ys}, $y \in Y$, $s \in S_y$. Denote this decomposition space by Z. $\mathcal{Z}$ is continuous by virtue of Problem 34. Let φ be the natural mapping of X onto Z. Since $\mathcal{Z}$ is continuous, $\varphi: X \to Z$ is continuous and closed and obviously compact, so that it is perfect. By our construction, φ is monotonic. Let ψ be the mapping, which associates to each continuum F_{ts}, i.e., point (F_{ts}) of Z the unique compactum of F_t, whose component is F_{ts}. Then $\psi: Z \to Y$, and obviously $f = \psi\varphi$. The mapping $\varphi: X \to Z$ is perfect and monotonic.

Consider $\psi: Z \to Y$. If $y_0 \in Y$, then $\psi^{-1}y_0$ is the decomposition space of $\mathcal{Z}_{y_0}$ of the compactum $f^{-1}y_0$ into its connected components. The decomposition $\mathcal{Z}_{y_0}$ is continuous (see Problem 365 of Chap. 3) and the decomposition $Z_{y_0} = \psi^{-1}y_0$ of $\mathcal{Z}_{y_0}$ is a 0-dimensional compactum (see Problem 366 of Chap. 3). Hence $\psi: Z \to Y$ is a compact 0-dimensional quotient mapping. Ths fact that it is closed then follows directly from the perfection of f and φ, and the fact that $f = \psi\varphi$ (see Problem 56). The theorem is proved.

36. Suppose that ω is an arbitrary open cover of X. Consider for each $y \in Y$ the cover $\omega_y = \{U \in \omega: U \cap f^{-1}y \neq \Lambda\}$ of the compact subspace $f^{-1}y$, from which we pick a finite subcover ω_y^0.

Consider the open set $Vy = Y \setminus f(X \setminus \tilde{\omega}_y^0)$, where $\tilde{\omega}_y^0 = \bigcup\{U: U \in \omega_y^0\}$; from the open cover $\{Vy: y \in Y\}$ of a compact space Y pick a finite subcover $\{Vy_1, \ldots, Vy_s\}$. Then

$$X = f^{-1}Y = f^{-1}\bigcup_{i=1}^{s} Vy_i = \bigcup_{i=1}^{s} f^{-1}Vy_i \subset \bigcup_{i=1}^{s} \tilde{\omega}_{y_i}^0.$$

Consequently,

$$\omega_0 = \bigcup_{i=1}^{s} \omega_{y_i}^0$$

is a finite subcover of ω.

38. Let $\gamma = \{U_\alpha: \alpha \in A\}$ be an open cover of X and let $|A| \leqslant \tau$. Let $\tilde{A}$ be the family of all finite subsets of A; then $|\tilde{A}| \leqslant \tau$. Since f^{-1} is τ-compact for every $y \in Y$, there is a finite set $\Gamma \subset A$ for which $f^{-1}y \subset \bigcup\{U_\alpha: \alpha \in \Gamma\}$. Put $V_\Gamma = Y \setminus f(X \setminus \bigcup\{U_\alpha: \alpha \in \Gamma\})$. V_Γ is open since f is closed, $y \in V_\Gamma$ and $f^{-1}(V_\Gamma) \subset \bigcup\{U_\alpha: \alpha \in \Gamma\}$. Consequently, $\lambda = \{V_\Gamma: \Gamma \in \tilde{A}\}$ is an open cover of Y of cardinality not exceeding τ. Since Y is τ-paracompact, there is a locally finite open cover $\{W_\beta: \beta \in B\}$ refining λ. For each $\beta \in B$, $\Gamma_\beta \in \tilde{A}$ can be chosen such that $f^{-1}(W_\beta) \subset \bigcup\{U_\alpha: \alpha \in \Gamma_\beta\}$. Then $\mu = \{f^{-1}(W_\beta) \cap U_\alpha: \beta \in B, \alpha \in \Gamma_\beta\}$ is a locally finite open cover of X, refining γ. In fact, $\{f^{-1}(W_\beta): \beta \in B\}$ is locally finite because $\{W_\beta: \beta \in B\}$ is locally finite in Y and f is continuous. Turning to μ, we replace every set $f^{-1}(W_\beta)$ by the finite family of open sets $\{f^{-1}(W_\beta) \cap U_\alpha: \alpha \in \Gamma_\beta\}$, whose union is precisely $f^{-1}(W_\beta)$. Hence μ is open and locally finite.

41. Since every perfect mapping is a biquotient mapping, the solution follows from Problem 20. Also give a direct proof.

43. The projection mapping $\pi: X \times Y \to X$ is perfect (Problem 33). Hence, since X is a paracompactum, $X \times Y$ is also a paracompactum (Problem 38).

44. *Necessity.* Suppose that f is a perfect mapping. If there is no adherent point for ξ in $f^{-1}y$, then each point $x \in f^{-1}y$ has a nhood Ox such that $Ox \cap Bx = \Lambda$ for some $Bx \in \xi$. From the cover $\{Ox: x \in f^{-1}y\}$ of $f^{-1}y$ pick a finite subcover $\{Ox_1, \dots, Ox_k\}$ and put $U = \bigcup\{Ox_i: i = 1, \dots, k\}$. From ξ choose an element B such that $B \subset \bigcap\{Bx_i: i = 1, \dots, k\}$. Then $U \supset f^{-1}y$ and $U \cap B = \Lambda$. Put $Oy = Y \setminus f(X \setminus U)$. Oy is open in Y, since f is closed, and $y \in Oy$ (since $U \supset f^{-1}y$). Furthermore, $f^{-1}Oy \subset U$ (this is obvious). Hence $Oy \cap fB = \Lambda$, i.e., $y \notin [fB]$; but this contradicts the fact that y is an adherent point of $f\xi \ni fB$.

Sufficiency. To show that f is closed, let $P \subset X$ be closed in X, $Q = fP \subset Y$ and $y \in [Q]_Y$.

Consider the prefilter $\eta = \{Q \cap OY: Oy \text{ is a nhood of } y \text{ in } Y\}$ and put $\xi = \{P \cap f^{-1}A: A \in \eta\}$. Obviously, ξ is a prefilter. Since $A \subset Q$ and $fP = Q$, $f(P \cap f^{-1}A) = A$ for all $A \in \eta$. Hence $f\xi = \eta$. Since y is an adherent point of η, ξ has adherent point $x \in f^{-1}y$. But then $x \in [P]$ (because all elements of ξ are subsets of P) and, consequently, $x \in P$, $y = f(x) \in fP = Q$, i.e., Q is closed. Thus f is a closed mapping. Now we verify that $f^{-1}y$ is a compact subspace of X for every point $y \in Y$. Assume the contrary. Then there is a prefilter ξ, consisting of subsets of $f^{-1}y$ such that $\bigcap\{[P]_X: P \in \xi\} = \Lambda$. Obviously, $\eta = f\xi$ is a prefilter on Y, whose unique element is the set $\{y\}$. Hence y is an adherent point of η. But, by the definition of ξ, no point of $f^{-1}y$ is an adherent point of ξ, a contradiction. Hence $f^{-1}y$ is compact for all $y \in Y$ and f is a perfect mapping.

45. *Sufficiency.* Let $\{\xi\} = \alpha X \setminus X$ and $\{\eta\} = \alpha Y \setminus Y$. We have $\tilde{f}(X) = Y$, $Y \subset \tilde{f}(\alpha X) \subset \alpha Y$. However, Y is not compact, but $\tilde{f}(\alpha X)$ is compact, being a continuous image of a compact space (Problem 91 of Chap. 3). Hence $\tilde{f}(\alpha X) = \alpha Y$, $\tilde{f}(X) \not\ni \eta$ and $\tilde{f}(\xi) = \eta$. Therefore, $\tilde{f}^{-1}(\eta) = \xi$ and $\tilde{f}^{-1}(Y) = X$. But αX and αY are compact and hence $\tilde{f}$ is a perfect mapping (Problem 92 of Chap. 3). Then the restriction $\tilde{f}$ to X is also a perfect mapping (since $\tilde{f}^{-1}(Y) = X$, see Problem 319 of Chap. 2). But $\tilde{f} \mid X = f$. Thus f is a perfect mapping.

Necessity. Put $\tilde{f}(x) = f(x)$ for all $x \in X$ and $\tilde{f}(\xi) = \eta$. Let $\Phi \subset X$, where Φ is a compactum. Put $O_\Phi(\xi) = \alpha X \setminus \Phi$. The family of all sets of the form $O_\Phi(\xi)$ forms a base for the topology of αX at the point ξ (see solution of Problem 176, Chap. 3). Similarly, sets of the form $O_F(\eta)$, where F is a compactum contained in Y, and $O_F(\eta) = \alpha Y \setminus F$, form a base for the topology of αY at η. But, obviously, $\tilde{f}^{-1}O_F(\eta) = O_{f^{-1}F}(\xi)$ and $f^{-1}F \subset X$, where $f^{-1}F$ is a compactum (since $f = \tilde{f} \mid X$ is a perfect mapping; see Problem 36). Consequently, $\tilde{f}$ is continuous at ξ. It is continuous at all points of X, because X is open in αX and $\tilde{f}$ coincides with the continuous mapping f on X.

46. From $\tilde{f}(bX) = bY$ and $\tilde{f}(bX \setminus X) \subset bY \setminus Y$ it follows that $\tilde{f}^{-1}Y = X$. But $\tilde{f}$ is a closed (and obviously a compact) mapping, since bX is a compactum and bY is Hausdorff. The restriction of a perfect mapping to a complete pre-image is a perfect mapping (Problem 319 of Chap. 2); hence $\tilde{f} \mid X$ is perfect.

47. It suffices to prove that $f^{-1}Y \subset X$, i.e., that $f^{-1}y \subset X$ for each point $y \in Y$. But $f^{-1}y = [f_0^{-1}y]_{bX} = f_0^{-1}y$, since $f_0^{-1}y$ is a compactum, $f_0^{-1}y \subset X$ and $f_0 = f \mid X$ is closed (in this connection see Problem 73 of the present chapter and Problem 166 of Chap. 4).

48. Consider the Stone-Čech compactifications βX and βY of X and Y respectively, and the extension $\bar{f}: \beta X \to \beta Y$ of f to these extensions (see Problem 31 of Chap. 4). By Problems 46 and 47, the perfection of f is equivalent to the fact that $\bar{f}(\beta X \setminus X) = \beta Y \setminus Y$ and hence also to the equality $\bar{f}^{-1}(\beta Y \setminus Y) = \beta X \setminus X$. On the other hand, the local compactness of X implies that the growth $\beta X \setminus X$ is a compactum, and the local compactness of Y implies the compactness of $\beta Y \setminus Y$ (see Problem 179 of Chap. 3). In the same manner, the Čech-compactness of X or Y means that $\beta X \setminus X$, or $\beta Y \setminus Y$ is σ-compact, i.e., is the union of a countable number of compacta. Hence, our proposition follows from the next statement.

A perfect image, as also a perfect pre-image of a compactum (resp. σ-compactum†) is also a compactum (resp. a σ-compactum) (see Problem 36 of the present chapter and Problem 91 of Chap. 3).

Notice that the mapping $\bar{f}: \beta X \setminus X \to \beta Y \setminus Y$ is perfect, because $\bar{f}: \beta X \to \beta Y$ is perfect (as a continuous mapping between compacta) and $\bar{f}^{-1}(\beta Y \setminus Y) = \beta X \setminus X$.

49. See Problem 8. Let us give another solution. Suppose that ρ is any metric on X generating the topology on X. For each $y \in Y$ and for every $n \in N^+$, put $O_n y = \{y' \in Y: f^{-1}(y') \subset O_{1/n}(f^{-1}y)\}$, where $O_{1/n}(f^{-1}y) = \{x \in X: \rho(x, f^{-1}y) < 1/n\}$. From Problem 321 of Chap. 2 and the perfection of f it follows that $O_n y$ is an open set. Set $\gamma_n = \{O_n y: y \in Y\}$ and show that $\gamma = \{\gamma_n: n \in N^+\}$ is a fundamental set of covers of X; this implies, by Problem 77 of Chap. 5, metrizability of X. Fix $y^* \in Y$ and a nhood Oy^* of y^* in Y. Then $\varepsilon^* = \rho\{f^{-1}y^*, X \setminus f^{-1}(Oy^*)\} > 0$, since $f^{-1}y^*$ is a compactum and $P^* = X \setminus f^{-1}(Oy^*)$ is a closed set not intersecting $f^{-1}y^*$ (see Problem 288 of Chap. 3). Put $V = O_{\varepsilon^*/2}(f^{-1}y^*)$ and pick a distinguished open set U in X, containing $f^{-1}y^*$ and contained in V. This is possible, because f is a closed mapping (see Problem 325 of Chap. 2). Recall that U is a distinguished set for f, if $f^{-1}fU = U$. Using the fact that $f^{-1}y^*$ is a compactum and the set $X \setminus U$ is closed, we conclude that $\rho(f^{-1}y^*, X \setminus U) = \delta^* > 0$. Hence $\delta^* \leqslant \varepsilon^*$. Now select $n^* \in N^+$ such that $1/n^* < \delta^*/2$ and consider a nhood $O_{n^*}y^*$ of y^* and the cover γ_{n^*}. It is to be shown that $\gamma_{n^*}(O_{n^*}y^*) \subset Oy^*$.

Consider any $y' \in Y$ for which $O_{n^*}y' \cap O_{n^*}y^* \neq \Lambda$. Fix $y'' \in O_{n^*}y' \cap O_{n^*}y^*$. Then $f^{-1}(y'') \subset O_{\delta^*/2}(f^{-1}y') \cap O_{\delta^*/2}(f^{-1}y^*)$, so that $\rho(f^{-1}y', f^{-1}y^*) < \delta^*$, i.e., $f^{-1}y' \cap$

---

†A σ-compactum is a space which is the union of countably many compact subspaces.

$O_{\delta*}(f^{-1}y^*) \neq \Lambda$. Then, by the choice of δ^*, we surely have $f^{-1}y' \cap U \neq \Lambda$ and, consequently, $f^{-1}y' \subset U \subset V$. Similarly, $f^{-1}(O_{\eta*}y') \subset O_{\delta*}(V) \subset O_{\varepsilon*/2}(V) \subset O_{\varepsilon*}(f^{-1}y^*) \subset f^{-1}(Oy^*)$. Hence $O_{\eta*}y' \subset Oy^*$, i.e., $\gamma_{\eta*}(O_{\eta*}y^*) \subset Oy^*$.

50. Consider any pointwise countable completely regular space X which is not a space of countable type (find such a space). Fix a compactum $F \subset X$ such that the character of each compactum Φ, contained in X and containing F, is uncountable.

Consider the natural projection mapping π of X onto its decomposition space Y, whose unique nontrivial element is F. The mapping π is perfect (see Problems 318 and 322 of Chap. 2). It is easy to show that Y is completely regular (do this yourself). Put $y^* = \pi F$. If there was a compactum $B \ni y^*$, $B \subset Y$ such that $\chi(B, Y) \leqslant \aleph_0$, then its pre-image $\pi^{-1}B$ would be a compactum (because π is a perfect mapping; see Problem 36) of countable character in X (see Problem 324 of Chap. 2)), containing F. This is a contradiction. Hence Y is not a point-countable space (at the point y^*).

51. See Filippov [2], where a stronger statement has been proved.

52. Suppose that X is an uncountable set, x^* is a fixed point of X and $\mathcal{F}$ is obtained by adding the empty set and the family $\{A \subset X: A \ni x^*\}$ to the family $\mathcal{A}$ of all countable subsets of $X \setminus \{x^*\}$. Obviously, $\mathcal{F}$ is the family of all closed sets with respect to some topology $\mathcal{T}$ on X. $(X, \mathcal{T})$ is Lindelöf because every nhood of x^* contains all points of X, except possibly some countable set. Evidently $(X, \mathcal{T})$ is a T_1-space, and since only one point of X is non-isolated (namely x^*), $(X, \mathcal{T})$ is a normal space (see Problem 15 of Chap. 3). Now suppose that there is a perfect mapping f of X onto some separable space Y. Let M be any countable dense set in Y. Fix for each $y \in M$ a point $x_y \in X$ such that $f(x_y) = y$. Then, $A = \{x_y: y \in M\}$ is countable. Consequently A is closed in X. But then $M = fA$ is closed in Y. Since M is dense in Y, we conclude that $M = Y$. Now observe that every compact subspace of X is finite (in fact, every countable set in X is closed). Hence $X = f^{-1}Y = f^{-1}M$ is countable, which contradicts the hypothesis.

53. From (*a*) it follows that $\tilde{f}^{-1}(bY \setminus Y) = bX \setminus X$ and $\tilde{f}^{-1}(Y) = X$. Hence $f = \tilde{f} \mid X$ is a perfect mapping (see Problem 46). Since $R(X)$ is closed in X, the objective will be achieved if it is shown firstly that $f(R(X)) = R(Y)$ and secondly that $f(x') \neq f(x'')$ for every two points $x' \in R(X)$, $x'' \in R(X)$, $x' \neq x''$. Using the condition (*a*) and also the continuity and closedness of $\tilde{f}$ we conclude that $\tilde{f}[bX \setminus X]_{bX} = [bY \setminus Y]_{bY}$. Since, in addition, $\tilde{f} \mid X = f$ and $f(X) = Y$, $\tilde{f}([bX \setminus X]_{bX} \setminus (bX \setminus X)) = [bY \setminus Y]_{bY} \setminus (bY \setminus Y)$. Consequently, $\tilde{f}(R(X)) = R(Y)$, because $R(X) = [bX \setminus X]_{bX} \setminus (bX \setminus X)$; $R(Y) = [bY \setminus Y]_{bY} \setminus (bY \setminus Y)$. Furthermore, since by hypothesis the restriction of f to the set $\varphi(Y)$ is a one-to-one mapping, the problem is solved if we can establish that $R(X) \subset \varphi(Y)$. We have $\tilde{f}([\varphi Y]_{bX}) = bY$, because $\tilde{f}([\varphi Y]_{bX})$ contains Y and is closed in bY. Since $\tilde{f}(\varphi Y) = Y$, $\tilde{f}([\varphi Y]_{bX} \setminus \varphi Y) \supset bY \setminus Y$. Hence $[\varphi Y]_{bX} \supset bX \setminus X$. Consequently, $R(X) \subset [bX \setminus X]_{bX} \subset [\varphi Y]_{bX}$. The proof is complete.

54. It is easy to construct a continuous mapping $\varphi: Y \to X \times Y$ such that $\pi\varphi: Y \to Y$ is an identity mapping. For example, fix $x^* \in X$ and put $\varphi(y) = (x^*, y)$ for all $y \in Y$. Arguing by contradiction, we can now apply the conclusion of Problem 53 and conclude that the restriction π to the set $R(X \times Y)$ of points of non-local compactness of the product $X \times Y$ is a homeomorphism. But in our case it is obvious that this is not so, for $R(X \times Y) = X \times R(Y)$ (see Problem 183 of Chap. 3) and $\pi \mid (X \times R(Y))$ is not one-to-one.

55. Denote by i_X and i_Z the identity mappings of X and Z into themselves, respectively. Obviously, the mapping $f\Delta g$ can be regarded as the composition of the diagonal product $i_X \Delta g$: $X \to X \times Z$ with the product $f \times i_Z$: $X \times Z \to Y \times Z$, i.e., $f\Delta g = (f \times i_Z)(i_X \Delta g)$. But $i_X \Delta g$ maps the space X homeomorphically onto the graph of the mapping g, which is a closed subspace of $X \times Z$, because Z is a Hausdorff space (see Problem 291 of Chap. 2). Also, $f \times i_Z$ is perfect, being the product of two perfect mappings (Problem 31). Hence $f\Delta g$ is a perfect mapping, as the composition of two perfect mappings (Problem 29).

56. Use the conclusion of the preceding problem. The diagonal product $f\Delta(gf)$: $X \to Y \times Z$ is a perfect mapping, with $(f\Delta gf)(X) \subset \Gamma_g = \{(y, g(y)) : y \in Y\}$, where Γ_g is the graph of g. As we know, the restriction $\pi \mid \Gamma_g$ of the projection π: $Y \times Z \to Y$ is a homeomorphism of Γ_g onto Y (see Problem 291 of Chap. 2). But obviously $\pi \cdot (f\Delta gf) = f$. Hence as a composition of a perfect mapping and a homeomorphism, f is a perfect mapping (Problem 29). The second assertion is trivial.

57. Put $f_1 = f \mid X_1$ and let i_1 be the identity mapping of X_1 into X. Then $f_1 = f \cdot i_1$ and we can apply Problem 56, by virtue of which i_1 is a perfect mapping. Hence, $X_1 = i_1(X_1)$ is closed in X.

58. We may assume that all factors are 'onto' mappings. Then, by applying Problems 55 of Chap. 3 and 325 of Chap. 2 it can be shown that the images are Hausdorff spaces. The rest follows directly from Problem 56.

Another solution of this problem for completely regular spaces, using Problems 46 and 47, is possible.

59. If f: $X \to Y$ is a perfect mapping and g: $X \to Z$ is a contraction, Z being Hausdorff, then the diagonal product $f\Delta g$ is a homeomorphism of X onto a closed subspace of $Y \times Z$ (see Problem 55 and the observation made in Problem 57).

60. Obviously, (b) implies (a); indeed, the projection π: $Y \times Z \to Y$ is a perfect mapping (see Problem 33) and the restriction of the latter to any closed subspace in $Y \times Z$ is also a perfect mapping (see Problem 30). Deduce (b) from (a). Since X is completely regular, there is a homeomorphic mapping i of X into some compactum Z (Problems 75, 38 of Chap. 3). By Problem 55, the diagonal product $f\Delta i$ is a perfect mapping of X into $Y \times Z$. But $f\Delta i$ is one-to-one. Hence $f\Delta i$ maps X homeomorphically onto the closed subspace $(f\Delta i)(X)$ of $Y \times Z$.

61. The space X, like every Hausdorff space with a net of cardinality not greater than τ, is contractible to a Hausdorff space of weight not greater than τ (see Problem 147 of Chap. 2). But the class of all spaces of weight not greater than τ is closed under finite products and taking subspaces. Hence applying the statement of Problem 59 we conclude that $wX \leqslant \tau$.

68. Clearly, every $U_{t\alpha}$, $\alpha \in A_t$, $t \in T$ is an open set in X, because X is locally connected, and $U_{t\alpha}$ is a connected component of the open set $f^{-1}V_t$ in X (see Problem 370 of Chap. 3). Let $x_0 \in X$ be arbitrary, where Ox_0 is any nhood of x_0. Consider a nhood $O'x_0$ of x_0 such that $O'x_0 \subset Ox_0$ and $F_0 = O'x_0 \cap f^{-1}fx_0$ is clopen in the subspace $f^{-1}fx_0$ (obviously, such a nhood $O'x_0$ of x_0 exists, because $f^{-1}fx_0$ is 0-dimensional, i.e., has a base of clopen subsets). Now consider the clopen set $F_1 = f^{-1}fx_0 \setminus F_0$ in

$f^{-1}fx_0$. Clearly, F_0 and F_1 are closed in X and $F_0 \cap F_1 = \Lambda$. Now use the normality of X. Let Γ_0 and Γ_1 be open sets in X, for which $F_0 \subset \Gamma_0$, $F_1 \subset \Gamma_1$, $\Gamma_0 \cap \Gamma_1 = \Lambda$ and, in addition, $\Gamma_0 \subset O'x_0$. Put $\Gamma = \Gamma_0 \cup \Gamma_1$. Obviously, Γ is open in X and $f^{-1}fx_0 \subset \Gamma$. Since $f: X \to Y$ is closed, $f^{\#}\Gamma = Y \setminus f(X \setminus \Gamma)$ is open in Y. It is clear that $fx_0 \in f^{\#}\Gamma$. Pick $V_{t_0} \in \theta$ such that $fx_0 \in V_{t_0} \subset f^{\#}\Gamma$ (recall that θ is a base for Y).

Consider now the unique connected component $U_{t_0\alpha_0}$ of $f^{-1}V_{t_0}$, for which $x_0 \in U_{t_0\alpha_0}$. Clearly, $U_{t_0\alpha_0} \subset \Gamma$. Recall that $\Gamma = \Gamma_0 \cup \Gamma_1$ and $\Gamma_0 \cap \Gamma_1 = \Lambda$. Then, $U_{t_0\alpha_0}$, because of its connectedness, is contained in Γ_0 and does not intersect Γ_1, since $x_0 \in U_{t_0\alpha_0}$, $x_0 \in \Gamma_0$ but $x_0 \notin \Gamma_1$ (obviously, it is relevant that $U_{t_0\alpha_0} \subset \Gamma_0 \cup \Gamma_1$). Thus $x_0 \in U_{t_0\alpha_0} \subset \Gamma_0 \subset O'x_0 \subset Ox_0$, where $U_{t_0\alpha_0} \in \sigma$. The proof is complete.

69. Let $\sigma_V = \{U_\alpha: \alpha \in A\}$. Since $f^{-1}V$ is open in X and X is locally connected, each U_α is open in X (see Problem 370 of Chap. 3). Put $f_V = f \mid f^{-1}V$. Since f is perfect and $f^{-1}V$ is a complete pre-image, $f_V: f^{-1}V \to V$ is a perfect mapping. For each $\alpha \in A$ fix a point $x_\alpha \in U_\alpha$ and put $F = \{x_\alpha: \alpha \in A\}$. Since σ_V is a disjoint family and $f^{-1}V = \bigcup\{U_\alpha: \alpha \in A\}$, F is a closed discrete subspace of $f^{-1}V$. Consider the restriction $f_F = f_V \mid F = f \mid F$. Since f_V is perfect and F is closed in $f^{-1}V$, the mapping $f_F: F \to fF$ is perfect, but since F is discrete, fF is also a discrete space, where $|fF| = |F|$. However, the cardinality of a discrete space always coincides with its weight. Thus $|F| = |fF| = w(fF) < wY < \tau$. Hence $|\sigma_V| \leqslant \tau$, completing the proof.

70. Since Y is metrizable, it has a σ-discrete base (see Problem 80 of Chap. 5): $\theta = \bigcup_{i=1}^{\infty} \theta_i$, where θ_i, $i \in N^+$ is a discrete family of open sets in Y. Let $\theta_i = \{V_{it}: t \in T_i\}$. For each $i \in N^+$, $t \in T_i$ consider the family $\sigma_{it} = \{U_{it\alpha}: \alpha \in A_{it}\}$ of connected components of $f^{-1}V_{it}$ open in X. Put $\sigma_i = \bigcup\{\sigma_{it}: t \in T_i\}$, $\sigma = \bigcup\{\sigma_i: i \in N^+\}$, $f^{-1}\theta_i = \{f^{-1}V_{it}: t \in T_i\}$. By using the perfect normality of Y we can pick the σ-discrete base $\theta = \bigcup_{i=1}^{\infty} \theta_i$ to be such that the family σ_i is σ-discrete in X for every $i \in N^+$. Then σ is also a σ-discrete family in X. For the rest refer to Problems 68 of the present chapter and 80 of Chap. 5.

71. Weight does not increase under a perfect mapping (see Problem 41); hence $wX > wY$. To prove the reverse inequality, use Problems 69 and 68.

72. By the preceding problem, a normal space X has a countable base. Hence X is metrizable (see Problem 79 of Chap. 5).

73. This is proved along similar lines as Problem 166 of Chap. 4.

75. If $f: X \to Y$, $f(X) = Y$ is a closed mapping and σ is a discrete family of closed sets in Y, then $f^{-1}\sigma = \{f^{-1}\Phi, \Phi \in \sigma\}$ is such a family in X. For the rest, use the definition of collectionwise normality (p. 271) and Problem 323 of Chap. 2.

76. The solution follows from Problem 150 of Chap. 5 and the following straightforward proposition: If $f: X \to Y$ is a closed mapping, $f(X) = Y$ and γ is a conservative (p. 271) family of sets in X, then $f\gamma = \{fA: A \in \gamma\}$ is a conservative family of sets in Y.

77. See Engelking [3].

78. This follows from Problem 155 of Chap. 5 and the preceding problem.

79. Put $A = f(F)$, where F is closed in $T(\omega_1)$. If $x \in [A]$, then there is a countable set $A' \subset A$, for which $x \in [A']$ (since X is a space of countable tightness). Obviously, there is a countable set $B' \subset F$ such that $f(B') = A'$. Put $B = [B']$. Then B is a compactum (Problem 84 of Chap. 3) and $B \subset F$. Since f is a continuous mapping into a Hausdorff space, $f(B)$ is compact and closed in X (Problems 56 and 91, Chap. 3). Furthermore, from $f(B) \supset f(B') = A'$ it now follows that $f(F) = A \supset fB = [fB] \supset [A'] \ni x$. Hence $f(F) = A$ is closed in X.

80. Assume the contrary. Then there is a sequence $\{x_n: n \in N^+\}$ of points of $\mathrm{Fr}\, f^{-1}y$, for which $|\varphi(x_{n+1})| > |\varphi(x_n)| + 1$ for all values of n. Put $V_i = \{x \in X: |\varphi(x) - \varphi(x_i)| < \frac{1}{3}\}$. The set V_i is open (since φ is continuous), $x_i \in V_i$ and the family $\gamma = \{V_i: i \in N^+\}$ is discrete in X (i.e., every point of X has a nhood intersecting at most one element of γ). Now pick $z_i \in V_i$, $i \in N^+$ such that (*a*) $z_i \neq x_i$, and (*b*) $f(z_i) \neq f(z_j)$ for $i \neq j$. This is easy to accomplish by induction: if for $i \leqslant k$ the points z_i are defined in accordance with the prescriptions (*a*) and (*b*), then we take z_{k+1} as an arbitrary point of the set

$$V_{k+1} \setminus (f^{-1}y \cup (\cup\{f^{-1}f(z_i): i = 1, \ldots, k\})). \tag{1}$$

Since x_{k+1} is a boundary point of $f^{-1}y$, and $f^{-1}f(z_i)$ is closed in X (in fact, Y is a T_1-space) and either does not intersect or coincide with $f^{-1}y$, z_{k+1} can indeed be so chosen.

As a result, the set $P = \{z_i: i \in N^+\}$ is closed in X. Moreover, every subset of P is also closed in X (we have already noted the discreteness of γ and the fact that $z_i \in V_i$, $i \in N^+$). From the fact that f is closed it now follows that every subset of fP is closed in Y. Hence, fP has no limit in Y (Problem 25 of Chap. 2). But, by (*b*), $|fP| = \aleph_0$. This contradicts the assumption that Y is countably compact.

81. If $\mathrm{Fr}\, f^{-1}y$ is not countably compact, then there is a closed discrete set $A \subset \mathrm{Fr}\, f^{-1}y$ of cardinality $\aleph_0$ (Problem 189 of Chap. 3). Enumerate the elements of A: $A = \{x_n: n \in N^+\}$ and the elements of any countable base $\mathcal{B}$ at the point y in Y: $\mathcal{B} = \{V_n: n \in N^+\}$. Since X is normal, there is in X a discrete family $\gamma = \{U_n: n \in N^+\}$ of sets U_n open in X, where $U_n \ni x_n$ (Problem 22 of Chap. 3). By reducing U_n, if necessary, we may assume that $fU_n \subset V_n$ for each value of n (since f is continuous). Since x_n lies on the frontier of $f^{-1}y$, $U \cap (X \setminus f^{-1}y) \neq \Lambda$. Pick for each $n \in N^+$ a point $x_n^* \in U_n \cap (X \setminus f^{-1}y)$. Then $f(x_n^*) \in V_n$ and, consequently, y belongs to the closure of the set $Q = \{fx_n^*: n \in N^+\}$, which is the image of the set $P = \{x_n^*: n \in N^+\}$ under f. But $x_n^* \notin f^{-1}y$, so that $y \notin Q$. Hence Q is not closed although P is closed (because the family $\{U_n: n \in N^+\}$ is discrete and $x_n^* \in U_n$). Since $fP = Q$, this contradicts the fact that f is closed.

82. This follows from the preceding problem and Problem 181 of Chap. 5.

83. Suppose that the frontier F of the pre-image of some point is not countably compact. The frontier of a closed set is always closed; hence F is a closed set in X. Let A be a closed discrete subset of F of cardinality $\aleph_0$ and consider an arbitrary unbounded real-valued function ψ on A. ψ is continuous on A, since A is a discrete space. Since X is normal, and A is closed in X, ψ extends to some continuous real-valued function φ on X (see Problem 34 of Chap. 3). This contradicts Problem 80.

84. To obtain the required conclusion, it suffices to carry out minor and obvious modifications to the proof of the assertions of Problem 81.

85. In a pointwise-countable space, every point has a countable family of nhoods satisfying the conditions of the preceding problem (see Problem 65 of Chap. 3). Reference to Problem 181 of Chap. 5 suffices for the rest.

86. (*a*). The space Z is obviously countably paracompact, hence its product with any metrizable compactum (and Y is such a compactum) is normal (Problem 222 of Chap. 5).

(*b*) Since the projection of X onto a countably compact space Z (see Problem 193 of Chap. 3) is a perfect mapping (see solution of Problem 33), X is countably compact (Problem 39).

(*c*) Every closed subspace F of X is countably compact. Hence its image fF under f is also countably compact. But $fF \subset Y$, and hence $|fF| \leqslant \aleph_0$. Consequently, fF is a compactum and fF is closed in Y.

Statement (*d*) is obvious.

87. For the proof, make use of Problem 82.

89. By Problems 85 and 88, Y is the image of some subspace of X under a perfect mapping. Hence Y is metrizable (see Problems 8 and 49).

90. Let $A \in \mathscr{Z}$ and $A \subset U \subset X$, where U is open in X. For $x \in A$ put $\varepsilon_x = \rho(x, X \setminus U)/2$. Then, $\varepsilon_x > 0$. Set $\mathscr{P} = \{B \in \mathscr{Z}: B \neq A \text{ and } \operatorname{diam} B > \varepsilon_x\}$ and $P = \bigcup \{B: B \in \mathscr{P}\}$. By hypothesis the family $\mathscr{P}$ is finite, consists of closed sets and its elements are disjoint from A (because all of them are distinct from A, and $\mathscr{Z}$ is a decomposition). Therefore, P is closed, $x \notin P$, and hence $\delta_x = \min \{\rho(x, P), \varepsilon_x\} > 0$. Put $Ox = \{y \in X: \rho(x, y) < \delta_x\}$. Then, obviously, if $B \in \mathscr{Z}$ and $B \cap Ox \neq \Lambda$, we have $\operatorname{diam} B \leqslant \varepsilon_x$ and $\rho(x, B) < \varepsilon_x$, so that $B \subset O_{2\varepsilon_x}x = \{y \in X: \rho(x, y) < 2\varepsilon_x\} \subset U$.

Consider the set $V = \bigcup \{Ox: x \in A\}$; V is an open set containing A; moreover, if $B \cap V \neq \Lambda$, then $B \subset U$. We conclude that $\mathscr{Z}$ is continuous (see p. 77).

91. Suppose that R is the set of all rational numbers of the interval [0, 1] of the real line, endowed with the usual metric and topology, and $Q = R^2 = \{(x, y): x, y \in R\}$ is the metric product $R \times R$. Enumerate the elements of R: $R = \{r_1, r_2, \ldots, r_n, \ldots\}$. For each $n \in N^+$ fix a finite disjoint cover γ_n of the space R by sets of diameter less than $1/n$, closed in R (non-compact). Put $\lambda_n = \{F \times \{r_n\}: F \in \gamma_n\}$ and $\lambda = \bigcup \{\lambda_n: n \in N^+\}$. Obviously, λ is a disjoint closed cover of Q and for each $\varepsilon > 0$ the diameter of at most finitely many elements of λ exceeds ε. Consequently (see the preceding problem), λ is a continuous decomposition of Q. Obviously, every element of λ is a non-compact set with empty interior.

92. Consider the decomposition λ of Q constructed in the preceding problem. It is continuous, hence the mapping $\pi: Q \to Z$ corresponding to this decomposition, where Z is the decomposition space of λ, is continuous and closed (Problem 322 of Chap. 2), and Z itself is a T_1-space since all elements of the decomposition under consideration are closed. But then Z is normal (Problem 329 of Chap. 2). Also $|Z| = \aleph_0$. Z is not first countable at any point: otherwise, the pre-image of this point would have a compact frontier (see Problem 82), but there is no element with a compact frontier in the given decomposition.

93. Assume the contrary. Suppose that D has a limit point $y^* \in Y$. Put $D_1 = D \setminus \{y^*\}$. Clearly, D_1 is not closed in Y. Since Y is a k-space, there is a compact subspace $F \subset Y$ for which the set $F \cap D_1$ is not closed in F. Consequently, $|F \cap D_1| \geqslant \aleph_0$. Pick a set $E' \subset F \cap D_1$, satisfying $|E'| = \aleph_0$. Since F is compact, some y^{**} is a limit point of E'. Put $E = E' \setminus \{y^{**}\}$. Then $y^{**} \in [E] \setminus E$, $E \subset D$, $|E| = \aleph_0$. Enumerate the elements of E: $E = \{y_n : n \in N^+\}$. Put $A_n = f^{-1}y_n$. Define by induction a sequence $\{x_n: n \in N^+\}$ in X. Take as x_1 any point of A_1. If the points x_n have been defined for all $n < k$, then take as x_{k+1} any point of the set $A_{k+1} \setminus (\bigcup \{\gamma(x_n): n \leqslant k\})$, where $\gamma(x) = \bigcup \{U: U \in \gamma$ and $U \ni x\}$. Since $y_{k+1} \in D$ and the cover γ is point-finite, $A_{k+1} \setminus (\bigcup \{\gamma(x_n): n \leqslant k\})$ is not empty and, consequently, x_{k+1} can be defined. Thus, the sequence $\{x_n: n \in N^+\}$ of points of X has been constructed.

It is to be shown that the set P of points of this sequence is closed in X (indeed, P is even discrete in X). Let $x \in X$. If $\gamma(x) \cap P = \Lambda$, then $x \notin [P]$ because γ is an open cover. If $x_{n'} \in \gamma(x)$, then $x \in \gamma(x_{n'})$ and $x_n \notin \gamma(x_{n'})$ for $n > n'$ (this is implied by the definition of x_n). Hence $|\gamma(x_{n'}) \cap P| < \aleph_0$. If $x \notin P$, then $x \notin P' = \gamma(x_{n'}) \cap P$. Using the fact that X is a T_1-space, we can then find a nhood Ox of x, which does not intersect the finite set P'. Then $O^*x = Ox \cap \gamma(x_{n'})$ is a nhood of x in X not intersecting P. Thus P is closed in X; hence $E = fP$ is closed in Y. But $y^{**} \in [E] \setminus E$. A contradiction. Hence D is discrete in Y.

94. X is a k-space (Problem 238 of Chap. 3). Since every closed continuous mapping is a quotient mapping and the image of a k-space under a quotient mapping is a k-space (Problem 246 of Chap. 3), Y is also a k-space. The cover λ of X by all open sets with compact closures in X is refinable to a point-finite open cover γ. Apply the result of the preceding problem to this γ and the mapping f, to conclude that the set A of all those $y \in Y$, whose pre-images are not covered by any finite collection of elements of γ, is discrete in Y. Then, if $y \in Y \setminus A$, we have $f^{-1}y \subset \bigcup \{U: U \in \gamma'\}$ where γ' is some finite subfamily of γ. This yields $f^{-1}y \subset \bigcup \{[U]: U \in \gamma'\} = \Phi$. For each $U \in \gamma'$ the set $[U]$ is compact (because γ is a refinement of λ, and λ is chosen in the appropriate manner). Then Φ is compact, as the union of a finite family of compacta, and, consequently, $f^{-1}y$ is also compact, as a closed subspace of the compact space Φ. Note that the closedness of $f^{-1}y$ follows from the continuity of f and the fact that Y is a T_1-space. (Y is a T_1-space since f is closed and X is, by hypothesis, a T_1-space.)

95. The restriction of a closed continuous mapping to a pre-image is always a closed continuous mapping (Problem 319 of Chap. 2). In the given case, it is necessary to take the pre-image of the set of all points whose pre-images are compact. The conclusion now follows from the preceding problem and the fact that the image of a locally compact space under a perfect mapping is locally compact (Problem 48).

96. The net $\mathcal{S}$ may be assumed to be such that $P \cap Q \in \mathcal{S}$ for each $P \in \mathcal{S}$ and $Q \in \mathcal{S}$. Put $\mathcal{T} = \{fP: P \in \mathcal{S}\}$ and let Y_0 be the set of all points $y \in Y$ such that there is a finite family $\mathcal{T}' \subset \mathcal{T}$, for which $\{y\} = \bigcap \{Q: Q \in \mathcal{T}'\}$. It is obvious that $|Y_0| \leqslant |\mathcal{T}| + \aleph_0 \leqslant |\mathcal{S}| + \aleph_0 \leqslant \aleph_0$. We must show that $f^{-1}y$ is a compact subspace of X for every point $y \in Y \setminus Y_0$. Fix a point $x \in f^{-1}y$. Further, let $\{P_n(x): n \in N^+\}$ be an enumreration of the elements of $\mathcal{S}$ containing this point x. Then every sequence $\{z_n: n \in N^+\}$, where $z_n \in P_n(x)$, has x as an adherent point, i.e., $x \in [\{z_n: n \in N^+\}]$. X, as a space with a countable net, is Lindelöf and, in addition, since it is regular, it is paracompact (see Problems 143 of Chap. 2 and 132 of Chap. 5). Every subspace $f^{-1}y$, $y \in Y$, is then also paracompact, because $f^{-1}y$ is closed in a paracompactum. Hence if the subspace $f^{-1}y$ is not compact, then it is also not countably compact (see (Problem 181 of Chap. 5). Then there is a sequence $A = \{x_n: n \in N^+\}$ of mutually distinct points of $f^{-1}y$, having no limit point in X (i.e., A is a closed discrete countably infinite set). Pick a discrete family $\{U_n: n \in N^+\}$ of open sets in X such that $x_n \in U_n$ for every $n \in N^+$ (one notes that X is a normal space, which is implied by its regularity and Lindelöfness; see Problem 19 of Chap. 3).

Construct by induction the three sequences

$$\xi_1 = \{x^1_n: n \in N^+\},$$
$$\xi_2 = \{x^n_n: n \in N^+\},$$
$$\xi_3 = \{O_n x: n \in N^+\}$$

in such a way that the following conditions are satisfied:

(*a*) $O_k x \ni x^1_n$ for all $n > k$ and $k \in N^+$;
(*b*) $X \setminus [O_k x] \supset f^{-1}fx^1_n$ for all $n \leqslant k$ and $k \in N^+$;
(*c*) $x^1_n \in P_n(x)$ for all $n \in N^+$;
(*d*) $x^n_n \in U_n$ for all $n \in N^+$;
(*e*) $fx^1_n = fx^n_n \neq y$ for all $n \in N^+$.

Note that $O_k x$, $k \in N^+$ is a nhood of x in X. The construction is carried out as follows.

Pick any point $x^1 \in X \setminus f^{-1}y$ and a nhood $O_1 x$ in accordance with condition (*b*). Suppose now that x^1_n, x^n_n and $O_n x$ are already defined for every $n < j$ in accordance with conditions (*a*) to (*e*). Since $\mathcal{S}$ is a net in X, there are $P, P' \in \mathcal{S}$, for which $x \in P \subset O_j x$ and $x_{j+1} \in P' \subset U_{j+1}$. Put $P^* = P \cap P_{j+1}(x)$. By hypothesis, $P^* \in \mathcal{S}$. We have: $(fP^* \cap fP') \setminus \{y\} \neq \Lambda$. Pick the points $x^1_{j+1} \in P^*$, $x^{j+1}_{j+1} \in P'$ in such a way that $fx^1_{j+1} = fx^{j+1}_{j+1} \neq y$. As $O_{j+1}x$ take any nhood of point x for which $x \in O_{j+1}x \subset [O_{j+1}x] \subset X \setminus \bigcup \{f^{-1}fx^1_i: i = 1, 2, \ldots, j+1\}$. The construction has been completed. Continuing in this way, we get the required sequences. Since $\{x^1_n: n \in N^+\}$ converges to x, $y = fx \in [\{fx^1_n: n \in N^+\}] = [\{fx^n_n: n \in N^+\}] = \{fx^n_n: n \in N^+\}$. This contradicts the fact that $y \neq fx^n_n$. We now note that f is a closed mapping and that $\{x^n_n: n \in N^+\}$ is a closed set (the latter is a consequence of the discreteness of $\{U_n: n \in N^+\}$ in X).

97. Let Y' be the set of those $y \in Y$ for which $f^{-1}y$ is not compact. Then $|Y'| \leqslant \aleph_0$ (see the preceding problem). Put $Y^* = Y \setminus Y'$, $X^* = f^{-1}Y^*$ and $f_* = f | X^*$. The mapping $f_*: X^* \to Y^*$ is perfect and $f(X^*) = Y^*$. Obviously, it suffices to show that Y^* decomposes into a countable family of subspaces homeomorphic to subspaces of X^*. Since X^* is a regular space with a countable net, X^* also has a countable net $\mathcal{S}^*$ consisting of closed sets in X^*. Every $f_*^{-1}y$ (where $y \in Y^*$) is therefore countably compact, and hence at least one point of $f_*^{-1}y$ is isolated in $f_*^{-1}y$. Keeping this remark in mind, consider for each $P \in \mathcal{S}^*$ the restriction $f_P = f | P$ of f to the subspace P. Since P is closed in X^*, f_P is a continuous closed mapping. Denote by $\tilde{P}$ the set of all $x \in P$, for which $f^{-1}f(x) = \{x\}$ (i.e., $\tilde{P}$ is a set of all points at which f_P is one-to-one). Obviously, $f_P^{-1} f_P \tilde{P} = \tilde{P}$, hence $f_{\tilde{P}} = f_P | \tilde{P}$ also is closed and continuous. But $f_{\tilde{P}}$ is, in addition, one-to-one; consequently, $f_{\tilde{P}}$ is a homeomorphism between $\tilde{P}$ and the subspace $f_{\tilde{P}}\tilde{P} = f\tilde{P}$ of Y. To show now that $Y^* = \bigcup \{f\tilde{P}: P \in \mathcal{S}^*\}$, let $y^* \in Y^*$. Choose a point x^* isolated in $f^{-1}y$. The set $U = X^* \setminus (f^{-1}y \setminus \{x^*\})$ is a nhood of x^* in X^*. Since $\mathcal{S}^*$ is a net in X^*, there is a $P^* \in \mathcal{S}^*$, for which $x^* \in P^* \subset U$. It is clear that then $f_{P^*}^{-1} f_{P^*}(x^*) = x^*$, i.e., $x^* \in \tilde{P}^*$ and $y^* = fx^* \in f(\tilde{P}^*) \subset Y^*$. Thus, $Y - (\bigcup\{\{y\}: y \in Y'\}) \bigcup (\bigcup\{\widetilde{fP}: P \in \mathcal{S}^*\})$ is the required representation.

98. If $\alpha_1, \alpha_2 \in M \setminus M_\xi$, $\alpha_1 \neq \alpha_2$, $\langle U_B \rangle \cap F_{\alpha_1} \neq \Lambda$ and $\langle U_\beta \rangle \cap F_{\alpha_2} \neq \Lambda$, then $U_\beta \supset \langle U_\beta \rangle \supset F_{\alpha_1}$ and $U_\beta \supset \langle U_\beta \rangle \supset F_{\alpha_2}$ (since $\langle U_\beta \rangle$ is a distinguished set), and hence either $\alpha_1 \neq \beta$, or $\alpha_2 \neq \beta$. For definiteness, let $\alpha_1 \neq \beta$. But then $\alpha_1 \in M_\xi$, a contradiction. The first statement is proved. The second follows from the first. Indeed, the image of $\langle U_\beta \rangle$ in the decomposition space Z is open (since the decomposition is continuous), contains $\{F_\beta\} \in Z$ and its intersection with the set $\{\{F_\alpha\}: \alpha \in M \setminus M_\xi\}$ contains at most one element. But $Z = \{\{F_\beta\}: \beta \in M\}$. Hence $Z \setminus Z_\xi$ is discrete in Z.

99. Put $\xi_i = \{O_{1/i} f^{-1}z: z \in Z\}, i \in N^+$. With respect to ξ_i, define Z_{ξ_i} as in the preceding problem and put $Y_i = Z \setminus Z_{\xi_i}$, $Y_0 = Z \setminus \bigcup \{Y_i: i \in N^+\}$. Then $Z = \bigcup \{Y_i: i \in N\}$ and $Y_0 = \bigcap \{Z_{\xi_i}: i \in N^+\}$. By the preceding problem, every Y_i for $i \geqslant 1$ is discrete in Z. It remains to show that $f^{-1}z$ is compact for every $z \in Y_0$. By the definition of Y_0, for each $i \in N^+$ there is a $z_i \in Z$, $z_i \neq z$ such that $O_{1/i} f^{-1}z_i \supset f^{-1}z$. Then, obviously, every subsequence of the sequence $\{z_i: i \in N^+\}$ has z as its limit point (because the closure of the union of the pre-images of elements of this sequence intersects $f^{-1}z$). Hence $\{z_i: i \in N^+\}$ converges to z. Let Z^* be the subspace of Z consisting of z and the points of the sequence $\{z_i: i \in N^+\}$. From what has been stated it is clear that Z^* is first countable. Put $X^* = f^{-1}Z^*$ and $f^* = f \mid X^*$. Then $f^*: X^* \to Z^*$ is a closed continuous mapping and $fX^* = Z^*$. By Problem 85, the frontier of $f^{-1}z$ in X^* is compact. But the relation $O_{1/i} f^{-1}z_i \supset f^{-1}z$ ensures that all points of $f^{-1}z$ are bounded with respect to X^*. Hence $f^{-1}z$ is a compactum.

100. By the preceding problem, there is a countable family γ of discrete closed subspaces of Y such that, for each point $y \in Y^* = Y \setminus \bigcup\{P: P \in \gamma\}$, the set $f^{-1}y$ is compact. Put $X^* = f^{-1}Y^*$ and $f^* = f \mid X^*$. Then f^* is a perfect mapping of the metric space X^* onto a T_1-space Y^*. Consequently, Y^* is metrizable (Problem 49). The desired expansion of Y is thus $Y = Y^* \cup (\bigcup \{P: P \in \gamma\})$.

101. By Problem 96, the set Y' of all points of Y, for which $f^{-1}y$ is not compact, is countable. Put $Y^* = Y \setminus Y'$, $X^* = f^{-1}Y^*$ and $f^* = f \mid X^*$. Then f^* is a perfect mapping of X^* onto Y^*. Hence Y^* has a countable base (Problem 41).

102. We shall assume that the weight of X is uncountable; otherwise, see Problem 96. Fix a base $\mathcal{B}$ for X, whose cardinality τ equals the weight of X. Furthermore, suppose that X is not compact; otherwise our problem is trivial. Then X is not countably compact (because it is paracompact, see Problem 181 of Chap. 5). Suppose that γ is any discrete (in X) countably infinite family of elements of $\mathcal{B}$: $\gamma = \{U_n: n \in N^+\}$.

Consider the set M_γ of points of Y, the pre-image of each of which intersects all elements of γ. We show that M_γ is discrete in Y. Assume the contrary. Choose a non-closed countable set $\{y_n: n \in N^+\} \subset M_\gamma$ and $x_n \in f^{-1}y_n \cap U_n: n \in N^+$. The set $\{x_n: n \in N^+\}$ is discrete in Y, because γ is discrete, but its image, the set $\{y_n: n \in N^+\}$ is not closed, contradicting the closedness of f. Thus M_γ is discrete in Y. But Y, as the continuous image of a space of weight τ, has a net of cardinality not greater than τ. Then M_γ also has such a net. But M_γ is a discrete space. Consequently, $|M_\gamma| \leqslant \tau$.

The cardinality of the family of all countable subsets of $\mathcal{B}$ does not exceed $\tau^{\aleph_0}$. Hence $|\bigcup\{M_\gamma: \gamma \subset \mathcal{B}, |\gamma| = \aleph_0$ and γ is discrete in $X\}| \leqslant \tau \cdot \tau^{\aleph_0} = \tau^{\aleph_0}$. It is to be shown that each point $y \in Y$, whose pre-image is not compact, is an element of some M_γ. If the space $f^{-1}y$ is not compact, then it is also not countably compact (since X, and hence also $f^{-1}y$ is paracompact). This permits us to pick out in $f^{-1}y$ a closed discrete set A of cardinality $\aleph_0$. Let $A = \{x_n: n \in N^+\}$. X is normal, hence in X there exist open sets $U_n \in \mathcal{B}$ such that $x_n \in U_n$ for all values of n and the family $\gamma = \{U_n: n \in N^+\}$ is discrete in X (see Problem 22 of Chap. 3). But then $y \in M_\gamma$.

103. Suppose that $f: X \to Y$ is a closed continuous mapping such that $f(X) = Y$. Let $\mathcal{S}$ be a net in X of cardinality not greater than τ. We may assume that the intersection of any finite set of elements of $\mathcal{S}$ belongs to $\mathcal{S}$. For each point $x \in X$ fix a sequence $\{P_n(x): n \in N^+\}$ of elements of $\mathcal{S}$, forming a net at this point. Such a sequence exists, because X is first countable every where. As in the solution of Problem 96, set $\mathcal{T} = \{fP: P \in \mathcal{S}\}$ and $Y_0 = \{y \in Y$: there is a finite set $\mathcal{T}' \subset \mathcal{T}$, for which $\{y\} = \bigcap\{Q: Q \in \mathcal{T}'\}\}$. Then $|\mathcal{T}| \leqslant \tau$ and hence $|Y_0| \leqslant \tau$. We show that for every $y \in Y \setminus Y_0$ the pre-image $f^{-1}y$ is a countably compact subspace of X. Since X is paracompact and $f^{-1}y$ is closed in X, we conclude from this that $f^{-1}y$ is compact. To prove that $f^{-1}y$ is countably compact, repeat the arguments in the solution of Problem 96.

108. X is metrizable. Hence we can apply the statement of Problem 99 (or Problem 96). By virtue of the latter, $Y = Y^* \cup Y'$, where $|Y'| \leqslant \aleph_0$ and $f^{-1}y$ is a compactum for every $y \in Y^*$, This and the inequality $|Y| > \aleph_0$ imply that $Y^* \neq \Lambda$. Fix a point $y^* \in Y^*$. The set $f^{-1}y^*$ is a compact subset of the metrizable space X. Hence $f^{-1}y^*$ has countable character in X. Since f is continuous and closed, we conclude that Y is first countable at y^* (see Problem 327 of Chap. 2). From the homogeneity of Y it now follows that Y is first countable at all points. However, we may now refer to Problem 89 and the required conclusion follows (noticing that Y is Lindelöf).

109. Since f is irreducible, $f^{\#}U$ is nonempty for every open set U in X, and since f is closed, $f^{\#}U$ is open in Y, because $f^{\#}U = Y \setminus f(X \setminus U)$ and U is open in X. Since f is continuous, $f^{-1}f^{\#}U$ is open in X. Thus, $f^{-1}f^{\#}U$ is nonempty and open in X. Clearly, $f^{-1}f^{\#}U \subset U$. It is to be shown that $f^{-1}f^{\#}U$ is dense in U. Suppose that U' is an open set in X and $U' \subset U$. Since f is irreducible, $f^{\#}U' \neq \Lambda$. Hence $U' \cap f^{-1}f^{\#}U \neq \Lambda$.

110. Because of the irreducibility and continuity of f, $f^{\#}U$ is nonempty and open in X, and $f^{-1}f^{\#}U$ is dense in U (see Problem 109). Then $[U] = [f^{-1}f^{\#}U]$ and, consequently, $f[U] = f[f^{-1}f^{\#}U] = [ff^{-1}f^{\#}U] = [f^{\#}U]$, completing the proof. (Here we have used the equality $f[A] = [fA]$, characterizing a closed continuous mapping.) The second statement is now immediate.

111. (*a*) Suppose that $x_0 \in X$ is isolated in X. Then the one-point subset $\{x_0\}$ is clopen in X. By Problem 109, the set $f^{\#}\{x_0\} = \{y \in Y: f^{-1}y \subset \{x_0\}\}$ is nonempty and open in Y. But, clearly, $f^{\#}\{x_0\} = \{fx_0\}$, whence fx_0 is isolated in Y.

(*b*). Suppose that fx_0 is isolated in Y. Then $f^{-1}fx_0$ is clopen in X. We show that $f^{-1}fx_0 = \{x_0\}$. Clearly, $x_0 \in f^{-1}fx_0$. Assume that the set $f^{-1}fx_0$ belongs to more than one point. Consider the set $X_0 = (X \setminus f^{-1}fx_0) \cup \{x_0\}$. Clearly, X_0 is closed in X and $f(X_0) = Y$. Moreover, X_0 is a proper subset of X. This is a contradiction.

112. Assume that $|f^{-1}y| > 1$ and that the interior Int $f^{-1}y$ of $f^{-1}y$ is nonempty—let x^* belong to it. Since $|f^{-1}y| > 1$, there is $x \in f^{-1}y$ such that $x^* \neq x$. Using the fact that X is a T_1-space, pick a nhood Ox^* of x^*, satisfying $Ox^* \subset \text{Int}\,(f^{-1}y)$ and $x \notin Ox^*$. Then $X^* = X \setminus Ox^*$ is a closed subspace of X, $f^{-1}y' \subset X^*$ for $y' \neq y$ and $f^{-1}y \cap X^* \supset \{x\} \neq \Lambda$. Hence $fX^* = Y$, which contradicts the irreducibility of f.

113. To prove the irreducibility of $\bar{f}$: $\beta X \to \beta Y$, it obviously suffices to show that for any open set Γ in βX there is a point $\eta \in \beta Y$, for which $\bar{f}^{-1}\eta \subset \Gamma$. Consider an open set Γ' in βX such that $[\Gamma']_{\beta X} \subset \Gamma$. Put $U' = \Gamma' \cap X$. Using the irreducibility of f: $X \to Y$, pick a point $y_0 \in Y$ for which $f^{-1}y_0 \subset U'$. Then $[f^{-1}y_0]_{\beta X} \subset [U']_{\beta X} = [\Gamma']_{\beta X} \subset \Gamma$. f being closed, we obtain $[f^{-1}y_0]_{\beta X} = \bar{f}^{-1}y_0$ (see Problem 73). Thus the point $y_0 \in Y \subset \beta Y$ satisfies $\bar{f}^{-1}y_0 \subset \Gamma$.

115. Suppose that $\gamma = \{U_\alpha : \alpha \in A\}$ is a base for X and suppose that $|\gamma| = wX$. Put $A_n = A$ for all $n \in N^+$ and regard each A_n as a discrete topological space. Consider the topological product $M(A) = \Pi\{A_n : n \in N^+\}$. Define the subspace S of $M(A)$ as follows: $S = \{\{\alpha_i : i \in N^+\} \in M(A) : \{U_{\alpha_i} : i \in N^+\}$ is a base at some point $x \in X$ and $U_{\alpha_{i+1}} \subset U_{\alpha_i}$ for all $i \in N^+\}$, and the point $\alpha = \{\alpha_i : i \in N^+\}$ of S is set in correspondence to the point $\varphi(\alpha) = \bigcap \{U_{\alpha_i} : i \in N^+\}$ of X. In the following, we shall denote by α, α', α^* etc., points of $M(A)$ and by α_i, α'_i, α^*_i etc., their i-th co-ordinates, elements of A_i. $M(A)$ and hence also S are metrizable, as the product of a countable set of metrizable (here, discrete) spaces. We show that φ: $S \to X$ is an open continuous mapping of S onto X.

Consider an arbitrary $\alpha \in S$, any nhood U of $x = \varphi(\alpha)$ in X and any nhood V of α in S. By the definition of φ, we have $U_{\alpha_{n'}} \subset U$ for some value n' of the index i. The sets $V_j = \{\alpha' \in S : \alpha'_i = \alpha_i$ for $1 \leqslant i \leqslant j\}$ form a base at the point α in S by the definition of the topology in $M(A)$. Hence there is a value n'' of the index j for which $V_{n''} \subset V$. Now put $n^* = \max\{n', n''\}$. Then $\alpha \in V_{n^*} \subset V$, $x \in U_{\alpha_{n^*}} \subset U$. We show that $\varphi(V_{n^*}) = U_{\alpha_{n^*}}$. If $\alpha' \in V_{n^*}$, then $\alpha'_i = \alpha_i$ for $1 \leqslant i \leqslant n^*$, and hence $U_{\alpha'_{n^*}} = U_{\alpha_{n^*}}$. From $\varphi(\alpha') \in U_{\alpha'_{n^*}} = U_{\alpha_{n^*}}$ we conclude that $\varphi(V_{n^*}) \subset U_{\alpha_{n^*}}$. If $x' \in U_{\alpha_{n^*}}$, then $x' \in U_{\alpha_i}$ for $1 \leqslant i \leqslant n^*$, and hence (using the fact that X is first countable) we can pick $\alpha'_i \in A_i$, $i \in N^+$ such that for $1 \leqslant i \leqslant n^*$ we have $\alpha_i = \alpha'_i$ (and, therefore, $U_{\alpha'_i} = U_{\alpha_i}$) and a family $\{U_{\alpha'_i} : i \in N^+\}$ forming a base at x', subject to the condition that $U_{\alpha'_{i+1}} \subset U_{\alpha'_i}$. Then $\alpha' \in V_{n^*}$ and $x' = \varphi(\alpha')$. Thus $\varphi(V_{n^*}) \supset U_{\alpha_{n^*}}$, and hence $\varphi(V_{n^*}) = U_{\alpha_{n^*}}$. Since α, U and V are arbitrary, we conclude that φ is a continuous open mapping. The relation $\varphi(S) = X$ is immediate from the fact that X is first countable at all points. The converse statement is obvious.

116. False. Every compactum satisfying the first axiom of countability (and among such there are non-metrizable compacta—see Problems 80, 82 and 87 of Chap. 3) is the image of a metric space under a continuous open mapping.

117. *Necessity*. Suppose that f: $X \to Y$ is the mapping given in the hypothesis. There is a base $\mathcal{B}$ for X, which is the union of families γ_i, $i = 1, 2, \ldots$, each of which is locally finite. Then every set $f^{-1}y$ intersects at most a countable set of elements of γ_i, $i = 1, 2, \ldots$ (since all elements of γ_i are open, and $f^{-1}y$ is a separable space). Hence $|\{U \in \mathcal{B} : U \cap f^{-1}y \neq \Lambda\}| \leqslant \aleph_0$. This yields $|\{fU : fU \ni y$ and $U \in \mathcal{B}\}| \leqslant$

$\aleph_0$, i.e., $\{fU: U \in \mathcal{B}\}$ is a pointwise countable family of sets. Since f is a continuous open mapping, $\{fU: U \in \mathcal{B}\}$ is a base for Y.

Sufficiency. The construction on which the solution of Problem 115 is based, need not start only from the family of all open sets in X: it is permissible to take as γ any pointwise countable base for the space in question, and assume that $|\gamma| = wX$. It should then be noted that the product of countably many spaces, each with a countable base, is always a space with a countable base.

118. *Sufficiency.* Suppose that $f: X \to Y$ is the mapping specified in the hypothesis, ρ is a metric on X and γ_i is some open locally finite cover of X by sets of diameter less than $1/i$, $i = 1, 2, \ldots$ We also assume that γ_{i+1} is always a refinement of γ_i. Put $\lambda_i = \{fU: U \in \gamma_i\}$. Since $f^{-1}y$ is a compactum, $|\{U: U \in \gamma_i \text{ and } U \cap f^{-1}y \neq \Lambda\}| < \aleph_0$. Hence $\{V: V \in \lambda_i \text{ and } V \ni y\}$ is finite for each $y \in Y$. Noticing also that f is an open mapping and $f(X) = Y$, we conclude that λ_i is an open point-finite covering of Y for each $i \in N^+$. We now show that $\{\lambda_i: i \in N^+\}$ is a refined family of covers. Fix $y \in Y$ and $V_i \in \lambda_i$ such that $V_i \ni y$ for all $i \in N^+$. Pick $U_i \in \gamma_i$ for which $fU_i = V_i$. Consider an arbitrary nhood Oy of y. Since f is continuous, there is a nhood G of $f^{-1}y$ in X, for which $fG \subset Oy$. But $f^{-1}y$ is compact and X is a metric space. Hence there is a natural number n, for which

$$O_{1/n}(f^{-1}y) = \{x \in X: \rho(x, f^{-1}y) < 1/n\} \subset G. \tag{1}$$

In addition, $U \in \gamma_n$ and $U \cap f^{-1}y \neq \Lambda$ imply (since the diameter of U does not exceed $1/n$) that $U \subset G$. Consequently, all elements of λ_n containing y, lie wholly in Oy; in particular, $V_n \subset Oy$. This shows that $\{\lambda_i: i \in N^+\}$ is a refined sequence of open point-finite covers of Y. Then $\bigcup \{\lambda_i: i \in N^+\}$ is a uniform base for Y (Problem 67 of Chap. 5).

Necessity. The proof is based on the fact that every uniform base $\mathcal{B}$ is representable in the form $\mathcal{B} = \bigcup \{\gamma_i: i \in N^+\}$, where each $\gamma_i = \{U_{\alpha_i}: \alpha_i \in A_i\}$ is a point-finite open cover of Y and the sequence $\{\gamma_i: i \in N^+\}$ is refined (Problem 66 of Chap. 5). The construction of the mapping (and then the investigation of its properties) is conducted in exactly the same way as in the solution of Problem 115 on the basis of sets A_i, $i \in N^+$ (though they do not now coincide, in general, with any one fixed set of A). The rest follows by noting that the product of a countable family of finite sets is compact.

119. By Problem 117, X has a point-countable base. It now remains to refer to Problem 87 of Chap. 5.

120. This is an open problem, as far as the authors are aware. However, if the requirement of complete regularity of X is weakened to Hausdorffness, the answer is negative. For example, let Q be an uncountable dense subset of the interval $[0, 1]$, in which every compactum is countable (see Problem 285 of Chap. 3). Put $T = [0, 1] \setminus Q$ and

$$X = \{(x, y): 0 \leqslant x \leqslant 1, y = 0\} \cup \{(x, y): x \in Q, y \in N\},$$

where N is the set of all natural numbers. As a base for the topology $\mathcal{T}$ on X, we take the collection of sets of the following three types:

(i) $\{(x, y)\}$ for all $x \in Q$ and $y \neq 0$;

(ii) $O_c(x) = \{(x, 0)\} \cup \{(x, y): y > c\}$ for all $x \in Q$ when $c \in N$;

(iii) $C_{\alpha c}(x) = \{(x, 0): x \in T \cap \alpha\} \cup \{(x, y): x \in Q \cap \alpha, y > c\}$, where α is an open set in $[0, 1]$ and $c \in N$.

The space X so constructed is obviously Hausdorff and has a uniform base.

By Problem 118, X is the image of some metric space M under a continuous open compact mapping, which we denote by f. Set in correspondence to each point (x, y) of X its first coordinate x, thus defining a mapping g of X onto the set of points of $[0, 1]$. Endow $[0, 1]$ with a non-standard topology $\mathcal{T}^*$ as follows: $A \in \mathcal{T}^*$ iff there is an open set $U \subset A$ in the usual topology of the interval such that $A \setminus U \subset Q$. Put $Y = ([0, 1], \mathcal{T}^*)$ (see Problem 209 of Chap. 3). Obviously, $g: X \to Y$ is a continuous open compact mapping. It now remains to refer to the properties of Y in Problem 209 of Chap. 3 and use Problem 118.

121. False. Furthermore, we can construct a continuous open mapping of finite multiplicity from a completely regular countable space, which does not contain any infinite compactum, onto an infinite compactum (Problem 140).

122. This is still an open problem, as far as the authors know. If the restriction of complete regularity of X is relaxed to Hausdorffness, then the answer is negative—the prescribed mapping $g: X \to Y$ is constructed in Problem 120.

123. Obviously, it suffices to show that f is one-to-one. Let $x_1, x_2 \in X$ and $x_1 \neq x_2$. Consider the nhoods Ux_1, Ux_2: $Ux_1 \cap Ux_2 = \Lambda$. Since f is irreducible, $f^{\#}Ux_1 = \{y \in Y: f^{-1}y \subset Ux_1\} \neq \Lambda$, and $f^{-1}f^{\#}Ux_1$ is dense in Ux_1. Then $f[Ux_1] = f[f^{-1}f^{\#}Ux_1] \subset [f^{\#}Ux_1]$ (here we have also used the continuity of f). Since f is open, fUx_2 is open in Y, and the set $Y \setminus fUx_2$ is closed. In addition, $f^{\#}Ux_1 \subset Y \setminus fUx_2$. Hence $[f^{\#}Ux_1] \subset Y \setminus fUx_2$. Consequently, $[f^{\#}Ux_1] \cap fUx_2 = \Lambda$. Moreover, since $fx_1 \in f[Ux_1] \subset [f^{\#}Ux_1]$, $fx_2 \in fUx_2$, we obtain $fx_1 \neq fx_2$.

124. Suppose that $f: X \to Y$ is the mapping in question, P is closed in X, $Q = fP$ and $y \notin fP$. Then $f^{-1}y \cap P = \Lambda$.

Consider the points $x_1, \ldots, x_k$ making up the set $f^{-1}y$. Pick for x_i a nhood Ox_i, not intersecting P and satisfying the restriction that $Ox_i \cap Ox_j = \Lambda$ for $i \neq j$ and f would map Ox_i homeomorphically onto an open set in Y. Then $V = \bigcap \{fOx_i: i = 1, \ldots, k\}$ is a nhood of y not intersecting Q. Hence $Q = fP$ is closed.

125. Suppose that $f: X \to Y$ is such a mapping. Fix x and put $y = f(x)$, $F = f^{-1}y$. Then $x \in F$ and $|F| = k$. Enumerate F as $\{x_1, \ldots, x_k\}$. For definiteness let $x = x_1$. Since X is a Hausdorff space, we can find mutually disjoint nhoods $Ox_1, \ldots, Ox_k$ of $x_1, \ldots, x_k$. Then $V = \bigcap \{fOx_i: i = 1, \ldots, k\}$ is an open set in Y, $V \ni y$. Put $U_i = Ox_i \cap f^{-1}V$. Obviously, $fU_i = V$, $i = 1, \ldots, k$, and $U_i \cap U_j = \Lambda$ for $i \neq j$. Hence $f^{-1}y' \cap U_i \neq \Lambda$ for each point $y' \in V$, and $|f^{-1}y'| = k$ implies that $f | U_i$ is a one-to-one mapping. But f is open and continuous. Hence, since U_i is open, the mapping $f | U_i$ is open (and continuous). Consequently, $f | U_i$ is a homeomorphism. Moreover, $f(U_i) = V$ is an open set in Y. For the rest see the preceding problem.

126. Put $Y_n = \{y \in Y: |f^{-1}y| = n\}$, $X_n = f^{-1}Y_n$, $f_n = f | X_n$, $f_n: X_n \to Y_n$, $n \in N^+$. Then f_n is open and continuous, since these properties are possessed by f, and X_n is a pre-image (see Problem 337 of Chap. 2). Moreover, f_n is precisely of multiplicity n. Hence (by the preceding problem) f_n is a local homeomorphism (and perfect). Therefore, X_n can be covered by sets open in X_n, the weight of each of which does not exceed τ. But f_n is closed (Problem 124) and is therefore perfect. Hence, from the above cover of X_n we can extract a cover of cardinality not greater than τ. Consequently, the weight of X_n does not exceed τ. But $X = \bigcup \{X_n: n \in N^+\}$. Therefore, X has a net of cardinality not greater than τ.

127. (*a*) Let τ be the weight of Y. From the preceding problem it follows that X has a net of cardinality not greater than τ. But then (see Problem 61) the weight of X also does not exceed τ. Obviously (by virtue of the continuity and openness of the mapping), the weight of Y does not exceed the weight of X. Hence, the weight of X equals the weight of Y. (*b*) follows from the preceding problem and Problem 227 of Chap. 5.

128. Consider the points $a, b \in I$, $a < b$. Put $L_b = \{x \in I: x < b\}$, $M_a = \{x \in I: a < x\}$. Clearly, L_b and M_a are open in I, and $U_{a,b} = L_b \cap M_a$ coincides with the interval (a, b). We have

$$\psi^{-1}U_{a,b} = \psi^{-1}L_b \cap \psi^{-1}M_a = B \cap A,$$

where $B = \psi^{-1}L_b = \{y: [\varphi(f^{-1}y)] \subset L_b\}$ and $A = \psi^{-1}M_a = \{y: \varphi(f^{-1}y) \cap M_a \neq \Lambda\}$.

Now fix a point $y^* \in B$. Because of the continuity of φ and the normality of I, there is an open set $V \subset X$ containing the set $f^{-1}y^*$ such that $[\varphi(V)] \subset L_b$. From the closedness of f it then follows that $f^{-1}(W) \subset V$ for some open set W in Y containing y^*. We have $[\varphi(f^{-1}(W))] \subset [\varphi(V)] \subset L_b$. Hence it follows that $W \subset B$. Therefore, B is open in Y. We now show that A is also open. Let $\tilde{y} \in A$ be arbitrary. Then there is $\tilde{x} \in X$ such that $f\tilde{x} = \tilde{y}$ and $\varphi(\tilde{x}) \in M_a$. Since φ is continuous, there is an open set $\tilde{V}$ in X such that $\tilde{x} \in \tilde{V}$ and $\varphi(\tilde{V}) \subset M_a$. Put $\tilde{W} = f\tilde{V}$. f is open, consequently, $\tilde{W}$ is open in Y. Moreover, $\tilde{y} \in \tilde{W}, f^{-1}y' \cap \tilde{V} \neq \Lambda$ for every point $y' \in \tilde{W}$ and $\varphi(f^{-1}y' \cap \tilde{V}) \subset \varphi(\tilde{V}) \subset M_a$, if $y' \in \tilde{W}$. Hence $\varphi(f^{-1}y') \cap M_a \neq \Lambda$ for each $y' \in \tilde{W}$, whence $\tilde{W} \subset A$. Thus A is open in Y. From the above it follows that the set $\psi^{-1}U_{a,b} = B \cap A$ is open in Y. Since the sets of the form $U_{a,b}$ form a base for the topology of I, the function ψ is continuous. The proof is complete.

129. The answer is negative. Not every such mapping is a c-covering. Denote by π the projection of the square $I \times I$ onto the interval I. The cardinality of the set $\mathcal{E}$ of all compact subspaces of $I \times I$ is equal to the cardinality of I (see Problem 226 of Chap. 3). Let φ be a one-to-one mapping of I onto $\mathcal{E}$. For each point $y \in I$ denote by $x(y)$ any element of $\varphi(y) \cap \pi^{-1}(y)$, provided $\varphi(y) \cap \pi^{-1}y \neq \Lambda$. If $\varphi(y) \cap \pi^{-1}(y) = \Lambda$, then pick $x(y) \in \pi^{-1}y$ arbitrarily. Put $A = \{x(y): y \in I\}$, $X' = X \setminus A$ and $\pi' = \pi \mid X'$. Then $\pi': X' \to I$ is an open continuous mapping (because π is open and $[(\pi')^{-1}(y)] = \pi^{-1}(y)$ for each $y \in I$, see Problem 340 of Chap. 2). Suppose there is a compactum $F \subset X'$, for which $\pi'(F) = I$. Then $F \in \mathcal{E}$; consider $y' = \varphi^{-1}(F)$ and $x(y')$. Since $\pi'(F) = I$, $F \cap \pi^{-1}(y') \neq \Lambda$. Consequently, $x(y') \in F$, but this contradicts the fact that $x(y') \in A$, $F \subset X \setminus A$.

130. Put $S_\infty = \{U \in S: |f^{-1}U| = \aleph_0\}$. From $|S_n| < \aleph_0$, $|S| = \aleph_0$ and condition (β) it follows that $S_\infty \cap S_n \neq \Lambda$, $n \in N^+$. Moreover, by virtue of (α), if $U \in S_n \cap S_\infty$, then there is a $V \in S_{n+1} \cap S_\infty$, for which $fV \ni U$. The latter remark enables us to construct immediately the required sequence $\{U_n: n \in N^+\}$ by induction.

131. We may assume that the sets S_i and S_j for $i \neq j$ are disjoint. Then $S = \bigcup \{S_n: n \in N^+\}$ is an infinite set. For $U \in S_{n+1}$ put $f(U) = \{f_n(U), f_{n-1}f_n(U), \ldots, f_1 \ldots$

$f_n(U)\}$, i.e., $V \in f(U)$, if there is a sequence $U_{n+1}, \ldots, U_1$, one of whose members is V, such that $U_{n+1} = U$ and $f_i(U_{i+1}) = U_i$ for $1 \leqslant i < n$. In addition, $f(U_{n+1}) = U_n$, where $U_{n+1} \in S_{n+1}$ and $U_n \in S_n$ iff $f_n(U_{n+1}) = U_n$. Obviously, the conditions (α) and (β) in the preceding problem are satisfied by f, S_n, $n \in N^+$. Making use of the preceding problem, we conclude that the required sequence $\{U_n: n \in N^+\}$ exists.

132. For each $U \in S_{n+1}$ fix an element of the family S_n containing U. Denote this by $f_n(U)$. Thus, we obtain a mapping $f_n: S_{n+1} \to S_n$.

By virtue of the preceding problem, we can pick $U_n \in S_n$ in order to have $f_n(U_{n+1}) = U_n$, $n \in N^+$. Then, by the definition of f_n, we have $U_{n+1} \subset U_n$.

133. Such a sequence is easy to construct by induction: suppose that Υ_n is already defined, where $f\Upsilon_n$ covers Y and all $U \in \Upsilon_n$ are open in X. Put $\lambda_{n+1} = \{V \subset X: V$ is open in X, V is contained in some element of Υ_n and $\operatorname{diam}(V) < 1/(n+1)\}$. Then $f\lambda_{n+1}$ is an open cover of Y. Consequently, there is a finite family Υ_{n+1} of elements of λ_{n+1} for which $f\Upsilon_{n+1}$ covers Y. Conditions (i), (ii) and (iii) are satisfied for Υ_{n+1} and γ_n.

134. There is (see the preceding problem) a sequence $\xi = \{\gamma_i: i \in N^+\}$ of finite families of closed sets in X such that for all $i \in N^+$:

(i) γ_{i+1} is a refinement of γ_i;
(ii) $f\gamma_i$ covers Y;
(iii) if $P \in \gamma_i$, then $\operatorname{diam} P \leqslant 1/i$.

Put $F_i = \bigcup\{P: P \in \gamma_i\}$. Each set F_i has a finite $1/i$-net (see p. 46). Hence it follows that the set $\Phi = \bigcap\{F_i: i \in N^+\}$ is totally bounded. It is also obvious that Φ is closed in X, because every set F_i, $i \in N^+$, is closed in X. We show that $f\Phi = Y$.

Let $y^* \in Y$ be arbitrary. For each $i \in N^+$ put $\gamma_i^* = \{P \in \gamma_i: P \cap f^{-1}y^* \neq \Lambda\}$. From the properties (i), (ii) and (iii) of the sequence ξ we obtain the following properties of the sequence $\xi^* = \{\gamma_i^*: i \in N^+\}$:

(i′) γ_{i+1}^* is a refinement of γ_i^*;
(ii′) $fP \ni y^*$ for each $P \in \gamma_i^*$ and every $i \in N^+$;
(iii′) if $P \in \gamma_i^*$, then $\operatorname{diam} P \leqslant 1/i$.

Moreover, recall that for every $i \in N^+$ each $P \in \gamma_i^* \subset \gamma_i$ is closed in X. Now set in correspondence to every element P of the family γ_{i+1}^* some element of γ_i^*, which contains P; we denote the latter by $\varphi(P)$. Now, apply the statement of Problem 131 to the choice function φ and to the sequence $\xi^* = \{\gamma_i^*: i \in N^+\}$. Then for each $i \in N^+$ (using Problem 131) one can select $P_i \in \gamma_i^*$ such that $\varphi(P_{i+1}) = P_i$ for all $i \in N^+$. From (ii′) we then find that $Q_{i+1} \subset Q_i$ for every $i \in N^+$, where $Q_i = f^{-1}y^* \cap P_i$. Clearly, every Q_i, $i \in N^+$, is closed in the complete metric space $(f^{-1}y^*, \rho)$.

In addition, from (iii′) and the definition of Q_i we get $\operatorname{diam} Q_i \leqslant 1/i$. Then from the completeness of $(f^{-1}y^*, \rho)$ it follows that $F = \bigcap\{Q_i: i \in N^+\} \neq \Lambda$ (Problem 187 of Chap. 2). But $F \subset \Phi$ and $f(F) \ni y^*$; consequently, $f\Phi \ni y^*$. Since $y^* \in Y$ was arbitrary, it follows that $f\Phi = Y$. Note that the set $f^{-1}y \cap \Phi$ is compact for every $y \in Y$. The proof is complete.

We give another proof of a special case of this statement based on Problem 141. By this problem, if (X, ρ) is complete (hence also Čech-complete, see Problem 92 of Chap.

5) and f is continuous and open, then there is a closed subspace $\Phi \subset X$, for which $f\Phi = Y$ and the restriction $f_0 = f \mid \Phi$ is perfect. Then Φ is a compactum, as the pre-image (under the mapping f_0) of the compact space Y (see Problem 36).

136. The metric space (X, ρ) has (Problem 134) a totally bounded subspace Φ such that $f\Phi = Y$. But (Problem 261 of Chap. 2) Φ is then separable, hence there exists a countable family $\mathcal{B}$ of open sets in X, containing some base at every point $x \in \Phi$. Then $f\mathcal{B} = \{fU: U \in \mathcal{B}\}$ (by the preceding problem) is a pseudo-base for Y. Obviously, $|f\mathcal{B}| \leqslant \aleph_0$. It remains to remark that a compactum with a countable pseudo-base also has a countable base (Problem 122 of Chap. 3).

We now construct another proof of the special case when X itself is complete; this is based on Problem 141. The space X, as a complete metric space, is Čech-complete (Problem 92 of Chap. 5). Using Problem 141, we take a closed subspace $\Phi \subset X$, for which $f\Phi = Y$ and the restriction $f \mid \Phi = f_0$ is perfect. $\Phi \subset X$ is compact, being a pre-image (under f_0) of the compact space Y (see Problem 36). Furthermore, Φ is a subspace with a countable base, since it is a metrizable compactum (see Problem 257 of Chap. 3). It now remains to note that the perfect mapping $f_0: \Phi \to Y$ (see Problem 41) does not increase weight.

139. False. As Φ take an infinite compactum. There is exactly one point of Φ which is not isolated, let us denote it by y^*. In addition, $|\Phi \setminus \{y^*\}| = \aleph_0$. We assume that the points of $\Phi \setminus \{y^*\}$ are enumerable as follows: $\Phi \setminus \{y^*\} = \{y_n: n \in N^+\}$. By the definition of the topology in Φ, the sequence $\{y_n: n \in N^+\}$ converges to y^*.

Suppose that X is a compactum in which there is no non-trivial convergent sequence, but there is a sequence $\eta = \{A_i: i \in N^+\}$ of nonempty mutually disjoint finite sets such that every nonempty open set intersects all the members of this sequence with sufficiently large index. Such a compactum is described in Problem 132 of Chap. 3.

Consider $Z = \Phi \times X$ (the Tychonoff product) and the continuous mapping $\pi: Z \to \Phi$ (projection onto the first factor). Put $S_n = \{y\} \times A_n$, $S^* = \{y^*\} \times X$, $F = \bigcup \{S_n: n \in N^+\} \cup S^*$ and $f = \pi \mid F, f: F \to \Phi$. Then F is closed in Z. Indeed, if $z \notin F$, then $z \notin S^*$, and hence $z \in \{y_{n^*}\} \times X$ for some $n^* \in N^+$. But the set $\{y_{n^*}\} \times X$ is open in Z, because $\{y_{n^*}\}$ is open in Φ and X is open in X. Then $(\{y_{n^*}\} \times X) \cap F = S_{n^*}$ is a finite set. Consequently, z is not a limit point of F, hence $z \notin F$ implies that $z \notin [F]$. The mapping f, being the restriction of a continuous mapping π, is continuous and hence also closed, because F is a compactum and Φ is a Hausdorff space. Finally, since A_n is nonempty for all $n \in N^+$, $fF = \Phi$.

To show that f is an open mapping, let W be open in F and $y \in fW$. If $y \neq y^*$, then the one-point set $\{y\}$ is open in Φ and $y \in \{y\} \subset fW$.

Now suppose that $y = y^*$. Then $W \cap S^* \neq \Lambda$. Pick $x^* \in X$ in such a way that $(y^*, x^*) \in W \cap S^*$. Fix an open set U in Z for which $U \cap F = W$. By virtue of the definition of the topology on a product, there exist $n^* \in N^+$ and an open set $V \ni x^*$ in X such that $(y_n, x) \in U$ for all $n > n^*$ and $x \in V$. But $V \supset U_{n_0}$ for some $n_0 \in N^+$. We obtain: if $n > n_0$, then $V \cap A_n \supset A_n \cap U_{n_0} \neq \Lambda$. For all values of n greater than $n_0 + n^*$, pick $x_n \in V \cap A_n$. Then $(y_n, x_n) \in U$ and $(y_n, x_n) \in F$. Consequently, $(y_n, x_n) \in U \cap F = W$. Hence $fW \ni f(y_n, x_n) = y_n$ for all $n > n_0 + n^*$, i.e., y^* is contained in fW together with some nhood of y^*. This shows that fW is an open set, and f is an open mapping.

We now show that there is no non-trivial convergent sequence in F. Since $S^* = \{y^*\} \times X$ is homeomorphic to X and there is no non-trivial convergent sequence in X, it suffices to consider an arbitrary sequence $\theta = \{(y_{n_i}, x_i): i \in N^+\}$ of points of $F \setminus S^*$, all of whose terms are mutually distinct and show that it has no limit in F. If $i' \neq i''$ but $n_{i'} = n_{i''}$, then $x_{i'} \neq x_{i''}$ since $y_{n_{i'}} = y_{n_{i''}}$ and $(y_{n_{i'}}, x_{i'}) \neq (y_{n_{i''}}, x_{i''})$. If $n_{i'} \neq n_{i''}$ then, since $x_{i'} \in A_{i'}$, $x_{i''} \in A_{i''}$ and $A_{i'} \cap A_{i''} = \Lambda$, we conclude that $x_{i'} \neq x_{i''}$. Consequently, $\lambda = \{x_i: i \in N^+\}$ is a non-trivial sequence of points of X.

Suppose that the sequence θ converges in F. Then the sequence λ obtained from θ by projection of the product $\Phi \times X$ onto X converges in X, which is impossible. Thus there is no non-trivial convergent sequence in F.

140. Consider ths mapping $f: F \to \Phi$ constructed in the preceding problem. In the set $S^* \subset F$ (see the preceding problem), pick any point x^* and put $X^* = (F \setminus S^*) \cup \{x^*\}$, $f^* = f \mid X^*$. Then $f^*: X^* \to \Phi$ is a continuous open mapping (obviously, of finite multiplicity) where $f^*X^* = \Phi$. There is no non-trivial convergent sequence in X^*, because there is no such sequence in $F \supset X^*$.

141. See Pas'nkov [2].

142. Relying on the preceding problem, we claim that there is a closed set $X_0 \subset X$ in X, for which $fX_0 = Y$ and the restriction $f_0 = f \mid X_0$ is perfect. The closed subspace X_0 of a Čech-complete space X is also Čech-complete. Then the space Y, as the perfect image (under f_0) of the Čech-complete space X_0, is also Čech-complete (see Problem 48).

143. For the proof it is first necessary to recall that a metrizable space is a metrizable space with a complete metric iff it is Čech-complete (Problem 92 of Chap. 5) and then that every metrizable space is paracompact (see Problem 45 of Chap. 5). For the rest, we need only use Problem 141 and recall that Čech-completeness is preserved under perfect mappings (see Problem 48).

144. Since X is a metrizable space with a complete metric, it is Čech-complete. Also Y, by hypothesis, is a paracompactum. Thus we arrive at the hypothesis of Problem 141. Consequently, we can find a closed set $F \subset X$ such that $fF = Y$, and the restriction $f_0 = f \mid F$ is perfect. The space Y, as the image of the metrizable space F under a perfect mapping, is metrizable (see Problem 49). For the rest, refer to the preceding problem.

145. Let $\Phi \subset Y$ be a compactum in Y. Consider the closed set $f^{-1}\Phi = X_0$ in X and the restriction $f_0 = f \mid X_0$ of the mapping f to the subspace $f^{-1}\Phi = X_0$. The mapping f_0 is open, because $f^{-1}\Phi$ is a pre-image of $\Phi \subset Y$ under an open mapping (see Problem 337 of Chap. 2). In addition, $X_0 = f^{-1}\Phi$ is a Čech-complete space, as a closed subspace of a Čech-complete space. Now use the statement of Problem 141, according to which there is a closed set $F \subset X_0$ in X_0 and hence also in X such that $f_0(F) = f(F) = \Phi$, and the restriction $f_0 \mid F = f \mid F = f_1$ is perfect. It now remains to note that the set F, as a pre-image of the compactum Φ under the perfect mapping f_1, is also a compactum (see Problem 36).

147. Recall the definition of the Baire space $B(\tau)$ of weight τ (see Problems 246-248 of Chap. 2). Suppose that A is a set of cardinality $|A| = \tau$. $B(\tau)$ has the following standard basis sets: $U_{\alpha_1 \ldots \alpha_k}$, $\alpha_1 \in A, \ldots, \alpha_k \in A$; $U_{\alpha_1 \ldots \alpha_k} = \{\xi = \{\alpha_i': i \in N^+\} \in B(\tau): \alpha_1' = \alpha_1, \ldots, \alpha_k' = \alpha_k\}$. Put $\theta_k = \{U_{\alpha_1 \ldots \alpha_k}: \alpha_1 \in A, \ldots, \alpha_k \in A\}$. For every

$k \in N^+$ the family θ_k is a disjoint open cover of $B(\tau)$ where diam $U_{\alpha_1} \ldots {}_{\alpha_k} < 1/k$ for every $U_{\alpha_1} \ldots {}_{\alpha_k}$. It is easy to see that the family $\theta = \{U_{\alpha_1} \ldots {}_{\alpha_k}: \alpha_1 \in A, \ldots, \alpha_k \in A, k \in N^+\}$ satisfies all the conditions of a completely refined A_τ-sequence of covers and, in addition, the disjointness condition. Furthermore, if $f: X \to Y$ is an open 'onto' mapping and $\theta = \{U_{\alpha_1} \ldots {}_{\alpha_k} : \alpha_1 \in A, \ldots, \alpha_k \in A, k \in N^+\}$ is a completely refined A_τ-sequence of open covers on X, then $f\theta = \{fU_{\alpha_1} \ldots {}_{\alpha_k}: \alpha_1 \in A, \ldots, \alpha_k \in A, k \in N^+\}$ is a completely refined A_τ-sequence of open covers of Y (verifying this assertion is trivial). Now let

$$\nu = \{V_{\alpha_1} \ldots {}_{\alpha_k} : \alpha_1 \in A, \ldots, \alpha_k \in A, k \in N^+\},$$

where A is a set of cardinality $|A| = \tau \geqslant \aleph_0$, be a completely refined A_τ-sequence of open covers of X. Put $\nu_k = \{V_{\alpha_1} \ldots {}_{\alpha_k}: \alpha_1 \in A, \ldots, \alpha_k \in A\}$. For each $k \in N^+$, the family ν_k is an open cover of X, and the sequence $\{\nu_k: k \in N^+\}$ of these covers is complete and refined. Let $\xi = \{\alpha_i: i \in N^+\}$ be an arbitrary point of $B(\tau)$ (see Problem 246 of Chap. 2), where $\alpha_i \in A$ for each $i \in N^+$. Consider the sets V_{α_1}, $V_{\alpha_1\alpha_2}, \ldots$, $V_{\alpha_1\alpha_2} \ldots {}_{\alpha_k}, \ldots$ in ν. Since ν is a completely refined A_τ-sequence, the intersection $\bigcap\{V_{\alpha_1} \ldots {}_{\alpha_k}: k \in N^+\}$ is nonempty and consists of exactly one point, which we denote by x. Put $f\xi = x$. Here $\xi = \{\alpha_i : i \in N^+\} \in B(\tau)$ and $\{x\} = \bigcap\{V_{\alpha_1} \ldots {}_{\alpha_i}: i \in N^+\}$. The mapping $f: B(\tau) \to X$ is thus defined. Consider in $B(\tau)$ an arbitrary standard basis set $U_{\alpha_1^0} \ldots {}_{\alpha_k^0} = \{\xi = \{\alpha_i: i \in N^+\} \in B(\tau): \alpha_1 = \alpha_1^0, \ldots, \alpha_k = \alpha_k^0\}$. From the definition of the mapping it follows that

$$fU_{\alpha_1^0} \ldots {}_{\alpha_k^0} = V_{\alpha_1^0} \ldots {}_{\alpha_k^0}.$$

Noting that ν is a base for X, and $\theta = \{U_{\alpha_1} \ldots {}_{\alpha_k}: \alpha_1 \in A, \ldots, \alpha_k \in A; k \in N^+\}$ is a base for $B(\tau)$, we infer that the mapping is continuous and open. It is routine to observe also that f is onto all of X.

148. It has already been shown in the preceding problem that there is a disjoint completely refined A_τ-sequence of open covers in $B(\tau)$. Suppose that $\nu = \{V_{\alpha_1} \ldots {}_{\alpha_k}: \alpha_1 \in A, \ldots, \alpha_k \in A; k \in N^+\}$, where $|A| = \tau$, is a disjoint completely refined A_τ-sequence of open covers of X. Consider the open mapping f of the Baire space $B(\tau)$ onto X, constructed in the preceding problem. From the disjointness of the A_τ-sequence ν it follows that $V_{\alpha_1} \ldots {}_{\alpha_k} \cap V_{\alpha_1'} \ldots {}_{\alpha_k'} = \Lambda$ if $\alpha_i \neq \alpha_i'$ for some $i \leqslant k$. Hence, if $\xi_1 = \{\alpha_i': i \in N^+\} \in B(\tau)$, $\xi_2 = \{\alpha_i'': i \in N^+\} \in B(\tau)$ and $\xi_1 \neq \xi_2$, then $\alpha_i' \neq \alpha_i''$ for some $i \in N^+$. Denote the first $i \in N^+$, for which $\alpha_i' \neq \alpha_i''$ by i_0. Consider standard basic sets $U_{\alpha_1'} \ldots {}_{\alpha_{i_0}'}$ and $U_{\alpha_1''} \ldots {}_{\alpha_{i_0}''}$ in $B(\tau)$. It is obvious that $U_{\alpha_1'} \ldots {}_{\alpha_{i_0}'} \cap U_{\alpha_1''} \ldots {}_{\alpha_{i_0}''} = \Lambda$, $\xi_1 \in U_{\alpha_1'} \ldots {}_{\alpha_{i_0}'}$, $\xi_2 \in U_{\alpha_1''} \ldots {}_{\alpha_{i_0}''}$. We have $fU_{\alpha_1'} \ldots {}_{\alpha_{i_0}'} = V_{\alpha_1'} \ldots {}_{\alpha_{i_0}'}$, $fU_{\alpha_1''} \ldots {}_{\alpha_{i_0}''} = V_{\alpha_1''} \ldots {}_{\alpha_{i_0}''}$, $V_{\alpha_1'} \ldots {}_{\alpha_{i_0}'} \cap V_{\alpha_1''} \ldots {}_{\alpha_{i_0}''} = \Lambda$, whence $f\xi_1 \neq f\xi_2$.

149. By the preceding problem, it suffices to construct a disjoint completely refined A_τ-sequence of open covers of X. Let A be a set of cardinality $|A| = \tau$. Since the weight of the metric space (X, ρ) is τ, X is not a compactum and $\dim X = 0$, using Problems 153 and 154 of Chap. 5 and 217 of Chap. 2 it is routine to construct a disjoint open cover ν_1 of cardinality τ. We may assume that diam $U < 1$ for every $U \in \nu_1$. Choose some one-to-one mapping of A onto ν_1. Then $\nu_1 = \{U_\alpha: \alpha \in A\}$; in addition, $U_{\alpha'} \cap U_{\alpha''} = \Lambda$ if $\alpha' \neq \alpha''$. Since X is homogeneous in weight, the weight of every subspace U_α, $\alpha \in A$, is also τ; hence (because U_α is not compact) we can construct for each $U_\alpha \in \nu_1$, $\alpha \in A$, a disjoint open cover θ_α of U_α of cardinality $|\theta_\alpha| = \tau$. It is easy to see that each U_α, $\alpha \in A$, is clopen in X, so that every $U \in \theta_\alpha$, $\alpha \in A$, is also clopen.

We can repeat the assumption that diam $U < \frac{1}{2}$ for every $U \in \theta_\alpha$ and $\alpha \in A$. Once again we effect some one-to-one mapping h_α of A onto θ_α for each $\alpha \in A$. Then every $U \in \theta_\alpha$ is some $U_{\alpha\alpha'}$, where $U_{\alpha\alpha'}$ is an image of $\alpha' \in A$ under the given one-to-one mapping of A onto θ_α. Put $\nu_2 = \bigcup\{\theta_\alpha: \alpha \in A\}$. Then $\nu_2 = \{U_{\alpha_1\alpha_2}: \alpha_1, \alpha_2 \in A\}$. It is also obvious that $|\nu_2| = \tau$. Note that $U_{\alpha_1\alpha_2} \cap U_{\alpha_1'\alpha_2'} = \Lambda$ if $\alpha_1 \neq \alpha_1'$ or if $\alpha_2 \neq \alpha_2'$. Moreover, ν_2 is a disjoint open cover of X. From the construction it also follows that $U_{\alpha_1} = \bigcup\{U_{\alpha_1\alpha_2}: \alpha_2 \in A\}$.

Now using the fact that $w(U_{\alpha_1\alpha_2}) = \tau$ and $U_{\alpha_1\alpha_2}$ is not a compactum for every subspace $U_{\alpha_1\alpha_2} \in \nu_2$, we carry out the same construction for each $U_{\alpha_1\alpha_2} \in \nu_2$, and so on. As a result, we obtain the family $\nu = \{U_{\alpha_1\alpha_2\ldots\alpha_k}: \alpha_1 \in A, \ldots, \alpha_k \in A;\ k \in N^+\}$ of clopen sets of X, for which the following conditions are satisfied:

(i) $\nu_k = \{U_{\alpha_1\alpha_2\ldots\alpha_k}: \alpha_1 \in A, \ldots, \alpha_k \in A\}$ is a disjoint open cover of X for each fixed $k \in N^+$;

(ii) $U_{\alpha_1\alpha_2\ldots\alpha_k} \cap U_{\alpha_1'\alpha_2'\ldots\alpha_k'} = \Lambda$ if $\alpha_i \neq \alpha_i'$ for some $i \leq k$;

(iii) $U_{\alpha_1\alpha_2\ldots\alpha_{k-1}} = \bigcup\{U_{\alpha_1\alpha_2\ldots\alpha_{k-1}\alpha_k}: \alpha_k \in A\}$;

(iv) diam $U_{\alpha_1\alpha_2\ldots\alpha_k} < 1/k$ for every $\alpha_1 \in A, \ldots, \alpha_k \in A$.

The conditions (i), (ii) and (iii) imply that ν is a disjoint A_τ-sequence of open covers of X. From the completeness of the metric space X and from condition (iv) it follows that ν is complete and refined. Thus we arrive at the conditions of the preceding problem, yielding the desired conclusion.

150. Let A be a set of cardinality $|A| = \tau$. Consider an open cover ω_1 of X such that diam $U < 1$ for every $U \in \omega_1$. Clearly, $|\omega_1| \leq \tau$. Fix some mapping g_1 of A onto ω_1. Put $U_\alpha = g_1\alpha$, where $\alpha \in A$. Thus, $\omega_1 = \{U_\alpha: \alpha \in A\}$, where it is possible to have $U_\alpha = U_{\alpha'}$ even when $\alpha \neq \alpha'$. For each $\alpha \in A$ consider an open cover θ_α of U_α such that diam $U < \frac{1}{2}$ for every $U \in \theta_\alpha$ and $\alpha \in A$. Fix for each α a mapping g_2^α of A onto θ_α. Put $U_{\alpha_1\alpha_2} = g_2^{\alpha_1}(\alpha_2)$, where $\alpha_1, \alpha_2 \in A$. Then $\theta_{\alpha_1} = \{U_{\alpha_1\alpha_2}: \alpha_2 \in A\}$ and $U_{\alpha_1} = \bigcup\{U_{\alpha_1\alpha_2}: \alpha_2 \in A\}$. Set $\nu_2 = \bigcup\{\theta_{\alpha_1}: \alpha_1 \in A\}$. Then $\nu_2 = \{U_{\alpha_1\alpha_2}: \alpha_1, \alpha_2 \in A\}$. We carry out the same construction for each $U_{\alpha_1\alpha_2}$, $\alpha_1, \alpha_2 \in A$, and so on. As a result, we obtain the family $\nu = \{U_{\alpha_1\alpha_2\ldots\alpha_k}: \alpha_1 \in A, \ldots, \alpha_k \in A, k \in N^+\}$. It is easy to see that ν is an A_τ-sequence of open covers of X. Since diam $U_{\alpha_1\alpha_2\ldots\alpha_k} < 1/k$ for every $\alpha_1 \in A, \alpha_2 \in A, \ldots, \alpha_k \in A$, and (X, ρ) is a complete metric space, the A_τ-sequence ν is complete and refined. Thus we arrive at the conditions of Problem 147, giving the required conclusion.

151. The solution follows from the preceding problem, Problem 144 and the completeness of the Baire spaces (see Problems 246 and 247 of Chap. 2).

152. In consequence of Problem 150, X is an image of the Baire space $B(\tau)$ under some open continuous mapping f. Since X is a paracompactum (as is every metric space) and $B(\tau)$ is a complete metric space (hence Čech-complete), in view of Problem 141 there is a closed set F in $B(\tau)$ such that $f(F) = X$ and the restriction $f_0 = f \mid F$ is a perfect mapping of F onto X. Recourse may now be had to Problem 78, which implies the existence of a closed continuous mapping g of the Baire space $B(\tau)$ onto its closed subspace F. The composition $f_0 \cdot g$ is a closed continuous mapping of $B(\tau)$ onto X. Another solution of this problem can be obtained from Problem 252 and, once again, Problem 78.

154. A regular extremally disconnected space obviously has a base of clopen sets. For the rest, see Problem 14 of Chap. 3.

155. (*a*) Suppose that X is extremally disconnected and $U_1, U_2 \subset X$ are two open sets in X such that $U_1 \cap U_2 = \Lambda$. Then (see Problem 37 of Chap. 2), $[U_1] \cap U_2 = \Lambda$. Since $[U_1]$ is open in X (because X is extremally disconnected), hence even $[U_1] \cap [U_2] = \Lambda$ (again see Problem 37 of Chap. 2).

(*b*) Suppose that any pair of disjoint open sets in X have disjoint closures. Let U be an arbitrary open set in X. Then $V = X \diagdown [U]$ is also open in X. In addition, $X = [U] \cup [V]$ and $U \cap V = \Lambda$. By hypothesis, $[U] \cap [V] = \Lambda$; consequently, $[U]$ and $[V]$ are clopen in X. The proof is complete.

157. Suppose that U_1 and U_2 are two disjoint open sets in X_0. Take open sets V_1 and V_2 in X for which $U_1 = V_1 \cap X_0$, $U_2 = V_2 \cap X_0$. Since X_0 is dense in X, $V_1 \cap V_2 = \Lambda$ and since X is extremally disconnected, we obtain (see Problem 155) $[V_1]_X \cap [V_2]_X = \Lambda$. We have $([V_1]_X \cap X_0) \cap ([V_2]_X \cap X_0) = \Lambda$, $[U_1]_{X_0} \subset [V_1]_X \cap X_0$, $[U_2]_{X_0} \subset [V_2]_X \cap X_0$. Consequently, $[U_1]_{X_0} \cap [U_2]_{X_0} = \Lambda$. Hence X_0 is extremally disconnected (see Problem 155).

158. Use the characteristic properties of the Stone-Čech compactification βX of a given space X, set forth in Problem 15 of Chap. 4. Suppose that Γ_1 and Γ_2 are two disjoint open sets in βX. Consider the open sets $U_1 = \Gamma_1 \cap X$ and $U_2 = \Gamma_2 \cap X$ in X. Obviously, $U_1 \cap U_2 = \Lambda$ but, since X is extremally disconnected, we get $[U_1]_X \cap [U_2]_X = \Lambda$. Since, in addition, $[U_1]_X$ and $[U_2]_X$ are clopen in X, they are functionally separable (see p. 120). Hence (see Problem 15 of Chap. 4), $[[U_1]_X]_{\beta X} \cap [[U_2]_X]_{\beta X} = \Lambda$. But, furthermore, since X is dense in βX, we have $[\Gamma_1]_{\beta X} = [\Gamma_1 \cap X]_{\beta X} = [[U_1]_X]_{\beta X}$, $[\Gamma_2]_{\beta X} = [\Gamma_2 \cap X]_{\beta X} = [[U_2]_X]_{\beta X}$. In view of all this we also obtain $[\Gamma_1]_{\beta X} \cap [\Gamma_2]_{\beta X} = \Lambda$. Thus βX is extremally disconnected (see Problem 155).

159. The space D, as a discrete space, is extremally disconnected. Then its Stone-Čech compactification is also extremally disconnected (see the preceding problem).

160. The extremal disconnectedness of X is a consequence of the fact that $f: X \to Y$ is a homeomorphism. Clearly, it suffices to show that f is one-to-one, because it is already known that f is continuous and closed. Thus, let $x_1, x_2 \in X$, $x_1 \neq x_2$. Consider nhoods Ux_1 and Ux_2 of x_1 and x_2 (X is a Hausdorff space), satisfying $Ux_1 \cap Ux_2 = \Lambda$. Then, obviously, $f^{\#}Ux_1 \cap f^{\#}Ux_2 = \Lambda$, where $f^{\#}Ux_i = \{y \in Y: f^{-1}y \subset Ux_i\} = Y \diagdown f(X \diagdown Ux_i)$, $i = 1, 2$. From Problem 109 it follows that these sets are open and nonempty in Y, because f is closed and irreducible. Then, since Y is extremally disconnected (by Problem 155), we obtain

$$[f^{\#}Ux_1] \cap [f^{\#}Ux_2] = \Lambda. \tag{*}$$

Also (see Problem 110), $fx_1 \in f[Ux_1] = [f^{\#}Ux_1]$, $fx_2 \in f[Ux_2] = [f^{\#}Ux_2]$. Taking note of the equality (*) we obtain $fx_1 \neq fx_2$, completing the proof.

161. There is an open set $\tilde{P}$ in X such that $\tilde{P} \cap Y = P$. P is closed in X (because Y is) and $\tilde{P} \supset P$. Since X is normal, there is an open set U in X for which $\tilde{P} \supset [U] \supset$

$U \supset P$. Put $G = [U]$. Thus G is open in X (because X is extremally disconnected), $G \supset P$ and $G \cap Y \subset \tilde{P} \cap Y \subset P$. Hence $G \cap Y = P$ and G is the desired clopen set.

162. X is a compactification bX_0 of X_0, because X_0 is dense in X. Then there is a natural continuous mapping h: $\beta X_0 \to bX_0$ (see Problem 22 of Chap. 4) of the Stone-Čech compactification βX_0 of X_0 onto the extremally disconnected compactum $bX_0 = X$. Obviously, h is an irreducible mapping because $hx = x$ and $h^{-1}hx = x$ for every $x \in X_0$ (see Problem 27 of Chap. 4). By Problem 160, h is then a homeomorphism; consequently, $\beta X_0 \equiv X$, completing the proof.

163. Since $D_1 \cap [D_2] = \Lambda$, $[D_1] \cap D_2 = \Lambda$ and X is regular, we can use Problem 19 of Chap. 3 to obtain open sets U_1 and U_2 in X such that $D_1 \subset U_1$, $D_2 \subset U_2$ and $U_1 \cap U_2 = \Lambda$. Then, the extremal disconnectedness of X implies that $[U_1] \cap [U_2] = \Lambda$. Consequently, $[D_1] \cap [D_2] = \Lambda$.

164. Since $|D_1| \leqslant \aleph_0$, $|D_2| \leqslant \aleph_0$, D_1 and D_2 are Lindelöf subspaces. Hence we arrive at the preceding problem.

165. It is clear that $[D]_X$ is a compactification bD of D. To show that $bD \equiv \beta D$, it suffices (see Problem 16 of Chap. 4) to show (by the normality of subspace D) that for every two disjoint closed subsets D_1 and D_2 in D, their closures are also disjoint in $bD = [D]_X$.

Thus suppose that $D_1 \subset D$, $D_2 \subset D$ are closed subsets of D (incidentally, every subset of D is clopen, because D is a discrete subspace) such that $D_1 \cap D_2 = \Lambda$. Obviously, D_1 and D_2 are discrete subspaces of the compactum X and $[D_1]_X \cap D_2 = \Lambda$, $[D_2]_X \cap D_1 = \Lambda$. Hence, by using the preceding problem we conclude that $[D_1]_X \cap [D_2]_X = \Lambda$. Moreover, obviously, $[D_1]_X \subset [D]_X = bD$ and $[D_2]_X \subset [D]_X = bD$. Hence $[D_1]_{bD} \cap [D_2]_{bD} = \Lambda$. Therefore, $[D]_X = bD \equiv \beta D \equiv \beta N$.

166. The preceding problem also implies the fact that an infinite regular space has an infinite countable discrete subspace.

167. This is direct from Problem 165.

168. The requirement here is to consider the Stone-Čech compactification βX of our extremally disconnected completely regular space X; for this use Problem 158 and then the preceding problem.

169. From condition (k) it follows, in particular, that D is discrete and, consequently, normal. Hence, if we can show that $D_1 \subset D$, $D_2 \subset D$ and $D_1 \cap D_2 = \Lambda$ imply $[D_1] \cap [D_2] = \Lambda$, then by Problem 16 of Chap. 4 it will follow that $[D]$ coincides with βD.

Put $\Gamma_1 = \bigcup\{Ud\colon d \in D_1\}$ and $\Gamma_2 = \bigcup\{Ud\colon d \in D_2\}$. Since $\gamma = \{Ud\colon d \in D\}$ is disjoint and $D_1 \cap D_2 = \Lambda$, $\Gamma_1 \cap \Gamma_2 = \Lambda$. Clearly, Γ_1 and Γ_2 are open in X. Hence, since X is extremally disconnected (see Problem 155), surely $[\Gamma_1] \cap [\Gamma_2] = \Lambda$. Consequently, $[D_1] \cap [D_2] = \Lambda$, because $[D_1] \subset [\Gamma_1]$, $[D_2] \subset [\Gamma_2]$.

170. The first part is obvious. The Cantor perfect set is an appropriate example (see Problem 185 of Chap. 2), because in an extremally disconnected compactum there is no non-trivial convergent sequence (Problem 167), but in the Cantor perfect set there is such a sequence.

171. This statement is a corollary to Problem 166 of the present chapter and Problem 53 of Chap. 4.

172. The answer is negative. Consider the Stone-Čech compactification βN of the discrete space of natural numbers. It is extremally disconnected (see Problem 159). However, $\beta N \setminus N$ is closed in βN but is not extremally disconnected; see Problem 178.

173. (*a*) If X is an extremally disconnected compactum and $X_0 \subset X$ is a dense subspace of X, then βX_0 coincides with X (see Problem 162).

(*b*) Suppose that the compactum X is the Stone-Čech compactification of each of its dense open subspaces. Let $U \subset X$ be an arbitrary open set in X. Consider the open set $V = X \setminus [U]$ in X. It is clear that $U \cap V = \Lambda$. Put $W = U \cup V$. From the definition of V it follows that W is open and dense in X. Also U and V are clopen in W, and since furthermore they are disjoint, they are functionally separable in W. Hence, since $\beta W \equiv X$, we have $[U]_X \cap [V]_X = \Lambda$ (see Problem 15 of Chap. 4). Therefore, from the obvious equality $X = [W]_X = [U \cup V]_X = [U]_X \cup [V]_X$ it also follows that $[U]_X$ is open, so that X is extremally disconnected.

174. Assume the contrary. Suppose that there is a compactum Y and a one-to-one continuous mapping $f: X_0 \to Y$. Since no point $d \in D$ is isolated in X, the open set $X \setminus \{d\}$ in X is dense in this compactum for each $d \in D$. Hence $X_0 = \bigcap \{(X \setminus \{d\}): d \in D\}$ is dense in X, as the intersection of a countable family of open dense sets in the compactum X (see Problem 88 of Chap. 3). Therefore, since X is extremally disconnected, from Problem 162 it follows that βX_0 coincides with X. On the other hand, since $\beta X_0 \equiv X$, there is an extension of the mapping $f: X_0 \to Y$ to a continuous mapping $\bar{f}: X \to Y$ (see Problem 31 of Chap. 4). Since X and Y are compacta and $\bar{f}$ is continuous, $\bar{f}$ is also perfect. Then, since D is discrete, the subspace $\bar{f}D \subset Y$ is also discrete. Furthermore, obviously, $|\bar{f}D| \leqslant \aleph_0$ because $|D| = \aleph_0$. For each point $p \in \bar{f}D$ fix a nhood Vp of p in Y in such a way that the family $\theta = \{Vp: p \in \bar{f}D\}$ is disjoint (this can be done since $\bar{f}D$ is a countable discrete subspace of the normal space Y; see Problem 22 of Chap. 3). The family $\bar{f}^{-1}\theta = \{\bar{f}^{-1}Vp: p \in \bar{f}D\}$ is also disjoint and consists of open sets in X.

Consider also the sets $A = \bar{f}^{-1}(\bar{f}(D))$ and $E = A \cap X_0$. Obviously, $|E| \leqslant \aleph_0$. E is discrete. Using the definitions of E and the family $\bar{f}^{-1}\theta$ and the extremal disconnectedness of X, it is routine to show that $[D] \cap [E] = \Lambda$. On the other hand, $\bar{f}[D] = [\bar{f}D], \bar{f}(E) = [\bar{f}E], \bar{f}D = \bar{f}E = fE$. Hence, there is a point $y_0 \in Y$, for which $\bar{f}^{-1}y_0 = f^{-1}y_0 = x_0$, $x_0 \in X$ and $x_0 \in [E]$, $x_0 \in [D]$, i.e., $[E] \cap [D] \neq \Lambda$. This is a contradiction.

175. Every infinite compactum X contains a countably infinite discrete subspace D. Enumerate D as a sequence: $D = \{d_i: i \in N^+\}$. For each point $d_i \in D$ fix a nhood U_i in X in such a way that the family $\gamma = \{U_i: i \in N^+\}$ is disjoint. Put $U_0 = X \setminus [\bigcup \{U_i: i \in N^+\}]$, $\gamma^* = \{U_i: i \in N\}$ and $W = \bigcup \{U_i: i \in N\}$ (we may assume that $U_0 \neq \Lambda$). Define a mapping f of the subspace W onto N as follows: put $fx = i$, where $i \in N$, if $x \in U_i$. It is obvious that $f: W \to N$ is continuous, where N is the space of natural numbers with the discrete topology. Furthermore, it is clear that W is dense in X. Since X is an extremally disconnected compactum, βW coincides with X (see Problem 162). Then there is an extension of f to a continuous mapping $\bar{f}: X \to \beta N$, $\bar{f}(X) = \beta N$.

176. The infinite extremally disconnected compactum X contains a topological image of βN (see Problem 166), but βN is not hereditarily normal (see Problem 56 of Chap. 4). Hence X is also not hereditartly normal.

177. As we know (see Problem 335 of Chap. 2), $f: X \to Y$ is an open mapping iff $f^{-1}[B] \subset [f^{-1}B]$ for every subset $B \subset Y$. Let $V \subset Y$ be an arbitrary open set in Y. $f^{-1}V$

is open in X, but $[f^{-1}V]$ is clopen in X (because X is extremally disconnected). Then $f^{-1}[V] \subset [f^{-1}V]$. Hence $ff^{-1}[V] = [V] \subset f[f^{-1}V] \subset [ff^{-1}V]$ (here we have again used the continuity of f). Since f is an open mapping and $[f^{-1}V]$ is an open set in X, $f[f^{-1}V]$ is open in Y. Then the inclusion $[V] \subset f[f^{-1}V] \subset [V]$ implies that $[V]$ coincides with the open set $f[f^{-1}V]$ in Y. Hence Y is extremally disconnected.

178. In the space X there is a family γ of cardinality $2^{\aleph_0}$ of nonempty mutually disjoint open (in X) sets (see Problem 52 of Chap. 4). Fix a point $x(U) \in U$ for each $U \in \gamma$. Put $D = \{x(U)\colon U \in \gamma\}$. Obviously, D is a discrete subspace. Suppose now that X is extremally disconnected (note also that X is a compactum). This takes us back to Problem 169, so that $[D]_X$ coincides with βD. Hence (see Problem 180) $|[D]_X| = |\beta D| = 2^{2^{|D|}} = 2^{2^{2^{\aleph_0}}}$. Since $|\beta N| = 2^{2^{\aleph_0}}$, $|[D]_X| > |\beta N|$, contradicting the inclusion $[D]_X \subset X \subset \beta N$.

179. The answer is negative.

180. Solve this on similar lines as Problem 53 of Chap. 4. It is non-trivial to show that $|\beta D| \geqslant 2^{2^\tau}$. For this, it is necessary to use the fact that the Tychonoff product of 2^τ copies of discrete double-points has a dense set of cardinality τ (see Problem 381 of Chap. 2). Hence the conclusion that βD is continuously mapped onto this product.

181. (*a*) Suppose that φ is any mapping of γ onto itself with no fixed point. Put $V(U) = U \cup (f^{-1}(\varphi U) \setminus Q)$, where $Q = \bigcup\{U\colon U \in \gamma\}$, $U \in \gamma$. Then $V(U) \supset U$, $P = \bigcup\{V(U)\colon U \in \gamma\} \supset Q$ and $f(V(U)) \subset fU \cup \varphi U \subset Q \subset P$. If $U \neq U'$, then $\varphi U \neq \varphi U'$, and hence $\varphi U \cap \varphi U' = \Lambda$, $f^{-1}(\varphi U) \cap f^{-1}(\varphi U') = \Lambda$. Hence, in view of $U \cap U' = \Lambda$, we deduce easily that $V(U) \cap V(U') = \Lambda$. The continuity of f implies that $V(U)$ is clopen (we have already noted that all the elements of γ are clopen). Obviously, $|\{V(U)\colon U \in \gamma\}| = |\gamma| = k$. In order to infer that $\lambda = \{V(U)\colon U \in \gamma\}$ is an (f, k)-family, it remains to verify that λ satisfies condition (i). We have $f(V(U)) \cap V(U) \subset (fU \cup \varphi U) \cap V(U) \subset (Q \setminus U) \cap (U \cup (X \setminus Q)) = \Lambda$. The relation $V(U) \supset U$ implies that γ is a refinement of $\lambda = \{V(U)\colon U \in \gamma\}$. Furthermore, $f^{-1}Q = \bigcup\{f^{-1}U\colon U \in \gamma\} = \bigcup\{f^{-1}(\varphi U)\colon U \in \gamma\} = \bigcup\{U \cup (f^{-1}\varphi(U) \setminus Q)\colon U \in \gamma\} = \bigcup\{V(U)\colon U \in \gamma\} = P$. Statement (*a*) is proved.

(*b*) Put $\gamma_1 = \gamma$ and assuming that the (f, k)-family γ_l is defined, take as γ_{l+1} any (f, k)-family such that $f^{-1}\langle \gamma_l \rangle = \langle \gamma_{l+1} \rangle$ and γ_l is a refinement of γ_{l+1} (from (*a*) it follows that such a γ_{l+1} exists). Consider the sequence $\{\gamma_n\colon n \in N^+\}$. Since, for each $n \in N^+$, γ_n is an (f, k)-family and refines γ_{n+1}, the elements of γ_n can be indexed as: $\gamma_n = \{U_i^n\colon i = 1, 2, \dots, k\}$, so that $U_i^{n_1} \subset U_i^{n_2}$ for $n_1 < n_2$ and every i and, in addition, $U_{i_1}^{n_1} \cap U_{i_2}^{n_2} = \Lambda$ for $i_1 \neq i_2$ and all n_1, n_2. Put $W_i = \bigcup\{U_i^n\colon n \in N^+\}$, $i = 1, 2, \dots, k$. Obviously, $\{W_i\colon i = 1, 2, \dots, k\}$ is not necessarily, in general, an (f, k)-family because W_i may not be closed. However, each W_i is open and $W_{i'} \cap W_{i''} = \Lambda$ if $i' \neq i''$. Put $V_i = [W_i]$, $i = 1, 2, \dots, k$ and consider the family $\mu = \{V_i\colon i = 1, 2, \dots, k\}$. Since X is extremally disconnected, every V_i, $i = 1, 2, \dots, k$ is a clopen set. Also, by extremal disconnectedness (see Problem 155), $V_{i'} \cap V_{i''} = \Lambda$ for $i' \neq i''$ since $W_{i'} \cap W_{i''} = \Lambda$. From the equality $fU_i^n \cap U_i^n = \Lambda$ it is easily inferred that $W_i \cap fW_i = \Lambda$. Since f is continuous, $fV_i = f[W_i] \subset \bigcup\{[fW_j]\colon j \neq i, j = 1, 2, \dots, k\}$. Hence μ is an (f, k)-family. Obviously, γ is a refinement of μ. To verify that $f^{-1}\langle \mu \rangle = \langle \mu \rangle$ put $G = \bigcup\{W_i\colon i = 1, 2, \dots, k\}$. We have $G = \bigcup\{\langle \gamma_n \rangle\colon n \in N^+\}$, $\langle \gamma_n \rangle \subset \langle \gamma_{n+1} \rangle$ and

$f^{-1}\langle\gamma_n\rangle = \langle\gamma_{n+1}\rangle$. Hence $f^{-1}G = G$. Clearly, Moreover, $\langle\mu\rangle = [G]$. Hence, by the continuity of f we obtain the inclusion $f^{-1}[G] \supset [G]$. To prove the converse inclusion, suppose that $y \in X$ and $f^{-1}y \notin [G]$. Then, since f is closed, for some nhood Oy of y we have $f^{-1}Oy \subset X \setminus [G]$ (see Problem 325 of Chap. 2). Consequently, $f^{-1}Oy \cap f^{-1}G = f^{-1}Oy \cap G = \Lambda$. Hence $Oy \cap G = \Lambda$ and $y \in [G]$. Therefore, $f^{-1}[G] \subset [G]$. Thus, finally, we have the equality $f^{-1}[G] = [G]$. Statement (*b*) is therefore proved.

(*c*) We have $\mathcal{S} = \{\langle\gamma_\alpha\rangle : \alpha \in A\}$, where $\langle\gamma_{\alpha'}\rangle \cap \langle\gamma_{\alpha''}\rangle = \Lambda$ for $\alpha' \neq \alpha''$ and $\gamma_\alpha = \{U_i^\alpha : i = 1, 2, \ldots, k)$ is an (f, k)-family. Put $W_i = \bigcup\{U_i^\alpha : \alpha \in A\}, i = 1, 2, \ldots, k$. Then $fW_i \cap W_i = \Lambda, i = 1, 2, \ldots, k$ and $f(\bigcup\{W_i : i = 1, 2, \ldots, k\}) \subset \bigcup\{W_i : i = 1, \ldots, k\}$. In exactly the same way as in the proof of (*b*) we show that $\mu = \{V_i : i = 1, 2, \ldots, k\}$, where $V_i = [W_i]$ is an (f, k)-family.

182. (i) Fix $x^* \in X$ to satisfy $y^* = f(x^*) \neq x^*$. Put $A = f(X)$. By hypothesis, A is closed in X. Select disjoint clopen sets U and V in X, satisfying $U \ni x^*$, $V \ni y^*$ and $fU \subset V$. fU is clopen in A; hence (Problem 161) there is a clopen set W in X such that $W \cap A = fU$. Also, we may suppose that $W \cap U = \Lambda$, replacing in the contrary case W by $W' = (X \setminus U) \cap W$. Now put $U_0 = U$, $U_1 = W$ and assume that for some integer $k \geqslant 1$ the disjoint clopen sets U_i, $i = 1, 2, \ldots, k$, in X are already defined to satisfy $f(U_{i-1}) \setminus U_0 = A \cap U_i$. Then $U_k \cap A = f(U_{k-1}) \setminus U_0$ and $fU_k \subset A$. Hence $U_i \cap fU_k \subset U_i \cap A \subset fU_{i-1}$ and $U_i \cap fU_k \subset fU_{i-1} \cap fU_k = \Lambda$ for all $i = 1, \ldots, k$ (we have already noted that $U_{i-1} \cap U_k = \Lambda$). Thus, fU_k is clopen in X and $fU_k \subset X \setminus \bigcup\{U_i : i = 1, \ldots, k\}$. Take as U_{k+1} any clopen set in X, satisfying $U_{k+1} \cap A = fU_k \setminus U_0$ and $U_{k+1} \subset X \setminus \bigcup\{U_i : i = 0, 1, \ldots, k\}$—such a set exists by virtue of Problem 161. Continuing in this manner we obtain an infinite sequence $U_0, U_1, \ldots, U_k, \ldots$ of disjoint clopen sets in X such that $fU_i \subset U_0 \cup U_{i+1}$ for all $i = 0, 1, \ldots, k, \ldots$ Some of the sets U_i may be empty. Put $G_0 = U_0$, $G_1 = \bigcup\{U_{2i+1} : i \in N\}$, $G_2 = \bigcup\{U_{2i} : i \in N^+\}$. Then, obviously, $f(G_0 \cup G_1 \cup G_2) = f(\bigcup\{U_i : i \in N\}) \subset \bigcup\{U_i : i \in N\} = G_0 \cup G_1 \cup G_2$ and $fG_1 \subset G_0 \cup G_2$, $fG_0 \subset G_1$, $fG_2 \subset G_0 \cup G_1$, i.e., $fG_0 \cap G_0 = \Lambda$, $fG_1 \cap G_1 = \Lambda$, $fG_2 \cap G_2 = \Lambda$, because G_0, G_1 and G_2 are disjoint. From the extremal disconnectedness of X it obviously follows that the sets $[G_0]$, $[G_1]$ and $[G_2]$ are also disjoint (see Problem 155). It is easy to see that $\gamma = \{[G_0], [G_1], [G_2]\}$ forms an $(f, 3)$-family. However, the set $G = [G_0] \cup [G_1] \cup [G_2]$ is nonempty because the points x^* and y^* a fortiori belong to it, and is $(f, 3)$-decomposable.

(ii) Fix in X any nonempty $(f, 3)$-decomposable set F. By Zorn's lemma (p. 6), the family $\{F\}$ can be extended to a maximal disjoint family γ of $(f, 3)$-decomposable sets. By (*c*) of the preceding problem we conclude that $\Phi = [\bigcup\{P \in \gamma\}]$, an $(f, 3)$-decomposable set, is nonempty. From (*b*) of the preceding problem it follows that there is an $(f, 3)$-decomposable set A, containing Φ, for which $f^{-1}A = A$.

Condition (II') is satisfied. If $A' \cap A = \Lambda$, $A' \neq \Lambda$ and A' is $(f, 3)$-separable, then $A' \cap \Phi = \Lambda$ and $\tilde{\gamma} = \{A'\} \cup \gamma$ is a disjoint family of $(f, 3)$-decomposable sets, strictly containing γ, which contradicts the maximality of γ. Hence condition (II'') is also satisfied.

(iii) Consider a set A satisfying the requirement (ii). It is clopen, as is every (f, k)-decomposable set. From $f^{-1}A = A$ it follows that $fB \subset B$, where $B = X \setminus A$. Consider the restriction f^* of the homeomorphism f to the clopen subspace B of X. Obviously, f^*B is closed in B. If f^* is not the identity, then it is possible, by applying

(i) of this problem to the homeomorphism $f^*: B \to B$, to find in B a nonempty $(f^*, 3)$-decomposable set C. Clearly, C is also $(f, 3)$-decomposable. On the other hand, $C \cap A = \Lambda$, because $C \subset B$ and $B = X \setminus A$. This leads to a contradiction with statement (II''). Thus $f^*x = x = fx$ for every point $x \in B$. The proof is complete.

183. This assertion follows directly from the definitions and statement (iii) of the preceding problem.

184. If the compactum X consists of three elements 1, 2 and 3 and the mapping f is such that $f(1) = 2$, $f(2) = 3$ and $f(3) = 1$, then it is easy to show that there is no nonempty $(f, 2)$-decomposabls set in X.

185. By Problem 183, the set U of all fixed points of the homeomorphism f is clopen in X. Clearly, $U \subset f(X)$. Since $f(X)$ is nowhere dense in X, we have $U = \Lambda$.

186. Fix some homeomorphism φ of X onto Y. If $F \supset Y$, then Y is also nowhere dense in X. Hence by the preceding problem φ has no fixed point. Pick a point $x \in F$ and consider $y = \varphi x$. Since, obviously, $\varphi(F) \subset Y \subset F$, $y \in Y \subset F$. Also, $x \neq y$ because φ has no fixed point. Now assume that there is a homeomorphism $f: F \to F$ such that $fy = x$. Consider the homeomorphism $g = f\varphi$, the composition of homeomorphisms φ and f. Clearly, g maps X into Y. We have: $g(x) = f(\varphi(x)) = fy = x$, i.e., the homeomorphism $g: X \to X$, where $g(X)$ is nowhere dense in X, has a fixed point (i.e., x), which contradicts the preceding problem. Thus, F is non-homogeneous.

187. Select a countably infinite discrete subspace $D \subset X$. But $X \subset \beta N$ is extremally disconnected (Problem 159); consequently (by Problem 165), $[D]_{\beta N} \equiv \beta D$, where βD is the Stone-Čech compactification of D. Clearly, βD is homeomorphic to βN, $\beta D \subset X$, and X is nowhere dense in βN. Hence we arrive at the hypothesis of the preceding problem, from which it follows that $X = \beta N \setminus N$ is non-homogeneous. Another solution rests on Problems 144 and 146 of Chap. 1.

188. Consider the set $P_\alpha(x_0) = \bigcup\{A \in \alpha: x_0 \notin A\}$. Since every $A \in \alpha$ is a closed set in X, and α is an open finite (or locally finite) cover, $P_\alpha(x_0)$ is closed in X. Then $x_0 \in X \setminus P_\alpha(x_0) \subset \alpha(x_0)$, so that x_0 is an internal point of $\alpha(x_0)$, i.e., $x_0 \in \operatorname{Int} \alpha(x_0)$.

189. (*a*) Suppose that for the element $A^{\alpha_2}_{i(\alpha_2)} \in \alpha_2$ there are two distinct elements $A^{\alpha_1}_{j_1} \in \alpha_1$, $A^{\alpha_1}_{j_2} \in \alpha_1$, for which $A^{\alpha_2}_{i(\alpha_2)} \subset A^{\alpha_1}_{j_1}$, $A^{\alpha_2}_{i(\alpha_2)} \subset A^{\alpha_1}_{j_2}$. Then, $\operatorname{Int} A^{\alpha_2}_{i(\alpha_2)} \subset \operatorname{Int} A^{\alpha_1}_{j_1}$, $\operatorname{Int} A^{\alpha_2}_{i(\alpha_2)} \subset \operatorname{Int} A^{\alpha_1}_{j_2}$; consequently, $\operatorname{Int} A^{\alpha_1}_{j_1} \cap \operatorname{Int} A^{\alpha_1}_{j_2} \neq \Lambda$, which contradicts that the cover (α_1) is regular.

(*b*) Pick an arbitrary element $A^{\alpha_1}_i \in \alpha_1$. Take a point $x_0 \in \operatorname{Int} A^{\alpha_1}_i$ and an element $A^{\alpha_2}_j$ for which $x_0 \in A^{\alpha_2}_j$. We first show that $A^{\alpha_2}_j \subset A^{\alpha_1}_i$. In the first place, $\operatorname{Int} A^{\alpha_1}_i \cap A^{\alpha_2}_j \neq \Lambda$. Consequently, $\operatorname{Int} A^{\alpha_1}_i \cap \operatorname{Int} A^{\alpha_2}_j \neq \Lambda$. Since $\alpha_2 > \alpha_1$, we have an element $A^{\alpha_1}_{i'} \in \alpha_1$ for which $A^{\alpha_2}_j \subset A^{\alpha_1}_{i'}$. Hence $\operatorname{Int} A^{\alpha_1}_i \cap \operatorname{Int} A^{\alpha_2}_j \neq \Lambda$ implies that $\operatorname{Int} A^{\alpha_1}_i \cap \operatorname{Int} A^{\alpha_1}_{i'} \neq \Lambda$ because $\operatorname{Int} A^{\alpha_2}_j \subset \operatorname{Int} A^{\alpha_1}_{i'}$. Consequently, since α_1 is a regular cover, $A^{\alpha_1}_{i'} = A^{\alpha_1}_i$. Thus, if $A^{\alpha_2}_j \in \alpha_2$ is such that $A^{\alpha_2}_j \cap \operatorname{Int} A^{\alpha_1}_i \neq \Lambda$, then $A^{\alpha_1}_i \subset A^{\alpha_2}_j$. This implies that $\operatorname{Int} A^{\alpha_1}_i \subset \bigcup\{A \in \alpha_2: A \cap \operatorname{Int} A^{\alpha_1}_i \neq \Lambda\}$ which, since the set $\bigcup\{A \in \alpha_2: A \cap \operatorname{Int} A^{\alpha_1}_i \neq \Lambda\}$ is closed, yields $A^{\alpha_1}_i = [\operatorname{Int} A^{\alpha_1}_i] \subset \bigcup\{A \in \alpha_2: A \cap \operatorname{Int} A^{\alpha_1}_i \neq \Lambda\}$. Since $A \in \alpha_2$ and $A \cap \operatorname{Int} A^{\alpha_1}_i \neq \Lambda$ implies $A \subset A^{\alpha_1}_i$, $A^{\alpha_1}_i = \bigcup\{A \in \alpha_2: A \cap \operatorname{Int} A^{\alpha_1}_i \neq \Lambda\}$. It remains to note that $A \in \alpha_2$, $A \subset A^{\alpha_1}_i$ implies that

$A \cap \operatorname{Int} A_i^{\alpha_1} \neq \Delta$, which enables us to write $A_i^{\alpha_1} = \bigcup \{A \in \alpha_2: A \subset A_i^{\alpha_1}\}$.

190. Put $\alpha^0 = \{\operatorname{Int} A: A \in \alpha\}$, $\beta^0 = \{\operatorname{Int} B: B \in \beta\}$. Clearly, α^0 and β^0 consist of disjoint open sets. Moreover, the sets $\tilde{\alpha}^0 = \bigcup \{\operatorname{Int} A : A \in \alpha\}$ and $\tilde{\beta}^0 = \bigcup \{\operatorname{Int} B: B \in \beta\}$ are dense in X. Consider the family $\gamma^0 = \{(\operatorname{Int} A \cap \operatorname{Int} B): A \in \alpha, B \in \beta, \operatorname{Int} A \cap \operatorname{Int} B \neq \Delta\}$. It is obvious that γ^0 consists of disjoint open sets, $\tilde{\gamma}^0 = \bigcup \{(\operatorname{Int} A \cap \operatorname{Int} B): A \in \alpha, B \in \beta\}$ being dense in X. Then $\gamma = \{[U]: U \in \gamma^0\}$ is a regular cover of X, and $\gamma = \alpha \wedge \beta$ (see the definition on p. 328). If $A \in \alpha$, $B \in \beta$, then obviously $[\operatorname{Int} A \cap \operatorname{Int} B] \subset [\operatorname{Int} A] = A$ and $[\operatorname{Int} A \cap \operatorname{Int} B] \subset [\operatorname{Int} B] = B$. Hence the regular cover $\gamma = \alpha \wedge \beta$ is a refinement of both α and β, i.e., $\gamma > \alpha$, $\gamma > \beta$.

191. The partial order axioms are easily verifiable. The directedness of the family $\mathfrak{U}(X)$ follows from the preceding problem.

192. (*a*) Suppose that X is regular. Let $x_0 \in X$ and let Ux_0 be any nhood of x_0. Since X is regular, we can select a nhood $U'x_0$ of x_0, for which $[U'x_0] \subset Ux_0$. Now consider the regular cover α_0, consisting of the two regular closed sets $[U'x_0]$ and $[X \setminus [U'x_0]]$. Obviously, the star of x_0 with respect to the cover α_0 coincides with $[U'x_0]$ and is, consequently, contained in Ux_0, so that the family $\mathfrak{U}(X)$ is refined.

(*b*) Conversely, suppose $\mathfrak{U}(X)$ is refined. Pick an arbitrary point $x_0 \in X$ and any nhood Ux_0 of it. Since the family $\mathfrak{U}(X)$ is refined, there is a regular cover $\alpha_0 \in \mathfrak{U}(X)$ such that $\alpha_0(x_0) \subset Ux_0$. From Problem 188 we find that $x_0 \in \operatorname{Int} \alpha_0(x_0)$. Put $U'x_0 = \operatorname{Int} \alpha_0(x_0)$. We have $x_0 \in U'x_0 \subset [U'x_0] = [\operatorname{Int} \alpha_0(x_0)] = \alpha_0(x_0) \subset Ux_0$.

193. Let ξ be a distinguished thread in $\mathfrak{U}(X)$, $x_1 \in \bigcap \{A: A \in \xi\}$ and $x_2 \in \bigcap \{A: A \in \xi\}$. Suppose that $x_1 \neq x_2$. Since X is a Hausdorff space, there is a nhood Ux_1 of x_1 such that $x_2 \notin [Ux_1]$. Consider the regular cover α_0, consisting of the two sets $[Ux_1]$ and $[X \setminus [Ux_1]]$. Then, by the definition of a thread, either $[Ux_1] \in \xi$, or $[X \setminus [Ux_1]] \in \xi$ (they cannot both belong to ξ). If $[Ux_1] \in \xi$, then $x_2 \notin \bigcap \{A: A \in \xi\}$, because $x_2 \notin [Ux_1]$. This is a contradiction. If $[X \setminus [UX_1]] \in \xi$, then $x_1 \notin \bigcap \{A: A \in \xi\}$ because $x_1 \notin [X \setminus [Ux_1]]$. Again we get a contradiction. Hence $\bigcap \{A: A \in \xi\}$ is a one-point set.

194. (*a*) Let η be a $\varkappa$-family (distinguished at the point x_0). Denote by $\mathcal{A}(\eta)$ all the $\varkappa$-families (distingushed at the point $x_0 \in X$), containing η. Let $\zeta_1, \zeta_2 \in \mathcal{A}(\eta)$. We shall assume that $\zeta_2 > \zeta_1$ if $\zeta_1 \subset \zeta_2$. Then $\mathcal{A}(\eta)$ is partially ordered. Suppose that $\mathcal{S} \subset \mathcal{A}(\eta)$ is a nest (see the definition on p. 6) in $\mathcal{A}(\eta)$. In order to show that $\mathcal{S}$ has a maximal element in $\mathcal{A}(\eta)$, it is required to establish that $\zeta^* = \bigcup \{\zeta: \zeta \in \mathcal{S}\}$ is a $\varkappa$-family (distinguished at $x_0 \in X$), which is easily verifiable. Now, from the maximal principle (see p. 6) it follows that $\mathcal{A}(\eta)$ has a maximal element, i.e., η is contained in some $\varkappa$-ultrafilter (distinguished at the point $x_0 \in X$).

(*b*) Now let η be a $\varkappa$-ultrafilter.

(i) If $A_1, A_2 \in \eta$, then by adding to the family η the set $[\operatorname{Int} A_1 \cap \operatorname{Int} A_2]$, we do not violate the $\varkappa$-centralization property. Hence, since η is maximal, surely $[\operatorname{Int} A_1 \cap \operatorname{Int} A_2] \in \eta$.

(ii) This statement is also trivial.

(iii) Let $B \in \mathfrak{B}(X)$ and $\operatorname{Int} B \cap \operatorname{Int} A \neq \Lambda$ for every $A \in \eta$. Put $\eta^* = \{B\} \cup \eta \cup \{[\operatorname{Int} B \cap \operatorname{Int} A]: A \in \eta\}$. It is easy to see that η^* is a $\varkappa$-family. Hence, by the maximality of the $\varkappa$-family η (in fact, η is a $\varkappa$-ultrafilter), we obtain $\eta = \eta^*$. Hence $B \in \eta$. Thus, every $\varkappa$-ultrafilter satisfies conditions (i), (ii) and (iii). Conversely, suppose the

$\varkappa$-family η satisfies these conditions. Suppose that η is contained in some $\varkappa$-family η^*. Then in every case we have $\text{Int } B \cap \text{Int } A \neq \Lambda$ if $B \in \eta^*$ and $A \in \eta \subset \eta^*$. However, η satisfies condition (iii). Hence $B \in \eta$, so that η^* necessarily coincides with η, i.e., η is a $\varkappa$-ultrafilter.

(*c*) This is implied by the preceding statement.

195. (*a*) Let ξ be a $\varkappa$-ultrafilter.

(i) Suppose that $\alpha \in \mathfrak{U}(X)$ is an arbitrary regular cover of X. It is required to show that there is an $A \in \alpha$ such that $A \in \xi$. Suppose that no such $A \in \alpha$ exists. Then by using (*c*) of the preceding problem, we have $[X \setminus A] \in \xi$ for all $A \in \alpha$. Hence (by (*b*) of the preceding problem) we have $[\bigcap \{\text{Int } [X \setminus A]: A \in \alpha\}] \in \xi$. But $\text{Int } [X \setminus A] = X \setminus A$ and $\bigcap \{(X \setminus A) : A \in \alpha\} = X \setminus \bigcup \{A : A \in \alpha\} = \Lambda$. Thus $A \in \xi$, a contradiction. Now, let $A_1, A_2 \in \alpha$ and $A_1, A_2 \in \xi$. Then $A_2 \subset [X \setminus A_1]$ and $[X \setminus A_1] \in \xi$ (see (*b*) of the preceding problem). This contradicts (*c*) of the preceding problem. Thus $\xi \cap \alpha$ is nonempty and consists of exactly one element for each $\alpha \in \mathfrak{U}(X)$.

(ii) Let $\alpha_1, \alpha_2 \in \mathfrak{U}(X)$ and $\alpha_2 > \alpha_1$. Let $\{A_1\} = \xi \cap \alpha_1$, $\{A_2\} = \xi \cap \alpha_2$, i.e., A_1 (resp. A_2) is the unique element of α_1 (resp. α_2), belonging to ξ. It is required to show that $A_2 \subset A_1$. This inclusion follows from Problem 189 and what has just been proved in (i).

(*b*) Suppose now that ξ is a thread in $\mathfrak{U}(X)$. It is required to show that ξ is a $\varkappa$-ultrafilter. Let $A_1, A_2 \in \xi$. By the definition of a thread, we have $\{A_1\} = \xi \cap \alpha_1$, $\{A_2\} = \xi \cap \alpha_2$ for some regular covers α_1 and α_2. Consider the regular cover $\alpha_3 = \alpha_1 \wedge \alpha_2$. It is shown in Problem 190 that $\alpha_3 > \alpha_1$, $\alpha_3 > \alpha_2$. Consider also the element $A_3 \in \xi \cap \alpha_3$. Then, by the definition of a thread, $A_3 \subset A_1$, $A_3 \subset A_2$. But $\alpha_3 = \alpha_1 \wedge \alpha_2$, so that $A_3 = [\text{Int } A' \cap \text{Int } A'']$, where $A' \in \alpha_1$, $A'' \subset \alpha_2$. Clearly, we must have $A_1 = A'$, $A_2 = A''$. Thus, $A_3 = [\text{Int } A_1 \cap \text{Int } A_2] \in \xi$. Hence, in particular, it follows that ξ is $\varkappa$-centralized, i.e., ξ is a $\varkappa$-family. Suppose that ξ is not a maximal $\varkappa$-family, i.e., ξ is not a $\varkappa$-ultrafilter. Then ξ is contained in some $\varkappa$-ultrafilter $\bar{\xi}$. From (*a*) of the present problem we find that $\bar{\xi} \cap \alpha$ is nonempty and consists of exactly one element for each $\alpha \in \mathfrak{U}(X)$. This contradicts the fact that this property is already possessed by a smaller subfamily of ξ. Thus ξ is a $\varkappa$-ultrafilter.

Now observe that if ξ is a distinguished thread, then, obviously, ξ is also a distinguished $\varkappa$-ultrafilter. The converse is also true. The proof is complete.

196. Consider the family ξ_0 where only element is the set A_0. Clearly, ξ_0 is a $\varkappa$-family distinguished at the point x_0 because $x_0 \in A_0$. By Problem 194(*a*), there is a $\varkappa$-ultrafilter ξ, which also is distinguished at x_0 and satisfies $\xi_0 \subset \xi$, i.e., ξ is a $\varkappa$-ultrafilter such that $A_0 \in \xi$ and $\{x_0\} = \bigcap \{A: A \in \xi\}$. But (Problem 195) every $\varkappa$-ultrafilter (distinguished at a given point) is at the same time a thread (distinguished at the same point). The proof is complete.

197. From the preceding problem it follows that for each point $x \in X$ there is a thread ξ distinguished at x, i.e., a thread such that $\{x\} = \bigcap \{A : A \in \xi\}$. By the definition of natural mapping, $\pi_X \xi = x$.

198. (*a*) Let ξ_0 be any point of $\Xi(X)$. By the definition of the Tychonoff topology in $P = \Pi \{\hat{\alpha}: \alpha \in \mathfrak{U}(X)\}$, sets of the form $O(A^0_{\alpha_1}, \ldots, A^0_{\alpha_s}) = \{\xi \in P: A^0_{\alpha_1} \in \xi, \ldots, A^0_{\alpha_s} \in \xi\}$, where $A^0_{\alpha_1} \in \alpha_1, \ldots, A^0_{\alpha_s} \in \alpha_s$, $\alpha_1 \in \mathfrak{U}(X), \ldots, \alpha_s \in \mathfrak{U}(X)$, form a standard base for the topology on $P = \Pi \{\hat{\alpha}: \alpha \in \mathfrak{U}(X)\}$. Suppose that $O(A^0_{\alpha_1}, \ldots, A^0_{\alpha_s}$ is an arbitrary nhood of the point $\xi_0 \in \Xi(X)$ in P in the standard base for P. Then it is obvious that $O(A^0_{\alpha_1}, \ldots, A^0_{\alpha_s}) \cap \Xi(X) = \langle \overline{A^0_{\alpha_1}, \ldots, A^0_{\alpha_s}} \rangle = \{\xi \in \Xi(X): A^0_{\alpha_1} \in \xi, \ldots, A^0_{\alpha_s} \in \xi\}$ is a nhood of $\xi_0 \in \Xi(X)$ in $\Xi(X)$.

Consider (using Problem 191) regular cover $\alpha_0 \in \mathfrak{U}(X)$, for which $\alpha_0 > \alpha_i$, $i = 1, 2, \ldots, s$. Pick an element $A^0_{\alpha_0} \in \alpha_0$, which is representable as a thread of ξ_0 in α_0, i.e., $\{A^0_{\alpha_0}\} = \xi_0 \cap \alpha_0$. Now consider the set $\langle \overline{A}^0_{\alpha_0} \rangle = \{\xi \in \Xi(X): A^0_{\alpha_0} \in \xi\}$. It is clear that $\langle \overline{A}^0_{\alpha_0} \rangle$ is a nhood of ξ_0 in $\Xi(X)$, because $\langle \overline{A}^0_{\alpha_0} \rangle = O(A^0_{\alpha_0}) \cap \Xi(X)$, where $O(A^0_{\alpha_0}) = \{\xi \in P: A^0_{\alpha_0} \in \xi\}$, $A^0_{\alpha_0} \in \alpha_0$ is a single-indexed standard nhood of ξ_0 in P. We show that $\langle \overline{A}^0_{\alpha_0} \rangle \subset \langle \overline{A^0_{\alpha_1}, \ldots, A^0_{\alpha_s}} \rangle$. Let $\xi \in \langle \overline{A}^0_{\alpha_0} \rangle$. Then $A^0_{\alpha_0} \in \xi$, where $A^0_{\alpha_0} \in \alpha_0$. Since $A^0_{\alpha_0} \in \xi_0$ (and $\langle \overline{A}^0_{\alpha_0} \rangle$ is a nhood of ξ_0), $A^0_{\alpha_1} \in \xi_0, \ldots, A^0_{\alpha_s} \in \xi_0$ and $\alpha_0 > \alpha_1, \ldots, \alpha_0 > \alpha_s$, we have $A^0_{\alpha_0} \subset A^0_{\alpha_i}$ for each $i = 1, 2, \ldots, s$. Hence, using the fact that $A^0_{\alpha_0} \in \xi$, it follows that $A^0_{\alpha_1} \in \xi$, $A^0_{\alpha_2} \in \xi, \ldots, A^0_{\alpha_s} \in \xi$, because ξ is a thread in the same $\varkappa$-ultrafilter (see Problems 195 and 194). Therefore, $\xi \in \langle \overline{A^0_{\alpha_1}, \ldots, A^0_{\alpha_s}} \rangle$.

(*b*) Let $\xi_0 \in P = \Pi\{\hat{\alpha}: \alpha \in \mathfrak{U}(X)\}$ and $\xi_0 \in [\Xi(X)]_P$. It is required to show that ξ_0 is a thread. Let $\alpha_1, \alpha_2 \in \mathfrak{U}(X)$ and $\alpha_2 > \alpha_1$. Consider the coordinates $A^0_{\alpha_1}$ and $A^0_{\alpha_2}$ of ξ_0 in $\hat{\alpha}_1$ and $\hat{\alpha}_2$, respectively, i.e., $\{A^0_{\alpha_1}\} = \xi_0 \cap \alpha_1$, $\{A^0_{\alpha_2}\} = \xi_0 \cap \alpha_2$. It is to be shown that $A^0_{\alpha_2} \subset A^0_{\alpha_1}$. Consider a standard nhood $O(A^0_{\alpha_1}, A^0_{\alpha_2})$ of ξ_0 in P. Then $O(A^0_{\alpha_1}, A^0_{\alpha_2}) \cap \Xi(X) \neq \Delta$ because $\xi_0 \in [\Xi(X)]_P$. Fix a point $\xi \in O(A^0_{\alpha_1}, A^0_{\alpha_2}) \cap \Xi(X)$. It follows that ξ is a thread and $A^0_{\alpha_1}, A^0_{\alpha_2} \in \xi$. Since $\alpha_2 > \alpha_1$, $A^0_{\alpha_2} \subset A^0_{\alpha_1}$, completing the proof.

(*c*) The compactness and Hausdorffness of the Tychonoff product $P = \Pi\{\hat{\alpha}: \alpha \in \mathfrak{U}(X)\}$ of the finite discrete spaces $\hat{\alpha}$, $\alpha \in \mathfrak{U}(X)$, is implied by (*b*).

(*d*) This is obvious because $\langle \overline{A}_0 \rangle \cap \dot{X} = \langle A_0 \rangle$.

199. (*a*) Let $\xi \in \langle A_0 \rangle$. Then $A_0 \in \xi$. Hence by the definition of π_X, $\pi_X\xi = \cap\{A: A \in \xi\} = \{x\} \subset A_0$, i.e., $\pi_X\xi \in A_0$. Thus, $\pi_X(\langle A_0 \rangle) \subset A_0$. Conversely, let $x_0 \in A_0$. Using Problem 196, consider a thread ξ_0 distinguished at x_0, for which $A_0 \in \xi_0$, i.e., $\xi_0 \in \dot{X}$, $A_0 \in \xi_0$ and $\{x_0\} = \cap\{A: A \in \xi_0\}$. Hence $\xi_0 \in \langle A_0 \rangle$ and $\pi_X\xi_0 = x_0$. Thus, $A_0 \subset \pi_X(\langle A_0 \rangle)$, and the statement (*a*) is proved.

(*b*) Let $x_0 \in \operatorname{Int} A_0$ and $A_0 \in \alpha_0$. Now consider any distinguished thread ξ, for which $\pi_X\xi = x_0$, i.e., $\{x_0\} = \cap\{A: A \in \xi\}$. The set A_0 in the cover α_0 is a representative of every such distinguished thread. Hence, every such thread belongs to $\langle A_0 \rangle$. But $\pi_X^{-1}x_0 = \{\xi \in \dot{X}: \cap\{A: A \in \xi\} = \{x_0\}\}$ so that $\pi_X^{-1}x_0 \subset \langle A_0 \rangle$. Hence $\pi_X^{\#}\langle A_0 \rangle \supset \operatorname{Int} A_0$. Conversely, let $x_0 \in \pi_X^{\#}\langle A_0 \rangle$, i.e., $\pi_X^{-1}x_0 = \{\xi \in \dot{X}: \cap\{A: A \in \xi\} = \{x_0\}\} \subset \langle A_0 \rangle$. Then $A_0 \in \xi$ for every $\xi \in \pi_X^{-1}x_0$. Suppose that $\alpha_0 \in \mathfrak{U}(X)$ is such that $A_0 \in \alpha_0$. If we had $x_0 \notin \operatorname{Int} A_0$, we would have one more element A_0' in α_0, $A_0' \neq A_0$, for which $x_0 \in A_0'$. This would yield a distinguished thread (see Problem 195) ξ_0', for which $A_0' \in \xi_0'$ and $\{x_0\} = \cap\{A': A' \in \xi_0'\}$. Then we would have $\xi_0' \in \pi_X^{-1}x_0$ and $A_0' \in \xi_0'$. On the other hand, since $\xi_0' \in \pi_X^{-1}x_0$, $A_0 \in \xi_0'$. We thus obtain $A_0 \in \xi_0' \cap \alpha_0$, $A_0' \in \xi_0' \cap \alpha_0$, $A_0 \neq A_0'$. This contradicts Problem 195. Thus, $\pi_X^{\#}\langle A_0 \rangle \subset \operatorname{Int} A_0$. Hence (*b*) is also proved.

200. To show that π_X is compact, let x_0 be any point of X. It is required to show that the subspace $\pi_X^{-1} x_0$ is compact. Since $\Xi(X)$ is compact, it suffices for this purpose to show that $\pi_X^{-1}x_0$ is closed in $\Xi(X)$ (hence compact). In the first place, it is obvious that $\pi_X^{-1}x_0 = \{\xi \in \dot{X}: \bigcap\{A: A \in \xi\} = \{x_0\}\}$. Let $\xi_0 \in [\pi_X^{-1}x_0]_{\Xi(X)}$. Hence for every single-indexed nhood $\langle \overline{A^0} \rangle$, $A^0 \in \xi_0$, of ξ_0 in $\Xi(X)$ we have $\langle \overline{A_0} \rangle \cap \pi_X^{-1}x_0 \neq \Lambda$. Let $\eta_0 \in \langle \overline{A_0} \rangle \cap \pi_X^{-1}x_0$, i.e., η_0 is a distinguished thread, $A^0 \in \eta_0$ and $\bigcap\{A: A \in \eta_0\} = \{x_0\}$. Hence in particular we have $x_0 \in A^0$. Thus $x_0 \in A^0$ for every $A^0 \in \xi_0$, i.e., ξ_0 is a distinguished thread and $\{x_0\} = \bigcap\{A^0: A^0 \in \xi_0\}$. Thus, $\xi_0 \in \pi_X^{-1}x_0$, so that $\pi_X^{-1}x_0$ is closed in $\Xi(X)$. The irreducibility of π_X is implied by Problems 199(*b*) and 198(*a*).

Suppose now that X is regular. Then the family $\mathfrak{A}(X)$ of all regular covers is refined (see Problem 192). Hence for each point $x_0 \in X$ and every nhood Ux_0 of x_0 there is a regular closed set A_0 such that $x_0 \in A_0 \subset Ux_0$. Hence, if $\xi_0 \in \dot{X}$ is such that $\pi_X\xi_0 = x_0$, then it is clear that $A_0 \in \xi_0$. Consequently, $\langle A_0 \rangle$ is a nhood of ξ_0, for which $\pi_X(\langle A_0 \rangle) = A_0 \subset Ux_0$ (here we have used Problem 199(*a*)). The proof is complete.

201. (i) follows from the obvious equality $\langle \overline{A} \rangle \cap \dot{X} = \langle A \rangle$ and the denseness of $\dot{X}$ in $\Xi(X)$ (see Problem 198).

(ii) The disjointness of the families $\langle \overline{\alpha} \rangle$ and $\langle \alpha \rangle$ follows from the fact that for every thread ξ the intersection $\xi \cap \alpha$ consists of just one element, and that $\langle \overline{\alpha} \rangle$ and $\langle \alpha \rangle$ are covers of $\Xi(X)$ and $\dot{X}$ respectively, which follows from the fact that $\xi \cap \alpha \neq \Lambda$ for every thread ξ. It is also obvious that every set $\langle \overline{A} \rangle$ (resp. $\langle A \rangle$) is clopen in $\Xi(X)$ (resp. in $\dot{X}$).

(iii) follows from the fact that every $\langle \overline{\alpha} \rangle$, $\alpha \in \mathfrak{A}(X)$ (resp. $\langle \alpha \rangle$) is disjoint, and the collection of sets of the form $\langle \overline{A} \rangle$, $A \in \mathfrak{P}(X)$ (resp. $\langle A \rangle$) forms a base for $\Xi(X)$ (resp. for $\dot{X}$).

(iv) is obvious.

202. The continuity, compactness and irreducibility of π_X have already been shown (see Problem 200). It has also been shown that π_X is an 'onto' mapping (see Problem 197). It remains to show that π_X is closed. Let $K \subset \dot{X}$ be an arbitrary closed set in $\dot{X}$. To prove that $\pi_X(K)$ is closed in X, it suffices to show that $\pi_X^{\#}(\dot{X} \setminus K) = X \setminus \pi_X(K)$ is open in X. Let $x_0 \in \pi_X^{\#}(\dot{X} \setminus K)$. Then, $\pi_X^{-1}x_0 \subset \dot{X} \setminus K$, and $\dot{X} \setminus K = O\pi_X^{-1}x_0$ is a nhood of the compact set $\pi_X^{-1}x_0$ in $\dot{X}$. For each point $\xi \in \pi_X^{-1}x_0$, consider a regular cover $\alpha_\xi \in \mathfrak{A}(X)$ such that

$$\langle \alpha_\xi \rangle(\xi) \subset O\pi_X^{-1} x_0, \tag{1}$$

where $\langle \alpha_\xi \rangle(\xi)$ is the star of ξ with respect to the cover $\langle \alpha_\xi \rangle$ of $\dot{X}$ (in this connection see the preceding problem). Since $\langle \alpha_\xi \rangle$ is a disjoint cover (see the preceding problem), $\langle \alpha_\xi \rangle(\xi)$ coincides with some set $\langle A(\xi) \rangle \in \langle \alpha_\xi \rangle$. Thus, for each point $\xi \in \pi_X^{-1}x_0$ we pick a cover $\alpha_\xi \in \mathfrak{A}(X)$ and an element $A(\xi) \in \alpha_\xi$ in such a manner that

$$\xi \in \langle A(\xi) \rangle \subset O\pi_X^{-1}x_0 \,. \tag{1'}$$

Consider the open cover

$$\omega = \{\langle A(\xi)\rangle : \xi \in \pi_X^{-1}x_0\} \tag{2}$$

of the compactum $\pi_X^{-1}x_0$ and extract from it the finite subcover

$$\omega_0 = \{\langle A(\xi_i)\rangle : i = 1, 2, \dots, s\}\cdot \tag{3}$$

Now take a regular cover $\alpha^* \in \mathfrak{A}(X)$ of X, satisfying $\alpha^* > \alpha_{\xi_i}$ for all $i = 1, 2, \dots, s$. Consider also a disjoint clopen cover $\langle \alpha^* \rangle$ of $\dot{X}$ (see the preceding problem). From the preceding problem we conclude that $\langle \alpha^* \rangle$ is a refinement of all the covers $\langle \alpha_{\xi_i} \rangle$, $i = 1, 2, \dots, s$. The inclusions (1), (1′) and the construction of $\langle \alpha^* \rangle$ lead to the inclusions

$$\langle \alpha^* \rangle (\pi_X^{-1}x_0) = \bigcup \{\langle A \rangle \in \langle \alpha^* \rangle : \langle A \rangle \cap \pi_X^{-1}x_0 \neq \Delta\} \subset O\pi_X^{-1}x_0 \cdot \tag{4}$$

Put

$$\dot{X} \setminus \langle \alpha^* \rangle (\pi_X^{-1}x_0) = H, \tag{5}$$

to obtain

$$H = \bigcup \{\langle A \rangle \in \langle \alpha^* \rangle : \langle A \rangle \cap \pi_X^{-1}x_0 = \Delta\}\cdot \tag{6}$$

The set H, as the union of finitely many clopen sets in $\dot{X}$, is itself clopen in $\dot{X}$. From (6), using Problem 199(*a*), we obtain $\pi_X(H) = \bigcup \{\pi_X(\langle A \rangle) : \langle A \rangle \in \alpha^*$ and $\langle A \rangle \cap \pi_X^{-1}x = \Delta\} = \bigcup \{A \in \alpha^* : x_0 \notin A\}$, so that the set $\pi_X(H)$, as the union of finitely many closed sets in X, is itself closed in X. Hence the set $X \setminus \pi_X(H)$ is open in X. Moreover, it contains x_0 and is contained in the set $\pi_X^{\#}(\dot{X} \setminus K)$. The proof is complete.

203. Since $A_1, A_2 \subset X$ are regular closed sets, $A_1 = [\mathrm{Int}\, A_1]$, $A_2 = [\mathrm{Int}\, A_2]$. Problem 110 then yields

$$fA_1 = [f^{\#}(\mathrm{Int}\, A_1)], \quad fA_2 = [f^{\#}(\mathrm{Int}\, A_2)]\cdot \tag{1}$$

It is easy to see that the inclusions

$$\begin{aligned} &f^{-1}(f^{\#}(\mathrm{Int}\, A_1)) \subset f^{-1}(\mathrm{Int}\, f(A_1)) \subset A_1, \\ &f^{-1}(f^{\#}(\mathrm{Int}\, A_2)) \subset f^{-1}(\mathrm{Int}\, f(A_2)) \subset A_2 \end{aligned} \tag{2}$$

hold. Moreover, since the set $f^{-1}(f^{\#}(\mathrm{Int}\, A_1))$ is dense in A_1 and $f^{-1}(f^{\#}(\mathrm{Int}\, A_2))$ is dense in A_2 (see Problem 109), we have

$$\begin{aligned} &f^{-1}(f^{\#}(\mathrm{Int}\, A_1)) = [f^{-1}(\mathrm{Int}\, f(A_1))] = A_1, \\ &f^{-1}(f^{\#}(\mathrm{Int}\, A_2)) = [f^{-1}(\mathrm{Int}\, f(A_2))] = A_2. \end{aligned} \tag{3}$$

But (1) and the equality $fA_1 = fA_2$ imply $\text{Int} f(A_1) = \text{Int} f(A_2)$. Hence, using this and (3), we obtain the desired equality $A_1 = A_2$.

204. (*a*) Let $\Gamma \subset \Xi(X)$ be an arbtitrary open set in $\Xi(X)$. It is required to show that $[\Gamma]_{\Xi(X)}$ is open in $\Xi(X)$. Since $\dot{X}$ is dense in $\Xi(X)$, we obtain the equality

$$[\Gamma]_{\Xi(X)} = [\Gamma \cap \dot{X}]_{\Xi(X)} = [[\Gamma \cap \dot{X}]_{\dot{X}}]_{\Xi(X)}. \tag{1}$$

Put

$$A_1 = \pi_X([\Gamma \cap \dot{X}]_{\dot{X}}), \quad A_2 = [X \setminus A_1)_X \tag{2}$$

and consider the cover $\alpha_0 = \{A_1, A_2\}$ of X. Since π_X is an irreducible perfect mapping (see Problem 202) and $[\Gamma \cap \dot{X}]_{\dot{X}}$ is a regular closed set in $\dot{X}$, A_1 is a regular closed set in X and, consequently, so is A_2. Hence α_0 is a regular cover of X.

Consider the clopen sets $\langle A_1 \rangle = \{\xi \in \dot{X}: A_1 \in \xi\}$, $\langle A_2 \rangle = \{\xi \in \dot{X}: A_2 \in \xi\}$ in $\dot{X}$. In Problem 199 we showed that $\pi_X(\langle A_1 \rangle) = A_1$. Hence, since $\langle A_1 \rangle$ is a regular closed set in $\dot{X}$, $[\Gamma \cap \dot{X}]_{\dot{X}}$ is a regular closed set in $\dot{X}$ and $\pi_X(\langle A_1 \rangle) = \pi_X([\Gamma \cap \dot{X}]_{\dot{X}})$, where π_X is perfect and irreducible, therefore $[\Gamma \cap \dot{X}]_{\dot{X}} = \langle A_1 \rangle$ (see the preceding problem). But $[\langle A_1 \rangle]_{\Xi(X)} = \langle \overline{A_1} \rangle$, $\langle \overline{A_1} \rangle$ is a clopen set in $\Xi(X)$; hence, taking note of (1), we obtain the equality $\langle \overline{A_1} \rangle = [\langle A_1 \rangle]_{\Xi(X)} = [[\Gamma \cap \dot{X}]_{\dot{X}}]_{\Xi(X)} = [\Gamma]_{\Xi(X)}$. Consequently, $[\Gamma]_{\Xi(X)}$ is clopen in $\Xi(X)$.

(*b*) Since $\Xi(X)$ is extremally disconnected, the dense subspace $\dot{X} \subset \Xi(X)$ is as well (see Problem 157). Moreover, $\Xi(X)$ is a compactum, as a closed subspace of the compactum $\Pi\{\hat{\alpha}: \alpha \in \mathfrak{U}(X)\}$, and $\dot{X}$ is completely regular. (*b'*) follows by the definition of an absolute (see p. 326), (*b*) and Problem 160.

(*c*) follows (by virtue of Problem 162) from the fact that $\Xi(X)$ is an extremally disconnected compactum and $\dot{X} \subset \Xi(X)$ is dense in $\Xi(X)$.

205. This follows from the preceding problem; the space $\dot{X}$ is also an absolute of X.

206. Pick an absolute $\dot{X}$ of X (see Problems 204 and 205) and a natural mapping $\pi_X: \dot{X} \to X$, which is perfect and irreducible (see Problem 202). $\dot{X}$ is completely regular and is contained in the 0-dimensional compactum $\Pi\{\hat{\alpha}: \alpha \in \mathfrak{U}(X)\}$. Hence $\dot{X}$ itself is also 0-dimensional (in the sense of p. 42).

207. Suppose that $\sigma = \{U_\lambda: \lambda \in T\}$ is a base for X such that $|\sigma| = w(X)$. For every $U_\lambda \in \sigma$ consider the regular cover $\lambda = \{[U_\lambda], [X \setminus [U_\lambda]]\}$ of X. For each $n \in N^+$ and each finite sequence $\alpha = (\lambda_1, \ldots, \lambda_n)$ consider the regular cover $\alpha = \lambda_1 \wedge \lambda_2 \wedge \ldots \wedge \lambda_n$ (see Problem 190). Denote by $\mathfrak{U}(\sigma, X)$ the family of regular covers of X constructed by the above prescribed method. Obviously, $|\mathfrak{U}(\sigma, X)| = \tau$. It is easy to see that $\mathfrak{U}(\sigma, X)$ is refined. The directedness of $\mathfrak{U}(\sigma, X)$ follows from its construction and Problem 190.

208. Suppose that σ is a base for X whose cardinality equals the weight of X. Consider a refined directed sequence $\mathfrak{U}(\sigma, X)$ of regular closed covers of X, constructed with respect to the given base σ (see the preceding problem). Further, consider the space $\dot{X}(\sigma)$ constructed exactly in the same way as the space $\dot{X}$ for the family $\mathfrak{U}(X)$ of regular covers of X, i.e., $\dot{X}(\sigma)$ is the space of all distinguished threads in $\mathfrak{U}(\sigma, X)$.

Exactly as before we construct a natural mapping $\pi_X: \dot{X}(\sigma) \to X$ and show that this is perfect and irreducible. Moreover, $\dot{X}(\sigma)$ is a subspace of the generalized discontinuum $\Pi\{\hat{\alpha}: \alpha \in \mathfrak{A}(\sigma, X)\}$ of weight $|\mathfrak{A}(\sigma, X)| = \tau$. Hence the weight of $\dot{X}(\sigma)$ does not exceed τ. Since perfect mappings do not increase weight (see Problem 41), the weight of $\dot{X}(\sigma)$ is τ. In addition, $\operatorname{ind}(\dot{X}(\sigma)) = 0$, because $\Pi\{\hat{\alpha}: \alpha \in \mathfrak{A}(\sigma, X)\}$ is a 0-dimensional compactum, containing a topological image of $\dot{X}(\sigma)$. The proof is complete.

213. Since ξ is a thread, it is also a $\varkappa$-ultrafilter. We shall show that $\tilde{f}\xi = \{fA: A \in \xi\}$ is a $\varkappa$-ultrafilter in the family $\mathfrak{P}(Y)$ of all regular closed sets in Y. First note that fA is a regular closed set in Y for every $A \in \xi$ (see Problem 110). Let $A_1, A_2 \in \xi$. Since ξ is a $\varkappa$-ultrafilter, $[\operatorname{Int} A_1 \cap \operatorname{Int} A_2] \neq \Lambda$ also belongs to ξ (see Problem 194). Since f is closed and irreducible, $[f^{\#}(\operatorname{Int} A_1)] = f[\operatorname{Int} A_1] = fA_1$ and $[f^{\#}(\operatorname{Int} A_2)] = f[\operatorname{Int} A_2] = fA_2$ (see Problem 110). On the other hand, $[\operatorname{Int} fA_1] = fA_1$, $[\operatorname{Int} fA_2] = fA_2$, where $\operatorname{Int} fA_1$ and $\operatorname{Int} fA_2$ are maximal open sets in Y with this property. This implies that the nonempty open set $f^{\#}[\operatorname{Int} A_1] \subset \operatorname{Int} fA_1$ is also dense in $\operatorname{Int} fA_1$. Similarly, $f^{\#}(\operatorname{Int} A_2) \subset \operatorname{Int} fA_2$ is also dense in $\operatorname{Int} fA_2$. We have $f^{\#}(\operatorname{Int} A_1 \cap \operatorname{Int} A_2) = f^{\#}(\operatorname{Int} A_1) \cap f^{\#}(\operatorname{Int} A_2) \subset \operatorname{Int} fA_1 \cap \operatorname{Int} fA_2$, where $f^{\#}(\operatorname{Int} A_1 \cap \operatorname{Int} A_2)$ is dense in the set $\operatorname{Int} fA_1 \cap \operatorname{Int} fA_2$. We obtain the equality $[f^{\#}(\operatorname{Int} A_1 \cap \operatorname{Int} A_2)] = f[\operatorname{Int} A_1 \cap \operatorname{Int} A_2] = [\operatorname{Int} fA_1 \cap \operatorname{Int} fA_2]$ (here we have also employed Problem 110). Since $[\operatorname{Int} A_1 \cap \operatorname{Int} A_2] \in \xi$, $[\operatorname{Int} fA_1 \cap \operatorname{Int} fA_2] \in \tilde{f}\xi$. Thus $\tilde{f}\xi$ is a $\varkappa$-family. To prove that $\tilde{f}\xi$ is a $\varkappa$-ultrafilter, it obviously suffices (see Problem 194) to verify that property (iii) of Problem 194 for $\varkappa$-ultrafilters is satisfied. Thus, let $B \in \mathfrak{P}(Y)$ be such that $\operatorname{Int} B \cap \operatorname{Int} fA \neq \Lambda$ for every $fA \in \tilde{f}\xi$. Again $f^{-1}(\operatorname{Int} fA)$ is open, nonempty, contained in A and dense in A. Hence $[f^{-1}(\operatorname{Int} fA)] = A$. $\operatorname{Int} B \cap \operatorname{Int} fA \neq \Lambda$ implies $f^{-1}(\operatorname{Int} B) \cap f^{-1}(\operatorname{Int} fA) \neq \Lambda$ and, consequently, $\operatorname{Int}[f^{-1}(\operatorname{Int} B)] \cap \operatorname{Int} A \neq \Lambda$ for every $A \in \xi$. Therefore, since ξ is a $\varkappa$-ultrafilter, we have $[f^{-1}(\operatorname{Int} B)] \in \xi$. Then, $f[f^{-1}(\operatorname{Int} B)] = [\operatorname{Int} B] = B$. Hence $B \in \tilde{f}\xi$. Thus, $\tilde{f}\xi$ is a $\varkappa$-ultrafilter in $\mathfrak{P}(Y)$ and is hence a thread in $\mathfrak{A}(Y)$ (see Problem 195).

(*b*) Let ξ be a distinguished thread, then $\bigcap\{A: A \in \xi\} \neq \Lambda$. Therefore, $f(\bigcap\{A: A \in \xi\}) \subset \bigcap\{fA: A \in \xi\}$; consequently, $\bigcap\{fA: A \in \xi\} \neq \Lambda$, i.e., $\tilde{f}\xi$ is a distinguished thread in $\mathfrak{A}(Y)$.

(*c*) If ξ_1 and ξ_2 are threads in $\mathfrak{A}(X)$ and $\xi_1 \neq \xi_2$, then there exist $A_1 \in \xi_1$, $A_2 \in \xi_2$ such that $A_1 \neq A_2$. Therefore, by Problem 203, $fA_1 \neq fA_2$. Hence $\tilde{f}\xi_1 \neq \tilde{f}\xi_2$.

214. By virtue of Problem 195, it suffices to show that ξ is a $\varkappa$-ultrafilter. For this it is necessary to verify conditions (i), (ii) and (iii) of Problem 194 for ξ (they are necessary and sufficient for a family of regular closed sets to be a $\varkappa$-ultrafilter). Note that our purpose will be adequately served by verifying that the properties (i) and (iii) are satisfied. It is obvious at the outset that ξ consists of nonempty regular closed sets in X.

(i) Let $[f^{-1}(\operatorname{Int} B_1)] \in \xi$, $[f^{-1}(\operatorname{Int} B_2)] \in \xi$, where $B_1, B_2 \in \eta$. We have

$$f[f^{-1}(\operatorname{Int} B_1) \cap f^{-1}(\operatorname{Int} B_2)] = [f(f^{-1}(\operatorname{Int} B_1) \cap f^{-1}(\operatorname{Int} B_2))] = [\operatorname{Int} B_1 \cap \operatorname{Int} B_2], \tag{1}$$

$$f[f^{-1}(\operatorname{Int}[\operatorname{Int} B_1 \cap \operatorname{Int} B_2])] = [\operatorname{Int}[\operatorname{Int} B_1 \cap \operatorname{Int} B_2]] = [\operatorname{Int} B_1 \cap \operatorname{Int} B_2].$$

Hence (by Problem 203) we obtain

$$[f^{-1}(\operatorname{Int} B_1) \cap f^{-1}(\operatorname{Int} B_2)] = [f^{-1}(\operatorname{Int} B_3)], \tag{2}$$

where $B_3 = [\text{Int } B_1 \cap \text{Int } B_2]$. But $B_1, B_2 \in \eta$; consequently, $[\text{Int } B_1 \cap \text{Int } B_2] = B_3 \in \eta$ (since η is a $\varkappa$-ultrafilter). Noting this and also equalities (2), (1) and the definition of ξ, we obtain

$$[\text{Int } [f^{-1}(\text{Int } B_1)] \cap \text{Int } [f^{-1}(\text{Int } B_2)]] \in \xi.$$

By the same token, property (i) of $\varkappa$-ultrafilters (see Problem 194) is proved. From what has been shown it follows in particular that ξ is a $\varkappa$-centralized family.

(iii) Suppose that $A \subset X$ is a regular closed set in X such that $\text{Int } A \cap \text{Int } [f^{-1}(\text{Int } B)] \neq \Lambda$ for every $B \in \eta$. Then $\text{Int } A \cap f^{-1}(\text{Int } B) \neq \Lambda$ for every $B \in \eta$ (because $f^{-1}(\text{Int } B)$ is contained in and is dense in $\text{Int } [f^{-1}(\text{Int } B)]$). Hence $f^{\#}(\text{Int } A) \cap \text{Int } B \neq \Lambda$ for every $B \in \eta$. Here, as usual, $f^{\#}(\text{Int } A) = \{y \in Y: f^{-1}y \subset \text{Int } A\}$. Hence we obtain the conclusion that $\text{Int } [f^{\#}(\text{Int } A)] \cap \text{Int } B \neq \Lambda$ for every $B \in \eta$. Then, since $fA = [f^{\#}(\text{Int } A)]$, we obtain $\text{Int } fA \cap \text{Int } B \neq \Lambda$ for every $B \in \eta$. The fact that η is a $\varkappa$-ultrafilter leads to the conclusion (see Problem 194) that $fA \in \eta$. But then $[f^{-1} \text{Int}(fA)] \in \xi$. Thus ξ is a $\varkappa$-ultrafilter and, consequently, a thread in $\mathfrak{U}(X)$.

215. The preceding problem shows that ξ is a thread in $\mathfrak{U}(X)$. It is now required to show just that $\bigcap \{[f^{-1}(\text{Int } B)]: B \in \eta\} \neq \Lambda$. Since η is a distinguished thread in $\mathfrak{U}(Y)$, $\bigcap \{B: B \in \eta\} \neq \Lambda$ consists of exactly one point, which we denote by y_0. Consider the compact set $f^{-1}y_0$. Since $y_0 \in B$ for every $B \in \eta$ and the equality $f[f^{-1}(\text{Int } B)] = B$ holds, $F_B = f^{-1}y_0 \cap [f^{-1}(\text{Int } B)] \neq \Lambda$ for every $B \in \eta$. It is easy to see that the family $\{F_B: B \in \eta\}$ is centralized and consists of compact sets. Hence $\bigcap \{F_B: B \in \eta\} \neq \Lambda$ and, consequently, $\bigcap \{[f^{-1}(\text{Int } B)]: B \in \eta\} \neq \Lambda$.

216. This follows from Problems 213 and 214.

217. This follows immediately from Problems 213-216.

218. (*a*) In the preceding problem it was shown that $\dot{f}$ is a one-to-one mapping of $\dot{X}$ onto $\dot{Y}$. Let $\langle A_0 \rangle = \{\xi \in \dot{X}: A_0 \in \xi\}$ be an arbitrary basic open set in $\dot{X}$ (see Problem 198), which is clopen in $\dot{X}$. The set fA_0 is closed in Y (see Problem 110). Consider the basic open set $\langle fA_0 \rangle = \{\eta \in \dot{Y}: fA_0 \in \eta\}$ in $\dot{Y}$. Then, we have $\dot{f}\langle A_0 \rangle = \langle fA_0 \rangle$. Indeed, if $\xi \in \langle A_0 \rangle$, then $A_0 \in \xi$; consequently, $fA_0 \in \dot{f}\xi$, i.e., $\dot{f}\xi \in \langle fA_0 \rangle$. Consequently, if $\eta \in \langle fA_0 \rangle$, then $fA_0 \in \eta$ and since $[f^{-1}(\text{Int } fA_0)] = A_0$, $A_0 \in \xi$, where $\xi = \{[f^{-1}(\text{Int } B)]: B \in \eta\}$. But $\xi \in \dot{X}$ and $\dot{f}\xi = \eta$. Hence $\xi = \dot{f}^{-1}\eta \in \langle A_0 \rangle$.

Since the sets of the form $\langle A \rangle$, $A \in \mathfrak{P}(X)$ form a base for $\dot{X}$, those of the form $\langle B \rangle$, $B \in \mathfrak{P}(Y)$ form a base for $\dot{Y}$, and each $B \in \mathfrak{P}(Y)$ is an image under f of a unique $A \in \mathfrak{P}(X)$; hence, taking note of the equality just proved, we infer that $\dot{f}$ is a topological mapping.

(*b*) Let $\xi \in \dot{X}$. Then $\dot{f}(\xi) = \{fA: A \in \xi\}$ and $\pi_Y(\dot{f}(\xi)) = \bigcap \{fA: A \in \xi\} = \{y_0\}$, where y_0 is a unique point of Y belonging to this intersection. On the other hand, $\pi_X(\xi) = \bigcap \{A: A \in \xi\}$ and $f(\pi_X(\xi)) = f(\bigcap\{A: A \in \xi\}) \subset \bigcap \{fA: A \in \xi\} = \{y_0\}$. However, since the intersection $\bigcap \{A: A \in \xi\}$ consists of exactly one point, $f(\pi_X(\xi)) = y_0$. Thus, $\pi_Y(\dot{f}(\xi)) = f(\pi_X(\xi))$, proving (*b*).

219. (*a*) is proved on precisely similar lines as (*b*) of the preceding problem. Statement (*b*) of the present problem is proved in the same way as (*a*) of the preceding problem. Note that $\dot{f}$ is in general not an 'onto' mapping, because f is generally not perfect. Statement (*c*) is also proved along similar lines as statement (*a*) of the preceding problem. The requirement is just to note first that $\tilde{f}$ is a one-to-one 'onto' mapping (see

Problems 216 and 217), and verify the equality $\tilde{f}\langle \overline{A} \rangle = \langle \overline{f(A)} \rangle$ for every regular closed set.

220. The Stone-Čech compactifications $\beta\dot{X}, \beta\dot{Y}$ of $\dot{X}$ and $\dot{Y}$ coincide with the spaces $\Xi(X)$ and $\Xi(Y)$, respectively (see Problem 204), but because of the preceding problem $\Xi(X)$ and $\Xi(Y)$ are homeomorphic.

221. The existence of an absolute of a regular space X has already been established (see Problems 204 and 205): an absolute of X is the space $\dot{X}$ of all distinguished threads in the family $\mathfrak{U}(X)$ of all regular covers of X. To show that $\dot{X}$ is the unique absolute of X suppose that Y is also an absolute of X. Then there is (by the definition of absolute) an irreducible perfect mapping f of Y onto X: f: $Y \to X$.

Consider the space $\dot{X}$ and the natural mapping π_X: $\dot{X} \to X$; also the family $\mathfrak{U}(Y)$ of all regular covers of Y, the space $\dot{Y}$ of all distinguished threads in $\mathfrak{U}(Y)$ and the natural mapping π_Y: $\dot{Y} \to Y$. Now, consider the homeomorphism $\dot{f}$: $\dot{Y} \to \dot{X}$ onto $\dot{X}$, generated by f: $Y \to X$ (see Problems 217 and 218). Then (see Problem 218), the diagram in Figure 4 is commutative.

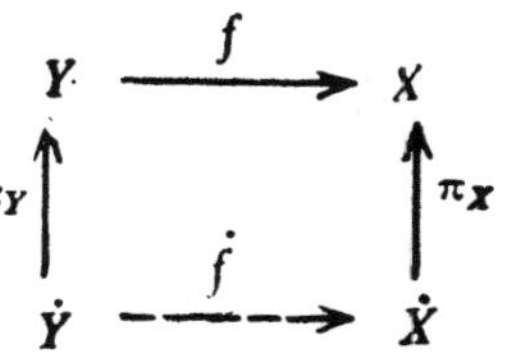

Figure 4.

Since Y is an absolute of X, and $\dot{Y}$ is the irreducible perfect pre-image of Y (for example, under the natural mapping π_Y), $\dot{Y}$ is homeomorphic to Y: $\dot{Y} \approx Y$ (by the definition of the absolute).

On the other hand, $\dot{f}$ is an 'onto' homeomorphism; consequently, $\dot{Y} \approx \dot{X}$. Hence $Y \approx \dot{X}$, completing the proof.

It remains to note from what has been just proved and Problems 204 and 205 that the absolute of a regular space is a completely regular extremally disconnected space.

224. If X and Y are co-absolute, then we consider their common absolute Z. By the definition of absolute, there exist irreducible perfect mappings p_X: $Z \to X$ and p_Y: $Z \to Y$. The converse follows from the preceding problem.

226. Consider the natural mappings $\pi_X: \dot{X} \to X$, π_Y: $\dot{Y} \to Y$ and $\pi_Z: \dot{Z} \to Z$; also the homeomorphisms $\dot{f}$: $\dot{X} \to \dot{Y}$, $\dot{g}: \dot{Z} \to \dot{Y}$, generated by f and g respectively. By Problem 218, the diagrams in Figures 5 and 6 are commutative.

Since Z is extremally disconnected, π_Z is a homeomorphism. Put $h = \pi_X \dot{f}^{-1} \dot{g} \pi_Z^{-1}$. Obviously, h is the required mapping.

227. Suppose that h_1: $\beta X \to b_1X$, h_2: $\beta X \to b_2X$ are natural mappings of the Stone-Čech compactification βX onto the extensions b_1X and b_2X, respectively (see Problem 22 of Chap. 4). It is clear that h_1 and h_2 are perfect irreducible mappings. Hence this implies that b_1X and b_2X are both co-absolute to βX (see Problem 223); consequently, b_1X and b_2X are co-absolute.

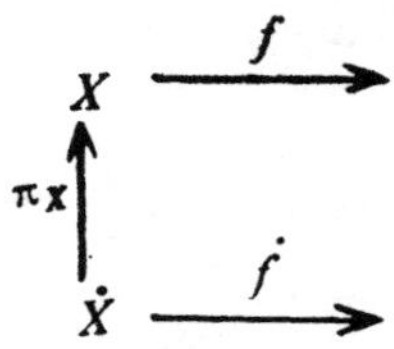

Figure 5.

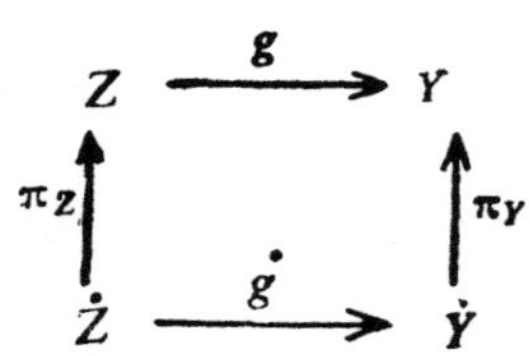

Figure 6.

228. By Problem 113, the extension $\tilde{f}$: $\beta X \to \beta Y$ of f: $X \to Y$ to the Stone-Čech compactifications of these spaces is irreducible. Then βX and βY are co-absolute (see Problem 223). Futhermore, βX is co-absolute to every compactification bX of X and in the same way βY is co-absolute to every compactification cY of Y (see the preceding problem). Sincc βX and βY are co-absolute, bX and cY are also co-absolute.

229. Suppose that π_X: $\dot{X} \to X$ is an irreducible perfect mapping of the absolute $\dot{X}$ of X onto X. $\dot{X}$ is extremally disconnected, hence so is its Stone-Čech compactification $\beta\dot{X}$ (see Problem 158).

Consider the extension $\bar{\pi}_X$: $\beta\dot{X} \to \beta X$ of π_X: $\dot{X} \to X$ to the Stone-Čech compactifications $\beta\dot{X}$, βX of $\dot{X}$ and X. $\bar{\pi}_X$ is an extension of an irreducible perfect mapping π_X, hence (see Problem 113) is itself irreducible (and, of course, perfect). Consequently, the extremally disconnected space $\beta\dot{X}$ admits an irreducible perfect mapping $\bar{\pi}_X$ onto βX, so that (see Problem 222) $\beta\dot{X}$ is the absolute of βX, i.e., $\beta\dot{X} = \dot{(\beta X)}$. The assertion is proved.

230. The two countable compacta X and Y are compactifications of the space N of natural numbers, N: $X = b_1N$, $Y = b_2N$. For the rest, see Problems 225, 229 and 158.

231. No, not necessarily. It is required to construct a regular non-normal space and a perfect irreducible mapping onto a normal space. See Mishenko [2].

232. No, in general, it is not necessary. Pick any regular, not completely regular space Y (see Problem 9 of Chap. 3). Then the absolute $\dot{Y}$ of the regular space Y must be a completely regular space (see Problem 205), i.e., every regular space is co-absolute with some completely regular space, namely with its absolute.

233. It follows from the fact that the absolutes $\dot{X}$ and $\dot{Y}$ of the spaces in question are homeomorphic, that $\dot{X}$ and $\dot{Y}$ are related to X and Y, respectively, by a perfect (even irreducible) mapping and finally that the topological properties in the hypothesis of the problem are invariant under a perfect mapping during the passage to both image and to pre-image (see Problems 48, 36, 37, 38 and 76 of the present chapter and Problems 91 and 203 of Chap. 3).

234. (*a*) The absolute $\dot{X}$ of an infinite compactum X is infinite, extremally disconnected and compact. Having noted this, refer to Problem 171 for the rest.

(*b*) Consider an arbitrary countable compactum X. Its absolute $\dot{X}$ (see Problem 230) is homeomorphic to the Stone-Čech compactfication βN of the natural numbers. For the rest, refer to Problem 53 of Chap. 4.

235. Suppose that σ is a π-base for X. Since f is closed and irreducible, then if U is open in X (see Problem 109), $f^{\#}U = \{y \in Y: f^{-1}y \subset U\} = Y \setminus f(X \setminus U)$ is nonempty and open in Y. It is easy to see that the family $f^{\#}\sigma = \{f^{\#}U: U \in \sigma\}$ is a π-base of Y. Hence, $\pi wY \leqslant \pi wX$.

Suppose now that σ is a π-base for Y. Consider the family $f^{-1}\sigma = \{f^{-1}V: V \in \sigma\}$. Every member of this family is open. Let $U \subset X$ be any open set in X. Consider the nonempty open set $f^{\#}U$ in Y (see Problem 109). We find a $V \in \sigma$ such that $V \subset f^{\#}U$ (since σ is a π-base for Y). Then $f^{-1}V \subset f^{-1}f^{\#}U \subset U$ and $f^{-1}V \in f^{-1}\sigma$. From all this it follows that $\pi wX \leqslant \pi wY$. By combining the two inequalities, we get the required equality $\pi wX = \pi wY$.

236. This follows from Problems 235 and 224.

237. (*a*) Suppose that the compactum X is co-absolute with the metrizable compactum Y. Then $\pi wX = \pi wY$ (see the preceding problem), but $\pi wY \leqslant wY \leqslant \aleph_0$ (even $\pi wY = wY$, because Y is a metrizable compactum, see Problems 216 of Chap. 2 and Problem 257 of Chap. 3), so that $\pi wX \leqslant \aleph_0$.

(*b*) X is a compactum and $\pi wX \leqslant \aleph_0$. Then (see Problem 326 of Chap. 3) there exist a metrizable compactum Y and an irreducible continuous mapping $f: X \to Y$ (which is automatically perfect). From Problem 223 it follows that X and Y are co-absolute.

238. A metrizable compactum is an irreducible perfect image of the Cantor perfect set C (see Problem 303 of Chap. 3). For the rest, see Problem 224.

239. Suppose that X is a compactum, $\pi wX \leqslant \aleph_0$ and X has no isolated points. It is required to show that X is co-absolute with the Cantor perfect set. Owing to the fact that $\pi wX \leqslant \aleph_0$, there exist a metrizable compactum Y and an irreducible perfect mapping $f: X \to Y$ (see Problem 326 of Chap. 3). Since X has no isolated points and f is perfect and irreducible, Y also does not have isolated points (see Problem 111). Problem 223 implies that X and Y are co-absolute and Problem 238 the same of Y and the Cantor perfect set C. Thus, X is co-absolute with the Cantor perfect set C.

Suppose that X is co-absolute with C. First note that $\pi wX = \pi wC = \aleph_0$ (see Problem 236). Next, make use of Problems 224 and 111.

240. Obviously, by virtue of Problem 224, it suffices to show that $s(X) = s(Y)$ and $c(X) = c(Y)$ in the case in which X and Y are related by an irreducible perfect mapping $f: X \to Y, f(X) = Y$. This follows from the following elementary propositions:

(i) if f is continuous and $A \subset X$ is dense in X, then fA is dense in Y;

(ii) let f be closed and irreducible, and let $B \subset Y$ be dense in Y. For each $y \in B$ fix $x_y \in f^{-1}y$. Then, the set $A = \{x_y: y \in B\}$ is dense in Y and $|A| = |B|$;

(iii) if f is continuous and $\gamma = \{V_t: t \in T\}$ is a disjoint family of open sets in Y, then the family $f^{-1}\gamma = \{f^{-1}V_t: t \in T\}$ is disjoint and consists of open sets in X;

(iv) If f is closed and irreducible and ω is a disjoint family of open sets in X, then $f^{\#}\omega = \{f^{\#}U: U \in \omega\}$ is a disjoint family of nonempty open sets in Y.

241. If a compactum X is co-absolute with some metrizable compactum, then $\pi wX \leqslant \aleph_0$ (see Problem 237). Consequently, $s(X) \leqslant \aleph_0$, i.e., X is separable. Conversely, suppose that X is a separable first countable compactum. Let $A \subset X, |A| \leqslant \aleph_0$ and $[A]_X = X$. For each point $a \in A$ fix, using the first countability of X, a countable base

σ_a at this point. Put $\sigma = \bigcup \{\sigma_a : a \in A\}$. Clearly, $|\sigma| \leqslant \aleph_0$ and σ is a family of open sets dense in X. Hence $\pi w X \leqslant \aleph_0$. Consequently (see Problem 237), X is co-absolute with some metrizable compactum.

242. This follows from the preceding problem. The requirement is just to note that a perfect normal compactum is first countable.

243. From Problem 237 it follows that the π-weight of X equals $\aleph_0$. Hence, by Problem 327 of Chap. 3, X has a countable base and is therefore metrizable. We shall now establish the second statement of the problem. For every dyadic compactum the Souslin number is $\aleph_0$ (see Problem 166 of Chap. 3), so that $c(bX) = \aleph_0$. Consequently, for the subspace X dense in bX, $c(X) = \aleph_0$. Using Problem 240, we then obtain $c(Y) = \aleph_0$ (because Y is co-absolute with X and $c(X) = \aleph_0$). Since Y is metrizable, $c(Y) = \pi w Y = wY = \aleph_0$ (see Problems 216 and 217 of Chap. 2). Hence Y has a countable base. Since X is co-absolute with Y, $\pi w X = \aleph_0$ (see Problem 236). But X is dense in bX, so that $\pi w(bX) = \aleph_0$. We conclude (see Problem 327 of Chap. 3) that $w(bX) = \aleph_0$, so that also $wX = \aleph_0$.

The third statement has been established in Problem 167 of Chap. 3. We shall attempt another proof. If X is a first countable compactum, then $|X| \leqslant 2^{\aleph_0}$ (see Problems 151 and 213 of Chap. 3), and hence $wX \leqslant 2^{\aleph_0}$. But the dyadic compactum X of weight $wX \leqslant 2^{\aleph_0}$ is a continuous image of the generalized Cantor discontinuum $D^{\mathfrak{c}}$ of weight $2^{\aleph_0}$ (see Problem 323 of Chap. 3). We know that $D^{\mathfrak{c}}$ has a countable dense subset (Problem 381 of Chap. 2). Then X is also separable (see Problem 282 of Chap. 2). By Problems 241 and 237, we have $\pi w X \leqslant \aleph_0$. Then $wX \leqslant \aleph_0$ (see Problem 327 of Chap. 3). The assertion is proved.

244. See Problems 326 of Chap. 3 and 223 of the present chapter.

245. (*a*) If every open cover of a regular space X has a locally finite regular refinement, hence a locally finite closed refinement, then X is paracompact (Problem 150 of Chap. 5).

(*b*) Suppose that X is a paracompactum and Ω is an arbitrary open cover of X. Refine Ω to some locally finite open cover ω. Let $(T, <)$ be a minimal well-ordered set of cardinality equal to the cardinality of ω. Enumerate the elements of ω by the elements $t \in T$; let $\omega = \{U_t : t \in T\}$. Using Problem 13 of Chap. 5, combinatorially refine ω to an open cover $\omega' = \{H_t : t \in T\}$ such that $[H_t] \subset U_t$ for every $t \in T$. It is clear that ω' and the closed cover $\overline{\omega'} = \{[H_t] : t \in T\}$ are locally finite in X. Put $V_0 = H_0$ and in general

$$V_t = H_t \setminus \bigcup \{[H_{t'}] : t' < t\}.$$

It is easy to see that $\alpha^0 = \{V_t : t \in T\}$ is a disjoint locally finite family of open sets in X, refining ω'. It is also easy to show that $\tilde{\alpha}^0 = \bigcup \{V_t : t \in T\}$ is a dense set in X. Then, $\alpha = \{[V_t] : t \in T\}$ is a locally finite regular cover of X, refining ω and hence also Ω. The proof is complete.

246. Suppose that ω is an arbitrary open cover of an extremally disconnected paracompactum X. Refine ω (using the preceding problem) to a locally finite regular cover α. Let $A_1, A_2 \in \alpha$ and $A_1 \neq A_2$. Then, Int $A_1 \cap$ Int $A_2 = \Lambda$ (α being regular); consequently (owing to the extremal disconnectedness of X, see Problem 155), [Int A_1] $\cap$ [Int A_2] $= \Lambda$, [Int A] is clopen in X for every $A \in \alpha$. It remains to note that [Int A] $= A$

since A is a regular closed set.

Thus, α is a disjoint cover of X by clopen sets, refining ω.

247. $\dot{X}$ being the absolute of a paracompactum X is paracompact (Problem 233) and extremally disconnected (Problem 222). Then, by the preceding problem, the absolute $\dot{X}$ is perfectly 0-dimensional. Every perfectly 0-dimensional space is strongly paracompact, however.

248. No, not necessarily. Pick a paracompactum X, which is not a strongly paracompact space (see Problem 163 of Chap. 5). Then, the absolute $\dot{X}$ of X is perfectly 0-dimensional, so it is also strongly paracompact (see the preceding problem). In addition, $\dot{X}$ and X are connected by a perfect (also irreducible) mapping $\pi_X: \dot{X} \to X$, $\pi_X(\dot{X}) = X$.

250. Suppose that ρ is a metric on X, and that it is also complete if X is homeomorphic to a complete metric space.

For each $i \in N^+$, consider the open cover ω_i of X such that diam $U^i < 1/i$ for every $U^i \in \omega_i$. Since the metrizable space X is paracompact (Problem 45 of Chap. 5), using Problem 245, pick for each $i \in N^+$ a locally finite regular cover φ_i refining ω_i. Put $\alpha_1 = \varphi_1$ and in general $\alpha_i = \varphi_1 \wedge \varphi_2 \wedge \ldots \wedge \varphi_i$, $i \in N^+$ (definition on p. 327). Then α_i is a locally finite regular cover for every $i \in N^+$, $\alpha_{i+1} > \alpha_i$ for every $i \in N^+$ and diam $A^i < 1/i$ for every $A^i \in \alpha_i$, $i \in N^+$. Hence $\{\alpha_i: i \in N^+\}$ is a countable refined sequence of locally finite regular covers, which is complete if ρ is a complete metric.

251. This is solved similarly to Problems 193, 196, 200 and 202.

252. By Problem 250, X has a refined sequence $\{\alpha_i: i \in N^+\}$ of locally finite regular covers, which is complete if the metric space X is complete. If the weight of X is τ, then we may assume that $|\alpha_i| = |\alpha_{i+1}| = \tau$ for every i. For the rest, refer to the preceding problem.

References

ALEXANDROFF, P. S.

[1] 'Compactifications of Topological Spaces', *Mat. Sbornik* **5** (47) (1939), 403-424 (with German summary).

[2] 'Notion of a Space in Topology' *Uspechi Mat. Nauk* **2** (1) (1947), 5-57.

[3] *Introduction to the General Theory of Sets and Functions*, Gostekhizdat, 1948.

[4] 'Some Results in the Theory of Topological Spaces, Obtained during the Last Twenty-five Years', *Russian Math. Surveys* **15** (2) (1960), 23-84.

[5] 'On Some Results Concerning Topological Spaces and their Continuous Mappings', in J. Novak (Ed.), *Proceedings of the Second Symposium on General Topology and its Relation to Modern Analysis and Algebra* (Prague Symposium, 1961, 41-54), Academic Press, New York, 1962.

[6] 'On Some Basic Directions in General Topology', *Russian Math. Surveys* **19**(6) (1964), 1-39.

ALEXANDROFF, P. S. and P. S. URYSOHN

[1] 'Mèmoir sur les espaces topologiques compacts', *Verh. Nederl. Akad. Wetensch. Afd. Naturk. Sect.* **14** (1929), 1-96.

ARKHANGEL'SKII, A. V.

[1] 'Compact Sets and the Topology of Spaces', *Trudy Moskov. mat. Obšč.* **13** (1965), 3-55.

[2] 'On a Class of Spaces containing all Metric and all Locally Compact Spaces', *Mat. Sbornik* **67**(109) (1965), 55-85.

[3] 'Closed Mappings, Compact Sets and a Problem of Alexandroff', *Mat. Sborink* **69(111)** (1966), 132-184.

[4] 'Open and Almost Open Mappings', *Trudy Moskov. mat. Obšč.* **15** (1966), 181-223.

[5] 'Mappings and Spaces', *Russian Math. Surveys* **21**(4) (1966), 115-162.

[6] 'Extremally Disconnected Compact Spaces of Nonhomogeneous Weight', *Doklady Akad. Nauk SSSR* **175**(4) (1967), 751-754.

[7] 'Approximation Theory for Dyadic Compact Spaces', *Doklady Akad. Nauk SSSR* **184**(4) (1969), 767-770.

[8] 'Cardinality of Compact First-Countable Spaces', *Doklady Akad. Nauk SSSR* **187**(5) (1969), 967-970.

[9] 'Souslin Number and Cardinality. Character of a Point in a Compact Space', *Doklady Acad. Nauk SSSR* **192**(2) (1970), 255-258.

[10] 'On Compacta Hereditarily Satisfying Souslin's Condition. Tightness and Free Sequences', *Soviet Math. Doklady* **12** (1971), 1153-1257.

[11] 'Čech-complete Topological Spaces', *Vestnik Moskov. Univ. Ser. Math.* **2** (1961), 37-40.

ARKHANGEL'SKII, A. V. and V. V. FILIPPOV

[1] 'Spaces with Bases of Finite Rank', *Mat. Sbornik* **87** (129) (1973), 147-158.

ARKHANGEL'SKII A. V. and V. I. PONOMAREV

[1] 'Dyadic Compact Spaces', *Doklady Acad. Nauk SSSR* **182** (5) (1968), 993-996.

AULL, C. E.

[1] 'Topological Spaces with σ-Finite Bases', *Proc. Amer. Math. Soc.* **29**(2) (1971), 411-416.

BEL'NOV, V. K.

[1] 'Metric Extensions', I and II, *Vestnik Moskov. Univ. Ser. Math. Meh.* **4**, 60-65 and **5**, 7-14 (1970).

[2] 'Metric Extensions', *Doklady Akad. Nauk SSSR* **207**(5) (1972), 202.

BELUGIN, V. I.

[1] 'Contractions of Compact Spaces', *Doklady Akad. Nauk SSSR* **207**(2) (1972), 259-261.

BEREZNITSKII, Y.

[1] 'Nonhomeomorphism of Two Compact Spaces', *Vestnik Moskov. Univ. Ser. Math.* **6** (1971), 8-10.

BING, R. H.

[1] 'Metrization of Topological Spaces', *Canad. J. Math.* **3**(2) (1951), 175-186.

[2] 'Challenging Conjectures', *Amer. Math. Monthly* **74** (1967), 56-64.

BORGES, C. J. R.

[1] 'On Function Spaces of Stratifiable Spaces and Compact Spaces', *Proc. Amer. Math. Soc.* **17**(5) (1966), 1074-1078.

BOURBAKI, N.

[1] *Theory of Sets*, Addison Wesley, Reading, 1966.

[2] *General Topology*, Parts 1 and 2, Addison-Wesley, Reading, 1966.

BURKE, D. K.

[1] 'On Subparacompact Spaces', *Proc. Amer. Math. Soc.* **23** (1969), 656-663.

ČECH, E.

[1] *Topological Spaces*, Rev. Ed. by M. Kalekov and Z. Frolik, Interscience, New York, 1966.

ČECH, E. and B. POSPIŠEL

[1] 'Sur les espaces compacts', *Publ. de la fac. de. univ. Masaryk, Brono* **258** (1938), 1-14.

CEDER, J. G.
[1] 'Some Generalizations of Metric Spaces', *Pacif. J. Math.* **11** (1961), 105-126.

CHOBAN, M. M.
[1] 'On the Behaviour of Metrizability under Quotient s-Mappings', *Doklady Akad. Nauk SSSR* **166** (1966), 562-565.
[2] 'Some Metrization Theorems for Feather Spaces', *Doklady Akad. Nauk SSSR* **173** (1967), 1270-1272.
[3] 'On σ-Paracompact Spaces', *Vestnik Moskov. Univ. Ser. Math. Meh.* **1** (1969), 20-27.
[4] 'On the Theory of p-Spaces', *Doklady Akad. Nauk SSSR* **194**(3) (1970), 528-531.
[5] 'Many-valued Functions and Borel Sets', *Trudy Moskov. mat. Obšč.* **22** (1971), 229-50; **23** (1972), 277-301.

COMFORT, W. W.
[1] 'Retractions and their Continuous Maps from βX onto $\beta X \setminus X$,' *Trans. Amer. Math. Soc.* **114**(1) (1965), 1-9.
[2] 'A Survey of Cardinal Invariants', *Gen. Topol. Appl.* **1** (1971), 163-199.

COMFORT, W. W. and S. NEGREPONTIS
[1] 'Homomorphs of Three Subspaces of $\beta N \setminus N$,' *Math. Zeitschr.* **107**(1) (1968), 53-58.
[2] 'Some Topological Properties Associated with Measurable Cardinals', *Fund. Math.* **69**(3) (1970), 191-205.

CORSON, H. H.
[1] 'Normality in Subsets of Product Spaces', *Amer. J. Math.* **81** (1959), 785-796.

DOWKER, C. H.
[1] 'On Countably Paracompact Spaces', *Canad. J. Math.* **3** (1951), 219-224.

EFIMOV, B. A.
[1] 'Dyadic Compacta', *Trudy Moskov. mat. Obšč.* **14** (1965), 211-247.
[2] 'On Extremally Disconnected Compacta of π-Weight $\mathfrak{c}$', *Doklady Akad. Nauk SSSR* **183** (3) (1968), 15-18.
[3] 'Extremally and Absolutely Disconnected Compacta', *Trudy Moskov. mat. Obšč.* **23** (1970), 235-276.

EL'KIN, A.
[1] 'On k-Decomposable Spaces which are not Maximally Decomposable', *Doklady Akad. Nauk SSSR* **195**(2) (1970), 274-277.

ENGELKING, R.
[1] 'Cartesion Products and Dyadic Spaces', *Fund. Math.* **58** (1965), 287-304.

[2] 'On Functions Defined on Cartesian Products', *Fund. Math.* **59** (1966), 221-231.
[3] 'On Closed Images of the Spaces of Irrationals', *Proc. Amer. Math. Soc.* **21**(3) (1969), 583-586.
[4] *Outline of General Topology*, North Holland, Amsterdam, 1968.

ERDÖS, P. and R. RADO
[1] 'A Partition Calculus in Set Theory', *Bull. Amer. Math. Soc.* **62** (1956), 427-489.

FEDORČUK, V. V.
[1] 'Irreducible Prefect Mappings and Generalized Proximities', *Mat. Sbornik* **76** (118) (1968), 513-536.
[2] 'Some Problems in the Theory of Ordered Spaces', *Cyb. matem. J.* **10**(1) (1969), 172-187.
[3] 'On *H*-Closed Extensions of θ-Proximity Spaces', *Mat. Sbornik* **89**(3) (1972), 400-412.

FILIPPOV, V. V.
[1] 'On a Complete Image of a Paracompact Space', *Doklady Akad. Nauk SSSR* **176**(3) (1967), 533-536.
[2] 'Factor Spaces and Multiplicity of Bases', *Mat. Sbornik* **80**(122) (1969), 521-532.
[3] 'On Perfectly Normal Compacta', *Doklady Akad. Nauk SSSR* **189** (1969), 36-39.
[4] 'On Paracompacta', *Doklady Akad. Nauk SSSR* **178**(3) (1968), 555-558.

FRAENKEL, A. A., Y. BAR-HILLEL and A. LEVY
[1] *Foundations of Set Theory*, Second Edn., North-Holland, Amsterdam, 1973.

FROLIK, Z.
[1] 'Applications of Complete Families of Continuous Functions to the Theory of *Q*-Spaces', *Zech. Math. J.* **11**(86) (1961), 115-133.
[2] 'Sums of Ultrafilters', *Bull. Amer. Math. Soc.* **73** (1967), 87-91.
[3] 'Homogeneity Problems for Extremally Disconnected Spaces', *Comment. Math. Univ. Carolinae* **8**(4) (1967), 757-763.

GEL'FAND, I. M., D. A. RAIKOV and G. E. SHILOV
[1] '*Commutative Normed Rings*', Chelsea, New York, 1964.

GILLMAN, L. and M. JERISON
[1] *Rings of Continuous Functions*, Van Nostrand-Reinhold, New York, 1960.

GRACE, E. E. and R. W. HEATH
[1] 'Separability and Metrizability in Pointwise Paracompact Moore Spaces', *Duke Math J.* **31**(4) (1964), 603-610.

HAJNAL, A. and I. JUHASZ

[1] 'Discrete Subspaces of Topological Spaces', *Indag. Math*, **29** (1967), 343-356.

[2] 'Discrete Subspaces of Topological Spaces', II, *Indag. Math.* **31** (1969), 18-30.

HAUSDORFF, F.

[1] *Set Theory*, Second Edition, Chelsea, New York, 1960.

[2] 'Erweiterung einer Homomöorphe', *Fund. Math.* **16** (1930), 353-360.

[3] 'Summen von $\aleph_1$ Mengen', *Fund. Math.* **26** (1936).

HEATH, R. W.

[1] Screenability, Pointwise Paracompactness, and Metrization of Moore Spaces', *Canad. J. Math.* **16** (1964), 763-770.

[2] 'On Spaces with Point-Countable Bases', *Bull. Acad. Polon. Sci. Ser. Math.* **13**(6) (1965), 393-395.

[3] 'Semi-metric Spaces and Related Space', in: *Proc. Topology Conf. Arizona State Univ.*, 1968, 153-161.

HENRIKSEN, M. and J. R. ISBELL

[1] 'Some Properties of Compactifications', *Duke Math. J.* **25**(1) (1958), 83-106.

HEWITT, E.

[1] 'A Note on 0-Dimensional Compact Groups', *Fund. Math.* **50**(1) (1961), 95-97.

ILIADIS, S. D. and S. V. FOMIN

[1] 'Centralized Families in the Theory of Topological Spaces', *Uspechi Mat. Nauk* **21**(4) (1966), 47-76.

ISBELL, J. R.

[1] *Uniform Spaces*, Amer. Math. Soc. Survey No. 12, American Mathematical Society, Providence, 1964.

ISIWATA, T.

[1] 'Mappings and Spaces', *Pacif. J. Math.* **20** (1967), 455-480.

JONES, F. B.

[1] 'Concerning Normal and Completely Normal Spaces', *Bull. Amer. Math. Soc.* **43** (1937), 671-679.

JUHASZ, I.

[1] *Cardinal Functions in Topology*, Math. Centre Tract No. 34, Amsterdam, 1971.

[2] 'On Two Problems of A. V. Arkhangel'skii', *Gen. Topol. Appl.* **2** (1978), 151-156.

KATĚKOV, M.

[1] 'Complete Normality of Cartesian Products', *Fund. Math.* **36** (1948), 271-274.

KEESLING, J.
[1] 'On the Equivalence of Normality and Compactness in Hyperspaces', *Pacif. J. Math.* **33** (1970), 657-667.

KEISLER, H. J. and A. TARSKI
[1] 'From Accessible to Inaccessible Cardinals', *Fund. Math.* **53** (1964), 225-308.

KELLEY, J. L.
[1] *General Topology*, Van Nostrand-Reinhold, New York, 1955.

KOFNER, A. Y.
[1] 'On Pseudocircular Spaces', *Fund. Math.* **70**(1) (1971), 25-47.

KOLMOGOROV, A. N. and S. V. FOMIN
[1] *Elements of the Theory of Functions and Functional Analysis*, Volumes 1 and 2, Graylock Press, Rochester, 1957, 1961.

KOMBAROV, A. P.
[1] 'On Σ-Product of Topological Spaces', *Doklady Akad. Nauk SSSR* **199**(3) (1971), 526-528.
[2] 'On the Product of Normal Spaces. Uniformity on Σ-Product', *Doklady Akad. Nauk SSSR* **205**(5) (1927), 1033-1035.

KURATOWSKI, C.
[1] *Topology*, Volumes 1 and 2, Academic Press, New York, 1966, 1969.

LASHNEV, N. S.
[1] 'Continuous Decompositions and Closed Mappings of Metric Spaces', *Soviet Math. Doklady* **6** (1965), 1504-1506.
[2] 'Closed Images of Metric Spaces', *Soviet Math. Doklady* **7** (1966), 1219-1221.

MAL'KHIN, V.
[1] 'On Countable Spaces not having a Compactification of Countable Tightness', *Doklady Akad. Nauk SSSR* **206**(6) (1972), 1293-1296.
[2] 'Products of Ultrafilters with Indecomposable Spaces', *Mat. Sbornik* **90**(1) (1973), 106-116.

MICHAEL, E.
[1] 'A Note on Paracompact Spaces', *Proc. Amer. Math. Soc.* **4** (1953), 831-838.
[2] 'Another Note on Paracompact Spaces', *Proc. Amer. Math. Soc.* **8** (1957), 822-828.
[3] $\aleph_0$-Spaces, *J. Math. Mech.* **15** (1966), 983-1002.
[4] 'Topologies on Spaces of Subsets', *Trans. Amer. Math. Soc.* **71** (1951), 152-182.
[5] 'Continuous Selections, I,' *Ann. Math.* **63** (1956), 361-382.

[6] 'Biquotient Maps and Cartesian Products of Quotient Maps', *Ann. Inst. Fourier/ Grebole* **18**(2) (1969), 287-302.
[7] 'A Quintuple Quotient Test', *Gen. Topol. Appl.* **2** (1972), 91-138.

MICHAEL, E. and A. H. STONE
[1] 'Quotients of the Space of Rationals', *Pacif. J. Math.*

MISHENKO, A. S.
[1] 'On Finally Compact (Lindelöf) Spaces', *Doklady Akad. Nauk SSSR* **145** (1962), 1224-1227.
[2] 'Examples of Irreducible Perfect Mappings of Nonnormal Spaces onto Normal Spaces', *Uspechi Mat. Nauk* **18**(5) (1963), 181-182.
[3] 'Some Theorems on Product Topological Spaces', *Fund. Math.* **58** (1966), 259-284.

MORITA, K.
[1] 'Paracompactness and Product Spaces', *Fund. Math.* **50**(3) (1962), 223-236.
[2] 'Products of Normal Spaces with Metric Spaces', *Math. Ann.* **154** (1964), 365-382.
[3] 'A Survey of the Theory of *M*-Spaces', *Gen. Topol. Appl.* **1** (1971), 49-55.

MROWKA, S.
[1] 'Some Properties of *Q*-Spaces', *Bull. Acad. Polon. Sci. Cl III* **5** (1957), 947-950.
[2] 'On *E*-Compact Spaces, II,' *Bull. Acad. Polon. Sci.* **14**(11) (1964), 597-605.

NAGAMI, K.
[1] 'Σ-Spaces', *Fund. Math.* **65** (1969), 169-192.

NAGATA, J.
[1] 'Mappings and *M*-Spaces', *Proc. Japan Acad.* **45**(1) (1969), 140-144.

NAIMARK, M. A.
[1] *Normed Algebras*, Noordhoff, Groningen, 1972.

NEDEV, S. I.
[1] 'Zero-dimensional Spaces', *Trudy Moskov. math. Obšč.* **24** (1971), 201-236.

NEGREPONTIS, S.
[1] 'The Existence of Certain Uniform Ultrafilters', *Ann. Math.* **76**(1) (1969), 23-32.

OKUYAMA, A.
[1] 'Some Generalizations of Metric Spaces, their Metrization Theorems and Product Spaces', *Sci. Rep. Tokyo Kyoiku Daigaku, Sec. A*, **9** (1967), 236-254.
[2] 'A Survey of Theory of σ-Spaces', *Gen. Topol. Appl.* **1** (1971), 57-63.

PAROVICHENKO, I. I.
[1] 'On Universal Compacta of Weight $\aleph_0$', *Doklady Akad. Nauk SSSR* **150** (1963), 36-39.

PAS'NKOV, B. A.

[1] 'On Almost Metrizable Groups', *Doklady Akad. Nauk SSSR* **161**(2) (1965), 281.

[2] 'On Open Mappings', *Doklady Akad. Nauk SSSR* **175**(2) (1967), 292-295.

[3] 'Partial Products', *Trudy Moskov. mat. Obšč.* **13**(1965), 136-245.

PEŁCZYŃSKI, A.

[1] 'A Remark on Spaces 2^X for Zero-dimensional X,' *Bull. Acad. Polon. Sci. Ser. Math.* **13**(2) (1965), 85-89.

PONOMAREV, V. I.

[1] 'New Spaces of Closed Sets and Many-Valued Continuous Mappings of Compacta', *Mat. Sbornik* **48**(2) (1959), 191-212.

[2] 'On Spaces Co-absolute with Metric Spaces', *Uspechi Mat. Nauk* **21**(4) (1966), 101-132.

[3] 'On Metrizability of Lindelöf p-Spaces with Precisely Countable Bases', *Doklady Akad. Nauk SSSR* **174**(6) (1967), 1274-1277.

[4] 'On Star-finite Covers and Clopen Sets', *Doklady Akad. Nauk SSSR* **186**(5) (1969), 1016-1019.

[5] 'On the Cardinality of a First Countable Compactum', *Doklady Akad. Nauk SSSR* **196**(2) (1971), 296-298.

POTOCZNY, H. B.

[1] 'A Nonparacompact Space which Admits a Closure-Preserving Cover of Compact Sets', *Proc. Amer. Math. Soc.* **32**(1) (1972), 309-311.

RAIKHBERG, A. I.

[1] 'An Extremally Disconnected Space with Unique Homeomorphism onto itself', *Doklady Akad. Nauk SSSR* **195**(3) (1970), 562-563.

ROY, P.

[1] 'Separability of Metric Spaces', *Trans. Amer. Math. Soc.* **149**(1) (1970), 19-43.

RUDIN, M. E.

[1] 'Partial Orders on the Types in βN,' *Trans. Amer. Math. Soc.* **155**(2) (1971), 353-367.

[2] 'A Normal Space X for which $X \times I$ is not Normal', *Fund. Math.* **73**(1971), 179-186.

RUDIN, W.

[1] 'Homogeneity Problems in the Theory of Čech Compactifications', *Duke Math. J.* **23**(3) 1956), 409-423.

SHANIN, N. A.

[1] 'On Product Topological Spaces', *Trudy Mat. Inst. Steklova* **24** (1948).

SHAPIROVSKII, B.

[1] 'On the Separability and Metrizability of Spaces with the Souslin Property', *Doklady Akad. Nauk SSSR* **207**(4) (1972).

[2] 'On Discrete Subspaces of Topological Spaces. Weight, Tightness and Souslin Number', *Doklady Akad. Nauk SSSR* **202**(4) (1972), 779-782.

[3] 'On Compactness of Topological Spaces', *Doklady Akad. Nauk SSSR* **206**(3) (1972), 559-562.

SIERPINSKI, W.

[1] *Cardinal and Ordinal Numbers*, Monografie Matematyizne, Vol. 34, Polska Akad. Nauk, Warsaw, 1958.

SIKORSKI, R.

[1] *Boolean Algebras*, Third Edn., Springer-Verlag, Heidelberg, 1969.

SIROTA, S.

[1] 'The Product Topological Groups and Extremal Disconnectedness', *Mat. Sbornik* **79**(2) (1969), 179-192.

SKLYARENKO, E. G.

[1] 'Some Problems in the Theory of Compactifications', *Izv. Akad. Nauk SSSR Ser. Math.* **26** (1962), 427-452.

SLOT, J. VAN DER

[1] *Universal Topological Properties*, Mathematical Centre Tract No. 20, Mathematisch Centrum, Amesterdam, 1966, pp. 1-9.

SMIRNOV, Y. M.

[1] 'On Strongly Paracompact Spaces', *Izv. Akad. Nauk SSSR Ser. Math.* **20** (1956), 253-274.

[2] 'Contractions of Compacta and Connexion with Compactifications and Retractions', *Fund. Math.* **63**(2) (1968), 199-211.

STONE, A. H.

[1] 'Paracompactness and Product Spaces', *Bull. Amer. Math. Soc.* **54** (1948), 977-982.

[2] 'Metrizability of Decomposition Spaces', *Proc. Amer. Math. Soc.* **7** (1956), 690-700.

[3] 'Non-separable Borel Sets, I,' *Rozprawy Mat.* **28** (1962), 3-40.

[4] 'Non-Separable Borel Sets, II,' *Gen. Topol. Appl.* 2 (1972), 249-270.

TAMANO, H.

[1] 'On Paracompactness', *Pacif. J. Math.* **10** (1960), 1043-1047.

URYSOHN, P. S.

[1] *Collected Works on Topology and other Branches of Mathematics*, Vols. 1 and 2, Izd. Akad. Nauk SSSR, 1951.

VLADIMOROV, D. A.
[1] *Boolean Algebras*, Nauka, 1969.

VLADIMOROV, D. A. and B. A. EFIMOV
[1] 'On the Cardinality of Extremally Disconnected Spaces and Complete Boolean Algebras', *Doklady Akad. Nauk SSSR* **194**(6) (1970), 1247-1250.

WICKE, H. H.
[1] 'On the Hausdorff Open Continuous Images of Hausdorff Paracompact p-spaces', *Proc. Amer. Math. Soc.* **22**(1) (1969), 136-140.

WOODS, R. GRANT
[1] 'Co-absolutes of Remainders of Stone-Čech Compactifications', *Pacif. J. Math.* **37**(2) (1971), 545-560

WORRELL, J. M.
[1] 'A Characterization of Metacompact Spaces'. *Portug. Math.* **25**(3) (1966), 171-174.
[2] 'The Closed Continuous Images of Metacompact Topological Spaces', *Portug. Math.* **25**(3) (1966), 176-179.

ZAITSEV, V. I.
[1] 'Finite Spectra of Topological Spaces and their Limit Spaces', *Math. Ann.* **179** (1969), 153-179.
[2] 'Projection Spectra', *Trudy Moskov. mat. Obšč.* **27**(1972), 129-193.

Index

Made in the USA
Monee, IL
13 September 2026

62976820R00240